T0323972

MODAL LOGIC

Cambridge Tracts in Theoretical Computer Science

Editorial Board

S. Abramsky, *Computer Laboratory, Oxford University*
P. H. Aczel, *Department of Computer Science, University of Manchester*
J. W. de Bakker, *Centrum voor Wiskunde en Informatica, Amsterdam*
Y. Gurevich, *Microsoft Research*
J. V. Tucker, *Department of Mathematics and Computer Science, University College of Swansea*

Titles in the series

1. G. Chaitin *Algorithmic Information Theory*
2. L. C. Paulson *Logic and Computation*
3. M. Spivey *Understanding Z*
5. A. Ramsey *Formal Methods in Artificial Intelligence*
6. S. Vickers *Topology via Logic*
7. J.-Y. Girard, Y. Lafont & P. Taylor *Proofs and Types*
8. J. Clifford *Formal Semantics & Progmatics for Natural Language Processing*
9. M. Winslett *Updating Logical Databases*
10. K. McEvoy & J. V. Tucker (eds) *Theoretical Foundations of VLSI Design*
11. T. H. Tse *A Unifying Framework for Structured Analysis and Design Models*
12. G. Brewka *Nonmonotonic Reasoning*
14. S. G. Hoggar *Mathematics for Computer Graphics*
15. S. Dasgupta *Design Theory and Computer Science*
17. J. C. M. Baeten (ed) *Applications of Process Algebra*
18. J. C. M. Baeten & W. P. Weijland *Process Algebra*
19. M. Manzano *Extensions of First Order Logic*
21. D. A. Wolfram *The Clausal Theory of Types*
22. V. Stoltenberg-Hansen, I. Lindström & E. Griffor *Mathematical Theory of Domains*
23. E.-R. Olderog *Nets, Terms and Formulas*
26. P. D. Mosses *Action Semantics*
27. W. H. Hesselink *Programs, Recursion and Unbounded Choice*
28. P. Padawitz *Deductive and Declarative Programming*
29. P. Gärdenfors (ed) *Belief Revision*
30. M. Anthony & N. Biggs *Computational Learning Theory*
31. T. F. Melham *Higher Order Logic and Hardware Verification*
32. R. L. Carpenter *The Logic of Typed Feature Structures*
33. E. G. Manes *Predicate Transformer Semantics*
34. F. Nielson & H. R. Nielson *Two Level Functional Languages*
35. L. Feijs & H. Jonkers *Formal Specification and Design*
36. S. Mauw & G. J. Veltink (eds) *Algebraic Specification of Communication Protocols*
37. V. Stavridou *Formal Methods in Circuit Design*
38. N. Shankar *Metamathematics, Machines and Gödel's Proof*
39. J. B. Paris *The Uncertain Reasoner's Companion*
40. J. Dessel & J. Esparza *Free Choice Petri Nets*
41. J.-J. Ch. Meyer & W. van der Hoek *Epistemic Logic for AI and Computer Science*
42. J. R. Hindley *Basic Simple Type Theory*
43. A. Troelstra & H. Schwichtenberg *Basic Proof Theory*
44. J. Barwise & J. Seligman *Information Flow*
45. A. Asperti & S. Guerrini *The Optimal Implementation of Functional Programming Languages*
46. R. M. Amadio & P.-L. Curien *Domains and Lambda-Calculi*
47. W.-P. de Roever & K. Engelhardt *Data Refinement*
48. H. Kleine Büning & T. Lettman *Propositional Logic*
49. L. Novak & A. Gibbons *Hybrid Graph Theory and Network Analysis*
51. H. Simmons *Derivation and Computation*
52. A. S. Troelstra & H. Schwictenberg *Basic Proof Theory* (Second Edition)
53. P. Blackburn, M. de Rijke & Y. Venema *Modal Logic*

Modal Logic

Patrick Blackburn
LORIA, Nancy

Maarten de Rijke
University of Amsterdam

Yde Venema
University of Amsterdam

CAMBRIDGE
UNIVERSITY PRESS

CAMBRIDGE UNIVERSITY PRESS
Cambridge, New York, Melbourne, Madrid, Cape Town,
Singapore, São Paulo, Delhi, Tokyo, Mexico City

Cambridge University Press
The Edinburgh Building, Cambridge CB2 8RU, UK

Published in the United States of America by
Cambridge University Press, New York

www.cambridge.org
Information on this title: www.cambridge.org/9780521527149

First published 2001
First paperback edition 2002
Fourth printing with corrections 2010

A catalogue record for this publication is available from the British Library

Library of Congress Cataloguing in Publication data
Blackburn, Patrick, 1959-
Modal logic / Patrick Blackburn, Maarten de Rijke, Yde Venema.
p. cm.
Includes bibliographical references and index.
ISBN 0 521 80200 8
1. Modality (Logic). I. Rijke, Maarten de. II. Venema, Yde, 1963- III. Title.
QA9.46.B58 2001
511.3 21-dc21 00-054667

ISBN 978-0-521-80200-0 Hardback
ISBN 978-0-521-52714-9 Paperback

For Johan

Contents

Preface   *page* xi

1	**Basic Concepts**	1
1.1	Relational Structures	2
1.2	Modal Languages	9
1.3	Models and Frames	16
1.4	General Frames	28
1.5	Modal Consequence Relations	31
1.6	Normal Modal Logics	33
1.7	Historical Overview	37
1.8	Summary of Chapter 1	48
2	**Models**	50
2.1	Invariance Results	51
2.2	Bisimulations	64
2.3	Finite Models	73
2.4	The Standard Translation	83
2.5	Modal Saturation via Ultrafilter Extensions	91
2.6	Characterization and Definability	100
2.7	Simulation and Safety	110
2.8	Summary of Chapter 2	117
Notes		118
3	**Frames**	123
3.1	Frame Definability	124
3.2	Frame Definability and Second-Order Logic	130
3.3	Definable and Undefinable Properties	138
3.4	Finite Frames	143
3.5	Automatic First-Order Correspondence	148
3.6	Sahlqvist Formulas	156
3.7	More about Sahlqvist Formulas	167

3.8	Advanced Frame Theory	178
3.9	Summary of Chapter 3	183
Notes		185
4	**Completeness**	**188**
4.1	Preliminaries	189
4.2	Canonical Models	196
4.3	Applications	201
4.4	Limitative Results	211
4.5	Transforming the Canonical Model	217
4.6	Step by Step	223
4.7	Rules for the Undefinable	229
4.8	Finitary Methods I	239
4.9	Finitary Methods II	247
4.10	Summary of Chapter 4	256
Notes		258
5	**Algebras and General Frames**	**261**
5.1	Logic as Algebra	262
5.2	Algebraizing Modal Logic	275
5.3	The Jónsson-Tarski Theorem	283
5.4	Duality Theory	294
5.5	General Frames	303
5.6	Persistence	318
5.7	Summary of Chapter 5	326
Notes		327
6	**Computability and Complexity**	**332**
6.1	Computing Satisfiability	333
6.2	Decidability via Finite Models	338
6.3	Decidability via Interpretations	347
6.4	Decidability via Quasi-models and Mosaics	356
6.5	Undecidability via Tiling	364
6.6	NP	373
6.7	PSPACE	381
6.8	EXPTIME	393
6.9	Summary of Chapter 6	406
Notes		407
7	**Extended Modal Logic**	**413**
7.1	Logical Modalities	414
7.2	Since and Until	426
7.3	Hybrid Logic	434
7.4	The Guarded Fragment	446

7.5   Multi-Dimensional Modal Logic   458
7.6   A Lindström Theorem for Modal Logic   470
7.7   Summary of Chapter 7   476
Notes   477

Appendix A   **A Logical Toolkit**   485

Appendix B   **An Algebraic Toolkit**   497

Appendix C   **A Computational Toolkit**   504

Appendix D   **A Guide to the Literature**   516

Bibliography   524
List of Notation   544

Index   547

Preface

Ask three modal logicians what modal logic is, and you are likely to get at least three different answers. The authors of this book are no exception, so we will not try to start off with a neat definition. Nonetheless, a number of general ideas guide our thinking about the subject, and we will present the most important right away as a series of three slogans. These are meant to be read now, and, perhaps more importantly, referred back to occasionally; doing so will help you obtain a firm grasp of the ideas and intuitions that have shaped this book. Following the slogans we will discuss the aims and content of the book in more detail.

Our first slogan is the simplest and most fundamental. It sets the basic theme on which the others elaborate:

Slogan 1: Modal languages are simple yet expressive languages for talking about relational structures.

In this book we will be examining various *propositional* modal languages: that is, the familiar language of propositional logic augmented by a collection of *modal operators*. Like the familiar boolean connectives ($\neg$, $\wedge$, $\vee$, $\rightarrow$, $\bot$, and $\top$), modal operators do *not* bind variables. Thus, as far as syntax is concerned, we will be working with the simplest non-trivial languages imaginable.

But in spite of their simplicity, propositional modal languages turn out to be an excellent way of talking about *relational structures*, and this book is essentially an attempt to map out some of the ramifications of this. For a start, it goes a long way towards explaining the recent popularity of modal languages in applied logic. Moreover, it introduces one of the fundamental themes in the mathematical study of modal logic: the use of relational structures (that is, *relational semantics*, or *Kripke semantics*) to explicate the logical structure of modal systems.

A *relational structure* is simply a set together with a collection of relations on that set. Given the broad nature of this definition, it is unsurprising that relational structures are to be found just about everywhere. Virtually all familiar mathematical structures can be thought of as relational structures. Moreover, the entities commonly used to model the phenomena of interest in various applications often turn out to be relational structures. For example, theoretical computer scientists use labeled transition systems to model program execution, but a labeled transition system is just a set (the states) together with a collection of binary relations (the transition relations) that model the behavior of programs. Moreover, relational structures play a fundamental modeling role in many other disciplines, including knowledge representation, computational linguistics, formal semantics, economics, and philosophy. As modal languages are the simplest languages in which relational structures can be described, constrained, and reasoned about, it is hardly surprising that applied modal logic has blossomed in recent years.

But relational structures have also played a fundamental role in the development of the mathematics of modal logic: their use turned modal logic from a rather esoteric branch of syntax manipulation into a concrete and intuitively compelling field. In fact, it is difficult to overstate the importance of relational models to modal logic: their discovery in the 1950s and early 1960s was the biggest single impetus to the development of the field. An early application was *completeness theory*, the classification of modal logics in relational terms. More recently, relational semantics has played an important role in mapping out the *computational complexity* of modal systems.

Modal languages may be simple – but what makes them special? Our next slogan tries to pin this down:

Slogan 2: Modal languages provide an internal, local perspective on relational structures.

That is, modal languages talk about relational structures in a special way: 'from the inside' and 'locally.' Rather than standing outside a relational structure and scanning the information it contains from some celestial vantage point, modal formulas are evaluated *inside* structures, *at a particular state*. The function of the modal operators is to permit the information stored at other states to be scanned – but, crucially, *only the states accessible from the current point via an appropriate transition may be accessed in this way*. This idea will be made precise in the following chapter when we define the satisfaction definition. In the meantime, the reader who pictures a modal formula as a little automaton standing at some state in a relational structure, and only permitted to explore the structure by making journeys to neighboring states, will have grasped one of the key intuitions of modal model theory.

The internal perspective modal languages offer makes them natural for many applications. For a start, the decidability of many important modal systems stems from the step-by-step way that modal formulas are evaluated. Moreover, in a number of disciplines, simple languages offering an internal perspective on relational structures have been devised; sometimes these (independently invented) systems turn out to be variants of well-known modal systems, and can be analyzed using modal techniques. For example, Kasper-Rounds logic (used in computational linguistics) is essentially a natural notation for a certain fragment of propositional dynamic logic with intersection, and many of the description logics used in knowledge representation can be usefully viewed as (fragments of) modal languages. Finally, it is also the stepwise way in which modal formulas are evaluated which explains why the notion of *bisimulation*, a crucial tool in the process theoretic study of labeled transition systems, unlocks the door to important characterizations of modal expressivity.

So far there have been only two characters in this discussion: modal languages and the structures which interpret them. Now it is certainly true that for much of its history modal logic was studied in isolation, but the true richness of the subject only becomes apparent when one adopts a broader perspective. Accordingly, the reader should bear in mind that:

Slogan 3: Modal languages are not isolated formal systems.

One of the key lessons to have emerged since about 1970 is that it is fruitful to systematically explore the way modal logic is related to other branches of mathematical logic. In the pair ⟨MODAL LANGUAGES, RELATIONAL STRUCTURES⟩, there are two obvious variations that should be considered: the relationships with other languages for describing relational structures, and the use of other kinds of structures for interpreting modal languages.

As regards the first option, there are many well-known alternative languages for talking about relational structure: most obviously, first- or second-order classical languages. And indeed, every modal language has *corresponding* classical languages that describe the same class of structures. But although both modal and classical languages talk about relational structures, they do so very differently. Whereas modal languages take an internal perspective, classical languages, with their quantifiers and variable binding, are the prime example of how to take an *external* perspective on relational structures. In spite of this, there is a *standard translation* of any modal language into its corresponding classical language. This translation provides a bridge between the worlds of modal and classical logic, enabling techniques and results to be imported and exported. The resultant study is called *correspondence theory*, and it is a cornerstone of modern modal logic.

In the most important example of the second variation, modal logic is linked up with universal algebra via the apparatus of *duality theory*. In this framework, modal formulas are viewed as algebraic terms which have a natural algebraic semantics in terms of *boolean algebras with operators*, and, from this perspective, modal logic is essentially the study of certain varieties of *equational logic*. Now, even in isolation, this algebraic perspective is of interest – but what makes it a truly formidable tool is the way it interacts with the perspective provided by relational structures. Roughly speaking, relational structures can be constructed out of algebras, and algebras can be constructed out of relational structures, and both constructions preserve essential logical properties. The key technical result that underlies this duality is the Jónsson-Tarski Theorem, a Stone-like representation theorem for boolean algebras with operators. This opens the door to the world of universal algebra and, as we will see, the powerful techniques to be found there lend themselves readily to the analysis of modal logic.

Slogan 3 is fundamental to the way the material in this book is developed: modal logic will be systematically linked to the wider logical world by both correspondence and duality theory. We do not view modal logic as a 'non-classical logic' that studies 'intensional phenomena' via 'possible world semantics.' This is one interpretation of the machinery we will discuss – but the real beauty of the subject lies deeper.

Let us try and summarize our discussion. Modal languages are syntactically simple languages that provide an internal perspective on relational structures. Because of their simplicity, they are becoming increasingly popular in a number of applications. Moreover, modal logic is surprisingly mathematically rich. This richness is due to the intricate interplay between modal languages and the relational structures that interpret them. At its most straightforward, the relational interpretation gives us a natural semantic perspective from which to attack problems directly. But the interplay runs deeper. By adopting the perspective of correspondence theory, modal logic can be regarded as a fragment of first- or second-order classical logic. Moreover, by adopting an algebraic perspective, we obtain a different (and no less classical) perspective: modal logic as equational logic. The fascination of modal logic ultimately stems from the (still not fully understood) links between these perspectives.

What this book is about

This book is a course in modal logic, intended for both novices and more experienced readers, that presents modal logic as a powerful and flexible tool for working with relational structures. It provides a thorough grounding in the basic relational perspective on modal logic, and applies this perspective to issues in completeness, computability, and complexity. In addition, it introduces and develops in some detail the perspectives provided by correspondence theory and algebra.

This much is predictable from our earlier discussion. However, three additional desiderata have helped shape the book. First, we have attempted to emphasize the *flexibility* of modal logic as a tool for working with relational structures. One still encounters with annoying frequency the view that modal logic amounts to rather simple-minded uses of two operators $\Diamond$ and $\Box$. This view has been out of date at least since the late 1960s (say, since Hans Kamp's expressive completeness result for since/until logic, to give a significant, if arbitrary, example), and in view of such developments as propositional dynamic logic and arrow logic it is now hopelessly anachronistic and unhelpful. We strongly advocate a liberal attitude in this book: we switch freely between various modal languages and in the final chapter we introduce a variety of further 'upgrades.' And as far as we are concerned, *it is all just modal logic*.

Second, two pedagogic goals have shaped the writing and selection of material: we want to explicate a range of *proof techniques* which we feel are significant and worth mastering, and, where appropriate, we want to draw attention to some important *general results*. These goals are pursued fairly single mindedly: on occasion, a single result may be proved by a variety of methods, and every chapter (except the following one) proves at least one very general and (we hope) very interesting result. The reader looking for a catalogue of facts about his or her favorite modal system probably will not find it here. But such a reader may well find the technique needed to algebraize it, to analyze its expressive power, to prove a completeness result, or to establish its decidability or undecidability – and may even discover that the relevant results are a special case of something known.

Finally, contemporary modal logic is profoundly influenced by its applications, particularly in theoretical computer science. Indeed, some of the most interesting advances in the subject (for example, the development of propositional dynamic logic, and the investigation of modal logic from a complexity-theoretic standpoint) were largely due to computer scientists, not modal logicians. Such influences must be acknowledged and incorporated, and we attempt to do so.

What this book is not about

Modal logic is a broad field, and inevitably we have had to leave out a lot of interesting material, indeed whole areas of active research. There are two principle omissions: there is no discussion of first-order modal systems or of non-Hilbert-style proof theory and automated reasoning techniques.

The first omission is relatively easy to justify. First-order modal logic is an enterprise quite distinct from the study of propositional systems: its principle concern is how best to graft together classical logic and propositional modal logic. It is an interesting field, and one in which there is much current activity, but its concerns lie outside the scope of this book.

The omission of proof theory and automated reasoning techniques calls for a little more explanation. A considerable proportion of this book is devoted to completeness theory and its algebraic ramifications; however, as is often the case in modal logic, the proof systems discussed are basically Hilbert-style axiomatic systems. There is no discussion of natural deduction, sequent calculi, labeled deductive systems, resolution, or display calculi. A (rather abstract) tableaux system is used once, but only as a tool to prove a complexity result. In short, there is little in this book that a proof theorist would regard as real proof theory, and nothing on implementation. Why is this? Essentially because modal proof theory and automated reasoning are still relatively youthful enterprises; they are exciting and active fields, but as yet there is little consensus about methods and few general results. Moreover, these fields are moving fast; much that is currently regarded as state of the art is likely to go rapidly out of date. For these reasons we have decided – rather reluctantly – not to discuss these topics.

In addition to these major areas, there are a host of more local omissions. One is provability and interpretability logic. While these are fascinating examples of how modal logical ideas can be applied in mathematics, the principle interest of these fields is not modal logic itself (which is simply used as a tool) but the formal study of arithmetic: a typical introduction to these topics (and several excellent ones exist, for example Boolos [68, 69], and Smoryński [416]) is typically about ten percent modal and ninety percent arithmetical. A second omission is a topic that is a traditional favorite of modal logicians: the fine structure of the lattice of normal modal logics in the basic $\Diamond$ and $\Box$ language; we confine ourselves in this book to the relatively easy case of logics extending **S4.3**. The reader interested in learning more about this type of work should consult Bull and Segerberg [75] or Chagrov and Zakharyaschev [88]. Other omissions we regret include: a discussion of meta-logical properties such as interpolation, a detailed examination of local versus global consequence, and an introduction to the modal μ-calculus and model checking. Restrictions of space and time made their inclusion impossible.

Audience and prerequisites

The book is aimed at people who use or study modal logic, and more generally, at people working with relational structures. We hope that the book will be of use to two distinct audiences: a less experienced audience, consisting of students of logic, computer science, artificial intelligence, philosophy, linguistics, and other fields where modal logic and relational structures are of importance, and a more experienced audience consisting of colleagues working in one or more of the above research areas who would like to learn and apply modal logic in their own area. To this end, there are two distinct tracks through this book: the basic track (this consists of selected sections from each chapter, and will be described shortly) and an advanced track (that is, the entire book).

The book starts at the beginning, and does not presuppose prior acquaintance with *modal* logic; but, even on the basic track, prior acquaintance with first-order logic and its semantics is essential. Furthermore, the development is essentially mathematical and assumes that the reader is comfortable with such things as sets, functions, relations and so on, and can follow mathematical argumentation, such as proofs by induction. In addition, although we have tried to make the basic track material as self-contained as possible, two of the later chapters probably require a little more background knowledge than this. In particular, a reader who has never encountered boolean (or some other) algebras before is likely to find Chapter 5 hard going, and the reader who has never encountered the concept of computable and uncomputable problems will find Chapter 6 demanding. That said, only a relatively modest background knowledge in these areas is required to follow the basic track material; certainly the main thrust of the development should be clear. The requisite background material in logic, algebra and computability can be found in Appendices A, B, and C.

Needless to say, we have also tried to make the advanced track material as readable and understandable as possible. However, largely because of the different kinds of background knowledge required in different places, advanced track readers may sometimes need to supplement this book with a background reading in model theory, universal algebra or computational complexity. Again, the required material is sketched in the appendices.

Contents

The chapter-by-chapter breakdown of the material is as follows.

Chapter 1. Basic Concepts. This chapter introduces a number of key modal languages (namely the *basic modal language, modal languages of arbitrary similarity type*, the *basic temporal language*, the language of *propositional dynamic logic*, and *arrow languages*), and shows how they are interpreted on various kinds of relational structures (namely *models, frames* and *general frames*). It also establishes notation, discusses some basic concepts such as *satisfaction, validity, logical consequence* and *normal modal logics*, and places them in historical perspective. The entire chapter is essentially introductory; all sections lie on the basic track.

Chapter 2. Models. This chapter examines modal languages as tools for talking about models. In the first five sections we prove some basic *invariance results*, introduce *bisimulations*, discuss the use of *finite models*, and, by describing the *standard translation*, initiate the study of *correspondence theory*. All five sections are fundamental to later developments – indeed the sections on bisimulations and the standard translation are among the most important in the entire book – and together they constitute the basic track selection. The remaining two sections are on the advanced track. They probe the expressive power of modal languages using ultrafilter extensions, ultraproducts, and saturated models; establish the fundamental role of bisimulations in correspondence theory; and introduce the concepts of simulation and safety.

Chapter 3. Frames. This chapter examines modal languages as tools for talking about frames; all sections, save the very last, lie on the basic track. The first three sections develop the basic theory of frame correspondence: we give examples of frame definability, show that relatively simple modal formulas can define frame conditions beyond the reach of any first-order formula (and explain why this happens), and introduce the concepts needed to state the celebrated *Goldblatt-Thomason* Theorem. After a short fourth section which discusses finite frames, we embark on the study of the *Sahlqvist fragment*. This is a large class of formulas, each of which corresponds to a first-order frame condition, and we devote three sections to it. In the final (advanced) section we introduce some further frame constructions and prove the Goldblatt-Thomason Theorem model theoretically.

Chapter 4. Completeness. This chapter has two parts; all sections, save the very last, lie on the basic track. The first part, consisting of the first four sections, is an introduction to basic completeness theory (including *canonical models*, *completeness-via-canonicity* proofs, *canonicity failure*, and *incompleteness*). The second part is a survey of methods that can be used to show completeness when canonicity fails. We discuss *transformation methods*, the *step-by-step* technique, the use of *rules for the undefinable*, and devote the final two sections to a discussion of *finitary methods*. The first of these sections proves the completeness of Propositional Dynamic Logic (**PDL**). The second (the only section on the advanced track) examines extensions of **S4.3**, proving (among other things) Bull's Theorem.

Chapter 5. Algebras and General Frames. The first three sections lie on the basic track: we discuss the role of algebra in logic, show how algebraic ideas can be applied to modal logic via *boolean algebras with operators*, and then prove the fundamental *Jónsson-Tarski Theorem*. With the basics thus laid we turn to *duality theory*, which soon leads us to an algebraic proof of the *Goldblatt-Thomason* Theorem (which was proved model theoretically in Chapter 3). In the two remaining sections (which lie on the advanced track) we discuss *general frames* from an algebraic perspective, introduce the concept of *persistence* (a generalization of the idea of canonicity) and use it to prove the *Sahlqvist Completeness Theorem*, the completeness-theoretic twin of the correspondence result proved in Chapter 3.

Chapter 6. Computability and Complexity. This chapter has two main parts. The first, comprising the first five sections, is an introduction to decidability and undecidability in modal logic. We introduce the basic ideas involved in computing modal satisfiability and validity problems, and then discuss three ways of proving decidability results: the use of *finite models*, the method of *interpretations*, and the use of *quasi-models* and *mosaics*. The fifth section gives two simple examples which illustrate how easily undecidable – and indeed, highly undecidable – modal logics can arise. All of the first part lies on the basic track. The remaining three sections examine modal logic from the perspective of computational complexity. In particular, the modal relevance of three central complexity classes (NP, PSPACE, and EXPTIME) is discussed in some detail. We pay particular attention to PSPACE, proving Ladner's general PSPACE-hardness result in detail. These sections lie on the advanced track, but this is partly because computational complexity is likely to be a new subject for some readers. The material is elegant and interesting, and we have tried to make these sections as self-contained and accessible as possible.

Chapter 7. Extended Modal Logic. This chapter has a quite different flavor from the others: it is essentially the party at the end of the book in which we talk about some of our favorite examples of extended modal systems. We will not offer any advice about what to read here – simply pick and choose and enjoy. The topics covered are: boosting the expressive power of modal languages with the aid of *logical modalities*, completeness-via-completeness proofs in *since/until logic*, naming states with the help of *hybrid logics*, and performing evaluation at sequences of states in *multi-dimensional modal logic*. We also show how to export modal ideas back to first-order logic by defining the *guarded fragment*, and conclude by proving a *Lindström Theorem* for modal logic.

Nearly all sections end with exercises. Each chapter starts with a chapter guide outlining the main themes of the sections that follow. Moreover, each chapter finishes with a summary, and – except the first – with a section entitled Notes. These give references for results discussed in the text. (In general we do not attribute results in the text, though where a name has become firmly attached – for example, Bull's Theorem or Lindenbaum's Lemma – we use it.) The Notes also give pointers to relevant work not covered in the text. The final section of Chapter 1 sketches the history of modal logic, and Appendix D gives a brief guide to textbooks, survey articles, and other material on modal logic.

Teaching the book

The book can be used as the basis for a number of different courses. Here are some suggestions.

Modal Logic and Relational Structures. (1 Semester, 2 hours a week)

All of Chapter 1, all the basic track sections in Chapter 2, and all the basic track sections in Chapter 3. This course introduces modal logic from a semantically oriented perspective. It is not particularly technical (in fact, only Section 2.5 is likely to cause any difficulties), and the student will come away with an appreciation of what modal languages are and the kind of expressivity they offer. It is deliberately one-sided – it is intended as an antidote to traditional introductions.

An Introduction to Modal Logic. (1 Semester, 4 hours a week)
All of Chapter 1, all the basic track material in Chapter 2, the first six or seven sections of Chapter 3, the first six or seven sections of Chapter 4, and the first four sections of Chapter 6. In essence, this course adds to the previous one the contents of a traditional introduction to modal logic (namely completeness-via-canonical models, and decidability-via-filtrations) and includes extra material on decidability which we believe *should* become traditional. This course gives a useful and fairly balanced picture of many aspects of modern modal logic.

Modal Logic for Computer Scientists. (1 Semester, 4 hours a week)
All of Chapter 1, the first four sections of Chapter 2, the first four sections of Chapter 3, the first four sections of Chapter 4 plus Section 4.8 (completeness of **PDL**), all of Chapter 6, and a selection of topics from Chapter 7. In our opinion, this course is more valuable than the previous one, and in spite of its title it is *not* just for computer science students. This course teaches basic notions of modal expressivity (bisimulation, the standard translation, and frame definability), key ideas on completeness (including incompleteness), covers both computability and complexity, and will give the student an impression of the wide variety of options available in modern modal logic. It comes close to our ideal of what a modern, well-rounded introduction to modal logic should look like.

Mathematical Aspects of Modal Logic. (1 Semester, 4 hours a week)
Chapters 1, 2, and 3, the first four sections of Chapter 4, and all of Chapter 5. If you are teaching logicians, this is probably the course to offer. It is a demanding course, and requires background knowledge in both model theory and algebra, but we think that students with this background will like the way the story unfolds.

Modal Logic. (2 Semesters, 4 hours a week)
But of course, there is another option: teach the whole book. Given enough background knowledge and commitment, this *is* do-able in 2 semesters. Though we should confess right away that the course's title is *highly* misleading: once you get to the end of the book, you will discover that far from having learned everything about modal logic, you have merely arrived at the beginning of an unending journey.

Hopefully these suggestions will spark further ideas. There is a lot of material here, and by mixing and matching, perhaps combined with judicious use of other sources (see Appendix D, the Guide to the Literature, for some suggestions), the instructor should be able to tailor courses for most needs. The dependency diagram (see Figure 1) will help your planning.

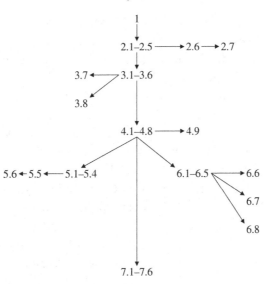

Fig. 1. Dependency Diagram

Electronic support

We have set up a home page for this book, where we welcome feedback, and where we will make selected solutions to the exercises and teaching materials available, as well as any corrections that may need to be made. The URL is

```
http://www.mlbook.org
```

Acknowledgments

We want to thank the following colleagues for their helpful comments and useful suggestions: Carlos Areces, Johan van Benthem, Giacomo Bonanno, Henry Chinaski, Jan van Eijck, Joeri Engelfriet, Paul Gochet, Rob Goldblatt, Val Goranko, Ramon Jansana, Theo Janssen, Tim Klinger, Johan W. Klüwer, Alexander Kurz, Holger Schlingloff, Moshe Vardi, and Rineke Verbrugge. Special thanks are due to Maarten Marx who worked through earlier incarnations of Chapter 6 in great detail; his comments transformed the chapter. We are grateful for the detailed comments Ian Hodkinson made on a later version of this chapter. We are extremely grateful to Costas Koutras for his extensive comments based on his experience of teaching the book.

We had the good fortune of being able to try out (parts of) the material on students in Amsterdam, Barcelona, Braga, Budapest, Cape Town, Chiba, Chia-Yi, Johannesburg, Lisbon, Saarbrücken, Utrecht and Warwick. We want to thank all our students, and in particular Maarten Stol, and the students of the University of Amsterdam classes on modal logic of the years 1999–2001.

We would like to thank our editor, David Tranah, for his support and advice, and the anonymous referees for their valuable feedback. We are also extremely grateful to our copy editor, who displayed amazing attention to detail in what (as we have discovered) is a very difficult task indeed.

We began the book when we were employed by the Netherlands Organization for Scientific Research (NWO), project 102/62-356 'Structural and Semantic Parallels in Natural Languages and Programming Languages.' We are grateful for the financial support by NWO. During the later stages of the writing we received support from a variety of sources, for which we are extremely grateful. Patrick Blackburn was based at the Department of Computational Linguistics at the University of Saarland. Maarten de Rijke was supported by the Spinoza project 'Logic in Action' at ILLC, the University of Amsterdam, and by NWO under project numbers 612-13-001, 365-20-005, 612.069.006, 612.000.106, 612.000.207, and 612.066.302. And Yde Venema was supported by a fellowship of the Royal Netherlands Academy of Arts and Sciences, and later, also by the Spinoza project 'Logic in Action.' We also want to thank the Department of Mathematics and Computer Science of the Free University in Amsterdam for the facilities they provided.

Concerning the second paperback printing

We made small corrections throughout the book, and more extensive changes to Example 3.57 and Section 6.4, but the page numbering is essentially unaltered. We are grateful to everyone who pointed out typos and errors in the first printing: Loredana Afanasiev, Fokko van de Bult, Seth Cable, Henry Chinaski, Rajeev Goré, Helle Hansen, Philip Hölzenspies, Tanja Hötte, Dick de Jongh, Suvi Karvonen, Clemens Kupke, Thomas Müller, Joshua Sacks, Thomas Schneider, Jerry Seligman, Dmitry Shkatov, and Zhou Chunlai. Special thanks are due to Bernhard Heinemann who went through the book very carefully and sent us long lists of typos.

Patrick Blackburn
Maarten de Rijke
Yde Venema

1

Basic Concepts

Languages of propositional modal logic are propositional languages to which sentential operators (usually called *modalities* or *modal operators*) have been added. In spite of their syntactic simplicity, such languages turn out to be useful tools for describing and reasoning about *relational structures*. A relational structure is a non-empty set on which a number of relations have been defined; they are widespread in mathematics, computer science, artificial intelligence and linguistics, and are also used to interpret first-order languages.

Now, when working with relational structures we are often interested in structures possessing certain properties. Perhaps a certain transitive binary relation is particularly important. Or perhaps we are interested in applications where 'dead ends,' 'loops,' and 'forkings' are crucial, or where each relation is a partial function. Wherever our interests lie, modal languages can be useful, for modal operators are essentially a simple way of accessing the information contained in relational structures. As we will see, the *local* and *internal* access method that modalities offer is strong enough to describe, constrain, and reason about many interesting and important aspects of relational structures.

Much of this book is essentially an exploration and elaboration of these remarks. The present chapter introduces the concepts and terminology we will need, and the concluding section places them in historical context.

Chapter guide

Section 1.1: Relational Structures. Relational structures are defined, and a number of examples are given.

Section 1.2: Modal Languages. We define the basic modal language and some of its extensions.

Section 1.3: Models and Frames. Here we link modal languages and relational structures. In fact, we introduce *two* levels at which modal languages can be used to talk about structures: the level of *models* (which we explore in Chapter 2) and the level of *frames* (which is examined in Chapter 3). This section contains the fundamental *satisfaction definition*, and defines the key logical notion of *validity*.

Section 1.4: General Frames. In this section we link modal languages and relational structures in yet another way: via *general frames*. Roughly speaking, general frames provide a third level at which modal languages can be used to talk about relational structures, a level intermediate between those provided by models and frames. We will make heavy use of general frames in Chapter 5.

Section 1.5: Modal Consequence Relations. Which conclusions do we wish to draw from a given set of modal premises? That is, which *consequence relations* are appropriate for modal languages? We opt for a *local* consequence relation, though we note that there is a *global* alternative.

Section 1.6: Normal Modal Logics. Both validity and local consequence are defined *semantically* (that is, in terms of relational structures). However, we want to be able to generate validities and draw conclusions *syntactically*. We take our first steps in modal proof theory and introduce Hilbert-style axiom systems for modal reasoning. This motivates a concept of central importance in Chapters 4 and 5: *normal modal logics*.

Section 1.7: Historical Overview. The ideas introduced in this chapter have a long and interesting history. Some knowledge of this will make it easier to understand developments in subsequent chapters, so we conclude with a historical overview that highlights a number of key themes.

1.1 Relational Structures

Definition 1.1 A *relational structure* is a tuple $\mathfrak{F}$ whose first component is a nonempty set W called the *universe* (or *domain*) of $\mathfrak{F}$, and whose remaining components are relations on W. We assume that every relational structure contains at least one relation. The elements of W have a variety of names in this book, including: *points*, *states*, *nodes*, *worlds*, *times*, *instants* and *situations*.   ⊣

An attractive feature of relational structures is that we can often display them as simple pictures, as the following examples show.

Example 1.2 *Strict partial orders* (SPOs) are an important type of relational structure. A *strict partial order* is a pair (W, R) such that R is *irreflexive* ($\forall x \, \neg Rxx$) and *transitive* ($\forall xyz \, (Rxy \wedge Ryz \rightarrow Rxz)$). A strict partial order R is a *linear order* (or a *total order*) if it also satisfies the *trichotomy* condition: $\forall xy \, (Rxy \vee x = y \vee Ryx)$.

An example of an SPO is given in Figure 1.1, where $W = \{1, 2, 3, 4, 6, 8, 12, 24\}$ and Rxy means 'x and y are different, and y can be divided by x.' Obviously this is

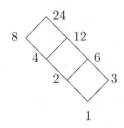

Fig. 1.1. A strict partial order

not a linear order. On the other hand, if we define Rxy by 'x is numerically smaller than y,' we obtain a linear order over the same universe W. Important examples of linear orders are $(\mathbb{N}, <)$, $(\mathbb{Z}, <)$, $(\mathbb{Q}, <)$ and $(\mathbb{R}, <)$, the *natural numbers, integers, rationals* and *reals* in their usual order. We sometimes use the notation $(\omega, <)$ for $(\mathbb{N}, <)$.

In many applications we want to work not with *strict* partial orders, but with plain old *partial orders* (POs). We can think of a partial order as the reflexive closure of a strict partial order; that is, if R is a strict partial order on W, then $R \cup \{(u, u) \mid u \in W\}$ is a partial order (for more on reflexive closures, see Exercise 1.1.3). Thus partial orders are transitive, *reflexive* ($\forall x\, Rxx$) and *antisymmetric* ($\forall xy\,(Rxy \wedge Ryx \rightarrow x = y)$). If a partial order is *connected* ($\forall xy\,(Rxy \vee Ryx)$) it is called a *reflexive linear order* (or a *reflexive total order*).

If we interpret the relation in Figure 1.1 reflexively (that is, if we take Rxy to mean 'x and y are equal, or y can be divided by x') we have a simple example of a partial order. Obviously, it is not a reflexive *linear* order. Important examples of reflexive linear orders include $(\mathbb{N}, \leq)$ (or $(\omega, \leq)$), $(\mathbb{Z}, \leq)$, $(\mathbb{Q}, \leq)$ and $(\mathbb{R}, \leq)$, the *natural numbers, integers, rationals* and *reals* under their respective 'less-than-or-equal-to' orderings. ⊣

Example 1.3 *Labeled Transition Systems* (LTSs), or more simply, *transition systems*, are a simple kind of relational structure widely used in computer science. An LTS is a pair $(W, \{R_a \mid a \in A\})$ where W is a non-empty set of states, A is a non-empty set (of *labels*), and for each $a \in A$, $R_a \subseteq W \times W$. Transition systems can be viewed as an abstract model of computation: the states are the possible states of a computer, the labels stand for programs, and $(u, v) \in R_a$ means that there is an execution of the program a that starts in state u and terminates in state v. It is natural to depict states as nodes and transitions R_a as directed arrows.

In Figure 1.2 a transition system with states w_1, w_2, w_3, w_4 and labels a, b, c is shown. Formally, $R_a = \{(w_1, w_2), (w_4, w_4)\}$, while $R_b = \{(w_2, w_3)\}$ and $R_c = \{(w_4, w_3)\}$. This transition system is actually rather special, for it is *deterministic*: if we are in a state where it is possible to make one of the three possible kinds of

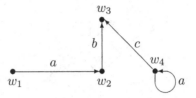

Fig. 1.2. A deterministic transition system

transition (for example, an a transition) then it is fixed which state that transition will take us to. In short, the relations R_a, R_b and R_c are all *partial functions*.

Deterministic transition systems are important, but in theoretical computer science it is more usual to take *non-deterministic* transition systems as the basic model of computation. A non-deterministic transition system is one in which the state we reach by making a particular kind of transition from a given state need not be fixed. That is, the transition relations do not have to be partial functions, but can be arbitrary relations.

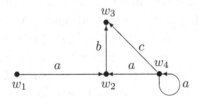

Fig. 1.3. A non-deterministic transition system

In Figure 1.3 a non-deterministic transition system is shown: a is now a non-deterministic program, for if we execute it in state w_4 there are two possibilities: either we loop back into w_4, or we move to w_2.

Transition systems play an important role in this book. This is not so much because of their computational interpretation (though that is interesting) but because of their sheer ubiquity. Sets equipped with collections of binary relations are one of the simplest types of mathematical structures imaginable, and they crop up just about everywhere.   ⊣

Example 1.4 For our next example we turn to the branch of artificial intelligence called knowledge representation. A central concern of knowledge representation is objects, their properties, their relations to other objects, and the conclusions one can draw about them. For example, Figure 1.4 represents some of the ways Mike relates to his surroundings.

One conclusion that can be drawn from this representation is that Sue has children. Others are not so clear. For example, does Mike love Sue, and does he

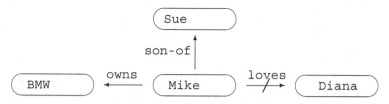

Fig. 1.4. Mike and others

love his BMW? Assuming that absence of a `not_loves` arc (like that connecting the Mike and the Diana nodes) means that the loves relation holds, this is a safe conclusion to draw. There are often such 'gaps' between pictures and relational structures, and to fill them correctly (that is, to know which relational structure the picture corresponds to) we have to know which diagrammatic conventions are being assumed.

Let us take the picture at face value. It gives us a set {BMW, Sue, Mike, Diana} together with binary relations `son-of`, `owns`, and `not_loves`. So we have here another labeled transition system. ⊣

Example 1.5 Finite trees are ubiquitous in linguistics. For example, the tree depicted in Figure 1.5 represents some simple facts about phrase-structure, namely that a sentence (S) can consist of a noun phrase (NP) and a verb phrase (VP); an NP can consist of a proper noun (PN); and VPs can consist of a transitive verb (TV) and an NP. ⊣

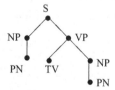

Fig. 1.5. A finite decorated tree

Trees play an important role in this book, so we will take this opportunity to define them. We first introduce the following important concepts.

Definition 1.6 Let W be a non-empty set and R a binary relation on W. Then R^+, the *transitive closure* of R, is the smallest transitive relation on W that contains R. That is,

$$R^+ = \bigcap \{R' \mid R' \text{ is a transitive binary relation on } W \ \& \ R \subseteq R'\}.$$

Furthermore, R^*, the *reflexive transitive closure* of R, is the smallest reflexive and transitive relation on W containing R. That is,

$$R^* = \bigcap \{R' \mid R' \text{ is a reflexive transitive binary relation on } W \ \& \ R \subseteq R'\}. \qquad \dashv$$

Note that R^+uv holds if and only if there is a sequence of elements $u = w_0, w_1, \ldots, w_n = v \ (n > 0)$ from W such that for each $i < n$ we have Rw_iw_{i+1}. That is, R^+uv means that v is reachable from u in a finite number of R-steps. Thus transitive closure is a natural and useful notion; see Exercise 1.1.3.

With these concepts at our disposal, it is easy to say what a tree is.

Definition 1.7 A *tree* $\mathfrak{T}$ is a relational structure (T, S) where:

(i) T, the set of nodes, contains a unique $r \in T$ (called the *root*) such that $\forall t \in T \ S^*rt$.

(ii) Every element of T distinct from r has a unique S-predecessor; that is, for every $t \neq r$ there is a unique $t' \in T$ such that $St't$.

(iii) S is *acyclic*; that is, $\forall t \neg S^+tt$. (It follows that S is irreflexive.) $\qquad \dashv$

Clearly, Figure 1.5 contains enough information to give us a tree (T, S) in the sense just defined: the nodes in T are the displayed points, and the relation S is indicated by means of a straight line segment drawn from a node to a node immediately below (that is, S is the obvious *successor* or *daughter-of* relation). The root of the tree is the topmost node (the one labeled S).

But the diagram also illustrates something else: often we need to work with structures consisting of not only a tree (T, S), but a whole lot else besides. For example, linguists would not be particularly interested in the bare tree (T, S) just defined, rather they would be interested in (at least) the structure

$$(T, S, \text{LEFT-OF}, S, \text{NP}, \text{VP}, \text{PN}, \text{TV}).$$

Here S, NP, VP, PN, and TV are *unary* relations on T (note that S and S are distinct symbols). These relations record the information attached to each node, namely the fact that some nodes are noun phrase nodes, while others are proper name nodes, sentential nodes, and so on. LEFT-OF is a binary relation which captures the left-to-right aspect of the above picture; the fact that the NP node is to the left of the VP node might be linguistically crucial.

Similar things happen in mathematical contexts. Sometimes we will need to work with relational structures which are much richer than the simple trees (T, S) just defined, but which, perhaps in an implicit form, contain a relation with all the properties required of S. It is useful to have a general term for such structures; we will call them *tree-like*. A formal definition here would do more harm than good, but in the text we will indicate, whenever we call a structure tree-like, where this implicit tree (T, S) can be found. That is, unless it is obvious, we will say *which* definable relation in the structure satisfies the conditions of Definition 1.7. One of the most important examples of tree-like structures is the Rabin structure, which we will meet in Section 6.3.

One often encounters the notion of a tree defined in terms of the (reflexive) transitive closure of the successor relation. Such trees we call *(reflexive and) transitive trees*, and they are dealt with in Exercises 1.1.4 and 1.1.5

Example 1.8 We have already seen that labeled transition systems can be regarded as a simple model of computation. Indeed, they can be thought of as models for practically any dynamic notion: each transition takes us from an input state to an output state. But this treatment of states and transitions is rather unbalanced: it is clear that transitions are second-class citizens. For example, if we talked about LTSs using a first-order language, we could not name transitions using constants (they would be talked about using relation symbols) but we could have constants for states. But there is a way to treat transitions as first-class citizens: we can work with *arrow structures*.

The *objects* of an arrow structure are things that can be pictured as *arrows*. As concrete examples, the mathematically inclined reader might think of vectors, or functions or morphisms in some category; the computer scientist of programs; the linguist of the context changing potential of a grammatically well-formed piece of text or discourse; the philosopher of some agent's cognitive actions; and so on. But note well: although arrows are the prime citizens of arrow structures, this does not mean that they should always be thought of as *primitive* entities. For example, in a *two-dimensional arrow structure*, an arrow a is thought of as a *pair* (a_0, a_1) of which a_0 represents the starting point of a, and a_1 its endpoint.

Having 'defined' the elements of arrow structures to be objects graphically representable as arrows, we should now ask: what are the basic *relations* which hold between arrows? The most obvious candidate is *composition*: vector spaces have an additive structure, functions can be composed, language fragments can be concatenated, and so on. So the central relation on arrows will be a ternary *composition relation* C, where $Cabc$ says that arrow a is the outcome of composing arrow b with arrow c (or conversely, that a can be decomposed into b and c). Note that in many concrete examples, C is actually a (partial) function; for example, in the two-dimensional framework we have

$$Cabc \text{ iff } a_0 = b_0, \ a_1 = c_1 \text{ and } b_1 = c_0. \tag{1.1}$$

What next? Well, in all the examples listed, the composition function has a neutral element; think of the identity function or the SKIP-program. So, arrow structures will contain degenerate arrows, transitions that do not lead to a different state. Formally, this means that arrow structures will contain a designated subset I of *identity arrows*; in the pair-representation, I will be (a subset of) the diagonal:

$$Ia \text{ iff } a_0 = a_1. \tag{1.2}$$

Another natural relation is *converse*. In linguistics and cognitive science we might view this as an 'undo' action (perhaps we have made a mistake and need to recover) and in many fields of mathematics arrow-like objects have converses (vectors) or inverses (bijective functions). So we will also give arrow structures a binary *reverse relation R*. Again, in many cases this relation will be a partial function. For example, in the two-dimensional picture, R is given by

$$Rab \text{ iff } a_0 = b_1 \text{ and } a_1 = b_0. \tag{1.3}$$

Although there are further natural candidates for arrow relations (notably some notion of *iteration*) we will leave it at this. And now for the formal definition: an *arrow frame* is a quadruple $\mathfrak{F} = (W, C, R, I)$ such that C, R and I are a ternary, a binary and a unary relation on W, respectively. Pictorially, we can think of them as follows:

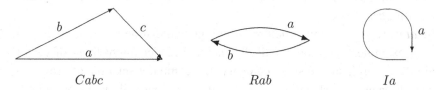

$$Cabc \qquad\qquad\qquad Rab \qquad\qquad\qquad Ia$$

The two-dimensional arrow structure, in which the universe consists of all pairs over the set U (and the relations C, R and I are given by (1.1), (1.3) and (1.2), respectively) is called the *square over U*, notation: $\mathfrak{S}_U$. The square arrow frame over U can be pictorially represented as a full graph over U: each arrow object (a_0, a_1) in $\mathfrak{S}_U$ can be represented as a 'real' arrow from a_0 to a_1; the relations are as pictured above. Alternatively, square arrow frames can be represented two-dimensionally; see the pictures in Example 1.27.   ⊣

Exercises for Section 1.1

1.1.1 Let (W, R) be a *quasi-order*; that is, assume that R is transitive and reflexive. Define the binary relation $\sim$ on W by putting $s \sim t$ iff Rst and Rts.

(a) Show that $\sim$ is an equivalence relation.

Let $[s]$ denote the equivalence class of s under this relation, and define the following relation on the collection of equivalence classes: $[s] \leq [t]$ iff Rst.

(b) Show that this is well defined.
(c) Show that $\leq$ is a partial order.

1.1.2 Let R be a transitive relation on a finite set W. Prove that R is well-founded iff R is irreflexive. (R is called *well-founded* if there are no infinite paths ... $Rs_2 Rs_1 Rs_0$.)

1.1.3 Let R be a binary relation on W. In Example 1.2 we defined the reflexive closure of R to be $R \cup \{(u, u) \mid u \in W\}$. But we can also give a definition analogous to those of R^+ and R^* in Definition 1.6, namely that it is the smallest reflexive relation on W that contains R:

$$R^{\mathbf{r}} = \bigcap \{R' \mid R' \text{ is a reflexive binary relation on } W \ \& \ R \subseteq R'\}.$$

Explain why this new definition (and the definitions of R^+ and R^*) are well defined. Show the equivalence of the two definitions of reflexive closure. Finally, show that R^+uv if and only if there is a sequence of elements $u = w_0, w_1, \ldots, w_n = v$ from W such that for each $i < n$ we have Rw_iw_{i+1}, and give an analogous sequence-based definition of *reflexive transitive closure*.

1.1.4 A *transitive tree* is an SPO $(T, <)$ such that (i) there is a *root* $r \in T$ satisfying $r < t$ for all $t \in T$ such that $r \neq t$, and (ii) for each $t \in T$, the set $\{s \in T \mid s < t\}$ of predecessors of t is finite and linearly ordered by $<$.

    (a) Prove that if (T, S) is a tree then (T, S^+) is a transitive tree.
    (b) Prove that $(T, <)$ is a transitive tree iff $(T, S_<)$ is a tree, where $S_<$ is the immediate successor relation given by $sS_<t$ iff $s < t$ and $s < v < t$ for no $v \in T$.
    (c) Under which conditions does the converse of (a) hold?

1.1.5 Define the notion of a reflexive and transitive tree, such that if (T, S) is a tree then (T, S^*) is a reflexive and transitive tree.

1.1.6 Show that the following formulas hold on square arrow frames:

    (a) $\forall xy \, (Rxy \rightarrow Ryx)$,
    (b) $\forall xyz \, ((Cxyz \wedge Iz) \leftrightarrow x = y)$,
    (c) $\forall xx_1x_2x_3 \, (\exists y \, (Cxx_1y \wedge Cyx_2x_3) \ \leftrightarrow \ \exists z \, (Cxzx_3 \wedge Czx_1x_2))$.

1.2 Modal Languages

It is now time to meet the modal languages we will be working with. First, we introduce the *basic modal language*. We then define *modal languages of arbitrary similarity type*. Finally we examine the following extensions of the basic modal language in more detail: the *basic temporal language*, the language of *propositional dynamic logic*, and a language of *arrow logic*.

Definition 1.9 The *basic modal language* is defined using a set of *proposition letters* (or *proposition symbols* or *propositional variables*) Φ whose elements are usually denoted p, q, r, and so on, and a unary modal operator $\Diamond$ ('diamond'). The well-formed *formulas* ϕ of the basic modal language are given by the rule

$$\phi ::= p \mid \bot \mid \neg \phi \mid \psi \vee \phi \mid \Diamond \phi,$$

where p ranges over elements of Φ. This definition means that a formula is either a proposition letter, the propositional constant falsum ('bottom'), a negated formula, a disjunction of formulas, or a formula prefixed by a diamond.

    Just as the familiar first-order existential and universal quantifiers are duals to each other (in the sense that $\forall x \, \alpha \leftrightarrow \neg \exists x \, \neg \alpha$), we have a dual operator $\Box$ ('box') for our diamond which is defined by $\Box\phi := \neg\Diamond\neg\phi$. We also make use of the classical abbreviations for conjunction, implication, bi-implication and the constant true ('top'): $\phi\wedge\psi := \neg(\neg\phi\vee\neg\psi)$, $\phi \to \psi := \neg\phi\vee\psi$, $\phi \leftrightarrow \psi := (\phi \to \psi)\wedge(\psi \to \phi)$ and $\top := \neg\perp$.   ⊣

Although we generally assume that the set Φ of proposition letters is a countably infinite set $\{p_0, p_1, \ldots\}$, occasionally we need to make other assumptions. For instance, when we are after decidability results, it may be useful to stipulate that Φ is finite, while doing model theory or frame theory we may need uncountably infinite languages. This is why we take Φ as an explicit parameter when defining the set of modal formulas.

Example 1.10 Three readings of diamond and box have been extremely influential. First, $\Diamond\phi$ can be read as 'it is *possibly* the case that ϕ.' Under this reading, $\Box\phi$ means 'it is not possible that not ϕ,' that is, '*necessarily* ϕ,' and examples of formulas we would probably regard as correct principles include all instances of $\Box\phi \to \Diamond\phi$ ('whatever is necessary is possible') and all instances of $\phi \to \Diamond\phi$ ('whatever is, is possible'). The status of other formulas is harder to decide. Should $\phi \to \Box\Diamond\phi$ ('whatever is, is *necessarily* possible') be regarded as a general truth about necessity and possibility? Should $\Diamond\phi \to \Box\Diamond\phi$ ('whatever is possible, is necessarily possible')? Are any of these formulas linked by a modal notion of logical consequence, or are they independent claims about necessity and possibility? These are difficult (and historically important) questions. The relational semantics defined in the following section offers a simple and intuitively compelling framework in which to discuss them.

Second, in *epistemic logic* the basic modal language is used to reason about knowledge, though instead of writing $\Box\phi$ for 'the agent knows that ϕ' it is usual to write $K\phi$. Given that we are talking about knowledge (as opposed to, say, belief or rumor), it seems natural to view all instances of $K\phi \to \phi$ as true: if the agent really *knows* that ϕ, then ϕ must hold. On the other hand (assuming that the agent is not omniscient) we would regard $\phi \to K\phi$ as false. But the legitimacy of other principles is harder to judge (if an agent knows that ϕ, does she know that she knows it?). Again, a precise semantics brings clarity.

Third, in *provability logic* $\Box\phi$ is read as 'it is *provable* (in some arithmetical theory) that ϕ.' A central theme in provability logic is the search for a complete axiomatization of the provability principles that are valid for various arithmetical theories (such as Peano Arithmetic). The *Löb* formula $\Box(\Box p \to p) \to \Box p$ plays a key role here. The arithmetical ramifications of this formula lie outside the scope of the book, but in Chapters 3 and 4 we will explore its modal content.   ⊣

That is the basic modal language. Let us now generalize it. There are two obvious ways to do so. First, there seems no good reason to restrict ourselves to languages with only one diamond. Second, there seems no good reason to restrict ourselves to modalities that take only a single formula as argument. Thus the general modal languages we will now define may contain many modalities, of arbitrary arities.

Definition 1.11 A *modal similarity type* is a pair $\tau = (O, \rho)$ where O is a non-empty set, and ρ is a function $O \to \mathbb{N}$. The elements of O are called *modal operators*; we use $\triangle$ ('*triangle*'), $\triangle_0$, $\triangle_1$, ..., to denote elements of O. The function ρ assigns to each operator $\triangle \in O$ a finite *arity*, indicating the number of arguments $\triangle$ can be applied to.

In line with Definition 1.9, we often refer to *unary* triangles as *diamonds*, and denote them by $\Diamond_a$ or $\langle a \rangle$, where a is taken from some index set. We often assume that the arity of operators is known, and do not distinguish between τ and O. ⊣

Definition 1.12 A *modal language* $ML(\tau, \Phi)$ is built up using a modal similarity type $\tau = (O, \rho)$ and a set of proposition letters Φ. The set $Form(\tau, \Phi)$ of *modal formulas* over τ and Φ is given by the rule

$$\phi ::= p \mid \bot \mid \neg\phi \mid \phi_1 \lor \phi_2 \mid \triangle(\phi_1, \ldots, \phi_{\rho(\triangle)}),$$

where p ranges over elements of Φ. ⊣

The similarity type of the basic modal language is called τ_0. In the sequel we sometimes state results for modal languages of arbitrary similarity types, give the proof for similarity types with diamonds only, and leave the general case as an exercise. For binary modal operators, we often use infix notation; that is, we usually write $\phi\triangle\psi$ instead of $\triangle(\phi, \psi)$. One other thing: note that our definition permits *nullary modalities* (or *modal constants*), triangles that take no arguments at all. Such modalities can be useful – we will see a natural example when we discuss arrow logic – but they play a relatively minor role in this book. Syntactically (and indeed, semantically) they are rather like propositional variables; in fact, they are best thought of as propositional *constants*.

Definition 1.13 We now define dual operators for non-nullary triangles. For each $\triangle \in O$ the *dual* $\triangledown$ of $\triangle$ is defined as $\triangledown(\phi_1, \ldots, \phi_n) := \neg\triangle(\neg\phi_1, \ldots, \neg\phi_n)$. The dual of a triangle of arity at least 2 is called a *nabla*. As in the basic modal language, the dual of a diamond is called a *box*, and is written $\Box_a$ or $[a]$. ⊣

Three extensions of the basic modal language deserve special attention. Two of these, the *basic temporal language* and the language of *propositional dynamic logic* will be frequently used in subsequent chapters. The third is a simple language of *arrow logic*; it will provide us with a natural example of a binary modality.

Example 1.14 (The Basic Temporal Language) The basic temporal language is built using a set of unary operators $O = \{\langle F \rangle, \langle P \rangle\}$. The intended interpretation of a formula $\langle F \rangle \phi$ is 'ϕ will be true at some *F*uture time,' and the intended interpretation of $\langle P \rangle \phi$ is 'ϕ was true at some *P*ast time.' This language is called the *basic temporal language*, and it is the core language underlying a branch of modal logic called *temporal logic*. It is traditional to write $\langle F \rangle$ as F and $\langle P \rangle$ as P, and their duals are written as G and H, respectively. (The mnemonics here are: 'it is always *G*oing to be the case' and 'it always *H*as been the case.')

We can express many interesting assertions about time with this language. For example, $P\phi \rightarrow GP\phi$, says 'whatever has happened will always have happened,' and this seems a plausible candidate for a general truth about time. On the other hand, if we insist that $F\phi \rightarrow FF\phi$ must always be true, it shows that we are thinking of time as *dense*: between any two instants there is always a third. And if we insist that $GFp \rightarrow FGp$ (the *McKinsey formula*) is true, for all propositional symbols p, we are insisting that atomic information true somewhere in the future eventually settles down to being always true. (We might think of this as reflecting a 'thermodynamic' view of information distribution.)

One final remark: computer scientists will have noticed that the binary *until modality* is conspicuous by its absence. As we will see in the following chapter, the basic temporal language is *not* strong enough to express until. We examine a language containing the until operator in Section 7.2.   ⊣

Example 1.15 (Propositional Dynamic Logic) Another important branch of modal logic, again involving only unary modalities, is *propositional dynamic logic*. PDL, the language of propositional dynamic logic, has an infinite collection of diamonds. Each of these diamonds has the form $\langle \pi \rangle$, where π denotes a (nondeterministic) *program*. The intended interpretation of $\langle \pi \rangle \phi$ is 'some terminating execution of π from the present state leads to a state bearing the information ϕ.' The dual assertion $[\pi]\phi$ states that 'every execution of π from the present state leads to a state bearing the information ϕ.'

So far, there is nothing really new – but a simple idea is going to ensure that PDL is highly expressive: we will make the inductive structure of the programs explicit in PDL's syntax. Complex programs are built out of basic programs using some repertoire of program constructors. By using diamonds which reflect this structure, we obtain a powerful and flexible language.

Let us examine the core language of PDL. Suppose we have fixed some set of basic programs a, b, c, and so on (thus we have basic modalities $\langle a \rangle$, $\langle b \rangle$, $\langle c \rangle$, ... at our disposal). Then we are allowed to define complex programs π (and hence, modal operators $\langle \pi \rangle$) over this base as follows:

(*choice*) if π_1 and π_2 are programs, then so is $\pi_1 \cup \pi_2$.

    The program $\pi_1 \cup \pi_2$ (non-deterministically) executes π_1 or π_2.

(*composition*) if π_1 and π_2 are programs, then so is π_1 ; π_2.

    This program first executes π_1 and then π_2.

(*iteration*) if π is a program, then so is π^*.

    π^* is a program that executes π a finite (possibly zero) number of times.

For the collection of diamonds this means that if $\langle \pi_1 \rangle$ and $\langle \pi_2 \rangle$ are modal operators, then so are $\langle \pi_1 \cup \pi_2 \rangle$, $\langle \pi_1$; $\pi_2 \rangle$ and $\langle \pi_1^* \rangle$. This notation makes it straightforward to describe properties of program execution. Here is a fairly straightforward example. The formula $\langle \pi^* \rangle \phi \leftrightarrow \phi \vee \langle \pi ; \pi^* \rangle \phi$ says that a state bearing the information ϕ can be reached by executing π a finite number of times if and only if either we already have the information ϕ in the current state, or we can execute π once and then find a state bearing the information ϕ after finitely many more iterations of π. Here is a far more demanding example:

$$[\pi^*](\phi \rightarrow [\pi]\phi) \rightarrow (\phi \rightarrow [\pi^*]\phi).$$

This is *Segerberg's axiom* (or the *induction axiom*) and the reader should try working out what exactly it is that this formula says. We discuss this formula further in Chapter 3; see Example 3.10.

    If we confine ourselves to these three constructors (and in this book for the most part we do) we are working with a version of PDL called *regular* PDL. (This is because the three constructors are the ones used in Kleene's well-known analysis of regular programs.) However, a wide range of other constructors have been studied. Here are two:

(*intersection*) if π_1 and π_2 are programs, then so is $\pi_1 \cap \pi_2$.

    The intended meaning of $\pi_1 \cap \pi_2$ is: execute both π_1 and π_2, in parallel.

(*test*) if ϕ is a formula, then ϕ? is a program.

    This program tests whether ϕ holds, and if so, continues; if not, it fails.

To flesh this out a little, the intended reading of $\langle \pi_1 \cap \pi_2 \rangle \phi$ is that if we execute both π_1 and π_2 in the present state, then there is at least one state reachable by both programs which bears the information ϕ. This is a natural constructor for a variety of purposes, and we will make use of it in Section 6.5.

    The key point to note about the test constructor is its unusual syntax: it allows us to make a modality out of a formula. Intuitively, this modality accesses the *current* state if the current state satisfies ϕ. On its own such a constructor is uninteresting ($\langle \phi? \rangle \psi$ simply means $\phi \wedge \psi$). However, when other constructors are present, it can be used to build interesting programs. For example, $(p? ; a) \cup (\neg p? ; b)$ is 'if p then a else b.'

    Nothing prevents us from viewing the basic programs as *deterministic*, and we will discuss a fragment of deterministic PDL (DPDL) in Section 6.5. ⊣

Example 1.16 (An Arrow Language) The type $\tau_\rightarrow$ of *arrow logic* is a similarity type with modal operators other than diamonds. The language of arrow logic is designed to talk about the objects in arrow structures (entities which can be pictured as arrows). The well-formed formulas ϕ of the arrow language are given by the rule

$$\phi ::= p \mid \perp \mid \neg\phi \mid \phi \vee \psi \mid \phi \circ \psi \mid \otimes\phi \mid 1'.$$

That is, 1' ('identity') is a nullary modality (a modal constant), the 'converse' operator $\otimes$ is a diamond, and the 'composition' operator $\circ$ is a dyadic operator. Possible readings of these operators are:

1'	identity	'skip',
$\otimes\phi$	converse	'ϕ conversely',
$\phi \circ \psi$	composition	'first ϕ, then ψ'.

$\dashv$

Example 1.17 (Feature Logic and Description Logic) As we mentioned in the Preface, researchers developing formalisms for describing graphs have sometimes (without intending to) come up with notational variants of modal logic. For example, computational linguists use *Attribute-Value Matrices* (AVMs) for describing *feature structures* (directed acyclic graphs that encode linguistic information). Here is a fairly typical AVM:

$$\begin{bmatrix} \text{AGREEMENT} & \begin{bmatrix} \text{PERSON} & \textit{1st} \\ \text{NUMBER} & \textit{plural} \end{bmatrix} \\ \text{CASE} & \textit{dative} \end{bmatrix}.$$

But this is just a two dimensional notation for the following modal formula:

$$\langle\text{AGREEMENT}\rangle(\langle\text{PERSON}\rangle\textit{1st} \wedge \langle\text{NUMBER}\rangle\textit{plural}) \wedge \langle\text{CASE}\rangle\textit{dative}.$$

Similarly, researchers in artificial intelligence needing a notation for describing and reasoning about ontologies developed *description logic*. For example, the concept of 'being a hired killer for the mob' is true of any individual who is a killer and is employed by a gangster. In description logic we can define this concept as follows:

```
killer ⊓ ∃employer.gangster.
```

But this is simply the following modal formula lightly disguised:

```
killer ∧ ⟨employer⟩gangster.
```

It turns out that the links between modal logic on the one hand, and feature and description logic on the other, are far more interesting than these rather simple examples might suggest. A modal perspective on feature or description logic capable of accounting for other important aspects of these systems (such as the ability to talk about re-entrancy in feature structures, or to perform ABox reasoning in description logic) must make use of the kinds of extended modal logics discussed in

Chapter 7 (in particular, logics containing the global modality, and hybrid logics). Furthermore, some versions of feature and description logic make use of ideas from PDL, and description logic makes heavy use of *counting modalities* (which say such things as 'at most 3 transitions lead to a ϕ state'). ⊣

Substitution

Throughout this book we will be working with the syntactic notion of one formula being a substitution instance of another. In order to define this notion we first introduce the concept of a substitution as a function mapping proposition letters to formulas.

Definition 1.18 Suppose we are working a modal similarity type τ and a set Φ of proposition letters. A *substitution* is a map $\sigma : \Phi \to Form(\tau, \Phi)$.

Now such a substitution σ induces a map $(\cdot)^\sigma : Form(\tau, \Phi) \to Form(\tau, \Phi)$ which we can recursively define as follows:

$$
\begin{aligned}
\bot^\sigma &= \bot, \\
p^\sigma &= \sigma(p), \\
(\neg\psi)^\sigma &= \neg\psi^\sigma, \\
(\psi \vee \theta)^\sigma &= \psi^\sigma \vee \theta^\sigma, \\
(\triangle(\psi_1, \ldots, \psi_n))^\sigma &= \triangle(\psi_1^\sigma, \ldots, \psi_n^\sigma).
\end{aligned}
$$

This definition spells out exactly what is meant by carrying out *uniform substitution*. Finally, we say that χ is a *substitution instance* of ψ if there is some substitution τ such that $\psi^\tau = \chi$. ⊣

To give an example, if σ is the substitution that maps p to $p \wedge \Box q$, q to $\Diamond\Diamond q \vee r$ and leaves all other proposition letters untouched, then we have

$$(p \wedge q \wedge r)^\sigma = ((p \wedge \Box q) \wedge (\Diamond\Diamond q \vee r) \wedge r).$$

Exercises for Section 1.2

1.2.1 Using $K\phi$ to mean 'the agent knows that ϕ' and $M\phi$ to mean 'it is consistent with what the agent knows that ϕ,' represent the following statements:

(a) If ϕ is true, then it is consistent with what the agent knows that she knows that ϕ.
(b) If it is consistent with what the agent knows that ϕ, and it is consistent with what the agent knows that ψ, then it is consistent with what the agent knows that $\phi \wedge \psi$.
(c) If the agent knows that ϕ, then it is consistent with what the agent knows that ϕ.
(d) If it is consistent with what the agent knows that it is consistent with what the agent knows that ϕ, then it is consistent with what the agent knows that ϕ.

Which of these seem plausible principles concerning knowledge and consistency?

1.2.2 Suppose $\Diamond\phi$ is interpreted as 'ϕ is permissible'; how should $\Box\phi$ be understood? List formulas which seem plausible under this interpretation. Should the Löb formula $\Box(\Box p \to p) \to \Box p$ be on your list? Why?

1.2.3 Explain how the program constructs 'while ϕ do π' and 'repeat π until ϕ' can be expressed in PDL.

1.2.4 Consider the following arrow formulas. Do you think they should be always true?

$$
\begin{aligned}
1' \circ p &\leftrightarrow p, \\
\otimes(p \circ q) &\leftrightarrow \otimes q \circ \otimes p, \\
p \circ (q \circ r) &\leftrightarrow (p \circ q) \circ r.
\end{aligned}
$$

1.2.5 Show that 'being-a-substitution-instance-of' is a transitive concept. That is, show that if χ is a substitution instance of ψ, and ψ is a substitution instance of ϕ, then χ is a substitution instance of ϕ.

1.3 Models and Frames

Although our discussion has contained many semantically suggestive phrases such as 'true' and 'intended interpretation,' as yet we have given them no mathematical content. The purpose of this (key) section is to put that right. We do so by interpreting our modal languages in relational structures. In fact, by the end of the section we will have done this in two distinct ways: at the level of *models* and at the level of *frames*. Both levels are important, though in different ways. The level of models is important because this is where the fundamental notion of *satisfaction* (or *truth*) is defined. The level of frames is important because it supports the key logical notion of *validity*.

Models and satisfaction

We start by defining frames, models, and the satisfaction relation for the basic modal language.

Definition 1.19 A *frame* for the basic modal language is a pair $\mathfrak{F} = (W, R)$ such that

(i) W is a non-empty set.
(ii) R is a binary relation on W.

That is, a frame for the basic modal language is simply a relational structure bearing a single binary relation. We remind the reader that we refer to the elements of W by many different names (see Definition 1.1).

A *model* for the basic modal language is a pair $\mathfrak{M} = (\mathfrak{F}, V)$, where $\mathfrak{F}$ is a frame for the basic modal language, and V is a function assigning to each proposition letter p in Φ a subset $V(p)$ of W. Formally, V is a map: $\Phi \rightarrow \mathcal{P}(W)$, where $\mathcal{P}(W)$ denotes the power set of W. Informally we think of $V(p)$ as the set of points in our model where p is true. The function V is called a *valuation*. Given a model $\mathfrak{M} = (\mathfrak{F}, V)$, we say that $\mathfrak{M}$ is *based on* the frame $\mathfrak{F}$, or that $\mathfrak{F}$ is the frame *underlying* $\mathfrak{M}$.   ⊣

Note that models for the basic modal language can be viewed as relational structures in a natural way, namely as structures of the form:

$$(W, R, V(p), V(q), V(r), \ldots).$$

That is, a model is a relational structure consisting of a domain, a single binary relation R, and the unary relations given to us by V. Thus, viewed from a purely structural perspective, a frame $\mathfrak{F}$ and a model $\mathfrak{M}$ based on $\mathfrak{F}$, are simply two relational models based on the same universe; indeed, a model is simply a frame enriched by a collection of unary relations.

But in spite of their mathematical kinship, frames and models are *used* very differently. Frames are essentially mathematical pictures of ontologies that we find interesting. For example, we may view time as a collection of points ordered by a strict partial order, or feel that a correct analysis of knowledge requires that we postulate the existence of situations linked by a relation of 'being an epistemic alternative to.' In short, we use the level of frames to make our fundamental assumptions mathematically precise.

The unary relations provided by valuations, on the other hand, are there to dress our frames with contingent information. Is it raining on Tuesday or not? Is the system write-enabled at time t_6? Is a situation where Janet does not love him an epistemic alternative for John? Such information is important, and we certainly need to be able to work with it – nonetheless, statements only deserve the description 'logical' if they are *invariant* under changes of contingent information. Because we have drawn a distinction between the fundamental information given by frames, and the additional descriptive content provided by models, it will be straightforward to define a modally reasonable notion of validity.

But this is jumping ahead. First we must learn how to interpret the basic modal language in models. This we do by means of the following satisfaction definition.

Definition 1.20 Suppose w is a state in a model $\mathfrak{M} = (W, R, V)$. Then we inductively define the notion of a formula ϕ being *satisfied* (or *true*) in $\mathfrak{M}$ at state w as follows:

$$\mathfrak{M}, w \Vdash p \quad \text{iff} \quad w \in V(p), \text{ where } p \in \Phi,$$

$$\mathfrak{M}, w \Vdash \bot \quad \quad \text{never},$$

$$\mathfrak{M}, w \Vdash \neg\phi \quad \text{iff} \quad \text{not } \mathfrak{M}, w \Vdash \phi,$$

$\mathfrak{M}, w \Vdash \phi \vee \psi$ iff $\mathfrak{M}, w \Vdash \phi$ or $\mathfrak{M}, w \Vdash \psi$,

$\mathfrak{M}, w \Vdash \Diamond\phi$ iff   for some $v \in W$ with Rwv we have $\mathfrak{M}, v \Vdash \phi$.   (1.4)

It follows from this definition that $\mathfrak{M}, w \Vdash \Box\phi$ if and only if for all $v \in W$ such that Rwv, we have $\mathfrak{M}, v \Vdash \phi$. Finally, we say that a *set* Σ of formulas is true at a state w of a model $\mathfrak{M}$, notation: $\mathfrak{M}, w \Vdash \Sigma$, if all members of Σ are true at w. $\dashv$

Note that this notion of satisfaction is intrinsically *internal* and *local*. We evaluate formulas *inside* models, at some particular state w (the *current state*). Moreover, $\Diamond$ works locally: the final clause (1.4) treats $\Diamond\phi$ as an instruction to scan states in search of one where ϕ is satisfied. Crucially, only *successors* of the current state (that is, states that are accessible from the current one by making one R-step) can be scanned by our operators. Much of the characteristic flavor of modal logic springs from the perspective on relational structures embodied in the satisfaction definition.

If $\mathfrak{M}$ does not satisfy ϕ at w we often write $\mathfrak{M}, w \nVdash \phi$, and say that ϕ is *false* or *refuted* at w. When $\mathfrak{M}$ is clear from the context, we write $w \Vdash \phi$ for $\mathfrak{M}, w \Vdash \phi$ and $w \nVdash \phi$ for $\mathfrak{M}, w \nVdash \phi$. It is convenient to extend the valuation V from proposition letters to arbitrary formulas so that $V(\phi)$ always denotes the set of states at which ϕ is true:

$$V(\phi) := \{w \mid \mathfrak{M}, w \Vdash \phi\}.$$

Definition 1.21 A formula ϕ is *globally* or *universally* *true* in a model $\mathfrak{M}$ (notation: $\mathfrak{M} \Vdash \phi$) if it is satisfied at all points in $\mathfrak{M}$ (that is, if $\mathfrak{M}, w \Vdash \phi$, for all $w \in W$). A formula ϕ is *satisfiable* in a model $\mathfrak{M}$ if there is *some* state in $\mathfrak{M}$ at which ϕ is true; a formula is *falsifiable* or *refutable* in a model if its negation is satisfiable.

A *set* Σ of formulas is globally true (satisfiable, respectively) in a model $\mathfrak{M}$ if $\mathfrak{M}, w \Vdash \Sigma$ for all states w in $\mathfrak{M}$ (some state w in $\mathfrak{M}$, respectively). $\dashv$

Example 1.22 (i) Consider the frame $\mathfrak{F} = (\{w_1, w_2, w_3, w_4, w_5\}, R)$, where Rw_iw_j iff $j = i + 1$:

If we choose a valuation V on $\mathfrak{F}$ such that $V(p) = \{w_2, w_3\}$, $V(q) = \{w_1, w_2, w_3, w_4, w_5\}$, and $V(r) = \varnothing$, then in the model $\mathfrak{M} = (\mathfrak{F}, V)$ we have that

- $\mathfrak{M}, w_1 \Vdash \Diamond\Box p$,
- $\mathfrak{M}, w_1 \nVdash \Diamond\Box p \rightarrow p$,
- $\mathfrak{M}, w_2 \Vdash \Diamond(p \wedge \neg r)$, and
- $\mathfrak{M}, w_1 \Vdash q \wedge \Diamond(q \wedge \Diamond(q \wedge \Diamond(q \wedge \Diamond q)))$.

Furthermore, $\mathfrak{M} \Vdash \Box q$. Now, it is clear that $\Box q$ is true at w_1, w_2, w_3 and w_4, but why is it true at w_5? Well, as w_5 has no successors at all (we often call such points '*dead ends*' or '*blind states*') it is vacuously true that q is true at all R-successors of w_5. Indeed, any 'boxed' formula $\Box \phi$ is true at any dead end in any model.

(ii) As a second example, let $\mathfrak{F}$ be the SPO given in Figure 1.1, where $W = \{1, 2, 3, 4, 6, 8, 12, 24\}$ and Rxy means 'x and y are different, and y can be divided by x.' Choose a valuation V on this frame such that $V(p) = \{4, 8, 12, 24\}$, and $V(q) = \{6\}$, and let $\mathfrak{M} = (\mathfrak{F}, V)$. Then we have that

- $\mathfrak{M}, 4 \Vdash \Box p$,
- $\mathfrak{M}, 6 \Vdash \Box p$,
- $\mathfrak{M}, 2 \not\Vdash \Box p$, and
- $\mathfrak{M}, 2 \Vdash \Diamond(q \wedge \Box p) \wedge \Diamond(\neg q \wedge \Box p)$.

(iii) Whereas a diamond $\Diamond$ corresponds to making a single R-step in a model, stacking diamonds one in front of the other corresponds to making a sequence of R-steps through the model. The following defined operators will sometimes be useful: we write $\Diamond^n \phi$ for ϕ preceded by n occurrences of $\Diamond$, and $\Box^n \phi$ for ϕ preceded by n occurrences of $\Box$. If we like, we can associate each of these defined operators with its own accessibility relation. We do so inductively: $R^0 xy$ is defined to hold if $x = y$, and $R^{n+1} xy$ is defined to hold if $\exists z \, (Rxz \wedge R^n zy)$. Under this definition, for any model $\mathfrak{M}$ and state w in $\mathfrak{M}$ we have $\mathfrak{M}, w \Vdash \Diamond^n \phi$ iff there exists a v such that $R^n wv$ and $\mathfrak{M}, v \Vdash \phi$.

(iv) The use of the word 'world' (or 'possible world') for the entities in W derives from the reading of the basic modal language in which $\Diamond \phi$ is taken to mean '*possibly ϕ*,' and $\Box \phi$ to mean '*necessarily ϕ*.' Given this reading, the machinery of frames, models, and satisfaction which we have defined is essentially an attempt to capture mathematically the view (often attributed to Leibniz) that *necessity* means *truth in all possible worlds*, and that *possibility* means *truth in some possible world*.

The satisfaction definition stipulates that $\Diamond$ and $\Box$ check for truth not at *all* possible worlds (that is, at all elements of W) but only at R-accessible possible worlds. At first sight this may seem a weakness of the satisfaction definition – but in fact, it is its greatest source of strength. The point is this: varying R is a mechanism which gives us a firm mathematical grip on the pre-theoretical notion of access between possible worlds. For example, by stipulating that $R = W \times W$ we can allow all worlds access to each other; this corresponds to the Leibnizian idea in its purest form. Going to the other extreme, we might stipulate that *no* world has access to any other. Between these extremes there is a wide range of options to explore. Should interworld access be reflexive? Should it be transitive? What impact do these choices have on the notions of necessity and possibility? For example, if we demand symmetry, does this justify certain principles, or rule others out?

(v) Recall from Example 1.10 that in epistemic logic $\Box$ is written as K and $K\phi$

is interpreted as 'the agent knows that ϕ.' Under this interpretation, the intuitive reading for the semantic clause governing K is: the agent knows ϕ in a situation w (that is, $w \Vdash K\phi$) iff ϕ is true in all situations v that are compatible with her knowledge (that is, if $v \Vdash \phi$ for all v such that Rwv). Thus, under this interpretation, W is to be thought of as a collection of situations, R is a relation which models the idea of one situation being epistemically accessible from another, and V governs the distribution of primitive information across situations. $\dashv$

We now define frames, models and satisfaction for modal languages of arbitrary similarity type.

Definition 1.23 Let τ be a modal similarity type. A τ-*frame* is a tuple $\mathfrak{F}$ consisting of the following ingredients:

(i) a non-empty set W,
(ii) for each $n \geq 0$, and each n-ary modal operator $\triangle$ in the similarity type τ, an $(n+1)$-ary relation $R_\triangle$.

So, again, frames are simply relational structures. If τ contains just a finite number of modal operators $\triangle_1, \ldots, \triangle_n$, we write $\mathfrak{F} = (W, R_{\triangle_1}, \ldots, R_{\triangle_n})$; otherwise we write $\mathfrak{F} = (W, R_\triangle)_{\triangle \in \tau}$ or $\mathfrak{F} = (W, \{R_\triangle \mid \triangle \in \tau\})$. We turn such a frame into a model in exactly the same way that we did for the basic modal language: by adding a valuation. That is, a τ-*model* is a pair $\mathfrak{M} = (\mathfrak{F}, V)$ where $\mathfrak{F}$ is a τ-frame, and V is a valuation with domain Φ and range $\mathcal{P}(W)$, where W is the universe of $\mathfrak{F}$.

The notion of a formula ϕ being *satisfied* (or *true*) at a state w in a model $\mathfrak{M} = (W, \{R_\triangle \mid \triangle \in \tau\}, V)$ (notation: $\mathfrak{M}, w \Vdash \phi$) is defined inductively. The clauses for the atomic and boolean cases are the same as for the basic modal language (see Definition 1.20). As for the modal case, when $\rho(\triangle) > 0$ we define

$$\mathfrak{M}, w \Vdash \triangle(\phi_1, \ldots, \phi_n) \quad \text{iff} \quad \text{for some } v_1, \ldots, v_n \in W \text{ with } R_\triangle wv_1 \ldots v_n$$
$$\text{we have, for each } i, \, \mathfrak{M}, v_i \Vdash \phi_i.$$

This is an obvious generalization of the way $\Diamond$ is handled in the basic modal language. Before going any further, the reader should formulate the satisfaction clause for $\triangledown(\phi_1, \ldots, \phi_n)$.

On the other hand, when $\rho(\triangle) = 0$ (that is, when $\triangle$ is a nullary modality) then $R_\triangle$ is a unary relation and we define

$$\mathfrak{M}, w \Vdash \triangle \quad \text{iff} \quad w \in R_\triangle.$$

That is, unlike other modalities, nullary modalities do not access other states. In fact, their semantics is identical to that of the propositional variables, save that the unary relations used to interpret them are *not* given by the valuation – rather, they are part of the underlying *frame*.

As before, we often write $w \Vdash \phi$ for $\mathfrak{M}, w \Vdash \phi$ where $\mathfrak{M}$ is clear from the context. The concept of *global truth* (or *universal truth*) in a model is defined as for the basic modal language: it simply means *truth at all states in the model*. And, as before, we sometimes extend the valuation V supplied by $\mathfrak{M}$ to arbitrary formulas. ⊣

Example 1.24 (i) Let τ be a similarity type with three unary operators $\langle a \rangle$, $\langle b \rangle$, and $\langle c \rangle$. Then a τ-frame has three binary relations R_a, R_b, and R_c (that is, it is a labeled transition system with three labels). To give an example, let W, R_a, R_b and R_c be as in Figure 1.2, and consider the formula $\langle a \rangle p \rightarrow \langle b \rangle p$. Informally, this formula is true at a state, if it has an R_a-successor satisfying p only if it has an R_b-successor satisfying p. Let V be a valuation with $V(p) = \{w_2\}$. Then the model $\mathfrak{M} = (W, R_a, R_b, R_c, V)$ has $\mathfrak{M}, w_1 \nVdash \langle a \rangle p \rightarrow \langle b \rangle p$.

(ii) Let τ be a similarity type with a binary modal operator $\triangle$ and a ternary operator $\bigcirc$. Frames for this τ contain a ternary relation $R_\triangle$ and a 4-ary relation $S_\bigcirc$. As an example, let $W = \{u, v, w, s\}$, $R_\triangle = \{(u, v, w)\}$, and $S_\bigcirc = \{(u, v, w, s)\}$ as in Figure 1.6, and consider a valuation V on this frame with $V(p_0) = \{v\}$, $V(p_1) = \{w\}$ and $V(p_2) = \{s\}$. Now, let ϕ be the formula

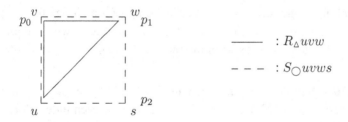

$$\begin{array}{ll} \text{———} & : R_\triangle uvw \\ \text{- - - -} & : S_\bigcirc uvws \end{array}$$

Fig. 1.6. A simple frame

$\triangle(p_0, p_1) \rightarrow \bigcirc(p_0, p_1, p_2)$. An informal reading of ϕ is 'any triangle of which the evaluation point is a vertex, and which has p_0 and p_1 true at the other two vertices, can be expanded to a rectangle with a fourth point at which p_2 is true.' The reader should be able to verify that ϕ is true at u, and indeed at all other points, and hence that it is globally true in the model. ⊣

Example 1.25 (Bidirectional Frames and Models) Recall from Example 1.14 that the basic temporal language has two unary operators F and P. Thus, according to Definition 1.23, models for this language consist of a set bearing two binary relations, R_F (the into-the-future relation) and R_P (the into-the-past relation), which are used to interpret F and P respectively. However, given the intended reading of the operators, most such models are inappropriate: clearly we ought to insist on working with models based on frames in which R_P is the *converse* of R_F (that is, frames in which $\forall xy\,(R_F xy \leftrightarrow R_P yx)$).

Let us denote the converse of a relation R by $\breve{R}$. We will call a frame of the form $(T, R, \breve{R})$ a *bidirectional frame*, and a model built over such a frame a *bidirectional model*. From now on, we will only interpret the basic temporal language in bidirectional models. That is, if $\mathfrak{M} = (T, R, \breve{R}, V)$ is a bidirectional model then:

$$\mathfrak{M}, t \Vdash F\phi \quad \text{iff} \quad \exists s\,(Rts \wedge \mathfrak{M}, s \Vdash \phi),$$
$$\mathfrak{M}, t \Vdash P\phi \quad \text{iff} \quad \exists s\,(\breve{R}ts \wedge \mathfrak{M}, s \Vdash \phi).$$

But of course, once we have made this restriction, we do not need to mention $\breve{R}$ explicitly any more: once R has been fixed, its converse is fixed too. That is, we are free to interpret the basic temporal languages on frames (T, R) for the basic modal language using the clauses

$$\mathfrak{M}, t \Vdash F\phi \quad \text{iff} \quad \exists s\,(Rts \wedge \mathfrak{M}, s \Vdash \phi),$$
$$\mathfrak{M}, t \Vdash P\phi \quad \text{iff} \quad \exists s\,(Rst \wedge \mathfrak{M}, s \Vdash \phi).$$

These clauses clearly capture a crucial part of the intended semantics: F looks forwards along R, and P looks backwards along R. Of course, our models will only start looking genuinely *temporal* when we insist that R has further properties (notably transitivity, to capture the flow of time), but at least we have pinned down the fundamental interaction between the two modalities.   ⊣

Example 1.26 (Regular Frames and Models) As explained in Example 1.15, the language of PDL has an infinite collection of diamonds, each indexed by a program π built from basic programs using the constructors $\cup$, ; and *. Now, according to Definition 1.23, a model for this language has the form

$$(W, \{R_\pi \mid \pi \text{ is a program }\}, V).$$

That is, a model is a labeled transition system together with a valuation. However, given our reading of the PDL operators, most of these models are uninteresting. As with the basic temporal language, we must insist on working with a class of models that does justice to our intentions.

Now, there is no problem with the interpretation of the basic programs: any binary relation can be regarded as a transition relation for a non-deterministic program. Of course, if we were particularly interested in *deterministic* programs we would insist that each basic program be interpreted by a partial function, but let us ignore this possibility and turn to the key question: which relations should interpret the structured modalities? Given our readings of $\cup$, ; and *, as choice, composition, and iteration, it is clear that we are only interested in relations constructed using the following inductive clauses:

$$R_{\pi_1 \cup \pi_2} = R_{\pi_1} \cup R_{\pi_2},$$
$$R_{\pi_1;\pi_2} = R_{\pi_1} \circ R_{\pi_2} \; (= \{(x,y) \mid \exists z \, (R_{\pi_1} xz \wedge R_{\pi_2} zy)\}),$$
$$R_{\pi_1^*} = (R_{\pi_1})^*, \text{ the reflexive transitive closure of } R_{\pi_1}.$$

These inductive clauses completely determine how each modality should be interpreted. Once the interpretation of the basic programs has been fixed, the relation corresponding to each complex program is fixed too. This leads to the following definition.

Suppose we have fixed a set of basic programs. Let Π be the smallest set of programs containing the basic programs and all the programs constructed over them using the regular constructors $\cup$, ; and *. Then a *regular frame for* Π is a labeled transition system $(W, \{R_\pi \mid \pi \in \Pi\})$ such that R_a is an arbitrary binary relation for each basic program a, and for all complex programs π, R_π is the binary relation inductively constructed in accordance with the previous clauses. A *regular model* for Π is a model built over a regular frame; that is, a regular model is a regular frame together with a valuation. When working with the language of PDL over the programs in Π, we will only be interested in regular models for Π, for these are the models that capture the intended interpretation.

What about the $\cap$ and ? constructors? Clearly the intended reading of $\cap$ demands that $R_{\pi_1 \cap \pi_2} = R_{\pi_1} \cap R_{\pi_2}$. As for ?, it is clear that we want the following definition:

$$R_{\phi?} = \{(x,y) \mid x = y \text{ and } y \Vdash \phi\}.$$

This is indeed the clause we want, but note that it is rather different from the others: it is not a *frame* condition. Rather, in order to determine the relation $R_{\phi?}$, we need information about the *truth* of the formula ϕ, and this can only be provided at the level of *models*. ⊣

Example 1.27 (Arrow Models) Arrow frames were defined in Example 1.8 and the arrow language in Example 1.16. Given these definitions, it is clear how the language of arrow logic should be interpreted. First, an *arrow model* is a structure $\mathfrak{M} = (\mathfrak{F}, V)$ such that $\mathfrak{F} = (W, C, R, I)$ is an arrow frame and V is a valuation. Then:

$$\mathfrak{M}, a \Vdash 1' \quad \text{iff} \quad Ia,$$
$$\mathfrak{M}, a \Vdash \otimes\phi \quad \text{iff} \quad \mathfrak{M}, b \Vdash \phi \text{ for some } b \text{ with } Rab,$$
$$\mathfrak{M}, a \Vdash \phi \circ \psi \quad \text{iff} \quad \mathfrak{M}, b \Vdash \phi \text{ and } \mathfrak{M}, c \Vdash \psi \text{ for some } b \text{ and } c \text{ with } Cabc.$$

When $\mathfrak{F}$ is a *square* frame $\mathfrak{S}_U$ (as defined in Example 1.8), this works out as follows. V now maps propositional variables to sets of *pairs* over U; that is, to binary relations. The truth definition can be rephrased as follows:

$$\mathfrak{M}, (a_0, a_1) \Vdash 1' \quad \text{iff} \quad a_0 = a_1,$$

$$\mathfrak{M}, (a_0, a_1) \Vdash \otimes\phi \quad \text{iff} \quad \mathfrak{M}, (a_1, a_0) \Vdash \phi,$$

$$\mathfrak{M}, (a_0, a_1) \Vdash \phi \circ \psi \quad \text{iff} \quad \mathfrak{M}, (a_0, u) \Vdash \phi \text{ and } \mathfrak{M}, (u, a_1) \Vdash \psi \text{ for some } u \in U.$$

Such situations can be represented pictorially in two ways. First, one could draw the graph-like structures as given in Example 1.8. Alternatively, one could draw a square model two-dimensionally, as in the picture below. It will be obvious that the modal constant 1' holds precisely at the *diagonal points* and that $\otimes\phi$ is true at a point iff ϕ holds at its *mirror image* with respect to the diagonal. The formula $\phi \circ \psi$ holds at a point a iff we can draw a rectangle $abcd$ such that: b lies on the vertical line through a, d lies on the horizontal line through a; and c lies on the diagonal.

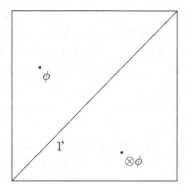 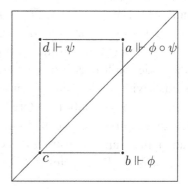

Frames and validity

It is time to define one of the key concepts in modal logic. So far we have been viewing modal languages as tools for talking about models. But models are composite entities consisting of a frame (our underlying ontology) and contingent information (the valuation). We often want to ignore the effects of the valuation and get a grip on the more fundamental level of frames. The concept of *validity* lets us do this. A formula is valid on a frame if it is true at every state in every model that can be built over the frame. In effect, this concept interprets modal formulas on frames by abstracting away from the effects of particular valuations.

Definition 1.28 A formula ϕ is *valid at a state w in a frame* $\mathfrak{F}$ (notation: $\mathfrak{F}, w \Vdash \phi$) if ϕ is true at w in every model $(\mathfrak{F}, V)$ based on $\mathfrak{F}$; ϕ is *valid in a frame* $\mathfrak{F}$ (notation: $\mathfrak{F} \Vdash \phi$) if it is valid at every state in $\mathfrak{F}$. A formula ϕ is *valid on a class of frames* F (notation: F $\Vdash \phi$) if it is valid on every frame $\mathfrak{F}$ in F; and it is *valid* (notation: $\Vdash \phi$) if it is valid on the class of all frames. The set of all formulas that are valid in a class of frames F is called the *logic* of F (notation: Λ_F).   ⊣

Our definition of the logic of a frame class F (as the set of 'all' formulas that are valid on F) is underspecified: we did not say which collection of proposition letters Φ should be used to build formulas. But usually the precise form of this collection is irrelevant for our purposes. On the few occasions in this book where more precision is required, we will explicitly deal with the issue. (If the reader is worried about this, he or she may just fix a countable set Φ of proposition letters and define Λ_F to be $\{\phi \in Form(\tau, \Phi) \mid F \Vdash \phi\}$.)

As will become abundantly clear in the course of the book, validity differs from truth in many ways. Here is a simple example. When a formula $\phi \vee \psi$ is true at a point w, this means that either ϕ or ψ is true at w (the satisfaction definition tells us so). On the other hand, if $\phi \vee \psi$ is valid on a frame $\mathfrak{F}$, this does *not* mean that either ϕ or ψ is valid on $\mathfrak{F}$ ($p \vee \neg p$ is a simple counterexample).

Example 1.29 (i) The formula $\Diamond(p \vee q) \rightarrow (\Diamond p \vee \Diamond q)$ is valid on all frames. To see this, take any frame $\mathfrak{F}$ and state w in $\mathfrak{F}$, and let V be a valuation on $\mathfrak{F}$. We have to show that if $(\mathfrak{F}, V), w \Vdash \Diamond(p \vee q)$, then $(\mathfrak{F}, V), w \Vdash \Diamond p \vee \Diamond q$. So assume that $(\mathfrak{F}, V), w \Vdash \Diamond(p \vee q)$. Then, by definition there is a state v such that Rwv and $(\mathfrak{F}, V), v \Vdash p \vee q$. But, if $v \Vdash p \vee q$ then either $v \Vdash p$ or $v \Vdash q$. Hence either $w \Vdash \Diamond p$ or $w \Vdash \Diamond q$. Either way, $w \Vdash \Diamond p \vee \Diamond q$.

(ii) The formula $\Diamond\Diamond p \rightarrow \Diamond p$ is not valid on all frames. To see this we need to find a frame $\mathfrak{F}$, a state w in $\mathfrak{F}$, and a valuation on $\mathfrak{F}$ that falsifies the formula at w. So let $\mathfrak{F}$ be a three-point frame with universe $\{0, 1, 2\}$ and relation $\{(0, 1), (1, 2)\}$. Let V be any valuation on $\mathfrak{F}$ such that $V(p) = \{2\}$. Then $(\mathfrak{F}, V), 0 \Vdash \Diamond\Diamond p$, but $(\mathfrak{F}, V), 0 \nVdash \Diamond p$ since 0 is not related to 2.

(iii) But there is a class of frames on which $\Diamond\Diamond p \rightarrow \Diamond p$ is valid: the class of *transitive* frames. To see this, take any transitive frame $\mathfrak{F}$ and state w in $\mathfrak{F}$, and let V be a valuation on $\mathfrak{F}$. We have to show that if $(\mathfrak{F}, V), w \Vdash \Diamond\Diamond p$, then $(\mathfrak{F}, V), w \Vdash \Diamond p$. So assume that $(\mathfrak{F}, V), w \Vdash \Diamond\Diamond p$. Then by definition there are states u and v such that Rwu and Ruv and $(\mathfrak{F}, V), v \Vdash p$. But as R is transitive, it follows that Rwv, hence $(\mathfrak{F}, V), w \Vdash \Diamond p$.

(iv) As the previous example suggests, when additional constraints are imposed on frames, more formulas may become valid. For example, consider the frame depicted in Figure 1.2. On this frame the formula $\langle a \rangle p \rightarrow \langle b \rangle p$ is not valid; a countermodel is obtained by putting $V(p) = \{w_2\}$. Now, consider a frame satisfying the condition $R_a \subseteq R_b$; an example is depicted in Figure 1.7.

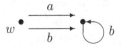

Fig. 1.7. A frame satisfying $R_a \subseteq R_b$

On this frame it is impossible to refute the formula $\langle a \rangle p \rightarrow \langle b \rangle p$ at w, because a refutation would require the existence of a point u with $R_a wu$ and p true at u, but not $R_b wu$; but such points are forbidden when we insist that $R_a \subseteq R_b$.

This is a completely general point: in *every* frame $\mathfrak{F}$ of the appropriate similarity type, if $\mathfrak{F}$ satisfies the condition $R_a \subseteq R_b$, then $\langle a \rangle p \rightarrow \langle b \rangle p$ is valid in $\mathfrak{F}$. Moreover, the converse to this statement also holds: whenever $\langle a \rangle p \rightarrow \langle b \rangle p$ is valid on a given frame $\mathfrak{F}$, then the frame must satisfy the condition $R_a \subseteq R_b$. To use the terminology we will introduce in Chapter 3, the formula $\langle a \rangle p \rightarrow \langle b \rangle p$ *defines* the property that $R_a \subseteq R_b$.

(v) When interpreting the basic temporal language (see Example 1.25) we observed that arbitrary frames of the form (W, R_P, R_F) were uninteresting given the intended interpretation of F and P, and we insisted on interpreting them using a relation R and its converse. Interestingly, there is a sense in which the basic temporal language itself is strong enough to enforce the condition that the relation R_P is the converse of the relation R_F: such frames are *precisely* the ones which validate both the formulas $p \rightarrow GPp$ and $p \rightarrow HFp$; see Exercise 3.1.1.

(vi) The formula $Fq \rightarrow FFq$ is not valid on all frames. To see this we need to find a frame $\mathfrak{T} = (T, R)$, a state t in $\mathfrak{T}$, and a valuation on $\mathfrak{T}$ that falsifies this formula at t. So let $T = \{0, 1\}$, and let R be the relation $\{(0, 1)\}$. Let V be a valuation such that $V(p) = \{1\}$. Then $(\mathfrak{T}, V), 0 \Vdash Fp$, but obviously $(\mathfrak{T}, V), 0 \nVdash FFp$.

(vii) But there is a frame on which $Fp \rightarrow FFp$ is valid. As the universe of the frame take the set of all rational numbers $\mathbb{Q}$, and let the frame relation be the usual $<$-ordering on $\mathbb{Q}$. To show that $Fp \rightarrow FFp$ is valid on this frame, take any point t in it, and any valuation V such that $(\mathbb{Q}, <, V), t \Vdash Fp$; we have to show that $t \Vdash FFp$. But this is easy: as $t \Vdash Fp$, there exists a t' such that $t < t'$ and $t' \Vdash p$. Because we are working on the rationals, there must be an s with $t < s$ and $s < t$ (for example, $(t + t')/2$). As $s \Vdash Fp$, it follows that $t \Vdash FFp$.

(viii) The special conditions demanded of PDL models also give rise to validities. For example,

$$\langle \pi_1 \,;\, \pi_2 \rangle p \leftrightarrow \langle \pi_1 \rangle \langle \pi_2 \rangle p$$

is valid on any frame such that $R_{\pi_1;\pi_2} = R_{\pi_1} \circ R_{\pi_2}$, and in fact the converse is also true. The reader is asked to prove this in Exercise 3.1.2.

(ix) In our last example we consider arrow logic. We claim that in any square arrow frame $\mathfrak{S}_U$, the formula $\otimes(p \circ q) \rightarrow \otimes q \circ \otimes p$ is valid. For, let V be a valuation on $\mathfrak{S}_U$, and suppose that for some pair of points u, v in U, we have $(\mathfrak{S}_U, V), (u, v) \Vdash \otimes(p \circ q)$. It follows that $(\mathfrak{S}_U, V), (v, u) \Vdash p \circ q$, and hence, there must be a $w \in U$ for which $(\mathfrak{S}_U, V), (v, w) \Vdash p$ and $(\mathfrak{S}_U, V), (w, u) \Vdash q$. But then we have $(\mathfrak{S}_U, V), (w, v) \Vdash \otimes p$ and $(\mathfrak{S}_U, V), (u, w) \Vdash \otimes q$. This in turn implies that $(\mathfrak{S}_U, V), (u, v) \Vdash \otimes q \circ \otimes p$. $\dashv$

Exercises for Section 1.3

1.3.1 Show that when evaluating a formula ϕ in a model, the only relevant information in the valuation is the assignments it makes to the propositional letters actually occurring in ϕ. More precisely, let $\mathfrak{F}$ be a frame, and V and V' be two valuations on $\mathfrak{F}$ such that $V(p) = V'(p)$ for all proposition letters p in ϕ. Show that $(\mathfrak{F}, V) \Vdash \phi$ iff $(\mathfrak{F}, V') \Vdash \phi$. Work in the basic modal language. Do this exercise by *induction on the number of connectives* in ϕ (or as we usually put it, by *induction on* ϕ). (If you are unsure how to do this, glance ahead to Proposition 2.3 where such a proof is given in detail.)

1.3.2 Let $\mathfrak{N} = (\mathbb{N}, S_1, S_2)$ and $\mathfrak{B} = (\mathbb{B}, R_1, R_2)$ be the following frames for a modal similarity type with two diamonds $\Diamond_1$ and $\Diamond_2$. Here $\mathbb{N}$ is the set of natural numbers, $\mathbb{B}$ is the set of strings of 0s and 1s, and the relations are defined by

$$
\begin{aligned}
mS_1 n &\quad \text{iff} \quad n = m + 1, \\
mS_2 n &\quad \text{iff} \quad m > n, \\
sR_1 t &\quad \text{iff} \quad t = s0 \text{ or } t = s1, \\
sR_2 t &\quad \text{iff} \quad t \text{ is a proper initial segment of } s.
\end{aligned}
$$

Which of the following formulas are valid on $\mathfrak{N}$ and $\mathfrak{B}$, respectively?

(a) $(\Diamond_1 p \wedge \Diamond_1 q) \to \Diamond_1 (p \wedge q)$,
(b) $(\Diamond_2 p \wedge \Diamond_2 q) \to \Diamond_2 (p \wedge q)$,
(c) $(\Diamond_1 p \wedge \Diamond_1 q \wedge \Diamond_1 r) \to (\Diamond_1 (p \wedge q) \vee \Diamond_1 (p \wedge r) \vee \Diamond_1 (q \wedge r))$,
(d) $p \to \Diamond_1 \Box_2 p$,
(e) $p \to \Diamond_2 \Box_1 p$,
(f) $p \to \Box_1 \Diamond_2 p$,
(g) $p \to \Box_2 \Diamond_1 p$.

1.3.3 Consider the basic temporal language and the frames $(\mathbb{Z}, <)$, $(\mathbb{Q}, <)$ and $(\mathbb{R}, <)$ (the integer, rational, and real numbers, respectively, all ordered by the usual less-than relation $<$). In this exercise we use $E\phi$ to abbreviate $P\phi \vee \phi \vee F\phi$, and $A\phi$ to abbreviate $H\phi \wedge \phi \wedge G\phi$. Which of the following formulas are valid on these frames?

(a) $GGp \to p$,
(b) $(p \wedge Hp) \to FHp$,
(c) $(Ep \wedge E\neg p \wedge A(p \to Hp) \wedge A(\neg p \to G\neg p)) \to E(Hp \wedge G\neg p)$.

1.3.4 Show that every formula that has the form of a propositional tautology is valid. Further, show that $\Box(p \to q) \to (\Box p \to \Box q)$ is valid.

1.3.5 Show that each of the following formulas is *not* valid by constructing a frame $\mathfrak{F} = (W, R)$ that refutes it.

(a) $\Box\bot$,
(b) $\Diamond p \to \Box p$,
(c) $p \to \Box\Diamond p$,
(d) $\Diamond\Box p \to \Box\Diamond p$.

Find, for each of these formulas, a non-empty class of frames on which it is valid.

1.3.6 Show that the arrow formulas $\phi \circ (\psi \circ \chi) \leftrightarrow (\phi \circ \psi) \circ \chi$ and $1' \circ \phi \leftrightarrow \phi$ are valid in any square.

1.4 General Frames

At the level of models the fundamental concept is satisfaction. This is a relatively simple concept involving only a frame and a *single* valuation. By ascending to the level of frames we get a deeper grip on relational structures – but there is a price to pay. Validity lacks the concrete character of satisfaction, for it is defined in terms of *all* valuations on a frame. However, there is an intermediate level: a *general frame* $(\mathfrak{F}, A)$ is a frame $\mathfrak{F}$ together with a restricted, but suitably well-behaved collection A of *admissible valuations*.

General frames are useful for at least two reasons. First, there may be application driven motivations to exclude certain valuations. For instance, if we were using $(\mathbb{N}, <)$ to model the temporal distribution of outputs from a computational device, it would be unreasonable to let valuations assign non-recursively enumerable sets to propositional variables. But perhaps the most important reason to work with general frames is that they support a notion of validity that is mathematically simpler than the frame-based one, without losing too many of the concrete properties that make models so easy to work with. This 'simpler behavior' will only really become apparent when we discuss the algebraic perspective on completeness theory in Chapter 5. It will turn out there is a fundamental and universal *completeness result* for general frame validity, something that the frame semantics lacks. Moreover, we will discover that general frames are essentially a set-theoretic representation of *boolean algebras with operators*. Thus, the A in (W, R, A) stands not only for *Admissible*, but also for *Algebra*.

So what is a 'suitably well-behaved collection of valuations'? It simply means a collection of valuations closed under the set-theoretic operations corresponding to our connectives and modal operators. Now, fairly obviously, the boolean connectives correspond to the boolean operations of union, relative complement, and so on – but what operations on sets do modalities correspond to? Here is the answer.

Let us first consider the basic modal similarity type with one diamond. Given a frame $\mathfrak{F} = (W, R)$, let m_R be the following operation on the power set of W:

$$m_R(X) = \{w \in W \mid Rwx \text{ for some } x \in X \}.$$

Think of $m_R(X)$ as the set of states that 'see' a state in X. This operation corresponds to the diamond in the sense that for any valuation V and any formula ϕ we have $V(\Diamond\phi) = m_R(V(\phi))$.

Moving to the general case, we obtain the following definition.

Definition 1.30 Given an $(n+1)$-ary relation R on a set W, we define the following n-ary operation m_R on the power set $\mathcal{P}(W)$ of W:

$$m_R(X_1, \ldots, X_n) =$$
$$\{w \in W \mid Rww_1 \ldots w_n \text{ for some } w_1 \in X_1, \ldots, w_n \in X_n\}. \quad \dashv$$

Example 1.31 Let $\otimes$ be the converse operator of arrow logic, and recall that we use the letter R to denote the accessibility relation for $\otimes$. Thus on a *square* frame $\mathfrak{S}_U$, by the rather special nature of R, we have that, for any subset X of U^2:

$$
m_R(X) \;=\; \{(a_0, a_1) \in U^2 \mid a_0 = x_1 \text{ and } a_1 = x_0 \text{ for some } (x_0, x_1) \in X \}
$$
$$
\;=\; \{(x_1, x_0) \in U^2 \mid (x_0, x_1) \in X\}.
$$

In other words, $m_R(X)$ is nothing but the *converse* of the binary relation X. $\dashv$

Definition 1.32 (General Frames) Let τ be a modal similarity type. A *general τ-frame* is a pair $(\mathfrak{F}, A)$ where $\mathfrak{F} = (W, R_\Delta)_{\Delta \in \tau}$ is a τ-frame, and A is a non-empty collection of *admissible* subsets of W closed under the following operations:

(i) union: if $X, Y \in A$, then $X \cup Y \in A$.
(ii) relative complement: if $X \in A$, then $W \setminus X \in A$.
(iii) modal operations: if $X_1, \ldots, X_n \in A$, then $m_{R_\Delta}(X_1, \ldots, X_n) \in A$ for all $\Delta \in \tau$.

A *model based on a general frame* is a triple $(\mathfrak{F}, A, V)$ where $(\mathfrak{F}, A)$ is a general frame and V is a valuation satisfying the constraint that for each proposition letter p, $V(p)$ is an element of A. Valuations satisfying this constraint are called *admissible* for $(\mathfrak{F}, A)$. $\dashv$

It follows immediately from the first two clauses of the definition that both the empty set and the universe of a general frame are always admissible. Note that an ordinary frame $\mathfrak{F} = (W, R_\Delta)_{\Delta \in \tau}$ can be regarded as a general frame where $A = \mathcal{P}(W)$ (that is, a general frame in which all valuations are admissible). Also, note that if a valuation V is admissible for a general frame $(\mathfrak{F}, A)$, then the closure conditions listed in Definition 1.32 guarantee that $V(\phi) \in A$, for all formulas ϕ. In short, a set of admissible valuations A is a 'logically closed' collection of information assignments.

Definition 1.33 A formula ϕ is *valid at a state w in a general frame* $(\mathfrak{F}, A)$ (notation: $(\mathfrak{F}, A), w \Vdash \phi$) if ϕ is true at w in every admissible model $(\mathfrak{F}, A, V)$ on $(\mathfrak{F}, A)$; and ϕ is *valid in a general frame* $(\mathfrak{F}, A)$ (notation: $(\mathfrak{F}, A) \Vdash \phi$) if ϕ is true at every state in every admissible model $(\mathfrak{F}, A, V)$ on $(\mathfrak{F}, A)$.

A formula ϕ is *valid on a class of general frames* G (notation: G $\Vdash \phi$) if it is valid on every general frame $(\mathfrak{F}, A)$ in G. Finally, if ϕ is valid on the class of all general frames we say that it is *g-valid* and write $\Vdash_g \phi$. We will learn in Chapter 4 (see Exercise 4.1.1) that a formula ϕ is valid if and only if it is g-valid. $\dashv$

Clearly, for any frame $\mathfrak{F}$, if $\mathfrak{F} \Vdash \phi$ then for any collection of admissible assignments A on $\mathfrak{F}$, we have $(\mathfrak{F}, A) \Vdash \phi$ too. The converse does not hold. Here is a counterexample that will be useful in Chapter 4.

Example 1.34 Consider the McKinsey formula, $\Box\Diamond p \to \Diamond\Box p$. It is easy to see that the McKinsey formula is *not* valid on the frame $(\mathbb{N}, <)$, for we obtain a countermodel by choosing a valuation for p that lets the truth value of p alternate infinitely often (for instance, by letting $V(p)$ be the collection of even numbers).

However, there is a general frame based on $(\mathbb{N}, <)$ in which the McKinsey formula *is* valid. First some terminology: a set is *co-finite* if its complement is finite. Now consider the general frame $\mathfrak{f} = (\mathbb{N}, <, A)$, where A is the collection of all finite and co-finite sets. We leave it as an exercise to show that $\mathfrak{f}$ satisfies all the constraints of Definition 1.32; see Exercise 1.4.5.

To see that the McKinsey formula is indeed valid on $\mathfrak{f}$, let V be an admissible valuation, and let $n \in \mathbb{N}$. If $(\mathfrak{f}, V), n \Vdash \Box\Diamond p$, then $V(p)$ must be co-finite (why?), hence for some k every state $l \geq k$ is in $V(p)$. But this means that $(\mathfrak{f}, V), n \Vdash \Diamond\Box p$, as required.   ⊣

Although we will make an important comment about general frames in Section 3.2, and use them to help prove an incompleteness result in Section 4.4, we will not really be in a position to grasp their significance until Chapter 5, when we introduce boolean algebras with operators. Until then, we will concentrate on modal languages as tools for talking about models and frames.

Exercises for Section 1.4

1.4.1 Define, analogous to m_R, an operation l_R on the power set of a frame such that for an arbitrary modal formula ϕ and an arbitrary valuation V we have that $l_R(V(\phi)) = V(\Box\phi)$. Extend this definition to the dual of a polyadic modal operator.

1.4.2 Consider the basic modal formula $\Diamond p \to \Box p$.

(a) Construct a frame $\mathfrak{F} = (W, R)$ and a general frame $\mathfrak{f} = (\mathfrak{F}, A)$ such that $\mathfrak{F} \not\Vdash \Diamond p \to \Box p$, but $\mathfrak{f} \Vdash \Diamond p \to \Box p$.
(b) Construct a general frame $(\mathfrak{F}, A)$ and a valuation V on $\mathfrak{F}$ such that $(\mathfrak{F}, A) \not\Vdash \Diamond p \to \Box p$, but $(\mathfrak{F}, V) \Vdash \Diamond p \to \Box p$.

1.4.3 Show that if B is any collection of valuations over some frame $\mathfrak{F}$, then there is a smallest general frame $(\mathfrak{F}, A)$ such that $B \subseteq A$. ('Smallest' means that for any general frame $(\mathfrak{F}, A')$ such that $B \subseteq A', A \subseteq A'$.)

1.4.4 Recall that in any arrow frame, we use C and I to denote the relations associated with the modalities $\circ$ and 1', respectively. Show that for *square* arrow frames, the operation m_C is nothing but *composition* of two binary relations. What is m_I?

1.4.5 Consider the basic modal language, and the general frame $\mathfrak{f} = (\mathbb{N}, <, A)$, where A is the collection of all finite and co-finite sets. Show that $\mathfrak{f}$ is a general frame.

1.4.6 Consider the structure $\mathfrak{g} = (\mathbb{N}, C, A)$ where A is the collection of finite and co-finite subsets of $\mathbb{N}$, and C is defined by

$$Cn_1n_2n_3 \text{ iff } n_1 \leq n_2 + n_3 \text{ and } n_2 \leq n_3 + n_1 \text{ and } n_3 \leq n_1 + n_2.$$

If C is the accessibility relation of a dyadic modal operator, show that $\mathfrak{g}$ is a general frame.

1.4.7 Let $\mathfrak{M} = (\mathfrak{F}, V)$ be some modal model. Prove that the structure

$$(\mathfrak{F}, \{V(\phi) \mid \phi \text{ is a formula }\})$$

is a general frame.

1.5 Modal Consequence Relations

While the idea of validity in frames (and indeed, validity in general frames) gives rise to logically interesting formulas, so far we have said nothing about what *logical consequence* might mean for modal languages. That is, we have not explained what it means for a set of modal formulas Σ to logically entail a modal formula ϕ.

This we will now do. In fact, we will introduce *two* families of consequence relations: a local one and a global one. Both families will be defined *semantically*; that is, in terms of classes of structures. We will define these relations for all three kinds of structures we have introduced, though in practice we will be primarily interested in semantic consequence over frames. Before going further, a piece of terminology. If S is a class of models, then *a model from* S is simply a model $\mathfrak{M}$ in S. On the other hand, if S is a class of frames (or a class of general frames) then a model from S is a model based on a frame (general frame) in S.

What is a modally reasonable notion of logical consequence? Two things are fairly clear. First, it seems sensible to hold on to the familiar idea that a relation of semantic consequence holds when the truth of the premises guarantees the truth of the conclusion. Second, it should be clear that the inferences we are entitled to draw will depend on the class of structures we are working with. (For example, different inferences will be legitimate on transitive and intransitive frames.) Thus our definition of consequence will have to be parametric: it must make reference to a class of structures S.

Here is the standard way of meeting these requirements. Suppose we are working with a class of structures S. Then, for a formula ϕ (the *conclusion*) to be a logical consequence of Σ (the *premises*) we should insist that whenever Σ is true at some point in some model from S, then ϕ should also be true in that same model *at the same point*. In short, this definition demands that the maintenance of truth should be guaranteed *point to point* or *locally*.

Definition 1.35 (Local Semantic Consequence) Let τ be a similarity type, and let S be a class of structures of type τ (that is a class of models, a class of frames, or a class of general frames of this type). Let Σ and ϕ be a set of formulas and a single formula from a language of type τ. We say that ϕ is a *local semantic consequence of Σ over* S (notation: $\Sigma \Vdash_S \phi$) if for all models $\mathfrak{M}$ from S, and all points w in $\mathfrak{M}$, if $\mathfrak{M}, w \Vdash \Sigma$ then $\mathfrak{M}, w \Vdash \phi$.   ⊣

Example 1.36 Suppose that we are working with Tran, the class of transitive frames. Then:

$$\{\Diamond\Diamond p\} \Vdash_{\mathsf{Tran}} \Diamond p.$$

On the other hand, $\Diamond p$ is *not* a local semantic consequence of $\{\Diamond\Diamond p\}$ over the class of *all* frames.   ⊣

Local consequence is the notion of logical entailment explored in this book, but it is by no means the only possibility. Here is an obvious variant.

Definition 1.37 (Global Semantic Consequence) Let τ, S, Σ and ϕ be as in Definition 1.35. We say that ϕ is a *global semantic consequence of Σ over* S (notation: $\Sigma \Vdash^g_S \phi$) if and only if for all structures $\mathfrak{G}$ in S, if $\mathfrak{G} \Vdash \Sigma$ then $\mathfrak{G} \Vdash \phi$. (Here, depending on the kind of structures S contains, $\Vdash$ denotes either validity in a frame, validity in a general frame, or global truth in a model.)   ⊣

Again, this definition hinges on the idea that premises guarantee conclusions, but here the guarantee covers *global* notions of correctness.

Example 1.38 The local and global consequence relations are different. Consider the formulas p and $\Box p$. It is easy to see that p does not locally imply $\Box p$ – indeed, that this entailment should *not* hold is pretty much the essence of locality. On the other hand, suppose that we consider a model $\mathfrak{M}$ where p is globally true. Then p certainly holds at all successors of all states, so $\mathfrak{M} \Vdash \Box p$, and so $p \Vdash^g \Box p$.   ⊣

Nonetheless, there is a systematic connection between the two consequence relations, as the reader is asked to show in Exercise 1.5.3.

Exercises for Section 1.5

1.5.1 Let K be a class of frames for the basic modal similarity type, and let M(K) denote the class of models based on a frame in K. Prove that $\Box p \Vdash^g_{\mathsf{M(K)}} p$ iff $\mathsf{K} \models \forall x \exists y\, Ryx$ (every point has a predecessor).
    Does this equivalence hold as well if we work with $\Vdash^g_{\mathsf{K}}$ instead?

1.5.2 Let M denote the class of all models, and F the class of all frames. Show that if $\Sigma \Vdash^g_{\mathsf{M}} \phi$ then $\Sigma \Vdash^g_{\mathsf{F}} \phi$, but that the converse is false.

1.5.3 Let Σ be a set of formulas in the basic modal language, and let M denote the class of all models. Show that $\Sigma \Vdash^g_{\mathsf{M}} \phi$ iff $\{\Box^n \sigma \mid \sigma \in \Sigma, n \in \omega\} \Vdash_{\mathsf{M}} \phi$.

1.5.4 Again, let M denote the class of all models. Show that the local consequence relation does have the deduction theorem: $\phi \Vdash_{\mathsf{M}} \psi$ iff $\Vdash \phi \to \psi$, but the global one does not. However, show that on the class Tran of transitive models we have that $\phi \Vdash^g_{\mathsf{Tran}} \psi$ iff $\Vdash^g_{\mathsf{Tran}} \Box\phi \to \psi$.

1.6 Normal Modal Logics

Till now our discussion has been largely *semantic*; but logic has an important *syntactic* dimension, and our discussion raises some obvious questions. Suppose we are interested in a certain class of frames F: are there syntactic mechanisms capable of generating Λ_F, the formulas valid on F? And are such mechanisms capable of coping with the associated semantic consequence relation? The modal logician's response to such questions is embodied in the concept of a *normal modal logic*.

A normal modal logic is simply a set of formulas satisfying certain syntactic closure conditions. Which conditions? We will work towards the answer by defining a Hilbert-style axiom system called **K**. **K** is the 'minimal' (or 'weakest') system for reasoning about frames; stronger systems are obtained by adding extra axioms. We discuss **K** in some detail, and then, at the end of the section, define normal modal logics. By then, the reader will be in a position to see that the definition is a more-or-less immediate abstraction from what is involved in Hilbert-style approaches to modal proof theory. We will work in the basic modal language.

Definition 1.39 A **K**-*proof* is a finite sequence of formulas, each of which is an *axiom*, or follows from one or more earlier items in the sequence by applying a *rule of proof*. The axioms of **K** are *all instances of propositional tautologies* plus:

(K) $\quad \Box(p \rightarrow q) \rightarrow (\Box p \rightarrow \Box q)$
(Dual) $\quad \Diamond p \leftrightarrow \neg\Box\neg p$.

The rules of proof of **K** are:

- *Modus ponens*: given ϕ and $\phi \rightarrow \psi$, prove ψ.
- *Uniform substitution*: given ϕ, prove θ, where θ is obtained from ϕ by uniformly replacing proposition letters in ϕ by arbitrary formulas.
- *Generalization*: given ϕ, prove $\Box\phi$.

A formula ϕ is **K**-*provable* if it occurs as the last item of some **K**-proof, and if this is the case we write $\vdash_\mathbf{K} \phi$. $\qquad\dashv$

Some comments. Tautologies may contain modalities (for example, $\Diamond q \vee \neg\Diamond q$ is a tautology, as it has the same form as $p \vee \neg p$). As tautologies are valid on all frames (Exercise 1.3.4), they are a safe starting point for modal reasoning. Our decision to add *all* propositional tautologies as axioms is an example of axiomatic overkill; we could have chosen a small set of tautologies capable of generating the rest via the rules of proof, but this refinement is of little interest for our purposes.

Modus ponens is probably familiar to all our readers, but there are two important points we should make. First, *modus ponens preserves validity*. That is, if $\Vdash \phi$ and $\Vdash \phi \rightarrow \psi$ then $\Vdash \psi$. Given that we want to reason about frames, this property is crucial. Note, however, that modus ponens also preserves two further properties, namely *global truth* (if $\mathfrak{M} \Vdash \phi$ and $\mathfrak{M} \Vdash \phi \rightarrow \psi$ then $\mathfrak{M} \Vdash \psi$) and *satisfiability* (if $\mathfrak{M}, w \Vdash \phi$ and $\mathfrak{M}, w \Vdash \phi \rightarrow \psi$ then $\mathfrak{M}, w \Vdash \psi$). That is, modus ponens is not only a correct rule for reasoning about frames, it is also a correct rule for reasoning about models, both globally and locally.

Uniform substitution should also be familiar. It mirrors the fact that validity abstracts away from the effects of particular assignments: if a formula is valid, this cannot be because of the particular value its propositional symbols have, thus we should be free to uniformly replace these symbols with any other formula whatsoever. And indeed, as the reader should check, *uniform substitution preserves validity*. Note, however, that it does *not* preserve either global truth or satisfiability. (For example, q is obtainable from p by uniform substitution, but just because p is globally true in some model, it does *not* follow that q is too!) In short, uniform substitution is strictly a tool for generating new validities from old.

That is the classical core of our Hilbert system, so let us turn to the the genuinely *modal* axioms and rules of proof. First the axioms. The K axiom is the fundamental one. It is clearly *valid* (as the reader who has not done Exercise 1.3.4 should now check) but why is it a useful addition to our Hilbert system?

K is sometimes called the *distribution axiom*, and is important because it lets us transform $\Box(\phi \rightarrow \psi)$ (a boxed formula) into $\Box\phi \rightarrow \Box\psi$ (an implication). This box-over-arrow distribution enables further purely propositional reasoning to take place. For example, suppose we are trying to prove $\Box\psi$, and have constructed a proof sequence containing both $\Box(\phi \rightarrow \psi)$ and $\Box\phi$. If we could apply modus ponens under the scope of the box, we would have proved $\Box\psi$. This is what distribution lets us do: as **K** contains the axiom $\Box(p \rightarrow q) \rightarrow (\Box p \rightarrow \Box q)$, by uniform substitution we can prove $\Box(\phi \rightarrow \psi) \rightarrow (\Box\phi \rightarrow \Box\psi)$. But then a first application of modus ponens proves $\Box\phi \rightarrow \Box\psi$, and a second proves $\Box\psi$ as desired.

The Dual axiom obviously reflects the duality of $\Diamond$ and $\Box$; nonetheless, readers familiar with other discussions of **K** (many of which have K as the sole modal axiom) may be surprised at its inclusion. Do we really need it? Yes, we do. In this book, $\Diamond$ is primitive and $\Box$ is an abbreviation. Thus our K axiom is really shorthand for $\neg\Diamond\neg(p \rightarrow q) \rightarrow (\neg\Diamond\neg p \rightarrow \neg\Diamond\neg q)$. We need a way to maneuver around these negations, and this is the *syntactic* role that Dual plays. (Incidentally had we chosen $\Box$ as our primitive operator, Dual would *not* have been required.) We prefer working with a primitive $\Diamond$ (apart from anything else, it is more convenient for the algebraic work of Chapter 5) and do not mind adding Dual as an extra axiom. Dual, of course, is valid.

It only remains to discuss the modal rule of proof: *generalization* (another common name for it is *necessitation*). Generalization 'modalizes' provable formulas by stacking boxes in front. Roughly speaking, while the K axiom lets us apply classical reasoning inside modal contexts, necessitation creates new modal contexts for us to work with; modal proofs arise from the interplay of these two mechanisms.

Note that generalization preserves validity: if it is impossible to falsify ϕ, then obviously we will never be able to falsify ϕ at any accessible state! Similarly, generalization preserves *global* truth. But it *does not* preserve satisfaction: just because p is true in some state, we cannot conclude that p is true at all accessible states.

K is the minimal modal Hilbert system in the following sense. As we have seen, its axioms are all valid, and all three rules of inference preserve validity, hence all **K**-provable formulas are valid. (To use the terminology introduced in Definition 4.9, **K** is *sound* with respect to the class of all frames.) Moreover, as we will prove in Theorem 4.23, the converse is also true: *if a basic modal formula is valid, then it is **K**-provable.* (That is, **K** is *complete* with respect to the class of all frames.) In short, **K** generates precisely the valid formulas.

Example 1.40 The formula $(\Box p \wedge \Box q) \to \Box(p \wedge q)$ is valid on any frame, so it should be **K**-provable. And indeed, it is. To see this, consider the following sequence of formulas:

1.	$\vdash p \to (q \to (p \wedge q))$	Tautology
2.	$\vdash \Box(p \to (q \to (p \wedge q)))$	Generalization: 1
3.	$\vdash \Box(p \to q) \to (\Box p \to \Box q)$	K axiom
4.	$\vdash \Box(p \to (q \to (p \wedge q))) \to (\Box p \to \Box(q \to (p \wedge q)))$	
		Uniform Substitution: 3
5.	$\vdash \Box p \to \Box(q \to (p \wedge q))$	Modus Ponens: 2, 4
6.	$\vdash \Box(q \to (p \wedge q)) \to (\Box q \to \Box(p \wedge q))$	Uniform Substitution: 3
7.	$\vdash \Box p \to (\Box q \to \Box(p \wedge q))$	Propositional Logic: 5, 6
8.	$\vdash (\Box p \wedge \Box q) \to \Box(p \wedge q)$	Propositional Logic: 7

Strictly speaking, this sequence is *not* a **K**-proof – it is a subsequence of the proof consisting of the most important items. The annotations in the right-hand column should be self-explanatory; for example 'Modus Ponens: 2, 4' labels the formula obtained from the second and fourth formulas in the sequence by applying modus ponens. To obtain the full proof, fill in the items that lead from line 6 to 8. ⊣

Remark 1.41 Warning: there is a pitfall that is *very* easy to fall into if you are used to working with natural deduction systems: we *cannot* freely make and discharge assumptions in the Hilbert system **K**. The following 'proof' shows what can go wrong if we do:

1.	p	Assumption
2.	$\Box p$	Generalization: 1
3.	$p \to \Box p$	Discharge assumption

So we have 'proved' $p \to \Box p$! This is obviously wrong: this formula is *not* valid, hence it is *not* **K**-provable. And it should be clear where we have gone wrong:

we *cannot* use assumptions as input to generalization, for, as we have already remarked, this rule does *not* preserve satisfiability. Generalization is there to enable
us to generate new validities from old. It is not a local rule of inference.   ⊣

For many purposes, **K** is too weak. If we are interested in transitive frames, we
would like a proof system which reflects this. For example, we know that $\Diamond\Diamond p \rightarrow$
$\Diamond p$ is valid on all transitive frames, so we would want a proof system that generates
this formula; **K** does not do this, for $\Diamond\Diamond p \rightarrow \Diamond p$ is not valid on all frames.

But we can extend **K** to cope with many such restrictions by adding extra ax-
ioms. For example, if we enrich **K** by adding $\Diamond\Diamond p \rightarrow \Diamond p$ as an axiom, we obtain
the Hilbert system called **K4**. As we will show in Theorem 4.27, **K4** is sound and
complete with respect to the class of all transitive frames (that is, it generates *precisely* the formulas valid on transitive frames). More generally, given any set of
modal formulas Γ, we are free to add them as extra axioms to **K**, thus forming the
axiom system **KΓ**. As we will learn in Chapter 4, in many important cases it is
possible to characterize such extensions in terms of frame validity.

One final issue remains to be discussed: do such axiomatic extensions of **K** give
us a grip on semantic consequence, and in particular, the local semantic consequence relation over classes of frames (see Definition 1.35)?

In many important cases they do. Here is the basic idea. Suppose we are interested in transitive frames, and are working with **K4**. We capture the notion of local
consequence over transitive frames in **K4** as follows. Let Σ be a set of formulas,
and ϕ a formula. Then we say that ϕ is a local *syntactic* consequence of Σ in **K4**
(notation: $\Sigma \vdash_{\mathbf{K4}} \phi$) if and only if there is some finite subset $\{\sigma_1, \ldots, \sigma_n\}$ of Σ
such that $\vdash_{\mathbf{K4}} \sigma_1 \wedge \cdots \wedge \sigma_n \rightarrow \phi$. In Theorem 4.27 we will show that

$$\Sigma \vdash_{\mathbf{K4}} \phi \ \text{ iff } \ \Sigma \Vdash_{\mathsf{Tran}} \phi,$$

where $\Vdash_{\mathsf{Tran}}$ denotes local semantic consequence over transitive frames. In short,
we have reduced the local *semantic* consequence relation over transitive frames to
provability in **K4**.

Definition 1.42 (Normal Modal Logics) A *normal modal logic* Λ is a set of for-
mulas that contains all tautologies, $\Box(p \rightarrow q) \rightarrow (\Box p \rightarrow \Box q)$, and $\Diamond p \leftrightarrow \neg\Box\neg p$,
and that is closed under *modus ponens*, *uniform substitution* and *generalization*.
We call the smallest normal modal logic **K**.   ⊣

This definition is a direct abstraction from the ideas underlying modal Hilbert systems. It throws away all talk of proof sequences and concentrates on what is really
essential: the presence of axioms and closure under the rules of proof.

We will rarely mention Hilbert systems again: we prefer to work with the more
abstract notion of normal modal logics. For a start, although the two approaches
are equivalent (see Exercise 1.6.6), it is simpler to work with the set-theoretical

notion of membership than with proof sequences. More importantly, in Chapters 4 and 5 we will prove results that link the semantic and syntactic perspectives on modal logic. These results will hold for *any* set of formulas fulfilling the normality requirements. Such a set might be the formulas generated by a Hilbert-style proof system – but it could just as well be the formulas provable in a natural-deduction system, a sequent system, a tableaux system, or a display calculus. Finally, the concept of a normal modal logic makes good semantic sense: for any class of frames F, we have that Λ_F, the set of formulas valid on F, is a normal modal logic; see Exercise 1.6.7.

Exercises for Section 1.6

1.6.1 Give **K**-proofs of $(\Box p \wedge \Diamond q) \to \Diamond(p \wedge q)$ and $\Diamond(p \vee q) \leftrightarrow (\Diamond p \vee \Diamond q)$.

1.6.2 Let ϕ^- be the 'demodalized' version of a modal formula ϕ; that is, ϕ^- is obtained from ϕ by simply erasing all diamonds. Prove that ϕ^- is a propositional tautology whenever ϕ is **K**-provable. Conclude that not every modal formula is **K**-provable.

1.6.3 The axiom system known as **S4** is obtained by adding the axiom $p \to \Diamond p$ to **K4**. Show that $\nvdash_{S4} p \to \Box \Diamond p$; that is, show that **S4** does *not* prove this formula. (Hint: find an appropriate class of frames for which **S4** is *sound*.) If we add this formula as an axiom to **S4** we obtain the system called **S5**. Give an **S5**-proof of $\Diamond \Box p \to \Box p$.

1.6.4 Try adapting **K** to obtain a minimal Hilbert system for the basic temporal language. Does your system cope with the fact that we only interpret this language on bidirectional frames? Then try and define a minimal Hilbert system for the language of propositional dynamic logic.

1.6.5 This exercise is only for readers who like syntactical manipulations and have a lot of time to spare. **KL** is the axiomatization obtained by adding the Löb formula $\Box(\Box p \to p) \to \Box p$ as an extra axiom to **K**. Try and find a **KL** proof of $\Box p \to \Box\Box p$. That is, show that **KL** = **KL4**.

1.6.6 In Chapter 4 we will use **K**Γ to denote the smallest normal modal logic containing Γ; the point of the present exercise is to relate this notation to our discussion of Hilbert systems. So (as discussed above) suppose we form the axiom system **K**Γ by adding as axioms all the formulas in Γ to **K**. Show that the *Hilbert system* **K**Γ proves precisely the formulas contained in the *normal modal logic* **K**Γ.

1.6.7 Let F be a class of frames. Show that Λ_F is a normal modal logic.

1.7 Historical Overview

The ideas introduced in this chapter have a long history. They evolved as responses to particular problems and challenges, and knowing something of the context in which they arose will make it easier to appreciate why they are considered important, and the way they will be developed in subsequent chapters. Some of the discussion that follows may not be completely accessible at this stage. If so, do not worry. Just note the main points, and try again once you have explored the chapters that follow.

We find it useful to distinguish three phases in the development of modal logic: the *syntactic* era, the *classical* era, and the *modern* era. Roughly speaking, most of the ideas introduced in this chapter stem from the classical era, and the remainder of the book will explore them from the point of view of the modern era.

The syntactic era (1918–1959)

We have opted for 1918, the year that C.I. Lewis published his *Survey of Symbolic Logic* [299], as the birth of modal logic as a mathematical discipline. Lewis was certainly not the first to consider modal reasoning, indeed he was not even the first to construct symbolic systems for this purpose: Hugh MacColl, who explored the consequences of enriching propositional logic with operators ϵ ('it is certain that') and η ('it is impossible that') seems to have been the first to do that (see his book *Symbolic Logic and its Applications* [305], and for an overview of his work, see [369]). But MacColl's work is firmly rooted in the nineteenth century algebraic tradition of logic (well-known names in this tradition include Boole, De Morgan, Jevons, Peirce, Schröder, and Venn), and linking MacColl's contributions to contemporary concerns is a non-trivial scholarly task. The link between Lewis's work and contemporary modal logic is more straightforward.

In his 1918 book, Lewis extended propositional calculus with a unary modality I ('it is impossible that') and defined the binary modality $\phi \prec \psi$ (ϕ *strictly implies* ψ) to be I($\phi \wedge \neg\psi$). Strict implication was meant to capture the notion of logical entailment, and Lewis presented a $\prec$-based axiom system. Lewis and Langford's joint book *Symbolic Logic* [300], published in 1932, contains a more detailed development of Lewis's ideas. Here $\Diamond$ ('it is possible that') is primitive and $\phi \prec \psi$ is defined to be $\neg\Diamond(\phi \wedge \neg\psi)$. Five axiom systems of ascending strength, **S1**–**S5**, are discussed; **S3** is equivalent to Lewis's system of 1918, and only **S4** and **S5** are normal modal logics. Lewis's work sparked interest in the idea of 'modalizing' propositional logic, and there were many attempts to axiomatize such concepts as obligation, belief and knowledge. Von Wright's monograph *An Essay in Modal Logic* [464] is an important example of this type of work.

But in important respects, Lewis's work seems strange to modern eyes. For a start, his axiomatic systems are not modular. Instead of extending a base system of propositional logic with specifically modal axioms (as we did in this chapter when we defined **K**), Lewis defines his axioms directly in terms of $\prec$. The modular approach to modal Hilbert systems is due to Kurt Gödel. Gödel [175] showed that (propositional) intuitionistic logic could be translated into **S4** in a theorem-preserving way. However instead of using the Lewis and Langford axiomatization, Gödel took □ as primitive and formulated **S4** in the way that has become standard: he enriched a standard system for classical propositional logic with the rule of generalization, the K axiom, and the additional axioms ($\Box p \rightarrow p$ and $\Box p \rightarrow \Box\Box p$).

But the fundamental difference between current modal logic and the work of Lewis and his contemporaries is that the latter is essentially *syntactic*. Propositional logic is enriched with some new modality. By considering various axioms, the logician tries to pin down the logic of the intended interpretation. This simple view of logical modeling has its attractions, but is open to serious objections. First, there are technical difficulties. Suppose we have several rival axiomatizations of some concept. Forget for now the problem of judging which is the best, for there is a more basic difficulty: how can we tell if they are really different? If we only have access to syntactic ideas, proving that two Hilbert systems generate different sets of formulas can be extremely difficult. Indeed, even showing syntactically that two Hilbert systems generate the *same* set of formulas can be highly non-trivial (recall Exercise 1.6.5).

Proving distinctness theorems was standard activity in the syntactic era; for instance, Parry [355] showed that **S2** and **S3** are distinct, and papers addressing such problems were common till the late 1950s. Algebraic methods were often used to prove distinctness. The propositional symbols would be viewed as denoting the elements of some algebra, and complex formulas would be interpreted using the algebraic operations. Indeed, algebras were the key tool driving the technical development of the period. For example, McKinsey [322] used them to analyze **S2** and **S4** and show their decidability; McKinsey and Tarski [324, 325] and McKinsey [323] extended this work in a variety of directions (giving, among other things, a topological interpretation of **S4**); while Dummett and Lemmon [117] built on this work to isolate and analyze **S4.2** and **S4.3**, two important normal logics between **S4** and **S5**. But for all their technical utility, algebraic methods seemed of limited help in providing reliable intuitions about modal languages and their associated logics. Sometimes algebraic elements were viewed as multiple truth values. But Dugundji [116] showed that no logic between **S1** and **S5** could be viewed as an n-valued logic for *finite* n, so the multi-valued perspective on modal logic was not suited as a reliable source of insight.

The lack of a natural semantics brings up a deeper problem facing the syntactic approach: how do we know we have considered all the relevant possibilities? Nowadays the normal logic **T** (that is, **K** enriched with the axiom $p \rightarrow \Diamond p$) would be considered a fundamental logic of possibility; but Lewis overlooked **T** (it is intermediate between **S2** and **S4** and neither contains nor is contained by **S3**). Moreover, although Lewis did isolate two logics still considered important (namely **S4** and **S5**), how could he claim that either system was, in any interesting sense, *complete*? Perhaps there are important axioms missing from both systems? The existence of so many competing logics should make us skeptical of claims that it is easy to find all the relevant axioms and rules; and without precise, intuitively acceptable, criteria of what the reasonable logics are (in short, the type of criteria a decent semantics provides us with) we have no reasonable basis for claiming success.

For further discussion of the work of this period, the reader should consult the historical section of Bull and Segerberg [75]). We close our discussion of the syntactic era by noting three lines of work that anticipate later developments: Carnap's state-description semantics, Prior's work on temporal logic, and the Jónsson and Tarski Representation Theorem for boolean algebras with operators.

A *state description* is simply a collection of propositional letters. (Actually, Carnap used state descriptions in his pioneering work on first-order modal logic, so a state for Carnap could be a set of first-order formulas.) If S is a collection of state descriptions, and $s \in S$, then a propositional symbol p is satisfied at s if and only if $p \in s$. Boolean operators are interpreted in the obvious way. Finally, $\Diamond \phi$ is satisfied at $s \in S$ if and only if there is some $s' \in S$ such that s' satisfies ϕ. (See, for example, Carnap [85, 86].)

Carnap's interpretation of $\Diamond \phi$ in state descriptions is strikingly close to the idea of satisfaction in models. However one crucial idea is missing: the use of an *explicit* relation R over state descriptions. In Carnap's semantics, satisfaction for $\Diamond$ is defined in terms of membership in S (in effect, R is taken to be $S \times S$). This implicit fixing of R reduces the utility of his semantics: it yields a semantics for one fixed interpretation of $\Diamond$, but deprives us of the vital parameter needed to map logical options.

Arthur Prior founded temporal logic (or as he called it, *tense logic*) in the early 1950s. He invented the basic temporal language and many other temporal languages, both modal and non-modal. Like most of his contemporaries, Prior viewed the axiomatic exploration of concepts as one of the logician's key tasks. But there the similarity ends: his writings are packed with an extraordinary number of semantic ideas and insights. By 1955 Prior had interpreted the basic modal language in models based on $(\omega, <)$ (see Prior [364], and Chapter 2 of Prior [365]), and used what would now be called soundness arguments to distinguish logics. Moreover, the relative expressivity of modal and classical languages (such as the Prior-Meredith U-calculus [327]) is a constant theme of his writings; indeed, much of his work anticipates later work in correspondence theory and extended modal logic. His work is hard to categorize, and impossible to summarize, but one thing is clear: because of his influence temporal logic was an essentially semantically driven enterprise. The best way into his work is via Prior [365].

With the work of Jónsson and Tarski [255, 256] we reach the most important (and puzzling) might-have-beens in the history of modal logic. Briefly, Jónsson and Tarski investigated the representation theory of boolean algebras with operators (that is, modal algebras). As we have remarked, while modal algebras were useful tools, they *seemed* of little help in guiding logical intuitions. The representation theory of Jónsson and Tarski should have swept this apparent shortcoming away for good, for in essence they showed how to represent modal algebras as the structures we now call models! In fact, they did a lot more than this. Their representation technique is essentially a model building technique, hence their work gave the technical tools needed to prove the completeness result that dominated the classical era (indeed, their approach is an algebraic analog of the canonical model technique that emerged 15 years later). Moreover, they provided all this for modal languages of arbitrary similarity type, not simply the basic modal language.

Unfortunately, their work was overlooked for 20 years; not until the start of the modern era was its significance appreciated. It is unclear to us why this happened. Certainly it did not help matters that Jónsson and Tarski do not mention modal logic in their classic article; this is curious since Tarski had already published joint papers with McKinsey on algebraic approaches to modal logic. Maybe Tarski did not see the connection at all: Copeland [97, page 13] writes that Tarski heard Kripke speak about relational semantics at a 1962 talk in Finland, a talk in which Kripke stressed the importance of the work by Jónsson and Tarski. According to Kripke, following the talk Tarski approached him and said he was unable to see any connection between the two lines of work.

Even if we admit that a connection which now seems obvious may not have been so at the time, a puzzle remains. Tarski was based in California, which in the 1960s was the leading center of research in modal logic, yet in all those years, the connection was never made. For example, in 1966 Lemmon (also based in California) published a two part paper on algebraic approaches to modal logic [295] which reinvented (some of) the ideas in Jónsson and Tarski (Lemmon attributes these ideas to Dana Scott), but only cites the earlier Tarski and McKinsey papers.

We present the work by Jónsson and Tarski in Chapter 5; their Representation Theorem underpins the work of the entire chapter.

The classical era (1959–1972)

'Revolutionary' is an overused word, but no other word adequately describes the impact relational semantics (that is, the concepts of frames, models, satisfaction, and validity presented in this chapter) had on the study of modal logic. Problems which had previously been difficult (for example, distinguishing Hilbert systems) suddenly yielded to straightforward semantic arguments. Moreover, like all revolutions worthy of the name, the new world view came bearing an ambitious research program. Much of this program revolved around the concept of completeness: at last is was possible to give a precise and natural meaning to claims that a logic generated everything it ought to. (For example, **K4** could now be claimed complete in a genuinely interesting sense: it generated *all* the formulas valid on transitive frames.) Such semantic characterizations are both simple and beautiful (especially when viewed against the complexities of the preceding era) and the hunt for such results was to dominate technical work for the next 15 years. The two outstanding monographs of the classical era – the existing fragment of Lemmon and Scott's *An Introduction to Modal Logic* [296], and Segerberg's *An Essay in Classical Modal Logic* [404] – are largely devoted to completeness issues.

Some controversy attaches to the birth of the classical era. Briefly, relational semantics is often called Kripke semantics, and Kripke [283] (in which **S5**-based modal predicate logic is proved complete with respect to models with an implicit global relation), Kripke [284] (which introduces an explicit accessibility relation R and gives semantic characterization of some propositional modal logics in terms of this relation) and Kripke [285] (in which relational semantics for first-order modal languages is defined) were crucial in establishing the relational approach: they are clear, precise, and ever alert to the possibilities inherent in the new framework: for example, Kripke [285] discusses provability interpretations of propositional modal languages. Nonetheless, Hintikka had already made use of relational semantics to analyze the concept of belief and distinguish logics, and Hintikka's ideas played an important role in establishing the new paradigm in philosophical circles; see, for example, [224]. Furthermore, it has since emerged that Kanger, in a series of papers and monographs published in 1957, had introduced the basic idea of relational semantics for propositional and first-order modal logic; see, for example, Kanger [261, 262]. And a number of other authors (such as Arthur Prior, and Richard Montague [335]) had either published or spoken about similar ideas earlier. Finally, the fact remains that Jónsson and Tarski had already presented and generalized the mathematical ideas needed to analyze propositional modal logics.

But disputes over priority should not distract the reader from the essential point: somewhere around 1960 modal logic was reborn as a new field, acquiring new questions, methods, and perspectives. The magnitude of the shift, not who did what when, is what is important here. (The reader interested in more detail on who did what when, should consult Goldblatt [182]. Incidentally, Goldblatt concludes that Kripke's contributions were the most significant.)

So by the early 1960s it was clear that relational semantics was an important tool for classifying modal logics. But how could its potential be unlocked? The key tool required – the *canonical models* we discuss in Chapter 4 – emerged with surprising speed. They seem to have first been used in Makinson [307] and in Cresswell [99] (although Cresswell's so-called subordination relation differs slightly from the canonical relation), and in Lemmon and Scott [296] they appear full-fledged in the form that has become standard.

Lemmon and Scott [296] is a fragment of an ambitious monograph that was intended to cover all then current branches of modal logic. At the time of Lemmon's death in 1966, however, only the historical introduction and the chapter on the basic modal languages had been completed. Nonetheless, it is a gem. Although for the next decade it circulated only in manuscript form (it was not published until 1977) it was enormously influential, setting much of the agenda for subsequent developments. It unequivocally established the power of the canonical model technique, using it to prove general results of a sort not hitherto seen. It also introduced *filtrations*, an important technique for building finite models we will discuss in Chapter 2, and used them to prove a number of decidability results.

While Lemmon and Scott showed how to exploit canonical models directly, many important normal logics (notably, **KL** and the modal and temporal logic of structures such as $(\mathbb{N}, <)$, $(\mathbb{Z}, <)$, $(\mathbb{Q}, <)$, and $(\mathbb{R}, <)$, and their reflexive counterparts) cannot be analyzed in this way. However, as Segerberg [403, 404] showed, it is possible to use canonical models indirectly: one can transform the canonical model into the required form and prove these (and a great many other) completeness results. Segerberg-style transformation proofs are discussed in Section 4.5.

But although completeness and canonical models were the dominant issues of the classical era, there is a small body of work which anticipates more recent themes. For example, Robert Bull, swimming against the tide of fashion, used *algebraic* arguments to prove a striking result: all normal extensions of **S4.3** are characterized by classes of finite *models* (see Bull [74]). Although model-theoretic proofs of Bull's Theorem were sought (see, for example, Segerberg [404, page 170]), not until Fine [128] did these efforts succeed. Kit Fine was shortly to play a key role in the birth of the modern era, and the technical sophistication which was to characterize his later work is already evident in this paper; we discuss Fine's proof in Theorem 4.96. As a second example, in his 1968 PhD thesis [258], Hans Kamp proved one of the few (and certainly the most interesting) *expressivity* results of the era. He defined two natural binary modalities, since and until (discussed in Chapter 7), showed that the standard temporal language was not strong enough to define them, and proved that over Dedekind continuous strict total orders (such as $(\mathbb{R}, <)$) his new modalities offered full first-order expressive power.

Summing up, the classical era supplied many of the fundamental concepts and methods used in contemporary modal logic. Nonetheless, viewed from a modern perspective, it is striking how differently these ideas were put to work then. For a start, the classical era took over many of the *goals* of the syntactic era. Modal investigations still revolved round much the same group of concepts: necessity, belief, obligation and time. Moreover, although modal research in the classical era was certainly not syntactical, it was, by and large, *syntactically driven*. That is – with the notable exception of the temporal tradition – relational semantics seems to have been largely viewed as a tool for analyzing logics: soundness results could distinguish logics, and completeness results could give them nice characterizations.

Relational structures, in short, were not really there to be *described* – they were there to fulfill an analytic role. (This goes a long way towards explaining the lack of expressivity results for the basic modal language; Kamp's result, significantly, was grounded in the Priorean tradition of temporal logic.) Moreover, it was a self-contained world in a way that modern modal logic is not. Modal languages and relational semantics: the connection between them seemed clear, adequate, and well understood. Surely nothing essential was missing from this paradise?

The modern era (1972–present)

Two forces gave rise to the modern era: the discovery of frame incompleteness results, and the adoption of modal languages in theoretical computer science. These unleashed a wealth of activity which profoundly changed the course of modal logic and continues to influence it till this day. The incompleteness results forced a fundamental reappraisal of what modal languages actually *are*, while the influence of theoretical computer science radically changed expectations of *what* they could be used for, and *how* they were to be applied.

Frame-based analyses of modal logic were revealing and intoxicatingly successful – but was *every* normal logic complete with respect to some class of frames? Lemmon and Scott knew that this was a difficult question; they had shown, for example, that there were obstacles to adapting the canonical model method to analyze the logic yielded by the McKinsey axiom. Nonetheless, they conjectured that the answer was *yes*:

> However, it seems reasonable to conjecture that, if a consistent normal K-system S is *closed with respect to substitution instances* ... then S determines a class Γ_S of world systems such that $\vdash_S \mathbf{A}$ iff $\models^{\Gamma_S} \mathbf{A}$. We have no proof of this conjecture. But to prove it would be to make a considerable difference to our theoretical understanding of the general situation. [296, page 76]

Other optimistic sentiments can be found in the literature of the period. Segerberg's thesis is more cautious, simply identifying it as 'probably the outstanding question in this area of modal logic at the present time' [404, page 29].

The question was soon resolved – *negatively*. In 1972, S.K. Thomason [433] showed that there were incomplete normal logics in the basic temporal language, and in 1974 Thomason [434] and Fine [129] both published examples of incomplete normal logics in the basic modal language. Moreover, in an important series of papers Thomason showed that these results were ineradicable: as tools for talking about frames, modal languages were essentially monadic second-order logic in disguise, and hence were intrinsically highly complex.

These results stimulated what remains some of the most interesting and innovative work in the history of the subject. For a start, it was now clear that it no longer sufficed to view modal logic as an isolated formal system; on the contrary, it was evident that a full understanding of what modal languages were, required that their position in the logical universe be located as accurately as possible. Over the next few years, modal languages were to be extensively mapped from the perspective of both *universal algebra* and *classical model theory*.

Thomason [433] had already adopted an algebraic perspective on the basic temporal language. Moreover, this paper introduced general frames, showed that they were equivalent to semantics based on boolean algebras with operators, and showed that these semantics were complete in a way that the frame-based semantics was not: every normal temporal logic was characterized by some algebra. Goldblatt introduced the universal algebraic approach towards modal logic and developed modal duality theory (the categorical study of the relation between relational structures endowed with topological structure on the one hand, and boolean algebras with operators on the other). This led to a belated appreciation of the fundamental contributions made in Jónsson and Tarski's pioneering work. Goldblatt and Thomason showed that the concepts and results of universal algebra could be applied to yield modally interesting results; the best known example of this is the Goldblatt-Thomason Theorem, a model theoretic characterization of modally definable frame classes obtained by applying the Birkhoff Variety Theorem to boolean algebras with operators. We discuss such work in Chapter 5 (and in Chapter 3 we discuss the Goldblatt-Thomason Theorem from the perspective of first-order model theory). Work by Blok made deeper use of algebras, and universal algebra became a key tool in the exploration of completeness theory. The revival of algebraic semantics – together with a genuine appreciation of *why* it was so important – is one of the most enduring legacies of this period.

But the modern period also firmly linked modal languages with classical model theory. One line of inquiry that led naturally in this direction was the following: given that modal logic was essentially second-order in nature, why was it so often first-order, and very simple first-order at that? That is, from the modern perspective, incomplete normal logics were to be expected – it was the elegant results of the classical period that now seemed in need of explanation. One type of answer was given in the work of Sahlqvist [396], who isolated a large set of axioms which guaranteed completeness with respect to first-order definable classes of frames. (We define the Sahlqvist fragment in Section 3.6, where we discuss the Sahlqvist Correspondence Theorem, an expressivity result. The twin Sahlqvist Completeness Theorem is proved algebraically in Theorem 5.91.) Another type of answer was developed in Fine [132] and van Benthem [40, 41]; we discuss this work (albeit from an algebraic perspective) in Chapter 5.

A different line of work also linked modal and classical languages: an investigation of modal languages viewed purely as *description languages*. The classical era largely ignored expressivity in favor of completeness, but the Sahlqvist Correspondence Theorem showed the narrowness of this perspective: a beautiful result about the basic modal language that did not even mention normal modal logics! Expressivity issues were subsequently studied by van Benthem, who developed the subject now known as *correspondence theory*, along two main lines; see [42, 43]. One views modal languages as tools for describing *frames* (that is, as second-order description languages) and probes their expressive power. This line of investigation, together with Sahlqvist's work, forms the basis of Chapter 3. The second line explores modal languages as tools for talking about *models*, an intrinsically first-order perspective. This lead van Benthem to isolate the concept of a *bisimulation*, and prove the fundamental Characterization Theorem: viewed as a tool for talking about models, modal languages are the bisimulation invariant fragment of the corresponding first-order language. Bisimulation driven investigations of modal expressivity are now standard, and much of Chapter 2 is devoted to such issues.

The impact of theoretical computer science was less dramatic than the discovery of the incompleteness results, but its influence has been equally profound. Burstall [82] already suggests using modal logic to reason about programs, but the birth of this line of work really dates from Pratt [362] (the paper which gave rise to PDL) and Pnueli [359] (which suggested using temporal logic to reason about execution-traces of programs). Computer scientists tended to develop powerful modal languages; PDL in its many variants is an obvious example (see Harel [209] for a detailed survey). And since the appearance of Gabbay *et al.* [160] the temporal languages used by computer scientists typically contain the until operator, and often additional operators which are evaluated with respect to *paths* (see Clarke and Emerson [94]). Gabbay also noted the significance of Rabin's Theorem [368] for modal decidability (we discuss this in Chapter 6), and applied it to a wide range of languages and logics; see Gabbay [145, 146, 147].

Computer scientists brought a new array of questions to the study of modal logic. For a start, they initiated the study of the computational complexity of normal logics. Already by 1977 Ladner [292] had showed that every normal logic between **K** and **S4** had a PSPACE-hard satisfiability problem, while the results of Fischer and Ladner [135] and Pratt [363] together show that PDL has an EXPTIME-complete satisfiability problem. (These results are proved in Chapter 6.) Moreover, the interest of the modal expressivity studies emerging in correspondence theory was reinforced by several lines of work in computer science. To give one particularly nice example, computer scientists studying concurrent systems independently isolated the notion of bisimulation (see Park [354]). This paved the way for the work of Hennessy and Milner [219] who showed that weak modal languages could be used to classify various notions of process invariance.

But one of the most significant endowments from computer science has actually been something quite simple: it has helped remove a lingering tendency to see modal languages as intrinsically 'intensional' formalisms, suitable only for analyzing such concepts as knowledge, obligation and belief. During the 1990s this point was strongly emphasized when connections were discovered between modal logic and knowledge representation formalisms. In particular, *description logics* are a family of languages that come equipped with effective reasoning methods, and a special focus on balancing expressive power and computational and algorithmic complexity; see Donini *et al.* [115]. The discovery of this connection has lead to a renewed focus on efficient reasoning methods, dedicated languages that are fine-tuned for specific modeling tasks, and a variety of novel uses of modal languages; see Schild [400] for the first paper to make the connection between the two fields, and De Giacomo [106], Areces [12], and Areces and de Rijke [15] for work exploiting the connection.

And this is but one example. Links with computer science and other disciplines have brought enormous richness and variety to modal logic. Computer science has seen a shift of emphasis from isolated programs to complex entities collaborating in heterogeneous environments; this gives rise to new challenges for the use of modal logic in theoretical computer science. For instance, agent-based theories require flexible modeling facilities together with efficient reasoning mechanisms; see Wooldridge and Jennings [463] for a discussion of the agent paradigm, and Bennet *et al.* [34] for the link with modal logic. More generally, complex computational architectures call for a variety of combinations of modal languages; see the proceedings of the *Frontiers of Combining Systems* workshop series for references [16, 152, 268].

Similar developments took place in foundational research in economics. Game theory (Osborne and Rubinstein [350]) also shows a nice interplay between the notions of action and knowledge; recent years have witnessed an increasing tendency to give a formal account of epistemic notions; see Battigalli and Bonanno [31] or Kaneko and Nagashima [260]. For modal logics that combine dynamic and epistemic notions to model games we refer to Baltag [21] and van Ditmarsch [109].

Further examples abound. Database theory continues to be a fruitful source of questions for logicians, modal or otherwise. For instance, developments in temporal databases have given rise to new challenges for temporal logicians (see Finger [134]), while description logicians have found new applications for their modeling and reasoning methods in the area of semistructured data (see Calvanese *et al.* [84]). In the related, but more philosophically oriented area of belief revision, Fuhrmann [144] has given a modal formalization of one of the most influential approaches in the area, the AGM approach [4]. Authors such as Friedman and Halpern [142], Gerbrandy and Groeneveld [170], de Rijke [384], and Segerberg [410] have discussed various alternative modal formalizations.

Cognitive phenomena have long been of interest to modal logicians. This is clear from examples such as belief revision, but perhaps even more so from language-related work in modal logic. The feature logic mentioned in Example 1.17 is but one example; authors such as Blackburn, Gardent, Meyer-Viol, and Spaan [61, 58], Kasper and Rounds [266, 394], Kracht [280], Kurtonina [287], and Reape [376] have offered a variety of modal logical perspectives on grammar formalisms. Others have analyzed the semantics of natural language by modal means; see Fernando [126] for a sample of modern work along these lines.

During the 1980s and 1990s a number of new themes on the interface of modal logic and mathematics received considerable attention. One of these themes concerns links between modal logic and non-wellfounded set theory; work that we should mention here includes Aczel [2], Barwise and Moss [27], and Baltag [20, 22]; see the Notes to Chapter 2 for further discussion. Non-well-founded sets and many other notions, such as automata and labeled transition systems, have been brought together under the umbrella of co-algebras (see Jacobs and Rutten [242]), which form a natural and elegant way to model state-based dynamic systems. Since it was discovered that modal logic is as closely related to co-algebras as equational logic is to algebras, there has been a wealth of results reporting on this connection; we only mention Jacobs [241], Kurz [290] and Rößiger [393] here.

Another 1990s theme on the interface of modal logic and mathematics concerns an old one: geometry. Work by Balbiani *et al.* [19], Stebletsova [423] and Venema [448] indicates that modal logic may have interesting things to say about geometry, while Aiello and van Benthem [3] and Lemon and Pratt [297] investigate the potential of modal logic as a tool for reasoning about space.

As should now be clear to all our readers, the simple question posed by the modal satisfaction definition – what happens at accessible states? – gives us a natural way of working with *any* relational structure. This has opened up a host of new applications for modal logic. Moreover, once the relational perspective has been fully assimilated, it opens up rich new approaches to traditional subjects: see van Benthem [45] and Fagin, Halpern, Moses, and Vardi [125] for thoroughly modern discussions of temporal logic and epistemic logic respectively.

1.8 Summary of Chapter 1

▶ *Relational Structures*: A relational structure is a set together with a collection of relations. Relational structures can be used to model key ideas from a wide range of disciplines.

▶ *Description Languages*: Modal languages are simple languages for describing relational structures.

▶ *Similarity Types*: The basic modal language contains a single primitive unary operator $\Diamond$. Modal languages of arbitrary similarity type may contain many modalities $\triangle$ of arbitrary arity.

▶ *Basic Temporal Language*: The basic temporal language has two operators F

and P whose intended interpretations are 'at some time in the future' and 'at some time in the past.'

▶ *Propositional Dynamic Logic*: The language of propositional dynamic logic has an infinite collection of modal operators indexed by programs π built up from atomic programs using union $\cup$, composition ; and iteration *; additional constructors such as intersection $\cap$ and test ? may also be used. The intended interpretation of $\langle\pi\rangle\phi$ is 'some terminating execution of program π leads to a state where ϕ holds.'

▶ *Arrow Logic*: The language of arrow logic is designed to talk about any object that may be represented by arrows; it has a modal constant 1' ('skip'), a unary operator $\otimes$ ('converse'), and a dyadic operator $\circ$ ('composition').

▶ *Satisfaction*: The *satisfaction definition* is used to interpret formulas inside models. This satisfaction definition has an obvious local flavor: modalities are interpreted as scanning the states accessible from the current state.

▶ *Validity*: A formula is *valid* on a frame when it is globally true, no matter what valuation is used. This concept allows modal languages to be viewed as languages for describing frames.

▶ *General Frames*: Modal languages can also be viewed as talking about general frames. A general frame is a frame together with a set of admissible valuations. General frames offer some of the advantages of both models and frames and are an important technical tool.

▶ *Semantic Consequence*: Semantic consequence relations for modal languages need to be relativized to classes of structures. The classical idea that the truth of the premises should guarantee the truth of the conclusion can be interpreted either locally or globally. In this book we almost exclusively use the local interpretation.

▶ *Normal Modal Logics*: Normal modal logics are the unifying concept in modal proof theory. Normal modal logics contain all tautologies, the K axiom and the Dual axiom; in addition they should be closed under modus ponens, uniform substitution and generalization.

2

Models

In Section 1.3 we defined what it means for a formula to be *satisfied* at a state in a model – but as yet we know virtually nothing about this fundamental semantic notion. What exactly can we say about models when we use modal languages to describe them? Which properties of models can modal languages express, and which lie beyond their reach?

In this chapter we examine such questions in detail. We introduce *disjoint unions, generated submodels, bounded morphisms*, and *ultrafilter extensions*, the 'big four' operations on models that leave modal satisfaction unaffected. We discuss two ways to obtain finite models and show that modal languages have the *finite model property*. Moreover, we define the *standard translation* of modal logic into first-order logic, thus opening the door to *correspondence theory*, the systematic study of the relationship between modal and classical logic. All this material plays a fundamental role in later work; indeed, the basic track sections in this chapter are among the most important in the book.

But the central concept of the chapter is that of a *bisimulation* between two models. Bisimulations reflect, in a particularly simple and direct way, the locality of the modal satisfaction definition. We introduce them early on, and they gradually come to dominate our discussion. By the end of the chapter we will have a good understanding of modal expressivity over models, and the most interesting results all hinge on bisimulations.

Chapter guide

Section 2.1: Invariance Results (Basic track). We introduce three classic ways of constructing new models from old ones that do not affect modal satisfaction: disjoint unions, generated submodels, and bounded morphisms. We also meet isomorphisms and embeddings.

Section 2.2: Bisimulations (Basic track). We introduce bisimulations and show that modal satisfaction is invariant under bisimulation. We will see that the model constructions introduced in the first section are all special cases of bisimulation, learn that modal equivalence does not always imply bisimilarity, and examine an important special case in which it does.

Section 2.3: Finite Models (Basic track). Here we show that modal languages enjoy the finite model property. We do so in two distinct ways: by the selection method (finitely approximating a bisimulation), and by filtration (collapsing a model into a finite number of equivalence classes).

Section 2.4: The Standard Translation (Basic track). We start our study of correspondence theory. By defining the standard translation, we link modal languages to first-order (and other classical) languages and raise the two central questions that dominate later sections: What part of first-order logic does modal logic correspond to? And which properties of models are definable by modal means?

Section 2.5: Modal Saturation via Ultrafilter Extensions (Basic track). The first step towards obtaining some answers is to introduce ultrafilter extensions, the last of the big four modal model constructions. We then show that although modal equivalence does not imply bisimilarity, it does imply bisimilarity somewhere else, namely in the ultrafilter extensions of the models concerned.

Section 2.6: Characterization and Definability (Advanced track). We prove the two main results of this chapter. First, we prove van Benthem's Theorem stating that modal languages are the bisimulation invariant fragments of first-order languages. Second, we show that modally definable classes of (pointed) models are those that are closed under bisimulations and ultraproducts and whose complements are closed under ultrapowers.

Section 2.7: Simulation and Safety (Advanced track). We prove two results that give the reader a glimpse of recent work in modal model theory. The first describes the properties that are preserved under simulations (a one-way version of bisimulation), the second characterizes the first-order definable operations on binary relations which respect bisimilarity.

2.1 Invariance Results

Mathematicians rarely study structures in isolation. They are usually interested in the relations between different structures, and in operations that build new structures from old. Questions that naturally arise in such contexts concern the structural properties that are *invariant* under, or are *preserved* by, such relations and operations. We will not give precise definitions of these notions, but roughly speaking, a property is preserved by a certain relation or operation if, whenever two structures are linked by the relation or operation, then the second structure has the property if the first one has it. We speak of invariance if the property is preserved in both directions.

Logicians add a descriptive twist to this. For example, modal logicians want to know when two structures, or perhaps two points in distinct structures, are indistinguishable by modal languages. That is, when do they satisfy exactly the same modal formulas?

Definition 2.1 Let $\mathfrak{M}$ and $\mathfrak{M}'$ be models of the same modal similarity type τ, and let w and w' be states in $\mathfrak{M}$ and $\mathfrak{M}'$ respectively. The τ-*theory* (or τ-*type*) *of* w is the set of all τ-formulas satisfied at w: that is, $\{\phi \mid \mathfrak{M}, w \Vdash \phi\}$. We say that w and w' are *(modally) equivalent* (notation: $w \leftrightsquigarrow w'$) if they have the same τ-theories.

The τ-*theory* of the model $\mathfrak{M}$ is the set of all τ-formulas satisfied by all states in $\mathfrak{M}$: that is, $\{\phi \mid \mathfrak{M} \Vdash \phi\}$. Models $\mathfrak{M}$ and $\mathfrak{M}'$ are called *(modally) equivalent* (notation: $\mathfrak{M} \leftrightsquigarrow \mathfrak{M}'$) if their theories are identical.   ⊣

We now introduce three important ways of constructing new models from old ones which leave the theories associated with states unchanged: *disjoint unions*, *generated submodels*, and *bounded morphisms*. These constructions (together with *ultrafilter extensions*, which we introduce in Section 2.5) play an important role throughout the book. For example, in the following chapter we will see that they lift to the level of frames (where they preserve validity), we will use them repeatedly in our work on completeness and complexity, and in Chapter 5 we will see that they have important algebraic analogs.

Disjoint Unions

Suppose we have the following two models:

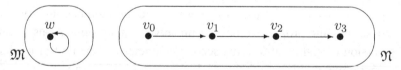

Do not worry that we have not specified the valuations – they are irrelevant here. All that matters is that $\mathfrak{M}$ and $\mathfrak{N}$ have disjoint domains, for we are now going to lump them together to form the model $\mathfrak{M} \uplus \mathfrak{N}$:

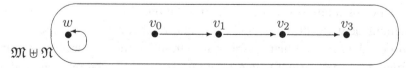

The model $\mathfrak{M} \uplus \mathfrak{N}$ is called the *disjoint union* of $\mathfrak{M}$ and $\mathfrak{N}$. It gathers together all the information in the two smaller models unchanged: we have not altered the way the points are related, nor the way atomic information is distributed. Suppose we are working in the basic modal language, and suppose that a formula ϕ is true at (say) v_1 in $\mathfrak{N}$: is ϕ still true at v_1 in $\mathfrak{M} \uplus \mathfrak{N}$? More generally, is modal satisfaction *preserved* from points in the original models to the points in the disjoint union? And what about the reverse direction: if a modal formula is true at some state in $\mathfrak{M} \uplus \mathfrak{N}$, is it also true at that same state in the smaller model it came from?

The answer to these questions is clearly *yes*: modal satisfaction must be *invariant* (that is, preserved in both directions) under the formation of disjoint unions. Modal satisfaction is intrinsically local: only the points accessible from the current state are relevant to truth or falsity. If we evaluate a formula ϕ at (say) w, it is completely irrelevant whether we perform the evaluation in $\mathfrak{M}$ or $\mathfrak{M} \uplus \mathfrak{N}$; ϕ simply cannot detect the presence or absence of states in other islands.

Definition 2.2 (Disjoint Unions) We first define disjoint unions for the basic modal language. We say that two models are *disjoint* if their domains contain no common elements. For disjoint models $\mathfrak{M}_i = (W_i, R_i, V_i)$ $(i \in I)$, their *disjoint union* is the structure $\uplus_i \mathfrak{M}_i = (W, R, V)$, where W is the union of the sets W_i, R is the union of the relations R_i, and for each proposition letter p, $V(p) = \bigcup_{i \in I} V_i(p)$.

Now for the general case. For disjoint τ-structures $\mathfrak{M}_i = (W_i, R_{\Delta i}, V_i)_{\Delta \in \tau}$ $(i \in I)$ of the same modal similarity type τ, their *disjoint union* is the structure $\uplus_i \mathfrak{M}_i = (W, R_\Delta, V)_{\Delta \in \tau}$ such that W is the union of the sets W_i; for each $\Delta \in \tau$, R_Δ is the union $\bigcup_{i \in I} R_{\Delta i}$; and V is defined as in the basic modal case.

If we want to put together a collection of models that are *not* disjoint, we first have to make them disjoint (say by indexing the domains of these models). To use the terminology introduced shortly, we simply take mutually disjoint isomorphic copies of the models we wish to combine, and combine the copies instead. $\dashv$

Proposition 2.3 *Let τ be a modal similarity type and, for all $i \in I$, let $\mathfrak{M}_i$ be a τ-model. Then, for each modal formula ϕ, for each $i \in I$, and each element w of $\mathfrak{M}_i$, we have $\mathfrak{M}_i, w \Vdash \phi$ iff $\uplus_{i \in I} \mathfrak{M}_i, w \Vdash \phi$. In words: modal satisfaction is invariant under disjoint unions.*

Proof. We will prove the result for the basic similarity type. The proof is by induction on ϕ. Let i be some index; we will prove, for each basic modal formula ϕ, and each element w of $\mathfrak{M}_i$, that $\mathfrak{M}_i, w \Vdash \phi$ iff $\mathfrak{M}, w \Vdash \phi$, where $\mathfrak{M}$ is the disjoint union $\uplus_{i \in I} \mathfrak{M}_i$.

First suppose that ϕ contains no connectives. Now, if ϕ is a proposition letter p, then we have $\mathfrak{M}_i, w \Vdash \phi$ iff $w \in V_i(p)$ iff (by definition of V) $w \in V(p)$ iff $\mathfrak{M}, w \Vdash \phi$. On the other hand, ϕ could be $\bot$ (for the purposes of inductive proofs it is convenient to regard $\bot$ as a propositional letter rather than as a logical connective). But trivially $\bot$ is false at w in both models, so we have the desired equivalence here too.

Our inductive hypothesis is that the desired equivalence holds for all formulas containing at most n connectives (where $n \geq 0$). We must now show that the equivalence holds for all formulas ϕ containing $n+1$ connectives. Now, if ϕ is of the form $\neg\psi$ or $\psi \vee \theta$ this is easily done – we will leave this to the reader – so as we are working with the basic similarity type, it only remains to establish the equivalence for formulas of the form $\Diamond\psi$. So assume that $\mathfrak{M}_i, w \Vdash \Diamond\psi$. Then there is a state v in $\mathfrak{M}_i$ with $R_i wv$ and $\mathfrak{M}_i, v \Vdash \psi$. By the inductive hypothesis, $\mathfrak{M}, v \Vdash \psi$. But by definition of $\mathfrak{M}$, we have Rwv, so $\mathfrak{M}, w \Vdash \Diamond\psi$.

For the other direction, assume that $\mathfrak{M}, w \Vdash \Diamond\psi$ holds for some w in $\mathfrak{M}_i$. Then there is a v with Rwv and $\mathfrak{M}, v \Vdash \psi$. It follows by the definition of R that $R_j wv$ for some j, and by the disjointness of the universes we must have that $j = i$. But then we find that v belongs to $\mathfrak{M}_i$ as well, so we may apply the inductive hypothesis; this yields $\mathfrak{M}_i, v \Vdash \psi$, so we find that $\mathfrak{M}_i, w \Vdash \Diamond\psi$.   ⊣

We will use Proposition 2.3 all through the book – here is a simple application which hints at the ideas we will explore in Chapter 7.

Example 2.4 Defined modalities are a convenient shorthand for concepts we find useful. We have already seen some examples. In this book $\Box$, the 'true at all accessible states modality,' is shorthand for $\neg\Diamond\neg$, and we have inductively defined a 'true somewhere n-steps from here' modality $\Diamond^n$ for each natural number n (see Example 1.22). But while it is usually easy to show that some modality *is* definable (we need simply write down its definition), how do we show that some proposed operator is *not* definable? Via invariance results! As an example, consider the *global modality*. The global diamond E has as its (intended) accessibility relation the relation $W \times W$ implicitly present in any model. That is:

$$\mathfrak{M}, w \Vdash \text{E}\phi \text{ iff } \mathfrak{M}, v \Vdash \phi \text{ for } some \text{ state } v \text{ in } \mathfrak{M}.$$

Its dual, A, the global box, thus has the following interpretation:

$$\mathfrak{M}, w \Vdash \text{A}\phi \text{ iff } \mathfrak{M}, v \Vdash \phi \text{ for } all \text{ states } v \text{ in } \mathfrak{M}.$$

Thus the global modality brings a genuinely global dimension to modal logic. But is it definable in the basic modal language? Intuitively, *no*: as $\Diamond$ and $\Box$ work locally, it seems unlikely that they can define a truly global modality over arbitrary structures. Fine – but how do we *prove* this?

With the help of the previous proposition. Suppose we could define A. Then we could write down an expression $\alpha(p)$ containing only symbols from the basic modal language such that for every model $\mathfrak{M}$, $\mathfrak{M}, w \Vdash \alpha(p)$ iff $\mathfrak{M} \Vdash p$. We now derive a contradiction from this supposition. Consider a model $\mathfrak{M}_1$ where p holds everywhere, and a model $\mathfrak{M}_2$ where p holds nowhere. Let w be some point in $\mathfrak{M}_1$. It follows that $\mathfrak{M}_1, w \Vdash \alpha(p)$, so as (by assumption) $\alpha(p)$ contains only symbols from the basic modal language, by Proposition 2.3 we have that $\mathfrak{M}_1 \uplus \mathfrak{M}_2, w \Vdash \alpha(p)$. But this implies that $\mathfrak{M}_1 \uplus \mathfrak{M}_2, v \Vdash p$ for every v in $\mathfrak{M}_2$, which, again by Proposition 2.3, in turn implies that $\mathfrak{M}_2 \Vdash p$: contradiction. We conclude that the global box (and hence the global diamond) is *not* definable in the basic modal language.

So, if we want the global modality, then we either have to introduce it as a primitive (we will do this in Section 7.1), or we have to work with restricted classes of models on which it *is* definable (in Exercise 1.3.3 we worked with a class of models in which we could define A in the basic temporal language). ⊣

Generated submodels

Disjoint unions are a useful way of making bigger models from smaller ones – but we also want methods for doing the reverse. That is, we would like to know when it is safe to throw points away from a model without affecting satisfiability. Disjoint unions tell us a little about this (if a model is a disjoint union of smaller models, we are free to work with the component models), but in practice we usually need something sharper: *generated submodels*.

Suppose we are using the basic modal language to talk about a model $\mathfrak{M}$ based on the frame $(\mathbb{Z}, <)$, the integers with their usual order. It does not matter what the valuation is – all that is important is that $\mathfrak{M}$ looks something like this:

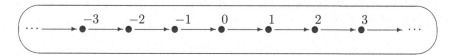

First suppose that we form a *submodel* $\mathfrak{M}^-$ of $\mathfrak{M}$ by throwing away all the positive numbers, and restricting the original valuation (whatever it was) to the remaining numbers. So $\mathfrak{M}^-$ looks something like this:

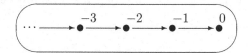

The basic modal language certainly *can* see that $\mathfrak{M}$ and $\mathfrak{M}^-$ are different. For example, it sees that 0 has successors in $\mathfrak{M}$ (note that $\mathfrak{M}, 0 \Vdash \Diamond\top$) but is a dead end in $\mathfrak{M}^-$ (note that $\mathfrak{M}^-, 0 \not\Vdash \Diamond\top$). So there is no invariance result for *arbitrary* submodels. But now consider the submodel $\mathfrak{M}^+$ of $\mathfrak{M}$ that is formed by omitting the negative numbers, and restricting the original valuation to the numbers that remain:

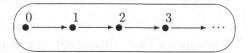

Suppose a basic modal formula ϕ is satisfied at some point n in $\mathfrak{M}$. Is ϕ also satisfied at the same point n in $\mathfrak{M}^+$? The answer must be *yes*. The only points that are relevant to ϕ's satisfiability are the points greater than n – *and all such points belong to* $\mathfrak{M}^+$. Similarly, it is clear that if $\mathfrak{M}^+$ satisfies a basic modal formula ϕ at m, then $\mathfrak{M}$ must too.

In short, it seems plausible that modal invariance holds for submodels which are closed under the accessibility relation of the original model. Such models are called *generated submodels*, and they do indeed give rise to the invariance result we are looking for.

Definition 2.5 (Generated Submodels) We first define generated submodels for the basic modal language. Let $\mathfrak{M} = (W, R, V)$ and $\mathfrak{M}' = (W', R', V')$ be two models; we say that $\mathfrak{M}'$ is a *submodel* of $\mathfrak{M}$ if $W' \subseteq W$, R' is the restriction of R to W' (that is: $R' = R \cap (W' \times W')$), and V' is the restriction of V to $\mathfrak{M}'$ (that is: for each p, $V'(p) = V(p) \cap W'$). We say that $\mathfrak{M}'$ is a *generated submodel* of $\mathfrak{M}$ (notation: $\mathfrak{M}' \rightarrowtail \mathfrak{M}$) if $\mathfrak{M}'$ is a submodel of $\mathfrak{M}$ and for all points w the following closure condition holds:

$$\text{if } w \text{ is in } \mathfrak{M}' \text{ and } Rwv, \text{ then } v \text{ is in } \mathfrak{M}'.$$

For the general case, we say that a model $\mathfrak{M}' = (W', R'_\triangle, V')_{\triangle \in \tau}$ is a *generated submodel* of the model $\mathfrak{M} = (W, R_\triangle, V)_{\triangle \in \tau}$ (notation: $\mathfrak{M}' \rightarrowtail \mathfrak{M}$) whenever $\mathfrak{M}'$ is a submodel of $\mathfrak{M}$ (with respect to $R_\triangle$ for all $\triangle \in \tau$), and the following closure condition is fulfilled for all $\triangle \in \tau$

$$\text{if } u \in W' \text{ and } R_\triangle u u_1 \ldots u_n, \text{ then } u_1, \ldots, u_n \in W'.$$

Let $\mathfrak{M}$ be a model, and X a subset of the domain of $\mathfrak{M}$; the *submodel generated by X* is the smallest generated submodel of $\mathfrak{M}$ whose domain contains X (such a model always exists: why?). Finally, a *rooted* or *point generated* model is a model that is generated by a singleton set, the element of which is called the *root* of the frame.   ⊣

Proposition 2.6 *Let τ be a modal similarity type and let $\mathfrak{M}$ and $\mathfrak{M}'$ be τ-models such that $\mathfrak{M}'$ is a generated submodel of $\mathfrak{M}$. Then, for each modal formula ϕ and each element w of $\mathfrak{M}'$ we have that $\mathfrak{M}, w \Vdash \phi$ iff $\mathfrak{M}', w \Vdash \phi$. In words: modal satisfaction is invariant under generated submodels.*

Proof. By induction on ϕ. The reader unused to such proofs should write out the proof in full. In Proposition 2.19 we provide an alternative proof based on the observation that generated submodels induce a bisimulation.   ⊣

Four remarks. First, note that the invariance result for disjoint unions (Proposition 2.3) is a special case of the result for generated submodels: any component of a disjoint union is a generated submodel of the disjoint union. Second, using an argument analogous to that used in Example 2.4 to show that the global box cannot be defined in the basic modal language, we can use Proposition 2.6 to show that we cannot define a backward looking modality in terms of $\diamond$; see Exercise 2.1.2. Thus if we want such a modality we have to add it as a primitive – which is exactly what we did, of course, when defining the basic temporal language. Third, although we have not explicitly discussed generated submodels for the basic temporal language, PDL, or arrow logic, the required concepts are all special cases of Definition 2.5, and thus the respective invariance results are special cases of Proposition 2.6. But it is worth making a brief comment about the basic temporal language. When we think explicitly in terms of bidirectional frames (see Example 1.25) it is obvious that we are interested in submodels closed under both R_F and R_P. But when working with the basic temporal language we usually leave R_P implicit: we work with ordinary models (W, R, V), and use $\breve{R}$, the converse of R, as R_P. Thus a *temporal* generated submodel of (W, R, V) is a submodel (W', R', V') that is closed under both R and $\breve{R}$. Finally, generated submodels are heavily used throughout the book: given a model $\mathfrak{M}$ that satisfies a formula ϕ at a state w, very often the first thing we will do is form the submodel of $\mathfrak{M}$ generated by w, thus trimming what may be a very unwieldy satisfying model down to a more manageable one.

Morphisms for modalities

In mathematics the idea of *morphisms* or *structure preserving maps* is of fundamental importance. What notions of morphism are appropriate for modal logic? That is, what kinds of morphism give rise to invariance results? We will approach the answer bit by bit, introducing a number of important concepts on the way. We will start by considering the general notion of *homomorphism* (this is too weak to yield invariance, but it is the starting point for better attempts), then we will define *strong homomorphisms*, *embeddings*, and *isomorphisms* (these do give us invariance, but are not particularly modal), and finally we will zero in on the answer: *bounded morphisms*.

Definition 2.7 (Homomorphisms) Let τ be a modal similarity type and let $\mathfrak{M}$ and $\mathfrak{M}'$ be τ-models. By a *homomorphism* f from $\mathfrak{M}$ to $\mathfrak{M}'$ (notation: $f : \mathfrak{M} \to \mathfrak{M}'$) we mean a function f from W to W' with the following properties:

(i) For each proposition letter p and each element w from $\mathfrak{M}$, if $w \in V(p)$, then $f(w) \in V'(p)$.

(ii) For each $n \geq 0$ and each n-ary $\triangle \in \tau$, and $(n + 1)$-tuple $\overline{w}$ from $\mathfrak{M}$, if $(w_0, \ldots, w_n) \in R_\triangle$ then $(f(w_0), \ldots, f(w_n)) \in R'_\triangle$ (the *homomorphic condition*).

We call $\mathfrak{M}$ the *source* and $\mathfrak{M}'$ the *target* of the homomorphism.   ⊣

Note that for the basic modal language, item (ii) is just this:

   if Rwu then $R'f(w)f(u)$.

Thus item (ii) simply says that homomorphisms preserve relational links.

   Are modal formulas invariant under homomorphisms? No: although homomorphisms reflect the structure of the source in the structure of the target, they do not reflect the structure of the target back in the source. It is easy to turn this observation into a counterexample, and we will leave this task to the reader as Exercise 2.1.3.

   So let us try and strengthen the definition. There is an obvious way of doing so: turn the conditionals into equivalences. This leads to a number of important concepts.

Definition 2.8 (Strong Homomorphisms, Embeddings and Isomorphisms) Let τ be a modal similarity type and let $\mathfrak{M}$ and $\mathfrak{M}'$ be τ-models. By a *strong homomorphism* of $\mathfrak{M}$ into $\mathfrak{M}'$ we mean a homomorphism $f : \mathfrak{M} \to \mathfrak{M}'$ which satisfies the following stronger version of the above items (i) and (ii):

(i) For each proposition letter p and element w from $\mathfrak{M}$, $w \in V(p)$ iff $f(w) \in V'(p)$.

(ii) For each $n \geq 0$ and each n-ary $\triangle$ in τ and $(n+1)$-tuple $\overline{w}$ from $\mathfrak{M}$, $(w_0, \ldots, w_n) \in R_\triangle$ iff $(f(w_0), \ldots, f(w_n)) \in R'_\triangle$ (the *strong homomorphic condition*).

An *embedding* of $\mathfrak{M}$ into $\mathfrak{M}'$ is a strong homomorphism $f : \mathfrak{M} \to \mathfrak{M}'$ which is injective. An *isomorphism* is a bijective strong homomorphism. We say that $\mathfrak{M}$ is *isomorphic* to $\mathfrak{M}'$, in symbols $\mathfrak{M} \cong \mathfrak{M}'$, if there is an isomorphism from $\mathfrak{M}$ to $\mathfrak{M}'$.   ⊣

Note that for the basic modal language, item (ii) is just:

   Rwu iff $R'f(w)f(u)$.

That is, item (ii) says that relational links are preserved from the source model to the target *and back again*. So it is not particularly surprising that we have a number of invariance results.

Proposition 2.9 *Let τ be a modal similarity type and let $\mathfrak{M}$ and $\mathfrak{M}'$ be τ-models. Then the following holds:*

(i) *For all elements w and w' of $\mathfrak{M}$ and $\mathfrak{M}'$, respectively, if there exists a surjective strong homomorphism $f : \mathfrak{M} \to \mathfrak{M}'$ with $f(w) = w'$, then w and w' are modally equivalent.*

(ii) *If* $\mathfrak{M} \cong \mathfrak{M}'$, *then* $\mathfrak{M} \rightsquigarrow \mathfrak{M}'$.

Proof. The first item follows by induction on ϕ; the second one is an immediate consequence. ⊣

None of the above results is particularly modal. For a start, as in all branches of mathematics, 'isomorphic' basically means 'mathematically identical.' Thus, we do not want to be able to distinguish isomorphic structures in modal (or indeed, any other) logic. Quite the contrary: we want to be free to work with structures 'up to isomorphism' – as we did, for example, in our discussion of disjoint union, when we talked of taking isomorphic copies. Item (ii) tells us that we can do this, but it is not a surprising result.

But why is item (i), the invariance result for strong homomorphisms, not 'genuinely modal'? Quite simply, because there are many morphisms which do give rise to invariance, but which fail to qualify as strong homomorphisms. To ensure modal invariance we need to ensure that some target structure is reflected back in the source, but strong morphisms do this in a much too heavy-handed way. The crucial concept is more subtle.

Definition 2.10 (Bounded Morphisms – the Basic Case) We first define bounded morphisms for the basic modal language. Let $\mathfrak{M}$ and $\mathfrak{M}'$ be models for the basic modal language. A mapping $f : \mathfrak{M} = (W, R, V) \rightarrow \mathfrak{M}' = (W', R', V')$ is a *bounded morphism* if it satisfies the following conditions:

(i) w and $f(w)$ satisfy the same proposition letters.
(ii) f is a homomorphism with respect to the relation R (that is, if Rwv then $R'f(w)f(v)$).
(iii) If $R'f(w)v'$ then there exists v such that Rwv and $f(v) = v'$ (the *back condition*).

If there is a *surjective* bounded morphism from $\mathfrak{M}$ to $\mathfrak{M}'$, then we say that $\mathfrak{M}'$ is a *bounded morphic image* of $\mathfrak{M}$, and write $\mathfrak{M} \twoheadrightarrow \mathfrak{M}'$. ⊣

The idea embodied in the back condition is utterly fundamental to modal logic – in fact, it is the idea that underlies the notion of bisimulation – so we need to get a good grasp of what it involves right away. Here is a useful example.

Example 2.11 Consider the models $\mathfrak{M} = (W, R, V)$ and $\mathfrak{M}' = (W', R', V')$, where

- $W = \mathbb{N}$ (the natural numbers), Rmn iff $n = m + 1$, and $V(p) = \{n \in \mathbb{N} \mid n \text{ is even}\}$,
- $W' = \{e, o\}$, $R' = \{(e, o), (o, e)\}$, and $V'(p) = \{e\}$.

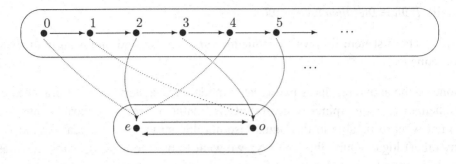

Fig. 2.1. A bounded morphism.

Now, let $f : W \to W'$ be the following map:

$$f(n) = \begin{cases} e & \text{if } n \text{ is even} \\ o & \text{if } n \text{ is odd} \end{cases}$$

Figure 2.1 sums this all up in a simple picture.

Now, f is *not* a strong homomorphism (why not?), but it *is* a (surjective) bounded morphism from $\mathfrak{M}$ to $\mathfrak{M}'$. Let us see why. Trivially f satisfies item (i) of the definition. As for the homomorphic condition consider an arbitrary pair $(n, n + 1)$ in R. There are two possibilities: n is either even or odd. Suppose n is even. Then $n + 1$ is odd, so $f(n) = e$ and $f(n + 1) = o$. But then we have $R'f(n)f(n + 1)$, as required. The argument for n odd is analogous.

And now for the interesting part: the back condition. Take an arbitrary element n of W and assume that $R'f(n)w'$. We have to find an $m \in W$ such that Rnm and $f(m) = w'$. Let us assume that n is odd (the case for even n is similar). As n is odd, $f(n) = o$, so by definition of R', we must have that $w' = e$. But then $f(n + 1) = w'$ since $n + 1$ is even, and by the definition of R we have that $n + 1$ is a successor of n. Hence, $n + 1$ is the m that we were looking for.   ⊣

Definition 2.12 (Bounded Morphisms – the General Case) The definition of a bounded morphism for general modal languages is obtained from the above by adapting the homomorphic and back conditions of Definition 2.10 as follows:

(ii)′ For all $\triangle \in \tau$, $R_\triangle wv_1 \ldots v_n$ implies $R'_\triangle f(w)f(v_1) \ldots f(v_n)$.

(iii)′ If $R'_\triangle f(w)v'_1 \ldots v'_n$ then there exist $v_1 \ldots v_n$ such that $R_\triangle wv_1 \ldots v_n$ and $f(v_i) = v'_i$ (for $1 \leq i \leq n$).   ⊣

Example 2.13 Suppose we are working in the modal similarity type of arrow logic; see Example 1.16 and 1.27. Recall that the language has a modal constant 1′, a unary operator $\otimes$ and a single dyadic operator $\circ$. Semantically, to these operators correspond a unary relation I, a binary R and a ternary C. We will define a bounded morphism from a square model to a model based on the addition of the integer numbers. We will use the following notation: if x is an element of $\mathbb{Z} \times \mathbb{Z}$, then x_0 denotes its first component, and x_1 its second component.

Consider the two models $\mathfrak{M} = (W, C, R, I, V)$ and $\mathfrak{M}' = (W', C', R', I', V')$ where

- $W = \mathbb{Z} \times \mathbb{Z}$, $Cxyz$ iff $x_0 = y_0$, $y_1 = z_0$ and $z_1 = x_1$, Rxy if $x_0 = y_1$ and $x_1 = y_0$, Ix iff $x_0 = x_1$, and finally, the valuation V is given by $V(p) = \{(x_0, x_1) \mid x_1 - x_0 \text{ is even }\}$,
- $W' = \mathbb{Z}$, $C'stu$ iff $s = t + u$, $R'st$ iff $s = -t$, $I's$ iff $s = 0$, and the valuation V' is given by $V'(p) = \{s \in \mathbb{Z} \mid s \text{ is even }\}$.

This example is best understood by looking at Figure 2.2. The left picture shows a fragment of the model $\mathfrak{M}$; the points of $\mathbb{Z} \times \mathbb{Z}$ are represented as disks or circles, depending on whether p is true or not. The diagonal is indicated by the dashed diagonal line.

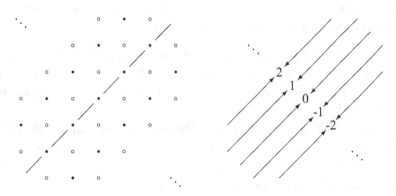

Fig. 2.2. Another bounded morphism.

The right-hand side of Figure 2.2 gives a pictorial representation of the function $f : \mathbb{Z} \times \mathbb{Z} \to \mathbb{Z}$ given by

$$f(z) = z_1 - z_0.$$

We claim that f is a bounded morphism for this similarity type. The clause for the propositional variables is trivial. For the unary relation I we only have to check that for any z in $\mathbb{Z} \times \mathbb{Z}$, $z_0 = z_1$ iff $z_1 - z_0 = 0$. This is obviously true. We leave the case of the binary relation R to the reader.

So let us turn to the clauses for the ternary relation C. To check item (ii) (the homomorphic condition), assume that $Cxyz$ holds for x, y and z in W. That is, we have that $x_0 = y_0$, $y_1 = z_0$ and $z_1 = x_1$. But then we find that

$$f(x) = x_1 - x_0 = z_1 - y_0 = z_1 - z_0 + y_1 - y_0 = f(z) + f(y),$$

so by definition of C' we do indeed find that $C'f(x)f(y)f(z)$.

For item (iii)' (the back condition) assume that we have $C'f(x)tu$ for some $x \in \mathbb{Z} \times \mathbb{Z}$ and $t, u \in \mathbb{Z}$. In other words, we have that $x_1 - x_0 = t + u$. Consider the pairs $y := (x_0, x_0 + t)$ and $z := (x_0 + t, x_1)$. It is obvious that $Cxyz$; we also find that $f(y) = t$ and $f(z) = x_1 - (x_0 + t) = (x_1 - x_0) - t = u$. Hence y and z are the elements of W that we need to satisfy item (iii)'. ⊣

Definition 2.12 covers the basic temporal language, PDL, and arrow logic, as special cases – but once more it is worth issuing a warning concerning the basic temporal language. Although R_P is usually presented implicitly (as the converse of the relation R in some model (W, R, V)) we certainly cannot ignore it. Thus a *temporal* bounded morphism from (W_1, R_1, V_1) to (W_2, R_2, V_2) is a bounded morphism from $(W_1, R_1, \breve{R}_1, V_1)$ to $(W_2, R_2, \breve{R}_2, V_2)$.

Proposition 2.14 *Let τ be a modal similarity type and let $\mathfrak{M}$ and $\mathfrak{M}'$ be τ-models such that $f : \mathfrak{M} \to \mathfrak{M}'$ is a bounded morphism. Then, for each modal formula ϕ, and each element w of $\mathfrak{M}$ we have $\mathfrak{M}, w \Vdash \phi$ iff $\mathfrak{M}', f(w) \Vdash \phi$. In words: modal satisfaction is invariant under bounded morphisms.*

Proof. Let $\mathfrak{M}$, $\mathfrak{M}'$ and f be as in the statement of the proposition. We will prove that for each formula ϕ and state w, $\mathfrak{M}, w \Vdash \phi$ iff $\mathfrak{M}', f(w) \Vdash \phi$. The proof is by induction on ϕ. We will assume that τ is the basic similarity type, leaving the general case to the reader.

The base step and the boolean cases are routine, so let us turn to the case where ϕ is of the form $\Diamond\psi$. Assume first that $\mathfrak{M}, w \Vdash \Diamond\psi$. This means there is a state v with Rwv and $\mathfrak{M}, v \Vdash \psi$. By the inductive hypothesis, $\mathfrak{M}', f(v) \Vdash \psi$. By the homomorphic condition, $R'f(w)f(v)$, so $\mathfrak{M}', f(w) \Vdash \Diamond\psi$.

For the other direction, assume that $\mathfrak{M}', f(w) \Vdash \Diamond\psi$. Thus there is a successor of $f(w)$ in $\mathfrak{M}'$, say v', such that $\mathfrak{M}', v' \Vdash \psi$. Now we use the back condition (of Definition 2.10). This yields a point v in $\mathfrak{M}$ such that Rwv and $f(v) = v'$. Applying the inductive hypothesis, we obtain $\mathfrak{M}, v \Vdash \psi$, so $\mathfrak{M}, w \Vdash \Diamond\psi$. ⊣

Here is a simple application: we will now show that any satisfiable formula can be satisfied in a *tree-like* model. To put it another way: modal logic has the *tree model property*.

Let τ be a modal similarity type containing only diamonds (thus if $\mathfrak{M}$ is a τ-model, it has the form $(W, R_1, R_2, \dots, V)$, where each R_i is a binary relation on W). In this context we will call a τ-model $\mathfrak{M}$ *tree-like* if the structure $(W, \bigcup_i R_i, V)$ is a tree in the sense of Definition 1.7.

Proposition 2.15 *Assume that τ is a modal similarity type containing only diamonds. Then, for any rooted τ-model $\mathfrak{M}$ there exists a tree-like τ-model $\mathfrak{M}'$ such that $\mathfrak{M}' \twoheadrightarrow \mathfrak{M}$. Hence any satisfiable τ-formula is satisfiable in a tree-like model.*

Proof. Let w be the root of $\mathfrak{M}$. Define the model $\mathfrak{M}'$ as follows. Its domain W' consists of all finite sequences $(w, u_1, \ldots, u_n)$ such that $n \geq 0$ and for some modal operators $\langle a_1 \rangle, \ldots, \langle a_n \rangle \in \tau$ there is a path $wR_{a_1}u_1 \cdots R_{a_n}u_n$ in $\mathfrak{M}$. Define $(w, u_1, \ldots, u_n)R'_a(w, v_1, \ldots, v_m)$ to hold if $m = n + 1$, $u_i = v_i$ for $i = 1, \ldots, n$, and $R_a u_n v_m$ holds in $\mathfrak{M}$. That is, R'_a relates two sequences iff the second is an extension of the first with a state from $\mathfrak{M}$ that is a successor of the last element of the first sequence. Finally, V' is defined by putting $(w, u_1, \ldots, u_n) \in V'(p)$ iff $u_n \in V(p)$. As the reader is asked to check in Exercise 2.1.4, the mapping $f : (w, u_1, \ldots, u_n) \mapsto u_n$ defines a surjective bounded morphism from $\mathfrak{M}'$ to $\mathfrak{M}$, thus $\mathfrak{M}'$ and $\mathfrak{M}$ are equivalent.

But then it follows that any satisfiable τ-formula is satisfiable in a tree-like model. For suppose ϕ is satisfiable in some τ-model at a point w. Let $\mathfrak{M}$ be the submodel generated by w. By Proposition 2.6, $\mathfrak{M}, w \Vdash \phi$, and as $\mathfrak{M}$ is rooted we can form an equivalent tree-like model $\mathfrak{M}'$ as just described. $\dashv$

The method used to construct $\mathfrak{M}'$ from $\mathfrak{M}$ is well known in both modal logic and computer science: it is called *unraveling* (or *unwinding*, or *unfolding*). In essence, we built $\mathfrak{M}'$ by treating the paths through $\mathfrak{M}$ as first class citizens: this untangles the (possibly very complex) way information is stored in $\mathfrak{M}$, and makes it possible to present it as a tree. We will make use of unraveling several times in later work; in the meantime, Exercise 2.1.7 asks the reader to extend the notion of 'tree-likeness' to arbitrary modal similarity types, and generalize Proposition 2.15.

Exercises for Section 2.1

2.1.1 Suppose we wanted an operator D with the following satisfaction definition: for any model $\mathfrak{M}$ and any formula ϕ, $\mathfrak{M}, w \Vdash D\phi$ iff there is a $u \neq w$ such that $\mathfrak{M}, u \Vdash \phi$. This operator is called the *difference operator* and we will discuss it further in Section 7.1. Is the difference operator definable in the basic modal language?

2.1.2 Use generated submodels to show that the backward looking modality (that is, the P of the basic temporal language) cannot be defined in terms of the forward looking operator $\Diamond$.

2.1.3 Give the simplest possible example which shows that the truth of modal formulas is *not* invariant under homomorphisms, even if condition (i) is strengthened to an equivalence. Is modal truth preserved under homomorphisms?

2.1.4 Show that the mapping f defined in the proof of Proposition 2.15 is indeed a surjective bounded morphism.

2.1.5 Let $\mathfrak{B} = (B, R)$ be the transitive binary tree; that is, B is the set of finite strings of 0s and 1s, and $R\sigma\tau$ holds if σ is a proper initial segment of τ. Let $\mathfrak{N} = (\mathbb{N}, <)$ be the frame of the natural numbers with the usual ordering.

(a) Let V_0 be the valuation on $\mathfrak{N}$ given by $V_0(p) = \{2n \mid n \in \mathbb{N}\}$ for each proposition letter p. Define a valuation U_0 on $\mathfrak{B}$ and a bounded morphism from $(\mathfrak{B}, U_0)$ to $(\mathfrak{N}, V_0)$.

(b) Let U_1 be the valuation on $\mathfrak{B}$ given by $U_1(p) = \{1\sigma \mid \sigma \in B\}$ for each proposition letter p. Give a valuation V_1 on $\mathfrak{N}$ and a homomorphism from $(\mathfrak{B}, U_1)$ to $(\mathfrak{N}, V_1)$.

(c) Can you also find *bounded* morphisms?

2.1.6 Show that every model is the bounded morphic image of the disjoint union of point-generated (that is: rooted) models. This exercise may look rather technical, but in fact it is very straightforward – think about it!

2.1.7 This exercise generalizes Proposition 2.15 to arbitrary modal similarity types.

(a) Define a suitable notion of tree-like model that works for arbitrary modal similarity types. (Hint: in case of $R_\triangle s_0 s_1 \ldots s_n$, think of s_0 as being the parent node and of $s_1, \ldots, s_n$ as the children.)

(b) Generalize Proposition 2.15 to arbitrary modal similarity types.

2.2 Bisimulations

What do the invariance results of the previous section have in common? They all deal with special sorts of *relations* between two models, namely relations with the following properties: related states carry identical atomic information, and whenever it is possible to make a transition in one model, it is possible to make a matching transition in the other. For example, with generated submodels the inter-model relation is identity, and every transition in one model is matched by an identical transition in the other. With bounded morphisms, the inter-model relation is a function, and the notion of matching involves both the homomorphic link from source to target, and the back condition which reflects target structure in the source.

This observation leads us to the central concept of the chapter: *bisimulations*. Quite simply, a bisimulation is a relation between two models in which related states have identical atomic information and matching transition possibilities. The interesting part of the definition is the way it makes the notion of 'matching transition possibilities' precise.

Definition 2.16 (Bisimulations – the Basic Case) We first give the definition for the basic modal language. Let $\mathfrak{M} = (W, R, V)$ and $\mathfrak{M}' = (W', R', V')$ be two models.

A non-empty binary relation $Z \subseteq W \times W'$ is called a *bisimulation between* $\mathfrak{M}$ and $\mathfrak{M}'$ (notation: $Z : \mathfrak{M} \underleftrightarrow{} \mathfrak{M}'$) if the following conditions are satisfied:

(i) If wZw' then w and w' satisfy the same proposition letters.

(ii) If wZw' and Rwv, then there exists v' (in $\mathfrak{M}'$) such that vZv' and $R'w'v'$ (the *forth condition*).

(iii) The converse of (ii): if wZw' and $R'w'v'$, then there exists v (in $\mathfrak{M}$) such that vZv' and Rwv (the *back condition*).

When Z is a bisimulation linking two states w in $\mathfrak{M}$ and w' in $\mathfrak{M}'$ we say that w and w' are *bisimilar*, and we write $Z : \mathfrak{M}, w \underline{\leftrightarrow} \mathfrak{M}', w'$. If there is a bisimulation Z such that $Z : \mathfrak{M}, w \underline{\leftrightarrow} \mathfrak{M}', w'$, we sometimes write $\mathfrak{M}, w \underline{\leftrightarrow} \mathfrak{M}', w'$, or $w \underline{\leftrightarrow} w'$ if the models are clear from the context. Finally, if $\mathfrak{M}$ and $\mathfrak{M}'$ are linked by some bisimulation, we write $\mathfrak{M} \underline{\leftrightarrow} \mathfrak{M}'$. ⊣

Think of Definition 2.16 pictorially. Figure 2.3 shows the content of the forth clause. Suppose we know that wZw' and Rwv (the solid arrow in $\mathfrak{M}$ and the Z-link at the bottom of the diagram display this information). Then the forth condition says that it is always possible to find a v' that 'completes the square' (this is shown by the dashed arrow in $\mathfrak{M}'$ and the dotted Z-link at the top of the diagram). Note the symmetry between the back and forth clauses: to visualize the back clause, simply reflect the picture through its vertical axis.

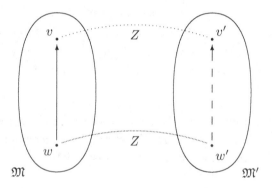

Fig. 2.3. The forth condition.

In effect, bisimulations are a relational generalization of bounded morphisms: we drop the directionality from source to target (and with it the homomorphic condition) and replace it with a back-and-forth system of matching moves between models.

Example 2.17 The models $\mathfrak{M}$ and $\mathfrak{M}'$ shown in Figure 2.4 are bisimilar. To see this, define the following relation Z between their states: $Z = \{(1, a), (2, b), (2, c), (3, d), (4, e), (5, e)\}$. Condition (i) of Definition 2.16 is obviously satisfied: Z-related states make the same propositional letters true. Moreover, the back and forth conditions are satisfied too: any move in $\mathfrak{M}$ can be matched by a similar move in $\mathfrak{M}'$, and conversely, as the reader should check.

This example also shows that bisimulation is a genuine generalization of the constructions discussed in the previous section. Although $\mathfrak{M}$ and $\mathfrak{M}'$ are bisimilar, neither is a generated submodel nor a bounded morphic image of the other. ⊣

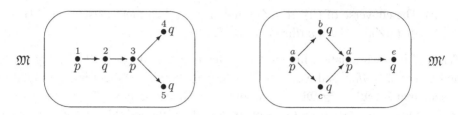

Fig. 2.4. Bisimilar models.

Definition 2.18 (Bisimulations – the General Case) Let τ be a modal similarity type, and let $\mathfrak{M} = (W, R_\triangle, V)_{\triangle \in \tau}$ and $\mathfrak{M}' = (W', R'_\triangle, V')_{\triangle \in \tau}$ be τ-models. A non-empty binary relation $Z \subseteq W \times W'$ is called a *bisimulation* between $\mathfrak{M}$ and $\mathfrak{M}'$ (notation: $Z : \mathfrak{M} \leftrightarrow \mathfrak{M}'$) if the above condition (i) from Definition 2.16 is satisfied (that is, Z-related states satisfy the same proposition letters) and in addition the following conditions (ii)′ and (iii)′ are satisfied:

(ii)′ If wZw' and $R_\triangle w v_1 \ldots v_n$ then there are v'_1, ..., v'_n (in W') such that $R'_\triangle w' v'_1 \ldots v'_n$ and for all i ($1 \leq i \leq n$) $v_i Z v'_i$ (the *forth* condition).

(iii)′ The converse of (ii)′: if wZw' and $R'_\triangle w' v'_1 \ldots v'_n$ then there are $v_1, \ldots, v_n$ (in W) such that $R_\triangle w v_1 \ldots v_n$ and for all i ($1 \leq i \leq n$) $v_i Z v'_i$ (the *back* condition). ⊣

Examples of bisimulations abound – indeed, as we have already mentioned, the constructions of the previous section (disjoint unions, generated submodels, isomorphisms, and bounded morphisms), are all bisimulations:

Proposition 2.19 *Let τ be a modal similarity type, and let $\mathfrak{M}$, $\mathfrak{M}'$ and $\mathfrak{M}_i$ ($i \in I$) be τ-models.*

(i) *If $\mathfrak{M} \cong \mathfrak{M}'$, then $\mathfrak{M} \leftrightarrow \mathfrak{M}'$.*

(ii) *For every $i \in I$ and every w in $\mathfrak{M}_i$, $\mathfrak{M}_i, w \leftrightarrow \biguplus_i \mathfrak{M}_i, w$.*

(iii) *If $\mathfrak{M}' \rightarrowtail \mathfrak{M}$, then $\mathfrak{M}', w \leftrightarrow \mathfrak{M}, w$ for all w in $\mathfrak{M}'$.*

(iv) *If $f : \mathfrak{M} \twoheadrightarrow \mathfrak{M}'$, then $\mathfrak{M}, w \leftrightarrow \mathfrak{M}', f(w)$ for all w in $\mathfrak{M}$.*

Proof. We only prove the second item, leaving the others as Exercise 2.2.2. Assume we are working in the basic modal language. Define a relation Z between $\mathfrak{M}_i$ and $\biguplus_i \mathfrak{M}_i$ by putting $Z = \{(w, w) \mid w \in \mathfrak{M}_i\}$. Then Z is a bisimulation. To see this, observe that clause (i) of Definition 2.16 is trivially fulfilled, and as to clauses (ii) and (iii), any R-step in $\mathfrak{M}_i$ is reproduced in $\biguplus_i \mathfrak{M}_i$, and by the disjointness condition every R-step in $\biguplus_i \mathfrak{M}_i$ that departs from a point that was originally in $\mathfrak{M}_i$, stems from a corresponding R-step in $\mathfrak{M}_i$. The reader should extend this argument to arbitrary similarity types. ⊣

We will now show that modal satisfiability is invariant under bisimulations (and hence, by Proposition 2.19, provide an alternative proof that modal satisfiability is invariant under disjoint unions, generated submodels, isomorphisms, and bounded morphisms). The key thing to note about the following proof is how straightforward it is – the back and forth clauses in the definition of bisimulation are *precisely* what is needed to push the induction through.

Theorem 2.20 *Let τ be a modal similarity type, and let $\mathfrak{M}$, $\mathfrak{M}'$ be τ-models. Then, for every $w \in W$ and $w' \in W'$, $w \underline{\leftrightarrow} w'$ implies that $w \rightsquigarrow w'$. In words, modal formulas are invariant under bisimulation.*

Proof. By induction on ϕ. The case where ϕ is a proposition letter follows from clause (i) of Definition 2.16, and the case where ϕ is $\bot$ is immediate. The boolean cases are immediate from the induction hypothesis.

As for formulas of the form $\Diamond\psi$, we have $\mathfrak{M}, w \Vdash \Diamond\psi$ iff there exists a v in $\mathfrak{M}$ such that Rwv and $\mathfrak{M}, v \Vdash \psi$. As $w \underline{\leftrightarrow} w'$ we find by clause (ii) of Definition 2.16 that there exists a v' in $\mathfrak{M}'$ such that $R'w'v'$ and $v \underline{\leftrightarrow} v'$. By the induction hypothesis, $\mathfrak{M}', v' \Vdash \psi$, hence $\mathfrak{M}', w' \Vdash \Diamond\psi$. For the converse direction use clause (iii) of Definition 2.16.

The argument for the general modal case, with triangles $\triangle$, is an easy extension of that just given, as the reader should check. $\dashv$

This finishes our discussion of the basics of bisimulation – so let us now try and understand the concept more deeply. Some of the remarks that follow are conceptual, and some are technical, but they all point to ideas that crop up throughout the book.

Remark 2.21 (Bisimulation, Locality, and Computation) In the Preface we suggested that the reader think of modal formulas as automata. Evaluating a modal formula amounts to running an automaton: we place it at some state inside a structure and let it search for information. The automaton is only permitted to explore by making transitions to neighboring states; that is, it works locally.

Suppose such an automaton is standing at a state w in a model $\mathfrak{M}$, and we pick it up and place it at a state w' in a different model $\mathfrak{M}'$; would it notice the switch? If w and w' are bisimilar, *no*. Our automaton cares only about the information at the current state and the information accessible by making a transition – it is indifferent to everything else. Thus the definition of bisimulation spells out exactly what we have to do if we want to fool such an automaton as to where it is being evaluated. Viewed this way, it is clear that the concept of bisimulation is a direct reflection of the locality of the modal satisfaction definition.

But there is a deeper link between bisimulation and computation than our informal talk of automata might suggest. As we discussed in Example 1.3, labeled

Fig. 2.5. Equivalent but not bisimilar.

transition systems (LTSs) are a standard way of thinking about computation: when we traverse an LTS we build a sequence of state transitions – or to put it another way, we compute. When are two LTSs computationally equivalent? More precisely, if we ignore practical issues (such as how long it takes to actually perform a computation) when can two different LTSs be treated as freely exchangeable ('observationally equivalent') black boxes? One natural answer is: when they are bisimilar. Bisimulation turns out to be a very natural notion of equivalence for both mathematical and computational investigations. For more on the history of bisimulation and the connection with computer science, see the Notes.   ⊣

Remark 2.22 (Bisimulation and First-Order Logic) According to Theorem 2.20 modal formulas cannot distinguish between bisimilar states or between bisimilar models, even though these states or models may be quite different. It follows that modal logic is very different from first-order logic, for arbitrary first-order formulas are certainly *not* invariant under bisimulations. For example, the model $\mathfrak{M}'$ of Example 2.17 satisfies the formula

$$\exists y_1 y_2 y_3 \, (y_1 \neq y_2 \wedge y_1 \neq y_3 \wedge y_2 \neq y_3 \wedge Rxy_1 \wedge Rxy_2 \wedge Ry_1 y_3 \wedge Ry_2 y_3),$$

if we assign the state a to the free variable x. This formula says that there is a diamond-shaped configuration of points, which is true of the point a in $\mathfrak{M}$, but not of the state 1 in $\mathfrak{M}$. But as far as modal logic is concerned, $\mathfrak{M}$ and $\mathfrak{M}$, being bisimilar, are indistinguishable. In Section 2.4 we will start examining the links between modal logic and first-order logic more systematically.   ⊣

Now for a fundamental question: is the converse of Theorem 2.20 true? That is, if two models are modally equivalent, must they be bisimilar? The answer is *no*.

Example 2.23 Consider the basic modal language. We may just as well work with an empty set of proposition letters here. Define models $\mathfrak{M}$ and $\mathfrak{N}$ as in Figure 2.5, where arrows denote R-transitions. Each of $\mathfrak{M}$ and $\mathfrak{N}$ has, for each $n > 0$, a finite branch of length n; the difference between the models is that, in addition, $\mathfrak{N}$ has an infinite branch.

One can show that for all modal formulas ϕ, $\mathfrak{M}, w \Vdash \phi$ iff $\mathfrak{N}, w' \Vdash \phi$ (this is easy if one is allowed to use some results that we will prove further on, namely Proposition 2.31 and Lemma 2.33, but it is not particularly hard to prove from first principles, and the reader may like to try this). But even though w and w' are modally equivalent, there is no bisimulation linking them. To see this, suppose that there was such a bisimulation Z: we will derive a contradiction from this supposition.

Since w and w' are linked by Z, there has to be a successor of w, say v_0, which is linked to the first point v_0' on the infinite path from w'. Suppose that n is the length of the (maximal) path leading from w through v_0, and let w, v_0, ..., v_{n-1} be the successive points on this path. Using the bisimulation conditions $n - 1$ times, we find points v_1', ..., v_{n-1}' on the infinite path emanating from w', such that $v_0' R' v_1' \ldots R' v_{n-1}'$ and $v_i Z v_i'$ for each i. Now v_{n-1}' has a successor, but v_{n-1} does not; hence, there is no way that these two points can be bisimilar. $\quad\dashv$

Nonetheless, it is possible to prove a restricted converse to Theorem 2.20, namely the Hennessy-Milner Theorem. Let τ be a modal similarity type, and $\mathfrak{M}$ a τ-model. $\mathfrak{M}$ is *image-finite* if for each state u in $\mathfrak{M}$ and each relation R in $\mathfrak{M}$, the set $\{\,(v_1, \ldots, v_n) \mid Ruv_1 \ldots v_n\,\}$ is finite; observe that we are *not* putting any restrictions on the total number of different relations R in the model $\mathfrak{M}$ – just that each of them is image-finite.

Theorem 2.24 (Hennessy-Milner Theorem) *Let τ be a modal similarity type, and let $\mathfrak{M}$ and $\mathfrak{M}'$ be two image-finite τ-models. Then, for every $w \in W$ and $w' \in W'$, $w \leftrightarrow w'$ iff $w \rightsquigarrow w'$.*

Proof. Assume that our similarity type τ only contains a single diamond (that is, we will work in the basic modal language). The direction from left to right follows from Theorem 2.20; for the other direction, we will prove that the relation $\rightsquigarrow$ of modal equivalence itself satisfies the conditions of Definition 2.16 – that is, we show that the relation of modal equivalence on these models is itself a bisimulation. (This is an important idea; we will return to it in Section 2.5.)

The first condition is immediate. For the second one, assume that $w \rightsquigarrow w'$ and Rwv. We will try to arrive at a contradiction by assuming that there is no v' in $\mathfrak{M}'$ with $R'w'v'$ and $v \rightsquigarrow v'$. Let $S' = \{u' \mid R'w'u'\}$. Note that S' must be non-empty, for otherwise $\mathfrak{M}', w' \Vdash \Box\bot$, which would contradict $w \rightsquigarrow w'$ since $\mathfrak{M}, w \Vdash \Diamond\top$. Furthermore, as $\mathfrak{M}'$ is image-finite, S' must be finite, say $S' = \{w_1', \ldots, w_n'\}$. By assumption, for every $w_i' \in S'$ there exists a formula ψ_i such that $\mathfrak{M}, v \Vdash \psi_i$ but $\mathfrak{M}', w_i' \nVdash \psi_i$. It follows that

$$\mathfrak{M}, w \Vdash \Diamond(\psi_1 \wedge \cdots \wedge \psi_n) \text{ and } \mathfrak{M}', w' \nVdash \Diamond(\psi_1 \wedge \cdots \wedge \psi_n),$$

which contradicts our assumption that $w \rightsquigarrow w'$. The third condition of Definition 2.16 may be checked in a similar way. Extending the proof to other similarity types is routine.   ⊣

Theorem 2.20 (together with the Hennessy-Milner Theorem) on the one hand, and Example 2.23 on the other, mark important boundaries. Clearly, bisimulations have something important to say about modal expressivity over models, but they do not tell us everything. Two pieces of the jigsaw puzzle are missing. For a start, we are still considering modal languages in isolation: as yet, we have made no attempt to systematically link them to first-order logic. We will remedy this in Section 2.4 and this will eventually lead us to a beautiful result, the van Benthem Characterization Theorem (Theorem 2.68): modal logic is the bisimulation invariant fragment of first-order logic.

The second missing piece is the notion of an *ultrafilter extension*. We will introduce this concept in Section 2.5, and this will eventually lead us to Theorem 2.62. Informally, this theorem says: modal equivalence implies bisimilarity-somewhere-else. Where is this mysterious 'somewhere else'? In the ultrafilter extension. As we will see, although modally equivalent models need not be bisimilar, they must have bisimilar ultrafilter extensions.

Remark 2.25 (Bisimulations for the Basic Temporal Language, PDL, and Arrow Logic) Although we have already said the most fundamental things that need to be said on this topic (Definition 2.18 and Theorem 2.20 covers these languages), a closer look reveals some interesting results for PDL and arrow logic. But let us first discuss the basic temporal language.

First we issue our (by now customary) warning. When working with the basic temporal language, we usually work with models (W, R, V) and implicitly take R_P to be $R^{\smile}$. Thus we need a notion of bisimulation which takes $\check{R}$ into account, and so we define a *temporal* bisimulation between models (W, R, V) and (W', R', V') to be a relation Z between the states of the two models that satisfies the clauses of Definition 2.16, and in addition the following two clauses (iv) and (v) requiring that backward steps in one model should be matched by similar steps in the other model:

(iv) If wZw' and Rvw, then there exists v' (in $\mathfrak{M}'$) such that vZv' and $R'v'w'$.

(v) The converse of (iv): if wZw' and $R'v'w'$, then there exists v (in $\mathfrak{M}$) such that vZv' and Rvw.

If we do not do this, we are in trouble. For example, if $\mathfrak{M}$ is a model whose underlying frame is the integers, and $\mathfrak{M}'$ is the submodel of $\mathfrak{M}$ generated by 0 (according to the definition for the basic modal language), then these two models are bisimilar in the sense of Definition 2.16, and hence equivalent as far as the basic *modal* language is concerned. But they are not equivalent as far as the basic *temporal* language is concerned: $\mathfrak{M}, 0 \Vdash P\top$, but $\mathfrak{M}', 0 \nVdash P\top$.

Given our previous discussion, this is unsurprising. What is (pleasantly) surprising is that things do not work this way in PDL. Suppose we are given two regular models. Checking that these models are bisimilar for the language of PDL means checking that bisimilarity holds for all the (infinitely many) relations that exist in regular models (see Example 1.26). But as it turns out, most of this work is unnecessary. Once we have checked that bisimilarity holds for all the relations which interpret the basic programs, we do not have to check anything else: the relations corresponding to complex programs will *automatically* be bisimilar. In Section 2.7 we will introduce some special terminology to describe this: the operations in regular PDL's modality building repertoire ($\cup$, ; and*) will be called *safe for bisimulation*. Note that taking the converse of a relation is *not* an operation that is safe for bisimulation (in effect, that is what we just noted when discussing the basic temporal language).

What about arrow logic? The required notion of bisimulation is given by Definition 2.18; note that the clause for 1' reads that for bisimilar points a and d we have Ia iff $I'a$.   ⊣

Remark 2.26 (The Algebra of Bisimulations) Bisimulations give rise to algebraic structure quite naturally. For instance, if Z_0 is a bisimulation between $\mathfrak{M}_0$ and $\mathfrak{M}_1$, and Z_1 a bisimulation between $\mathfrak{M}_1$ and $\mathfrak{M}_2$, then the composition of Z_0 and Z_1 is a bisimulation linking $\mathfrak{M}_0$ and $\mathfrak{M}_2$. It is also a rather easy observation that the set of bisimulations between two models is closed under taking arbitrary (finite or infinite) unions. This shows that if two points are bisimilar, there is always a *maximal* bisimulation linking them; see Exercise 2.2.8. Further information on closure properties of the set of bisimulations between two models can be found in Section 2.7.   ⊣

Exercises for Section 2.2

2.2.1 Consider a modal similarity type with two diamonds $\langle a \rangle$ and $\langle b \rangle$, and with $\Phi = \{p\}$. Show that the following two models are bisimilar:

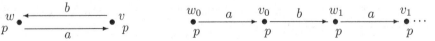

2.2.2 This exercise asks the reader to complete in detail the proof of Proposition 2.19, which links bisimulations and the model constructions discussed in the previous section. You should prove these results for arbitrary similarity types.

(a) Show that if $\mathfrak{M} \cong \mathfrak{M}'$, then $\mathfrak{M} \underline{\leftrightarrow} \mathfrak{M}'$.
(b) Show that if $\biguplus_i \mathfrak{M}_i$ is the disjoint union of the models $\mathfrak{M}_i$ ($i \in I$), then, for each i, $\mathfrak{M}_i \underline{\leftrightarrow} \biguplus_i \mathfrak{M}_i$.
(c) Show that if $\mathfrak{M}'$ is a generated submodel of $\mathfrak{M}$, then $\mathfrak{M}' \underline{\leftrightarrow} \mathfrak{M}$.
(d) Show that if $\mathfrak{M}'$ is a bounded morphic image of $\mathfrak{M}$, then $\mathfrak{M}' \underline{\leftrightarrow} \mathfrak{M}$.

2.2.3 This exercise is about *temporal* bisimulations.

- (a) Show *from first principles* that the truth of basic temporal formulas is invariant under temporal bisimulations. (That is, do not appeal to any of the results proved in this section.)
- (b) Let $\mathfrak{M}$ and $\mathfrak{M}'$ be *finite* rooted models for basic temporal logic with F and P. Let w and w' be the roots of $\mathfrak{M}$ and $\mathfrak{M}'$, respectively. Prove that if w and w' satisfy the same basic temporal formulas with F and P, then there exists a basic temporal bisimulation that relates w and w'.

2.2.4 Consider the binary until operator U. In a model $\mathfrak{M} = (W, R, V)$ its truth definition reads:

$$\mathfrak{M}, t \Vdash U(\phi, \psi) \quad \text{iff} \quad \text{there is a } v \text{ such that } Rtv \text{ and } v \Vdash \phi, \text{ and}$$
$$\text{for all } u \text{ such that } Rtu \text{ and } Ruv: u \Vdash \psi.$$

Prove that U is not definable in the basic modal language. Hint: think about the following two models, but with arrows added to make sure that the relations are transitive:

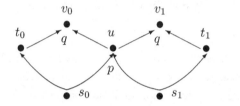 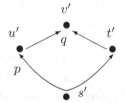

2.2.5 Consider the following two models, which we are going to use to interpret the basic *temporal* language: $\mathfrak{M}_0 = (\mathbb{R}, <, V_0)$ and $\mathfrak{M}_1 = (\mathbb{R}, <, V_1)$, where V_0 makes q true at all non-zero integers and V_1 in addition makes q true at all points of the form $1/z$ with z a non-zero integer number.

- (a) Prove that there is a temporal bisimulation between $\mathfrak{M}_0$ and $\mathfrak{M}_1$, linking 0 (in the one model) to 0 (in the other model).
- (b) Let Π be the *progressive* operator defined by the following truth table:

$$\mathfrak{M}, s \Vdash \Pi\phi \quad \text{iff} \quad \text{there are } t \text{ and } u \text{ such that } t < s < u \text{ and}$$
$$\mathfrak{M}, x \Vdash \phi \text{ for all } x \text{ between } t \text{ and } u.$$

Prove that this operator is not definable in the basic temporal language.

2.2.6 Suppose we have two bisimilar LTSs. Show that bisimilar states in these LTSs satisfy exactly the same formulas of PDL.

2.2.7 Prove that two square arrow models $\mathfrak{M} = (\mathfrak{S}_U, V)$ and $\mathfrak{M}' = (\mathfrak{S}_{U'}, V')$ are bisimilar if and only if there is a relation Z between *pairs* over U and *pairs* over U' such that

- (i) if $(u, v)Z(u', v')$, then $(u, v) \in V(p)$ iff $(u', v') \in V'(p)$,
- (ii) if $(u, v)Z(u', v')$, then $u = v$ iff $u' = v'$,
- (iii) if $(u, v)Z(u', v')$, then $(v, u)Z(v', u')$,
- (iv) if $(u, v)Z(u', v')$, then for any $w \in U$ there exists a $w' \in U'$ such that both $(u, w)Z(u', w')$ and $(w, v)Z(w', v')$,
- (v) the converse of (iv): if $(u, v)Z(u', v')$, then for any $w' \in U'$ there exists a $w \in U$ such that both $(u, w)Z(u', w')$ and $(w, v)Z(w', v')$,

Must any two bisimilar square arrow models be isomorphic? (Hint: think of $V(p)$ and $V'(p)$ as the natural ordering relations of the rational and the real numbers, respectively.)

2.2.8 Suppose that $\{Z_i \mid i \in I\}$ is a non-empty collection of bisimulations between $\mathfrak{M}$ and $\mathfrak{M}'$. Prove that the relation $\bigcup_{i \in I} Z_i$ is also a bisimulation between $\mathfrak{M}$ and $\mathfrak{M}'$. Conclude that if $\mathfrak{M}$ and $\mathfrak{M}'$ are bisimilar, then there is a maximal bisimulation between $\mathfrak{M}$ and $\mathfrak{M}'$; that is, a bisimulation Z_m such that for any bisimulation $Z : \mathfrak{M} \leftrightarrow \mathfrak{M}'$ we have $Z \subseteq Z_m$.

2.3 Finite Models

Preservation and invariance results can be viewed either positively or negatively. Viewed negatively, they map the limits of modal expressivity: they tell us, for example, that modal languages are incapable of distinguishing a model from its generated submodels. Viewed positively, they are a toolkit for transforming models into more desirable forms without affecting satisfiability. Proposition 2.15 has already given us a taste of this perspective (we showed that modal languages have the tree model property) and it will play an important role when we discuss completeness in Chapter 4.

The results of this section are similarly double-edged. We are going to investigate modal expressivity over finite models, and the basic result we will prove is that modal languages have the *finite model property*: if a modal formula is satisfiable on an arbitrary model, then it is satisfiable on a finite model.

Definition 2.27 (Finite Model Property) Let τ be a modal similarity type, and let M be a class of τ-models. We say that τ has the *finite model property with respect to* M if the following holds: if ϕ is a formula of similarity type τ, and ϕ is satisfiable in some model in M, then ϕ is satisfiable in a *finite* model in M.   ⊣

In this section we will mostly be concerned with the special case in which M in Definition 2.27 is the collection of *all* τ-models, so to simplify terminology here we will use the term 'finite model property' for this special case. The fact that modal languages have the finite model property (in this sense) can be viewed as a limitative result: modal languages simply lack the expressive strength to force the existence of infinite models. (By way of contrast, it is easy to write down first-order formulas which can only be satisfied on infinite models.) On the other hand, the result is a source of strength: we do not need to bother about (arbitrary) infinite models, for we can always find an equivalent finite one. This opens the door to the decidability results of Chapter 6. (The satisfiability problem for first-order logic, as the reader probably knows, is undecidable over arbitrary models.)

We will discuss two methods for building finite models for satisfiable modal formulas. The first is to (carefully!) select a finite *submodel* of the satisfying model, the second (called the filtration method) is to define a suitable *quotient* structure.

Selecting a finite submodel

The selection method draws together four observations. Here is the first. We know that modal satisfaction is intrinsically *local*: modalities scan the states accessible from the current state. How much of the model can a modal formula see from the current state? That obviously depends on how deeply the modalities it contains are nested.

Definition 2.28 (Degree) We define the *degree* of modal formulas as follows:

$$\begin{aligned}
\deg(p) &= 0, \\
\deg(\bot) &= 0, \\
\deg(\neg\phi) &= \deg(\phi), \\
\deg(\phi \vee \psi) &= \max\{\deg(\phi), \deg(\psi)\}, \\
\deg(\triangle(\phi_1, \ldots, \phi_n)) &= 1 + \max\{\deg(\phi_1), \ldots, \deg(\phi_n)\}.
\end{aligned}$$

In particular, the degree of a basic modal formula $\Diamond\phi$ is $1 + \deg(\phi)$.   ⊣

Second, we observe the following:

Proposition 2.29 *Let τ be a finite modal similarity type, and assume that our collection of proposition letters is finite as well.*

  (i) *For all n, up to logical equivalence there are only finitely many formulas of degree at most n.*
  (ii) *For all n, and every τ-model $\mathfrak{M}$ and state w of $\mathfrak{M}$, the set of all τ-formulas of degree at most n that are satisfied by w, is equivalent to a single formula.*

Proof. We prove the first item by induction on n. The case $n = 0$ is obvious. As for the case $n+1$, observe that every formula of degree $\leq n+1$ is a boolean combination of proposition letters and formulas of the form $\Diamond\psi$, where $\deg(\psi) \leq n$. By the induction hypothesis there can only be finitely many non-equivalent such formulas ψ. Thus there are only finitely many non-equivalent boolean combinations of proposition letters and formulas $\Diamond\psi$, where ψ has degree at most n. Hence, there are only finitely many non-equivalent formulas of degree at most $n + 1$.

Item (ii) is immediate from item (i).   ⊣

Third, we observe that there is a natural way of finitely approximating a bisimulation. These finite approximations will prove crucial in our search for finite models.

Definition 2.30 (n-Bisimulations) Here we define n-bisimulations for modal similarity types containing only diamonds, leaving the definition of the general case as part of Exercise 2.3.2. Let $\mathfrak{M}$ and $\mathfrak{M}'$ be models, and let w and w' be states of $\mathfrak{M}$ and $\mathfrak{M}'$, respectively. We say that w and w' are *n-bisimilar* (notation:

$w \mathbin{\underline{\leftrightarrow}}_n w'$) if there exists a sequence of binary relations $Z_n \subseteq \cdots \subseteq Z_0$ with the following properties (for $i + 1 \leq n$):

    (i) wZ_nw'.
    (ii) If vZ_0v' then v and v' agree on all proposition letters.
    (iii) If $vZ_{i+1}v'$ and Rvu, then there exists u' with $R'v'u'$ and uZ_iu'.
    (iv) If $vZ_{i+1}v'$ and $R'v'u'$, then there exists u with Rvu and uZ_iu'.   ⊣

The intuition is that if $w \mathbin{\underline{\leftrightarrow}}_n w'$, then w and w' *bisimulate up to depth* n. Clearly, if $w \mathbin{\underline{\leftrightarrow}} w'$, then $w \mathbin{\underline{\leftrightarrow}}_n w'$ for all n – but the converse need not hold; see Exercise 2.3.1.

Fourth, we observe that for languages containing only finitely many proposition letters, there is an *exact* match between modal equivalence and n-bisimilarity for all n. That is, for such languages not only does n-bisimilarity for all n imply modal equivalence, but the converse holds as well.

Proposition 2.31 *Let τ be a finite modal similarity type, Φ a finite set of proposition letters, and let $\mathfrak{M}$ and $\mathfrak{M}'$ be models for this language. Then for every w in $\mathfrak{M}$ and w' in $\mathfrak{M}'$, the following are equivalent:*

    (i) $w \mathbin{\underline{\leftrightarrow}}_n w'$.
    (ii) w *and* w' *agree on all modal formulas of degree at most* n.

It follows that 'n-bisimilarity for all n' and modal equivalence coincide as relations between states.

Proof. The implication (i) $\Rightarrow$ (ii) may be proved by induction on n. For the converse implication one can use an argument similar to the one used in the proof of Theorem 2.24; we leave the proof as part of Exercise 2.3.2.   ⊣

It is time to draw these observations together. The following definition and lemma, which are about *rooted* models, give us half of what we need to build finite models.

Definition 2.32 Let τ be a modal similarity type containing only diamonds. Let $\mathfrak{M} = (W, R_1, \ldots, R_n, \ldots, V)$ be a rooted τ-model with root w. The notion of the *height* of states in $\mathfrak{M}$ is defined by induction. The only element of height 0 is the root of the model; the states of height $n + 1$ are those immediate successors of elements of height n that have not yet been assigned a height smaller than $n + 1$. The *height of a model* $\mathfrak{M}$ is the maximum n such that there is a state of height n in $\mathfrak{M}$, if such a maximum exists; otherwise the height of $\mathfrak{M}$ is infinite.

For a natural number k, the *restriction* of $\mathfrak{M}$ to k (notation: $\mathfrak{M} \restriction k$) is defined as the submodel containing only states whose height is at most k. More precisely, $(\mathfrak{M} \restriction k) = (W_k, R_{1k}, \ldots, R_{nk}, \ldots, V_k)$, where $W_k = \{v \mid \operatorname{height}(v) \leq k\}$, $R_{nk} = R_n \cap (W_k \times W_k)$, and for each p, $V_k(p) = V(p) \cap W_k$.   ⊣

In words: the restriction of $\mathfrak{M}$ to k contains all states that can be reached from the root in at most k steps along the accessibility relations. Typically, this will not give a *generated* submodel, so why does it interest us? Because, as we can now show, given a formula ϕ of degree k that is satisfiable in some rooted model $\mathfrak{M}$, the restriction of $\mathfrak{M}$ to k contains all the states we need to satisfy ϕ. To put it another way: we are free to simply delete all states that lie beyond the 'k-horizon.'

Lemma 2.33 *Let τ be a modal similarity type that contains only diamonds. Let $\mathfrak{M}$ be a rooted τ-model, and let k be a natural number. Then, for every state w of $(\mathfrak{M} \restriction k)$, we have $(\mathfrak{M} \restriction k), w \underline{\leftrightarrow}_l \mathfrak{M}, w$, where $l = k - \text{height}(w)$.*

Proof. Take the identity relation on $(\mathfrak{M} \restriction k)$. We leave the reader to work out the details as Exercise 2.3.3. The following comment may be helpful: in essence this lemma tells us that if we are only interested in the satisfiability of modal formulas of degree at most k, then generating submodels of height k suffices to maintain satisfiability.   ⊣

Putting together Proposition 2.31 and Lemma 2.33, we conclude that every satisfiable modal formula can be satisfied on a model of finite *height*. This is clearly useful, but we are only halfway to our goal: the resulting model may still be infinite, as it may be infinitely branching. We obtain the finite model we are looking for by a further selection of points; in effect this discards unwanted branches and leads to the desired finite model.

Theorem 2.34 (Finite Model Property – via Selection) *Let τ be a modal similarity type containing only diamonds, and let ϕ be a τ-formula. If ϕ is satisfiable, then it is satisfiable on a finite model.*

Proof. Fix a modal formula ϕ with $\deg(\phi) = k$. We restrict our modal similarity type τ and our collection of proposition letters to the modal operators and proposition letters actually occurring in ϕ. Let $\mathfrak{M}_1, w_1$ be such that $\mathfrak{M}_1, w_1 \Vdash \phi$. By Proposition 2.15, there exists a tree-like model $\mathfrak{M}_2$ with root w_2 such that $\mathfrak{M}_2, w_2 \Vdash \phi$. Let $\mathfrak{M}_3 := (\mathfrak{M}_2 \restriction k)$. By Lemma 2.33 we have $\mathfrak{M}_2, w_2 \underline{\leftrightarrow}_k \mathfrak{M}_3, w_2$, and by Proposition 2.31 it follows that $\mathfrak{M}_3, w_2 \Vdash \phi$.

By induction on $n \leq k$ we define finite sets of states $S_0, \ldots, S_k$ and a (final) model $\mathfrak{M}_4$ with domain $S_0 \cup \cdots \cup S_k$; the points in each S_n will have height n.

Define S_0 to be the singleton $\{w_2\}$. Next, assume that $S_0, \ldots, S_n$ have already been defined. Fix an element v of S_n. By Proposition 2.29 there are only finitely many non-equivalent modal formulas whose degree is at most $k - n$, say $\psi_1, \ldots, \psi_m$. For each such formula that is of the form $\langle a \rangle \chi$ and holds in $\mathfrak{M}_3$ at v, select a state u from $\mathfrak{M}_3$ such that $R_a v u$ and $\mathfrak{M}_3, u \Vdash \chi$. Add all these us to S_{n+1}, and repeat this selection process for every state in S_n. S_{n+1} is defined as the set of all points that have been selected in this way.

Finally, define $\mathfrak{M}_4$ as follows. Its domain is $S_0 \cup \cdots \cup S_k$; as each S_i is finite, $\mathfrak{M}_4$ is finite. The relations and valuation are obtained by restricting the relations and valuation of $\mathfrak{M}_3$ to the domain of $\mathfrak{M}_4$. By Exercise 2.3.4 we have that $\mathfrak{M}_4, w_2 \rightleftharpoons_k \mathfrak{M}_3, w_2$, and hence $\mathfrak{M}_4, w_2 \Vdash \phi$, as required. ⊣

How well does the selection method generalize to other modal languages? For certain purposes it is fine. For example, to deal with arbitrary modal similarity types, the notion of a tree-like model needs to be adapted (in fact, we explained how to do this in Exercise 2.1.7), but once this has been done we can prove a general version of Proposition 2.15. Next, the notion of n-bisimilarity needs to be adapted to other similarity types, but that too is straightforward (it is part of Exercise 2.3.2). Finally, the selection process in the proof of Theorem 2.34 needs adaptation, but this is unproblematic. In short, we can show that the finite model property holds for arbitrary similarity types using the selection method.

The method has a drawback: the input model for our construction may satisfy important relational properties (such as being symmetric), but the end result is always a finite tree-like model, and the desired relational properties may be (and often are) lost. So if we want to establish the finite model property with respect to a class of models satisfying additional properties – something that is very important in practice – we may have to do additional work once we have obtained our finite tree-like model. In such cases, the selection method tends to be harder to use than the filtration method (which we discuss next). Nonetheless, the idea of (intelligently!) selecting points to build submodels is important, and (as we will see in Section 6.6 when we discuss NP-completeness) the idea really comes into its own when the model we start with is already finite.

Finite models via filtrations

We now examine the classic modal method for building finite models: filtration. Whereas the selection method builds finite models by *deleting* superfluous material from large, possibly infinite models, the filtration method produces finite models by taking a large, possibly infinite model and *identifying* as many states as possible. We first present the filtration method for the basic modal language.

Definition 2.35 A set of formulas Σ is *closed under subformulas* (or: *subformula closed*) if for all formulas ϕ, ϕ': if $\phi \vee \phi' \in \Sigma$ then so are ϕ and ϕ'; if $\neg\phi \in \Sigma$ then so is ϕ; and if $\triangle(\phi_1, \ldots, \phi_n) \in \Sigma$ then so are $\phi_1, \ldots, \phi_n$. (For the basic modal language, this means that if $\Diamond\phi \in \Sigma$, then so is ϕ.) ⊣

Definition 2.36 (Filtrations) We work in the basic modal language. Let $\mathfrak{M} = (W, R, V)$ be a model and Σ a subformula closed set of formulas. Let $\leftrightsquigarrow_\Sigma$ be the

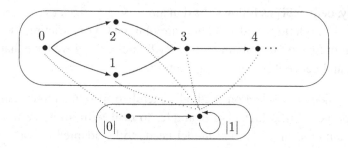

Fig. 2.6. A model and its filtration.

relation on the states of $\mathfrak{M}$ defined by:

$$w \leftrightsquigarrow_\Sigma v \text{ iff for all } \phi \text{ in } \Sigma: (\mathfrak{M}, w \Vdash \phi \text{ iff } \mathfrak{M}, v \Vdash \phi).$$

Note that $\leftrightsquigarrow_\Sigma$ is an equivalence relation. We denote the equivalence class of a state w of $\mathfrak{M}$ with respect to $\leftrightsquigarrow_\Sigma$ by $|w|_\Sigma$, or simply by $|w|$ if no confusion will arise. The mapping $w \mapsto |w|$ that sends a state to its equivalence class is called the *natural map*.

Let $W_\Sigma = \{|w|_\Sigma \mid w \in W\}$. Suppose $\mathfrak{M}^f_\Sigma$ is any model (W^f, R^f, V^f) such that:

(i) $W^f = W_\Sigma$.
(ii) If Rwv then $R^f |w||v|$.
(iii) If $R^f |w||v|$ then for all $\Diamond\phi \in \Sigma$, if $\mathfrak{M}, v \Vdash \phi$ then $\mathfrak{M}, w \Vdash \Diamond\phi$.
(iv) $V^f(p) = \{|w| \mid \mathfrak{M}, w \Vdash p\}$, for all proposition letters p in Σ.

Then $\mathfrak{M}^f_\Sigma$ is called a *filtration of* $\mathfrak{M}$ *through* Σ; we will often suppress subscripts and write $\mathfrak{M}^f$ instead of $\mathfrak{M}^f_\Sigma$.   ⊣

Because of item (ii), the natural map associated with any filtration is guaranteed to be a homomorphism (see Definition 2.7). And at first glance it may seem that it is even guaranteed to be a bounded morphism (see Definition 2.10), for item (iii) seems reminiscent of the back condition. Unfortunately, this is *not* the case, as the following example shows.

Example 2.37 Let $\mathfrak{M}$ be the model $(\mathbb{N}, R, V)$, where $R = \{(0, 1), (0, 2), (1, 3)\} \cup \{(n, n + 1) \mid n \geq 2\}$, and V has $V(p) = \mathbb{N} \setminus \{0\}$ and $V(q) = \{2\}$.

Further, assume that $\Sigma = \{\Diamond p, p\}$. Clearly Σ is subformula closed. Then, the model $\mathfrak{N} = (\{|0|, |1|\}, \{(|0|, |1|), (|1|, |1|)\}, V')$, where $V'(p) = \{|1|\}$, is a filtration of $\mathfrak{M}$ through Σ. See Figure 2.6.

Clearly, $\mathfrak{N}$ can *not* be a bounded morphic image of $\mathfrak{M}$: any bounded morphism would have to preserve the formula q, and the natural map does not preserve q, and need not, because q is not an element of our subformula closed set Σ.   ⊣

But in many other respects filtrations are well-behaved. For a start, the method gives us a bound (albeit an exponential one) on the size of the resulting finite model:

Proposition 2.38 *Let Σ be a finite subformula closed set of basic modal formulas. For any model $\mathfrak{M}$, if $\mathfrak{M}^f$ is a filtration of $\mathfrak{M}$ through a subformula closed set Σ, then $\mathfrak{M}^f$ contains at most 2^n nodes (where n denotes the size of Σ).*

Proof. The states of $\mathfrak{M}^f$ are the equivalence classes in W_Σ. Let g be the function with domain W_Σ and range $\mathcal{P}(\Sigma)$ defined by $g(|w|) = \{\phi \in \Sigma \mid \mathfrak{M}, w \Vdash \phi\}$. It follows from the definition of $\leftrightsquigarrow_\Sigma$ that g is well defined and injective. Thus the size of W_Σ is at most 2^n, where n is the size of Σ. $\dashv$

Moreover – crucially – filtrations preserve satisfaction in the following sense.

Theorem 2.39 (Filtration Theorem) *Consider the basic modal language. Let $\mathfrak{M}^f$ $(= (W_\Sigma, R^f, V^f))$ be a filtration of $\mathfrak{M}$ through a subformula closed set Σ. Then for all formulas $\phi \in \Sigma$, and all nodes w in $\mathfrak{M}$, we have $\mathfrak{M}, w \Vdash \phi$ iff $\mathfrak{M}^f, |w| \Vdash \phi$.*

Proof. By induction on ϕ. The base case is immediate from the definition of V^f. The boolean cases are straightforward; the fact that Σ is closed under subformulas allows us to apply the inductive hypothesis.

So suppose $\Diamond\phi \in \Sigma$ and $\mathfrak{M}, w \Vdash \Diamond\phi$. Then there is a v such that Rwv and $\mathfrak{M}, v \Vdash \phi$. As $\mathfrak{M}^f$ is a filtration, $R^f|w||v|$. As Σ is subformula closed, $\phi \in \Sigma$, thus by the inductive hypothesis $\mathfrak{M}^f, |v| \Vdash \phi$. Hence $\mathfrak{M}^f, |w| \Vdash \Diamond\phi$.

Conversely, suppose $\Diamond\phi \in \Sigma$ and $\mathfrak{M}^f, |w| \Vdash \Diamond\phi$. Thus there is a state $|v|$ in $\mathfrak{M}^f$ such that $R^f|w||v|$ and $\mathfrak{M}^f, |v| \Vdash \phi$. As $\phi \in \Sigma$, by the inductive hypothesis $\mathfrak{M}, v \Vdash \phi$. So the third clause in Definition 2.36 is applicable, and we conclude that $\mathfrak{M}, w \Vdash \Diamond\phi$. $\dashv$

Observe that clauses (ii) and (iii) of Definition 2.36 are designed to make the modal case of the induction step go through in the proof above.

But we still have not done one vital thing: we have not actually shown that filtrations exist! Observe that the clauses (ii) and (iii) in Definition 2.36 only impose conditions on candidate relations R^f – but we have not yet shown that a suitable R^f can always be found. In fact, there are always at least two ways to define binary relations that fulfill the required conditions. Define R^s and R^l as follows:

(i) $R^s|w||v|$ iff $\exists w' \in |w| \exists v' \in |v| \, Rw'v'$.

(ii) $R^l|w||v|$ iff for all formulas $\Diamond\phi$ in Σ: $\mathfrak{M}, v \Vdash \phi$ implies $\mathfrak{M}, w \Vdash \Diamond\phi$.

These relations – which are not necessarily distinct – give rise to the *smallest* and *largest* filtrations respectively.

Lemma 2.40 *Consider the basic modal language. Let $\mathfrak{M}$ be any model, Σ any subformula closed set of formulas, W_Σ the set of equivalence classes induced by $\leftrightsquigarrow_\Sigma$, and V^f the standard valuation on W_Σ. Then both (W_Σ, R^s, V^f) and (W_Σ, R^l, V^f) are filtrations of $\mathfrak{M}$ through Σ. Furthermore, if (W_Σ, R^f, V^f) is any filtration of $\mathfrak{M}$ through Σ then $R^s \subseteq R^f \subseteq R^l$.*

Proof. We show that (W_Σ, R^s, V^f) is a filtration; the rest is left as an exercise. It suffices to show that R^s fulfills clauses (ii) and (iii) of Definition 2.36. But R^s satisfies clause (ii) by definition, so it remains to check clause (iii). Suppose $R^s|w||v|$, and further suppose that $\Diamond\phi \in \Sigma$ and $\mathfrak{M}, v \Vdash \phi$. As $R^s|w||v|$, there exist $w' \in |w|$ and $v' \in |v|$ such that $Rw'v'$. As $\phi \in \Sigma$ and $\mathfrak{M}, v \Vdash \phi$, then because $v \leftrightsquigarrow_\Sigma v'$, we get $\mathfrak{M}, v' \Vdash \phi$. But $Rw'v'$, so $\mathfrak{M}, w' \Vdash \Diamond\phi$. But $\Diamond\phi \in \Sigma$, thus as $w' \leftrightsquigarrow_\Sigma w$ it follows that $\mathfrak{M}, w \Vdash \Diamond\phi$.   ⊣

Theorem 2.41 (Finite Model Property – via Filtrations) *Let ϕ be a basic modal formula. If ϕ is satisfiable, then it is satisfiable on a finite model. Indeed, it is satisfiable on a finite model containing at most 2^m nodes, where m is the number of subformulas of ϕ.*

Proof. Assume that ϕ is satisfiable on a model $\mathfrak{M}$; take any filtration of $\mathfrak{M}$ through the set of subformulas of ϕ. That ϕ is satisfied in the filtration is immediate from Theorem 2.39. The bound on the size of the filtration is immediate from Proposition 2.38.   ⊣

There are several points worth making about filtrations. The first has to do with the possible loss of properties when moving from a model to one of its filtrations. As we have already discussed, a drawback of the selection method is that it can be hard to preserve such properties. Filtrations are far better in this respect – but they certainly are not perfect. Let us consider the matter more closely.

Suppose (W_Σ, R^f, V^f) is a filtration of (W, R, V). Now, clause (ii) of Definition 2.36 means that the natural map from $\mathfrak{M}$ to $\mathfrak{M}^f$ is a surjective homomorphism with respect to the accessibility relation R. Thus any property of relations which is preserved under such maps will automatically be inherited by any filtration. Obvious examples include reflexivity and right unboundedness ($\forall x \exists y\, Rxy$).

However, many interesting relational properties are *not* preserved under homomorphisms: transitivity and symmetry are obvious counterexamples. Thus we need to find special filtrations which preserve these properties. Sometimes this is easy; for example, the smallest filtration preserves symmetry. Sometimes we need new ideas to find a good filtration; the classic example involves transitivity. Let us see what this involves.

Lemma 2.42 *Let $\mathfrak{M}$ be a model, Σ a subformula closed set of formulas, and W_Σ*

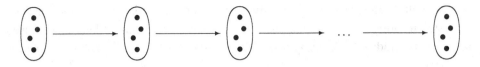

Fig. 2.7. Filtrating a model based on $(\mathbb{Q}, <)$.

the set of equivalence classes induced on $\mathfrak{M}$ by $\leftrightsquigarrow_\Sigma$. Let R^t be the binary relation on W_Σ defined by:

$$R^t|w||v| \; \textit{iff for all } \phi, \textit{ if } \Diamond\phi \in \Sigma \textit{ and } \mathfrak{M}, v \Vdash \phi \vee \Diamond\phi \textit{ then } \mathfrak{M}, w \Vdash \Diamond\phi.$$

If R is transitive then (W_Σ, R^t, V^f) is a filtration and R^t is transitive.

Proof. Left as Exercise 2.3.5. ⊣

In short, filtrations are flexible – but it is not a matter of 'plug and play.' Creativity is often required to exploit them.

The second point worth making is that filtrations of an infinite model through a finite set manage to represent an infinite amount of information in a finitary manner. It seems obvious, at least from an intuitive point of view, that this can only be achieved by *identifying* lots of points. As we have seen in Example 2.37, an infinite chain may be collapsed onto a single reflexive point by a filtration. An even more informative example is provided by models based on the rationals. For instance, what happens to the density condition in the filtration? Let $\mathfrak{M} = (\mathbb{Q}, <, V)$; then any (finite) filtration of $\mathfrak{M}$ has the form displayed in Figure 2.7. What is going on here? Instead of viewing models as structures made up of states and relations between them, in the case of filtrations it can be useful to view them as *sets* of states (namely, the sets of identified states) and relations between those sets. The following definition captures this idea.

Definition 2.43 Let (W, R, V) be a transitive frame. A *cluster* on (W, R, V) is a maximal, nonempty equivalence class under R. That is, $C \subseteq W$ is a cluster if the restriction of R to C is an equivalence relation, and this is *not* the case for any subset D properly extending C.

A cluster is *simple* if it consists of a single reflexive point, and *proper* if it contains more than one point. ⊣

As Figure 2.7 shows, a (finite) filtration of $(\mathbb{Q}, <)$ can be thought of as resulting in a finite linear sequence of clusters, perhaps interspersed with singleton irreflexive points, no two of which can be adjacent. (Note: the displayed model is transitive, even though we haven't drawn in the arrows needed to indicate this.) The reader is asked to check this claim in Exercise 2.3.9. Clusters will play an important role in Section 4.5.

To conclude this section we briefly indicate how the filtration method can be extended to other modal languages. Let us first consider modal languages based on arbitrary modal similarity types τ. Fix a τ-model $\mathfrak{M} = (W, R_\triangle, V)_{\triangle \in \tau}$ and a subformula closed set Σ as in Definition 2.36. Suppose $\mathfrak{M}_\Sigma^f = (W_\Sigma, R_\triangle^f, V^f)_{\triangle \in \tau}$ is a τ-model where W_Σ and V^f are as in Definition 2.36, and for $\triangle \in \tau$, $R_\triangle^f$ satisfy

(ii)′ if $R_\triangle w v_1 \ldots v_n$ then $R^f |w| |v_1| \ldots |v_n|$,

(iii)′ if $R^f |w| |v_1| \ldots |v_n|$, then for all $\phi_1, \ldots, \phi_n \in \Sigma$, if $\triangle(\phi_1, \ldots, \phi_n) \in \Sigma$ and $\mathfrak{M}, v_1 \Vdash \phi_1, \ldots, \mathfrak{M}, v_n \Vdash \phi_n$, then $\mathfrak{M}, w \Vdash \triangle(\phi_1, \ldots, \phi_n)$.

Then $\mathfrak{M}_\Sigma^f$ is a *τ-filtration of* $\mathfrak{M}$ *through* Σ.

With this definition at hand, Proposition 2.38 and Theorem 2.39 can be reformulated and proved for τ-filtrations, and suitable versions of the smallest and largest filtrations can also be defined, resulting in a general modal analog of Theorem 2.41, the Finite Model Property.

What about basic temporal logic, PDL, and arrow logic? It turns out that the filtration method works well for all of these. For basic temporal logic we need to issue the customary warning (we need to be explicit about what the filtration does to $R^{\smallsmile}$), but with this observed, matters are straightforward. Exercise 2.3.7 asks the reader to define transitive filtrations for the basic temporal language.

Matters are far more interesting (and difficult) with PDL – but here too, by making use of a clever idea called the Fischer-Ladner closure, it is possible to use a filtration style argument to show that PDL has the finite model property; we will do this in Section 4.8 as part of a completeness proof (Theorem 4.91). Exercise 2.3.10 deals with the finite model property for arrow logic.

Exercises for Section 2.3

2.3.1 Find two models $\mathfrak{M}$ and $\mathfrak{M}'$ and states w and w' in these models such that $w \underline{\leftrightarrow}_n w'$ for all n, but it is *not* the case that $w \underline{\leftrightarrow} w'$ are bisimilar. (Hint: we drew a picture of such a pair of models in the previous section.)

2.3.2 Generalize the definition of n-bisimulations (Definition 2.30) from diamond-only to arbitrary modal languages. Then prove Proposition 2.31 (that n-bisimilarity for all n implies modal equivalence and conversely) for arbitrary modal languages.

2.3.3 Lemma 2.33 tells us that if we are only interested in the satisfiability of modal formulas of degree at most k, we can delete all states that lie beyond the k-horizon without affecting satisfiability. Prove this.

2.3.4 The proof of Theorem 2.34 uses a selection-of-points argument to establish the finite model property. But no proof details were given for the last (crucial) claim in the proof, namely that $\mathfrak{M}_4, w_2$ is k-bisimilar to $\mathfrak{M}_3, w_2$. Fill in this gap.

2.3.5 First show that not every filtration of a transitive model is transitive. Then prove Lemma 2.42. That is, show that the relation R^t defined there is indeed a filtration, and that any filtration of a transitive model that makes use of R^t is guaranteed to be transitive.

2.3.6 Finish the proof of Lemma 2.40. That is, prove that the filtrations R^s and R^l are indeed the smallest and the largest filtration, respectively. In addition, give an example of a model and a set of formulas for which R^s and R^l coincide.

2.3.7 Show that every transitive model (W, R, V) has a transitive *temporal* filtration. (Take care to specify what the filtration does to $R^{\smallsmile}$.)

2.3.8 Call a frame or model *euclidean* if it satisfies $\forall xyz\,((Rxy \wedge Rxz) \rightarrow Ryz)$, and let E be the class of euclidean models. Fix a formula ξ, and let Σ be the smallest subformula closed set of formulas containing ξ that satisfies, for all formulas ψ: if $\Diamond\psi \in \Sigma$, then $\Box\Diamond\psi \in \Sigma$. (Recall that $\Box$ is an abbreviation of $\neg\Diamond\neg$.) Note that in general, Σ will be infinite.

(a) Prove that $E \Vdash \Diamond\psi \rightarrow \Box\Diamond\psi$.
(b) Prove that every euclidean model can be filtrated through Σ to a euclidean model.
(c) Show that every euclidean model satisfies the following modal reduction principles:
$\Diamond\Diamond\Diamond \leftrightarrow \Diamond\Diamond$, $\Diamond\Diamond\Box \leftrightarrow \Diamond\Box$, $\Diamond\Box\Diamond \leftrightarrow \Diamond\Diamond$ and $\Diamond\Box\Box \leftrightarrow \Diamond\Box$. That is, prove that the formulas $\Diamond\Diamond\Diamond\phi \leftrightarrow \Diamond\Diamond\phi$, ... are true throughout every euclidean model. Conclude that Σ is finite modulo equivalence on euclidean models.
(d) Prove that the basic modal similarity type has the finite model property with respect to the class of euclidean models. Can you prove this result simply by filtrating through any subformula closed set of formulas containing ξ?

2.3.9 Let $\mathfrak{M}^f$ be a finite, transitive filtration of a model based on the rationals with their usual ordering. Describe the possible shape of $\mathfrak{M}^f$ in terms of clusters and sets consisting of a single, irreflexive point. In particular, show that there is a natural way to impose a linear order on this collection of subsets of $\mathbb{Q}$. Can $\mathfrak{M}^f$ have two adjacent singleton clusters? Two adjacent singleton sets each consisting of an irreflexive point?

2.3.10 Consider the similarity type $\tau_{\rightarrow}$ of arrow logic.

(i) Show that $\tau_{\rightarrow}$ has the finite model property with respect to the class of all arrow models.
(ii) Consider the class of arrow models based on arrow frames $\mathfrak{F} = (W, C, R, I)$ such that for all s, t and u in W we have (i) $Cstu$ iff $Csut$ iff $Ctus$ and (ii) $Cstu$ and Iu iff $s = t$. Prove that arrow formulas have the finite model property with respect to this class of arrow models.
(iii) Prove that $\tau_{\rightarrow}$ does not have the finite model property with respect to the class of all square models. (Hint: try to express that the extension of the propositional variable p is a dense, linear ordering.)

2.4 The Standard Translation

In the Preface we warned the reader against viewing modal logic as an isolated formal system (remember Slogan 3?), yet here we are, halfway through Chapter 2, and we still have not linked modal logic with the wider logical world. We now put this right. We define a link called the *standard translation*. This paves the way for the results on modal expressivity in the sections that follow, for the study of frames in the following chapter, and for the introduction of the guarded fragment in Section 7.4.

We first specify our *correspondence languages* – that is, the languages we will translate modal formulas into.

Definition 2.44 For τ a modal similarity type and Φ a collection of proposition letters, let $\mathcal{L}^1_\tau(\Phi)$ be the first-order language (with equality) which has unary predicates P_0, P_1, P_2, ... corresponding to the proposition letters p_0, p_1, p_2, ... in Φ, and an $(n+1)$-ary relation symbol $R_\triangle$ for each (n-ary) modal operator $\triangle$ in our similarity type. We write $\alpha(x)$ to denote a first-order formula α with one free variable, x.   ⊣

We are now ready to define the standard translation.

Definition 2.45 (Standard Translation) Let x be a first-order variable. The *standard translation* ST_x taking modal formulas to first-order formulas in $\mathcal{L}^1_\tau(\Phi)$ is defined as follows:

$$
\begin{aligned}
ST_x(p) &= Px, \\
ST_x(\bot) &= x \neq x, \\
ST_x(\neg\phi) &= \neg ST_x(\phi), \\
ST_x(\phi \vee \psi) &= ST_x(\phi) \vee ST_x(\psi), \\
ST_x(\triangle(\phi_1, \ldots, \phi_n)) &= \exists y_1 \ldots \exists y_n \, (R_\triangle x y_1 \ldots y_n \wedge \\
&\qquad ST_{y_1}(\phi_1) \wedge \cdots \wedge ST_{y_n}(\phi_n)),
\end{aligned}
$$

where $y_1, \ldots, y_n$ are fresh variables (that is, variables that have not been used so far in the translation). When working with the basic modal language, the last clause boils down to:

$$
ST_x(\Diamond\phi) = \exists y \, (Rxy \wedge ST_y(\phi)).
$$

Note that (to keep notation simple) we prefer to use R rather than $R_\Diamond$, and we will continue to do this. We leave to the reader the task of working out what $ST_x(\triangledown(\phi_1, \ldots, \phi_n))$ is, but we will point out that for the basic modal language the required clause is:

$$
ST_x(\Box\phi) = \forall y \, (Rxy \rightarrow ST_y(\phi)).
$$ ⊣

Example 2.46 Let us see how this works. Consider the formula $\Diamond(\Box p \rightarrow q)$.

$$
ST_x(\Diamond(\Box p \rightarrow q)) = \exists y_1 \, (Rxy_1 \wedge ST_{y_1}(\Box p \rightarrow q))
$$

$$= \exists y_1 \left(Rxy_1 \wedge \left(ST_{y_1}(\Box p) \to ST_{y_1}(q) \right) \right)$$
$$= \exists y_1 \left(Rxy_1 \wedge \left(\forall y_2 \left(Ry_1y_2 \to ST_{y_2}(p) \right) \to Qy_1 \right) \right)$$
$$= \exists y_1 \left(Rxy_1 \wedge \left(\forall y_2 \left(Ry_1y_2 \to Py_2 \right) \to Qy_1 \right) \right).$$

Note that (this version of) the standard translation leaves the choice of fresh variables unspecified. For example, $\exists y_{256} \left(Rxy_{256} \wedge \left(\forall y_{14} \left(Ry_{256}y_{14} \to Py_{14} \right) \to Qy_{256} \right) \right)$ is a legitimate translation of $\Diamond(\Box p \to q)$, and indeed there are infinitely many others, all differing only in the bound variables they contain. Later in the section we remove this indeterminacy – elegantly. ⊣

It should be clear that the standard translation makes good sense: it is essentially a first-order reformulation of the modal satisfaction definition. For any modal formula ϕ, $ST_x(\phi)$ will contain exactly one free variable (namely x); the role of this free variable is to mark the current state; this use of a free variable makes it possible for the global notion of first-order satisfaction to mimic the local notion of modal satisfaction. Furthermore, observe that modalities are translated as *bounded quantifiers*, and in particular, quantifiers bounded to act only on related states; this is the obvious way of mimicking the local action of the modalities in first-order logic. Because of its importance it is worth pinning down just why the standard translation works.

Models for modal languages based on a modal similarity type τ and a collection of proposition letters Φ can also be viewed as models for $\mathcal{L}^1_\tau(\Phi)$. For example, if τ contains just a single diamond $\Diamond$, then the corresponding first-order language $\mathcal{L}^1_\tau(\Phi)$ has a binary relation symbol R and a unary predicate symbol corresponding to each proposition letter in Φ – and a first-order model for this language needs to provide an interpretation for these symbols. But a (modal) model $\mathfrak{M} = (W, R, V)$ supplies precisely what is required: the binary relation R can be used to interpret the relation symbol R, and the set $V(p_i)$ can be used to interpret the unary predicate P_i. This should *not* come as a surprise. As we emphasized in Chapter 1 (especially Sections 1.1 and 1.3) there is no mathematical distinction between modal and first-order models – both modal and first-order models are simply relational structures. Thus it makes perfect sense to write things like $\mathfrak{M} \models ST_x(\phi)[w]$, which means that the first-order formula $ST_x(\phi)$ is satisfied (in the usual sense of first-order logic) in the model $\mathfrak{M}$ when w is assigned to the free variable x.

Proposition 2.47 (Local and Global Correspondence on Models) *Fix a modal similarity type τ, and let ϕ be a τ-formula. Then:*

(i) *For all $\mathfrak{M}$ and all states w of $\mathfrak{M}$: $\mathfrak{M}, w \Vdash \phi$ iff $\mathfrak{M} \models ST_x(\phi)[w]$.*
(ii) *For all $\mathfrak{M}$: $\mathfrak{M} \Vdash \phi$ iff $\mathfrak{M} \models \forall x \, ST_x(\phi)$.*

Proof. By induction on ϕ. We leave this to the reader as Exercise 2.4.1. ⊣

Summing up: when interpreted on models, modal formulas are equivalent to first-order formulas in one free variable. Fine – but what does that give us? Lots! Proposition 2.47 is a bridge between modal and first-order logic – and we can use this bridge to import results, ideas, and proof techniques from one to the other.

Example 2.48 First-order logic has the compactness property: if Θ is a set of first-order formulas, and every finite subset of Θ is satisfiable, then so is Θ itself. It also has the downward Löwenheim-Skolem property: if a set of first-order formulas has an infinite model, then it has a countably infinite model.

It follows that modal logic must have both these properties (over models) too. Consider compactness. Suppose Σ is a set of modal formulas every finite subset of which is satisfiable – is Σ itself satisfiable? Yes. Consider the set $\{ST_x(\phi) \mid \phi \in \Sigma\}$. As every finite subset of Σ has a model it follows (reading item (i) of Proposition 2.47 left to right) that every finite subset of $\{ST_x(\phi) \mid \phi \in \Sigma\}$ does too, and hence (by first-order compactness) that this whole set is satisfiable in some model, say $\mathfrak{M}$. But then it follows (this time reading item (i) of Proposition 2.47 right to left) that Σ is satisfiable in $\mathfrak{M}$, hence modal satisfiability over models is compact.

And there is interesting traffic from modal logic to first-order logic too. For example, a significant difference between modal and first-order logic is that modal logic is decidable (over arbitrary models) but first-order logic is not. By using our understanding of modal decidability, it is possible to locate novel decidable fragments of first-order logic, a theme we will return to in Section 7.4 when we discuss the guarded fragment.   ⊣

Just as importantly, the standard translation gives us a new research agenda for investigating modal expressivity: *correspondence theory*. The central aim of this chapter is to explore the expressivity of modal logic over models – but how is expressivity to be measured? Proposition 2.47 suggests an interesting strategy: try to characterize the fragment of first-order logic picked out by the standard translation.

It is obvious on purely syntactic grounds that the standard translation is not surjective (standard translations of modal formulas contain only bounded quantifiers) – but could every first-order formula (in the appropriate correspondence language) be *equivalent* to the translation of a modal formula? No. This is very easy to see: whereas modal formulas are invariant under bisimulations, first-order formulas need not be; thus any first-order formula which is not invariant under bisimulations cannot be equivalent to the translation of a modal formula. We have seen such a formula in Section 2.2 (namely $\exists y_1 y_2 y_3 \, (y_1 \neq y_2 \wedge y_1 \neq y_3 \wedge y_2 \neq y_3 \wedge Rxy_1 \wedge Rxy_2 \wedge Ry_1 y_3 \wedge Ry_2 y_3))$, and it is easy to find simpler examples.

Thus the (first-order formulas equivalent to) standard translations of model formulas are a proper subset of the correspondence language. Which subset? Here is a nice observation. The standard translation can be reformulated so that it maps every modal formula into a very small fragment of $\mathcal{L}^1_\tau(\Phi)$, namely a certain *finite-variable fragment*. Suppose the variables of $\mathcal{L}^1_\tau(\Phi)$ have been ordered in some way. Then the *n*-variable fragment of $\mathcal{L}^1_\tau(\Phi)$ is the set of $\mathcal{L}^1_\tau(\Phi)$ formulas that contain only the first n variables. As we will now see, by judicious reuse of variables, a modal language with operators of arity at most n can be translated into the $(n+1)$-variable fragment of $\mathcal{L}^1_\tau(\Phi)$. (Reuse of variables is the name of the game when working with finite variable fragments. For example, we can express the existence of *three* different points in a linear ordering using only *two* variables as follows: $\exists xy\,(x < y \wedge \exists x\,(y < x))$.)

Proposition 2.49   (i) *Let τ be a modal similarity type that only contains diamonds. Then, every τ-formula ϕ is equivalent to a first-order formula containing at most two variables.*

   (ii) *More generally, if τ does not contain modal operators $\triangle$ whose arity exceeds n, all τ-formulas are equivalent to first-order formulas containing at most $(n + 1)$ variables.*

Proof. Assume τ contains only diamonds $\langle a \rangle$, $\langle b \rangle$, ...; proving the general case is left as Exercise 2.4.2. Fix two distinct individual variables x and y. Define two variants ST_x and ST_y of the standard translation as follows:

$$
\begin{array}{ll}
ST_x(p) = Px & ST_y(p) = Py \\
ST_x(\bot) = x \neq x & ST_y(\bot) = y \neq y \\
ST_x(\neg\phi) = \neg ST_x(\phi) & ST_y(\neg\phi) = \neg ST_y(\phi) \\
ST_x(\phi \vee \psi) = ST_x(\phi) \vee ST_x(\psi) & ST_y(\phi \vee \psi) = ST_y(\phi) \vee ST_y(\psi) \\
ST_x(\langle a \rangle \phi) = \exists y\,(R_a xy \wedge ST_y(\phi)) & ST_y(\langle a \rangle \phi) = \exists x\,(R_a yx \wedge ST_x(\phi)).
\end{array}
$$

Then, for any τ-formula ϕ, its ST_x-translation contains at most the two variables x and y, and $ST_x(\phi)$ is equivalent to the original standard translation of ϕ.   ⊣

Example 2.50 Let us see how this modified standard translation works. Consider again the formula $\Diamond(\Box p \rightarrow q)$.

$$
\begin{aligned}
ST_x(\Diamond(\Box p \rightarrow q)) &= \exists y\,(Rxy \wedge ST_y(\Box p \rightarrow q)) \\
&= \exists y\,(Rxy \wedge (\forall x\,(Ryx \rightarrow ST_x(p)) \rightarrow Qy)) \\
&= \exists y\,(Rxy \wedge (\forall x\,(Ryx \rightarrow Px) \rightarrow Qy)).
\end{aligned}
$$

That is, we just keep flipping between the two variables x and y. The result is a translation containing only two variables (instead of the three used in Example 2.46). As a side effect, the indeterminacy associated with the original version of the standard translation has disappeared.   ⊣

This raises another question: is every first-order formula $\alpha(x)$ in two variables equivalent to the translation of a basic modal formula? Again the answer is *no*. There is even a first-order formula in a single variable x which is not equivalent to any modal formula, namely Rxx. To see this, assume for the sake of a contradiction that ϕ is a modal formula such that $ST_x(\phi)$ is equivalent to Rxx. Let $\mathfrak{M}$ be a singleton reflexive model and let w be the unique state in $\mathfrak{M}$; obviously (irrespective of the valuation) $\mathfrak{M} \models Rxx[w]$. Let $\mathfrak{N}$ be a model based on the strict ordering of the integers; obviously (again, irrespective of the valuation), for every integer v, $\mathfrak{N} \models \neg Rxx[v]$. Let Z be the relation which links every integer with the unique state in $\mathfrak{M}$, and assume that the valuations in $\mathfrak{N}$ and $\mathfrak{M}$ are such that Z is a bisimulation (for example, make all proposition letters true at all points in both models). As $\mathfrak{M} \models Rxx[w]$, it follows by Proposition 2.47 that $\mathfrak{M}, w \Vdash \phi$ (after all, by assumption Rxx is equivalent to $ST_x(\phi)$). But for any integer v, we have that $w \leftrightarrow v$, hence $\mathfrak{N}, v \Vdash \phi$. Hence (again by Proposition 2.47 and our assumption that $ST_x(\phi)$ is equivalent to Rxx) we have that $\mathfrak{N} \models Rxx[v]$, contradicting the fact that $\mathfrak{N} \models \neg Rxx[v]$.

We will not discuss correspondence theory any further here, but in Section 2.6 we will prove one of its central results, the van Benthem Characterization Theorem: a first-order formula is equivalent to the translation of a modal formula if and only if it is invariant under bisimulations.

Proposition 2.47 is also going to help us investigate modal expressivity in other ways, notably via the concept of definability.

Definition 2.51 Let τ be a modal similarity type, C a class of τ-models, and Γ a set of formulas over τ. We say that Γ *defines* or *characterizes* a class K of models *within* C if for all models $\mathfrak{M}$ in C we have that $\mathfrak{M}$ is in K iff $\mathfrak{M} \Vdash \Gamma$. If C is the class of all τ-models, we simply say that Γ defines or characterizes K; we omit brackets whenever Γ is a singleton. We will say that a formula ϕ defines a *property* whenever ϕ defines the class of models satisfying that property.   ⊣

It is immediate from Proposition 2.47 that if a class of models is definable by a set of modal formulas, then it is also definable by a set of first-order formulas – but this is too obvious to be interesting. The important way in which Proposition 2.47 helps, is by making it possible to exploit standard model construction techniques from first-order model theory. For example, in Section 2.6 we will prove Theorem 2.75 which says that a class of (pointed) models is modally definable if and only if it is closed under bisimulations and ultraproducts (an important construction known from first-order model theory; see Appendix A), and its complement is closed under ultrapowers (another standard model theoretic construction). It would be difficult to overemphasize the importance of the standard translation; it is remarkable that such a simple idea can lead to so much.

To conclude this section, let us see how to adapt these ideas to the basic temporal language, PDL, and arrow logic. The case of basic temporal logic is easy: all we have to do is add a clause for translating the backward looking operator P:

$$ST_x(P\phi) = \exists y\,(Ryx \wedge ST_y(\phi)).$$

Note that we are using the more sophisticated approach introduced in the proof of Proposition 2.49: flipping between two translations ST_x and ST_y. (Thus we really need to add a mirror clause which flips the variables back.) So, just like the basic modal language, the basic temporal language can be mapped into a two variable fragment of the correspondence language. Moreover (again, as with the basic modal language) not every first-order formula in two variables is equivalent to (the translation of) a basic temporal formula (see Exercise 2.4.3).

Propositional dynamic logic calls for more drastic changes. Let us first look at the $*$-free fragment – that is, at PDL formulas without occurrences of the Kleene star. In PDL both formulas and modalities are recursively structured, so we are going to need two interacting translation functions: one to handle the formulas, the other to handle the modalities. The only interesting clause in the formula translation is the following:

$$ST_x(\langle\pi\rangle\phi) = \exists y\,(ST_{xy}(\pi) \wedge ST_y(\phi)).$$

That is, instead of returning a fixed relation symbol (say R), the formula translation ST_x calls on ST_{xy} to start recursively decomposing the program π. Why does this part of the translation require two free variables? Because its task is to define a binary relation.

$$
\begin{aligned}
ST_{xy}(a) &= R_a xy \text{ (and similarly for other pairs of variables),}\\
ST_{xy}(\pi_1 \cup \pi_2) &= ST_{xy}(\pi_1) \vee ST_{xy}(\pi_2),\\
ST_{xy}(\pi_1 \,;\, \pi_2) &= \exists z\,(ST_{xz}(\pi_1) \wedge ST_{zy}(\pi_2)).
\end{aligned}
$$

It follows that we can translate the $*$-free fragment of PDL into a *three* variable fragment of the correspondence language. The details are worth checking; see Exercise 2.4.4.

But the really drastic change comes when we consider the full language of PDL (that is, with Kleene star). Recall that a program α^* is interpreted using the reflexive, transitive closure of R_α. But the reflexive, transitive closure of an arbitrary relation is *not* a first-order definable relation (see Exercise 2.4.5). So the standard translation for PDL needs to take us to a richer background logic than first-order logic, one that can express this concept. Which one should we use? There are many options here, but to motivate our actual choice recall the definition of the meaning of a PDL program α^*:

$$R_{\alpha^*} = \bigcup_n (R_\alpha)^n,$$

where R_α^n is defined by

$$R^0 xy \text{ iff } x = y \quad \text{and} \quad R^{n+1} xy \text{ iff } \exists z \, (R^n xz \wedge Rzy).$$

Thus, if we were allowed to write infinitely long disjunctions, it would be easy to capture the meaning of an iterated program α^*:

$$(R_\alpha)^* xy \text{ iff } (x = y) \vee R_\alpha xy \vee \bigvee_{n \geq 1} \exists z_1 \ldots z_n \, (R_\alpha xz_1 \wedge \cdots \wedge R_\alpha z_n y).$$

In *infinitary logic* we can do this. More precisely, in $\mathcal{L}_{\omega_1 \omega}$ we are allowed to form formulas as in first-order logic, and, in addition, to build countably infinite disjunctions and conjunctions. We will take $\mathcal{L}_{\omega_1 \omega}$ as the target logic for the standard translation of PDL. We have seen most of the clauses we need: we use the clauses for the $*$-free fragment given above, and in addition the following clause to cater for the Kleene star:

$$ST_{xy}(\alpha^*) =$$
$$(x = y) \vee ST_{xy}(\alpha) \vee \bigvee_{n \geq 1} \exists z_1 \ldots z_n \, (ST_{xz_1}(\alpha) \wedge \cdots \wedge ST_{z_n y}(\alpha)).$$

This example of PDL makes an important point vividly: we cannot always hope to embed modal logic into first-order logic. Indeed in the following chapter we will see that when it comes to analyzing the expressive power of modal logic at the level of *frames*, the natural correspondence language (even for the basic modal language) is second-order logic.

There is nothing particularly interesting concerning the standard translation for the arrow language of Example 1.16. However, this changes when we turn to *square* models: in Exercise 2.4.6 the reader is asked to prove that on this class of models, the arrow language corresponds to a first-order language with *binary* predicate symbols, and that, in fact, it is expressively *equivalent* to the three variable fragment of such a language.

Exercises for Section 2.4

2.4.1 Prove Proposition 2.47. That is, check that the standard translation really is correct.

2.4.2 Prove Proposition 2.49 for arbitrary modal languages. That is, show that if τ does not contain modal operators $\triangle$ whose arity exceeds n, all τ-formulas are equivalent to first-order formulas containing at most $(n + 1)$ variables.

2.4.3 Show that there are first-order formulas $\alpha(x)$ using at most two variables that are not equivalent to the standard translation of a basic temporal formula.

2.4.4 In this exercise you should fill in some of the details for the standard translation for PDL.

(a) Check that the translation for the $*$-free fragment of PDL really does map all such formulas into the three variable fragment of the corresponding first-order language.
(b) Show that in fact, there is a translation into the *two* variable fragment of this corresponding first-order language.

2.4.5 The aim of this exercise is to show that taking the reflexive, transitive closure of a binary relation is not a first-order definable operation.

(a) Show that the class of connected graphs is not first-order definable:
  (i) For $l \in \mathbb{N}$, let $\mathfrak{C}_l$ be the graph given by a cycle of length $l + 1$:
$$\mathfrak{C}_l = (\{0, \ldots, l\}, \{(i, i + 1), (i + 1, i) \mid 0 \le i < l\} \cup \{(0, l), (l, 0)\})$$
    Show that for every $k \in \mathbb{N}$ and $l \ge 2^k$ the graph $\mathfrak{C}_l$ satisfies the same first-order sentences of quantifier rank at most k as the disjoint union $\mathfrak{C}_l \uplus \mathfrak{C}_l$.
  (ii) Conclude that the class of connected graphs is not first-order definable.
(b) Use item (a) to conclude that the reflexive transitive closure of a relation is not first-order definable.

2.4.6 Consider the class of square models for arrow logic. Observe that a square model $\mathfrak{M} = (\mathfrak{S}_U, V)$ can be seen as a first-order model $\mathfrak{M}^* = (U, V(p))_{p \in \Phi}$ if we let each propositional variable $p \in \Phi$ correspond to a *dyadic* relation symbol P.

(a) Work out this observation in the following sense. Define a suitable translation $(\cdot)^*$ mapping an arrow formula ϕ to a formula $\phi^*(x_0, x_1)$ in this 'dyadic correspondence language.' Prove that this translation has the property that for all arrow formulas ϕ and all square models $\mathfrak{M}$ the following correspondence holds:
$$\mathfrak{M}, (a_0, a_1) \Vdash \phi \text{ iff } \mathfrak{M}^* \models \phi^*(x_0, x_1)[a_0, a_1].$$
(b) Show that this translation can be done within the three variable fragment of first-order logic.
(c) Prove that conversely, every formula $\alpha(x_0, x_1)$ that uses only three variables, in a first-order language with binary predicates only, is equivalent to the translation of an arrow formula on the class of square models.

2.5 Modal Saturation via Ultrafilter Extensions

Bisimulations and the standard translation are two of the tools we need to understand modal expressivity over models. This section introduces the third: *ultrafilter extensions*. To motivate their introduction, we will first discuss *Hennessy-Milner model classes* and *modally saturated models*; both generalize ideas met in our earlier discussion of bisimulations. We will then introduce ultrafilter extensions as a way of building modally saturated models, and this will lead us to an elegant result: modal equivalence implies bisimilarity-somewhere-else.

M-saturation

Theorem 2.20 tells us that bisimilarity implies modal equivalence, but we have already seen that the converse does not hold in general (recall Figure 2.5). The

Hennessy-Milner Theorem shows that the converse does hold in the special case of image-finite models. Let us try and generalize this theorem.

First, when proving Theorem 2.24, we exploited the fact that, between image-finite models, the relation of modal equivalence *itself* is a bisimulation. Classes of models for which this holds are evidently worth closer study.

Definition 2.52 (Hennessy-Milner Classes) Let τ be a modal similarity type, and K a class of τ-models. K is a *Hennessy-Milner* class, or *has the Hennessy-Milner property*, if for every two models $\mathfrak{M}$ and $\mathfrak{M}'$ in K and any two states w, w' of $\mathfrak{M}$ and $\mathfrak{M}'$, respectively, $w \rightsquigarrow w'$ implies $\mathfrak{M}, w \leftrightarrow \mathfrak{M}', w'$.   ⊣

For example, by Theorem 2.24, the class of image-finite models has the Hennessy-Milner property. On the other hand, no class of models containing the two models in Figure 2.5 has the Hennessy-Milner property.

We generalize the notion of image-finiteness; doing so leads us to the concept of *modally-saturated* or (briefly) *m-saturated* models. Suppose we are working in the basic modal language. Let $\mathfrak{M} = (W, R, V)$ be a model, let w be a state in W, and let $\Sigma = \{\phi_0, \phi_1, \ldots\}$ be an infinite set of formulas. Suppose that w has successors $v_0, v_1, v_2, \ldots$ where (respectively) ϕ_0, $\phi_0 \wedge \phi_1$, $\phi_0 \wedge \phi_1 \wedge \phi_2, \ldots$ hold. If there is no successor v of w where *all* formulas from Σ hold *at the same time*, then the model is in some sense incomplete. A model is called m-saturated if incompleteness of this kind does not occur.

To put it another way: suppose that we are looking for a successor of w at which every formula ϕ_i of the infinite set of formulas $\Sigma = \{\phi_0, \phi_1, \ldots\}$ holds. M-saturation is a kind of compactness property, according to which it suffices to find satisfying successors of w for arbitrary finite approximations of Σ.

Definition 2.53 (M-saturation) Let $\mathfrak{M} = (W, R, V)$ be a model of the basic modal similarity type, X a subset of W and Σ a set of modal formulas. Σ is *satisfiable* in the set X if there is a state $x \in X$ such that $\mathfrak{M}, x \Vdash \phi$ for all ϕ in Σ; Σ is *finitely satisfiable* in X if every finite subset of Σ is satisfiable in X.

The model $\mathfrak{M}$ is called *m-saturated* if it satisfies the following condition for every state $w \in W$ and every set Σ of modal formulas:

If Σ is finitely satisfiable in the set of successors of w,
then Σ is satisfiable in the set of successors of w.

The definition of m-saturation for arbitrary modal similarity types runs as follows. Let τ be a modal similarity type, and let $\mathfrak{M}$ be a τ-model. $\mathfrak{M}$ is called *m-saturated* if, for every state w of $\mathfrak{M}$ and every (n-ary) modal operator $\triangle \in \tau$ and sequence $\Sigma_1, \ldots, \Sigma_n$ of sets of modal formulas, we have the following:

If for every sequence of finite subsets $\Delta_1 \subseteq \Sigma_1, \ldots, \Delta_n \subseteq \Sigma_n$ there are states $v_1, \ldots, v_n$ such that $Rwv_1 \ldots v_n$ and $v_1 \Vdash \Delta_1, \ldots, v_n \Vdash \Delta_n$, *then* there are states $v_1, \ldots, v_n$ in $\mathfrak{M}$ such that $Rwv_1 \ldots v_n$ and $v_1 \Vdash \Sigma_1$, $\ldots, v_n \Vdash \Sigma_n$. ⊣

Proposition 2.54 *Let τ be a modal similarity type. Then the class of m-saturated τ-models has the Hennessy-Milner property.*

Proof. We only prove the proposition for the basic modal language. Let $\mathfrak{M} = (W, R, V)$ and $\mathfrak{M}' = (W', R', V')$ be two m-saturated models. It suffices to prove that the relation $\leftrightsquigarrow$ of modal equivalence between states in $\mathfrak{M}$ and states in $\mathfrak{M}'$ is a bisimulation. We confine ourselves to a proof of the forth condition of a bisimulation, since the condition concerning the propositional variables is trivially satisfied, and the back condition is completely analogous to the case we prove.

So, assume that $w, v \in W$ and $w' \in W'$ are such that Rwv and $w \leftrightsquigarrow w'$. Let Σ be the set of formulas true at v. It is clear that for every finite subset Δ of Σ we have $\mathfrak{M}, v \Vdash \bigwedge \Delta$, hence $\mathfrak{M}, w \Vdash \Diamond \bigwedge \Delta$. As $w \leftrightsquigarrow w'$, it follows that $\mathfrak{M}', w' \Vdash \Diamond \bigwedge \Delta$, so w' has an R'-successor v_Δ such that $\mathfrak{M}', v_\Delta \Vdash \bigwedge \Delta$. In other words, Σ is finitely satisfiable in the set of successors of w'; but, then, by m-saturation, Σ itself is satisfiable in a successor v' of w'. Thus $v \leftrightsquigarrow v'$. ⊣

Ultrafilter extensions

So the class of m-saturated models satisfies the Hennessy-Milner property – but how do we actually *build* m-saturated models? To this end, we will now introduce the last of the 'big four' model constructions: *ultrafilter extensions*. The ultrafilter extension of a structure (model or frame) is a kind of *completion* of the original structure. The construction adds states to a model in order to make it m-saturated. Sometimes the result is a model isomorphic to the original (for example, when the original model is finite) but when working with infinite models, the ultrafilter extension always adds lots of new points. In the definition of ultrafilter extension we need operations on the power set algebra of a frame; we have already met one of these operations in Section 1.4 when we introduced general frames.

Definition 2.55 Given an $(n+1)$-ary relation R on a set W, we define the following two n-ary operations m_R and l_R on the power set $\mathcal{P}(W)$ of W:

$$m_R(X_1, \ldots, X_n) := \{w \in W \mid \text{there exist } w_1, \ldots, w_n \text{ such that}$$
$$Rww_1 \ldots w_n \text{ and } w_i \in X_i \text{ for all } i\},$$
$$l_R(X_1, \ldots, X_n) := \{w \in W \mid \text{for all } w_1, \ldots, w_n: \text{ if } Rww_1 \ldots w_n,$$
$$\text{then there is an } i \text{ with } w_i \in X_i\}. \quad \dashv$$

For a binary relation R, $m_R(X)$ is the set of points that 'can see' a state in X, and $l_R(X)$ is the set of points that 'only see' states in X. It follows that for any model $\mathfrak{M} = (W, R, V)$ we have

$$V(\Diamond\phi) = m_R(V(\phi)) \text{ and } V(\Box\phi) = l_R(V(\phi)).$$

Similar identities hold for modal operators of higher arity. Furthermore, for any relation R, m_R and l_R are each other's dual, in the following sense:

Proposition 2.56 *Let R be a relation of arity $n + 1$ on the set W. Then, for every n-tuple $X_1, \ldots, X_n$ of subsets of W we have*

$$l_R(X_1, \ldots, X_n) = W \setminus m_R(W \setminus X_1, \ldots, W \setminus X_n).$$

Proof. Left to the reader.   ⊣

We are ready to define ultrafilter extensions. As the name is meant to suggest, the states of the ultrafilter extension of a model $\mathfrak{M}$ are the ultrafilters over the universe of $\mathfrak{M}$. Filters and ultrafilters are discussed in Appendix A. Readers encountering this notion for the first time, are advised to do Exercises 2.5.1–2.5.4.

Definition 2.57 (Ultrafilter Extension) Let τ be a modal similarity type, and $\mathfrak{F} = (W, R_\Delta)_{\Delta \in \tau}$ a τ-frame. The *ultrafilter extension* $\mathfrak{ue}\,\mathfrak{F}$ of $\mathfrak{F}$ is defined as the frame $(Uf(W), R_\Delta^{ue})_{\Delta \in \tau}$. Here $Uf(W)$ is the set of ultrafilters over W and $R_\Delta^{ue} u_0 u_1 \ldots u_n$ holds for a tuple $u_0, \ldots, u_n$ of ultrafilters over W if we have that $m_{R_\Delta}(X_1, \ldots, X_n) \in u_0$ whenever $X_i \in u_i$ for all i with $1 \leq i \leq n$.

The *ultrafilter extension* of a τ-model $\mathfrak{M} = (\mathfrak{F}, V)$ is the model $\mathfrak{ue}\,\mathfrak{M} = (\mathfrak{ue}\,\mathfrak{F}, V^{ue})$ where $V^{ue}(p_i)$ is the set of ultrafilters of which $V(p_i)$ is a member.   ⊣

What are the intuitions behind this definition? First, note that the main ingredients have a logical interpretation. Any subset of a frame can, in principle, be viewed as (the extension or interpretation of) a *proposition*. A filter over the universe of the frame can thus be seen as a *theory*, in fact as a logically closed theory, since filters are both closed under intersection (conjunction) and upward closed (entailment). Viewed this way, a proper filter is a *consistent* theory, for it does not contain the empty set (falsum). Finally, an ultrafilter is a *complete* theory, or, as we will call it, a *state of affairs*: for each proposition (subset of the universe) an ultrafilter chooses between it and its negation (between the subset and its complementation).

How does this relate to ultrafilter extensions? In a given frame $\mathfrak{F}$ not every state of affairs needs to be 'realized', in the sense that there is a state satisfying all and only the propositions belonging to the state of affairs; only the states of affairs that correspond to the *principal* ultrafilters are realized, namely, as the points of the frame. We build $\mathfrak{ue}\,\mathfrak{F}$ by adding every state of affairs for $\mathfrak{F}$ as a new element of the domain – that is, $\mathfrak{ue}\,\mathfrak{F}$ realizes every proposition in $\mathfrak{F}$.

How should we relate these new elements in $\mathfrak{ue}\,\mathfrak{F}$ to each other and to the original elements from $\mathfrak{F}$? The obvious choice is to stipulate that $R^{ue}_{\Delta} u_0 u_1 \ldots u_n$ if u_0 'sees' the n-tuple $u_1, \ldots, u_n$. That is, whenever $X_1, \ldots, X_n$ are propositions of $u_1, \ldots, u_n$ respectively, then u_0 'sees' this combination: that is, the proposition $m_{R_{\Delta}}(X_1, \ldots, X_n)$ is a member of u_0. The definition of the valuation V^{ue} is self-explanatory.

One final comment: a special role in this section is played by the so-called *principal* ultrafilters over W. Recall that, given an element $w \in W$, the *principal ultrafilter* generated by w is the set $\pi_w = \{ X \subseteq W \mid w \in X \}$. By identifying a state w of a frame $\mathfrak{F}$ with the principal ultrafilter π_w, it is easily seen that any frame $\mathfrak{F}$ is (isomorphic to) a *submodel* (but in general not a *generated* submodel) of its ultrafilter extension. For we have the following equivalences, here proved for the basic modal similarity type:

$$
\begin{aligned}
Rwv \quad &\text{iff} \quad w \in m_R(X) \text{ for all } X \subseteq W \text{ such that } v \in X \\
&\text{iff} \quad m_R(X) \in \pi_w \text{ for all } X \subseteq W \text{ such that } X \in \pi_v \qquad (2.1) \\
&\text{iff} \quad R^{ue} \pi_w \pi_v.
\end{aligned}
$$

Let us make our discussion more concrete by considering an example.

Example 2.58 Consider the frame $\mathfrak{N} = (\mathbb{N}, <)$, the natural numbers in their usual ordering (a transitive model, though we haven't drawn in arrows to indicate this):

What is the ultrafilter extension of $\mathfrak{N}$? There are two kinds of ultrafilters over an infinite set: the principal ultrafilters that are in one-to-one correspondence with the points of the set, and the non-principal ones which contain all co-finite sets, and only infinite sets, cf. Exercise 2.5.4. We have just remarked (see (2.1)) that the principal ultrafilters form an isomorphic copy of the frame $\mathfrak{N}$ inside $\mathfrak{ue}\,\mathfrak{N}$. So where are the non-principal ultrafilters situated? The key fact here is that for any pair u, u' of ultrafilters, if u' is non-principal, then $R^{ue} u u'$. To see this, let u' be a non-principal ultrafilter, and let $X \in u'$. As X is infinite, for any $n \in \mathbb{N}$ there is an m such that $n < m$ and $m \in X$. This shows that $m_<(X) = \mathbb{N}$. But $\mathbb{N}$ is an element of every ultrafilter u.

This shows that the ultrafilter extension of $\mathfrak{N}$ looks like a gigantic balloon at the end of an infinite string: it consists of a copy of $\mathfrak{N}$, followed by a large (uncountable) cluster consisting of all the non-principal ultrafilters (again, the following diagram represents a transitive model):

We will prove two results concerning ultrafilter extensions. The first one, Proposition 2.59, is an invariance result: any state in the original model is modally equivalent to the corresponding principal ultrafilter in the ultrafilter extension. Then, in Proposition 2.61 we show that ultrafilter extensions are m-saturated. Putting these two facts together leads us to the main result of this section: two states are modally equivalent iff their representatives in the ultrafilter extensions are bisimilar.

Proposition 2.59 *Let τ be a modal similarity type, and $\mathfrak{M}$ a τ-model. Then, for any formula ϕ and any ultrafilter u over W, $V(\phi) \in u$ iff $ue\, \mathfrak{M}, u \Vdash \phi$. Hence, for every state w of $\mathfrak{M}$ we have $w \leftrightsquigarrow \pi_w$.*

Proof. The second claim of the proposition is immediate from the first one by the observation that $w \Vdash \phi$ iff $w \in V(\phi)$ iff $V(\phi) \in \pi_w$.

The proof of the first claim is by induction on ϕ. The basic case is immediate from the definition of V^{ue}. The proofs of the boolean cases are straightforward consequences of the defining properties of ultrafilters. As an example, we treat negation; suppose that ϕ is of the form $\neg\psi$, then

$$
\begin{aligned}
V(\neg\psi) \in u \quad &\text{iff} \quad W \setminus V(\psi) \in u \\
&\text{iff} \quad V(\psi) \notin u \\
&\text{iff} \quad ue\, \mathfrak{M}, u \not\Vdash \psi \quad \text{(induction hypothesis)} \\
&\text{iff} \quad ue\, \mathfrak{M}, u \Vdash \neg\psi.
\end{aligned}
$$

Next, consider the case where ϕ is of the form $\Diamond\psi$ (we only treat the basic modal similarity type, leaving the general case as an exercise to the reader). Assume first that $ue\, \mathfrak{M}, u \Vdash \Diamond\psi$. Then, there is an ultrafilter u' such that $R^{ue}uu'$ and $ue\, \mathfrak{M}, u' \Vdash \psi$. The induction hypothesis implies that $V(\psi) \in u'$, so by the definition of R^{ue}, $m_R(V(\psi)) \in u$. Now the result follows immediately from the observation that $m_R(V(\psi)) = V(\Diamond\psi)$.

The left-to-right implication requires a bit more work. Assume that $V(\Diamond\psi) \in u$. We have to find an ultrafilter u' such that $V(\psi) \in u'$ and $R^{ue}uu'$. The latter constraint reduces to the condition that $m_R(X) \in u$ whenever $X \in u'$, or equivalently (see Exercise 2.5.5):

$$u_0' := \{Y \mid l_R(Y) \in u\} \subseteq u'.$$

We will first show that u_0' is closed under intersection. Let Y, Z be members of u_0'. By definition, $l_R(Y)$ and $l_R(Z)$ are in u. But then $l_R(Y \cap Z) \in u$, as $l_R(Y \cap Z) = l_R(Y) \cap l_R(Z)$, as a straightforward proof shows. This proves that $Y \cap Z \in u_0'$.

Next we make sure that for any $Y \in u_0', Y \cap V(\psi) \neq \varnothing$. Let Y be an arbitrary element of u_0', then by definition of $u_0', l_R(Y) \in u$. As u is closed under intersection and does not contain the empty set, there must be an element x in $l_R(Y) \cap V(\Diamond\psi)$. But then x must have a successor y in $V(\psi)$. Finally, $x \in l_R(Y)$ implies $y \in Y$.

From the fact that u'_0 is closed under intersection, and the fact that for any $Y \in u'_0$, $Y \cap V(\psi) \neq \varnothing$, it follows that the set $u'_0 \cup \{V(\psi)\}$ has the finite intersection property. So the Ultrafilter Theorem (Fact A.14 in the Appendix) provides us with an ultrafilter u' such that $u'_0 \cup \{V(\psi)\} \subseteq u'$. This ultrafilter u' has the desired properties: it is clearly a successor of u, and the fact that $\mathfrak{ue}\,\mathfrak{M}, u' \Vdash \psi$ follows from $V(\psi) \in u'$ and the induction hypothesis. ⊣

Example 2.60 As with the invariance results of Section 2.1 (disjoint unions, generated submodels, and bounded morphisms), our new invariance result can be used to compare the relative expressive power of modal languages. Consider the modal constant $\circlearrowleft$ whose truth definition in a model for the basic modal language is

$$\mathfrak{M}, w \Vdash \circlearrowleft \text{ iff } \mathfrak{M} \models Rxx[v] \text{ for some } v \text{ in } \mathfrak{M}.$$

Can such a modality be defined in the basic modal language? No – a bisimulation based argument given at the end of the previous section already establishes this. Alternatively, we can see this by comparing the pictures of the frame $(\mathbb{N}, <)$ and its ultrafilter extension given in Example 2.58. The former is loop-free (thus in any model over this frame, $\mathfrak{M}, 0 \not\Vdash \circlearrowleft$), but the latter contains uncountably many loops (thus $\mathfrak{ue}\,\mathfrak{M}, \pi_0 \Vdash \circlearrowleft$). So if we want $\circlearrowleft$ we have to add it as a primitive. ⊣

Proposition 2.61 *Let τ be a modal similarity type, and let $\mathfrak{M}$ be a τ-model. Then $\mathfrak{ue}\,\mathfrak{M}$ is m-saturated.*

Proof. We only prove the proposition for the basic modal similarity type. Let $\mathfrak{M} = (W, R, V)$ be a model; we will show that its ultrafilter extension $\mathfrak{ue}\,\mathfrak{M}$ is m-saturated. Consider an ultrafilter u over W, and a set Σ of modal formulas which is finitely satisfiable in the set of successors of u. We have to find an ultrafilter u' such that $R^{ue}uu'$ and $\mathfrak{ue}\,\mathfrak{M}, u' \Vdash \Sigma$. Define

$$\Delta = \{V(\phi) \mid \phi \in \Sigma'\} \cup \{Y \mid l_R(Y) \in u\},$$

where Σ' is the set of (finite) conjunctions of formulas in Σ. We claim that the set Δ has the finite intersection property. Since both $\{V(\phi) \mid \phi \in \Sigma'\}$ and $\{Y \mid l_R(Y) \in u\}$ are closed under taking intersections, it suffices to prove that for an arbitrary $\phi \in \Sigma'$ and an arbitrary set $Y \subseteq W$ for which $l_R(Y) \in u$, we have $V(\phi) \cap Y \neq \varnothing$. But if $\phi \in \Sigma'$, then by assumption, there is a successor u'' of u such that $\mathfrak{ue}\,\mathfrak{M}, u'' \Vdash \phi$, or, in other words, $V(\phi) \in u''$. Then, $l_R(Y) \in u$ implies $Y \in u''$ by Exercise 2.5.5. Hence, $V(\phi) \cap Y$ is an element of the ultrafilter u'' and, therefore, cannot be identical to the empty set.

It follows by the Ultrafilter Theorem that Δ can be extended to an ultrafilter u'. Clearly, u' is the required successor of u in which Σ is satisfied. ⊣

We have finally arrived at the main result of this section: a characterization of modal equivalence as bisimilarity-somewhere-else – namely, between ultrafilter extensions.

Theorem 2.62 *Let τ be a modal similarity type, and let $\mathfrak{M}$ and $\mathfrak{M}'$ be τ-models, and w, w' two states in $\mathfrak{M}$ and $\mathfrak{M}'$, respectively. Then*

$$\mathfrak{M}, w \leftrightsquigarrow \mathfrak{M}', w' \text{ iff } ue\,\mathfrak{M}, \pi_w \underline{\leftrightarrow} ue\,\mathfrak{M}', \pi_{w'}.$$

Proof. Immediate by Propositions 2.59, 2.61 and 2.54.   ⊣

Three remarks. First, it is easy to define ultrafilter extensions and prove an analog of Theorem 2.62 for the basic temporal logic and arrow logic; see Exercises 2.5.8 and 2.5.9. With PDL the situation is a bit more complex; see Exercise 2.5.11. (The problem is that the property of one relation being the reflexive transitive closure of another is not preserved under taking ultrafilter extensions.) Second, we have not seen the last of ultrafilter extensions. Like disjoint unions, generated submodels, and bounded morphisms, ultrafilter extensions are a fundamental modal model construction technique, and we will make use of them when we discuss frames (in Chapter 3) and algebras (in Chapter 5). We will shortly see that ultrafilter extensions tie in neatly with ideas from first-order model theory – and we will use this to prove a second bisimilarity-somewhere-else result, Lemma 2.66. Finally, some readers may still have the feeling that taking the ultrafilter extension of a model is a far less natural construction than the other model operations that we have met. These readers are advised to hold on until (or take a peek ahead towards) Chapter 5, where we will see that ultrafilter extensions are indeed a very natural byproduct of modal logic's duality theory.

Exercises for Section 2.5

2.5.1 Let E be any subset of $\mathcal{P}(W)$, and let F be the filter generated by E.

(a) Prove that indeed, F is a filter over W. (Show that in general, the intersection of a collection of filters is again a filter.)

(b) Show that F is the set of all $X \in \mathcal{P}(W)$ such that either $X = W$ or for some $Y_1, \ldots, Y_n \in E$,

$$Y_1 \cap \cdots \cap Y_n \subseteq X.$$

(c) Prove that F is proper (that is: it does not coincide with $\mathcal{P}(W)$) iff E has the finite intersection property.

2.5.2 Let W be a non-empty set, and let w be an element of W. Show that the principal ultrafilter generated by w, that is, the set $\{X \in \mathcal{P}(W) \mid w \in X\}$, is indeed an ultrafilter over W.

2.5.3 Let F be a filter over W.

(a) Prove that F is an ultrafilter if and only if it is proper and maximal, that is, it has no proper extensions.

(b) Prove that F is an ultrafilter if and only if it is proper and for each pair of subsets X, Y of W we have that $X \cup Y \in F$ iff $X \in F$ or $Y \in F$.

2.5.4 Let W be an infinite set. Recall that $X \subseteq W$ is *co-finite* if $W \setminus X$ is finite.

(a) Prove that the collection of co-finite subsets of W has the finite intersection property.

(b) Show that there are ultrafilters over W that do not contain any finite set.

(c) Prove that an ultrafilter is non-principal if and only if it contains only infinite sets if and only if it contains all co-finite sets.

(d) Prove that any ultrafilter over W has uncountably many elements.

2.5.5 Given a model $\mathfrak{M} = (W, R, V)$ and two ultrafilters u and v over W, show that $R^{ue}uv$ if and only if $\{Y \mid l_R(Y) \in u\} \subseteq v$.

2.5.6 Let $\mathfrak{B} = (B, R)$ be the transitive binary tree; that is, B is the set of finite strings of 0s and 1s, and $R\sigma\tau$ holds if σ is a proper initial segment of τ. The aim of this exercise is to prove that any non-principal ultrafilter over B determines an *infinite* string of 0s and 1s.

More precisely, let B^ω be the set of finite and infinite strings of 0s and 1s, and R^ω the relation on B^ω given by $R^\omega\sigma\tau$ if either σ is finite and a proper initial segment of τ, or else $\sigma = \tau$. Define a bounded morphism $f : \mathfrak{ue}\,\mathfrak{B} \to \mathfrak{B}^\omega$.

2.5.7 Give an example of a model $\mathfrak{M}$ which is point-generated while its ultrafilter extension is not.

2.5.8 Develop a notion of ultrafilter extension for basic temporal logic, and establish an analog of Theorem 2.62 for basic temporal logic.

2.5.9 Develop a notion of ultrafilter extension for the arrow language introduced in Example 1.14, and establish an analog of Theorem 2.62 for this language.

2.5.10 Show that, in general, first-order formulas are not preserved under ultrafilter extensions. That is, give a model $\mathfrak{M}$, a state w, and a first-order formula $\alpha(x)$ such that $\mathfrak{M} \models \alpha(x)[w]$, but $\mathfrak{ue}\,\mathfrak{M} \not\models \alpha(x)[\pi_w]$, where π_w is the principal ultrafilter generated by w.

2.5.11 Consider a modal similarity type with two diamonds, $\Diamond$ and $\langle * \rangle$, and take any model $\mathfrak{M} = (S, R, R_*, V)$ with

$$
\begin{aligned}
S &= \mathbb{N} \cup \{\infty\}, \\
R &= \{(n+1, n), (\infty, n) \mid n \in \mathbb{N}\}, \\
R_* &= \{(m, n) \mid m, n \in \mathbb{N}, m \geq n\} \cup (\{\infty\} \times S).
\end{aligned}
$$

Note that R_* is the reflexive transitive closure of R.

(a) Show that $\mathfrak{M}, \infty \Vdash \Box \langle * \rangle \Box \bot$.

(b) Let u be an arbitrary non-principal ultrafilter over S. Prove that $R^{ue}\pi_\infty u$.

(c) Let u be an arbitrary non-principal ultrafilter over S. Prove that u has an R^{ue}-successor in $\mathfrak{ue}\,\mathfrak{M}$, and that each of its R^{ue}-successors is again a non-principal ultrafilter.

(d) Now suppose that we add a new diamond $\langle \star \rangle$ to the language, and that in the model $\mathfrak{ue}\, \mathfrak{M}$ we take $R_\star$ to be the reflexive transitive closure of R^{ue}. Show that $\mathfrak{ue}\, \mathfrak{M}, \pi_\infty \Vdash \Diamond[\star]\Diamond\top$.

(e) Prove that $R_\star^{ue} \neq R_\star$. (Hint: use Proposition 2.59, and conclude that the ultrafilter extension of a regular PDL-model need not be a regular PDL-model.)

(f) Prove that every non-principal ultrafilter over S has a *unique* R^{ue}-successor.

2.6 Characterization and Definability

In Section 2.3 we posed two important questions about modal expressivity:

(i) What is the modal fragment of first-order logic? That is, which first-order formulas are equivalent to the standard translation of a modal formula?

(ii) Which properties of models are definable by means of modal formulas?

In this, the first advanced track section of the book, we answer both questions. Our main tool will be a second characterization of modal equivalence as bisimilarity-somewhere-else, the Detour Lemma. Unlike the characterization just proved (Theorem 2.62), the Detour Lemma rests on a number of non-modal concepts and results, all of which are centered on *saturated* models (a standard concept of first-order model theory). We start by introducing saturated models and use them to describe the modal fragment of first-order logic. After that we show how to build saturated models. As corollaries we obtain results on modally definable properties of models. For background information on first-order model theory, see Appendix A.

The van Benthem Characterization Theorem

To define the notion of saturated models, we need the concept of α-saturation, but before giving a formal definition of the latter, we provide an informal description, which the reader may want to use as a 'working' definition.

Informally, then, the notion of α-saturation can be explained as follows. First of all, let $\Gamma(x)$ be a set of first-order formulas in which a single individual variable x may occur free – such a set of formulas is called a *type*. A first-order model $\mathfrak{M}$ *realizes* $\Gamma(x)$ if there is an element w in $\mathfrak{M}$ such that for all $\gamma \in \Gamma$, $\mathfrak{M} \models \gamma[w]$.

Next, let $\mathfrak{M}$ be a model for a given first-order language $\mathcal{L}^1$ with domain W. For a subset $A \subseteq W$, $\mathcal{L}^1[A]$ is the language obtained by extending $\mathcal{L}^1$ with new constants $\underline{a}$ for all elements $a \in A$. $\mathfrak{M}_A$ is the expansion of $\mathfrak{M}$ to a structure for $\mathcal{L}^1[A]$ in which each $\underline{a}$ is interpreted as a.

Assume that A is of size at most α. For the sake of our informal definition of α-saturation, assume that $\alpha = 3$ and $A = \{a_1, a_2\}$. Let $\Gamma(\underline{a}_1, \underline{a}_2, x)$ be a type of the language $\mathcal{L}^1[A]$; it is not difficult to see that $\Gamma(\underline{a}_1, \underline{a}_2, x)$ is consistent with the first-order theory of $\mathfrak{M}_A$ iff $\Gamma(\underline{a}_1, \underline{a}_2, x)$ is finitely realizable in $\mathfrak{M}_A$, (that is, $\mathfrak{M}_A$ realizes every *finite* subset Δ of $\Gamma(\underline{a}_1, \underline{a}_2, x)$). So, for this particular set $\Gamma(\underline{a}_1, \underline{a}_2, x)$, 3-saturation of $\mathfrak{M}$ means that if $\Gamma(\underline{a}_1, \underline{a}_2, x)$ is finitely realizable in $\mathfrak{M}_A$, then $\Gamma(\underline{a}_1, \underline{a}_2, x)$ is realizable in $\mathfrak{M}_A$.

Yet another way of looking at 3-saturation for this particular set of formulas is the following. Consider a formula $\gamma(\underline{a}_1, \underline{a}_2, x)$, and let $\gamma(x_1, x_2, x)$ be the formula with the fresh variables x_1 and x_2 replacing each occurrence in γ of $\underline{a}_1$ and $\underline{a}_2$, respectively. Then we have the following equivalence:

$\mathfrak{M}_A$ realizes $\{\gamma(\underline{a}_1, \underline{a}_2, x)\}$ iff there is a b such that $\mathfrak{M} \models \gamma(x_1, x_2, x)[a_1, a_2, b]$.

So, a model is α-saturated iff the following holds for every $n < \alpha$, and every set Γ of formulas of the form $\gamma(x_1, \ldots, x_n, x)$.

> If $(a_1, \ldots, a_n)$ is an n-tuple such that for every *finite* $\Delta \subseteq \Gamma$ there is a b_Δ such that $\mathfrak{M} \models \gamma(x_1, \ldots, x_n, x)[a_1, \ldots, a_n, b_\Delta]$ for every $\gamma \in \Delta$,
> *then* we have that there is a b such that $\mathfrak{M} \models \gamma(x_1, \ldots, x_n, x)[a_1, \ldots, a_n, b]$ for every $\gamma \in \Gamma$.

This way of looking at α-saturation is useful, for it makes the analogy with m-saturation of the previous section clear. Both m-saturated and countably saturated models are rich in the number of types $\Gamma(x)$ they realize, but the latter are far richer than the former: they realize the maximum number of types.

Now, for the 'official' definition of α-saturation.

Definition 2.63 Let α be a natural number, or ω. A model $\mathfrak{M}$ is *α-saturated* if for every subset $A \subseteq W$ of size less than α, the expansion $\mathfrak{M}_A$ realizes every set $\Gamma(x)$ of $\mathcal{L}^1[A]$-formulas (with only x occurring free) that is consistent with the first-order theory of $\mathfrak{M}_A$. An ω-saturated model is usually called *countably saturated*. $\dashv$

Example 2.64 (i) Every finite model is countably saturated. For, if $\mathfrak{M}$ is finite, and $\Gamma(x)$ is a set of first-order formulas consistent with the first-order theory of $\mathfrak{M}$, there exists a model $\mathfrak{N}$ that is elementarily equivalent to $\mathfrak{M}$ and that realizes $\Gamma(x)$. But, as $\mathfrak{M}$ and $\mathfrak{N}$ are finite, elementary equivalence implies isomorphism, and hence $\Gamma(x)$ is realized in $\mathfrak{M}$.

(ii) The ordering of the rational numbers $(\mathbb{Q}, <)$ is countably saturated as well. The relevant first-order language $\mathcal{L}^1$ has $<$ and $=$. Take a subset A of $\mathbb{Q}$ and let $\Gamma(x)$ be a set of formulas in the resulting expansion $\mathcal{L}^1[A]$ of this first-order language that is consistent with the theory of $(\mathbb{Q}, <, a)_{a \in A}$. Then there exists a model $\mathfrak{N}$ of the theory of $(\mathbb{Q}, <, a)_{a \in A}$ that realizes $\Gamma(x)$. Now take a countable elementary submodel $\mathfrak{N}'$ of $\mathfrak{N}$ that contains at least one object realizing $\Gamma(x)$. Then $\mathfrak{N}'$ is a countable dense linear ordering without endpoints, and hence the ordering of $\mathfrak{N}'$ is isomorphic to $(\mathbb{Q}, <)$. The interpretations (in $\mathfrak{N}$) of the constants $\underline{a}$ for elements a in A may be copied across to $\mathfrak{N}$. Hence, as $\mathfrak{N}$ realizes $\Gamma(x)$, so does $\mathfrak{N}'$, and hence, so does $(\mathbb{Q}, <)$, as required.

(iii) The ordering of the natural numbers $(\mathbb{N}, <)$ is not countably saturated. To see this, consider the following set of formulas:

$$\Gamma(x) \; := \; \{\exists y_1 \, (y_1 < x), \ldots, \exists y_1 \ldots y_n \, (y_1 < \cdots < y_n < x), \ldots\}.$$

$\Gamma(x)$ is clearly consistent with the theory of $(\mathbb{N}, <)$ as each of its finite subsets is realizable in $(\mathbb{N}, <)$. Yet $\Gamma(x)$ is clearly not realizable in $(\mathbb{N}, <)$.   ⊣

The following result explains why countably saturated models matter to us.

Theorem 2.65 *Let τ be a modal similarity type. Any countably saturated τ-model is m-saturated. It follows that the class of countably saturated τ-models has the Hennessy-Milner property.*

Proof. We only consider the basic modal language. Assume that $\mathfrak{M} = (W, R, V)$, viewed as a first-order model, is countably saturated. Let a be a state in W, and consider a set Σ of modal formulas which is finitely satisfiable in the successor set of a. Define Σ' to be the set

$$\Sigma' = \{R\underline{a}x\} \cup ST_x(\Sigma),$$

where $ST_x(\Sigma)$ is the set $\{ST_x(\phi) \mid \phi \in \Sigma\}$ of standard translations of formulas in Σ. Clearly, Σ' is consistent with the first-order theory of $\mathfrak{M}_a$: $\mathfrak{M}_a$ realizes every finite subset of Σ', namely in some successor of a. So, by the countable saturation of $\mathfrak{M}$, Σ' itself is realized in some state b. By $\mathfrak{M}_a \models R\underline{a}x[b]$ it follows that b is a successor of a. Then, by Proposition 2.47 and the fact that $\mathfrak{M}_a \models ST_x(\phi)[b]$ for all $\phi \in \Sigma$, it follows that $\mathfrak{M}, b \Vdash \Sigma$. Thus Σ is satisfiable in a successor of a.   ⊣

In fact, we only need 2-saturation for the proof of Theorem 2.65 to go through. This is because we restricted ourselves to the *basic* modal similarity type. We leave it to the reader to check to which extent the 'amount of saturation' needed to make the proof of Theorem 2.65 go through depends on the rank of the operators of the similarity type.

We have yet to show that countably saturated models actually exist; this issue will be addressed below (see Theorem 2.74). For now, we merely want to record the following important use of saturated models; you may want to recall the definition of an elementary embedding before reading the result (see Appendix A).

Lemma 2.66 (Detour Lemma) *Let τ be a modal similarity type, and let $\mathfrak{M}$ and $\mathfrak{N}$ be τ-models, and w and v states in $\mathfrak{M}$ and $\mathfrak{N}$, respectively. Then the following are equivalent:*

(i) *For all modal formulas ϕ: $\mathfrak{M}, w \Vdash \phi$ iff $\mathfrak{N}, v \Vdash \phi$.*

(ii) *There exists a bisimulation* $Z : \mathfrak{ue}\, \mathfrak{M}, \pi_w \; \underline{\leftrightarrow}\; \mathfrak{ue}\, \mathfrak{N}, \pi_v.$

(iii) *There exist countably saturated models* $\mathfrak{M}^*, w^*$ *and* $\mathfrak{N}^*, v^*$ *and elementary embeddings* $f : \mathfrak{M} \preccurlyeq \mathfrak{M}^*$ *and* $g : \mathfrak{N} \preccurlyeq \mathfrak{N}^*$ *such that*

    (a) $f(w) = w^*$ *and* $g(v) = v^*$,

    (b) $\mathfrak{M}^*, w^* \; \underline{\leftrightarrow}\; \mathfrak{N}^*, v^*.$

What does the Detour Lemma say in words? Obviously (i) $\Rightarrow$ (ii) is just our old bisimulation-somewhere-else result (Theorem 2.62). The key new part is the implication (i) $\Rightarrow$ (iii). This says that if $\mathfrak{M}, w$ and $\mathfrak{N}, v$ are modally equivalent, then both can be extended – more accurately: elementarily extended – to countably saturated models $\mathfrak{M}^*, w^*$ and $\mathfrak{N}^*, v^*$. As $\mathfrak{M}, w$ and $\mathfrak{N}, v$ were modally equivalent, so are $\mathfrak{M}^*, w^*$ and $\mathfrak{N}^*, v^*$; it follows by Theorem 2.65 that the latter two models are bisimilar. In short, this is a second 'bisimilarity-somewhere-else' result, this time the 'somewhere else' being 'in some suitable ultrapower'. Notice that in order to prove the Detour Lemma all we need to establish is that every model can be elementarily embedded in a countably saturated model. There are standard first-order techniques for doing this, and we will introduce one in the second half of this section.

With the help of the Detour Lemma, we can now precisely characterize the relation between first-order logic, modal logic, and bisimulations. To prove the theorem we need to explicitly define a concept which we have already invoked informally on several occasions.

Definition 2.67 A first-order formula $\alpha(x)$ in $\mathcal{L}^1_\tau$ is *invariant for bisimulations* if for all models $\mathfrak{M}$ and $\mathfrak{N}$, and all states w in $\mathfrak{M}$, v in $\mathfrak{N}$, and all bisimulations Z between $\mathfrak{M}$ and $\mathfrak{N}$ such that wZv, we have $\mathfrak{M} \models \alpha(x)[w]$ iff $\mathfrak{N} \models \alpha(x)[v]$. $\dashv$

Theorem 2.68 (van Benthem Characterization Theorem) *Let* $\alpha(x)$ *be a first-order formula in* $\mathcal{L}^1_\tau$. *Then* $\alpha(x)$ *is invariant for bisimulations iff it is equivalent to the standard translation of a modal τ-formula.*

Proof. The direction from right to left is a consequence of Theorem 2.20. To prove the direction from left to right, assume that $\alpha(x)$ is invariant for bisimulations and consider the set of modal consequences of α:

$$\mathrm{MOC}(\alpha) = \{ST_x(\phi) \mid \phi \text{ is a modal formula, and } \alpha(x) \models ST_x(\phi)\}.$$

Our first claim is that if $\mathrm{MOC}(\alpha) \models \alpha(x)$, then $\alpha(x)$ is equivalent to the translation of a modal formula. To see why this is so, assume that $\mathrm{MOC}(\alpha) \models \alpha(x)$; then, by the Compactness Theorem for first-order logic, for some finite subset $X \subseteq \mathrm{MOC}(\alpha)$ we have $X \models \alpha(x)$. So $\models \bigwedge X \rightarrow \alpha(x)$. Trivially $\models \alpha(x) \rightarrow \bigwedge X$, thus $\models \alpha(x) \leftrightarrow \bigwedge X$. And as every $\beta \in X$ is the translation of a modal formula, so is $\bigwedge X$. This proves our claim.

So it suffices to show that $\text{MOC}(\alpha) \models \alpha(x)$. Assume $\mathfrak{M} \models \text{MOC}(\alpha)[w]$; we need to show that $\mathfrak{M} \models \alpha(x)[w]$. Let

$$T(x) = \{ST_x(\phi) \mid \mathfrak{M} \models ST_x(\phi)[w]\}.$$

We claim that $T(x) \cup \{\alpha(x)\}$ is consistent. Why? Assume, for the sake of contradiction, that $T(x) \cup \{\alpha(x)\}$ is inconsistent. Then, by compactness, for some finite subset $T_0(x) \subseteq T(x)$ we have $\models \alpha(x) \rightarrow \neg \bigwedge T_0(x)$. Hence $\neg \bigwedge T_0(x) \in \text{MOC}(\alpha)$. But this implies $\mathfrak{M} \models \neg \bigwedge T_0(x)[w]$, which contradicts $T_0(x) \subseteq T(x)$ and $\mathfrak{M} \models T(x)[w]$.

So, let $\mathfrak{N}, v$ be such that $\mathfrak{N} \models T(x) \cup \{\alpha(x)\}[v]$. Observe that w and v are modally equivalent: $\mathfrak{M}, w \Vdash \phi$ implies $ST_x(\phi) \in T(x)$, which implies $\mathfrak{N}, v \Vdash \phi$; and likewise, if $\mathfrak{M}, w \not\Vdash \phi$ then $\mathfrak{M}, w \Vdash \neg\phi$, and $\mathfrak{N}, v \Vdash \neg\phi$. If modal equivalence implied bisimilarity we would be done, because then $\mathfrak{M}, w$ and $\mathfrak{N}, v$ would be bisimilar, and from this we would be able to deduce the desired conclusion $\mathfrak{M} \models \alpha(x)[w]$ by invariance under bisimulation. But, in general, modal equivalence does not imply bisimilarity, so this is not a sound argument.

However, we can use the Detour Lemma and make a detour through a Hennessy-Milner class where modal equivalence and bisimilarity do coincide! More precisely, the Detour Lemma yields two countably saturated models $\mathfrak{M}^*, w^* \succcurlyeq \mathfrak{M}, w$ and $\mathfrak{N}^*, v^* \succcurlyeq \mathfrak{N}, v$ such that $\mathfrak{M}^*, w^* \underline{\leftrightarrow} \mathfrak{N}^*, v^*$:

$$\begin{array}{ccc} \mathfrak{M}, w & & \mathfrak{N}, v \\ \preccurlyeq \Big| & & \Big| \succcurlyeq \\ \mathfrak{M}^*, w^* & \underline{\leftrightarrow} & \mathfrak{N}^*, v^*. \end{array}$$

This is where we really need the new characterization of modal equivalence in terms of bisimulation-somewhere-else that Theorem 2.74 gives us. We need to 'lift' the first-order formula $\alpha(x)$ from the model $\mathfrak{N}, v$ to the model $\mathfrak{N}^*, v^*$. By definition, the truth of first-order formulas is preserved under elementary embeddings, so that this can indeed be done. However, first-order formulas need not be preserved under ultrafilter extensions (see Exercise 2.5.10), and for that reason we cannot use the ultrafilter extension $\mathfrak{ue} \, \mathfrak{N}, \pi_v$ instead of $\mathfrak{N}^*, v^*$.

Returning to the main argument, $\mathfrak{N} \models \alpha(x)[v]$ implies $\mathfrak{N}^* \models \alpha(x)[v^*]$. As $\alpha(x)$ is invariant for bisimulations, we get $\mathfrak{M}^* \models \alpha(x)[w^*]$. By invariance under elementary embeddings, we have $\mathfrak{M} \models \alpha(x)[w]$. This proves the theorem. $\dashv$

Ultraproducts

The preceding discussion left us with an important technical question: how do we get countably saturated models? Our next aim is to answer this question and thereby prove the Detour Lemma.

The fundamental construction underlying our proof is that of an ultraproduct. Here we briefly recall the basic ideas; further details may be found in Appendix A.

We first apply the construction to sets, and then to models. Suppose $I \neq \varnothing$, U is an ultrafilter over I, and for each $i \in I$, W_i is a non-empty set. Let $C = \prod_{i \in I} W_i$ be the cartesian product of those sets. That is: C is the set of all functions f with domain I such that for each $i \in I$, $f(i) \in W_i$. For two functions $f, g \in C$ we say that f and g are *U-equivalent* (notation $f \sim_U g$) if $\{i \in I \mid f(i) = g(i)\} \in U$. The result is that $\sim_U$ is an equivalence relation on the set C.

Definition 2.69 (Ultraproduct of Sets) Let f_U be the equivalence class of f modulo $\sim_U$, that is: $f_U = \{g \in C \mid g \sim_U f\}$. The *ultraproduct of W_i modulo U*, denoted as $\prod_U W_i$, is the set of all equivalence classes of $\sim_U$. So

$$\prod_U W_i = \{f_U \mid f \in \prod_{i \in I} W_i\}.$$

In the case where all the sets are the same, say $W_i = W$ for all i, the ultraproduct is called the *ultrapower of W modulo U*, and written $\prod_U W$. ⊣

Following the general definition of the ultraproduct of first-order models (Definition A.17), we now define the ultraproduct of modal models.

Definition 2.70 (Ultraproduct of Models) Fix a modal similarity type τ, and let $\mathfrak{M}_i$ ($i \in I$) be τ-models. The *ultraproduct $\prod_U \mathfrak{M}_i$ of $\mathfrak{M}_i$ modulo U* is the model described as follows:

(i) The universe W_U of $\prod_U \mathfrak{M}_i$ is the set $\prod_U W_i$, where W_i is the universe of $\mathfrak{M}_i$.

(ii) Let V_i be the valuation of $\mathfrak{M}_i$. Then the valuation V_U of $\prod_U \mathfrak{M}_i$ is defined by

$$f_U \in V_U(p) \text{ iff } \{i \in I \mid f(i) \in V_i(p)\} \in U.$$

(iii) Let $\triangle$ be a modal operator in τ, and $R_{\triangle i}$ its associated relation in the model $\mathfrak{M}_i$. The relation $R_{\triangle U}$ in $\prod_U \mathfrak{M}_i$ is given by

$$R_{\triangle U} f_U^1 \ldots f_U^{n+1} \text{ iff } \{i \in I \mid R_{\triangle i} f^1(i) \ldots f^{n+1}(i)\} \in U.$$

In particular, for a diamond, item (iii) boils down to

$$R_{\diamond U} f_U g_U \text{ iff } \{i \in I \mid R_{\diamond i} f(i) g(i)\} \in U. \qquad ⊣$$

To show that the above definition is consistent, we should check that V_U and R_U depend only on the equivalence classes $f_U^1, \ldots, f_U^{n+1}$.

Proposition 2.71 *Let $\prod_U \mathfrak{M}$ be an ultrapower of $\mathfrak{M}$. Then, for all modal formulas ϕ we have $\mathfrak{M}, w \Vdash \phi$ iff $\prod_U \mathfrak{M}, (f_w)_U \Vdash \phi$, where f_w is the constant function such that $f_w(i) = w$, for all $i \in I$.*

Proof. This is left as Exercise 2.6.1.　　　　　　　　　　　　　　　　⊣

To build countably saturated models, we use ultraproducts based on a special kind of ultrafilter. An ultrafilter is *countably incomplete* if it is not closed under countable intersections (of course, it will be closed under finite intersections).

Example 2.72 Consider the set of natural numbers $\mathbb{N}$. Let U be an ultrafilter over $\mathbb{N}$ that does not contain any singletons $\{n\}$. (The reader is asked to prove that such ultrafilters exist in Exercise 2.5.4.) Then, for all n, $(\mathbb{N} \setminus \{n\}) \in U$. But

$$\varnothing = \bigcap_{n \in \mathbb{N}} (\mathbb{N} \setminus \{n\}) \notin U.$$

So U is countably incomplete.　　　　　　　　　　　　　　　　　　　⊣

Lemma 2.73 *Let $\mathcal{L}$ be a countable first-order language, U a countably incomplete ultrafilter over a non-empty set I, and $\mathfrak{M}$ an $\mathcal{L}$-model. The ultrapower $\prod_U \mathfrak{M}$ is countably saturated.*

Proof. A standard result. See Appendix A for a proof reference.　　　　⊣

We are now ready to prove the Detour Lemma. In Theorem 2.62 we showed that 'bisimulation-somewhere-else' can mean 'in the ultrafilter extension.' Now we will show that it can also mean: 'in a suitable ultrapower of the original models.'

Theorem 2.74 *Let τ be a modal similarity type, and let $\mathfrak{M}$ and $\mathfrak{N}$ be τ-models, and w and v states in $\mathfrak{M}$ and $\mathfrak{N}$, respectively. Then the following are equivalent:*

(i) *For all modal formulas ϕ: $\mathfrak{M}, w \Vdash \phi$ iff $\mathfrak{N}, v \Vdash \phi$.*

(ii) *There exist ultrapowers $\prod_U \mathfrak{M}$ and $\prod_U \mathfrak{N}$ as well as a bisimulation Z : $\prod_U \mathfrak{M}, (f_w)_U \underline{\leftrightarrow} \prod_U \mathfrak{N}, (f_v)_U$ linking $(f_w)_U$ and $(f_v)_U$, where f_w (f_v) is the constant function mapping every index to w (v).*

Proof. It is easy to see that (ii) implies (i). By Proposition 2.71 $\mathfrak{M}, w \Vdash \phi$ iff $\prod_U \mathfrak{M}, (f_w)_U \Vdash \phi$. By assumption this is equivalent to $\prod_U \mathfrak{N}, (f_v)_U \Vdash \phi$, and the latter is equivalent to $\mathfrak{N}, v \Vdash \phi$.

To prove the implication from (i) to (ii) we have to do some more work. Assume that for all modal formulas ϕ we have $\mathfrak{M}, w \Vdash \phi$ iff $\mathfrak{N}, v \Vdash \phi$. We need to create bisimilar ultrapowers of $\mathfrak{M}$ and $\mathfrak{N}$.

Take the set of natural numbers $\mathbb{N}$ as our index set, and let U be a countably incomplete ultrafilter over $\mathbb{N}$ (cf. Example 2.72). By Lemma 2.73 the ultrapowers $\prod_U \mathfrak{M}$ and $\prod_U \mathfrak{N}$ are countably saturated. Now $(f_w)_U$ and $(f_v)_U$ are modally equivalent: for all modal formulas ϕ, $\prod_U \mathfrak{M}, (f_w)_U \Vdash \phi$ iff $\prod_U \mathfrak{N}, (f_v)_U \Vdash \phi$. This claim follows from the assumption that w and v are modally equivalent together with Proposition 2.71. Next, apply Theorem 2.65: as $(f_w)_U$ and $(f_v)_U$ are modally equivalent and $\prod_U \mathfrak{M}$ and $\prod_U \mathfrak{N}$ are countably saturated, there exists a bisimulation $Z : \prod_U \mathfrak{M}, (f_w)_U \underline{\leftrightarrow} \prod_U \mathfrak{N}, (f_v)_U$. This proves the theorem. ⊣

We obtain the Detour Lemma as an immediate corollary of Theorem 2.74 and Theorem 2.62.

Definability

Our next aim is to answer the second of the two questions posed at the start of this section: which properties of models are definable by means of modal formulas? Like the Detour Lemma, the answer is a corollary of Theorem 2.74. We formulate the result in terms of *pointed models*. Given a modal similarity type τ, a pointed model is a pair $(\mathfrak{M}, w)$ where $\mathfrak{M}$ is a τ-model and w is a state of $\mathfrak{M}$. Although the results below can also be given for models, the use of pointed models allows for a smoother formulation, mainly because pointed models reflect the local way in which modal formulas are evaluated.

We need some further definitions. A class of pointed models K is said to be *closed under bisimulations* if $(\mathfrak{M}, w)$ in K and $\mathfrak{M}, w \underline{\leftrightarrow} \mathfrak{N}, v$ implies $(\mathfrak{N}, v)$ in K. K is *closed under ultraproducts* if any ultraproduct $\prod_U (\mathfrak{M}_i, w_i)$ of a family of pointed models $(\mathfrak{M}_i, w_i)$ in K belongs to K. If K is a class of pointed τ-models, $\overline{\mathsf{K}}$ denotes the complement of K within the class of all pointed τ-models. Finally, K is *definable by a set of modal formulas* if there is a set of modal formulas Γ such that for any pointed model $(\mathfrak{M}, w)$ we have $(\mathfrak{M}, w)$ in K iff for all $\gamma \in \Gamma$, $\mathfrak{M}, w \Vdash \gamma$; K is definable by a single modal formula iff it is definable by a singleton set.

By Theorem 2.20 definable classes of pointed models must be closed under bisimulations, and by Proposition 2.47 and Corollary A.20 they must be closed under ultraproducts as well. Theorems 2.75 and 2.76 below show that these two closure conditions suffice to completely describe the classes of pointed models that are definable by means of modal formulas.

Theorem 2.75 *Let τ be a modal similarity type, and K a class of pointed τ-models. Then the following are equivalent:*

(i) K *is definable by a set of modal formulas.*
(ii) K *is closed under bisimulations and ultraproducts, and $\overline{\mathsf{K}}$ is closed under ultrapowers.*

Proof. The implication from (i) to (ii) is easy. For the converse, assume K and $\overline{\mathsf{K}}$ satisfy the stated closure conditions. Observe that $\overline{\mathsf{K}}$ is closed under bisimulations, as K is. Define T as the set of modal formulas holding in K:

$$T = \{\phi \mid \text{for all } (\mathfrak{M}, w) \text{ in K: } \mathfrak{M}, w \Vdash \phi\}.$$

We will show that T defines the class K. First of all, by definition every pointed model $(\mathfrak{M}, w)$ in K is a model satisfying T in the sense that $\mathfrak{M}, w \Vdash T$. Second, assume that $\mathfrak{M}, w \Vdash T$; to complete the proof of the theorem we show that $(\mathfrak{M}, w)$ must be in K.

Define Σ to be the modal theory of w; that is, $\Sigma = \{\phi \mid \mathfrak{M}, w \Vdash \phi\}$. It is obvious that Σ is finitely satisfiable in K; for suppose that the set $\{\sigma_1, \ldots, \sigma_n\} \subseteq \Sigma$ is not satisfiable in K. Then the formula $\neg(\sigma_1 \wedge \cdots \wedge \sigma_n)$ would be true on all pointed models in K, so it would belong to T, yet be false in $\mathfrak{M}, w$. But then the following claim shows that Σ is satisfiable in the ultraproduct of pointed models in K.

Claim 1 *Let Σ be a set of modal formulas, and K a class of pointed models in which Σ is finitely satisfiable. Then Σ is satisfiable in some ultraproduct of models in K.*

Proof of Claim. Define an index set I as the collection of all finite subsets of Σ:

$$I = \{\Sigma_0 \subseteq \Sigma \mid \Sigma_0 \text{ is finite}\}.$$

By assumption, for each $i \in I$ there is a pointed model $(\mathfrak{N}_i, v_i)$ in K such that $\mathfrak{N}_i, v_i \Vdash i$. We now construct an ultrafilter U over I such that the ultraproduct $\prod_U \mathfrak{N}_i$ has a state f_U with $\prod_U \mathfrak{N}_i, f_U \Vdash \Sigma$.

For each $\sigma \in \Sigma$, let $\hat{\sigma}$ be the set of all $i \in I$ such that $\sigma \in i$. Then the set $E = \{\hat{\sigma} \mid \sigma \in \Sigma\}$ has the finite intersection property because

$$\{\sigma_1, \ldots, \sigma_n\} \in \hat{\sigma}_1 \cap \cdots \cap \hat{\sigma}_n.$$

So, by Fact A.14, E can be extended to an ultrafilter U over I. This defines $\prod_U \mathfrak{N}_i$; for the definition of f_U, let W_i denote the universe of the model $\mathfrak{N}_i$ and consider the function $f \in \prod_{i \in I} W_i$ such that $f(i) = v_i$.

It is left to prove that

$$\prod_U \mathfrak{N}_i, f_U \Vdash \Sigma. \tag{2.2}$$

To prove (2.2), observe that for $i \in \hat{\sigma}$ we have $\sigma \in i$, and so $\mathfrak{N}_i, v_i \Vdash \sigma$. Therefore, for each $\sigma \in \Sigma$

$$\{i \in I \mid \mathfrak{N}_i, v_i \Vdash \sigma\} \supseteq \hat{\sigma} \text{ and } \hat{\sigma} \in U.$$

It follows that $\{i \in I \mid \mathfrak{N}_i, v_i \Vdash \sigma\} \in U$, so by Theorem A.19, $\prod_U \mathfrak{N}_i, f_U \Vdash \sigma$. This proves (2.2), and, hence, Claim 1. $\dashv$

It follows from Claim 1 and the closure of K under taking ultraproducts that Σ is satisfiable in some pointed model $(\mathfrak{N}, v)$ in K. But $\mathfrak{N}, v \Vdash \Sigma$ implies that v and the state w from our original pointed model $(\mathfrak{M}, w)$ are modally equivalent. So by Theorem 2.74 there exists an ultrafilter U' such that

$$\prod_{U'}(\mathfrak{N}, v), (f_v)_U \rightleftarrows \prod_{U'}(\mathfrak{M}, w), (f_w)_U.$$

By closure under ultraproducts, the pointed model $(\prod_{U'}(\mathfrak{N}, v), (f_v)_U)$ belongs to K. Hence by closure under bisimulations, $(\prod_{U'}(\mathfrak{M}, w), (f_w)_U)$ is in K as well. By closure of $\overline{K}$ under ultrapowers it follows that $(\mathfrak{M}, w)$ is in K. This completes the proof. ⊣

Theorem 2.76 *Let τ be a modal similarity type, and* K *a class of pointed τ-models. Then the following are equivalent:*

(i) K *is definable by means of a single modal formula.*

(ii) *Both* K *and* $\overline{K}$ *are closed under bisimulations and ultraproducts.*

Proof. The direction from (i) to (ii) is easy. For the converse we assume that K, $\overline{K}$ satisfy the stated closure conditions. Then both are closed under ultraproducts, hence by Theorem 2.75 there are sets of modal formulas T_1, T_2 defining K and $\overline{K}$, respectively. Obviously their union is inconsistent in the sense that there is no pointed model $(\mathfrak{M}, w)$ such that $(\mathfrak{M}, w) \Vdash T_1 \cup T_2$. So then, by compactness, there exist $\phi_1, \ldots, \phi_n \in T_1$ and $\psi_1, \ldots, \psi_m \in T_2$ such that for all pointed models $(\mathfrak{M}, w)$

$$\mathfrak{M}, w \Vdash \phi_1 \wedge \cdots \wedge \phi_n \rightarrow \neg\psi_1 \vee \cdots \vee \neg\psi_m. \tag{2.3}$$

To complete the proof we show that K is in fact defined by the conjunction $\phi_1 \wedge \cdots \wedge \phi_n$. By definition, for any $(\mathfrak{M}, w)$ in K we have $\mathfrak{M}, w \Vdash \phi_1 \wedge \cdots \wedge \phi_n$. Conversely, if $\mathfrak{M}, w \Vdash \phi_1 \wedge \cdots \wedge \phi_n$, then, by (2.3), $\mathfrak{M}, w \Vdash \neg\psi_1 \vee \cdots \vee \neg\psi_m$. Hence, $\mathfrak{M}, w \nVdash T_2$. Therefore, $(\mathfrak{M}, w)$ does not belong to $\overline{K}$, whence $(\mathfrak{M}, w)$ belongs to K. ⊣

Theorems 2.75 and 2.76 correspond to analogous definability results in first-order logic: to get the analogous first-order results, simply replace closure under bisimulations in 2.75 and 2.76 by closure under isomorphisms; see the Notes at the end of the chapter for further details. This close connection to first-order logic may explain why the results of this section seem to generalize to any modal logic that has a standard translation into first-order logic. For example, all of the results of this section can also be obtained for basic temporal logic.

Exercises for Section 2.6

2.6.1 Prove Proposition 2.71: Let $\prod_U \mathfrak{M}$ be an ultrapower of $\mathfrak{M}$. Then, for all modal formulas ϕ we have $\mathfrak{M}, w \Vdash \phi$ iff $\prod_U \mathfrak{M}, (f_w)_U \Vdash \phi$, where f_w is the constant function such that $f_w(i) = w$, for all $i \in I$.

2.6.2 Give simple proofs of Theorem 2.75 and Theorem 2.76 using the analogous proof for first-order logic (see Theorem A.23).

2.6.3 Let I be an index set, and let $\{\mathfrak{M}_i\}_{i\in I}$ and $\{\mathfrak{N}_i\}_{i\in I}$ be two collections of models such that for each $i \in I$, $\mathfrak{M}_i \leftrightarrow \mathfrak{N}_i$. Show that for any ultrafilter U over I, the ultraproducts of the two collections are bisimilar: $\prod_U \mathfrak{M}_i \leftrightarrow \prod_U \mathfrak{N}_i$.

2.6.4   (a) Show that the ultraproduct of point-generated models need not be point-generated.
   (b) How is this for transitive models?

2.7 Simulation and Safety

Theorem 2.68 provided a result characterizing the modal fragment of first-order logic as the class of formulas invariant for bisimulations. In this section we present two further results in the same spirit; we focus on these results not just because they are interesting and typical of current work in modal model theory, but also because they provide instructive examples of how to apply the tools and proof strategies we have discussed. We first look at a notion of simulation that has been introduced in various settings, and characterize the modal formulas preserved by simulations. We then examine a question that arises in the setting of dynamic logic and process algebra: which operations on models preserve bisimulation? That is, if we have the back and forth clauses holding for R, and we apply an operation O to R which returns a new relation $O(R)$, then under which conditions do we also have the back and forth clauses for $O(R)$?

Simulations

A simulation is simply a bisimulation from which half of the atomic clause and the back clause have been omitted.

Definition 2.77 (Simulations) Let τ be a modal similarity type. Let $\mathfrak{M} = (W, R_\triangle, V)_{\triangle \in \tau}$ and $\mathfrak{M}' = (W', R'_\triangle, V')_{\triangle \in \tau}$ be τ-models. A non-empty binary relation $Z \subseteq W \times W'$ is called a τ-*simulation* from $\mathfrak{M}$ to $\mathfrak{M}'$ if the following conditions are satisfied:

(i) If wZw' and $w \in V(p)$, then $w' \in V'(p)$.

(ii) If wZw' and $R_\triangle w v_1 \ldots v_n$ then there are $v'_1, \ldots, v'_n$ (in W') such that $R'_\triangle w' v'_1 \ldots v'_n$ and for all i $(1 \leq i \leq n)$ $v_i Z v'_i$.

Thus, simulations only require that atomic information is preserved and that the forth condition holds.

If Z is a simulation from w in $\mathfrak{M}$ to w' in $\mathfrak{M}'$, we write $Z : \mathfrak{M}, w \rightrightarrows \mathfrak{M}', w'$; if there is a simulation Z such that $Z : \mathfrak{M}, w \rightrightarrows \mathfrak{M}', w'$, we sometimes write $\mathfrak{M}, w \rightrightarrows \mathfrak{M}', w'$.

A modal formula ϕ is *preserved under* simulations if for all models $\mathfrak{M}$ and $\mathfrak{M}'$, and all states w and w' in $\mathfrak{M}$ and $\mathfrak{M}'$, respectively, $\mathfrak{M}, w \Vdash \phi$ implies $\mathfrak{M}', w' \Vdash \phi$, whenever it is the case that $\mathfrak{M}, w \rightrightarrows \mathfrak{M}', w'$. ⊣

In various forms and under various names simulations have been considered in theoretical computer science. In the study of refinement, $\rightrightarrows$ is interpreted as follows: if $\mathfrak{M}, w \rightrightarrows \mathfrak{M}', w'$ then (the system modeled by) $\mathfrak{M}', w'$ refines or implements (the system modeled by) $\mathfrak{M}, w$. And in the database world one looks at simulations the other way around: if $\mathfrak{M}, w \rightrightarrows \mathfrak{M}', w'$, then $\mathfrak{M}', w'$ *constrains* the structure of $\mathfrak{M}, w$ by only allowing those relational patterns that are present in $\mathfrak{M}', w'$ itself. Note that if $\mathfrak{M}, w \rightrightarrows \mathfrak{M}', w'$ then $\mathfrak{M}', w'$ cannot enforce the presence of patterns. (See the Notes for references.) The following question naturally arises: which formulas are preserved when passing from $\mathfrak{M}, w$ to $\mathfrak{M}', w'$ along a simulation? Or, dually, which constraints on $\mathfrak{M}, w$ can be expressed by requiring that $\mathfrak{M}, w \rightrightarrows \mathfrak{M}', w'$?

Clearly simulations do not preserve the truth of all modal formulas. In particular, let $\mathfrak{M}$ be a one-point model with domain $\{w\}$ and empty relation; then, there is a simulation from $\mathfrak{M}, w$ to any state with the same valuation, no matter which model it lives in. Using this observation it is easy to show that universal modal formulas of the form $\Box(\cdots)$ or $\nabla(\cdots)$ are not preserved under simulations. On the other hand, by clause (ii) of Definition 2.77 existential modal formulas of the form $\Diamond(\cdots)$ or $\triangle(\cdots)$ are preserved under simulations. This leads to the conjecture that a modal formula is preserved under simulations if and only if it is equivalent to a formula that has been built from proposition letters, using only $\wedge$, $\vee$ and existential modal operators, that is, diamonds or triangles. Below we will prove this conjecture; our proof follows the proof of Theorem 2.68 to a large extent but there is an important difference. Since we are working *within* a modal language, and not in first-order logic, we can make do with a detour via (m-saturated) ultrafilter extensions rather than the (countably saturated) ultrapowers needed in the proof of Theorem 2.68.

Call a modal formula *positive existential* if it has been built up from proposition letters, using only $\wedge$, $\vee$ and existential modal operators $\Diamond$ and $\triangle$.

Theorem 2.78 *Let τ be a modal similarity type, and let ϕ be a τ-formula. Then ϕ is preserved under simulations iff it is equivalent to a positive existential formula.*

Proof. The easy inductive proof that positive existential formulas are preserved under simulations is left to the reader. For the converse, assume that ϕ is preserved under simulations, and consider the set of positive existential consequences of ϕ:

$$\text{PEC}(\phi) = \{\psi \mid \psi \text{ is positive existential and } \phi \models \psi\}.$$

We will show that $\text{PEC}(\phi) \models \phi$; then, by compactness, ϕ is equivalent to a positive existential modal formula. Assume that $\mathfrak{M}, w \Vdash \text{PEC}(\phi)$; we need to show that $\mathfrak{M}, w \Vdash \phi$. Let $\Gamma = \{\neg\psi \mid \psi \text{ is positive existential and } \mathfrak{M}, w \not\Vdash \psi\}$.

Our first claim is that the set $\{\phi\} \cup \Gamma$ is consistent. For, suppose otherwise. Then there are formulas $\neg\psi_1, \ldots, \neg\psi_n \in \Gamma$ such that $\phi \models \psi_1 \vee \cdots \vee \psi_n$. By definition each formula ψ_i is a positive existential formula, hence, so is $\psi_1 \vee \cdots \vee \psi_n$. But then $\mathfrak{M}, w \Vdash \psi_1 \vee \cdots \vee \psi_n$, by assumption; from this it follows that $\mathfrak{M}, w \Vdash \psi_i$ for some i ($1 \leq i \leq n$). This contradicts $\neg\psi_i \in \Gamma$.

As a corollary we find a model $\mathfrak{N}$ and a state v of $\mathfrak{N}$ such that $\mathfrak{N}, v \Vdash \phi \wedge \bigwedge \Gamma$. Clearly, for every positive existential formula ψ, if $\mathfrak{N}, v \Vdash \psi$, then $\mathfrak{M}, w \Vdash \psi$. It follows from Proposition 2.59 that for the ultrafilter extensions $\mathfrak{ue}\,\mathfrak{M}$ and $\mathfrak{ue}\,\mathfrak{N}$ we have the same relation: for every positive existential formula ψ, if $\mathfrak{ue}\,\mathfrak{N}, \pi_v \Vdash \psi$, then $\mathfrak{ue}\,\mathfrak{M}, \pi_w \Vdash \psi$. By exploiting the fact that ultrafilter extensions are m-saturated (Proposition 2.61), it can be shown that this relation is in fact a simulation from $\mathfrak{ue}\,\mathfrak{N}, \pi_v$ to $\mathfrak{ue}\,\mathfrak{M}, \pi_w$; see Exercise 2.7.1.

In a diagram we have now the following situation.

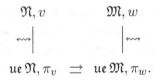

$$\mathfrak{N}, v \qquad\qquad \mathfrak{M}, w$$

$$\mathfrak{ue}\,\mathfrak{N}, \pi_v \quad \rightrightarrows \quad \mathfrak{ue}\,\mathfrak{M}, \pi_w.$$

We can carry ϕ around the diagram from $\mathfrak{N}, v$ to $\mathfrak{M}, w$ as follows. $\mathfrak{N}, v \Vdash \phi$ implies $\mathfrak{ue}\,\mathfrak{N}, \pi_v \Vdash \phi$ by Proposition 2.59. Since ϕ is preserved under simulations, we get $\mathfrak{ue}\,\mathfrak{M}, \pi_w \Vdash \phi$. By Proposition 2.59 again we conclude $\mathfrak{M}, w \Vdash \phi$. $\dashv$

Using Theorem 2.78 we can also answer the second of the two questions raised above. Call a constraint ϕ *expressible* if whenever $\mathfrak{M}, w$ satisfies ϕ and $\mathfrak{N}, v \rightrightarrows \mathfrak{M}, w$, then $\mathfrak{N}, v$ also satisfies ϕ. By Theorem 2.78 the expressible constraints (in first-order logic) are precisely the ones that are (equivalent to) the standard translations of negative universal modal formulas, that is, translations of modal formulas built up from negated proposition letters using only $\vee$, $\wedge$ and universal modal operators $\square$ and $\triangledown$.

Safety

Recall from Exercise 2.2.6 that bisimulations preserve the truth of formulas from propositional dynamic logic. This result hinges on the fact that bisimulations not only preserve the relations R_a corresponding to atomic programs, but also relations that are definable from these using PDL's relational repertoire $\cup$, ; and *. Put differently, if the back and forth conditions in the definition of a bisimulation hold for each relation R_a then they also hold for any relation that is definable from the basic ones using $\cup$, ; and *; these operations are 'safe' for bisimulation.

In this part of the section we work with a modal similarity type τ having diamonds only.

Definition 2.79 Let τ be a modal similarity type having diamonds only, and let $\alpha(x, y)$ denote an $\mathcal{L}^1_\tau(\Phi)$-formula with at most two free variables. Then $\alpha(x, y)$ is called *safe for bisimulations* if the following holds, for any bisimulation $Z : \mathfrak{M} \underline{\leftrightarrow} \mathfrak{M}'$:

> *if* wZw' and $\mathfrak{M} \models \alpha(x, y)[wv]$ for some state v of $\mathfrak{M}$,
> *then* there is a state v' of $\mathfrak{M}'$ such that $\mathfrak{M}' \models \alpha(x, y)[w'v']$ and vZv'.

In words, $\alpha(x, y)$ is safe if the back and forth clauses hold for $\alpha(x, y)$ whenever they hold for the atomic relations. ⊣

Example 2.80 (i) All PDL program constructors ($\cup$, ; and *) are safe for bisimulations (where we stretch the definition of safety to program constructors in an obvious way). For instance, assume that wZw', where Z is a bisimulation, and $(w, v) \in (R ; S)$ in $\mathfrak{M}$. Then, there exists u with Rwu and Suv in $\mathfrak{M}$; hence by the back and forth conditions for R and S, we find u' with uZu' and $R'w'u'$ in $\mathfrak{M}'$, and a state v' with vZv' and $S'u'v'$ in $\mathfrak{M}'$. Then v' is the required $(R;S)$-successor of w' in $\mathfrak{M}'$.

(ii) Atomic tests $P?$, defined by $P? := \{(x, y) \mid x = y \wedge Py\}$, are safe. For, assume that wZw', where Z is a bisimulation, and $(w, v) \in P?$. Then $w = v$ and $\mathfrak{M} \models Px[w]$. By the atomic clause in the definition of bisimulation, this implies $\mathfrak{M}' \models Px[w']$. Hence, $(w', w') \in P?$, as required.

(iii) Dynamic negation $\sim R$, defined by $\sim R = \{(x, y) \mid x = y \wedge \neg \exists z \, Rxz\}$, is safe. For, assume that wZw', where Z is a bisimulation, and $(w, v) \in \sim R$ in $\mathfrak{M}$. Then, $w = v$ and w has no R-successors in $\mathfrak{M}$. Now, suppose that w' did have an R'-successor in $\mathfrak{M}'$; then, by the back and forth conditions, w would have to have an R-successor in $\mathfrak{M}$ – a contradiction.

(iv) Intersection of relations is not safe; see Exercise 2.7.2. ⊣

Which operations are safe for bisimulations? Below, we give a complete answer for the restricted case where we consider first-order definable operations and languages with diamonds only. We need some preparations before we can prove this result.

First, in the remainder of this section we will use the term *labeled tree models* for τ-models of the form $(W, R_a, V)_{a \in \tau}$ such that $(W, \bigcup_a R_a, V)$ is a tree in the sense of Definition 1.7.

Second, let p be a fixed proposition letter. We write $\underline{\leftrightarrow}^-$ to denote the existence of a bisimulation for the modal language without the proposition letter p (exactly which proposition letter is meant will always be clear from the context).

Third, we define a modal formula ϕ to be *completely additive in the proposition letter* p if it satisfies the following:

For every family of non-empty sets $\{X_i\}_{i \in I}$ such that $V(p) = \bigcup_i X_i$ we have $(W, R_a, V)_{a \in \tau}, w \Vdash \phi$ iff, for some i, $(W, R_a, V_i)_{a \in \tau}, w \Vdash \phi$, where $V_i(p) = X_i$ and $V_i(q) = V(q)$ for $q \neq p$.

Completely additive formulas have a nice syntactic characterization.

Lemma 2.81 *A modal formula is completely additive in p iff it is equivalent to a disjunction of path formulas, that is, formulas of the form*

$$\psi_0 \wedge \langle a_1 \rangle (\psi_1 \wedge \cdots \wedge \langle a_n \rangle (\psi_n \wedge p) \cdots), \tag{2.4}$$

where p occurs in none of the formulas ψ_i.

Proof. We only prove the hard direction. Assume that ϕ is completely additive in p. Define

$$\mathrm{COC}(\phi) := \bigvee \{ \psi \mid \psi \text{ is of the form (2.4) and } \psi \Vdash \phi \},$$

that is, $\mathrm{COC}(\phi)$ is an infinite disjunction of modal formulas. We will show that $\phi \Vdash \mathrm{COC}(\phi)$; then, by compactness, ϕ is equivalent to a finite disjunction of formulas of the form specified in (2.4), and this proves the lemma.

So, assume that $\mathfrak{M}, w_0 \Vdash \phi$; we need to show that $\mathfrak{M}, w_0 \Vdash \mathrm{COC}(\phi)$. As the reader may verify by doing Exercise 2.7.3, we may assume that $\mathfrak{M}$ is an m-saturated, labeled tree model with root w_0. As ϕ is completely additive in p, we may also assume that $V(p)$ is just a singleton w_n; note that still, we may assume $\mathfrak{M}$ to be m-saturated with respect to the p free language. Since $\mathfrak{M}$ is a labeled tree with root w_0, there is a path $w_0 R_{a_1} \cdots R_{a_n} w_n$ from w_0 to w_n, see Figure 2.8.

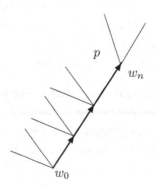

Fig. 2.8. True at only one state

For $0 \leq i \leq n$, let Ψ_i be the set of formulas in the p free language that hold at w_i, and consider the following description of the above path leading up to w_n:

$$\Psi \;=\; \{ \psi \mid \psi \text{ is of the form (2.4), with for all } i: \psi_i \in \Psi_i \text{ and } R_{a_i} w_i w_{i+1} \}.$$

The remainder of the proof is devoted to showing that $\Psi \Vdash \phi$, and this will do to prove the lemma. For if $\Psi \Vdash \phi$, then, for some finite subset $\Psi' \subseteq \Psi$ we have $\bigwedge \Psi' \Vdash \phi$, by compactness. It is not difficult to show that Ψ is closed (modulo equivalence) under taking finite conjunctions, so $\bigwedge \Psi'$ is equivalent to a formula $\psi \in \Psi$. Hence, we have found our path formula satisfying $\mathfrak{M}, w_0 \Vdash \psi$ and $\psi \Vdash \phi$.

To show that $\Psi \Vdash \phi$ we proceed as follows. Take a model $\mathfrak{N}$ with $\mathfrak{N}, v_0 \Vdash \Psi$; we need to show that $\mathfrak{N}, v_0 \Vdash \phi$. Again, we may assume that $\mathfrak{N}$ is an m-saturated, labeled tree with root v_0. By m-saturation, there are points $v_1, \ldots, v_n$ such that for each $0 \le i \le n-1$ we have $R_{a_i} v_i v_{i+1}$ and $\mathfrak{N}, v_i \Vdash \Psi_i$.

It follows from the definition of the Ψ_i that each w_i and v_i agree on all p free modal formulas. So by m-saturation, there is a bisimulation $Z : \mathfrak{M}, w_0 \underline{\leftrightarrow}^- \mathfrak{N}, v_0$. It follows from Exercise 2.7.4 that there are extensions $\mathfrak{M}'$ and $\mathfrak{N}'$ of $\mathfrak{M}$ and $\mathfrak{N}$ respectively, such that $(\mathfrak{M}, w_0) \underline{\leftrightarrow} (\mathfrak{M}', w_0)$ and $(\mathfrak{N}, v_0) \underline{\leftrightarrow} (\mathfrak{N}', v_0)$; we may also conclude from this exercise that there is a bisimulation Z' between $\mathfrak{M}'$ and $\mathfrak{N}'$ such that for all i, the points w_i and v_i are only related to each other; see Figure 2.9.

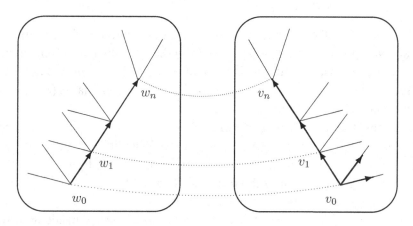

Fig. 2.9. Linking w_i only to v_i $(1 \le i \le n)$

Now we amend the models $\mathfrak{M}'$ and $\mathfrak{N}'$ as follows: We shrink the interpretations of p so that p only holds at w_n and v_n, respectively. Then Z' is a bisimulation between the resulting models $\mathfrak{M}''$ and $\mathfrak{N}''$ with respect to the *full* language, and further relations between the models are as indicated in (2.5) below:

$$
\begin{array}{ccc}
(\mathfrak{M}, w_0) & Z : \underline{\leftrightarrow}^- & (\mathfrak{N}, v_0) \\[4pt]
\underline{\leftrightarrow} \Big| & & \Big| \underline{\leftrightarrow} \\[4pt]
(\mathfrak{M}', w_0) & Z' : \underline{\leftrightarrow}^- & (\mathfrak{N}', v_0) \\[4pt]
\text{shrink } V(p) \Big| & & \Big| \text{shrink } V(p) \\[4pt]
(\mathfrak{M}'', w_0) & Z' : \underline{\leftrightarrow} & (\mathfrak{N}'', v_0),
\end{array}
\qquad (2.5)
$$

We can chase ϕ around this diagram, from $(\mathfrak{M}, u_0)$ to $(\mathfrak{N}, v_0)$; see Exercise 2.7.5. This proves the lemma.   ⊣

Lemma 2.82 *For any program a and any formulas ϕ and ψ, the following identities hold in any model:*

(i) $(\neg\phi)? = \sim(\phi?)$,

(ii) $(\phi \wedge \psi)? = (\phi)? ; (\psi)?$,

(iii) $(\langle a \rangle \phi)? = \sim\sim(a ; \phi?)$.

The proof of this lemma is left as Exercise 2.7.6.

Theorem 2.83 (Safety Theorem) *Let τ be a modal similarity type containing only diamonds, and let $\alpha(x, y)$ be a first-order formula in $\mathcal{L}^1_\tau(\Phi)$. Then $\alpha(x, y)$ is safe for bisimulations iff it can be defined from atomic formulas $R_a xy$ and atomic tests P? using only $\cup$, ; and $\sim$.*

Proof. To see that the constructions mentioned are indeed safe, consult Example 2.80. Now, to prove the converse, let $\alpha(x, y)$ be a safe first-order operation, and choose a *new* proposition letter p. Our first observation is that $\exists y\, (\alpha(x, y) \wedge Py)$ is preserved under bisimulations. So by Theorem 2.68, the formula $\exists y\, (\alpha(x, y) \wedge Py)$ is equivalent to a modal formula ϕ.

Next we exploit special properties of ϕ to arrive at our conclusion. First, because of its special form, $\exists y\, (\alpha(x, y) \wedge Py)$ is completely additive in P, in the obvious sense, and hence, ϕ is completely additive in p. Therefore, by Lemma 2.81 it is (equivalent to) a disjunction of the form specified in (2.4). Then, $\alpha(x, y)$ must be definable using the corresponding union of relations $(\psi_0)?; a_1; (\psi_1)?; \cdots; a_n; (\psi_n)?$. Finally, by using Lemma 2.82 all complex tests can be pushed inside until we get a formula of the required form, involving only $\cup$, ;, $\sim$ and ?.   ⊣

Exercises for Section 2.7

2.7.1 Assume that $\mathfrak{M}$ and $\mathfrak{M}'$ are m-saturated models and suppose that for every positive existential formula ϕ it holds that $\mathfrak{M}, w \Vdash \phi$ only if $\mathfrak{M}', w' \Vdash \phi$ for some w and w'. Prove that $\mathfrak{M}, w \rightleftharpoons \mathfrak{M}', w'$.

2.7.2 Prove that intersection of relations is not an operation that is safe for bisimulations (see Example 2.80).

2.7.3   (a) Suppose that Z is a bisimulation linking the models $\mathfrak{M}$ and $\mathfrak{M}'$. Suppose further that $\mathfrak{M}$ is m-saturated and that Z is surjective (that is, every point in $\mathfrak{M}'$ is linked by Z to some point in $\mathfrak{M}$). Prove that $\mathfrak{M}'$ is m-saturated as well.

(b) Let Σ be a set of formulas, and ϕ a formula. Prove that ϕ is a consequence of Σ if and only if for every m-saturated, labeled tree model $\mathfrak{M}$ with root w it holds that $\mathfrak{M}, w \Vdash \Sigma$ only if $\mathfrak{M}, w \Vdash \phi$.

2.7.4 Assume that $Z : \mathfrak{M}, w_0 \; \underline{\leftrightarrow}^- \; \mathfrak{N}, v_0$, where $\mathfrak{M}$ and $\mathfrak{N}$ are labeled tree models with roots w_0 and v_0, respectively. Assume further that in $\mathfrak{M}$ we have $w_0 R_{a_1} \cdots R_{a_n} w_n$ and in $\mathfrak{N}$, $v_0 R_{a_1} \cdots R_{a_n} v_n$, while Z links w_i (in $\mathfrak{M}$) to v_i (in $\mathfrak{N}$) for each i ($1 \le i \le n$).

In this exercise the reader is asked to show that there are extensions $(\mathfrak{M}', w_0)$ of $(\mathfrak{M}, w_0)$ and $(\mathfrak{N}', v_0)$ of $(\mathfrak{N}, v_0)$ (i.e., the universe of $\mathfrak{M}$ is a subset of the universe of $\mathfrak{M}'$, and likewise for $\mathfrak{N}$ and $\mathfrak{N}'$) such that

$$
\begin{array}{ccc}
(\mathfrak{M}, w_0) & Z : \underline{\leftrightarrow}^- & (\mathfrak{N}, v_0) \\
\underline{\leftrightarrow}\Big| & & \Big|\underline{\leftrightarrow} \\
(\mathfrak{M}', w_0) & Z' : \underline{\leftrightarrow}^- & (\mathfrak{N}', v_0),
\end{array}
$$

where Z' is such that for any i ($1 \le i \le n$) we have that w_i and v_i are only related to each other.

(a) Explain why we may assume that all bisimulation links (between $\mathfrak{M}$ and $\mathfrak{N}$) occur between states at the same height in the trees. (Hint: remove all links between points of different heights and prove that the remaining links still form a bisimulation.)

(b) Next, work your way up along the branch $w_0 R_{a_1} \cdots R_{a_n} w_n$ and remove any double bisimulation links involving the w_i. More precisely, and starting at height 1, do the following for each double link $w_1 Z v$. Add a copy of the submodel generated by w_1 to $\mathfrak{M}$, connect w_0 to the copy w_1^v of w_1 by R_{a_1}, and 'divert' the bisimulation link $w_1 Z v$ to $w_1^v Z v$. Show that the resulting model $\mathfrak{M}^v$ satisfies $\mathfrak{M} \; \underline{\leftrightarrow} \; \mathfrak{M}^v$ and $\mathfrak{M}^v \underline{\leftrightarrow}^- \mathfrak{N}$. Proceed in a similar way with points of larger height.

(c) Similar to the previous item, but now working up the branch $v_0 R_{a_1} \cdots R_n v_{a_n}$ in $\mathfrak{N}$ to eliminate any double bisimulation links ending in one of the v_is ($1 \le i \le n$).

(d) Prove the existence of the desired $\mathfrak{M}'$, $\mathfrak{N}'$ and Z' by putting together the previous items.

2.7.5 Explain why we can chase ϕ around the diagram displayed in (2.5) to infer $\mathfrak{N}, v_0 \Vdash \phi$ from $\mathfrak{M}, w_0 \Vdash \phi$.

2.7.6 Prove Lemma 2.82.

2.8 Summary of Chapter 2

▶ *New Models from Old Ones*: Taking disjoint unions, generated submodels, and bounded morphic images are three important ways of building new models from old that leave the truth values of modal formulas invariant.

▶ *Bisimulations*: Bisimulations offer a unifying perspective on model invariance, and each of the constructions just mentioned is a kind of bisimulation. Bisimilarity implies modal equivalence, but the converse does not hold in general. On image-finite models, however, bisimilarity and modal equivalence coincide.

▶ *Using Bisimulations*: Bisimulations can be used to establish non-definability results (for example, to show that the global modality is not definable in the basic modal language), or to create models satisfying special relational properties (for example, to show that every satisfiable formula is satisfiable in a tree-like model).

▶ *Finite Model Property*: Modal languages have the finite model property (f.m.p.). One technique for establishing the f.m.p. is by a selection-of-states argument involving finite approximations to bisimulations. Another, the filtration method, works by collapsing as many states as possible.

▶ *Standard Translation*: The standard translation maps modal languages into classical languages (such as the language of first-order logic) in a way that reflects the satisfaction definition. Every modal formula is equivalent to a first-order formula in one free variable; if the similarity type is finite, finitely many variables suffice to translate all modal formulas. Propositional dynamic logic has to be mapped into a richer classical logic capable of expressing transitive closure.

▶ *Ultrafilter Extensions*: Ultrafilter extensions are built by using the ultrafilters over a given model as the states of a new model, and defining an appropriate relation between them. This leads to the first bisimilarity-somewhere-else result: two states in two models are modally equivalent if and only if (their counterparts in) the ultrafilter extensions of the two models are bisimilar.

▶ *van Benthem Characterization Theorem*: The Detour Lemma – a bisimilarity-somewhere-else result in terms of ultrapowers – can be used to prove the van Benthem Characterization Theorem: the modal fragment of first-order logic is the set of formulas in one free variable that are invariant for bisimulations.

▶ *Definability*: The Detour Lemma also leads to the following result: the modally definable classes of (pointed) models are those that are closed under bisimulations and ultraproducts, while their complements are closed under ultrapowers.

▶ *Simulation*: The modal formulas preserved under simulations are precisely the positive existential ones.

▶ *Safety*: An operation on relations is safe for bisimulations if whenever the back and forth conditions hold for the base relations, they also hold for the result of applying the operation to the relations. The first-order operations safe for bisimulations are the ones that can be defined from atoms and atomic tests, using only composition, union, and dynamic negation.

Notes

Kripke, Kanger, Hintikka, and others introduced models to modal logic in the late 1950s and early 1960s, and relational semantics (or Kripke semantics as it was usually called) swiftly became the standard way of thinking about modal logic. In spite of this, much of the material discussed in this chapter dates not from the 1960s, or even the 1970s, but from the late 1980s and 1990s. Why? Because relational semantics was not initially regarded as of independent interest, rather it was thought of as a tool that lead to interesting modal completeness theory and decidability results. Only in the early 1970s (with the discovery of the frame incompleteness results) did modal expressivity become an active topic of research

– and even then, such investigations were initially confined to expressivity at the level of frames rather than at the level of models. Thus the most fundamental level of modal semantics was actually the last to be explored mathematically.

Generated submodels and bounded morphisms arose as tools for manipulating the canonical models used in modal completeness theory (we discuss canonical models in Chapter 4). *Point*-generated submodels, however, were already mentioned, under the name of connected model structures, in Kripke [284]. Bounded morphisms go back to at least Segerberg [404] (a very similar, earlier, notion can be found in de Jongh and Troelstra [249]) where they are called *pseudo epimorphisms*; this soon got shortened down to *p-morphism*, which remains the most widely used terminology. The name *bounded morphism* stems from Goldblatt [186]. Disjoint unions and ultrafilter extensions seem to have first been isolated when modal logicians started investigating modal expressivity over frames in the 1970s (along with generated submodels and bounded morphisms they are the four constructions needed in the Goldblatt-Thomason Theorem, which we discuss in the following chapter). Neither construction is as useful as generated submodels and bounded morphisms when it comes to proving completeness results, which is probably why they were not noted earlier. However, both arise naturally in the context of modal duality theory; see Goldblatt [184, 185]. Ultrafilter extensions independently came about in the model-theoretic analysis of modal logic, see Fine [132]; the name seems to be due to van Benthem. The unraveling construction (that is, unwinding arbitrary models into trees; see Proposition 2.15) is helpful in many situations. Surprisingly, it was first used as early as in 1959, by Dummett and Lemmon [117], but the method seems to have become widely known because of Sahlqvist's use of it in his classic 1975 paper [396].

Vardi [441] has stressed the importance of the *tree model property* of modal logic: the property that a formula is satisfiable iff it is satisfiable at the root of a tree-like model. The tree model property paves the way for the use of automata-theoretic tools and tableaux-based proof methods. Moreover, it is essential for explaining the so-called robust decidability of modal logic – the phenomenon that the basic modal logic is decidable itself, and of reasonably low complexity, and that these features are preserved when the basic modal logic is extended by a variety of additional constructions, including counting, transitive closure, and least fixed points.

We discussed two ways of building finite models: the selection method and filtration. However, the use of finite *algebras* predates the use of finite models: they were first used in 1941 by McKinsey [322]; Lemmon [295] used and extended this method in 1966. The use of model-theoretic filtration dates back to Lemmon and Scott's long unpublished monograph *An Introduction to Modal Logic* [296] (which began circulating in the mid 1960s); it was further developed in Segerberg's *An Essay in Classical Modal Logic* [404], which also seems to have given the method its name (see also Segerberg [402]). We introduced the selection method via the notion of finitely approximating a bisimulation, an idea which seems to have first appeared in 1985 in Hennessy and Milner [219].

The standard translation, in various forms, can be found in the work of a number of writers on modal and tense logic in the 1960s – but its importance only became fully apparent when the first frame incompleteness results were proved. Thomason [433], the paper in which frame incompleteness results was first established, uses the standard translation – and shows why the move to frames and validities requires a *second*-order perspective (something we will discuss in the following chapter). Thus the need became clear for a thorough investigation of the relation between modal and classical logic, and correspondence theory was born. But although other authors (notably Sahlqvist [396]) helped pioneer correspondence theory, it was the work of van Benthem [36] which made clear the importance of systematic use of the standard translation to access results and techniques from classical modal theory. The observation that at most two variables are needed to translate basic modal formulas into first-order logic is due to Gabbay [149]. The earliest systematic study of finite variable fragments seems to be due to Henkin [216] in the setting of algebraic logic, and Immerman and Kozen [240] study the link with complexity and database theory. Consult Otto [351] for more on finite variable logics. Keisler [267] is still a valuable reference for infinitary logic. A variety of other translations from modal to classical logic have been studied, and for a wide variety of purposes. For example, simply standardly translating modal logics into first-order logic and then feeding the result to a theorem prover is not an efficient way of automating modal theorem proving. But the idea of automating modal reasoning via translation is interesting, and a variety of translations more suitable for this purpose have been devised; see Ohlbach *et al.* [345] for a survey.

Under the name of p-relations, bisimulations were introduced by Johan van Benthem in the course of his work on correspondence theory. Key references here are van Benthem's 1976 PhD thesis [36]; his 1983 book based on the thesis [36]; and [43], his 1984 survey article on correspondence theory. In keeping with the spirit of the times, most of van Benthem's early work on correspondence theory dealt with frame definability (in fact he devotes only 6 of the 227 pages in his book to expressivity over models). Nonetheless, much of this chapter has its roots in this early work, for in his thesis van Benthem introduced the concept of a bisimulation (he used the name *p-relation* in [36, 42], and the name *zigzag relation* in [43]) and proved the Characterization Theorem. His original proof differs from the one given in the text: instead of appealing to saturated models, he employs an elementary chains argument. Explicitly isolating the Detour Lemma (which brings out the importance of ultrapowers) opens the way to Theorems 2.75 and 2.76 on definability and makes explicit the interesting analogies with first-order model theory discussed below. On the other hand, the original proof is more concrete. Both are worth knowing. The first published proof using saturated models seems to be due to Rodenburg [390], who used it to characterize the first-order fragment corresponding to intuitionistic logic.

The back and forth clauses of a bisimulation can be adapted to analyze the expressivity of a wide range of extended modal logics (such as those studied in Chapter 7), and such analyses are now commonplace. Bisimulation based characterizations have been given for the modal mu-calculus by Janin and Walukiewicz [243], for temporal logics with since and until by Kurtonina and de Rijke [288], for subboolean fragments of knowledge representation languages by Kurtonina and de Rijke [289], and for CTL* by Moller and Rabinovich [333]. Related model-theoretic characterizations can be found in Immerman and Kozen [240] (for finite variable logics) and Toman and Niwiński [438] (for temporal query languages). Rosen [392] presents a version of the Characterization Theorem that also works for the case of finite models; the proof given in the text breaks down in the finite case as it relies on compactness and saturated models.

But bisimulations did not just arise in modal logic – they were independently invented in computer science as an equivalence relation on process graphs. Park [354] seems to have been the first author to have used bisimulations in this way. The classic paper on the subject is Hennessy and Milner [219], the key reference for the Hennessy-Milner Theorem. The reader should be warned, however, that just as the notion of bisimulation can be adapted to cover many different modal systems, the notion of bisimulation can be adapted to cover many different concepts of process – in fact, a survey of bisimulation in process algebra in the early 1990s lists over 155 variants of the notion [173]! Our definitions do not exclude bisimulations between a model and itself (*auto-bisimulations*); the quotient of a model with respect to its largest auto-bisimulation can be regarded as a minimal representation of this model. The standard method for computing the largest auto-bisimulation is the so-called Paige-Tarjan algorithm; see the contributions to Ponse, de Rijke and Venema [360] for relevant pointers and surveys.

More recently, bisimulations have become fundamental in a third area, non-well-founded set theory. In such theories, the axiom of foundation is dropped, and sets are allowed to be members of themselves. Sets are thought of as graphs, and two sets are considered identical if and only if they are bisimilar. The classic source for this approach is Aczel [2], who explicitly draws on ideas from process theory. A recent text on the subject is Barwise and Moss [27], who link their work with the modal tradition. For recent work on modal logic and non-well-founded set theory, see Baltag [20].

The name 'm-saturation' stems from Visser [450], but the notion is older: its first occurrence in the literature seems to be in Fine [132] (under the name 'modally saturated$_2$'). The concept of a Hennessy-Milner class is from Goldblatt [179] and Hollenberg [233]. Theorem 2.62, that equivalence of models implies bisimilarity between their ultrafilter extensions, is due to [233]. Chang and Keisler [91, Chapters 4 and 6] is the classic reference for the ultraproduct construction; their Chapters 2 and 5 also contain valuable material on saturated models. Doets and van Benthem [112] give an intuitive explanation of the ultraproduct construction.

The results proved in this chapter are often analogs of standard results in first-order model theory, with bisimulations replacing partial isomorphisms. The Keisler-Shelah Theorem (see Chang and Keisler [91, Theorem 6.1.15]) states that two models are elementarily equivalent iff they have isomorphic ultrapowers; a weakened form, due to Doets and van Benthem [112], replaces 'isomorphic' with 'partially isomorphic'. Theorem 2.74, which is due to de Rijke [380], is a modal analog of this weakened characterization theorem. Proposition 2.31 is similar to characterizations of logical equivalence for first-order logic due to Ehrenfeucht [119] and Fraïssé [141]; in fact, bisimulations can be regarded as the modal cousins of the model theoretic Ehrenfeucht-Fraïssé games. We will return to the theme of analogies between first-order and modal model theory in Section 7.6 when we prove a Lindström theorem for modal logic. See de Rijke [380] and Sturm [426] for further work on modal model theory; de Rijke and Sturm [386] provide global counterparts for the local definability results presented in Section 2.6. One can also characterize modal definability of model classes using 'modal' structural operations only, that is, bisimulations, disjoint unions and ultrafilter extensions; see Venema [447].

Sources for the use of simulations in refinement are Henzinger *et al.* [221] and He Jifeng [246], and for their use in a database setting, consult Buneman *et al.* [76]; see de Rijke [380] for Theorem 2.78. The Safety Theorem 2.83 is due to van Benthem [48]. The text follows the original proof fairly closely; an alternative proof has been given by Hollenberg [232], who also proves generalizations.

One final remark. Given the importance of *finite* model theory, the reader may be surprised to find so little in this chapter on the topic. But we do not neglect finite model theory in this book: virtually all the results proved in Chapter 6 revolve around finite models and the way they are structured. That said, the topic of finite modal model theory has received less attention from modal logicians than it deserves. In spite of Rosen's [392] proof of the van Benthem characterization theorem for finite models, and in spite of work on modal 0-1 laws (Halpern and Kapron [204], Goranko and Kapron [191], and Grove *et al.* [200, 199]), finite modal model theory is an area where many interesting questions remain.

3

Frames

As we saw in Section 1.3, the concept of *validity*, which abstracts away from the effects of particular valuations, allows modal languages to get to grips with frame structure. As we will now see, this makes it possible for modal languages to *define* classes of frames, and most of the chapter is devoted to exploring this idea.

The following picture will emerge. Viewed as tools for defining frames, every modal formula corresponds to a second-order formula. Although this second-order formula sometimes has a first-order equivalent, even quite simple modal formulas can define classes of frames that no first-order formula can. In spite of this, there are extremely simple first-order definable frame classes which no modal formula can define. In short, viewed as frame description languages, modal languages exhibit an unusual blend of first- and second-order expressive powers.

The chapter has three main parts. The first, consisting of the first four sections, introduces frame definability, explains why it is intrinsically second-order, presents the four fundamental frame constructions and states the *Goldblatt-Thomason Theorem*, and discusses finite frames. The second part, consisting of the next three sections, is essentially a detailed exposition of the *Sahlqvist Correspondence Theorem*, which identifies a large class of modal formulas which correspond to first-order formulas. The final part, consisting of the last section, studies further frame constructions and gives a model-theoretic proof of the Goldblatt-Thomason Theorem. With the exception of the last two sections, all the material in this chapter lies on the basic track.

Chapter guide

Section 3.1: Frame Definability (Basic track). This section introduces frame definability, and gives several examples of modally definable frame classes.

Section 3.2: Frame Definability and Second-Order Logic (Basic Track). We explain why frame definability is intrinsically second-order, and give examples of frame classes that are modally definable but not first-order definable.

Section 3.3: Definable and Undefinable Properties (Basic track). We first show that validity is preserved under the formation of *disjoint unions*, *generated subframes* and *bounded morphic images*, and anti-preserved under *ultrafilter extensions*. We then use these constructions to give examples of frame classes that are *not* modally definable, and state the Goldblatt-Thomason Theorem.

Section 3.4: Finite Frames (Basic track). Finite frames enjoy a number of pleasant properties. We first prove a simple analog of the Goldblatt-Thomason Theorem for finite transitive frames. We then introduce the *finite frame property*, and show that a normal modal logic has the finite frame property if and only if it has the finite model property.

Section 3.5: Automatic First-Order Correspondence (Basic track). Here we prepare for the proof of the *Sahlqvist Correspondence Theorem* in the following section. We introduce positive and negative formulas, and show that their monotonicity properties can help eliminate second-order quantifiers.

Section 3.6: Sahlqvist Formulas (Basic track). In this section we prove the Sahlqvist Correspondence Theorem. Our approach is incremental. We first explore the key ideas in the setting of two smaller fragments, and then state and prove the main result.

Section 3.7: More About Sahlqvist Formulas (Advanced track). We first discuss the limitations of the Sahlqvist Correspondence Theorem. We then prove Kracht's Theorem, which provides a syntactic description of the first-order formulas that can be obtained as translations of Sahlqvist formulas.

Section 3.8: Advanced Frame Theory (Advanced track). We finish off the chapter with some advanced material on frame constructions, and prove the Goldblatt-Thomason Theorem model-theoretically.

3.1 Frame Definability

This chapter is mostly about using modal formulas to define classes of frames. In this section we introduce the basic ideas (*definability*, and *first- and second-order frame languages*), and give a number of examples of modally definable frames classes. Most of these examples – and, indeed, most of the examples given in this chapter – are important in their own right and will be used in later chapters.

Frame definability rests on the notion of a formula being *valid* on a frame, a concept which was discussed in Section 1.3 (see in particular Definition 1.28). We first recall and extend this definition.

Definition 3.1 (Validity) Let τ be a modal similarity type. A formula ϕ (of this similarity type) is *valid at a state* w *in a frame* $\mathfrak{F}$ (notation: $\mathfrak{F}, w \Vdash \phi$; here, of course, $\mathfrak{F}$ is a frame of type τ) if ϕ is true at w in every model $(\mathfrak{F}, V)$ based on $\mathfrak{F}$; ϕ is *valid on a frame* $\mathfrak{F}$ (notation: $\mathfrak{F} \Vdash \phi$) if it is valid at every state in $\mathfrak{F}$. A formula ϕ is *valid on a class of frames* K (notation: K $\Vdash \phi$) if it is valid on every frame $\mathfrak{F}$ in K. We denote the class of frames where ϕ is valid by Fr_ϕ.

These concepts can be extended to sets of formulas in the obvious way. In particular, a set Γ of modal formulas (of type τ) is *valid on a frame* $\mathfrak{F}$ (also of type τ) if every formula in Γ is valid on $\mathfrak{F}$; and Γ is *valid on a class* K *of frames* if Γ is valid on every member of K. We denote the class of frames where Γ is valid by Fr_Γ. ⊣

Now for the concept underlying most of our work in this chapter:

Definition 3.2 (Definability) Let τ be a modal similarity type, ϕ a modal formula of this type, and K a class of τ-frames. We say that ϕ *defines* (or *characterizes*) K if for all frames $\mathfrak{F}$, $\mathfrak{F}$ is in K if and only if $\mathfrak{F} \Vdash \phi$. Similarly, if Γ is a set of modal formulas of this type, we say that Γ *defines* K if $\mathfrak{F}$ is in K if and only if $\mathfrak{F} \Vdash \Gamma$.

A class of frames is *(modally) definable* if there is some set of modal formulas that defines it. ⊣

In short, a modal formula defines a class of frames if the formula pins down precisely the frames that are in that class via the concept of validity. The following generalization of this concept is sometimes useful:

Definition 3.3 (Relative Definability) Let τ be a modal similarity type, ϕ a modal formula of this type, and C a class of τ-frames. We say that ϕ *defines* (or *characterizes*) a class K of frames *within* C (or *relative to* C) if for all frames $\mathfrak{F}$ in C we have that $\mathfrak{F}$ is in K if and only if $\mathfrak{F} \Vdash \phi$.

Similarly, if Γ is a set of modal formulas of this type, we say that Γ *defines* a class K of frames *within* C (or *relative to* C) if for all frames $\mathfrak{F}$ in C we have that $\mathfrak{F}$ is in K if and only if $\mathfrak{F} \Vdash \Gamma$. ⊣

Note that when C is the class of *all* τ-frames, definability within C is our original notion of definability. In Section 3.4 we will investigate which frames are definable within the class of finite transitive frames, but for the most part we will work with the 'absolute' notion of definability given in Definition 3.2.

We often say that a formula ϕ (or a set of formulas Γ) defines a *property* (for example, reflexivity) if it defines the class of frames satisfying that property. For example, we will shortly see that $p \to \Diamond p$ defines the class of reflexive frames; in practice, we would often simply say that $p \to \Diamond p$ defines reflexivity.

Up till now our discussion has been purely modal – but, of course, as frames are just relational structures, we are free to define frame classes using a wide variety of *non-modal* languages. For example, the class of reflexive frames is simply the class of all frames that make $\forall x \, Rxx$ true. In this chapter, we are interested in comparing modal languages with the following classical languages as tools for defining frame classes:

Definition 3.4 (Frame Languages) For any modal similarity type τ, the *first-order frame language* of τ is the first-order language that has the identity symbol $=$ together with an $(n+1)$-ary relation symbol $R_\triangle$ for each n-ary modal operator $\triangle$ in τ. We denote this language by $\mathcal{L}_\tau^1$. We often call it the *first-order correspondence language* (for τ).

Let Φ be any set of proposition letters. The *monadic second-order frame language* of τ over Φ is the monadic second-order language obtained by augmenting $\mathcal{L}_\tau^1$ with a Φ-indexed collection of monadic predicate variables. (That is, this language has all the resources of $\mathcal{L}_\tau^1$, and in addition is capable of quantifying over subsets of frames.) We denote this language by $\mathcal{L}_\tau^2(\Phi)$, though sometimes we suppress reference to Φ and write $\mathcal{L}_\tau^2$. Moreover, we often simply call it the *second-order frame language* or the *second-order correspondence language* (for τ), taking it for granted that only monadic second-order quantification is permitted.   ⊣

Note that the second-order frame language is extremely powerful, even for the basic modal similarity type. For example, if R is interpreted as the relation of set membership, second-order Zermelo-Fraenkel (ZF) set theory can be axiomatized by a single sentence of this language.

Definition 3.5 (Frame Correspondence) If a class of frames (or more informally, a property) can be defined by a modal formula ϕ and by a formula α from one of these frame languages, then we say that ϕ and α are each others (global) frame *correspondents*.   ⊣

For example, the basic modal formula $p \to \Diamond p$ and the first-order sentence $\forall x \, Rxx$ are correspondents, for we will shortly see that $p \to \Diamond p$ defines reflexivity. Later in this chapter we will show how to systematically find correspondents of modal formulas by adopting a slightly different perspective on the standard translation introduced in Section 2.4.

In Definition 3.5 we did not mention the possibility that modal formulas correspond to a *set* of first-order formulas. Why not? The reason is that this situation simply cannot occur, as we ask the reader to show in Exercise 3.8.3.

There are a number of practical reasons for being interested in frame definability. First, some applications of modal logic are essentially *syntactically* driven; their starting point is some collection of modal formulas expressing axioms, laws, or principles which for some reason we find interesting or significant. Frame definability can be an invaluable tool in such work, for by determining which frame classes these formulas define we obtain a mathematical perspective on their content. On the other hand, some applications of modal logic are essentially *semantically* driven; their starting point is some class of frames of interest. But here too definability is a useful concept. For a start, can the modal language distinguish the 'good' frames from the 'bad' ones? And which properties can the modal language express *within* the class of 'good' frames? Finally, many applied modal languages contain several modalities, whose intended meanings are interrelated. Sometimes it is clear that these relationships should validate certain formulas, and we want to extract the frame-theoretic property they correspond to. On the other hand it may be clear what the relevant frame-theoretic property is (for example, in the basic temporal language we want the P and F operators to scan backwards and forwards along the *same* relation) and we want to see whether there is a modal formula that defines this property. In short, thinking in terms of frame definability can be useful for a variety of reasons – and as the following examples will make clear, modal languages can define some very interesting frame classes indeed.

Example 3.6 In Example 1.10 in Section 1.2 we mentioned the following reading of the modalities: read $\Diamond\phi$ as 'it is *possibly* the case that ϕ' and $\Box\phi$ as '*necessarily* ϕ.' We also mentioned that a number of interesting looking principles concerning necessity and possibility could be stated in the basic modal language. Here are three important examples, together with their traditional names:

(T) $p \to \Diamond p$
(4) $\Diamond\Diamond p \to \Diamond p$
(5) $\Diamond p \to \Box\Diamond p$

But now the problems start. While the status of T seems secure (if p holds here-and-now, p must be *possible*) but what about 4 and 5? When we have to deal with embedded modalities, our intuitions tend to fade, even for such simple formulas as 4 and 5; it is not easy to say whether they should be accepted, and if we only have our everyday understanding of the words 'necessarily' and 'possibly' to guide us, it is difficult to determine whether these principles are interrelated. What we need is a *mathematical* perspective on their content, and that is what the frame definability offers. So let us see what frame conditions these principles define.

Our first claim is that for any frame $\mathfrak{F} = (W, R)$, the axiom T corresponds to *reflexivity* of the relation R:

$$\mathfrak{F} \Vdash \text{T iff } \mathfrak{F} \models \forall x\, Rxx. \tag{3.1}$$

The proof of the right to left direction of (3.1) is easy: let $\mathfrak{F}$ be a reflexive frame, and take an arbitrary valuation V on $\mathfrak{F}$, and a state w in $\mathfrak{F}$ such that $(\mathfrak{F}, V), w \Vdash p$. We need to show that $\Diamond p$ holds at some state that is accessible from w – but as R is reflexive, w is accessible from itself, and $w \Vdash \Diamond p$.

For the other direction, we use contraposition: suppose that R is *not* reflexive, that is, there exists a state w which is not accessible from itself. To falsify T in $\mathfrak{F}$, it suffices to find a valuation V and a state v such that p holds at v, but $\Diamond p$ does not. It is pretty obvious that we should choose v to be our irreflexive state w. Now the valuation V has to satisfy two conditions: (1) $w \in V(p)$ and (2) $\{x \in W \mid Rwx\} \cap V(p) = \varnothing$. Consider the *minimal* valuation V satisfying condition (1), that is, take

$$V(p) = \{w\}.$$

Then it is immediate that $(\mathfrak{F}, V), w \Vdash p$. Now let v be an R-successor of w. As Rww does not hold in $\mathfrak{F}$, v must be distinct from w, so $v \not\Vdash p$. As v was arbitrary, $w \not\Vdash \Diamond p$. This proves (3.1).

Likewise, one can prove that for any frame $\mathfrak{F} = (W, R)$

$$\mathfrak{F} \Vdash 4 \quad \text{iff} \quad R \text{ is transitive, and} \tag{3.2}$$

$$\mathfrak{F} \Vdash 5 \quad \text{iff} \quad R \text{ is euclidean,} \tag{3.3}$$

where a relation is *euclidean* if it satisfies $\forall xyz\,((Rxy \wedge Rxz) \rightarrow Ryz)$. We leave the proofs of (3.2) and the easy (right to left) direction of (3.3) to the reader. For the left to right direction of (3.3), we again argue by contraposition. Assume that $\mathfrak{F}$ is a non-euclidean frame; then there must be states u, v and w such that Ruv, Ruw, but not Rvw:

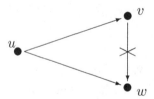

We will try to falsify 5 in u; for this purpose we have to find a valuation V such that $(\mathfrak{F}, V), u \Vdash \Diamond p$ and $(\mathfrak{F}, V), u \not\Vdash \Box\Diamond p$. In other words, we have to make p *true* at some R-successor x of u, and *false* at all R-successors of some R-successor y of u. Some reflection shows that appropriate candidates for x and y are w and v, respectively. Note that again the constraints on V are twofold: (1) $w \in V(p)$ and (2) $\{z \mid Rvz\} \cap V(p) = \varnothing$.

Let us take a *maximal* V satisfying condition (2), that is, define

$$V(p) = \{z \in W \mid \text{it is not the case that } Rvz\}.$$

Now clearly $v \not\Vdash \Diamond p$, so $u \not\Vdash \Box\Diamond p$. On the other hand we have $w \Vdash p$, since w is in the set $\{z \in W \mid \text{it is not the case that } Rvz\}$. So $u \Vdash \Diamond p$. In other words, we have indeed found a valuation V and a state u such that 5 does not hold in u. Therefore, 5 is not valid in $\mathfrak{F}$. This proves (3.3).   ⊣

Example 3.7 Suppose that we are working with the basic temporal language (see Section 1.3 and in particular Example 1.25) and that we are interested in *dense* bidirectional frames (that is, structures in which between every two points there is a third). This property can be defined using a first-order sentence (namely $\forall xy \, (x < y \rightarrow \exists z \, (x < z \land z < y))$) but can the basic temporal language define it too?

It can. The following simple formula suffices: $Fp \rightarrow FFp$. To see this, let $\mathfrak{T} = (T, <)$ be a frame such that $\mathfrak{T} \Vdash Fp \rightarrow FFp$. Suppose that a point $t \in T$ has a $<$-successor t'. To show that t and t' satisfy the density condition, consider the following *minimal* valuation V_m guaranteeing that $(\mathfrak{T}, V_m), t \Vdash Fp$:

$$V_m(p) = \{t'\}.$$

Now, under this valuation $t \Vdash Fp$, and by assumption $\mathfrak{T} \Vdash Fp \rightarrow FFp$, hence $t \Vdash FFp$. This means there is a point s such that $t < s$ and $s \Vdash Fp$. But as t' is the *only* state where p holds, this implies that $s < t'$, so s is the intermediate point we were looking for.

Conversely, let $\mathfrak{T} = (T, <)$ be a dense frame, and assume that under some valuation V, Fp holds at some $t \in T$. Then there is a point t' such that $t < t'$ and $t' \Vdash p$. But as $\mathfrak{T}$ is dense, there is a point s such that $t < s < t'$, hence $s \Vdash Fp$ and hence $t \Vdash FFp$.

Note that nothing in the previous argument depended on the fact that we were working with the basic temporal language; the previous argument also shows that density is definable in the basic modal language using the formula $\Diamond p \rightarrow \Diamond\Diamond p$. Note that this is the converse of the 4 axiom that defines transitivity. ⊣

Example 3.8 Here is a more abstract example. Suppose we are working with a similarity type with three binary operators $\triangle_1$, $\triangle_2$ and $\triangle_3$, and that we are interested in the class of frames in which the three ternary accessibility relations (denoted by R_1, R_2 and R_3, respectively), offer, so to speak, three 'perspectives' on the same relation. To put this precisely, suppose we want the condition

$$R_1 stu \text{ iff } R_2 tus \text{ iff } R_3 ust$$

to hold for all s, t and u in such frames. Can we define this class of frames?

We can. We will show that for all frames $\mathfrak{F} = (W, R_1, R_2, R_3)$ we have

$$\mathfrak{F} \Vdash p \land (q \triangle_1 r) \rightarrow (q \land r \triangle_2 p) \triangle_1 r \text{ iff } \mathfrak{F} \models \forall xyz \, (R_1 xyz \rightarrow R_2 yzx). \quad (3.4)$$

(Recall that we use infix notation for dyadic operation symbols.) The easy direction is from right to left. Let $\mathfrak{F}$ be a frame satisfying $\forall xyz \, (R_1 xyz \rightarrow R_2 yzx)$. Consider an arbitrary valuation V on $\mathfrak{F}$ and an arbitrary state s such that $(\mathfrak{F}, V), s \Vdash p \land (q \triangle_1 r)$. Then, $s \Vdash p$ and there are states t and u with $R_1 stu$, $t \Vdash q$ and $u \Vdash r$. From $R_1 stu$ we derive $R_2 tus$. But then $t \Vdash q \land r \triangle_2 p$, so by $R_1 stu$ we have $s \Vdash (q \land r \triangle_2 p) \triangle_1 r$.

For the other direction, suppose that the modal formula $p \wedge (q \triangle_1 r) \rightarrow (q \wedge r \triangle_2 p) \triangle_1 r$ is valid in $\mathfrak{F}$, and consider states s, t and u in $\mathfrak{F}$ with $R_1 stu$. We will show that $R_2 tus$. Consider a valuation V with $V(p) = \{s\}$, $V(q) = \{t\}$ and $V(r) = \{u\}$. Then $(\mathfrak{F}, V)$, $s \Vdash p \wedge q \triangle_1 r$, so by our assumption, $s \Vdash (q \wedge r \triangle_2 p) \triangle_1 r$. Hence, there must be states t', u' with $R_1 st'u'$, $t' \Vdash q \wedge r \triangle_2 p$ and $u' \Vdash r$. From $t' \Vdash q$ it follows that $t = t'$, so we have $t \Vdash r \triangle_2 p$. Again, using the truth definition we find states s'', u'' with $R_2 tu''s''$, $u'' \Vdash r$ and $s'' \Vdash p$. The latter two facts imply that $u'' = u$ and $s'' = s$. But then we have $R_2 tus$, as required.   ⊣

From these examples the reader could easily get the impression that modal formulas always correspond to frame properties that are definable in first-order logic. This impression is wrong, and in the next section we will see why.

Exercises for Section 3.1

3.1.1 Consider a language with two diamonds $\langle 1 \rangle$ and $\langle 2 \rangle$. Show that $p \rightarrow [2]\langle 1 \rangle p$ is valid on precisely those frames for the language that satisfy the condition $\forall xy\,(R_2 xy \rightarrow R_1 yx)$. What sort of frames does $p \rightarrow [1]\langle 1 \rangle p$ define?

3.1.2 Consider a language with three diamonds $\langle 1 \rangle$, $\langle 2 \rangle$, and $\langle 3 \rangle$. Show that the modal formula $\langle 3 \rangle p \leftrightarrow \langle 1 \rangle \langle 2 \rangle p$ is valid on a frame for this language if and only if the frame satisfies the condition $\forall xy\,(R_3 xy \leftrightarrow \exists z\,(R_1 xz \wedge R_2 zy))$.

3.2 Frame Definability and Second-Order Logic

In this section we show that modal languages can get to grips with notions that exceed the expressive power of first-order logic, and explain why. We start by presenting three well-known examples of modal formulas that define frame properties which cannot be expressed in first-order logic. Then, drawing on our discussion of the standard translation in Section 2.4, we show that such results are to be expected: as we will see, modal formulas standardly correspond to *second-order* frame conditions. Indeed, the real mystery is not why they do so (this turns out to be rather obvious), but why they sometimes correspond to simple *first-order* conditions such as reflexivity or transitivity (we discuss this more difficult issue in Sections 3.5–3.7).

Example 3.9 Consider the Löb formula $\square(\square p \rightarrow p) \rightarrow \square p$, which we will call L for brevity. This formula plays an essential role in *provability* logic, a branch of modal logic where $\square \phi$ is read as 'it is *provable* (in some formal system) that ϕ'. The formula L is named after Löb, who proved L as a theorem of the provability logic of Peano Arithmetic. We will first show that L defines the class of frames (W, R) such that R is transitive and R's converse is well-founded. (A relation R is *well-founded* if there is no infinite sequence $\ldots Rw_2 Rw_1 Rw_0$; hence, R's converse is well-founded if there is no infinite R-path emanating from any state. In particular, this excludes cycles and loops.)

We will then show that this is a class of frames that first-order frame languages *cannot* define; that is, we will show that this class is not *elementary*.

To see that L defines the stated property, assume that $\mathfrak{F} = (W, R)$ is a frame with a transitive and conversely well-founded relation R, and then suppose for the sake of a contradiction that L is not valid in $\mathfrak{F}$. This means that there is a valuation V and a state w such that $(\mathfrak{F}, V), w \not\Vdash \Box(\Box p \rightarrow p) \rightarrow \Box p$. In other words, $w \Vdash \Box(\Box p \rightarrow p)$, but $w \not\Vdash \Box p$. Then w must have a successor w_1 such that $w_1 \not\Vdash p$, and as $\Box p \rightarrow p$ holds at all successors of w, we have that $w_1 \not\Vdash \Box p$. This in turn implies that w_1 must have a successor w_2 where p is false; note that by the transitivity of R, w_2 is also a successor of w. But now, simply by repeating our argument, we see that w_2 must have a p-falsifying successor w_3 (which by transitivity must be a successor of w_1), that w_3 has a successor w_4 (which by transitivity must be a successor of w_1), and so on. In short, we have found an infinite path $w R w_1 R w_2 R w_3 R \ldots$, contradicting the converse well-foundedness of R. (Note that the points w_1, w_2, ... need not all be distinct.)

For the other direction, we use contraposition. That is, we assume that either R is not transitive or its converse is not well-founded; in both cases we have to find a valuation V and a state w such that $(\mathfrak{F}, V), w \not\Vdash L$. We leave the case where R is not transitive to the reader (hint: instead of L, consider the frame equivalent formula $\Diamond p \rightarrow \Diamond(p \wedge \neg \Diamond p)$) and only consider the second case. So assume that R is transitive, but not conversely well-founded. In other words, suppose we have a transitive frame containing an infinite sequence $w_0 R w_1 R w_2 R \ldots$. We exploit the presence of this sequence by defining the following valuation V:

$$V(p) = W \setminus \{x \in W \mid \text{there is an infinite path starting from } x\}.$$

We leave it to the reader to verify that under this valuation, $\Box p \rightarrow p$ is true *everywhere* in the model, whence, certainly, $(\mathfrak{F}, V), w_0 \Vdash \Box(\Box p \rightarrow p)$. The claim then follows from the fact that $(\mathfrak{F}, V), w_0 \not\Vdash \Box p$.

Finally, to show that the class of frames defined by L is not elementary, an easy compactness argument suffices. Suppose for the sake of a contradiction that there is a first-order formula equivalent to L; call this formula λ. As λ is equivalent to L, any model making λ true must be transitive. Let $\sigma_n(x_0, \ldots, x_n)$ be the first-order formula stating that there is an R-path of length n through $x_0, \ldots, x_n$:

$$\sigma_n(x_0, \ldots, x_n) = \bigwedge_{0 \leq i < n} R x_i x_{i+1}.$$

Obviously, every *finite* subset of

$$\Sigma = \{\lambda\} \cup \{\forall xyz\,((Rxy \wedge Ryz) \rightarrow Rxz)\} \cup \{\sigma_n \mid n \in \omega\}$$

is satisfiable in a finite linear order, and hence in the class of transitive, conversely well-founded frames. Thus by the Compactness Theorem, Σ itself must have a model. But it is clear that Σ is *not* satisfiable in any conversely well-founded frame – and λ, being equivalent to L, is supposed to define the class of transitive, conversely well-founded frames. From this contradiction we conclude that L cannot be equivalent to any first-order formula.

Could L then perhaps be equivalent to an (infinite) *set* of first-order formulas? No – we already mentioned (right after Definition 3.5) that this kind of correspondence never occurs.   ⊣

Our next example concerns *propositional dynamic logic* (PDL). Recall that this language contains a family of diamonds $\{\langle\pi\rangle \mid \pi \in \Pi\}$ (where Π is a collection of programs) and the program constructors $\cup$, ; and *. In the intended frames for this language (that is, the *regular* frames; see Example 1.26) we want the accessibility relations for diamonds built using these constructors to reflect choice, composition, and iteration of programs, respectively. Now, to reflect iteration we demanded that the relation R_{π^*} used for the program π^* be the reflexive, transitive closure of the relation R_π used for π. But it is well known that this constraint *cannot* be expressed in first-order logic (as with the Löb example, this can be shown using a compactness argument, and the reader was asked to do this in Exercise 2.4.5). Because of this, when we discussed PDL at the level of models in Section 2.4 we used the *infinitary* language $\mathcal{L}_{\omega_1\omega}$ as the correspondence language for PDL; using infinite disjunctions enabled us to capture the 'keep looking!' force of $*$ that eludes first-order logic. But although first-order logic cannot get to grips with $*$, PDL itself can – via the concept of frame definability.

Example 3.10 PDL can be interpreted on any transition system of the form $\mathfrak{F} = (W, R_\pi)_{\pi \in \Pi}$. Let us call such a frame $*$-*proper* if the transition relation R_{π^*} of each program π^* is the reflexive and transitive closure of the transition relation R_π of π. Can we single out, by modal means, the $*$-proper frames within the class of all transition systems of the form $(W, R_\pi)_{\pi \in \Pi}$? And can we then go on to single out the class of all regular frames?

The answer to both questions is *yes*. Consider the following set of formulas

$$\Delta = \{[\pi^*](p \rightarrow [\pi]p) \rightarrow (p \rightarrow [\pi^*]p),\ \langle\pi^*\rangle p \leftrightarrow (p \vee \langle\pi\rangle\langle\pi^*\rangle p) \mid \pi \in \Pi\}.$$

As we mentioned in Example 1.15, $[\pi^*](p \rightarrow [\pi]p) \rightarrow (p \rightarrow [\pi^*]p)$ is called *Segerberg's axiom*, or the *induction axiom*. We claim that for any PDL-frame $\mathfrak{F}$:

$$\mathfrak{F} \Vdash \Delta \text{ iff } \mathfrak{F} \text{ is } *\text{-proper.} \tag{3.5}$$

The reader is asked to supply a proof of this in Exercise 3.2.1.

A straightforward consequence is that PDL is strong enough to define the class of regular frames. The constraints on the relations interpreting $\cup$ and ; are simple first-order conditions, and

$$\Gamma = \{\langle \pi_1; \pi_2 \rangle p \leftrightarrow \langle \pi_1 \rangle \langle \pi_2 \rangle p, \langle \pi_1 \cup \pi_2 \rangle p \leftrightarrow \langle \pi_1 \rangle p \vee \langle \pi_2 \rangle p \mid \pi \in \Pi \}.$$

pins down down what is required. So $\Delta \cup \Gamma$ defines the regular frames. ⊣

In the previous two examples we encountered modal formulas that expressed frame properties that were, although not elementary, still relatively easy to understand. (Note however that in order to formally express (converse) well-foundedness in a classical language, one needs heavy machinery – the infinitary language $\mathcal{L}_{\omega_1 \omega}$ does not suffice!) The next example shows that extremely simple modal formulas can define second-order frame conditions that are not easy to understand at all.

Example 3.11 We will show that the McKinsey formula (M) $\Box \Diamond p \to \Diamond \Box p$ does not correspond to a first-order condition by showing that it violates the Löwenheim-Skolem Theorem.

Consider the frame $\mathfrak{F} = (W, R)$, where

$$W = \{w\} \cup \{v_n, v_{(n,i)} \mid n \in \mathbb{N}, i \in \{0, 1\}\} \cup \{z_f \mid f : \mathbb{N} \to \{0, 1\}\},$$

and

$$\begin{aligned} R = \ & \{(w, v_n), (v_n, v_{(n,i)}), (v_{(n,i)}, v_{(n,i)}) \mid n \in \mathbb{N}, i \in \{0, 1\}\} \cup \\ & \{(w, z_f), (z_f, v_{(n, f(n))}) \mid n \in \mathbb{N}, f : \mathbb{N} \to \{0, 1\}\}. \end{aligned}$$

In a picture:

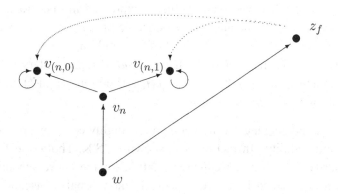

Note that W contains uncountably many points, for the set of functions indexing the z points is uncountable.

Our first observation is that $\mathfrak{F} \Vdash \Box \Diamond p \to \Diamond \Box p$. We leave it to the reader to verify that for all u different from w, $\mathfrak{F}, u \Vdash \Box \Diamond p \to \Diamond \Box p$. As to showing that

$\mathfrak{F}, w \Vdash \square\lozenge p \rightarrow \lozenge\square p$, suppose that $(\mathfrak{F}, V), w \Vdash \square\lozenge p$. Then, for each $n \in \mathbb{N}$, $(\mathfrak{F}, V), v_n \Vdash \lozenge p$. From this we get either $(\mathfrak{F}, V), v_{(n,0)} \Vdash p$ or $(\mathfrak{F}, V), v_{(n,1)} \Vdash p$. Choose $f : \mathbb{N} \rightarrow \{0, 1\}$ such that $(\mathfrak{F}, V), v_{(n,f(n))} \Vdash p$, for each $n \in \mathbb{N}$. Then clearly, $(\mathfrak{F}, V), z_f \Vdash \square p$, and so $(\mathfrak{F}, V), w \Vdash \lozenge\square p$.

In order to show that $\square\lozenge p \rightarrow \lozenge\square p$ does not define a first-order frame condition, let us view the frame $\mathfrak{F}$ as a first-order model with domain W. By the downward Löwenheim-Skolem Theorem (here we need the strong version of Theorem A.11) there must be a countable elementary submodel $\mathfrak{F}'$ of $\mathfrak{F}$ whose domain W' contains w, and each v_n, $v_{(n,0)}$ and $v_{(n,1)}$. As W is uncountable and W' countable, there must be a mapping $f : \mathbb{N} \rightarrow \{0, 1\}$ such that z_f does not belong to W'. Now, if the McKinsey formula was equivalent to a first-order formula it would be valid on $\mathfrak{F}'$ (the Löwenheim-Skolem Theorem tells us that $\mathfrak{F}$ and $\mathfrak{F}'$ are elementarily equivalent). But we will show that the McKinsey formula is *not* valid on $\mathfrak{F}'$, hence it cannot be equivalent to a first-order formula.

Let V' be a valuation on $\mathfrak{F}'$ such that $V'(p) = \{v_{(n,f(n))} \mid n \in \mathbb{N}\}$; here f is a mapping such that z_f does not belong to W'. We will show that under V', $\square\lozenge p$ is true at w, but $\lozenge\square p$ is not.

It is easy to see that $(\mathfrak{F}', V'), w \nVdash \lozenge\square p$. For a start, since p holds at exactly one of $v_{(n,0)}$ and $v_{(n,1)}$, $\square p$ is false at each v_n. Now consider an arbitrary element z_g in W'. Then g is distinct from f, so there must be an element $n \in \mathbb{N}$ such that $g(n) \neq f(n)$. So p is true at $v_{(n,f(n))}$ and false at $v_{(n,g(n))}$; this means that z_g has a successor where p is false, so $(\mathfrak{F}', V'), z_g \nVdash \square p$. Hence, we have not been able to find a successor for w where $\square p$ holds, so $(\mathfrak{F}', V'), w \nVdash \lozenge\square p$.

In order to show that $(\mathfrak{F}', V'), w \Vdash \square\lozenge p$ we reason as follows. Note first that $(\mathfrak{F}', V'), v_n \Vdash \lozenge p$, for each state v_n. Now consider an arbitrary element z_g of W'. Call two states z_h and z_k of $\mathfrak{F}$ *complementary* if for all n, $h(n) = 1 - k(n)$; the reader should verify that this relation can be expressed in first-order logic. Now suppose that z_g is complementary to z_f; since complementary states are unique, the fact that $\mathfrak{F}'$ is an elementary submodel of $\mathfrak{F}$ would imply that z_f exists in $\mathfrak{F}'$ as well. Clearly then, we may conclude that z_g is *not* complementary to z_f. Hence, there exists some $n \in \mathbb{N}$ such that $g(n) = f(n)$. Therefore, $(\mathfrak{F}', V'), z_g \Vdash \lozenge p$. But then $\lozenge p$ holds at every successor of w. $\dashv$

Clearly then, modal languages can express many highly complex properties via the notion of frame validity. In fact, as was shown by S.K. Thomason for the basic modal similarity type, the consequence relation for the entire second-order language $\mathcal{L}_\tau^2$ can be reduced in a certain sense to the (global) consequence relation over frames. More precisely, Thomason showed that there is a computable translation f taking $\mathcal{L}^2$ sentences α to modal formulas $f(\alpha)$, and a special fixed modal formula δ, such that for all sets of $\mathcal{L}^2$ sentences Σ, we have that

$$\Sigma \models \alpha \text{ iff } \{\delta\} \cup \{f(\sigma) \mid \sigma \in \Sigma\} \Vdash^g f(\alpha).$$

On the frame level, propositional modal logic must be understood as a rather strong fragment of classical monadic second-order logic. We now face the question: *why*?

The answer turns out to be surprisingly simple. Recall from Definition 3.1 that validity is defined by quantifying over all states of the universe and all possible valuations. But a valuation assigns a *subset* of a frame to each proposition letter, and this means that when we quantify across all valuations we are implicitly quantifying across all subsets of the frame. In short, monadic second-order quantification is hard-wired into the very definition of validity; it is hardly surprising that frame-definability is such a powerful concept.

Let us make this answer more precise. In the previous chapter, we saw that at the level of models, the modal language $ML(\tau, \Phi)$ can be translated in a truth-preserving way into the first-order language $\mathcal{L}^1_\tau(\Phi)$ (see Proposition 2.47). Let us adopt a slightly different perspective:

View the predicate symbol P that corresponds to the proposition letter p as a monadic second-order variable that we can quantify over.

If we do this, we are in effect viewing the standard translation as a way of translating into the second-order frame language $\mathcal{L}^2_\tau(\Phi)$ introduced in Definition 3.4. And if we view the standard translation this way we are led, virtually immediately, to the following result:

Proposition 3.12 *Let τ be a modal similarity type, and ϕ a τ-formula. Then for any τ-frame $\mathfrak{F}$ and any state w in $\mathfrak{F}$:*

$$\mathfrak{F}, w \Vdash \phi \quad \text{iff} \quad \mathfrak{F} \models \forall P_1 \ldots \forall P_n \, ST_x(\phi)[w],$$
$$\mathfrak{F} \Vdash \phi \quad \text{iff} \quad \mathfrak{F} \models \forall P_1 \ldots \forall P_n \forall x \, ST_x(\phi).$$

Here, the second-order quantifiers bind second-order variables P_i corresponding to the proposition letters p_i occurring in ϕ.

Proof. Let $\mathfrak{M} = (\mathfrak{F}, V)$ be any model based on $\mathfrak{F}$, and let w be any state in $\mathfrak{F}$. Then we have that

$$(\mathfrak{F}, V), w \Vdash \phi \quad \text{iff} \quad \mathfrak{F} \models ST_x(\phi)[w, P_1, \ldots, P_n],$$

where the notation $[w, P_1, \ldots, P_n]$ means 'assign w to the free first-order variable x in $ST_x(\phi)$, and $V(p_1), \ldots, V(p_n)$ to the free monadic second-order variables.' Note that this equivalence is nothing new; it is simply a restatement of Proposition 2.47 in second-order terms. But then we obtain the first part of the theorem simply by universally quantifying over the free variables $P_1, \ldots, P_n$. The second part follows from the first by universally quantifying over the states of the frame (as in Proposition 3.30). ⊣

It is fairly common to refer to the $\mathcal{L}^2_\tau(\Phi)$ formula $\forall P_1 \dots \forall P_n \forall x \, ST_x(\phi)$ as the standard translation of ϕ, since it is usually clear whether we are working at the level of models or the level of frames. Nonetheless, we will try and reserve the term standard translation to mean the $\mathcal{L}^1_\tau(\Phi)$ formula produced by the translation process, and refer to $\forall P_1 \dots \forall P_n \forall x \, ST_x(\phi)$ as the *second-order translation* of ϕ.

Let us sum up what we have learned. That modal formulas can define second-order properties of frames is neither mysterious nor surprising: because modal validity is defined in terms of quantification over subsets of frames, it is intrinsically second-order, hence so is the notion of frame definability. Indeed, the real mystery lies not with such honest, hard-working, formulas as Löb and McKinsey, but with such lazy formulas as T, 4 and 5 discussed in the previous section. For example, if we apply the second-order translation to T (that is, $p \to \Diamond p$) we obtain

$$\forall P \forall x \, (Px \to \exists y (Rxy \wedge Py)).$$

We already know that T defines reflexivity, so this must be a (somewhat baroque) second-order way of expressing reflexivity – and it is fairly easy to see that this is so. But this sort of thing happens a lot: 4 and 5 give rise to (fairly complex) second-order expressions, yet the complexity melts away leaving a simple first-order equivalent behind. The contrast with the McKinsey formula is striking: what *is* going on? This is an interesting question, and we discuss it in detail in Sections 3.5–3.7.

Another point is worth making: our discussion throws light on the somewhat mysterious *general frames* introduced in Section 1.4. Recall that a general frame is a frame together with a collection of valuations A satisfying certain modally natural closure conditions. We claimed that general frames combined the key advantage of frames (namely, that they support the key logical notion of validity) with the advantage of models (namely, that they are concrete and easy to work with). The work of this section helps explain why.

The key point is this. A general frame can be viewed as a *generalized model* for (monadic) second-order logic. A generalized model for second-order logic is a model in which the second-order quantifiers are viewed as ranging not over *all* subsets, but only over a pre-selected sub-collection of subsets. And of course, the collection of valuations A in a general frame is essentially such a sub-collection of subsets. This means that the following equivalence holds:

$$(\mathfrak{F}, A) \Vdash \phi \ \text{ iff } \ (\mathfrak{F}, A) \models \forall P_1 \dots \forall P_n \forall x \, ST_x(\phi).$$

Here the block of quantifiers $\forall P_1 \dots \forall P_n$ denotes not genuine second-order quantification, but generalized second-order quantification (that is, quantification over the subsets in A). Generalized second-order quantification is essentially a first-order 'approximation' of second-order quantification that possesses many properties that genuine second-order quantification lacks (such as Completeness, Compactness, and Löwenheim-Skolem). In short, one of the reasons general frames are so useful is that they offer a first-order perspective (via generalized models) on what is essentially a second-order phenomenon (frame validity). This is not the full story – the algebraic perspective on general frames is vital to modal logic – but it should make clear that these unusual looking structures fill an important logical niche.

Exercises for Section 3.2

3.2.1   (a) Consider a modal language with two diamonds $\langle 1 \rangle$ and $\langle 2 \rangle$. Prove that the class of frames in which R_1 is the reflexive transitive closure of R_2 is defined by the conjunction of the formulas $\langle 1 \rangle p \rightarrow (p \vee \langle 1 \rangle (\neg p \wedge \langle 2 \rangle p))$ and $\langle 1 \rangle p \leftrightarrow (p \vee \langle 2 \rangle \langle 1 \rangle p)$.

(b) Conclude that in the similarity type of PDL, the set Δ as defined in Example 3.10 defines the class of ∗-proper frames.

(c) Consider the example of multi-agent epistemic logic; let $\{1, \ldots, n\}$ be the set of agents. Suppose that one is interested in the operators E ($E\phi$ stands for 'everybody knows ϕ') and C ($C\phi$ meaning that 'it is common knowledge that ϕ'). The intended relations modeling E and C are given by:

$$R_E uv \quad \text{iff} \quad \bigwedge_{1 \leq i \leq n} R_i uv,$$
$$R_C uv \quad \text{iff} \quad \text{there is a path } u = x_0 R_E x_1 R_E \ldots x_{n-1} R_E x_n = v.$$

Write down a set of (epistemic) formulas that characterizes the class of epistemic frames where these conditions are met.

3.2.2 Show that Grzegorczyk's formula, $\Box(\Box(p \rightarrow \Box p) \rightarrow p) \rightarrow p$, characterizes the class of frames $\mathfrak{F} = (W, R)$ satisfying (i) R is reflexive, (ii) R is transitive and (iii) there are no infinite paths $x_0 R x_1 R x_2 R \ldots$ such that for all i, $x_i \neq x_{i+1}$.

3.2.3 Consider the basic temporal language (see Example 1.25). Recall that a frame $\mathfrak{F} = (W, R_F, R_P)$ for this language is called *bidirectional* if R_P is the converse of R_F.

(a) Prove that among the finite bidirectional frames, the formula $G(Gp \rightarrow p) \rightarrow Gp$ together with its converse, $H(Hp \rightarrow p) \rightarrow Hp$ defines the transitive and irreflexive frames.

(b) Prove that among the bidirectional frames that are transitive, irreflexive, and satisfy $\forall xy\,(R_F xy \vee x = y \vee R_P xy)$, this same set defines the finite frames.

(c) Is there a finite set of formulas in the *basic* modal language that has these same definability properties?

3.2.4 Consider the following formula in the basic similarity type:

$$\psi := \Diamond \Box p \rightarrow \Diamond (\Box(p \wedge q) \vee \Box(p \wedge \neg q)).$$

The aim of this exercise is to show that ψ does not define a first-order condition on frames.

(a) To obtain some intuitions about the meaning of ψ, let us first give a relatively simple first-order condition *implying* the validity of ψ:

$$\alpha := \forall xy\,(Rxy \rightarrow \exists z\,(Rxz \wedge \forall uv\,((Rzu \wedge Rzv) \rightarrow (u = v \wedge Ryu)))),$$

stating (in words) that for every pair (x, y) in R, x has a successor z which itself has at most one successor, this point being also a successor of y.

Show that ψ is valid in any frame satisfying α.

(b) Consider the frame $\mathfrak{F} = (W, R)$ which we define as follows. Let u be a non-principal ultrafilter over the set $\mathbb{N}$ of the natural numbers. Then $W := \{u\} \cup u \cup \mathbb{N}$, that is, the states of W are u itself, each subset of $\mathbb{N}$ that is a member of u and each natural number. The relation R is the converse of the membership relation, that is, Rst iff $t \in s$. Show that $\mathfrak{F} \not\Vdash \alpha$ and $\mathfrak{F} \Vdash \psi$.

(c) Prove that ψ does not have a first-order correspondent by showing that ψ is invalid on all *countable* structures that are elementarily equivalent to $\mathfrak{F}$ (that is, all countable structures satisfying the same first-order formulas as $\mathfrak{F}$).

3.3 Definable and Undefinable Properties

We have seen that modal languages are a powerful tool for defining frames: we have seen examples of modally definable frame classes that are not first-order definable, and it is clear that validity is an inherently second-order concept. But what are the limits of modal definability? For example, can modal languages define all first-order frame classes (the answer is *no*, as we will shortly see)? And anyway, how should we go about showing that a class of frames is *not* modally definable? After all, we cannot try out all possible formulas; something more sophisticated is needed.

In this section we will answer these question by introducing four fundamental frame constructions: *disjoint unions, generated subframes, bounded morphic images*, and *ultrafilter extensions*. The names should be familiar: these are the frame theoretic analogs of the model-theoretic constructions studied in the previous chapter, and they are going to do a lot of work for us, both here and in later chapters. For a start, it is a more-or-less immediate consequence of the previous chapter's work that the first three constructions preserve modal validity, while the fourth anti-preserves it. But this means that these constructions provide powerful tests for modal definability: by showing that some class of frames is *not* closed under one of these constructions, we will be able to show that it *cannot* be modally definable.

Definition 3.13 The definitions of the disjoint union of a family of frames, a generated subframe of a frame, and a bounded morphism from one frame to another, are obtained by deleting the clauses concerning valuations from Definitions 2.2, 2.5 and 2.10.

That is, for disjoint τ-frames $\mathfrak{F}_i = (W_i, R_{\Delta i})_{\Delta \in \tau}$ ($i \in I$), their *disjoint union* is the structure $\biguplus_i \mathfrak{F}_i = (W, R_\Delta)_{\Delta \in \tau}$ such that W is the union of the sets W_i and for each $\Delta \in \tau$, R_Δ is the union $\bigcup_{i \in I} R_{\Delta i}$. With frames that are not disjoint, proceed as in Definition 2.2.

We say that a τ-frame $\mathfrak{F}' = (W', R'_\Delta)_{\Delta \in \tau}$ is a *generated subframe* of the frame $\mathfrak{F} = (W, R_\Delta)_{\Delta \in \tau}$ (notation: $\mathfrak{F}' \rightarrowtail \mathfrak{F}$) whenever $\mathfrak{F}'$ is a subframe of $\mathfrak{F}$ (with respect to R_Δ for all $\Delta \in \tau$), and the following heredity condition is fulfilled for all $\Delta \in \tau$

if $u \in W'$ and $R_\triangle uu_1 \ldots u_n$, then $u_1, \ldots, u_n \in W'$.

Let X be a subset of the universe of a frame $\mathfrak{F}$; we denote by $\mathfrak{F}_X$ *the subframe generated by X*, that is, the generated subframe of $\mathfrak{F}$ that is based on the smallest set W' that contains X and satisfies the above heredity condition. If X is a singleton $\{w\}$, we write $\mathfrak{F}_w$ for the *subframe generated by w*; if a frame $\mathfrak{F}$ is generated by a singleton subset of its universe, we call it *rooted* or *point-generated*.

And finally, a bounded morphism from a τ-frame $\mathfrak{F} = (W, R_\triangle)_{\triangle \in \tau}$ to a τ-frame $\mathfrak{F}' = (W', R'_\triangle)_{\triangle \in \tau}$ is a function from W to W' satisfying the following two conditions:

(*forth*) For all $\triangle \in \tau$, $R_\triangle wv_1 \ldots v_n$ implies $R'_\triangle f(w)f(v_1) \ldots f(v_n)$.

(*back*) If $R'_\triangle f(w)v'_1 \ldots v'_n$ then there exist $v_1 \ldots v_n$ such that $R_\triangle wv_1 \ldots v_n$ and $f(v_i) = v'_i$ (for $1 \le i \le n$).

We say that $\mathfrak{F}'$ is a bounded morphic image of $\mathfrak{F}$, notation: $\mathfrak{F} \twoheadrightarrow \mathfrak{F}'$, if there is a surjective bounded morphism from $\mathfrak{F}$ onto $\mathfrak{F}'$. ⊣

It is an essential characteristic of modal formulas that their validity is preserved under the structural operations just defined:

Theorem 3.14 *Let τ be a modal similarity type, and ϕ a τ-formula.*

(i) *Let $\{\mathfrak{F}_i \mid i \in I\}$ be a family of frames. Then $\biguplus \mathfrak{F}_i \Vdash \phi$ if $\mathfrak{F}_i \Vdash \phi$ for every i in I.*

(ii) *Assume that $\mathfrak{F}' \rightarrowtail \mathfrak{F}$. Then $\mathfrak{F}' \Vdash \phi$ if $\mathfrak{F} \Vdash \phi$.*

(iii) *Assume that $\mathfrak{F} \twoheadrightarrow \mathfrak{F}'$. Then $\mathfrak{F}' \Vdash \phi$ if $\mathfrak{F} \Vdash \phi$.*

Proof. We only prove (iii), the preservation result for taking bounded morphic images, and leave the other cases to the reader as Exercise 3.3.1. So, assume that f is a surjective bounded morphism from $\mathfrak{F}$ onto $\mathfrak{F}'$, and that $\mathfrak{F} \Vdash \phi$. We have to show that $\mathfrak{F}' \Vdash \phi$. So suppose that ϕ is *not* valid in $\mathfrak{F}'$. Then there must be a valuation V' and a state w' such that $(\mathfrak{F}', V'), w' \not\Vdash \phi$. Define the following valuation V on $\mathfrak{F}$:

$$V(p_i) = \{x \in W \mid f(x) \in V'(p_i)\}.$$

This definition is tailored to make f a bounded morphism between the models $(\mathfrak{F}, V)$ and $(\mathfrak{F}', V')$ – the reader is asked to verify the details. Now we use the fact that f is surjective to find a w such that $f(w) = w'$. It follows from Proposition 2.14 that $(\mathfrak{F}, V), w \not\Vdash \phi$. In other words, we have falsified ϕ in the frame $\mathfrak{F}$, and shown the contrapositive of the desired result. ⊣

Think of these frame constructions as *test criteria* for the definability of frame properties: if a property is not preserved under one (or more) of these frame constructions, then it cannot be modally definable. Let us consider some examples of such testing.

Example 3.15 The class of finite frames is not modally definable. For suppose there was a set of formulas Δ (in the basic modal similarity type) characterizing the finite frames. Then Δ would be valid in every one-point frame $\mathfrak{F}_i = (\{w_i\}, \{(w_i, w_i)\})$ $(i < \omega)$. By Theorem 3.14(i) this would imply that Δ was also valid in the disjoint union $\biguplus_i \mathfrak{F}_i$:

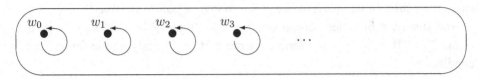

But clearly this cannot be the case, for $\biguplus_i \mathfrak{F}_i$ is infinite.

The class of frames having a reflexive point ($\exists x\, Rxx$) does not have a modal characterization either (again we work with the basic modal similarity type). For suppose that the set Δ characterized this class. Consider the following frame $\mathfrak{F}$:

As w is a reflexive state, $\mathfrak{F} \Vdash \Delta$. Now consider the generated subframe $\mathfrak{F}_v$ of $\mathfrak{F}$. Clearly, Δ cannot be valid in $\mathfrak{F}_v$, since neither v nor u is reflexive. But this contradicts the fact that validity of modal formulas is preserved under taking generated subframes (Theorem 3.14(ii)).

The two final examples involve the use of bounded morphisms. First, irreflexivity is not definable. To see this, simply note that the function which collapses the set of natural numbers in their usual order to a single reflexive point is a surjective bounded morphism. As the former frame is irreflexive, while the latter is not, irreflexivity cannot be modally definable.

Actually, a more sophisticated variant of this example lets us prove even more. Consider the following two frames: $\mathfrak{F} = (\omega, S)$, the natural numbers with the successor relation (Smn iff $n = m + 1$), and $\mathfrak{G} = (\{e, o\}, \{(e, o), (o, e)\})$ as depicted below.

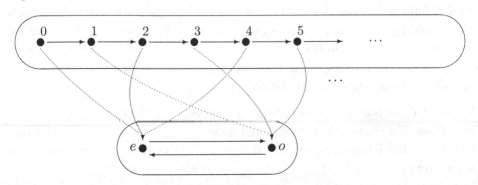

In Example 2.11 we saw that the map f sending even numbers to e and odd numbers to o is a surjective bounded morphism. By the same style of reasoning as in the earlier examples, it follows that no property P is modally definable if $\mathfrak{F}$ has P and $\mathfrak{G}$ lacks it. This shows, for example, that there is no set of formulas characterizing the asymmetric frames ($\forall xy\,(Rxy \rightarrow \neg Ryx)$). ⊣

Now for the fourth frame construction. Recall that in Section 2.5 we introduced the idea of *ultrafilter extensions*; see Definition 2.57 and Proposition 2.59. Once again, simply by ignoring the parts of the definition that deal with valuations, we can lift this concept to the level of frames, and this immediately provides us with the following *anti*-preservation result:

Corollary 3.16 *Let τ be a modal similarity type, $\mathfrak{F}$ a τ-frame, and ϕ a τ-formula. Then $\mathfrak{F} \Vdash \phi$ if $\mathfrak{ue}\,\mathfrak{F} \Vdash \phi$.*

Proof. Assume that ϕ is not valid in $\mathfrak{F}$. That is, there is a valuation V and a state w such that $(\mathfrak{F}, V), w \Vdash \neg\phi$. By Proposition 2.59, $\neg\phi$ is true at u_w in the ultrafilter extension of $\mathfrak{M}$. But then we have refuted ϕ in $\mathfrak{ue}\,\mathfrak{F}$. ⊣

Once again, we can use this result to show that some frame properties are not modally definable. For example, working in the basic modal similarity type, consider the property that every state has a reflexive successor: $\forall x\exists y\,(Rxy \wedge Ryy)$. We claim that this property is *not* modally definable, even though it is preserved under taking disjoint unions, generated subframes and bounded morphic images. To verify our claim, the reader is asked to consider the frame in Example 2.58. It is easy to see that every state of $\mathfrak{ue}\,\mathfrak{F}$ has a reflexive successor – take any non-principal ultrafilter. But $\mathfrak{F}$ itself clearly does not satisfy the property, as $\mathfrak{F}$ has *no* reflexive states. Now suppose that the property were modally definable, say by the set of formulas Δ. Then we would have $\mathfrak{ue}\,\mathfrak{F} \Vdash \Delta$, but $\mathfrak{F} \not\Vdash \Delta$ – a clear violation of Corollary 3.16.

Note the direction of the preservation result in Corollary 3.16. It states that modal validity is *anti*-preserved under taking ultrafilter extensions. This naturally raises the question whether the other direction holds as well, that is, whether $\mathfrak{F} \Vdash \phi$ implies $\mathfrak{ue}\,\mathfrak{F} \Vdash \phi$. For a partial answer to this question, we need the following theorem:

Theorem 3.17 *Let τ be a modal similarity type, and $\mathfrak{F}$ a τ-frame. Then $\mathfrak{F}$ has an ultrapower $\prod_U \mathfrak{F}$ such that $\prod_U \mathfrak{F} \twoheadrightarrow \mathfrak{ue}\,\mathfrak{F}$. In a diagram:*

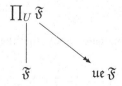

Proof. Advanced track readers will be asked to supply a proof of this theorem in Exercise 3.8.1 below.   ⊣

And now we have the following partial converse to Corollary 3.16:

Corollary 3.18 *Let τ be a modal similarity type, and ϕ a τ-formula. If ϕ defines a first-order property of frames, then frame validity of ϕ is preserved under taking ultrafilter extensions.*

Proof. Let ϕ be a modal formula which defines a first-order property of frames, and let $\mathfrak{F}$ be a frame such that $\mathfrak{F} \Vdash \phi$. By the previous theorem, there is an ultrapower $\prod_U \mathfrak{F}$ of $\mathfrak{F}$ such that $\prod_U \mathfrak{F} \twoheadrightarrow \mathfrak{ue}\,\mathfrak{F}$. As first-order properties are preserved under taking ultrapowers, $\prod_U \mathfrak{F} \Vdash \phi$. But then $\mathfrak{ue}\,\mathfrak{F} \Vdash \phi$ by Theorem 3.14.   ⊣

We are on the verge of one of the best-known results in modal logic: the *Goldblatt-Thomason Theorem*. This result tells us that – at least as far as first-order definable frame classes are concerned – the four frame constructions we have discussed constitute necessary *and sufficient* conditions for a class of frames to be modally definable. We are not going to prove this important result right away, but we will take this opportunity to state it precisely. We use the following terminology: a class of frames K *reflects* ultrafilter extensions if $\mathfrak{ue}\,\mathfrak{F} \in$ K implies $\mathfrak{F} \in$ K.

Theorem 3.19 (Goldblatt-Thomason Theorem) *Let τ be a modal similarity type. A first-order definable class* K *of τ-frames is modally definable if and only if it is closed under taking bounded morphic images, generated subframes, disjoint unions and reflects ultrafilter extensions.*

Proof. A model-theoretic proof will be given in Section 3.8 below; this proof lies on the advanced track. An algebraic proof will be given in Chapter 5; this proof lies on the basic track. In addition, a simple special case which holds for finite transitive frames is proved in the following section.   ⊣

In fact, we can weaken the condition of first-order definability to closure under ultrapowers, cf. Exercise 3.8.4 or Theorem 5.54.

Exercises for Section 3.3

3.3.1   (a) Prove that frame validity is preserved under taking generated subframes and disjoint unions.
  (b) Which of the implications in Theorem 3.14 can be replaced with an equivalence?
  (c) Is frame validity preserved under taking ultraproducts?

3.3.2 Consider the basic modal language. Show that the following properties of frames are not modally definable:

  (a) antisymmetry ($\forall xy(Rxy \wedge Ryx \rightarrow x = y)$),

(b) $|W| > 23$,
(c) $|W| < 23$,
(d) acyclicity (there is no path from any x to itself),
(e) every state has at most one predecessor,
(f) every state has at least two successors.

3.3.3 Consider a language with three diamonds, $\Diamond_1$, $\Diamond_2$ and $\Diamond_3$. For each of the frame conditions on the corresponding accessibility relations below, find out whether it is modally definable or not:

(a) R_1 is the union of R_2 and R_3,
(b) R_1 is the intersection of R_2 and R_3,
(c) R_1 is the complement of R_2,
(d) R_1 is the composition of R_2 and R_3,
(e) R_1 is the identity relation,
(f) R_1 is the complement of the identity relation.

3.3.4 Show that any frame is a bounded morphic image of the disjoint union of its rooted generated subframes.

3.4 Finite Frames

In this section we prove two simple results about finite frames. First we state and prove a version of the Goldblatt-Thomason Theorem for finite transitive frames. Next we introduce the finite frame property, and show that a normal modal logic has the finite frame property if and only if it has the finite model property.

Finite transitive frames

An elegant analog of the Goldblatt-Thomason Theorem holds for finite transitive frames: within this class, closure under the three structural operations of (finite) disjoint unions, generated subframes, and bounded morphisms is a necessary *and sufficient* condition for a class of frames to be modally definable. The proof is straightforward and makes use of *Jankov-Fine formulas*.

Let $\mathfrak{F} = (W, R)$ be a point-generated finite transitive frame for the basic modal similarity type, and let w be a root of $\mathfrak{F}$. The Jankov-Fine formula $\phi_{\mathfrak{F},w}$ is essentially a description of $\mathfrak{F}$ that has the following property: it is satisfiable on a frame $\mathfrak{G}$ if and only if $\mathfrak{F}$ is a bounded morphic image of a generated subframe of $\mathfrak{G}$.

We build Jankov-Fine formulas as follows. Enumerate the states of $\mathfrak{F}$ as $w_0, \dots, w_n$, where $w = w_0$. Associate each state w_i with a distinct proposition letter p_i. Let $\phi_{\mathfrak{F},w}$ be the conjunction of the following formulas:

(i) p_0
(ii) $\Box(p_0 \lor \cdots \lor p_n)$
(iii) $(p_i \to \neg p_j) \land \Box(p_i \to \neg p_j)$, for each i, j with $i \neq j \leq n$
(iv) $(p_i \to \Diamond p_j) \land \Box(p_i \to \Diamond p_j)$, for each i, j with $Rw_i w_j$

(v) $(p_i \rightarrow \neg \Diamond p_j) \wedge \Box(p_i \rightarrow \neg \Diamond p_j)$, for each i, j with $\neg R w_i w_j$

Note that as R is transitive, each node in $\mathfrak{F}$ is accessible in one step from w. It follows that when formulas of the form $\psi \wedge \Box \psi$ are satisfied at w, ψ is true throughout $\mathfrak{F}$. With this observed, the content of Jankov-Fine formulas should be clear: the first three conjuncts state that each node in $\mathfrak{F}$ is uniquely labeled by some p (with p_0 labelling w_0) while the last two conjuncts use this labeling to describe the frame structure.

Lemma 3.20 *Let $\mathfrak{F}$ be a transitive, finite, point-generated frame, let w be a root of $\mathfrak{F}$, and let $\phi_{\mathfrak{F},w}$ be the Jankov-Fine formula for $\mathfrak{F}$ and w. Then for any transitive frame $\mathfrak{G}$ we have the following equivalence: there is a valuation V and a node v such that $(\mathfrak{G}, V), v \Vdash \phi_{\mathfrak{F},w}$ if and only if there exists a bounded morphism from $\mathfrak{G}_v$ onto $\mathfrak{F}$.*

Proof. Left to the reader as Exercise 3.4.1.   ⊣

With the help of this lemma, it is easy to prove the following Goldblatt-Thomason analog:

Theorem 3.21 *Recall that τ_0 denotes the basic modal similarity type. Let K be a class of finite, transitive τ_0-frames. Then K is definable within the class of finite, transitive τ_0-frames if and only if it is closed under taking (finite) disjoint unions, generated subframes, and bounded morphic images.*

Proof. The right to left direction is immediate: we know from the previous section that any modally definable frame class is closed under these operations. So let us consider the more interesting converse.

Assume that K is a class of finite, transitive τ_0-frames and satisfies the stated closure conditions. Let Λ_K be the logic of K; that is, $\Lambda_K = \{\phi \mid \mathfrak{F} \Vdash \phi$, for all $\mathfrak{F} \in K\}$. We will show that Λ_K defines K. Clearly Λ_K is valid on every frame in K, so to complete the proof we need to show that if $\mathfrak{F} \Vdash \Lambda_K$, where $\mathfrak{F}$ is finite and transitive, then $\mathfrak{F} \in K$. We split the proof into two cases.

First suppose that $\mathfrak{F}$ is point-generated with root w. Consider the Jankov-Fine formula $\phi_{\mathfrak{F},w}$ for $\mathfrak{F}$ and w. Clearly $\phi_{\mathfrak{F},w}$ is satisfiable in $\mathfrak{F}$ at w, so $\neg \phi_{\mathfrak{F},w} \notin \Lambda_K$. Hence there is some $\mathfrak{G} \in K$ such that $\mathfrak{G} \nVdash \neg \phi_{\mathfrak{F},w}$; in other words, for some valuation V and state v we have $(\mathfrak{G}, V), v \Vdash \phi_{\mathfrak{F},w}$. Thus by the previous lemma, $\mathfrak{F}$ is a bounded morphic image of the point-generated subframe $\mathfrak{G}_v$ of $\mathfrak{G}$. By the closure conditions on K, it follows that $\mathfrak{F} \in K$.

So suppose that $\mathfrak{F}$ is *not* point-generated. But then as $\mathfrak{F} \Vdash \Lambda_K$, so does each point-generated subframe of $\mathfrak{F}$, hence by the work of the previous paragraph all these subframes belong to K. But by Exercise 3.3.4, $\mathfrak{F}$ is a bounded morphic image of the disjoint union of its rooted generated subframes, so $\mathfrak{F}$ belongs to K too.   ⊣

The finite frame property

Our next result deals not with frame definability, but with the relationship between normal modal logics and finite frames. Normal modal logics were introduced in Section 1.6 (see in particular Definition 1.42). Recall that normal modal logics are sets of formulas (containing certain axioms) that are closed under three simple conditions (modus ponens, uniform substitution, and generalization). They are the standard tool for capturing the notion of validity *syntactically*.

Now, in Section 2.3 we introduced the finite *model* property. We did not apply the concept to normal modal logics – but as a normal logic is simply a set of formulas, we can easily extend the definition to permit this:

Definition 3.22 A normal modal logic Λ has the finite model property with respect to some class of models M if M $\Vdash \Lambda$ and every formula *not* in Λ is refuted in a *finite* model $\mathfrak{M}$ in M. Λ has the finite model property if it has the finite model property with respect to some class of models. ⊣

Informally, if a normal modal logic has the finite model property, it has a finite *semantic* characterization: *it is precisely the set of formulas that some collection of finite models makes globally true.* This is an attractive property, and as we will see in Chapter 6 when we discuss the decidability of normal logics, a useful one too.

But something seems wrong. It is the level of *frames*, rather than the level of models, which supports the key logical concept of validity. It certainly seems sensible to try and semantically characterize normal logics in terms of finite structures – but it seems we should do so using finite *frames*, not finite models. That is, the following property seems more appropriate:

Definition 3.23 (Finite Frame Property) Let Λ be a normal modal logic and F a class of finite frames. We say Λ has the *finite frame property with respect to* F if and only if F $\Vdash \Lambda$, and for every formula ϕ such that $\phi \notin \Lambda$ there is some $\mathfrak{F} \in$ F such that ϕ is falsifiable on $\mathfrak{F}$. We say Λ has the *finite frame property* if and only if it has the finite frame property with respect to some class of finite frames. ⊣

Note that to establish the finite frame property of a normal modal logic Λ, it is not sufficient to prove that any formula $\phi \notin \Lambda$ can be refuted on a model where Λ is globally true: in addition one has to ensure that the underlying *frame* of the model validates Λ. If a logic has the finite frame property (and many important ones do, as we will learn in Chapter 6) then clearly there is no room for argument: it really can be characterized semantically in terms of finite structures.

But now for a surprising result. The finite frame property is *not* stronger than the finite model property: we will show that a normal modal logic has the finite frame property if and only if it has the finite model property. This result will prove useful at a number of places in Chapters 4 and 6. Moreover, while proving it we will meet some other concepts, notably *definable variants* and *distinguishing models*, which will be useful when proving Bull's Theorem in Section 4.9.

Definition 3.24 (Definable Variant) Let $\mathfrak{M} = (W, R, V)$ be a model and $U \subseteq W$. We say U is *definable in* $\mathfrak{M}$ if and only if there is a formula ϕ_U such that for all states $w \in W$, $\mathfrak{M}, w \Vdash \phi_U$ iff $w \in U$.

Any model $\mathfrak{M}'$ based on the frame (W, R) is called a *variant* of $\mathfrak{M}$. A variant (W, R, V') of $\mathfrak{M}$ is *definable in* $\mathfrak{M}$ if and only if for all proposition symbols p, $V'(p)$ is definable in $\mathfrak{M}$. If $\mathfrak{M}'$ is a variant of $\mathfrak{M}$ that is definable in $\mathfrak{M}$, we call $\mathfrak{M}'$ a *definable variant* of $\mathfrak{M}$.   ⊣

Recall that normal modal logics are closed under uniform substitution, the process of uniformly replacing propositional symbols with arbitrary formulas (see Section 1.6), and that a formula obtained from ϕ by uniform substitution is called a *substitution instance* of ϕ. Our intuitive understanding of uniform substitution suffices for most purposes, but in order to prove the following lemma we need to refer to the precise concepts of Definition 1.18.

Lemma 3.25 *Let $\mathfrak{M} = (\mathfrak{F}, V)$ be a model and $\mathfrak{M}' = (\mathfrak{F}, V')$ be a definable variant of $\mathfrak{M}$. For any formula ϕ, let ϕ' be the result of uniformly replacing each atomic symbol p in ϕ by $\phi_{V'(p)}$, where $\phi_{V'(p)}$ defines $V'(p)$ in $\mathfrak{M}$. Then for all formulas ϕ, and all normal modal logics Λ:*

  (i) *$\mathfrak{M}', w \Vdash \phi$ iff $\mathfrak{M}, w \Vdash \phi'$.*
 (ii) *If every substitution instance of ϕ is true in $\mathfrak{M}$, then every substitution instance of ϕ is true in $\mathfrak{M}'$.*
(iii) *If $\mathfrak{M} \Vdash \Lambda$ then $\mathfrak{M}' \Vdash \Lambda$.*

Proof. Item (i) follows by induction on ϕ. For the base case we have $\mathfrak{M}', w \Vdash p$ iff $\mathfrak{M}, w \Vdash \phi_{V'(p)}$. As $(\neg\phi)' = \neg\phi'$, $(\phi \vee \psi)' = \phi' \vee \psi'$, and $(\Diamond\phi)' = \Diamond\phi'$ (cf. Definition 1.18) the inductive steps are immediate.

For item (ii), we show the contrapositive. Let ψ be a substitution instance of ϕ and suppose that $\mathfrak{M}' \not\Vdash \psi$. Thus there is some w in $\mathfrak{M}$ such that $\mathfrak{M}', w \not\Vdash \psi$. By item (i), $\mathfrak{M}, w \not\Vdash \psi'$, which means that $\mathfrak{M} \not\Vdash \psi'$. But as ψ' is a substitution instance of ψ, and ψ is a substitution instance of ϕ, we have that ψ' is a substitution instance of ϕ (see Exercise 1.2.5) and the result follows.

Item (iii) is an immediate consequence of item (ii), for normal modal logics are closed under uniform substitution.   ⊣

We now isolate a type of model capable of defining *all* its variants:

Definition 3.26 (Distinguishing Model) A model $\mathfrak{M}$ is *distinguishing* if the relation $\leftrightsquigarrow$ of modal equivalence between states of $\mathfrak{M}$ is the identity relation.   ⊣

In other words, a model $\mathfrak{M}$ is *distinguishing* if and only if for all states w and u in $\mathfrak{M}$, if $w \neq u$, then there is a formula ϕ such that $\mathfrak{M}, w \Vdash \phi$ and $\mathfrak{M}, u \nVdash \phi$. Many important models are distinguishing. For example, all filtrations (see Definition 2.36) are distinguishing. Moreover, the canonical models introduced in Section 4.2 are distinguishing too. And, and as we will now see, when a distinguishing model is *finite*, it can define all its variants.

Lemma 3.27 *Let* $\mathfrak{M} = (\mathfrak{F}, V)$ *be a finite distinguishing model. Then:*

  (i) *For every state w in $\mathfrak{M}$ there is a formula ϕ_w that is true at, and only at, w.*
  (ii) $\mathfrak{M}$ *can define any subset of* $\mathfrak{F}$. *Hence* $\mathfrak{M}$ *can define all its variants.*
  (iii) *If* $\mathfrak{M} \Vdash \phi$ *then* $\mathfrak{F} \Vdash \phi$.

Proof. For item (i), suppose that $\mathfrak{F} = (W, R)$, and enumerate the states in W as $w_1, \ldots, w_n$. For all pairs (i, j) such that $1 \leq i, j \leq n$ and $i \neq j$, choose $\phi_{i,j}$ to be a formula such that $\mathfrak{M}, w_i \Vdash \phi_{i,j}$ and $\mathfrak{M}, w_j \nVdash \phi_{i,j}$ (such a formula exists, for $\mathfrak{M}$ is distinguishing) and define ϕ_{w_i} to be $\phi_{i,1} \wedge \cdots \wedge \phi_{i,n}$. Clearly ϕ_{w_i} is true at w_i and false everywhere else.

Item (ii) is an easy consequence. For let U be any subset of W. Then $\bigvee_{w \in U} \phi_w$ defines U. Hence as $\mathfrak{M}$ can define all subsets of W, it can define $V'(p)$, for any valuation V' on $\mathfrak{F}$ and propositional symbol p.

As for item (iii), suppose $\mathfrak{M} \Vdash \phi$. By item (iii) of the previous lemma we have that $\mathfrak{M}' \Vdash \phi$, where $\mathfrak{M}'$ is any definable variant of $\mathfrak{M}$. But we have just seen that $\mathfrak{M}$ can define all its variants, hence $\mathfrak{F} \Vdash \phi$. $\dashv$

Lemmas 3.25 and 3.27 will be important in their own right when we prove Bull's Theorem in Section 4.9. And with the help of a neat filtration argument, they yield the main result:

Theorem 3.28 *A normal modal logic has the finite frame property iff it has the finite model property.*

Proof. The left to right direction is immediate. For the converse, suppose that Λ is a normal modal logic with the finite model property. Since we will need to take a filtration through Λ, we have to be explicit about the set of proposition letters of the formulas in Λ, so assume that $\Lambda \subseteq Form(\tau, \Phi)$.

Take a formula in the language $ML(\tau, \Phi)$ that does not belong to Λ. We will show that ϕ can be refuted on a finite frame $\mathfrak{F}$ such that $\mathfrak{F} \Vdash \Lambda$.

As Λ has the finite model property, there is a finite model $\mathfrak{M}$ such that $\mathfrak{M} \Vdash \Lambda$ and $\mathfrak{M}, w \nVdash \phi$ for some state w in $\mathfrak{M}$. Let Σ be the set of all subformulas of formulas in $\{\phi\} \cup \Lambda$, and let $\mathfrak{M}^f$ be any filtration of $\mathfrak{M}$ through Σ. As $\mathfrak{M}$ is finite, so is $\mathfrak{M}^f$. As $\mathfrak{M}^f$ is a filtration, it is a distinguishing model. By the Filtration Theorem (Theorem 2.39), $\mathfrak{M}^f, |w| \nVdash \phi$. Moreover $\mathfrak{M}^f \Vdash \Lambda$, for as every state in $\mathfrak{M}$ satisfies all formulas in Λ, so does every state in $\mathfrak{M}^f$ (again, this follows from the Filtration Theorem). Let $\mathfrak{F}$ be the (finite) frame underlying $\mathfrak{M}^f$. By Lemma 3.27 item (iii), $\mathfrak{F} \Vdash \Lambda$, and we have proved the theorem.   ⊣

Note the somewhat unusual use of filtrations in this proof. Normally we filtrate infinite models through finite sets of formulas. Here we filtrated a *finite* model through an *infinite* sets of formulas to guarantee that an entire logic remained true.

This result shows that the concepts of normal modal logics and frame validity fit together well in the finite domain: if a normal logic has a finite semantic characterization in terms of models, then it is guaranteed to have a finite *frame-based* semantic characterization as well. But be warned: one of the most striking results of the following chapter is that logics and frame validity do not always fit together so neatly. In fact, the *frame incompleteness results* will eventually lead us (in Chapter 5) to the use of new semantic structures, namely modal algebras, to analyze normal modal logics. But this is jumping ahead. It is time to revert to our discussion of frame definability – but from a rather different perspective. So far, our approach has been firmly *semantical*. This has taught us a lot: in particular, the Goldblatt-Thomason Theorem has given us a model-theoretic characterization of the elementary frame classes that are modally definable. Moreover, we will see in Chapter 5 that the semantic approach has an important algebraic dimension. But it is also possible to approach frame definability from a more *syntactic* perspective, and that is what we are going to do now. This will lead us to the other main result of the chapter: the Sahlqvist Correspondence Theorem.

Exercises for Section 3.4

3.4.1 Prove Lemma 3.20. That is, suppose that $\phi_{\mathfrak{F},w}$ is the Jankov-Fine formula for a transitive finite frame $\mathfrak{F}$ with root w. Show that for any frame $\mathfrak{G}$, $\phi_{\mathfrak{F},w}$ is satisfiable on $\mathfrak{G}$ at a node v iff $\mathfrak{F}$ is a bounded morphic image of $\mathfrak{G}_v$.

3.4.2 Let $\mathfrak{M}$ be a model, let $\mathfrak{M}^f$ be any filtration of $\mathfrak{M}$ through some finite set of formulas Σ, and let f be the natural map associated with the filtration. If u is a point in the filtration, show that $f^{-1}[u]$ is definable in $\mathfrak{M}$.

3.5 Automatic First-Order Correspondence

We have learned a lot about frame definability in the previous sections. In particular, we have learned that frame definability is a second-order notion, and that the second-order correspondent of any modal formula can be straightforwardly computed using the second-order translation. Moreover, we know that many modal formulas have first-order correspondents, and that the Goldblatt-Thomason Theorem gives us a model-theoretic characterization of the frame classes they define.

Nonetheless, there remains a gap in our understanding: although many modal formulas define first-order conditions on frames, it is not really clear *why* they do so. To put it another way, in many cases the (often difficult to decipher) second-order condition yielded by the second-order translation is equivalent to a much simpler first-order condition. Is there any system to this? Better, are there algorithms that enable us to compute first-order correspondents automatically, and if so, how general are these algorithms? This section, and the two that follow, develop some answers.

A large part of this work centers on a beautiful positive result: there is a large class of formulas, the *Sahlqvist formulas*, each of which defines a first-order condition on frames which is effectively calculable using the *Sahlqvist-van Benthem algorithm*; this is the celebrated *Sahlqvist Correspondence Theorem*, which we will state and prove in the following section. The proof of this theorem sheds light on why so many second-order correspondents turn out to be equivalent to a first-order condition. Moreover each Sahlqvist formula is *complete* with respect to the class of first-order frames it defines; this is the *Sahlqvist Completeness Theorem*, which we will formulate more precisely in Theorem 4.42 and prove in Section 5.6. All in all, the Sahlqvist fragment is interesting from both theoretical and practical perspectives, and we devote a lot of attention to it.

In this section we lay the groundwork for the proof of the Sahlqvist Correspondence Theorem. We are going to introduce two simple classes of modal formulas, the *closed* formulas and the *uniform* formulas, and show that they define first-order conditions on frames. Along the way we are going to learn about positive and negative formulas, what they have to do with monotonicity, and how they can help us get rid of second-order quantifiers. These ideas will be put to work, in a more sophisticated way, in the following section.

One other thing: in what follows we are going to work with a stronger notion of correspondence. The concept of correspondence given in Definition 3.5 is *global*: a modal and a (first- or second-order) frame formula are called correspondents if they are valid on precisely the same frames. But it is natural to demand that validity matches *locally*:

Definition 3.29 (Local Frame Correspondence) Let ϕ be a modal formula in some similarity type, and $\alpha(x)$ a formula in the corresponding first- or second-order frame language (x is supposed to be the only free variable of α). Then we say that ϕ and $\alpha(x)$ are *local frame correspondents* of each other if the following holds, for any frame $\mathfrak{F}$ and any state w of $\mathfrak{F}$:

$$\mathfrak{F}, w \Vdash \phi \text{ iff } \mathfrak{F} \models \alpha[w]. \qquad \dashv$$

In fact, we have been implicitly using local correspondence all along. In Example 3.6 we showed that $p \rightarrow \Diamond p$ corresponds to $\forall x Rxx$ – but inspection of the proof reveals we did so by showing that $p \rightarrow \Diamond p$ *locally* corresponds to Rxx.

Similarly, in Example 3.7 we showed that $\Diamond p \rightarrow \Diamond\Diamond p$ corresponds to density by showing that $\Diamond p \rightarrow \Diamond\Diamond p$ *locally* corresponds to $\forall y\,(Rxy \rightarrow \exists z\,(Rxz \wedge Rzy))$. It should be clear from these examples that the local notion of correspondence is fundamental, and that the following connection holds between the local and global notions:

Proposition 3.30 *If $\alpha(x)$ is a local correspondent of the modal formula ϕ, then $\forall x\,\alpha(x)$ is a global correspondent of ϕ. So if ϕ has a first-order local correspondent, then it also has a first-order global correspondent.*

Proof. Trivial.   ⊣

What about the converse? In particular, suppose that the modal formula ϕ has a *first order* global correspondent; will it also have a first-order local correspondent? Intriguingly, the answer to this question is negative, as we will see in Example 3.57.

But until we come to this result, we will not mention global correspondence much: it is simpler to state and prove results in terms of local correspondence, relying on the previous lemma to guarantee correspondence in the global sense. With this point settled, it is time to start thinking about correspondence theory systematically.

Closed formulas

There is one obvious class of modal formulas guaranteed to correspond to first-order frame conditions: formulas which contain no proposition letters.

Example 3.31 Consider the basic temporal language. The formula $P\top$ defines the property that there is no first point of time. More precisely, $P\top$ is valid on precisely those frames such that every point has a predecessor.

Now, obviously it is easy to prove this directly, but for present purposes the following argument is more interesting. By Proposition 3.12, for any bidirectional frame $\mathfrak{F}$ and any point w in $\mathfrak{F}$ we have that:

$$\mathfrak{F}, w \Vdash P\top \quad \text{iff} \quad \mathfrak{F} \models \forall P_1 \ldots \forall P_n\, ST_x(P\top)[w],$$

where $P_1, \ldots, P_n$ are the unary predicate variables corresponding to the proposition letters $p_1, \ldots, p_n$ occurring in $P\top$. But $P\top$ contains *no* propositional variables, hence there are no second-order quantifiers, and hence:

$$\mathfrak{F}, w \Vdash P\top \quad \text{iff} \quad \mathfrak{F} \models ST_x(P\top)[w].$$

But $ST_x(P\top)$ is $\exists y\,(Ryx \wedge y = y)$, which is equivalent to $\exists y\, Ryx$. So $P\top$ *locally* corresponds to $\exists y\, Ryx$ (and thus *globally* corresponds to $\forall x \exists y\, Ryx$).   ⊣

The argument used in this example is extremely simple, and obviously generalizes. We will state and prove the required generalization, and then move on to richer pastures.

Definition 3.32 A modal formula ϕ is *closed* if and only if it contains no proposition letters. Thus closed formulas are built up from $\top$, $\bot$, and any nullary modalities (or modal constants) the signature may contain. ⊣

Proposition 3.33 *Let ϕ be a closed formula. Then ϕ locally corresponds to a first-order formula $c_\phi(x)$ which is effectively computable from ϕ.*

Proof. By Proposition 3.12 and the fact that ϕ contains no propositional variables we have:

$$\mathfrak{F}, w \Vdash \phi \text{ iff } \mathfrak{F} \models ST_x(\phi)[w].$$

As it is easy to write a program that computes $ST_x(\phi)$, the claim follows immediately. ⊣

Closed formulas arise naturally in some applications (a noteworthy example is provability logic), thus the preceding result is quite useful in practice.

Uniform formulas

Although the previous proposition was extremely simple, it does point the way to the strategy followed in our approach to the Sahlqvist Correspondence Theorem: we are going to look for ways of stripping off the initial block of monadic second-order universal quantifiers in $\forall P_1 \ldots \forall P_n ST_x(\phi)$, thus reducing the translation to $ST_x(\phi)$. The obvious way of getting rid of universal quantifiers is to perform universal instantiation, and this is exactly what we will do. Both here, and in the work of the next section, we will look for simple instantiations for the $P_1, \ldots, P_n$, which result in first-order formulas equivalent to the original. We will be able to make this strategy work because of the syntactic restrictions placed on ϕ.

One of the restrictions imposed on Sahlqvist formulas invokes the idea of *positive* and *negative* occurrences of proposition letters. We now introduce this idea, study its semantic significance, and then, as an introduction to the techniques of the following section, use a simple instantiation argument to show that the second-order translations of *uniform* formulas are effectively reducible to first-order conditions on frames.

Definition 3.34 An occurrence of a proposition letter p is a *positive* occurrence if it is in the scope of an even number of negation signs; it is a *negative* occurrence if it is in the scope of an odd number of negation signs. (This is one of the few places in the book where it is important to think in terms of the primitive connectives.

For example, the occurrence of p in $\Diamond(p \rightarrow q)$ is *negative*, for this formula is shorthand for $\Diamond(\neg p \vee q)$.) A modal formula ϕ is *positive in p* (*negative in p*) if all occurrences of p in ϕ are positive (negative). A formula is called *positive* (*negative*) if it is positive (negative) in all proposition letters occurring in it.

Analogous concepts are defined for the corresponding second-order language. That is, an occurrence of a unary predicate variable P in a second-order formula is *positive* (*negative*) if it is in the scope of an even (odd) number of negation signs. A second-order formula ϕ is *positive in P* (*negative in P*) if all occurrences of P in ϕ are positive (negative), and it is called *positive* (*negative*) if it is positive (negative) in all unary predicate variables occurring in it. ⊣

Lemma 3.35 *Let ϕ be a modal formula.*

(i) *ϕ is positive in p iff $ST_x(\phi)$ is positive in the corresponding unary predicate P.*

(ii) *If ϕ is positive (negative) in p, then $\neg\phi$ is negative (positive) in p.*

Proof. Virtually immediate. ⊣

Positive and negative formulas are important because of their special semantic properties. In particular, they exhibit a useful form of *monotonicity*.

Definition 3.36 Fix a modal language $ML(\tau, \Phi)$, and let $p \in \Phi$. A modal formula ϕ is *upward monotone in p* if its truth is preserved under extensions of the interpretation of p. More precisely, ϕ is upward monotone in p if for every model $(W, R_\Delta, V)_{\Delta \in \tau}$, every state $w \in W$, and every valuation V' such that $V(p) \subseteq V'(p)$ and for all $q \neq p$, $V(q) = V'(q)$, the following holds:

$$\text{if } (W, R_\Delta, V)_{\Delta \in \tau}, w \Vdash \phi, \text{ then } (W, R_\Delta, V')_{\Delta \in \tau}, w \Vdash \phi.$$

In short, extending $V(p)$ (while leaving the interpretation of other proposition letters unchanged) has the effect of extending $V(\phi)$ (or keeping it the same).

Likewise, a formula ϕ is *downward monotone in p* if its truth is preserved under shrinkings of the interpretation of p. That is, for every model $(W, R_\Delta, V)_{\Delta \in \tau}$, every state $w \in W$, and every valuation V' such that $V'(p) \subseteq V(p)$ and for all $q \neq p$, $V(q) = V'(q)$, the following holds:

$$\text{if } (W, R_\Delta, V)_{\Delta \in \tau}, w \Vdash \phi, \text{ then } (W, R_\Delta, V')_{\Delta \in \tau}, w \Vdash \phi.$$

The notions of a second-order formula being *upward* and *downward monotone in* a unary predicate variable P are defined analogously; we leave this task to the reader. ⊣

Lemma 3.37 *Let ϕ be a modal formula.*

(i) *If ϕ is positive in p, then it is upward monotone in p.*

(ii) *If ϕ is negative in p, then it is downward monotone in p.*

Proof. Prove both parts simultaneously by induction on ϕ; see Exercise 3.5.3. ⊣

But what do upward and downward monotonicity have to do with frame definability? The following example is instructive:

Example 3.38 The formula $\Diamond \Box p$ locally corresponds to a first-order formula. For suppose $\mathfrak{F}, w \Vdash \Diamond \Box p$. Regardless of the valuation, the formula $\Diamond \Box p$ holds at w. So consider a *minimal* valuation (for p) on $\mathfrak{F}$; that is, choose any V_m such that $V_m(p) = \varnothing$. Then as $w \Vdash \Diamond \Box p$, there must be a successor v of w such that $\Box p$ holds at v. However, there are no p-states, so v must be blind (that is, without successors). In other words, we have shown that

$$(\mathfrak{F}, V_m), w \Vdash \Diamond \Box p \text{ only if } \mathfrak{F} \models \exists y \, (Rxy \wedge \neg \exists z \, Ryz)[w].$$

Now for the interesting direction: assume that the state w in the frame $\mathfrak{F}$ has a blind successor. It follows immediately that $(\mathfrak{F}, V_m), w \Vdash \Diamond \Box p$, where V_m is any minimal valuation (for p). We claim that the formula $\Diamond \Box p$ is valid at w. To see this, consider an arbitrary valuation V and a point w of $\mathfrak{F}$. By item (i) of Lemma 3.37, $\Diamond \Box p$ is upward monotone in p. Hence it follows from the fact that $V_m(p) \subseteq V(p)$ that $(\mathfrak{F}, V), w \Vdash \Diamond \Box p$. As V was arbitrary, $\Diamond \Box p$ is valid on $\mathfrak{F}$ at w. ⊣

The key point is the last part of the argument: the use of a minimal valuation followed by an appeal to monotonicity to establish a result about *all* valuations. But now think about this argument from the perspective of the second-order correspondence language: in effect, we *instantiated* the predicate variable corresponding to p with the smallest subset of the frame possible, and then used a monotonicity argument to establish a result about *all* assignments to P.

This simple idea lies behind much of our work on the Sahlqvist fragment. To illustrate the style of argumentation it leads to, we will now use an instantiation argument to show that all *uniform* modal formulas define first-order conditions on frames.

Definition 3.39 A proposition letter p occurs *uniformly* in a modal formula if it occurs only positively, or only negatively. A predicate variable P occurs uniformly in a second-order formula if it occurs only positively, or only negatively. A modal formula is *uniform* if all the proposition letters it contains occur uniformly. A second-order formula is uniform if all the unary predicate variables it contains occur uniformly. ⊣

Theorem 3.40 *If ϕ is a uniform modal formula, then ϕ locally corresponds to a first-order formula $c_\phi(x)$ on frames. Moreover, c_ϕ is effectively computable from ϕ.*

Proof. Consider the universally quantified second-order equivalent of ϕ:

$$\forall P_1 \ldots \forall P_n \, ST_x(\phi), \tag{3.6}$$

where $P_1, \ldots, P_n$ are second-order variables corresponding to the proposition letters in ϕ. Our aim is to show that (3.6) is equivalent to a first-order formula by performing appropriate instantiations for the universally quantified monadic second-order variables $P_1, \ldots, P_n$.

As ϕ is uniform, by Lemma 3.35 so is $ST_x(\phi)$. We will instantiate the unary predicates that occur positively with a predicate denoting as small a set as possible (that is, the empty set), and the unary predicates that occur negatively with a predicate denoting as large a set as possible (that is, all the states in the frame). We will use Church's λ-notation for the required substitution instance providing the formulas that define these predicates. For every P occurring in $ST_x(\phi)$, define

$$\sigma(P) \equiv \begin{cases} \lambda u.\, u \neq u, & \text{if } ST_x(\phi) \text{ is positive in } P, \\ \lambda u.\, u = u, & \text{if } ST_x(\phi) \text{ is negative in } P. \end{cases}$$

Of course, the idea is that instantiating a universal second-order formula according to this substitution σ simply means (i) removing the second-order quantifiers and (ii) replacing every atomic subformula Py with the formula $\sigma(P)(y)$, that is, with either $y \neq y$ or $y = y$ (as given by the definition).[1]

Now consider the following instance of (3.6) in which every unary predicate P has been replaced by $\sigma(P)$:

$$[\sigma(P_1)/P_1, \ldots, \sigma(P_n)/P_n] \, ST_x(\phi). \tag{3.7}$$

We will show that (3.7) is equivalent to (3.6). It is immediate that (3.6) implies (3.7), for the latter is an instantiation of the former. For the converse implication we assume that

$$\mathfrak{M} \models [\sigma(P_1)/P_1, \ldots, \sigma(P_n)/P_n] \, ST_x(\phi)[w], \tag{3.8}$$

and we have to show that

$$\mathfrak{M} \models \forall P_1 \ldots \forall P_n \, ST_x(\phi)[w].$$

By the choice of $\sigma(P)$, for predicates P that occur only positively in $ST_x(\phi)$ we have that $\mathfrak{M} \models \forall y\, (\sigma(P)(y) \to P(y))$, and for predicates P that occur only negatively in $ST_x(\phi)$, we have that $\mathfrak{M} \models \forall y\, (P(y) \to \sigma(P)(y))$. (Readers familiar with λ-notation will realize that we have implicitly appealed to β-conversion here. Readers unfamiliar with λ-notation should simply note that when $\sigma(P)$ is a predicate denoting the empty set, then $\sigma(P)(y)$ is false no matter what y denotes, while

---

[1] If you are unfamiliar with λ-notation, all you really need to know to follow the proof is that $\lambda u.\, u \neq u$ and $\lambda u.\, u = u$ are predicates denoting the empty set and the set of all states respectively. Some explanatory remarks on λ-notation are given following the proof.

if $\sigma(P)$ denotes the set of all states, $\sigma(P)(y)$ is guaranteed to be true.) Hence, as $ST_x(\phi)$ is positive or negative in all unary predicates P occurring in it, (3.8) together with Lemma 3.37 imply that for *any* choice of $P_1, \ldots, P_n$,

$$(\mathfrak{M}, P_1, \ldots, P_n) \models ST_x(\phi)[w],$$

which means that $\mathfrak{M} \models \forall P_1 \ldots \forall P_n \, ST_x(\phi)$ as required. Finally, in any programming language with decent symbol manipulation facilities it is straightforward to write a program which, when given a uniform formula ϕ, produces $ST_x(\phi)$ and carries out the required instantiations. Hence the first-order correspondents of uniform formulas are computable. ⊣

On λ-notation

Although it is not essential to use λ-notation, it *is* convenient and we will apply it in the following section. For readers unfamiliar with it, here is a quick introduction to the fundamental ideas.

We have used Church's λ-notation as a way of writing predicates, that is, entities which denote subsets. But lambda expressions do not denote subsets directly; rather they denote their *characteristic functions*. Suppose we are working with a frame (W, R). Let $S \subseteq W$. Then the characteristic function of S (with respect to W) is the function χ_S with domain W and range $\{0, 1\}$ such that $\chi_S(s) = 1$ if $s \in S$ and $\chi_S(s) = 0$ otherwise. Reading 1 as true and 0 as false, χ_S is simply the function that says truthfully of each element of W whether it belongs to S or not.

Lambda expressions pick out characteristic functions in the obvious way. For example, when working with a frame (W, R), $\lambda u. u \neq u$ denotes the function from W to $\{0, 1\}$ that assigns 1 to every element $w \in W$ that satisfies $u \neq u$ and 0 to everything else. But for *no* choice of w is it the case that $w \neq w$; hence, as we stated in the previous proof, $\lambda u. u \neq u$ denotes the characteristic function of the empty set. Similarly, $\lambda u. u = u$ denotes the characteristic function of W, for $w = w$ for every $w \in W$.

Lambda expressions take the drudgery out of dealing with substitutions. Consider the second-order formula Px. This is satisfied in a model if and only if the element assigned to x belongs to the subset assigned to P. For example, if P is assigned the empty set, Px will be false no matter what x is assigned. Now suppose we substitute $(\lambda u. u \neq u)$ for P in Px. This yields the expression $(\lambda u. u \neq u)x$. Read this as 'apply the function denoted by $\lambda u. u \neq u$ to the state denoted by x.' Clearly this yields the value 0 (that is, *false*). The process of β-conversion mentioned in the proof is essentially a way of rewriting such functional applications to simpler but equivalent forms; for more details, consult one of the introductions cited in the Notes. Newcomers to λ-notation should try Exercise 3.5.1 right away.

Exercises for Section 3.5

3.5.1 Explain why we could have used the following predicate definitions in the proof of Example 3.38: for every P occurring in $ST_x(\phi)$, define

$$\sigma(P) \equiv \begin{cases} \lambda u.\perp, & \text{if } ST_x(\phi) \text{ is positive in } P, \\ \lambda u.\top, & \text{if } ST_x(\phi) \text{ is negative in } P. \end{cases}$$

If you have difficulties with this, consult one of the introductions to λ-calculus cited in the notes before proceeding further.

3.5.2 Let ϕ be a modal formula which is positive in all proposition letters. Prove that ϕ can be rewritten into a normal form which is built up from proposition letters, using $\wedge$, $\vee$, $\diamond$ and $\square$ only.

3.5.3 Prove Lemma 3.37. That is, show that if a modal formula ϕ is positive in p, then it is upward monotone in p, and that if it is negative in p, then it is downward monotone in p.

3.6 Sahlqvist Formulas

In the proof of Theorem 3.40 we showed that uniform formulas correspond to first-order conditions by finding a suitable *instantiation* for the universally quantified monadic second-order variables in their second-order translation and appealing to *monotonicity*. This is an important idea, and the rest of this section is devoted to extending it: the Sahlqvist fragment is essentially a large class of formulas to which this style of argument can be applied.

Very simple Sahlqvist formulas

Roughly speaking, Sahlqvist formulas are built up from implications $\phi \rightarrow \psi$, where ψ is positive and ϕ is of a restricted form (to be specified below) from which the required instantiations can be read off. We now define a limited version of the Sahlqvist fragment for the basic modal language; generalizations and extensions will be discussed shortly.

Definition 3.41 We will work in the basic modal language. A *very simple Sahlqvist antecedent* over this language is a formula built up from $\top$, $\perp$ and proposition letters, using only $\wedge$ and $\diamond$. A *very simple Sahlqvist formula* is an implication $\phi \rightarrow \psi$ in which ψ is positive and ϕ is a very simple Sahlqvist antecedent. $\dashv$

Examples of very simple Sahlqvist formulas include $p \rightarrow \diamond p$ and $(p \wedge \diamond\diamond q) \rightarrow \square\diamond(p \wedge q)$.

The following theorem is central for understanding what Sahlqvist correspondence is all about. Its proof describes and justifies an algorithm for converting very simple Sahlqvist formulas into first-order formulas; the algorithms given later for richer Sahlqvist fragments elaborate on ideas introduced here. Examples of the algorithm in action are given below; it is a good idea to refer to these while studying the proof.

Theorem 3.42 *Let $\chi = \phi \rightarrow \psi$ be a very simple Sahlqvist formula in the basic modal language $ML(\tau_0, \Phi)$. Then χ locally corresponds to a first-order formula $c_\chi(x)$ on frames. Moreover, c_χ is effectively computable from χ.*

Proof. Our starting point is the formula $\forall P_1 \ldots \forall P_n \left(ST_x(\phi) \rightarrow ST_x(\psi) \right)$, which is the local second-order translation of χ. We assume that this translation has undergone a pre-processing step to ensure that no two quantifiers bind the same variable, and no quantifier binds x. Let us denote $ST_x(\psi)$ by POS; that is, we have a translation of the form:

$$\forall P_1 \ldots \forall P_n \left(ST_x(\phi) \rightarrow \text{POS} \right). \tag{3.9}$$

We will now rewrite (3.9) to a form from which we can read off the instantiations that will yield its first-order equivalent.

Step 1. Pull out diamonds.
Use equivalences of the form

$$\left(\exists x_i\, \alpha(x_i) \land \beta \right) \leftrightarrow \exists x_i \left(\alpha(x_i) \land \beta \right),$$

and

$$\left(\exists x_i\, \alpha(x_i) \rightarrow \beta \right) \leftrightarrow \forall x_i \left(\alpha(x_i) \rightarrow \beta \right),$$

(in that order) to move all existential quantifiers in the antecedent $ST_x(\phi)$ of (3.9) to the front of the implication. Note that by our definition of Sahlqvist antecedents, the existential quantifiers only have to cross conjunctions before they reach the main implication. Of course, the above equivalences are not valid if the variable x_i occurs freely in β, but by our assumption on the pre-processing of the formula, this problem does not arise.

Step 1 results in a formula of the form

$$\forall P_1 \ldots \forall P_n \forall x_1 \ldots \forall x_m \left(\text{REL} \land \text{AT} \rightarrow \text{POS} \right), \tag{3.10}$$

where REL is a conjunction of atomic first-order statements of the form Rx_ix_j corresponding to occurrences of diamonds, and AT is a conjunction of (translations of) proposition letters. It may be helpful at this point to look at the concrete examples given below.

Step 2. Read off instances.
We can assume that every unary predicate P that occurs in the consequent of the matrix of (3.10), also occurs in the antecedent of the matrix of (3.10): otherwise (3.10) is positive in P and we can substitute $\lambda u.\, u \neq u$ for P (that is, make use of

the substitution used in the proof of Theorem 3.40) to obtain an equivalent formula without occurrences of P.

Let P_i be a unary predicate occurring in (3.10), and let $P_i x_{i_1}, \ldots, P_i x_{i_k}$ be all the occurrences of the predicate P_i in the antecedent of (3.10). Define

$$\sigma(P_i) \equiv \lambda u. (u = x_{i_1} \lor \cdots \lor u = x_{i_k}).$$

Note that $\sigma(P_i)$ is the *minimal* instance making the antecedent REL $\land$ AT true; this lambda expression says that if a node u has property P_i, then u must be one of the nodes $x_{i_1}, x_{i_2}, \ldots, x_{i_k}$ explicitly stated to have property P_i in the antecedent. But this is nothing else than saying that if some model $\mathfrak{M}$ makes the formula AT true under some assignment, then the interpretation of the predicate P must *extend* the set of points where $\sigma(P)$ holds:

$$\mathfrak{M} \models \text{AT}[ww_1 \ldots w_m] \text{ implies } \mathfrak{M} \models \forall y \, (\sigma(P_i)(y) \to P_i y)[ww_1 \ldots w_m] \quad (3.11)$$

This observation, in combination with the positivity of the consequent of the Sahlqvist formula, forms the key to understanding why Sahlqvist formulas have first-order correspondents.

Step 3. Instantiating.
We now use the formulas of the form $\sigma(P_i)$ found in Step 2 as instantiations; we substitute $\sigma(P_i)$ for each occurrence of P_i in the first-order matrix of (3.10). This results in a formula of the form

$$[\sigma(P_1)/P_1, \ldots, \sigma(P_n)/P_n] \forall x_1 \ldots \forall x_m \, (\text{REL} \land \text{AT} \to \text{POS}).$$

Now, there are no occurrences of monadic second-order variables in REL. Furthermore, observe that by our choice of the substitution instances $\sigma(P)$, the formula $[\sigma(P_1)/P_1, \ldots, \sigma(P_n)/P_n]$AT will be trivially true. So after carrying out these substitutions we end up with a formula that is equivalent to one of the form

$$\forall x_1 \ldots \forall x_m \, (\text{REL} \to [\sigma(P_1)/P_1, \ldots, \sigma(P_n)/P_n]\text{POS}). \quad (3.12)$$

As we assumed that every unary predicate occurring in the consequent of (3.10) also occurs in its antecedent, (3.12) must be a first-order formula involving only $=$ and the relation symbol R. So, to complete the proof of the theorem it suffices to show that (3.12) is equivalent to (3.10). The implication from (3.10) to (3.12) is simply an instantiation. To prove the other implication, assume that (3.12) and the antecedent of (3.10) are true. That is, assume that

$$\mathfrak{M} \models \forall x_1 \ldots \forall x_m (\text{REL} \to [\sigma(P_1)/P_1, \ldots, \sigma(P_n)/P_n]\text{POS})$$

and

$$\mathfrak{M} \models \text{REL} \land \text{AT}[ww_1 \ldots w_m].$$

We need to show that $\mathfrak{M} \models \text{POS}[ww_1 \ldots w_m]$. First of all, it follows from the above assumptions that

$$\mathfrak{M} \models [\sigma(P_1)/P_1, \ldots, \sigma(P_n)/P_n]\text{POS}[ww_1 \ldots w_m].$$

As POS is positive, it is upwards monotone in all unary predicates occurring in it, so it suffices to show that $\mathfrak{M} \models \forall y\,(\sigma(P_i)(y) \to P_i y)[ww_1 \ldots w_m]$. But, by the essential observation (3.11) in Step 2, this is precisely what the assumption $\mathfrak{M} \models \text{AT}[ww_1 \ldots w_m]$ amounts to. ⊣

Example 3.43 First consider the formula $p \to \Diamond p$. Its second-order translation is the formula

$$\forall P\,(\underbrace{Px}_{\text{AT}} \to \exists z\,(Rxz \wedge Pz)).$$

There are no diamonds to be pulled out here, so we can read off the minimal instance $\sigma(P) \equiv \lambda u.\, u = x$ immediately. Instantiation gives

$$(\lambda u.\, u = x)x \to \exists z\,(Rxz \wedge (\lambda u.\, u = x)z),$$

which (either by β-conversion or semantic reasoning) yields the following first-order formula.

$$x = x \to \exists z\,(Rxz \wedge z = x).$$

Note that this is equivalent to Rxx.

Our second example is the density formula $\Diamond p \to \Diamond\Diamond p$, which has

$$\forall P\,(\exists x_1\,(Rxx_1 \wedge Px_1) \to \exists z_0\,(Rxz_0 \wedge \exists z_1\,(Rz_0z_1 \wedge Pz_1))).$$

as its second-order translation. Here we can pull out the diamond $\exists x_1$:

$$\forall P \forall x_1\,(\underbrace{Rxx_1}_{\text{REL}} \wedge \underbrace{Px_1}_{\text{AT}} \to \exists z_0\,(Rxz_0 \wedge \exists z_1\,(Rz_0z_1 \wedge Pz_1))).$$

Instantiating with $\sigma(P) \equiv \lambda u.\, u = x_1$ gives

$$\forall x_1\,(Rxx_1 \wedge x_1 = x_1 \to \exists z_0\,(Rxz_0 \wedge \exists z_1\,(Rz_0z_1 \wedge z_1 = x_1))),$$

which can be simplified to $\forall x_1\,(Rxx_1 \to \exists z_0\,(Rxz_0 \wedge Rz_0x_1))$.

Our last example of a very simple Sahlqvist formula is $(p \wedge \Diamond\Diamond p) \to \Diamond p$. Its second-order translation is

$$\forall P\,(Px \wedge \exists x_1\,(Rxx_1 \wedge \exists x_2\,(Rx_1x_2 \wedge Px_2)) \to \exists z_0\,(Rxz_0 \wedge Pz_0)).$$

Pulling out the diamonds $\exists x_1$ and $\exists x_2$ results in

$$\forall P \forall x_1 \forall x_2\,(\underbrace{Rxx_1 \wedge Rx_1x_2}_{\text{REL}} \wedge \underbrace{Px \wedge Px_2}_{\text{AT}} \to \exists z_0\,(Rxz_0 \wedge Pz_0)).$$

Our minimal instantiation here is: $\sigma(P) \equiv \lambda u. (u = x \vee u = x_2)$. After instantiating we obtain

$$\forall x_1 \forall x_2 \, (Rxx_1 \wedge Rx_1x_2 \wedge (x = x \vee x = x_2) \wedge (x_2 = x \vee x_2 = x_2) \rightarrow \\ \exists z_0 \, (Rxz_0 \wedge (z_0 = x \vee z_0 = x_2))).$$

This formula simplifies to $\forall x_1 \forall x_2 \, (Rxx_1 \wedge Rx_1x_2 \rightarrow (Rxx \vee Rxx_2))$.   ⊣

Simple Sahlqvist formulas

What is the crucial observation we need to make about the preceding proof? Simply this: the algorithm for very simple Sahlqvist formulas worked because we were able to find a minimal instantiation for their antecedents. We now show that minimal instantiations can be found for more complex Sahlqvist antecedents. First a motivating example.

Example 3.44 Consider the formula $\Diamond_1 \Box_2 p \rightarrow \Box_2 \Diamond_1 p$; we will show that this formula locally corresponds to a kind of local *confluence* (or *Church-Rosser*) property of R_1 and R_2:

$$\forall x_1 z_0 \, (R_1 x x_1 \wedge R_2 x z_0 \rightarrow \exists z_1 (R_2 x_1 z_1 \wedge R_1 z_0 z_1)).$$

The reason for the apparently unnatural choice of variable names will soon become clear, as will the somewhat roundabout approach to the proof that we take. The name 'confluence' is explained by the following picture:

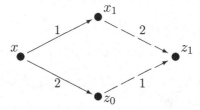

Let $\mathfrak{F} = (W, R_1, R_2)$ be a frame and w a state in $\mathfrak{F}$ such that $\mathfrak{F}, w \Vdash \Diamond_1 \Box_2 p \rightarrow \Box_2 \Diamond_1 p$, and let v be a state in $\mathfrak{F}$ such that $R_1 wv$. A sufficient condition for a valuation to make $\Diamond_1 \Box_2 p$ true at w would be that p holds at all R_2-successors of v. So a *minimal* such valuation can be defined as

$$V_m(p) = \{x \in W \mid R_2 vx\}.$$

That is, V_m makes p true at *precisely* the R_2-successors of v. As $\mathfrak{F}, w \Vdash \Diamond_1 \Box_2 p \rightarrow \Box_2 \Diamond_1 p$, we have $(\mathfrak{F}, V_m), w \Vdash \Box_2 \Diamond_1 p$, but what does this tell us about the (first-order) properties of $\mathfrak{F}$? The crucial observation is that by the choice of V_m:

$$(\mathfrak{F}, V_m), w \Vdash \Box_2 \Diamond_1 p \text{ iff}$$
$$(\mathfrak{F}, V_m) \models \forall z_0 \, (R_2 x z_0 \rightarrow \exists z_1 \, (R_2 x_1 z_1 \wedge R_1 z_0 z_1))[wv], \quad (3.13)$$

which yields that $\mathfrak{F} \models \forall x_1 z_0 (R_1 x x_1 \wedge R_2 x z_0 \rightarrow \exists z_1 (R_2 x_1 z_1 \wedge R_1 z_0 z_1))[w]$.

Conversely, assume that $\mathfrak{F}$ has the confluence property at w. In order to show that $\mathfrak{F}, w \Vdash \Diamond_1 \Box_2 p \rightarrow \Box_2 \Diamond_1 p$, let V be a valuation on $\mathfrak{F}$ such that $(\mathfrak{F}, V), w \Vdash \Diamond_1 \Box_2 p$. We have to prove that $w \Vdash \Box_2 \Diamond_1 p$. By the truth definition of $\Diamond_1$, w has an R_1-successor v satisfying $v \Vdash \Box_2 p$. Now we use the minimal valuation V_m again; first note that by the definition of V_m, we have $V_m(p) \subseteq V(p)$. Therefore, Lemma 3.37 ensures that it suffices to show that $\Box_2 \Diamond_1 p$ holds at w *under the valuation* V_m. But this is immediate by the assumption that $\mathfrak{F}$ is confluent and (3.13). $\dashv$

This example inspires the following definitions:

Definition 3.45 Let τ be a modal similarity type. A *boxed atom* is a formula of the form $\Box_{i_1} \cdots \Box_{i_k} p$ ($k \geq 0$), where $\Box_{i_1}, \ldots, \Box_{i_k}$ are (not necessarily distinct) boxes of the language. In the case where $k = 0$, the boxed atom $\Box_{i_1} \cdots \Box_{i_k} p$ is just the proposition letter p. $\dashv$

Convention 3.46 In the sequel, it will be convenient to treat sequences of boxes as single boxes. We will therefore denote the formula $\Box_{i_1} \cdots \Box_{i_k} p$ by $\Box_\beta p$, where β is the sequence $i_1 \ldots i_k$ of indices. Analogously, we will pretend to have a corresponding binary relation symbol R_β in the frame language $\mathcal{L}^1_\tau$. Thus the expression $R_\beta xy$ abbreviates the formula

$$\exists y_1 (R_{i_1} x y_1 \wedge \exists y_2 (R_{i_2} y_1 y_2 \wedge \cdots \wedge \exists y_{k-1} (R_{i_{k-1}} y_{k-2} y_{k-1} \wedge R_{i_k} y_{k-1} y) \ldots)).$$

Note that this convention allows us to write the second-order translation of the boxed atom $\Box_\beta p$ as $\forall y (R_\beta xy \rightarrow Py)$.

If $k = 0$, β is the empty sequence ϵ; in this case the formula $R_\epsilon xy$ should be read as $x = y$. Note that the second-order translation of $\Box_\epsilon p$ (that is, of the proposition letter p) can indeed be written as $\forall y (R_\epsilon xy \rightarrow Py)$.

Definition 3.47 Let τ be a modal similarity type. A *simple Sahlqvist antecedent* over this similarity type is a formula built up from $\top$, $\bot$ and boxed atoms, using only $\wedge$ and existential modal operators ($\Diamond$ and $\triangle$). A *simple Sahlqvist formula* is an implication $\phi \rightarrow \psi$ in which ψ is positive (as before) and ϕ is a simple Sahlqvist antecedent. $\dashv$

Example 3.48 Typical examples of simple Sahlqvist formulas are $\Diamond p \rightarrow \Diamond \Diamond p$, $\Box p \rightarrow \Box \Box p$, $\Diamond_1 \Box_2 p \rightarrow \Box_3 p$, $\Diamond_1 \Box_2 p \rightarrow \Box_2 \Diamond_1 p$ and $(\Box_1 \Box_2 p) \triangle (\Diamond_3 p \wedge \Box_2 \Box_1 q) \rightarrow \Diamond_3 (q \triangle p)$.

Typically *forbidden* in a simple Sahlqvist antecedent are:

(i) boxes over disjunctions, as in $H(r \vee Fq) \rightarrow G(Pr \wedge Pq)$,
(ii) boxes over diamonds, as in $\Box \Diamond p \rightarrow \Diamond \Box p$,

(iii) dual-triangled atoms, as in $p \bigtriangledown p \to p$.   ⊣

Theorem 3.49 *Let τ be a modal similarity type, and let $\chi = \phi \to \psi$ be a simple Sahlqvist formula over τ. Then χ locally corresponds to a first-order formula $c_\chi(x)$ on frames. Moreover, c_χ is effectively computable from χ.*

Proof. The proof of this theorem is an adaptation of the proof of Theorem 3.42. Consider the universally quantified second-order transcription of χ:

$$\forall P_1 \ldots \forall P_n \, (ST_x(\phi) \to ST_x(\psi)). \tag{3.14}$$

Again, we first make sure that no two quantifiers bind the same variable, and that no quantifier binds x. As before, the idea of the algorithm is to rewrite (3.14) to a formula from which we can easily read off instantiations which yield a first-order equivalent of (3.14).

Step 1. Pull out diamonds.
This is the same as before. This process results in a formula of the form

$$\forall P_1 \ldots \forall P_n \forall x_1 \ldots \forall x_m \, (\text{REL} \wedge \text{BOX-AT} \to ST_x(\psi)), \tag{3.15}$$

where REL is a conjunction of atomic first-order statements of the form $Rx_i x_j$ corresponding to occurrences of diamonds, and BOX-AT is a conjunction of (translations of) boxed atoms, that is, formulas of the form $\forall y \, (R_\beta x_i y \to Py)$.

Step 2. Read off instances.
Let P be a unary predicate occurring in (3.15), and let $\pi_1(x_{i_1}), \ldots, \pi_k(x_{i_k})$ be all the (translations of the) boxed atoms in the antecedent of (3.10) in which the predicate P occurs. Observe that every π_j is of the form $\forall y \, (R_{\beta_j} x_{i_j} y \to Py)$, where β_j is a sequence of diamond indices (recall Convention 3.46). Define

$$\sigma(P) \equiv \lambda u. \, (R_{\beta_1} x_{i_1} u \vee \cdots \vee R_{\beta_k} x_{i_k} u).$$

Again, $\sigma(P_1), \ldots, \sigma(P_n)$ form the *minimal* instances making the antecedent REL$\wedge$ BOX-AT true.

The remainder of the proof is the same as the proof of Theorem 3.42, with the proviso that all occurrences of 'AT' should be replaced by 'BOX-AT'.   ⊣

As in the case of very simple Sahlqvist formulas, the algorithm is best understood by inspecting some examples:

Example 3.50 Let us investigate some of the formulas given in Example 3.48. The simple Sahlqvist formula $\Box_1 \Box_2 p \to \Box_3 p$ has the following second-order translation:

$$\forall P \, (\underbrace{\forall y \, (R_{12} xy \to Py)}_{\text{BOX-AT}} \to \forall z \, (R_3 xz \to Pz)).$$

There are no diamonds to be pulled out here, so we can read off the required substitution instance $\sigma(P) \equiv \lambda u. R_{12}xu$ immediately. Carrying out the substitution we obtain

$$\forall y\, (R_{12}xy \rightarrow R_{12}xy) \rightarrow \forall z\, (R_3xz \rightarrow R_{12}xz),$$

which is equivalent to $\forall z\, (R_3xz \rightarrow R_{12}xz)$.

Next we consider the confluence formula $\Diamond_1 \Box_2 p \rightarrow \Box_2 \Diamond_1 p$, whose second-order translation is

$$\forall P(\exists x_1\, (R_1xx_1 \wedge \forall y\, (R_2x_1y \rightarrow Py)) \rightarrow \forall z_0\, (R_2xz_0 \rightarrow \exists z_1\, (R_1z_0z_1 \wedge Pz_1))).$$

Pulling out the existential quantification $\exists x_1$ yields

$$\forall P \forall x_1\, (\underbrace{R_1xx_1}_{\text{REL}} \wedge \underbrace{\forall y\, (R_2x_1y \rightarrow Py)}_{\text{BOX-AT}} \rightarrow \forall z_0\, (R_2xz_0 \rightarrow \exists z_1\, (R_1z_0z_1 \wedge Pz_1))).$$

The minimal instance making BOX-AT true is $\sigma(P) \equiv \lambda u. R_2x_1u$. After instantiating we obtain

$$\forall x_1\, (R_1xx_1 \wedge \forall y\, (R_2x_1y \rightarrow R_2x_1y) \rightarrow \forall z_0\, (R_2xz_0 \rightarrow \exists z_1\, (R_1z_0z_1 \wedge R_2x_1z_1))),$$

which can be simplified to

$$\forall x_1 \forall z_0\, (R_1xx_1 \wedge R_2xz_0 \rightarrow \exists z_1\, (R_1z_0z_1 \wedge R_2x_1z_1)).$$

As our final example, let us treat a formula using a dyadic modality $\triangle$:

$$(\Box_1 \Box_2 p) \triangle (\Diamond_3 p \wedge \Box_2 \Box_1 q) \rightarrow \Diamond_3 (q \triangle p).$$

We use a ternary relation symbol T for the triangle $\triangle$. Its second-order translation is the rather formidable looking

$$\forall P \forall Q\, (\exists x_1 x_2\, (Txx_1x_2 \wedge \forall y\, (R_{12}x_1y \rightarrow Py) \wedge$$
$$\exists x_3\, (R_3x_2x_3 \wedge Px_3) \wedge \forall y\, (R_{21}x_2y \rightarrow Qy))$$
$$\rightarrow \exists z\, (R_3xz \wedge \exists z_1 z_2\, (Tzz_1z_2 \wedge Qz_1 \wedge Pz_2))),$$

from which we can pull out the diamonds $\exists x_1$, $\exists x_2$ and $\exists x_3$. This leads to

$$\forall P \forall Q \forall x_1 x_2 \forall x_3\, (\overbrace{Txx_1x_2 \wedge R_3x_2x_3}^{\text{REL}} \wedge$$
$$\overbrace{\forall y\, (R_{12}x_1y \rightarrow Py) \wedge Px_3 \wedge \forall y\, (R_{21}x_2y \rightarrow Qy)}^{\text{BOX-AT}} \rightarrow$$
$$\exists z\, (R_3xz \wedge \exists z_1 z_2\, (Tzz_1z_2 \wedge Qz_1 \wedge Pz_2))).$$

Now we can easily read off the required instantiations:

$$\sigma(P) \equiv \lambda u.\, (R_{12}x_1u \vee u = x_3),$$
$$\sigma(Q) \equiv \lambda u.\, (R_{21}x_2u).$$

Performing the substitution $[\sigma(P)/P, \sigma(Q)/Q]$ and deleting the tautological parts from the antecedent gives

$$\forall x_1 x_2 \forall x_3 (T x x_1 x_2 \wedge R_3 x_2 x_3 \rightarrow$$
$$\exists z (R_3 x z \wedge \exists z_1 z_2 (T z z_1 z_2 \wedge R_{21} x_2 z_1 \wedge (R_{12} x_1 z_2 \vee z_2 = x_3))). \qquad \dashv$$

Sahlqvist formulas

We are now ready to introduce the full Sahlqvist fragment and the full version of the Sahlqvist-van Benthem algorithm.

Definition 3.51 Let τ be a modal similarity type. A *Sahlqvist antecedent* over τ is a formula built up from $\top$, $\bot$, boxed atoms, and negative formulas, using $\wedge$, $\vee$ and existential modal operators ($\Diamond$ and $\triangle$). A *Sahlqvist implication* is an implication $\phi \rightarrow \psi$ in which ψ is positive and ϕ is a Sahlqvist antecedent.

A *Sahlqvist* formula is a formula that is built up from Sahlqvist implications by freely applying boxes and conjunctions, and by applying disjunctions only between formulas that do not share any proposition letters. $\dashv$

Example 3.52 Both simple and very simple Sahlqvist formulas are examples of Sahlqvist formulas, as are $\Box(p \rightarrow \Diamond p)$, $p \wedge \Diamond \neg p \rightarrow \Diamond p$, and $\Box(\Diamond_1 \Box_2 p \rightarrow \Box_2 \Diamond_1 p) \wedge \Box_1(p \rightarrow \Diamond_2 p)$. As with simple Sahlqvist formulas, typically forbidden combinations in Sahlqvist antecedent are 'boxes over disjunctions,' 'boxes over diamonds,' and 'dual-triangled atoms' as in $p \triangledown p \rightarrow p$ (see Example 3.48). $\dashv$

The following lemma is instrumental in reducing the correspondence problem for arbitrary Sahlqvist formulas, first to that of Sahlqvist implications, and then to that of simple Sahlqvist formulas.

Lemma 3.53 *Let τ be a modal similarity type, and let ϕ and ψ be τ-formulas.*

(i) *If ϕ and $\alpha(x)$ are local correspondents, then so are $\Box_\beta \phi$ and $\forall y (R_\beta xy \rightarrow [y/x]\alpha)$.*

(ii) *If ϕ (locally) corresponds to α, and ψ (locally) corresponds to β, then $\phi \wedge \psi$ (locally) corresponds to $\alpha \wedge \beta$.*

(iii) *If ϕ locally corresponds to α, ψ locally corresponds to β, and ϕ and ψ have no proposition letters in common, then $\phi \vee \psi$ locally corresponds to $\alpha \vee \beta$.*

Proof. Left as Exercise 3.6.3. $\dashv$

The local perspective in parts one and three of the lemma is essential. For instance, one can find a modal formula ϕ that globally corresponds to a first-order condition $\forall x \, \alpha(x)$ without $\Box \phi$ globally corresponding to the formula $\forall x \forall y (Rxy \rightarrow \alpha(y))$; see Exercise 3.6.3.

Theorem 3.54 *Let τ be a modal similarity type, and let χ be a Sahlqvist formula over τ. Then χ locally corresponds to a first-order formula $c_\chi(x)$ on frames. Moreover, c_χ is effectively computable from χ.*

Proof. The proof of the theorem is virtually the same as the proof of Theorem 3.49, with the exception of the use of Lemma 3.53 and of the fact that we have to do some pre-processing of the formula χ.

By Lemma 3.53 it suffices to show that the theorem holds for all Sahlqvist implications. So assume that χ has the form $\phi \to \psi$ where ϕ is a Sahlqvist antecedent and ψ a positive formula. Proceed as follows.

Step 1. Pull out diamonds and pre-process.
Using the same strategy as in the proof of Theorem 3.49 together with equivalences of the form

$$((\alpha \vee \beta) \to \gamma) \leftrightarrow ((\alpha \to \gamma) \wedge (\beta \to \gamma))$$

and

$$\forall \ldots (\alpha \wedge \beta) \leftrightarrow (\forall \ldots \alpha \wedge \forall \ldots \beta),$$

we can rewrite the second-order translation of $\phi \to \psi$ into a conjunction of formulas of the form

$$\forall P_1 \ldots \forall P_n \forall x_1 \ldots \forall x_m \, (\text{REL} \wedge \text{BOX-AT} \wedge \text{NEG} \to ST_x(\psi)), \qquad (3.16)$$

where REL is a conjunction of atomic first-order statements of the form $R_\triangle \vec{x}$ corresponding to occurrences of diamonds and triangles, BOX-AT is a conjunction of (translations of) boxed atoms, and NEG is a conjunction of (translations of) negative formulas. By Lemma 3.53(ii) it suffices to show that each formula of the form displayed in (3.16) has a first-order equivalent. This is done by using the equivalence

$$(\alpha \wedge \text{NEG} \to \beta) \leftrightarrow (\alpha \to \beta \vee \neg\text{NEG}),$$

where $\neg$NEG is the *positive* formula that arises by negating the negative formula NEG. Using this equivalence we can rewrite (3.16) to obtain a formula of the form

$$\forall P_1 \ldots \forall P_n \forall x_1 \ldots \forall x_m \, (\text{REL} \wedge \text{BOX-AT} \to \text{POS}),$$

and from here on we can proceed as in Step 2 of the proof of Theorem 3.49. ⊣

Example 3.55 By way of example we determine the local first-order correspondents of two of the modal formulas given in Example 3.52. To determine the first-order correspondent of the Sahlqvist formula $\square(p \to \Diamond p)$ we first recall that the local first-order correspondent of $p \to \Diamond p$ is Rxx. So, by Lemma 3.53(i) $\square(p \to \Diamond p)$ locally corresponds to $\forall y \, (Rxy \to Ryy)$.

Next we consider the Sahlqvist formula $(p \wedge \Diamond \neg p) \rightarrow \Diamond p$. Its translation is

$$\forall P \left(Px \wedge \exists y \left(Rxy \wedge \neg Py \right) \rightarrow \exists z \left(Rxz \wedge Pz \right) \right).$$

Pulling out the diamond produces

$$\forall P \forall y \left(\underbrace{Px}_{\text{BOX-AT}} \wedge \underbrace{Rxy}_{\text{REL}} \wedge \underbrace{\neg Py}_{\text{NEG}} \rightarrow \underbrace{\exists z \left(Rxz \wedge Pz \right)}_{\text{POS}} \right),$$

and moving the negative part $\neg Py$ to the consequent we get

$$\forall P \forall y \left(\underbrace{Px}_{\text{BOX-AT}} \wedge \underbrace{Rxy}_{\text{REL}} \rightarrow \underbrace{Py \vee \exists z \left(Rxz \wedge Pz \right)}_{\text{POS}} \right).$$

The minimal instantiation to make Px true is $\lambda u.\, u = x$. After instantiation we obtain

$$\forall y \left(Rxy \rightarrow y = x \vee \exists z \left(Rxz \wedge z = x \right) \right),$$

which can be simplified to $\forall y \left(Rxy \wedge x \neq y \rightarrow Rxx \right)$. $\dashv$

Exercises for Section 3.6

3.6.1 Compute the first-order formulas locally corresponding to the following Sahlqvist formulas:

(a) $\Diamond_1 \Diamond_2 p \rightarrow \Diamond_2 \Diamond_1 p$,
(b) $(p \wedge \Box p \wedge \Box \Box p) \rightarrow \Diamond p$,
(c) $\Diamond^k \Box^l p \rightarrow \Box^m \Diamond^n p$, for arbitrary natural numbers k, l, m and n,
(d) $(\Box p) \triangle (\Box p) \rightarrow p \triangledown p$,
(e) $\Diamond (\neg p \wedge \Diamond (p \wedge q)) \rightarrow \Diamond (p \wedge q)$,
(f) $\Box ((p \wedge \Box \neg p \wedge q) \rightarrow \Diamond q))$.

3.6.2   (a) Show that the formula $\Box (p \vee q) \rightarrow \Diamond (\Box p \vee \Box q)$ does not locally correspond to a first-order formula on frames. (Hint: modify the frame of Example 3.11.)
(b) Use this example to show that dual-*triangled* atoms cannot be allowed in Sahlqvist antecedents.

3.6.3 Prove Lemma 3.53:

(a) Show that if ϕ and $\alpha(x)$ locally correspond, so do $\Box_\beta \phi$ and $\forall y \left(R_\beta xy \rightarrow \alpha(y) \right)$.
(b) Prove that if ϕ (locally) corresponds to $\alpha(x)$, and ψ (locally) corresponds to $\beta(x)$, then $\phi \wedge \psi$ (locally) corresponds to $\alpha(x) \wedge \beta(x)$.
(c) Show that if ϕ locally corresponds to $\alpha(x)$, ψ locally corresponds to $\beta(x)$, and ϕ and ψ have no proposition letters in common, then $\phi \vee \psi$ locally corresponds to $\alpha(x) \vee \beta(x)$.
(d) Prove that (a) and (c) do not hold for global correspondence, and that the condition on the proposition letters in (c) is necessary as well. (Hint: for (a), think of the modal formula $\Diamond \Diamond p \rightarrow \Diamond p$ and the first-order formula $\forall xyz \left(Ryz \wedge Rzx \rightarrow Ryx \right)$.)

3.7 More about Sahlqvist Formulas

It is time to step back and think more systematically about the Sahlqvist fragment, for a number of questions need addressing. For a start, does this fragment contain *all* modal formulas with first-order correspondents? And why did we forbid disjunctions in the scope of boxes, and occurrences of nested duals of triangles in Sahlqvist antecedents, while we allowed boxed atoms? Most interesting of all, *which* first-order conditions are expressible by means of Sahlqvist formulas? That is, is it possible to prove some sort of converse to the Sahlqvist Correspondence Theorem?

Limitative results

To set the stage for our discussion, we first state (without proof) the principal limitative result in this area: Chagrova's Theorem. Good presentations of the proof are available in the literature; see the Notes for references.

Theorem 3.56 (Chagrova's Theorem) *It is undecidable whether an arbitrary basic modal formula has a first-order correspondent.*

This implies that, even for the basic modal language, it is *not* possible to write a computer program which, when presented with an arbitrary modal formula as input, will terminate after finitely many steps, returning the required first-order correspondent (if there is one) or saying 'No!' (if there is not).

Quite apart from its intrinsic interest, this result immediately tells us that the Sahlqvist fragment cannot possibly contain *all* modal formulas with first-order correspondents. For it is straightforward to decide whether a modal formula is a Sahlqvist formula, and to compute the first-order correspondents of Sahlqvist formulas. Hence if all modal formulas with first-order correspondents were Sahlqvist, this would contradict Chagrova's Theorem.

But a further question immediately presents itself: is every modal formula with a first-order correspondent *equivalent* to a Sahlqvist formula? (The preceding argument does not rule this out.) The answer is *no*: there are modal formulas corresponding to first-order frame conditions which are not equivalent to any Sahlqvist formula.

Example 3.57 Consider the conjunction of the following two formulas:

(M) $\Box\Diamond p \rightarrow \Diamond\Box p,$

(4) $\Diamond\Diamond q \rightarrow \Diamond q.$

(M) is the McKinsey formula we discussed in Example 3.11, and (4) is the transitivity axiom. It is obvious that M itself is not a Sahlqvist axiom, and by Example 3.11 it does not express a first-order condition.

It requires a little argument to show that the *conjunction* M ∧ 4 is not *equivalent* to a Sahlqvist formula. One way to do so is by proving that M ∧ 4 does not have a *local* first-order correspondent (cf. Exercise 3.7.1).

Nevertheless, the conjunction M∧4 does have a global first-order correspondent, as we can prove the following equivalence for all transitive frames $\mathfrak{F}$:

$$\mathfrak{F} \Vdash M \text{ iff } \mathfrak{F} \models \forall x \exists y \, (Rxy \land \forall z \, (Ryz \to z = y)). \tag{3.17}$$

We leave the right to left direction as an exercise to the reader. To prove the other direction, assume for contradiction that there is a transitive frame $\mathfrak{F} = (W, R)$ on which the McKinsey formula is valid, but which does *not* satisfy the first-order formula given in (3.17). Let r be a state witnessing that the first-order formula in (3.17) does not hold in $\mathfrak{F}$. That is, assume that each successor s of r has a successor distinct from it. We may assume that the frame is generated from r, so that $\mathfrak{F} \models \forall y \exists z \, (Ryz \land y \neq z)$.

In order to derive a contradiction from this, we need to introduce some terminology. Call a subset X of W *cofinal in* W if for all $w \in W$ there is an $x \in X$ such that Rwx. We now claim that

$$W \text{ has a subset } X \text{ such that both } X \text{ and } W \setminus X \text{ are cofinal in } W. \tag{3.18}$$

From (3.18) we can immediately derive a contradiction by considering the valuation V given by $V(p) = X$. For, cofinality of X implies that $(\mathfrak{F}, V), r \Vdash \Box \Diamond p$, while cofinality of $W \setminus X$ likewise gives $(\mathfrak{F}, V), r \Vdash \Box \Diamond \neg p$. But then $(\mathfrak{F}, V), r \not\Vdash$ M.

To prove (3.18), consider the collection C of all pairs of disjoint non-empty subsets $Y, Z \subset W$ satisfying $\forall y \in Y \exists z \in Z \, Ryz$ and $\forall z \in Z \exists y \in Y \, Rzy$. We first prove that this collection is non-empty. From $\mathfrak{F} \models \forall y \exists z \, (Ryz \land y \neq z)$ it follows that there is a path $w_0 R w_1 R w_2 R \ldots$ such that $w_i \neq w_{i+1}$ for all i. Let n be the first index such that $w_n \in \{w_0, \ldots, w_{n-1}\}$, or $n = \omega$ if there is no such index. It is then an easy exercise to verify that the pair consisting of $Y_0 := \{w_{2i} \mid 2i < n\}$ and $Y_1 := \{w_{2i+1} \mid 2i + 1 < n\}$ belongs to C.

Order C under coordinate-wise inclusion. It is obvious that every chain in this partial ordering is bounded above; hence, we may apply Zorn's Lemma and obtain a *maximal* such pair Z_0, Z_1. We claim that

$$Z_0 \cup Z_1 = W. \tag{3.19}$$

Since Z_0 and Z_1 are disjoint, this implies that $Z_1 = W \setminus Z_0$ and thus proves (3.18).

Suppose that (3.19) does *not* hold. Then there is an element $w \in W$ which belongs neither to Z_0 nor to Z_1. If w has successor in $Z_0 \cup Z_1$ then (by transitivity of R) the pair $(Z_0 \cup \{w\}, Z_1)$ would belong to C, contradicting the maximality of (Z_0, Z_1) in the ordering of C. But if w has no successor in $Z_0 \cup Z_1$ then by transitivity of R, the subframe $\mathfrak{F}_w$ of $\mathfrak{F}$ is disjoint from $Z_0 \cup Z_1$. With the argument used above to show that $C \neq \varnothing$ we can construct a pair $(Y_0, Y_1) \in C$ such that Y_0 and Y_1 fall entirely within $\mathfrak{F}_w$, and thus satisfy $(Y_0 \cup Y_1) \cap (Z_0 \cup Z_1) = \varnothing$. It is then straightforward to check that the pair $(Y_0 \cup Y_1, Z_0 \cup Z_1)$ belongs to C; but this membership would contradict the maximality of (Z_0, Z_1). This proves (3.19). ⊣

Of course, this example begs the question whether there is a modal formula that *locally* corresponds to a first-order formula without being equivalent to a Sahlqvist formula. However, the answer to this is also affirmative: the formula $\Box M \wedge 4$ is an example. In Exercise 3.7.1 the reader is asked to show that it has a local first-order correspondent; in Chapter 5 we will develop the techniques needed to prove that the formula is not equivalent to a Sahlqvist formula, see Exercise 5.6.2.

Thus the Sahlqvist fragment does not contain all modal formulas with first-order correspondents. So the next question is: can the Sahlqvist fragment be further extended? The answer is *yes* – but we should reflect a little on what we hope to achieve through such extensions. The Sahlqvist fragment is essentially a good compromise between the demands of generality and simplicity. By adding further restrictions it is possible to extend it further, but it is not obvious that the resulting loss of simplicity is really worth it. Moreover, the Sahlqvist fragment also gives rise to a matching completeness theorem; we would like proposed extensions to do so as well. We do not know of simple generalizations of the Sahlqvist fragment which manage to do this. In short, while there is certainly room for experiment here, it is unclear whether anything interesting is likely to emerge.

However, one point is worth stressing once more: the Sahlqvist fragment *cannot* be further extended simply by dropping some of the restrictions in the definition of a Sahlqvist formula. We forbid disjunctions in the scope of boxes and nested duals of triangles in Sahlqvist antecedents for a very good reason: these forbidden combinations easily lead to modal formulas that have no first-order correspondent, as we have seen in Example 3.11 and Exercise 3.6.2.

Kracht's Theorem

Let us turn to a nice positive result. As has already been mentioned, not only does each Sahlqvist formula define a first-order class of frames, but when we use one as an axiom in a normal modal logic, that logic is guaranteed to be complete with respect to the elementary class of frames the axiom defines. (This is the content of the Sahlqvist Completeness Theorem; see Theorem 4.42 for a precise statement.) So it would be very pleasant to know *which* first-order conditions are the correspondents of Sahlqvist formulas. Kracht's Theorem is a sort of converse to the Sahlqvist Correspondence Theorem which gives us this information.

Before we can define the fragment of first-order logic corresponding to Sahlqvist formulas we need some auxiliary definitions; we also introduce some helpful notation. For reasons of notational simplicity, we work in the basic modal similarity type. First of all, we will abbreviate the first-order formula $\forall y \, (Rxy \rightarrow \alpha(y))$ to $(\forall y \triangleright x)\alpha(y)$, speaking of *restricted quantification* and calling x the *restrictor* of y. Likewise $\exists y \, (Rxy \wedge \alpha(y))$ is abbreviated to $(\exists y \triangleright x)\alpha(y)$. We will call the constructs $(\forall y \triangleright x)$ and $(\exists y \triangleright x)$ *restricted quantifiers*. If we wish not to specify the restrictor of a restricted quantifier we will write $\forall^r y$ or $\exists^r y$. Moreover, if we do not wish to specify whether a quantifier is existential or universal we denote it by Q (Q^r in the restricted case). Second, for the duration of this subsection it will be convenient for us to consider formulas of the form $u \neq u$ as atomic. Third, in this subsection we will work exclusively with formulas in which no variable occurs both free and bound, and in which no two distinct (occurrences of) quantifiers bind the same variable; we will call such formulas *clean*.

Now we call a formula *restrictedly positive* if it is built up from atomic formulas, using $\wedge$, $\vee$ and restricted quantifiers only; observe that monadic predicates occur positively in restrictedly positive formulas. Finally, we assume that the reader knows how to rewrite an arbitrary positive propositional formula to a *disjunctive normal form* or DNF (that is, to an equivalent disjunction of conjunctions of atomic formulas) and to a *conjunctive normal form* or CNF (that is, to an equivalent conjunction of disjunctions of atomic formulas).

The crucial notion in this subsection is that of a variable occurring *inherently universally* in a first-order formula.

Definition 3.58 We say that an occurrence of the variable y in the (clean!) formula α is inherently universal if either y is free, or else y is bound by a restricted quantifier of the form $(\forall y \triangleright x)\beta$ which is not in the scope of an existential quantifier. A formula $\alpha(x)$ in the basic first-order frame language is called a *Kracht formula* if α is clean, restrictedly positive and, furthermore, every atomic formula is either of the form $u = u$ or $u \neq u$, or else it contains at least one inherently universal variable.   ⊣

Restricted quantification is obviously the modal face of quantification in first-order logic; indeed, we could have defined the standard translation of a modal formula using this notion. As for Kracht formulas, first observe that every *universal* restricted first-order formula satisfies the definition. A second example of a Kracht formula is $(\forall w \triangleright v)(\forall x \triangleright v)(\exists y \triangleright w)Rxy$: note that it does not matter that the 'x' in Rxy falls within the scope of an existential quantifier; what matters is that the universal *quantifier* that *binds* x does not occur within the scope of any existential quantification. On the other hand, the formula $(\exists w \triangleright v)(\forall x \triangleright v)w = x$ is not a Kracht formula since the occurrence of neither w nor x in $w = x$ is inherently universal: w is disqualified because it is bound by an existential quantifier and x because it is bound within the scope of the existential quantifier $(\exists w \triangleright v)$.

The following result states that Kracht formulas are the first-order counterparts of Sahlqvist formulas – but not only that. As will become apparent from its proof, from a given Kracht formula we can *compute* a Sahlqvist formula locally corresponding to it. The reader is advised to glance at the examples provided below while reading the proof.

Theorem 3.59 *Any Sahlqvist formula locally corresponds to a Kracht formula; and conversely, every Kracht formula is a local first-order correspondent of some Sahlqvist formula which can be effectively obtained from the Kracht formula.*

Proof. For the left to right direction, we leave it as an exercise to the reader to show that the algorithm discussed in the sections 3.5 and 3.6 in fact produces, given a Sahlqvist formula, a first-order correspondent *within* the Kracht fragment. We will give the proof of the other direction: we will show how rewrite a given Kracht formula to an equivalent Sahlqvist formula.

Our first step is to provide special prenex formulas as normal forms for Kracht formulas. Define a *type 1* formula to be of the form

$$\forall^r x_1 \ldots \forall^r x_n Q_1^r y_1 \ldots Q_m^r y_m \, \beta(x_0, \ldots, x_n, y_1, \ldots, y_m)$$

such that $n, m \geq 0$ and each variable is restricted by an earlier variable (that is, the restrictor of any x_i is some x_j with $j < i$ and the restrictor of any y_i is either some x_k or some y_j with $j < i$. Furthermore we require that β is a DNF of formulas $u = u$, $u \neq u$, Rux, $u = x$ and Rxu (that is, we allow all atomic formulas that are *not* of the form Ryy' or $y = y'$). Here and in the remainder of this proof we use the convention that u and z denote arbitrary variables in $\{x_0, \ldots, x_n, y_1, \ldots, y_m\}$ and x an arbitrary variable in $\{x_0, \ldots, x_n\}$.

Clearly then, type 1 formulas form a special class of Kracht formulas. This inclusion is not proper (modulo equivalence), since we can prove the following claim.

Claim 1 *Every Kracht formula can be effectively rewritten into an equivalent type 1 formula.*

Proof of Claim. Let $\alpha(x_0)$ be a Kracht formula. By definition it is built up from atomic formulas using $\wedge$, $\vee$ and restricted quantifiers. Furthermore, since $\alpha(x_0)$ is clean, in a subformula of the form $Q^r v \, \beta$ the variable v may not occur outside of β. Hence, we may use the equivalences

$$(Q^r v \, \beta) \heartsuit \gamma \leftrightarrow Q^r v \, (\beta \heartsuit \gamma) \tag{3.20}$$

(where $\heartsuit$ uniformly denotes either $\wedge$ or $\vee$) to pull out quantifiers to the front. However, if we want to remain within the Kracht fragment we have to take care about the *order* in which we pull out quantifiers.

Without loss of generality we may assume that each inherently universal variable is named x_i for some i, while each of the remaining variables is named y_j for some j. This ensures that no atomic subformula of $\alpha(x_0)$ is of the form Ryy' or $y = y'$ (with distinct variables y and y').

Observe also that in every subformula of the form $((\forall x \triangleright u)\beta)\heartsuit\gamma$, the variable u occurs free. If this u is not the variable x_0 then it is a bound variable of α; hence, the mentioned subformula must occur in the scope of a quantifier $(Q'u \triangleright x')$. This quantification must have been universal, for, otherwise, the variable x could not have been among the inherently universal ones. But this means that the variable u itself must be inherently universal as well, so u is some x_i. This shows that by successively pulling out restricted universal quantifiers $\forall x$ we end up with a Kracht formula of the form

$$\forall^r x_1 \ldots \forall^r x_n \, \alpha'(x_0, \ldots, x_n, y_1, \ldots, y_m),$$

such that each atomic formula of α' is of the form $u = u$ or $u \neq u$, or else it contains some occurrence of a variable x_i. Furthermore, the restrictor of each x_i is some x_j with $j < i$.

It remains to pull out the other restricted quantifiers from α'. But this can easily be done using the equivalences of (3.20), since we do not have to worry anymore about the order in which we pull out the quantifiers. In the end, we arrive at a formula of the form

$$\forall^r x_1 \ldots \forall^r x_n \, Q_1^r y_1 \ldots Q_m^r y_m \, \alpha''(x_0, \ldots, x_n, y_1, \ldots, y_m)$$

such that the atomic subformulas of α'' satisfy the same condition of those in α' (in fact, they are the very same formulas), while in addition, α'' is quantifier free. Hence, if we rewrite α'' into disjunctive normal form, we are finished.   ⊣

Enter diamonds and boxes. A *type 2* formula is a formula in the second-order frame language of the form

$$\tilde{\forall} P_0 \ldots \tilde{\forall} P_n \tilde{\forall} Q_0 \ldots \tilde{\forall} Q_n \forall^r x_1 \ldots \forall^r x_n \left(\bigwedge_{0 \leq i \leq n} ST_{x_i}(\sigma_i) \rightarrow \beta \right)$$

such that each σ_i is a conjunction of boxed atoms in p_i and q_i, whereas β is a DNF of formulas $ST_x(\psi)$, with ψ some modal formula which is positive in each p_i, q_j.

Claim 2 *Every type 1 formula can be effectively rewritten into an equivalent type 2 formula.*

Proof of Claim. Now the prominent role of the inherently universal formulas will come out: they determine the proposition letters of the Sahlqvist formula and the

'BOX-AT' part of its antecedent. Consider the type 1 formula

$$\forall^r x_1 \ldots \forall^r x_n Q_1^r y_1 \ldots Q_m^r y_m \, \beta(x_0, \ldots, x_n, y_1, \ldots, y_m).$$

We abbreviate the sequence $\forall^r x_1 \ldots \forall^r x_n$ by $\forall^r \bar{x}$, and use similar abbreviations for other sequences of quantifiers. Recall that β is a DNF of formulas $u = u$, $u \neq u$, $u = x_i$, $Ruox_i$ and $Rx_i u$. Our first move is to replace such subformulas with the formulas $ST_u(\top)$, $ST_u(\bot)$, $ST_u(p_i)$, $ST_u(\Diamond p_i)$ and $ST_u(q_i)$, respectively; call the resulting formula β'.

Our first claim is that

$$\forall^r \bar{x} \, Q^r \bar{y} \, \beta \quad \text{is equivalent to}$$

$$\tilde{\forall} \bar{P} \bar{Q} \, \forall^r \bar{x} \left(\bigwedge_{0 \le i \le n} ST_{x_i}(p_i \wedge \Box q_i) \to Q^r \bar{y} \, \beta' \right). \tag{3.21}$$

Forbidding as (3.21) may look, its proof is completely analogous to proofs in Sections 3.5 and 3.6: the direction from right to left is immediate by instantiation, while the other direction simply follows from the fact that β is monotone in each predicate symbol P_i and Q_i.

Two remarks are in order here. First, since β may contain atomic formulas of the form $Rx_i x_j$ and $x_i = x_j$ (that is, with *both* variables being inherently universal), there is some *choice* here. For instance, the formula $Rx_i x_j$ may be replaced with either $ST_{x_i}(\Diamond p_j)$ or with $ST_{x_j}(q_j)$. Having this choice can sometimes be of use if one wants to find Sahlqvist correspondents satisfying some additional constraints.

Related to this is our second remark: we do not need to introduce *both* proposition letters p_i and q_i for *each* x_i. We can do with any supply of proposition letters that is sufficient to replace all atomic formulas of β with the standard translation of either $ST_u(p_i)$, $ST_u(\Diamond p_i)$ or $ST_u(q_i)$. A glance at the examples below will make this point clear.

We are now halfway through the proof of Claim 2: observe that β is already a DNF of formulas $ST_u(\psi)$ with ψ positive in each p_i, q_j. It remains to eliminate the quantifier sequence $Q^r \bar{y}$. This will be done step by step, using the following procedure:

Consider the formula

$$(\exists y_{i+1} \rhd z) \left(\bigvee_{k \le K} \bigwedge_{l \le L_k} ST_{u_{kl}}(\psi_{kl}) \right), \tag{3.22}$$

where each modal formula ψ_{kl} is positive in all variables p_i, q_j; z is either an x or a y_j with $j \le i$; and each u is either an x or a y_j with $j \le i + 1$. We first distribute the existential quantifier over the disjunction, yielding a disjunction of formulas

$$(\exists y_{i+1} \triangleright z) \bigwedge_{l \leq L_k} ST_{u_{kl}}(\psi_{kl}). \tag{3.23}$$

We may assume all these variables u to be distinct (otherwise, replace $ST_u(\psi') \wedge ST_u(\psi'')$ with $ST_u(\psi' \wedge \psi'')$); we may also assume that y_{i+1} is the variable u_{lL_k} (if y_{i+1} does not occur among the us, add a conjunct $ST_{y_{i+1}}(\top)$). But then (3.23) is equivalent to the formula

$$ST_z(\Diamond\psi_{kL_k}) \wedge \bigwedge_{l < L_k} ST_{u_{kl}}(\psi_{kl}),$$

whence (3.22) is equivalent to a disjunction of such formulas. Observe further that y_{i+1} does not occur in these formulas.

This shows how to get rid of an existential innermost restricted quantifier of the prenex $K^r \bar{y}$. A universal innermost restricted quantifier can be removed dually, by first converting the matrix β' into a *conjunctive* normal form; details are left to the reader. In any case, it will be clear that by this procedure we can rewrite any type 1 formula into an equivalent type 2 formula.   ⊣

We are now almost through with the proof of Theorem 3.59. All we have to do now is show how to massage arbitrary type 2 formulas into Sahlqvist shape.

Claim 3 *Any type 2 formula can be effectively rewritten into an equivalent Sahlqvist formula.*

Proof of Claim. Let

$$\tilde{\forall}\bar{P}\bar{Q}\,\forall^r\bar{x} \left(\bigwedge_{0 \leq i \leq n} ST_{x_i}(\sigma_i) \; \rightarrow \; \beta \right) \tag{3.24}$$

be an arbitrary type 2 formula.

First we rewrite β into conjunctive normal form, and we distribute the implication and the prenex of universal quantifiers over the conjunctions. Thus we obtain a conjunction of formulas of the form

$$\tilde{\forall}\bar{P}\bar{Q}\forall^r\bar{x} \left(\bigwedge_{0 \leq i \leq n} ST_{x_i}(\sigma_i) \; \rightarrow \; \beta' \right), \tag{3.25}$$

where β' is a disjunction of formulas of the form $ST_x(\psi)$ with each ψ positive in all p_i and q_j. As before, we may assume that each x_i occurs in exactly one disjunct of β', so (3.25) is equivalent to a formula

$$\tilde{\forall}\bar{P}\bar{Q}\,\forall^r\bar{x}\left(\bigwedge_{0\le i\le n} ST_{x_i}(\sigma_i) \;\rightarrow\; \bigvee_{0\le i\le n} ST_{x_i}(\psi_i)\right),$$

where each σ_i is a Sahlqvist antecedent and each ψ_i is positive. But clearly then, (3.25) is equivalent to the formula

$$\tilde{\forall}\bar{P}\bar{Q}\neg\exists^r\bar{x}\bigwedge_{0\le i\le n} ST_{x_i}(\sigma_i \wedge \neg\psi_i).$$

Observe that each modal formula $\sigma_i \wedge \neg\psi_i$ is a Sahlqvist antecedent.

But now, as before, working inside out we may eliminate all remaining restricted quantifiers, step by step. For, observe that the formula

$$\exists^r x_1 \ldots \exists^r x_{k-1}(\exists x_k \rhd x_j)\bigwedge_{0\le i\le k} ST_{x_i}(\chi_i)$$

is equivalent to

$$\exists^r x_1 \ldots \exists^r x_{k-1}\left(ST_{x_j}(\chi_j \wedge \Diamond\chi_k) \wedge \bigwedge_{0\le i< k, i\ne j} ST_{x_i}(\chi_i)\right).$$

Note that $\chi_j \wedge \Diamond\chi_{k+1}$ is a Sahlqvist antecedent if χ_j and χ_{k+1} are.

It turns out that for some Sahlqvist antecedent ϕ, (3.25) is equivalent to the second-order formula

$$\tilde{\forall}\bar{P}\bar{Q}\;\neg ST_{x_0}(\phi).$$

But then (3.24) is equivalent to a conjunction of such formulas, and thus equivalent to a formula

$$\tilde{\forall}\bar{P}\bar{Q}\;ST_{x_0}\left(\bigvee_l \phi_l \rightarrow \bot\right),$$

which is the local second-order frame correspondent of the formula $\bigvee_l \phi_l \rightarrow \bot$, which is obviously in Sahlqvist form. $\dashv$

This completes the proof of the third claim, and hence of the theorem. $\dashv$

Example 3.60 Consider the formula

$$\alpha(x_0) \;\equiv\; (\forall x_1 \rhd x_0)(\exists y_1 \rhd x_0)(\exists y_2 \rhd y_1)\, Rx_1 y_2.$$

This is already a type 1 Kracht formula, so we proceed by the procedure described in the proof of Claim 2 in the proof of Theorem 3.59. According to (3.21), $\alpha(x_0)$ is equivalent to the second order formula

$$\tilde{\forall}Q_1(\forall x_1 \rhd x_0)\,(ST_{x_1}(\Box q_1) \rightarrow (\exists y_1 \rhd x_0)(\exists y_2 \rhd y_1)ST_{y_2}(q_1)).$$

Then, using the equivalences described further on in the proof of Claim 2 we obtain the following sequences of formulas that are equivalent to $\alpha(x_0)$:

$$\tilde{\forall}Q_1(\forall x_1 \triangleright x_0)\, (ST_{x_1}(\Box q_1) \to (\exists y_1 \triangleright x_0)(\exists y_2 \triangleright y_1)ST_{y_2}(q_1))$$
$$\Leftrightarrow\quad \tilde{\forall}Q_1(\forall x_1 \triangleright x_0)\, (ST_{x_1}(\Box q_1) \to (\exists y_1 \triangleright x_0)ST_{y_1}(\Diamond q_1))$$
$$\Leftrightarrow\quad \tilde{\forall}Q_1(\forall x_1 \triangleright x_0)\, (ST_{x_1}(\Box q_1) \to ST_{x_0}(\Diamond\Diamond q_1)).$$

The last formula is a type 2 formula. Hence, the only thing left to do is to rewrite it to an equivalent Sahlqvist formula; this we do via the sequence of equivalent formulas below, following the pattern of the proof of Claim 3:

$$\tilde{\forall}Q_1(\,(\forall x_1 \triangleright x_0)\, (ST_{x_1}(\Box q_1) \to ST_{x_0}(\Diamond\Diamond q_1))\,)$$
$$\Leftrightarrow\quad \tilde{\forall}Q_1(\,(\forall x_1 \triangleright x_0)\, \neg(ST_{x_1}(\Box q_1) \wedge \neg ST_{x_0}(\Diamond\Diamond q_1))\,)$$
$$\Leftrightarrow\quad \tilde{\forall}Q_1(\,(\forall x_1 \triangleright x_0)\, \neg(ST_{x_1}(\Box q_1) \wedge ST_{x_0}(\neg\Diamond\Diamond q_1))\,)$$
$$\Leftrightarrow\quad \tilde{\forall}Q_1(\,\neg(\exists x_1 \triangleright x_0)\, (ST_{x_1}(\Box q_1) \wedge ST_{x_0}(\neg\Diamond\Diamond q_1))\,)$$
$$\Leftrightarrow\quad \tilde{\forall}Q_1(\,\neg((\exists x_1 \triangleright x_0)ST_{x_1}(\Box q_1) \wedge ST_{x_0}(\neg\Diamond\Diamond q_1))\,)$$
$$\Leftrightarrow\quad \tilde{\forall}Q_1(\,\neg(ST_{x_0}(\Diamond\Box q_1) \wedge ST_{x_0}(\neg\Diamond\Diamond q_1))\,)$$
$$\Leftrightarrow\quad \tilde{\forall}Q_1(\,\neg ST_{x_0}(\Diamond\Box q_1 \wedge \neg\Diamond\Diamond q_1)\,)$$
$$\Leftrightarrow\quad \tilde{\forall}Q_1(\, ST_{x_0}((\Diamond\Box q_1 \wedge \neg\Diamond\Diamond q_1) \to \bot)\,).$$

This means that $\alpha(x_0)$ locally corresponds to the Sahlqvist formula $(\Diamond\Box q_1 \wedge \neg\Diamond\Diamond q_1) \to \bot$, or to the equivalent formula $\Diamond\Box q_1 \to \Diamond\Diamond q_1$. $\dashv$

Example 3.61 Consider the Kracht formula

$$\alpha(x_0) \equiv (\forall x_1 \triangleright x_0)(\forall x_2 \triangleright x_0)\, (Rx_1 x_2 \vee Rx_2 x_1 \vee x_1 = x_2).$$

According to (3.21), $\alpha(x_0)$ is equivalent to

$$\tilde{\forall}P_1\tilde{\forall}Q_1(\forall x_1 \triangleright x_0)(\forall x_2 \triangleright x_0)\, (ST_{x_1}(p_1 \wedge \Box q_1)$$
$$\to (ST_{x_2}(q_1) \vee ST_{x_2}(\Diamond p_1) \vee ST_{x_2}(p_1)))$$

and to

$$\tilde{\forall}P_1\tilde{\forall}Q_1(\forall x_1 \triangleright x_0)(\forall x_2 \triangleright x_0)\, (ST_{x_1}(p_1 \wedge \Box q_1) \to ST_{x_2}(q_1 \vee \Diamond p_1 \vee p_1)).$$

The latter is a type 2 formula; in order to find a Sahlqvist equivalent for it, we proceed as follows:

$$\tilde{\forall}P_1\tilde{\forall}Q_1(\forall x_1 \triangleright x_0)(\forall x_2 \triangleright x_0)\, (ST_{x_1}(p_1 \wedge \Box q_1) \to ST_{x_2}(q_1 \vee \Diamond p_1 \vee p_1))$$
$$\Leftrightarrow\quad \tilde{\forall}P_1\tilde{\forall}Q_1(\forall x_1 \triangleright x_0)(\forall x_2 \triangleright x_0)\, \neg(ST_{x_1}(p_1 \wedge \Box q_1) \wedge$$
$$\neg ST_{x_2}(q_1 \vee \Diamond p_1 \vee p_1))$$
$$\Leftrightarrow\quad \tilde{\forall}P_1\tilde{\forall}Q_1(\forall x_1 \triangleright x_0)(\forall x_2 \triangleright x_0)\, \neg(ST_{x_1}(p_1 \wedge \Box q_1) \wedge$$

$$ST_{x_2}(\neg(q_1 \vee \Diamond p_1 \vee p_1)))$$

$$\Leftrightarrow \quad \tilde{\forall} P_1 \tilde{\forall} Q_1 \neg(\exists x_1 \triangleright x_0)(\exists x_2 \triangleright x_0)\,(ST_{x_1}(p_1 \wedge \Box q_1) \wedge$$
$$ST_{x_2}(\neg(q_1 \vee \Diamond p_1 \vee p_1)))$$

$$\Leftrightarrow \quad \tilde{\forall} P_1 \tilde{\forall} Q_1 \neg(\exists x_1 \triangleright x_0)\,(ST_{x_1}(p_1 \wedge \Box q_1) \wedge$$
$$(\exists x_2 \triangleright x_0)ST_{x_2}(\neg(q_1 \vee \Diamond p_1 \vee p_1)))$$

$$\Leftrightarrow \quad \tilde{\forall} P_1 \tilde{\forall} Q_1 \neg(\exists x_1 \triangleright x_0)\,(ST_{x_1}(p_1 \wedge \Box q_1) \wedge ST_{x_0}(\Diamond \neg(q_1 \vee \Diamond p_1 \vee p_1)))$$

$$\Leftrightarrow \quad \tilde{\forall} P_1 \tilde{\forall} Q_1 \neg((\exists x_1 \triangleright x_0)ST_{x_1}(p_1 \wedge \Box q_1) \wedge ST_{x_0}(\Diamond \neg(q_1 \vee \Diamond p_1 \vee p_1)))$$

$$\Leftrightarrow \quad \tilde{\forall} P_1 \tilde{\forall} Q_1 \neg(ST_{x_0}(\Diamond(p_1 \wedge \Box q_1)) \wedge ST_{x_0}(\Diamond \neg(q_1 \vee \Diamond p_1 \vee p_1)))$$

$$\Leftrightarrow \quad \tilde{\forall} P_1 \tilde{\forall} Q_1 \neg(ST_{x_0}(\Diamond(p_1 \wedge \Box q_1) \wedge \Diamond \neg(q_1 \vee \Diamond p_1 \vee p_1))).$$

From this, the fastest way to proceed is by observing that the last formula is equivalent to

$$\tilde{\forall} P_1 \tilde{\forall} Q_1 \,(ST_{x_0}(\Diamond(p_1 \wedge \Box q_1) \rightarrow \neg\Diamond\neg(q_1 \vee \Diamond p_1 \vee p_1))),$$

and hence, to the Sahlqvist formula

$$\Diamond(p_1 \wedge \Box q_1) \rightarrow \Box(q_1 \vee \Diamond p_1 \vee p_1). \qquad \dashv$$

Example 3.62 Consider the type 1 Kracht formula

$$\alpha(x_0) \equiv (\forall x_1 \triangleright x_0)(\exists y_1 \triangleright x_1)\, y_1 \neq y_1.$$

According to (3.21), we can rewrite $\alpha(x_0)$ into the equivalent

$$\tilde{\forall} P_0(\forall x_1 \triangleright x_0)\,(ST_{x_0}(p_0) \rightarrow (\exists y_1 \triangleright x_1)ST_{y_1}(\bot))$$

and, hence, to

$$\tilde{\forall} P_0(\forall x_1 \triangleright x_0)\,(ST_{x_0}(p_0) \rightarrow ST_{x_1}(\Diamond\bot)).$$

This is a type 2 formula for which we can find a Sahlqvist equivalent as follows:

$$\tilde{\forall} P_0(\forall x_1 \triangleright x_0)\,(ST_{x_0}(p_0) \rightarrow ST_{x_1}(\Diamond\bot))$$
$$\Leftrightarrow \quad \tilde{\forall} P_0(\forall x_1 \triangleright x_0)\,\neg(ST_{x_0}(p_0) \wedge \neg ST_{x_1}(\Diamond\bot))$$
$$\Leftrightarrow \quad \tilde{\forall} P_0 \neg(\exists x_1 \triangleright x_0)\,(ST_{x_0}(p_0) \wedge ST_{x_1}(\neg\Diamond\bot))$$
$$\Leftrightarrow \quad \tilde{\forall} P_0 \neg\,(ST_{x_0}(p_0) \wedge (\exists x_1 \triangleright x_0)ST_{x_1}(\neg\Diamond\bot))$$
$$\Leftrightarrow \quad \tilde{\forall} P_0 \neg\,(ST_{x_0}(p_0) \wedge ST_{x_0}(\Diamond\neg\Diamond\bot))$$
$$\Leftrightarrow \quad \tilde{\forall} P_0 \,(ST_{x_0}(\neg(p_0 \wedge \Diamond\neg\Diamond\bot)).$$

The latter formula is equivalent to the Sahlqvist formula $p_0 \rightarrow \Box\Diamond\bot$. (Obviously, the latter formula is equivalent to $\Box\Diamond\bot$ and, hence, to $\Box\bot$. Our algorithm will not always provide the simplest correspondents!) $\qquad \dashv$

This finishes our discussion of Sahlqvist correspondence. In the next chapter we will see that Sahlqvist formulas also have very nice completeness properties, in that any modal logic axiomatized by Sahlqvist formulas is complete with respect to the class of frames defined by (the global first-order correspondents of) the formulas. Here Kracht's Theorem can be useful: if we want to axiomatize a class of frames defined by formulas of the form $\forall x\, \alpha(x)$ with $\alpha(x)$ a Kracht formula, then it suffices to compute the Sahlqvist correspondents of these formulas and add these as axioms to the basic modal logic.

Exercises for Section 3.7

3.7.1   (a) Prove that the conjunction $M \wedge 4$ of McKinsey's formula $\square\Diamond p \to \Diamond\square p$ and the transitivity formula $\Diamond p \to \Diamond\Diamond p$ does not have a *local* first-order correspondent. Conclude that this conjunction is not equivalent to a Sahlqvist formula.
   (b) Show that on the other hand, the formula $\square M \wedge 4$ *does* have a local first-order correspondent.

3.7.2 Prove that the local correspondent of a Sahlqvist formula is a Kracht formula.

3.7.3 Find Sahlqvist formulas that locally correspond to the following formulas:

(a) $(\forall y \triangleright x)\, Ryy$,
(b) $(\forall y_1 \triangleright x)(\forall y_2 \triangleright x)(\forall y_3 \triangleright x)\, (y_1 = y_2 \vee y_1 = y_3 \vee y_2 = y_3)$,
(c) $(\forall y_1 \triangleright x)(\forall y_2 \triangleright y_1)\, (y_1 = y_2 \vee \exists z\, (Rxz \vee (Ry_1 z \wedge Ry_2 z)))$,
(d) $(\forall x_1 \triangleright x)(\exists y_1 \triangleright x)(\forall y_2 \triangleright y_1)\, (Ry_1 x_1 \vee (Rxy_2 \wedge Ry_2 x_1))$.

3.7.4 Prove that if $\phi \to \psi$ is a simple Sahlqvist formula, then $\square(\phi \to \psi)$ is equivalent to a simple Sahlqvist formula.

3.7.5 Consdier the basic temporal similarity type. Show that over the class of bidirectional frames, every simple Sahlqvist formula is equivalent to a very simple Sahlqvist formula. (Hint: first find a very simple Sahlqvist formula that is equivalent to the formula $FGp \to GFp$.)

3.8 Advanced Frame Theory

The main aim of this section is to prove Theorem 3.19, the Goldblatt-Thomason Theorem, characterizing the elementary frame classes that are modally definable. We will also prove a rather technical result needed in our later work on algebras. We will start by proving the Goldblatt-Thomason Theorem.

Theorem 3.19 *Let τ be a modal similarity type. A first-order definable class K of τ-frames is modally definable if and only if it is closed under taking bounded morphic images, generated subframes, disjoint unions and reflects ultrafilter extensions.*

Proof. The preservation direction follows from earlier results. For the other direction let K be a class of frames which is elementary (hence, closed under taking ultraproducts), closed under taking bounded morphic images, generated subframes and disjoint unions, and reflecting ultrafilter extensions. Let Λ_K be the logic of K; that is, $\Lambda_K = \{\phi \mid \mathfrak{F} \Vdash \phi, \text{ for all } \mathfrak{F} \in K\}$. We will show that Λ_K defines K. In order to avoid cumbersome notation we restrict ourselves to the basic modal similarity type.

Let $\mathfrak{F} = (W, R)$ be a frame such that $\mathfrak{F} \Vdash \Lambda_K$. We need to show that $\mathfrak{F}$ is a member of K. This we will do by moving around lots of structures; here is a map of where we are heading for in the proof:

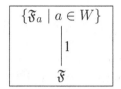

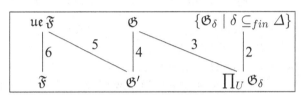

First, we can assume without loss of generality that $\mathfrak{F}$ is point-generated. For if $\mathfrak{F}$ validates Λ_K, then each of its point-generated subframes does so as well. And if we can prove that each point-generated subframe of $\mathfrak{F}$ is in K, then the membership in K of $\mathfrak{F}$ itself follows immediately from the closure properties of K and the fact that any frame is a bounded morphic image of the disjoint union of its point-generated subframes (as the reader was asked to show in Exercise 3.3.4). So from now on we assume that $\mathfrak{F}$ is generated by the point w.

Now for (one of) the main idea(s) of the proof. Let Φ be a set of proposition letters containing a proposition letter p_A for each subset A of W. This may be a huge language: if W is infinite, then Φ will be uncountable. We will look at the model $\mathfrak{M} = (\mathfrak{F}, V)$ where V is the natural valuation given by $V(p_A) = A$. Now let Δ be the modal type of w; that is, $\Delta = \{\phi \in ML(\tau, \Phi) \mid \mathfrak{M}, w \Vdash \phi\}$. We claim that

$$\Delta \text{ is satisfiable in K.} \tag{3.26}$$

In order to prove this, we first show that Δ is finitely satisfiable in K. Let δ be a finite subset of Δ. It is easy to see that δ is satisfiable in K: if it were not, then $\neg \bigwedge \delta$ would belong to Λ_K whence we would have $\mathfrak{F} \Vdash \neg \bigwedge \delta$. (Note that whereas Δ is written in a *particular* language, namely, the one having a proposition letter for each subset of W, when we are talking about Λ_K we are not really interested in a specific language. This is why we simply assume that '$\neg \bigwedge \delta$ would belong to Λ_K' even though we have not verified that this formula uses only proposition letters that occur in Λ_K.) But $\mathfrak{F} \Vdash \neg \bigwedge \delta$ would contradict that $\mathfrak{M}, w \Vdash \bigwedge \delta$. Thus each finite $\delta \subset \Delta$ is finitely satisfiable in some frame $\mathfrak{G}_\delta$ in K, so Δ is satisfiable in some ultraproduct of these frames (the reader is asked to supply a proof of this in Exercise 3.8.2 below). Since K is closed under ultraproducts by assumption, this proves (3.26).

But to say that Δ is satisfiable in K amounts to the following. There is a model $\mathfrak{N} = (X, S, U)$ and a point b in X such that the underlying frame $\mathfrak{G} = (X, S)$ is in K and $\mathfrak{N}, b \Vdash \Delta$. Since K is closed under (point-)generated subframes and modal truth is preserved under taking generated subframes, we may assume that the frame $\mathfrak{G}$ is generated from b.

The only thing left to do is to link up $\mathfrak{G}$ with our original frame $\mathfrak{F}$. This link is as follows:

$$\mathfrak{ue}\,\mathfrak{F} \text{ is a bounded morphic image of some ultrapower of } \mathfrak{G}. \tag{3.27}$$

We first ensure the existence of an m-saturated ultrapower of $\mathfrak{N}$. Note that we may view $\mathfrak{N}$ as a first-order structure for the language $\mathcal{L}^1_\tau(\Phi)$, analogous to the perspective in the previous chapter. Now consider a countably saturated ultrapower of this first-order structure, which we see again as a modal model $\mathfrak{N}' = (X', S', U')$. Note that the existence of such an ultrapower is not guaranteed by Lemma 2.73, since the first-order language $\mathcal{L}^1_{\tau, \mathfrak{F}}$ may not be countable. We need some heavier model-theoretic equipment here; the reader is referred to Theorems 6.1.4 and 6.1.8 in [91]. In any case, $\mathfrak{N}'$ is m-saturated and also has the property that every set Σ that is finitely satisfiable in $\mathfrak{N}'$ is satisfiable in $\mathfrak{N}'$.

How are we going to define the bounded morphism? That is, given an element s of X', which ultrafilter over W (the universe of our original frame $\mathfrak{F}$) are we going to assign to it? Recall that an ultrafilter over W is some collection of subsets of W; this means that given s, we have to decide for each subset of W whether to put it in $f(s)$ or not. But now it will become clear that there is only one natural choice for $f(s)$: simply put a subset A of W in $f(s)$ if p_A is true at s in the model $\mathfrak{N}'$:

$$f(s) = \{ A \subseteq W \mid \mathfrak{N}', s \Vdash p_A \}.$$

We will now show that f indeed maps points in $\mathfrak{N}'$ to ultrafilters over W, that f is a bounded morphism, and that f is onto $\mathfrak{ue}\,\mathfrak{F}$. In these proofs, the following equivalence comes in handy:

$$\text{for all formulas } \phi \in ML(\tau, \Phi), \; \mathfrak{M} \Vdash \phi \text{ iff } \mathfrak{N}' \Vdash \phi. \tag{3.28}$$

The proof of (3.28) is by the following chain of equivalences:

$$
\begin{array}{llll}
\mathfrak{M} \Vdash \phi & \Leftrightarrow & \mathfrak{M}, w \Vdash \Box^n \phi \text{ for all } n \in \mathbb{N} & (\mathfrak{M} \text{ is generated from } w) \\
& \Leftrightarrow & \Box^n \phi \in \Delta \text{ for all } n \in \mathbb{N} & (\text{definition of } \Delta) \\
& \Leftrightarrow & \mathfrak{N}, b \Vdash \Box^n \phi \text{ for all } n \in \mathbb{N} & (\text{definition of } \mathfrak{N} \text{ and } b) \\
& \Leftrightarrow & \mathfrak{N} \Vdash \phi & (\mathfrak{N} \text{ is generated from } b) \\
& \Leftrightarrow & \mathfrak{N}' \Vdash \phi & (\mathfrak{N}' \text{ is an ultrapower of } \mathfrak{N}).
\end{array}
$$

This proves (3.28).

Let us now first check that for all $s \in X'$, $f(s)$ is indeed an ultrafilter over W. We will only check the condition that $f(s)$ is closed under intersection, leaving the other conditions as exercises for the reader. Suppose that A and B are subsets of W that both belong to $f(s)$. Hence, by the definition of $f(s)$ we have that $\mathfrak{N}, s \Vdash p_A$ and $\mathfrak{N}', s \Vdash p_B$. It is easy to see that the formula $p_A \wedge p_B \leftrightarrow p_{A \cap B}$ holds throughout the original model $\mathfrak{M}$. It then follows from (3.28) that $\mathfrak{N} \Vdash p_A \wedge p_B \leftrightarrow p_{A \cap B}$. In particular, this formula is true at s, so we find that $\mathfrak{N}, s \Vdash p_{A \cap B}$. Hence, by the definition of f, $A \cap B$ belongs to $f(s)$.

In order to show that f is a bounded morphism, we will prove that for all ultrafilters u over W and all points s in X', we have that $u = f(s)$ if and only if u (in $\mathfrak{ue}\,\mathfrak{M}$) and s (in $\mathfrak{N}'$) satisfy the same formulas. This suffices, by Proposition 2.54 and the m-saturation of $\mathfrak{ue}\,\mathfrak{M}$ and $\mathfrak{N}$. The right to left direction of the equivalence is easy to prove. If the same formulas hold in s and u, then in particular we have for each $A \subseteq W$ that $\mathfrak{N}', s \Vdash p_A$ iff $\mathfrak{ue}\,\mathfrak{M}, u \Vdash p_A$. But by definition of the valuation on $\mathfrak{ue}\,\mathfrak{M}$ we have that $\mathfrak{ue}\,\mathfrak{M}, u \Vdash p_A$ iff $A = V(p_A) \in u$. Hence, we find that $\mathfrak{N}', s \Vdash p_A$ iff $A \in u$. This immediately yields $u = f(s)$.

For the other direction, it suffices to show that for each formula $\phi \in ML(\tau, \Phi)$ and each point s in $\mathfrak{N}$, $\mathfrak{ue}\,\mathfrak{M}, f(s) \Vdash \phi$ only if $\mathfrak{N}', s \Vdash \phi$. Suppose that ϕ holds at $f(s)$ in $\mathfrak{ue}\,\mathfrak{M}$. By Proposition 2.59 we have that $V(\phi) \in f(s)$. Thus by definition of f we obtain that $\mathfrak{N}', s \Vdash p_{V(\phi)}$. It follows easily from the definition of V that $\mathfrak{M} \Vdash \phi \leftrightarrow p_{V(\phi)}$, so by (3.28) we have that $\mathfrak{N} \Vdash \phi \leftrightarrow p_{V(\phi)}$. But then we may immediately infer that $\mathfrak{N}', s \Vdash \phi$.

Finally, we have to show that f is surjective; that is, each ultrafilter over W should belong to its range. Let u be such an ultrafilter; we claim that the set $\Sigma = \{p_A \mid A \in u\}$ is finitely satisfiable in $\mathfrak{N}$. Let σ be a finite subset of Σ. To start with, σ is satisfiable in $\mathfrak{M}$. Since $\mathfrak{M}$ is generated from w, this shows that $\mathfrak{M}, w \Vdash \Diamond^n \bigwedge \sigma$ for some natural number n. From the definition of $\mathfrak{N}$ and b it follows that $\mathfrak{N}, b \Vdash \Diamond^n \bigwedge \sigma$, so from the fact that $\mathfrak{N}$ is point-generated from b we obtain that $\bigwedge \sigma$ is satisfiable in $\mathfrak{N}$. Now $\mathfrak{N}'$ is an ultrapower of $\mathfrak{N}$, so we have that $\bigwedge \sigma$ is also satisfiable in $\mathfrak{N}'$. But $\mathfrak{N}'$ is countably saturated; so Σ, being finitely satisfiable in $\mathfrak{N}'$, is satisfiable in some point s of $\mathfrak{N}'$. It is then immediate that $f(s) = u$.

This proves (3.27), but why does that mean that $\mathfrak{F}$ belongs to K? Here we use the closure properties of K. Recall that $\mathfrak{G}$ is the underlying frame of the model $\mathfrak{N}$ in which we assumed that the set Δ is satisfiable. Since $\mathfrak{G}$ is in K by assumption, $\mathfrak{G}'$ belongs to K by closure under ultraproducts; $\mathfrak{ue}\,\mathfrak{F}$ is in K as it is a bounded morphic image of $\mathfrak{G}'$; and finally, $\mathfrak{F}$ is in K since K reflects ultrafilter extensions. $\dashv$

The following proposition, which is of a rather technical nature, will be put to good use in Chapter 5.

Proposition 3.63 *Let τ be a modal similarity type, and* K *a class of τ-frames. Suppose that $\mathfrak{G}$ is an ultrapower of the disjoint union $\biguplus_{i \in I} \mathfrak{F}_i$, where $\{\mathfrak{F}_i \mid i \in I\}$ is a family of frames in* K*. Then $\mathfrak{G}$ is a bounded morphic image of a disjoint union of ultraproducts of frames in* K*.*

Proof. Let $\mathfrak{F} = (W, R)$ denote the disjoint union $\biguplus_{i \in I} \mathfrak{F}_i$, and assume that $\mathfrak{G}$ is some ultrapower of $\mathfrak{F}$, say $\mathfrak{G} = \prod_U \mathfrak{F}$, where U is an ultrafilter over some index set J. We assume that τ contains only one operator $\triangle$, of arity n. This allows us to write $\mathfrak{F} = (W, R)$ and $\mathfrak{F}_i = (W_i, R_i)$ (that is, the subscript i refers to an index element of I, not to an operator from the similarity type).

Consider an arbitrary state t of $\mathfrak{G}$. By the definition of ultrapowers, there exists a sequence $f_t \in \prod_{j \in J} W$ such that

$$t = (f_t)_U = \{g \in \textstyle\prod_{j \in J} W \mid f_t \sim_U g\}.$$

As W is the disjoint union of the universes W_i, for each $j \in J$ there exists an element $i_j \in I$ such that $f_t(j)$ is an element of W_{i_j}. Form the ultraproduct

$$\mathfrak{F}_t := \textstyle\prod_U \mathfrak{F}_{i_j}.$$

Clearly this frame is an ultraproduct of frames in K.

We will now define a map θ_t sending states of the frame $\mathfrak{F}_t$ to states of the frame $\mathfrak{G}$, and show that θ_t is a bounded morphism with t in its range. From this it easily follows that $\mathfrak{G}$ is a bounded morphic image of the disjoint union $\biguplus_{t \in X} \mathfrak{F}_t$, where X is the universe of $\mathfrak{G}$. Observe that a typical element of $\mathfrak{F}_t$ has the form

$$g^U := \{h \in \textstyle\prod_{j \in J} W_{i_j} \mid g \sim_U h\}$$

for some $g \in \prod_{j \in J} W_{i_j}$. Since $\prod_{j \in J} W_{i_j} \subseteq \prod_{j \in J} W$, we have that $g^U \subseteq g_U$. Note that in general these two equivalence classes will not be identical, since g_J may contain elements h for which $h(j') \in W \setminus W_{i_{j'}}$ for some index j'. However, it is evident that *if* both g and h are in $\prod_{j \in J} W_{i_j}$, then we find that $g^U = h^U$ iff $g \sim_U h$ iff $g_U = h_U$. This means that if we put

$$\theta_t(g^U) := g_U,$$

we have found a well-defined map from the universe of $\mathfrak{F}_t$ to the universe X of $\mathfrak{G}$ (in fact, this map is injective).

Now consider the element $f_t \in \prod_{j \in J} W$. By definition of the indices i_j, we must have $f_t \in \prod_{j \in J} W_{i_j}$. It follows that f_t^U is in the domain of θ_t. Now

$$\theta_t(f_t^U) = (f_t)_U = t.$$

It remains to be proved that θ_t is a bounded morphism. However, this follows by a straightforward argument using standard properties of ultrafilters. $\dashv$

Exercises for Section 3.8

3.8.1 Let τ be an arbitrary modal similarity type and $\mathfrak{F}$ a τ-frame. Prove that the ultrafilter extension of $\mathfrak{F}$ is the bounded morphic image of some ω-saturated ultrapower of $\mathfrak{F}$; in other words, supply a proof for Theorem 3.17. (Hint: use an argument analogous to one in the proof of Theorem 3.19. That is, consider a language having a proposition letter p_A for each subset A of the universe of $\mathfrak{F}$, and take a countably saturated ultrapower of the model $\mathfrak{M} = (\mathfrak{F}, V)$, where V is the natural valuation mapping p_A to A for each variable p_A.)

3.8.2 Let K be some class of frames, and Δ a set of formulas which is finitely satisfiable in K. Show that Δ is satisfiable in an ultraproduct of frames in K.

3.8.3 (a) Show that the complement of a modally definable class is closed under taking ultrapowers.

Now suppose that the class K of frames is definable by a *single* formula ϕ.

   (b) Show that the complement of K is closed under taking ultraproducts. (Hint: let $\Gamma(\phi)$ be the set of first-order sentences that are semantic consequences of ϕ, in the sense that for any frame $\mathfrak{F}$ we have that $\mathfrak{F} \Vdash \phi$ only if $\mathfrak{F} \models \Gamma(\phi)$. In other words, $\Gamma(\phi)$ is the first-order theory of K.)
   (c) Prove that ϕ is a semantic consequence of $\Gamma(\phi)$. (Hint: reason by contraposition and use (b).)
   (d) Prove that ϕ is a semantic consequence of a finite subset of $\Gamma(\phi)$. (Hint: prove that $\Gamma(\phi) \models \forall x \, ST_x(\phi)$, and use compactness.)
   (e) Conclude that if a modal formula ϕ defines an elementary frame class, then ϕ corresponds to a (single) first-order formula.

3.8.4 Prove the strong version of the Goldblatt-Thomason Theorem which applies to any frame class that is closed under taking ultrapowers. (Hint: strengthen the result of Exercise 3.8.2 by showing that any set of modal formulas that is finitely satisfiable in a frame class K is itself satisfiable in an ultrapower of a disjoint union of frames in K.)

3.8.5 Point out where, in the picture summarizing the proof of Theorem 3.19, we use which closure conditions on K. (For instance: in step 2 we need the fact that K is closed under taking ultraproducts.)

3.9 Summary of Chapter 3

▶ *Frame Definability*: A modal formula is valid on a frame if and only if it is satisfied at every point in the frame, no matter which valuation is used. A modal formula defines a class of frames if and only if it is valid on precisely the frames in that class.

▶ *Frame Definability is Second-Order*: Because the definition of validity quantifies across all possible valuations, and because valuations are assignments of *subsets* of frames, the concept of validity, and hence frame definability, is intrinsically second-order.

▶ *Frame Languages*: Every modal formula can be translated into the appropriate second-order frame language. Such languages have an $(n + 1)$-place relation symbol for every n-place modality. Proposition letters correspond to unary predicate variables. The required translation is called the second-order translation. This is simply the standard translation modified to send proposition letters to (unary) predicate *variables* rather than predicate *constants*.

▶ *Correspondence*: Sometimes the second-order formulas obtained using this translation are equivalent to first-order formulas. But often they correspond to genuinely second-order formulas. This can sometimes be shown by exhibiting a failure of Compactness or the Löwenhein-Skolem property.

▶ *Frame Constructions*: The four fundamental model constructions discussed in the previous chapter have obvious frame-theoretic counterparts. Moreover, *validity* is preserved under the formation of disjoint unions, generated subframes and bounded morphic images, and anti-preserved under ultrafilter extensions.

▶ *Goldblatt-Thomason Theorem*: A first-order definable frame class is modally definable if and only if it is closed under disjoint unions, generated subframes and bounded morphic images, and reflects ultrafilter extensions.

▶ *Modal Definability on Finite Transitive Frames*: A class of finite transitive frames is modally definable if and only if it is preserved under (finite) disjoint unions, generated subframes and bounded morphic images.

▶ *The Finite Frame Property*: A normal modal logic Λ has the finite frame property if and only if any formula that does not belong to Λ can be falsified on a finite frame that validates all the formulas in Λ. A normal logic has the finite frame property if and only if it has the finite model property.

▶ *The Sahlqvist Fragment*: Formulas in the Sahlqvist fragment have the property that the second-order formula obtained via the second-order translation can be reduced to an equivalent first-order formula. The Sahlqvist-van Benthem algorithm is an effective procedure for carrying out such reductions.

▶ *Why Sahlqvist Formulas have First-Order Correspondents*: Syntactically, the Sahlqvist fragment forbids universal operators to take scope over existential or disjunctive connectives in the antecedent. Semantically, this guarantees that we will always be able to find a unique minimal valuation that makes the antecedent true. This ensures that Sahlqvist formulas have first-order correspondents.

▶ *Negative Results*: There are non-Sahlqvist formulas that define first-order conditions. Moreover, Chagrova's Theorem tells us that it is undecidable whether a modal formula has a first-order equivalent.

▶ *Kracht's Theorem*: Kracht's Theorem takes us back from first-order languages to modal languages. It identifies a class of first-order formulas that are the first-order correspondents of Sahlqvist formulas.

▶ *Frames and their Ultrafilter Extensions*: The ultrafilter extension of a frame may be obtained as a bounded morphic image of an ultrapower of the frame.

▶ *Ultrapowers of Disjoint Unions*: Ultrapowers of a disjoint union may be obtained as bounded morphic images of disjoint unions of ultraproducts.

Notes

The study of frames has been central to modal logic since the dawn of the classical era (see the Historical Overview in Chapter 1), but the way frames have been studied has changed dramatically over this period. The insight that gave birth to the classical era was that simple properties of frames (such as transitivity and reflexivity) could be used to characterize normal modal logics, and most of the 1960s were devoted to exploring this topic. It is certainly an important topic. For example, in the first half of the following chapter we will see that most commonly encountered modal logics can be given simple, intuitively appealing, frame-based characterizations. But the very success of this line of work meant that for a decade modal logicians paid little attention to modal languages as tools for *describing* frame structure. Frames were simply tools for analyzing normal logics. The notion of frame definability, and the systematic study of modal expressivity over frames, only emerged as a research theme after the frame incompleteness results showed that not all normal logics could be given frame-based characterizations. The first incompleteness result (shown for the basic temporal language) was published in 1972 by S.K. Thomason [433]. The first incompleteness results for the basic modal language were published in 1974 by S.K. Thomason [434] and Kit Fine [129].

The frame incompleteness theorems and the results which accompanied them decisively changed the research agenda of modal logic, essentially because they made it clear that the modal perspective on frames was intrinsically second-order. We have seen ample evidence for this in this chapter: as we saw in Example 3.11 a formula as innocuous looking as McKinsey's $\Box\Diamond p \rightarrow \Diamond\Box p$ defines a non-elementary class of frames. This was proved independently by Goldblatt [183] and van Benthem [35]. The proof given in the text is from Theorem 10.2 of van Benthem [42]. It was shown by S.K. Thomason [436] that on the level of frames, modal logic is expressive enough to capture the semantic consequence relation for $\mathcal{L}^2$. Moreover, in unpublished work, Doets showed that modal formulas can act as a reduction class for the theory of finite types; see van Benthem [42, pages 23–24] for further discussion.

So by the mid 1970s it was clear that modal logic embodied a substantial fragment of second-order logic, and a radically different research program was well under way. One strand of this program was algebraic: these years saw the (re)-emergence of algebraic semantics together with a belated appreciation of the work of Jónsson and Tarski [255, 256]; this line of work is treated in Chapter 5. The other strand was the emergence of correspondence theory.

Given that modal logic over frames is essentially second-order logic in disguise, it may seem that the most obvious way to develop correspondence theory would be to chart the second-order powers of modal logic. In fact, examples of modal formulas that define second-order classes of frames were known by the early 1970s (for example, Johan van Benthem proved that the Löb formula defined the class of transitive and converse well-founded frames using the argument given in Example 3.9). And there is interesting work on more general results on second-order frame definability, much of which may be found in Chapters XVII–XIX of van Benthem [42]. Nonetheless, most work on correspondence theory for frames has concentrated on its *first-order* aspects. There are two main reasons for this. First, second-order model theory is less well understood than first-order model theory, so investigations of second-order correspondences have fewer useful results to draw on. Second, there is a clear sense that it is the first-order aspects of frame definability which are truly mysterious (this has long been emphasized by Johan van Benthem). With the benefit of hindsight, the second-order nature of validity is obvious; understanding when – and why – it is sometimes first-order is far harder.

In this chapter we examined the two main strands in first-order correspondence theory (for frames): the *semantic*, exemplified by the Goldblatt-Thomason Theorem, and the *syntactic*, exemplified by the Sahlqvist Correspondence Theorem. (Incidentally, as we will learn in Chapter 5, both results have a substantial algebraic dimension.)

What we call the Goldblatt-Thomason Theorem was actually proved by Goldblatt. His result was in fact stronger than our Theorem 3.19, applying to any frame class that is closed under elementary equivalence. This theorem was published in a joint paper [188] with S.K. Thomason, who added a more general result which applies to all definable frame classes but has a less appealing frame construction. The model-theoretic proof of the theorem that we supplied in this chapter is due to van Benthem [46], who also proved the finite transitive version we recorded as Theorem 3.21. Barwise and Moss [28] obtain correspondence results for *models* as opposed to frames; their main result is that if a modal formula ϕ has a first-order frame correspondent c_ϕ, then for all models $\mathfrak{M}$, $\mathfrak{M}$ satisfies all substitution instances of ϕ in infinitary modal logic iff a certain frame underlying $\mathfrak{M}$ satisfies c_ϕ.

Concerning the identification of syntactic classes of modal formulas that correspond to first-order formulas, Sahlqvist's result was not the first. As early as in the Jónsson-Tarski papers [255, 256] particular examples such as reflexivity and transitivity were known. And an article by Fitch [136] was a stimulus for van Benthem's investigations in this area, which lead to van Benthem (unaware of Sahlqvist's earlier work) proving what is now known as Sahlqvist's Theorem. But Sahlqvist's paper [396] (essentially a presentation of results contained in his Masters thesis) remains the classic reference in the area. It greatly generalized all previous known results in the area and drew a beautiful link between definability and completeness.

Kracht isolated the first-order formulas that are the correspondents of Sahlqvist formulas in [276], as an application of his so-called calculus of internal describability. This calculus relates modal and first-order formulas on the level of general frames; see also [279].

During the 1990s a number of alternative correspondence languages have been considered for the basic modal language. In the so-called functional translation the accessibility relations are replaced by certain terms which can be seen as functions mapping worlds to accessible worlds. From a certain point of view this functional language is more expressive than the relational language, and certain second-order frame properties can be mapped to formulas expressed in the functional language – but this is not too surprising: in the functional language one can quantify over functions; this additional expressive power allows one to do without quantification over unary predicate variables; see Ohlbach and Schmidt [346] and Ohlbach *et al.* [344] and Simmons [414].

As with finite model theory, the theory of finite frames is rather underdeveloped. However, some of the basic results have been known a long time. We showed in Theorem 3.28 that a normal logic has the finite model property if and only if it has the finite frame property. This result is due to Segerberg [404, Corollary 3.8, page 33]. For some interesting results concerning frame correspondence theory over the class of finite frames the reader should consult the dissertation of Doets [110].

To conclude these Notes, we will tidy up a few loose ends. Exercise 3.6.2 is due to van Benthem [42, Theorem 10.4]. Exercise 3.2.4 is based on a result in Fine [132]. Second, we mentioned Chagrova's Theorem [89] that it is undecidable whether a modal formula has a first-order equivalent. For pointers to, and a brief discussion of, extensions of this line of work, see Chagrov and Zakharyaschev [88, Chapter 17]. At the end of Section 3.2 we remarked that general frames can be seen as a model version of the *generalized models* or *Henkin models* for second-order logic. Henkin [217] introduced such models, and good discussions of them can be found in Doets and van Benthem [112] or Manzano [314]. Finally, for more on the lambda calculus see Barendregt [24] or Hindley and Seldin [223].

4

Completeness

This chapter is about the completeness – and incompleteness – of normal modal logics. As we saw in Section 1.6, normal modal logics are collections of formulas satisfying certain simple closure conditions. They can be specified either syntactically or semantically, and this gives rise to the questions which dominate the chapter: Given a semantically specified logic, can we give it a syntactic characterization, and if so, how? And: Given a syntactically specified logic, can we give it a semantic characterization (and in particular, a characterization in terms of frames), and if so, how? To answer either type of question we need to know how to prove (soundness and) *completeness* theorems, and the bulk of the chapter is devoted to developing techniques for doing so.

The chapter has two major parts. The first, comprising the first four sections, is an introduction to basic completeness theory. It introduces canonical models, explains and applies the completeness-via-canonicity proof technique, discusses the Sahlqvist Completeness Theorem, and proves two fundamental limitative results. The material introduced in these sections (which are all on the basic track) is needed to follow the second part and the algebraic investigations of Chapter 5.

In the second part of the chapter we turn to the following question: what are we to do when canonicity fails? (As will become clear, canonicity failure is a fact of life for temporal logic, propositional dynamic logic, and other applied modal languages.) This part of the chapter is technique oriented: it introduces five important ways of dealing with such difficulties.

Chapter guide

Section 4.1: Preliminaries (Basic track). This section introduces the fundamental concepts: normal modal logics, soundness, and completeness.

Section 4.2: Canonical Models (Basic track). Canonical models are introduced, and the fundamental Canonical Model Theorem is proved.

Section 4.3: Applications (Basic track). This section discusses the key concept of canonicity, and uses completeness-via-canonicity arguments to put canonical models to work. We prove completeness results for a number of modal and temporal logics, and finish with a discussion of the *Sahlqvist Completeness Theorem*.

Section 4.4: Limitative Results (Basic track). We prove two fundamental limitative results: not all normal logics are canonical, and not all normal logics are characterized by some class of frames. This section concludes our introduction to basic completeness theory.

Section 4.5: Transforming the Canonical Model (Basic track). Often we need to build models with properties for which we lack a canonical formula. What are we to do in such cases? This section introduces one approach: use transformation methods to try and massage the 'faulty' canonical model into the required shape.

Section 4.6: Step by Step (Basic track). Sometimes we can cope with canonicity failure using the step-by-step method. This is a technique for building models with special properties inductively.

Section 4.7: Rules for the Undefinable (Basic track). Special proof rules (that in a certain sense manage to express undefinable properties of models and frames) sometimes allow us to construct special canonical models containing submodels with undefinable properties.

Section 4.8: Finitary Methods I (Basic track). We discuss a method for proving weak completeness results for non-compact logics: *finite* canonical models. We use such models to prove the completeness of propositional dynamic logic.

Section 4.9: Finitary Methods II (Advanced track). This section further explores finitary methods, this time the direct use of filtrations. We illustrate this with an analysis of the normal logics extending **S4.3**.

4.1 Preliminaries

In this section we introduce some of the fundamental concepts that we will use throughout the chapter. We begin by defining *modal logics* – these could be described as propositional logics in a modal language.

Throughout the chapter we assume we are working with a fixed countable language of proposition letters.

Definition 4.1 (Modal Logics) A *modal logic* Λ is a set of modal formulas that contains all propositional tautologies and is closed under *modus ponens* (that is, if $\phi \in \Lambda$ and $\phi \to \psi \in \Lambda$ then $\psi \in \Lambda$) and *uniform substitution* (that is, if ϕ belongs to Λ then so do all of its substitution instances). If $\phi \in \Lambda$ we say that ϕ is a *theorem* of Λ and write $\vdash_\Lambda \phi$; if not, we write $\nvdash_\Lambda \phi$. If Λ_1 and Λ_2 are modal logics such that $\Lambda_1 \subseteq \Lambda_2$, we say that Λ_2 is an *extension* of Λ_1. In what follows, we usually drop the word 'modal' and talk simply of 'logics.'   ⊣

Note that modal logics contain all substitution instances of the propositional tautologies: for example, $\Diamond p \vee \neg \Diamond p$, belongs to every modal logic. Even though such substitution instances may contain occurrences of $\Diamond$ and $\Box$, we still call them *tautologies*. Clearly tautologies are valid in every class of models.

Example 4.2   (i) The collection of all formulas is a logic, the *inconsistent logic*.

(ii) If $\{\Lambda_i \mid i \in I\}$ is a collection of logics, then $\bigcap_{i \in I} \Lambda_i$ is a logic.

(iii) Define Λ_{S} to be $\{\phi \mid \mathfrak{S} \Vdash \phi, \text{ for all structures } \mathfrak{S} \in \mathsf{S}\}$, where S is any class of frames or any class of general frames. Λ_{S} is a logic. If S is the singleton class $\{\mathfrak{S}\}$, we usually call this logic $\Lambda_{\mathfrak{S}}$, rather than $\Lambda_{\{\mathfrak{S}\}}$.

(iv) If M is a class of models, then Λ_{M} need *not* be a logic. Consider a model $\mathfrak{M}$ in which p is true at all nodes but q is not. Then $p \in \Lambda_{\mathfrak{M}}$, but $q \notin \Lambda_{\mathfrak{M}}$. But q is obtainable from p by uniform substitution.   ⊣

It follows from Examples 4.2(i) and 4.2(ii) that there is a smallest logic containing any set of formulas Γ; we call this *the logic generated by* Γ. For example, the logic generated by the empty set contains all the tautologies and nothing else; we call it **PC** and it is a subset of every logic. This generative perspective is essentially *syntactic*. However, as Example 4.2(iii) shows, there is a natural *semantic* perspective on logics: both frames and general frames give rise to logics in an obvious way. Even the empty class of frames gives rise to a logic, namely the inconsistent logic. Finally, Example 4.2(iv) shows that models may fail to give rise to logics. This 'failure' is actually the behavior we should expect: as we discussed in Section 1.6, genuine logics arise at the level of *frames*, via the concept of *validity*.

Definition 4.3 Let $\psi_1, \ldots, \psi_n, \phi$ be modal formulas. We say that ϕ is *deducible in propositional calculus from assumptions* $\psi_1, \ldots, \psi_n$ if $(\psi_1 \wedge \cdots \wedge \psi_n) \to \phi$ is a tautology.   ⊣

All logics are closed under deduction in propositional calculus: if ϕ is deducible in propositional calculus from assumptions $\psi_1, \ldots, \psi_n$, then $\vdash_\Lambda \psi_1, \ldots, \vdash_\Lambda \psi_n$ implies $\vdash_\Lambda \phi$.

Definition 4.4 If $\Gamma \cup \{\phi\}$ is a set of formulas then ϕ *is deducible in Λ from* Γ (or: ϕ *is Λ-deducible from* Γ) if $\vdash_\Lambda \phi$ or there are formulas $\psi_1, \ldots, \psi_n \in \Gamma$ such that

$$\vdash_\Lambda (\psi_1 \wedge \cdots \wedge \psi_n) \to \phi.$$

If this is the case we write $\Gamma \vdash_\Lambda \phi$, if not, $\Gamma \nvdash_\Lambda \phi$. A set of formulas Γ is Λ-*consistent* if $\Gamma \nvdash_\Lambda \bot$, and Λ-*inconsistent* otherwise. A formula ϕ is Λ-consistent if $\{\phi\}$ is Λ-consistent; otherwise ϕ is Λ-inconsistent. ⊣

It is a simple exercise in propositional logic to check that a set of formulas Γ is Λ-inconsistent if and only if there is a formula ϕ such that $\Gamma \vdash_\Lambda \phi \wedge \neg\phi$ if and only if for all formulas ψ, $\Gamma \vdash_\Lambda \psi$. Moreover, Γ is Λ-consistent if and only if every finite subset of Γ is. (That is, our notion of deducibility has the *compactness* property.) From now on, when Λ is clear from context or irrelevant, we drop explicit references to it and talk simply of 'theorems', 'deducibility', 'consistency' and 'inconsistency', and use the notation $\vdash \phi$, $\Gamma \vdash \phi$, and so on.

The preceding definitions merely generalize basic ideas of propositional calculus to modal languages. Now we come to a genuinely *modal* concept: *normal modal logics*. These logics are the focus of this chapter's investigations. We initially restrict our discussion to the basic modal language; the full definition is given at the end of the section. As we discussed in Section 1.6, the following definition is essentially an abstraction from Hilbert-style approaches to modal proof theory.

Definition 4.5 A modal logic Λ is *normal* if it contains the formulas:

(K) $\Box(p \rightarrow q) \rightarrow (\Box p \rightarrow \Box q)$,
(Dual) $\Diamond p \leftrightarrow \neg\Box\neg p$,

and is closed under *generalization* (that is, if $\vdash_\Lambda \phi$ then $\vdash_\Lambda \Box\phi$).   ⊣

Syntactic issues do not play a large role in this book; nonetheless, readers new to modal logic should study the following lemma and attempt Exercise 4.1.2.

Lemma 4.6 *For any normal logic Λ, if $\vdash_\Lambda \phi \leftrightarrow \psi$ then $\vdash_\Lambda \Diamond\phi \leftrightarrow \Diamond\psi$.*

Proof. Suppose $\vdash_\Lambda \phi \leftrightarrow \psi$. Then $\vdash_\Lambda \phi \rightarrow \psi$ and $\vdash_\Lambda \psi \rightarrow \phi$. If we can show that $\vdash_\Lambda \Diamond\phi \rightarrow \Diamond\psi$ and $\vdash_\Lambda \Diamond\psi \rightarrow \Diamond\phi$, the desired result follows. Now, as $\vdash_\Lambda \psi \rightarrow \psi$, we have $\vdash_\Lambda \neg\psi \rightarrow \neg\phi$, hence by generalization $\vdash_\Lambda \Box(\neg\psi \rightarrow \neg\phi)$. By uniform substitution into the K axiom we obtain $\vdash_\Lambda \Box(\neg\psi \rightarrow \neg\phi) \rightarrow (\Box\neg\psi \rightarrow \Box\neg\phi)$. It follows by modus ponens that $\vdash_\Lambda \Box\neg\psi \rightarrow \Box\neg\phi$. Therefore, $\vdash_\Lambda \neg\Box\neg\phi \rightarrow \neg\Box\neg\psi$, and two uses of Dual yield $\vdash_\Lambda \Diamond\phi \rightarrow \Diamond\psi$, as desired. As $\vdash \psi \rightarrow \phi$, an analogous argument shows that $\vdash_\Lambda \Diamond\psi \rightarrow \Diamond\phi$, and the result follows.   ⊣

Remark 4.7 The above definition of normal logics (with or without Dual, depending on the choice of primitive operators) is probably the most popular way of stipulating what normal logics are. But it is not the only way. Here, for example, is a simple diamond-based formulation of the concept, which will be useful in our later algebraic work: a logic Λ is normal if it contains the axioms $\Diamond\bot \leftrightarrow \bot$ and $\Diamond(p \vee q) \leftrightarrow \Diamond p \vee \Diamond q$, and is closed under the following rule: $\vdash_\Lambda \phi \rightarrow \psi$ implies

$\vdash_\Lambda \Diamond\phi \to \Diamond\psi$. This formulation is equivalent to Definition 4.5, as the reader is asked to show in Exercise 4.1.2.   ⊣

Example 4.8   (i) The inconsistent logic is a normal logic.

  (ii) **PC** is not a normal logic.

  (iii) If $\{\Lambda_i \mid i \in I\}$ is a collection of normal logics, then $\bigcap_{i\in I} \Lambda_i$ is a normal logic.

  (iv) If F is any class of frames, then Λ_{F} is a normal logic.

  (v) If G is any class of general frames, then Λ_{G} is a normal logic. (The reader is asked to prove this in Exercise 4.1.1.)   ⊣

Examples 4.8(i) and 4.8(iii) guarantee that there is a smallest normal modal logic containing any set of formulas Γ. We call this the normal modal logic *generated* or *axiomatized* by Γ. The normal modal logic generated by the empty set is called **K**, and it is the smallest (or minimal) normal modal logic: for any normal modal logic Λ, **K** $\subseteq \Lambda$. If Γ is a non-empty set of formulas we usually denote the normal logic generated by Γ by **KΓ**. Moreover, we often make use of Hilbert axiomatization terminology, referring to Γ as *axioms* of this logic, and say that the logic was generated using the *rules of proof* modus ponens, uniform substitution, and generalization. We justified this terminology in Section 1.6, and also asked the reader to prove that the logic **KΓ** consists of precisely those formulas that can be proved in a Hilbert-style derivation from the axioms in Γ using the standard modal proof rules (see Exercise 1.6.6).

Defining a logic by stating which formulas generate it (that is, extending the minimal normal logic **K** with certain axioms of interest) is the usual way of syntactically specifying normal logics. Much of this chapter explores such axiomatic extensions. Here are some of the better known axioms, together with their traditional names:

(4) $\Diamond\Diamond p \to \Diamond p$

(T) $p \to \Diamond p$

(B) $p \to \Box\Diamond p$

(D) $\Box p \to \Diamond p$

(.3) $\Diamond p \wedge \Diamond q \to \Diamond(p \wedge \Diamond q) \vee \Diamond(p \wedge q) \vee \Diamond(q \wedge \Diamond p)$

(L) $\Box(\Box p \to p) \to \Box p$.

There is a convention for talking about the logics generated by such axioms: if $A_1, \ldots, A_n$ are axioms then **K$A_1 \ldots A_n$** is the normal logic generated by $A_1, \ldots, A_n$. But irregularities abound. Many historical names are firmly entrenched, thus modal logicians talk of **T, B, S4, S4.3**, and **S5** instead of **KT, KB, KT4, KT4.3**, and **KT4B** respectively. Moreover, many axioms have multiple names. For example,

K	the class of all frames
K4	the class of transitive frames
T	the class of reflexive frames
B	the class of symmetric frames
KD	the class of right-unbounded frames
S4	the class of reflexive, transitive frames
S5	the class of frames whose relation is an equivalence relation
K4.3	the class of transitive frames with no branching to the right
S4.3	the class of reflexive, transitive frames with no branching to the right
KL	the class of finite transitive trees (*weak* completeness only)

Table 4.1. *Some Soundness and Completeness Results*

the axiom we call L (for Löb) is also known as G (for Gödel) and W (for well-founded); and the axiom we call .3 has also been called H (for Hintikka). We adopt a fairly relaxed attitude towards naming logics, and use the familiar names as much as possible.

Now that we know what normal modal logics are, we are ready to introduce the two fundamental concepts linking the syntactic and semantic perspectives: *soundness* and *completeness*.

Definition 4.9 (Soundness) Let S be a class of frames (or models, or general frames). A normal modal logic Λ is *sound* with respect to S if $\Lambda \subseteq \Lambda_S$. (Equivalently: Λ is *sound* with respect to S if for all formulas ϕ, and all structures $\mathfrak{S} \in S$, $\vdash_\Lambda \phi$ implies $\mathfrak{S} \Vdash \phi$.) If Λ is sound with respect to S we say that S *is a class of frames* (or models, or general frames) *for* Λ. ⊣

Table 4.1 lists a number of well-known logics together with classes of frames for which they are sound. Recall that a *right-unboundedness* frame (W, R) is a frame such that $\forall x \exists y \, Rxy$. Also, a frame (W, R) satisfying $\forall x \forall y \forall z \, (Rxy \wedge Rxz \rightarrow (Ryz \vee y = z \vee Rzy))$ is said to have *no branching to the right*.

The *soundness* claims made in Table 4.1 (with the exception of the last one, which was shown in Example 3.9) are easily demonstrated. In all cases one shows that the axioms are valid, and that the three rules of proof (modus ponens, generalization, and uniform substitution) preserve validity on the class of frames in question. In fact, the proof rules preserve validity on *any* class of frames or general frames (see Exercise 4.1.1), so proving soundness boils down to checking the validity of the axioms. Soundness proofs are often routine, and when this is the case we rarely bother to explicitly state or prove them. But the concept of *completeness* leads to the problems that will occupy us for the remainder of the chapter.

Definition 4.10 (Completeness) Let S be a class of frames (or models, or general frames). A logic Λ is *strongly complete* with respect to S if for any set of formulas $\Gamma \cup \{\phi\}$, if $\Gamma \Vdash_S \phi$ then $\Gamma \vdash_\Lambda \phi$. That is, if Γ semantically entails ϕ on S (recall Definition 1.35) then ϕ is Λ-deducible from Γ.

A logic Λ is *weakly complete* with respect to S if for any formula ϕ, if S $\Vdash \phi$ then $\vdash_\Lambda \phi$. Λ is strongly complete (weakly complete) with respect to a single structure $\mathfrak{S}$ if Λ is strongly complete (weakly complete) with respect to $\{\mathfrak{S}\}$.   ⊣

Note that weak completeness is the special case of strong completeness in which Γ is empty, thus strong completeness with respect to some class of structures implies weak completeness with respect to that same class. (The converse does *not* hold, as we will later see.) Note that the definition of weak completeness can be reformulated to parallel the definition of soundness: Λ is weakly complete with respect to S if $\Lambda_S \subseteq \Lambda$. Thus, if we prove that a syntactically specified logic Λ is both sound and weakly complete with respect to some class of structures S, we have established a perfect match between the syntactical and semantical perspectives: $\Lambda = \Lambda_S$. Given a semantically specified logic Λ_S (that is, the logic of some class of structures S of interest) we often want to find a simple collection of formulas Γ such that Λ_S is the logic generated by Γ; in such a case we sometimes say that Γ *axiomatizes* S.

Example 4.11 With the exception of **KL**, all the logics mentioned in Table 4.1 are strongly complete with respect to the corresponding classes of frames. However, **KL** is only weakly complete with respect to the class of finite transitive trees. As we will learn in section 4.4, **KL** is not strongly complete with respect to this class of frames, or indeed with respect to any class of frames whatsoever.   ⊣

These completeness results are among the best known in modal logic, and we will soon be able to prove them. Together with their soundness counterparts, they constitute perspicuous semantic characterizations of important logics. **K4**, for example, is not just the logic obtained by enriching **K** with some particular axiom: it is precisely the set of formulas valid on all transitive frames. There is always something arbitrary about syntactic presentations; it is pleasant (and useful) to have these semantic characterizations at our disposal.

We make heavy use, usually without explicit comment, of the following result:

Proposition 4.12 *A logic Λ is strongly complete with respect to a class of structures* S *iff every Λ-consistent set of formulas is satisfiable on some $\mathfrak{S} \in$ S. Λ is weakly complete with respect to a class of structures* S *iff every Λ-consistent formula is satisfiable on some $\mathfrak{S} \in$ S.*

Proof. The result for weak completeness follows from the one for strong completeness, so we examine only the latter. To prove the right to left implication we argue by contraposition. Suppose Λ is not strongly complete with respect to S. Thus there is a set of formulas $\Gamma \cup \{\phi\}$ such that $\Gamma \Vdash_S \phi$ but $\Gamma \nvdash_\Lambda \phi$. Then $\Gamma \cup \{\neg\phi\}$ is Λ-consistent, but not satisfiable on any structure in S. The left to right direction is left to the reader. ⊣

To conclude this section, we extend the definition of normal modal logics to arbitrary similarity types.

Definition 4.13 Assume we are working with a modal language of similarity type τ. A *modal logic* in this language is (as before) a set of formulas containing all tautologies that is closed under modus ponens and uniform substitution. A modal logic Λ is *normal* if for every operator $\triangledown$ it contains the axiom $K_\triangledown^i$ (for all i such that $1 \leq i \leq \rho(\triangledown)$) and the axiom $Dual_\triangledown$, and is closed under the generalization rules described below.

The required axioms are obvious polyadic analogs of the earlier K and Dual axioms:

$(K_\triangledown^i)$ $\quad \triangledown(r_1, \ldots, p \to q, \ldots, r_{\rho(\triangledown)}) \to$
$$\to \left(\triangledown(r_1, \ldots, p, \ldots, r_{\rho(\triangledown)}) \to \triangledown(r_1, \ldots, q, \ldots, r_{\rho(\triangledown)}) \right).$$
$(Dual_\triangledown)$ $\quad \triangle(r_1, \ldots, r_{\rho(\triangledown)}) \leftrightarrow \neg\triangledown(\neg r_1, \ldots, \neg r_{\rho(\triangledown)}).$

(Here $p, q, r_1, \ldots, r_{\rho(\triangledown)}$ are distinct propositional variables, and the occurrences in $K_\triangledown^i$ of p and q occur in the i-th argument place of $\triangledown$.) Finally, for a polyadic operator $\triangledown$, generalization takes the following form:

$$\vdash_\Lambda \sigma \text{ implies } \vdash_\Lambda \triangledown(\bot, \ldots, \sigma, \ldots, \bot).$$

That is, an n-place operator $\triangledown$ is associated with n generalization rules, one for each of its n argument positions.

Note that these axioms and rules do not apply to *nullary* modalities. Nullary modalities are rather like propositional variables and – as far as the minimal logic is concerned – they do not give rise to any axioms or rules. ⊣

Definition 4.14 Let τ be a modal similarity type. Given a set of τ-formulas Γ, we define $\mathbf{K_\tau\Gamma}$, the normal modal logic *axiomatized* or *generated* by Γ, to be the smallest normal modal τ-logic containing all formulas in Γ. Formulas in Γ are called *axioms* of this logic, and Γ may be called an *axiomatization* of $\mathbf{K_\tau\Gamma}$. The normal modal logic generated by the empty set is denoted by $\mathbf{K_\tau}$. ⊣

Exercises for Section 4.1

4.1.1 Show that if G is any class of general frames, then Λ_G is a normal logic. (To prove this, you will have to show that the modal proof rules preserve validity on any general frame.)

4.1.2 First, show that the diamond-based definition of normal modal logics given in Remark 4.7 is equivalent to the box-based definition. Then, for languages of arbitrary similarity type, formulate a $\triangle$-based definition of normal modal logics, and prove it equivalent to the $\triangledown$-based one given in Definition 4.13.

4.1.3 Show that the set of all normal modal logics (in some fixed language) ordered by set theoretic inclusion forms a complete lattice. That is, prove that every family $\{\Lambda_i \mid i \in I\}$ of logics has both an infimum and a supremum. (An infimum is a logic Λ such that $\Lambda \subseteq \Lambda_i$ for all $i \in I$, and for any other logic Λ' that has this property, it holds that $\Lambda' \subseteq \Lambda$; the concept of a supremum is defined analogously, with '$\supseteq$' replacing '$\subseteq$.')

4.1.4 Show that the normal logic generated by $\Box(p \wedge \Box p \to q) \vee \Box(q \wedge \Box q \to p)$ is sound with respect to the class of **K4.3** frames (see Table 4.1). Further, show that the normal modal logic generated by $\Box(\Box p \to q) \vee \Box(\Box q \to p)$ is *not* sound with respect to this class of frames, but that it is sound with respect to the class of **S4.3** frames.

4.2 Canonical Models

Completeness theorems are essentially model existence theorems – that is the content of Proposition 4.12. Given a normal logic Λ, we prove its strong completeness with respect to some class of structures by showing that every Λ-consistent set of formulas can be satisfied in some suitable model. Thus the fundamental question we need to address is: *how do we build (suitable) satisfying models?*

This section introduces the single most important answer: build models out of *maximal consistent sets of formulas*, and in particular, build *canonical models*. It is difficult to overstress the importance of this idea. In one form or another it underlies almost every modal completeness result the reader is likely to encounter. Moreover, as we will learn in Chapter 5, the idea has substantial algebraic content.

Definition 4.15 (Λ-MCSs) A set of formulas Γ is *maximal Λ-consistent* if Γ is Λ-consistent, and any set of formulas properly containing Γ is Λ-inconsistent. If Γ is a maximal Λ-consistent set of formulas then we say it is a Λ-MCS. $\dashv$

Why use MCSs in completeness proofs? To see this, first note that every point w in every model $\mathfrak{M}$ for a logic Λ is associated with a set of formulas, namely $\{\phi \mid \mathfrak{M}, w \Vdash \phi\}$. It is easy to check (and the reader should do so) that this set of formulas is actually a Λ-MCS. That is: if ϕ is true in some model for Λ, then ϕ belongs to a Λ-MCS. Second, if w is related to w' in some model $\mathfrak{M}$, then it is clear that the information embodied in the MCSs associated with w and w' is 'coherently related'. Thus our second observation is: models give rise to collections of coherently related MCSs.

The idea behind the canonical model construction is to try and turn these observations around: that is, to work backwards from collections of coherently related MCSs to the desired model. The goal is to prove a Truth Lemma which tells us that

'ϕ belongs to an MCS' is actually *equivalent* to 'ϕ is true in some model.' How will we do this? By building a special model – the *canonical model* – whose points are all MCSs of the logic of interest. We will pin down what it means for the information in MCSs to be 'coherently related,' and use this notion to define the required accessibility relations. Crucially, we will be able to prove an Existence Lemma which states that there are enough coherently related MCSs to ensure the success of the construction, and this will enable us to prove the desired Truth Lemma.

To carry out this plan, we need to learn a little more about MCSs.

Proposition 4.16 (Properties of MCSs) *If Λ is a logic and Γ is a Λ-MCS then:*

   (i) *Γ is closed under modus ponens: if ϕ, $\phi \to \psi \in \Gamma$, then $\psi \in \Gamma$;*
   (ii) *$\Lambda \subseteq \Gamma$;*
   (iii) *for all formulas ϕ: $\phi \in \Gamma$ or $\neg\phi \in \Gamma$;*
   (iv) *for all formulas ϕ, ψ: $\phi \vee \psi \in \Gamma$ iff $\phi \in \Gamma$ or $\psi \in \Gamma$.*

Proof. Exercise 4.2.1.   ⊣

As MCSs are to be our building blocks, it is vital that we have enough of them. In fact, any consistent set of formulas can be extended to a maximal consistent one.

Lemma 4.17 (Lindenbaum's Lemma) *If Σ is a Λ-consistent set of formulas then there is a Λ-MCS Σ^+ such that $\Sigma \subseteq \Sigma^+$.*

Proof. Let ϕ_0, ϕ_1, ϕ_2, ... be an enumeration of the formulas of our language. We define the set Σ^+ as the union of a chain of Λ-consistent sets as follows:

$$
\begin{aligned}
\Sigma_0 &= \Sigma, \\
\Sigma_{n+1} &= \begin{cases} \Sigma_n \cup \{\phi_n\}, & \text{if this is } \Lambda\text{-consistent} \\ \Sigma_n \cup \{\neg\phi_n\}, & \text{otherwise} \end{cases} \\
\Sigma^+ &= \bigcup_{n \geq 0} \Sigma_n.
\end{aligned}
$$

The proof of the following properties of Σ^+ is left as Exercise 4.2.2: (i) Σ_n is Λ-consistent, for all n; (ii) exactly one of ϕ and $\neg\phi$ is in Σ^+, for every formula ϕ; (iii) if $\Sigma^+ \vdash_\Lambda \phi$, then $\phi \in \Sigma^+$; and finally (iv) Σ^+ is a Λ-MCS.   ⊣

We are now ready to build models out of MCSs, and in particular, to build the very special models known as canonical models. With the help of these structures we will be able to prove the Canonical Model Theorem, a universal completeness result for normal logics. We first define canonical models and prove this result for the basic modal language; at the end of the section we generalize our discussion to languages of arbitrary similarity type.

Definition 4.18 The *canonical model* $\mathfrak{M}^\Lambda$ for a normal modal logic Λ (in the basic language) is the triple $(W^\Lambda, R^\Lambda, V^\Lambda)$ where:

(i) W^Λ is the set of all Λ-MCSs;

(ii) R^Λ is the binary relation on W^Λ defined by $R^\Lambda wu$ if for all formulas ψ, $\psi \in u$ implies $\Diamond\psi \in w$. R^Λ is called the *canonical relation*;

(iii) V^Λ is the valuation defined by $V^\Lambda(p) = \{w \in W^\Lambda \mid p \in w\}$. V^Λ is called the *canonical* (or *natural*) *valuation*.

The pair $\mathfrak{F}^\Lambda = (W^\Lambda, R^\Lambda)$ is called the *canonical frame* for Λ.   ⊣

All three clauses deserve comment. First, the canonical valuation equates the truth of a propositional symbol at w with its membership in w. Our ultimate goal is to prove a Truth Lemma which will lift this 'truth = membership' equation to arbitrary formulas.

Second, note that the states of $\mathfrak{M}^\Lambda$ consist of *all* Λ-consistent MCSs. The significance of this is that, by Lindenbaum's Lemma, *any* Λ-consistent set of formulas is a subset of some point in $\mathfrak{M}^\Lambda$ – hence, by the Truth Lemma proved below, any Λ-consistent set of formulas is true at some point in this model. In short, the single structure $\mathfrak{M}^\Lambda$ is a 'universal model' for the logic Λ, which is why it is called 'canonical.'

Finally, consider the canonical relation: a state w is related to a state u precisely when for each formula ψ in u, w contains the information $\Diamond\psi$. Intuitively, this captures what we mean by MCSs being 'coherently related.' The reader should compare the present discussion with the account of ultrafilter extensions in Chapter 2 – in Chapter 5 we will discuss a unifying framework. In the meantime, the following lemma shows that we are getting things right:

Lemma 4.19 *For any normal logic Λ, $R^\Lambda wv$ iff for all formulas ψ, $\Box\psi \in w$ implies $\psi \in v$.*

Proof. For the left to right direction, suppose $R^\Lambda wv$. Further suppose $\psi \notin v$. As v is an MCS, by Proposition 4.16 $\neg\psi \in v$. As $R^\Lambda wv$, $\Diamond\neg\psi \in w$. As w is consistent, $\neg\Diamond\neg\psi \notin w$. That is, $\Box\psi \notin w$ and we have established the contrapositive. We leave the right to left direction to the reader.   ⊣

In fact, the definition of R^Λ is exactly what we require; all that remains to be checked is that enough 'coherently related' MCSs exist for our purposes.

Lemma 4.20 (Existence Lemma) *For any normal modal logic Λ and any state $w \in W^\Lambda$, if $\Diamond\phi \in w$ then there is a state $v \in W^\Lambda$ such that $R^\Lambda wv$ and $\phi \in v$.*

Proof. Suppose $\Diamond\phi \in w$. We will construct a state v such that $R^\Lambda wv$ and $\phi \in v$. Let v^- be $\{\phi\} \cup \{\psi \mid \Box\psi \in w\}$. Then v^- is consistent. For suppose not. Then there are $\psi_1, \ldots, \psi_n$ such that $\vdash_\Lambda (\psi_1 \wedge \cdots \wedge \psi_n) \to \neg\phi$, and it follows by an easy argument that $\vdash_\Lambda \Box(\psi_1 \wedge \cdots \wedge \psi_n) \to \Box\neg\phi$. As the reader should check, the formula $(\Box\psi_1 \wedge \cdots \wedge \Box\psi_n) \rightarrow \Box(\psi_1 \wedge \cdots \wedge \psi_n)$ is a theorem of every normal modal logic, hence by propositional calculus, $\vdash_\Lambda (\Box\psi_1 \wedge \cdots \wedge \Box\psi_n) \rightarrow \Box\neg\phi$. Now, $\Box\psi_1 \wedge \cdots \wedge \Box\psi_n \in w$ (for $\Box\psi_1, \ldots, \Box\psi_n \in w$, and w is an MCS) thus it follows that $\Box\neg\phi \in w$. Using Dual, it follows that $\neg\Diamond\phi \in w$. But this is impossible: w is an MCS containing $\Diamond\phi$. We conclude that v^- is consistent after all.

Let v be any MCS extending v^-; such extensions exist by Lindenbaum's Lemma. By construction $\phi \in v$. Furthermore, for all formulas ψ, $\Box\psi \in w$ implies $\psi \in v$. Hence by Lemma 4.19, $R^\Lambda wv$.   ⊣

With this established, the rest is easy. First we lift the 'truth = membership' equation to arbitrary formulas:

Lemma 4.21 (Truth Lemma) *For any normal modal logic Λ and any formula ϕ, $\mathfrak{M}^\Lambda, w \Vdash \phi$ iff $\phi \in w$.*

Proof. By induction on the degree of ϕ. The base case follows from the definition of V^Λ. The boolean cases follow from Proposition 4.16. It remains to deal with the modalities. The left to right direction is more or less immediate from the definition of R^Λ:

$$\mathfrak{M}^\Lambda, w \Vdash \Diamond\phi \quad \text{iff} \quad \exists v\,(R^\Lambda wv \wedge \mathfrak{M}^\Lambda, v \Vdash \phi)$$

$$\text{iff} \quad \exists v\,(R^\Lambda wv \wedge \phi \in v) \qquad \text{(Induction Hypothesis)}$$

$$\text{only if} \quad \Diamond\phi \in w \qquad \text{(Definition } R^\Lambda\text{)}.$$

For the right to left direction, suppose $\Diamond\phi \in w$. By the equivalences above, it suffices to find an MCS v such that $R^\Lambda wv$ and $\phi \in v$ – and this is precisely what the Existence Lemma guarantees.   ⊣

Theorem 4.22 (Canonical Model Theorem) *Any normal modal logic is strongly complete with respect to its canonical model.*

Proof. Suppose Σ is a consistent set of the normal modal logic Λ. By Lindenbaum's Lemma there is a Λ-MCS Σ^+ extending Σ. By the previous lemma, $\mathfrak{M}^\Lambda, \Sigma^+ \Vdash \Sigma$.   ⊣

At first glance, the Canonical Model Theorem may seem rather abstract. It is a completeness result with respect to a class of *models*, not frames, and a rather abstract class at that. (That **K4** is complete with respect to the class of transitive frames is interesting; that it is complete with respect to the singleton class containing only its canonical model seems rather dull.) But appearances are misleading: canonical models are by far the most important tool used in the present chapter. For a start, the Canonical Model Theorem immediately yields the following result:

Theorem 4.23 **K** *is strongly complete with respect to the class of all frames.*

Proof. By Proposition 4.12, to prove this result it suffices to find, for any **K**-consistent set of formulas Γ, a model $\mathfrak{M}$ (based on any frame whatsoever) and a state w in $\mathfrak{M}$ such that $\mathfrak{M}, w \Vdash \Gamma$. This is easy: simply choose $\mathfrak{M}$ to be $(\mathfrak{F}^{\mathbf{K}}, V^{\mathbf{K}})$, the canonical model for **K**, and let Γ^+ be any **K**-MCS extending Γ. By the previous lemma, $(\mathfrak{F}^{\mathbf{K}}, V^{\mathbf{K}}), \Gamma^+ \Vdash \Gamma$. ⊣

More importantly, it is often easy to get useful information about the structure of canonical frames. For example, as we will learn in the next section, the canonical frame for **K4** is transitive – and this immediately yields the (more interesting) result that **K4** is complete with respect to the class of transitive frames. Even when a canonical model is not as cleanly structured as we would like, it still embodies a vast amount of information about its associated logic; one of the important themes pursued later in the chapter is how to make use of this information indirectly. Furthermore, canonical models are mathematically natural. As we will learn in Chapter 5, from an algebraic perspective canonical models are not abstract oddities: indeed, they are precisely the structures one is lead to by considering the ideas underlying the Stone Representation Theorem.

To conclude this section we sketch the generalizations required to extend the results obtained so far to languages of arbitrary similarity types.

Definition 4.24 Let τ be a modal similarity type, and Λ a normal modal logic in the language over τ. The *canonical model* $\mathfrak{M}^\Lambda = (W^\Lambda, R_\triangle^\Lambda, V^\Lambda)_{\triangle \in \tau}$ for Λ has W^Λ and V^Λ as defined in Definition 4.18, while for an n-ary operator $\triangle \in \tau$ the relation $R_\triangle^\Lambda \subseteq (W^\Lambda)^{n+1}$ is defined by $R_\triangle^\Lambda w u_1 \ldots u_n$ if for all formulas $\psi_1 \in u_1$, $\ldots, \psi_n \in u_n$ we have $\triangle(\psi_1, \ldots, \psi_n) \in w$. ⊣

There is an analog of Lemma 4.19.

Lemma 4.25 *For any normal modal logic Λ, $R_\triangle^\Lambda w u_1 \ldots u_n$ iff for all formulas $\psi_1, \ldots, \psi_n$, $\triangledown(\psi_1, \ldots, \psi_n) \in w$ implies that for some i such that $1 \le i \le n$, $\psi_i \in u_i$.*

Proof. See Exercise 4.2.3. ⊣

Now for the crucial lemma – we must show that enough coherently related MCSs exist. This requires a more delicate approach than was needed for Lemma 4.20.

Lemma 4.26 (Existence Lemma) *Suppose $\triangle(\psi_1, \ldots, \psi_n) \in w$. Then there are $u_1, \ldots, u_n$ such that $\psi_1 \in u_1, \ldots, \psi_n \in u_n$ and $R^\Lambda w u_1 \ldots u_n$.*

Proof. The proof of Lemma 4.20 establishes the result for any unary operators in the language, so it only remains to prove the (trickier) case for modalities of higher arity. To keep matters simple, assume that $\triangle$ is binary; this illustrates the key new idea needed.

So, suppose $\triangle(\psi_1, \psi_2) \in w$. Let $\phi_0, \phi_1, \ldots$ enumerate all formulas. We construct two sequences of sets of formulas

$$\{\psi_1\} = \Pi_0 \subseteq \Pi_1 \subseteq \cdots \quad \text{and} \quad \{\psi_2\} = \Sigma_0 \subseteq \Sigma_1 \subseteq \cdots$$

such that all Π_i and Σ_i are finite and consistent, Π_{i+1} is either $\Pi_i \cup \{\phi_i\}$ or $\Pi_i \cup \{\neg\phi_i\}$, and similarly for Σ_{i+1}. Moreover, putting $\pi_i := \bigwedge \Pi_i$ and $\sigma_i := \bigwedge \Sigma_i$, we will have that $\triangle(\pi_i, \sigma_i) \in w$.

The key step in the inductive construction is

$$\triangle(\pi_i, \sigma_i) \in w \;\Rightarrow\; \triangle\left(\pi_i \wedge (\phi_i \vee \neg\phi_i), \sigma_i \wedge (\phi_i \vee \neg\phi_i)\right) \in w$$
$$\Rightarrow\; \triangle\left((\pi_i \wedge \phi_i) \vee (\pi_i \wedge \neg\phi_i), (\sigma_i \wedge \phi_i) \vee (\sigma_i \wedge \neg\phi_i)\right) \in w$$
$$\Rightarrow\; \text{one of the formulas } \triangle(\pi_i \wedge [\neg]\phi_i, \sigma_i \wedge [\neg]\phi_i) \text{ is in } w.$$

If, for example, $\triangle(\pi_i \wedge \phi_i, \sigma_i \wedge \neg\phi_i) \in w$, we take $\Pi_{i+1} := \Pi_i \cup \{\phi_i\}$, $\Sigma_{i+1} := \Sigma_i \cup \{\neg\phi_i\}$. Under this definition, all Π_i and Σ_i have the required properties. Finally, let $u_1 = \bigcup_i \Pi_i$ and $u_2 = \bigcup_i \Sigma_i$. It is easy to see that u_1, u_2 are Λ-MCSs and $R_\triangle^\Lambda w u_1 u_2$, as required. ⊣

With this lemma established, the real work has been done. The Truth Lemma and the Canonical Model Theorem for general modal languages are now obvious analogs of Lemma 4.21 and Theorem 4.22. The reader is asked to state and prove them in Exercise 4.2.4.

Exercises for Section 4.2

4.2.1 Show that all MCSs have the properties stated in Proposition 4.16. In addition, show that if Σ and Γ are distinct MCSs, then there is at least one formula ϕ such that $\phi \in \Sigma$ and $\neg\phi \in \Gamma$.

4.2.2 Lindenbaum's Lemma is not fully proved in the text. Give proofs of the four claims made at the end of our proof sketch.

4.2.3 Prove Lemma 4.25. (This is a good way of getting to grips with the definition of normality for modal languages of arbitrary similarity type.)

4.2.4 State and prove the Truth Lemma and the Canonical Model Theorem for languages of arbitrary similarity type. Make sure you understand the special case for nullary modalities (recall that we have no special axioms or rules of proof for these).

4.3 Applications

In this section we put canonical models to work. First we show how to prove the frame completeness results noted in Example 4.11 using a simple and uniform method of argument. This leads us to isolate one of most important concepts of modal completeness theory: *canonicity*. We then switch to the basic temporal language and use similar arguments to prove two important temporal completeness results. We conclude with a statement of the *Sahlqvist Completeness Theorem*, which we will prove in Chapter 5.

Suppose we suspect that a normal modal logic Λ is strongly complete with respect to a class of frames F; how should we go about proving it? Actually, there is no infallible strategy. (Indeed, as we will learn in the following section, many normal modal logics are not complete with respect to any class of frames whatsoever.) Nonetheless, a very simple technique works in a large number of interesting cases: simply show that the canonical frame for Λ belongs to F. We call such proofs *completeness-via-canonicity* arguments, for reasons which will soon become clear. Let us consider some examples.

Theorem 4.27 *The logic* **K4** *is strongly complete with respect to the class of transitive frames.*

Proof. Given a **K4**-consistent set of formulas Γ, it suffices to find a model $(\mathfrak{F}, V)$ and a state w in this model such that (1) $(\mathfrak{F}, V), w \Vdash \Gamma$, and (2) $\mathfrak{F}$ is transitive. Let (W^{K4}, R^{K4}, V^{K4}) be the canonical model for **K4** and let Γ^+ be any **K4**-MCS extending Γ. By Lemma 4.21, $(W^{K4}, R^{K4}, V^{K4}), \Gamma^+ \Vdash \Gamma$ so step (1) is established. It remains to show that (W^{K4}, R^{K4}) is transitive. So suppose w, v and u are points in this frame such that $R^{K4}wv$ and $R^{K4}vu$. We wish to show that $R^{K4}wu$. Suppose $\phi \in u$. As $R^{K4}vu$, $\Diamond\phi \in v$, so as $R^{K4}wv$, $\Diamond\Diamond\phi \in w$. But w is a **K4**-MCS, hence it contains $\Diamond\Diamond\phi \rightarrow \Diamond\phi$, thus by modus ponens it contains $\Diamond\phi$. Thus $R^{K4}wu$.   ⊣

In spite of its simplicity, the preceding result is well worth reflecting on. Two important observations should be made.

First, the proof actually establishes something more general than the theorem claims: namely, that the canonical frame of *any* normal logic Λ containing $\Diamond\Diamond p \rightarrow \Diamond p$ is transitive. The proof works because all MCSs in the canonical frame contain the 4 axiom; it follows that the canonical frame of any extension of **K4** is transitive, for all such extensions contain the 4 axiom.

Second, the result suggests that there may be a connection between the structure of canonical frames and the frame correspondences studied in Chapter 3. We know from our previous work that $\Diamond\Diamond p \rightarrow \Diamond p$ *defines* transitivity – and now we know that it imposes this property on canonical frames as well.

Theorem 4.28 **T**, **KB** *and* **KD** *are strongly complete with respect to the classes of reflexive frames, of symmetric frames, and of right-unbounded frames, respectively.*

Proof. For the first claim, it suffices to show that the canonical model for **T** is reflexive. Let w be a point in this model, and suppose $\phi \in w$. As w is a **T**-MCS, $\phi \rightarrow \Diamond\phi \in w$, thus by modus ponens, $\Diamond\phi \in w$. Thus $R^T ww$.

For the second claim, it suffices to show that the canonical model for **KB** is symmetric. Let w and v be points in this model such that $R^{\mathbf{KB}}wv$, and suppose that $\phi \in w$. As w is a **KB**-MCS, $\phi \to \Box\Diamond\phi \in w$, thus by modus ponens $\Box\Diamond\phi \in w$. Hence by Lemma 4.19, $\Diamond\phi \in v$. But this means that $R^{\mathbf{KB}}vw$, as required.

For the third claim, it suffices to show that the canonical model for **KD** is right-unbounded. (This is slightly less obvious than the previous claims since it requires an existence proof.) Let w be any point in the canonical model for **KD**. We must show that there exists a v in this model such that $R^{\mathbf{KD}}wv$. As w is a **KD**-MCS it contains $\Box p \to \Diamond p$, thus by closure under uniform substitution it contains $\Box\top \to \Diamond\top$. Moreover, as $\top$ belongs to all normal modal logics, by generalization $\Box\top$ does too; so $\Box\top$ belongs to **KD**, hence by modus ponens $\Diamond\top \in w$. Hence, by the Existence Lemma, w has an $R^{\mathbf{KD}}$ successor v. ⊣

Once again, these results hint at a link between definability and the structure of canonical frames: after all, T defines reflexivity, B defines symmetry, and D right unboundedness. And yet again, the proofs actually establish something more general than the theorem states: the canonical frame of *any* normal logic containing T is reflexive, the canonical frame of *any* normal logic containing B is symmetric, and the canonical frame of *any* normal logic containing D is right unbounded. This allows us to 'add together' our results. Here are two examples:

Theorem 4.29 **S4** *is strongly complete with respect to the class of reflexive, transitive frames.* **S5** *is strongly complete with respect to the class of frames whose relation is an equivalence relation.*

Proof. The proof of Theorem 4.27 shows that the canonical frame of *any* normal logic containing the 4 axiom is transitive, while the proof of the first clause of Theorem 4.28 shows that the canonical frame of *any* normal logic containing the T axiom is reflexive. As **S4** contains both axioms, its canonical frame has both properties, thus the completeness result for **S4** follows.

As **S5** contains both the 4 and the T axioms, it also has a reflexive, transitive canonical frame. As it also contains the B axiom (which by the proof of the second clause of Theorem 4.28 means that its canonical frame is symmetric), its canonical relation is an equivalence relation. The desired completeness result follows. ⊣

As these examples suggest, canonical models are an important tool for proving frame completeness results. Moreover, their utility evidently hinges on some sort of connection between the properties of canonical frames and the frame correspondences studied earlier. Let us introduce some terminology to describe this important phenomenon.

Definition 4.30 (Canonicity) A formula ϕ is *canonical* if, for any normal logic Λ, $\phi \in \Lambda$ implies that ϕ is valid on the canonical frame for Λ. A normal logic Λ is *canonical* if its canonical frame is a frame for Λ. (That is, Λ is canonical if for all ϕ such that $\vdash_\Lambda \phi$, ϕ is valid on the canonical frame for Λ.)   ⊣

Clearly 4, T, B and D axioms are all canonical formulas. For example, any normal logic Λ containing the 4 axiom has a transitive canonical frame, and the 4 axiom is valid on transitive frames. Similarly, any modal logic containing the B axiom has a symmetric canonical frame, and the B axiom is valid on symmetric frames.

Moreover **K4, T, KB, KD, S4** and **S5** are all canonical logics. Our previous work has established that all the axioms involved are valid on the relevant canonical frames. But (see Exercise 4.1.1) modus ponens, uniform substitution, and generalization preserve frame validity. It follows that *every* formula in each of these logics is valid on that logic's canonical frame. In general, to show that $\mathbf{KA_1 \ldots A_n}$ is a canonical logic it suffices to show that $A_1, \ldots, A_n$ are canonical formulas.

Definition 4.31 (Canonicity for a Property) Let ϕ be a formula, and P be a property. If the canonical frame for any normal logic Λ containing ϕ has property P, and ϕ is valid on any class of frames with property P, then ϕ is *canonical for P*. For example, we say that the 4 axiom is canonical for transitivity, because the presence of 4 forces canonical frames to be transitive, and 4 is valid on all transitive frames.   ⊣

Let us sum up the discussion so far. Many important frame completeness results can be proved straightforwardly using canonical models. The key idea in such proofs is to show that the relevant canonical frame has the required properties. Such proofs boil down to the following task: showing that the axioms of the logic are canonical for the properties we want (which is why we call them completeness-via-canonicity arguments).

Now for some rather different application of completeness-via-canonicity arguments. The theorems just proved were *syntactically* driven: we began with syntactically specified logics (for example **K4** and **T**) and showed that they could be semantically characterized as the logics of certain frame classes. Canonical models are clearly useful for such proofs – but how do they fare when proving *semantically* driven results? That is, suppose F is a class of frames we find interesting, and we have isolated a set of axioms which we hope generates Λ_F. Can completeness-via-canonicity arguments help establish their adequacy?

As such semantically driven questions are typical of temporal logic, let us switch to the basic temporal language. Recall from Example 1.14 that this language has two diamonds, F and P, whose respective duals are G and H. The F operator looks forwards along the flow of time, and P looks backwards. Furthermore, recall from Example 1.25 that we are only interested in the frames for this language in which the relations corresponding to F and P are mutually converse. That is, a bidirectional frame is a triple $(W, \{R_P, R_F\})$ such that

$$R_P = \{(y, x) \mid (x, y) \in R_F\}.$$

Recall that by convention we present bidirectional frames as unimodal frames (T, R); in such presentations we understand that $R_F = R$ and $R_P = R^{\smallsmile}$. The class of all bidirectional frames is denoted by F_t, and a bidirectional model is a model whose underlying frame belongs to F_t.

So, what is a temporal logic? As a first step towards answering this we define:

Definition 4.32 The *minimal temporal logic* $\varLambda_{\mathsf{F}_t}$ is $\{\phi \mid \mathsf{F}_t \Vdash \phi\}$. $\quad\dashv$

That is, the minimal temporal logic contains precisely the formulas valid on all bidirectional frames. This is a semantic definition, and, given our interest in frames, a sensible one. But can we axiomatize $\varLambda_{\mathsf{F}_t}$? That is, can we give $\varLambda_{\mathsf{F}_t}$ a simple *syntactic* characterization? First, note that $\varLambda_{\mathsf{F}_t}$ is *not* identical to the minimal normal logic in the basic temporal language. As we noted in Example 1.29(v), for any frame $\mathfrak{F} = (W, \{R_F, R_P\})$ we have that

$$\mathfrak{F} \Vdash (q \to HFq) \wedge (q \to GPq) \text{ iff } \mathfrak{F} \in \mathsf{F}_t.$$

The two conjuncts define the 'mutually converse' property enjoyed by R_F and R_P. Clearly, both conjuncts belong to $\varLambda_{\mathsf{F}_t}$. Equally clearly, they do *not* belong to the minimal normal logic in the basic temporal language. Nonetheless, although $\varLambda_{\mathsf{F}_t}$ is stronger, it is not much stronger: the only axioms we need to add are these converse-defining conjuncts.

Definition 4.33 A *normal temporal logic* $\varLambda$ is a normal modal logic (in the basic temporal language) that contains $p \to GPp$ and $p \to HFp$ (the *converse axioms*). The smallest normal temporal logic is called $\mathbf{K}_t$. We usually call normal temporal logics *tense logics*.

Note that in the basic temporal language the K axioms are $G(p \to q) \to (Gp \to Gq)$ and $H(p \to q) \to (Hp \to Hq)$, and the Dual axioms are $Fp \leftrightarrow \neg G \neg p$ and $Pp \leftrightarrow \neg H \neg p$. Closure under generalization means that if $\vdash_\varLambda \phi$ then $\vdash_\varLambda G\phi$ and $\vdash_\varLambda H\phi$. $\quad\dashv$

We want to show that $\mathbf{K}_t$ generates exactly the formulas in $\varLambda_{\mathsf{F}_t}$. Soundness is immediate: clearly $\mathbf{K}_t \subseteq \varLambda_{\mathsf{F}_t}$. We show completeness using a canonicity argument. So, what are canonical models for tense logics? Nothing new: simply the following instance of Definition 4.24:

Definition 4.34 The *canonical model for a tense logic* $\varLambda$ is the structure $\mathfrak{M}^\varLambda = (T^\varLambda, \{R_P^\varLambda, R_F^\varLambda\}, V^\varLambda)$ where:

(i) $T^\varLambda$ is the set of all $\varLambda$-MCSs.

(ii) R_P^Λ is the binary relation on T^Λ defined by $R_P^\Lambda ts$ if for all formulas ϕ, $\phi \in s$ implies $P\phi \in t$.

(iii) R_F^Λ is the binary relation on T^Λ defined by $R_F^\Lambda ts$ if for all formulas ϕ, $\phi \in s$ implies $F\phi \in t$.

(iv) V^Λ is the valuation defined by $V^\Lambda(p) = \{t \in T^\Lambda \mid p \in t\}$.   ⊣

We immediately inherit a number of results from the previous section, such as an Existence Lemma, a Truth Lemma, and a Canonical Model Theorem telling us that each tense logic is complete with respect to its canonical model. This is very useful – but it is not quite enough. We want to show that $\mathbf{K}_t$ generates all the *temporal* validities. None of the results just mentioned allow us to conclude this, and for a very obvious reason: we do not yet know whether canonical frames for tense logics are bidirectional frames! In fact they are, and this is where the converse axioms come into play. As the next lemma shows, these axioms are canonical; they force R_P^Λ and R_F^Λ to be mutually converse.

Lemma 4.35 *For any tense logic Λ, if $R_P^\Lambda ts$ then $R_F^\Lambda st$, and if $R_F^\Lambda ts$ then $R_P^\Lambda st$.*

Proof. Rather like the proof that B is canonical for symmetry (see Theorem 4.28). We leave it to the reader as Exercise 4.3.2.   ⊣

Thus canonical frames of tense logics are bidirectional frames, so from now on we present them as pairs (T^Λ, R^Λ). Moreover, we now have the desired result:

Corollary 4.36 $\mathbf{K}_t$ *is strongly complete with respect to the class of all bidirectional frames, and* $\mathbf{K}_t = \Lambda_{\mathsf{F}_t}$.

Proof. $\mathbf{K}_t$ is strongly complete with respect to its canonical model. As we have just seen, this model is based on a *bidirectional* frame, so the strong frame completeness result follows. Strong completeness implies weak completeness, so $\Lambda_{\mathsf{F}_t} \subseteq \mathbf{K}_t$. The inclusion $\mathbf{K}_t \subseteq \Lambda_{\mathsf{F}_t}$ has already been noted.   ⊣

With this basic result established, we are ready to start a semantically driven exploration of tense logic. That is, we can now attempt to capture the logics of 'time-like' classes of frames as axiomatic extensions of $\mathbf{K}_t$. Here we limit ourselves to the following question: how can the temporal logic of *dense unbounded weak total orders* be axiomatized? From the point of view of tense logic, this is an interesting problem: dense frames and totally ordered frames both play an important role in modeling temporal phenomena. Moreover, as we will see, there is an instructive problem that must be overcome if we build totally ordered models. This will give us a gentle initiation to the fundamental difficulty faced by semantically driven completeness results, a difficulty which we will explore in more detail later in the chapter.

Definition 4.37 A bidirectional frame (T, R) is *dense* if there is a point between any two related points $(\forall xy\,(Rxy \rightarrow \exists z\,(Rxz \wedge Rzy)))$. It is *right-unbounded* if every point has a successor, *left-unbounded* if every point has a predecessor, and *unbounded* if it is both right- and left-unbounded. It is *trichotomous* if any two points are equal or are related one way or the other $(\forall xy\,(Rxy \vee x = y \vee Ryx))$, and a *weak total order* (or *weakly linear*) if it is both transitive and trichotomous. We call a frame with all these properties a DUWTO-frame. $\dashv$

Note that weakly linear frames are allowed to contain both reflexive and irreflexive points. Indeed, they are allowed to contain non-empty subsets S such that for all $s, s' \in S$, Rss'. Thus they do not fully model the idea of linearity. Linearity is better captured by the class of *strict* total orders, which are transitive, trichotomous and *irreflexive*. Building strictly totally ordered models is harder than building weakly totally ordered models; we examine the problem in detail later in the chapter.

Our first task is to select suitable axioms. Three of the choices are fairly obvious:

(4)　　$FFp \rightarrow Fp$
(D_r)　$Gp \rightarrow Fp$
(D_l)　$Hp \rightarrow Pp$

Note that $FFp \rightarrow Fp$ is simply the 4 axiom in tense logical notation. We know (by the proof of Theorem 4.27) that it is canonical for transitivity, hence choosing it as an axiom ensures the transitive canonical frame we want. Next, D_r (a tense logical analog of the D axiom) is (by the proof of the third claim of Theorem 4.28) canonical for right-unboundedness. Similarly, its backwards-looking companion $Hp \rightarrow Pp$ is canonical for *left*-unboundedness, so we obtain an unbounded canonical frame without difficulty.

What about density? Here we are in luck. The following formula is canonical for density:

(*den*)　$Fp \rightarrow FFp$

This is worth a lemma, since the proof is not trivial. (Note that density is a universal-existential property, rather than a universal property like transitivity or reflexivity. This means that proving canonicity requires establishing the *existence* of certain MCSs.)

Lemma 4.38 $Fp \rightarrow FFp$ *is canonical for density.*

Proof. Let Λ be any tense logic containing $Fp \rightarrow FFp$, let (T^Λ, R^Λ) be its canonical frame, and let t and t' be points in this frame such that $R^\Lambda tt'$. We have to show that there is a Λ-MCS s such that $R^\Lambda ts$ and $R^\Lambda st'$. If we could show that $\{\phi \mid G\phi \in t\} \cup \{F\psi \mid \psi \in t'\}$ was consistent we would have the desired result (for by the Lemmas 4.19 and 4.35, any MCS extending this set would be a suitable choice for s).

So suppose for the sake of contradiction that this set is not consistent. Then, for some finite set of formulas $\phi_1, \ldots, \phi_m, \psi_1, \ldots, \psi_n$ from this set,

$$\vdash_\Lambda (\phi_1 \wedge \cdots \wedge \phi_m \wedge F\psi_1 \wedge \cdots \wedge F\psi_n) \to \bot .$$

Define $\widehat{\phi}$ to be $\phi_1 \wedge \cdots \wedge \phi_m$ and $\widehat{\psi}$ to be $\psi_1 \wedge \cdots \wedge \psi_n$. Note that $\widehat{\psi} \in t'$.

Now, $\vdash_\Lambda F\widehat{\psi} \to F\psi_1 \wedge \cdots \wedge F\psi_n$, hence $\vdash_\Lambda \widehat{\phi} \wedge F\widehat{\psi} \to \bot$, hence $\vdash_\Lambda \widehat{\phi} \to \neg F\widehat{\psi}$, and hence $\vdash_\Lambda G\widehat{\phi} \to G\neg F\widehat{\psi}$. Because $G\phi_1, \ldots, G\phi_m \in t$, we have that $G\widehat{\phi} \in t$ too, hence $G\neg F\widehat{\psi} \in t$, and hence $\neg G\neg F\widehat{\psi} \notin t$. That is, $FF\widehat{\psi} \notin t$. But this means that $F\widehat{\psi} \notin t$, as (by uniform substitution in *den*) $F\widehat{\psi} \to FF\widehat{\psi} \in t$. But now we have a contradiction: as $\widehat{\psi} \in t'$ and $R^\Lambda tt'$, $F\widehat{\psi}$ must be in t. We conclude that $\{\phi \mid G\phi \in t\} \cup \{F\psi \mid \psi \in t'\}$ is consistent after all. (Note that this proof makes no use of the converse axioms, thus we have also proved that $\Diamond p \to \Diamond\Diamond p$ is canonical for density.)   ⊣

So it only remains to ensure trichotomy – but here we encounter an instructive difficulty. Because modal (and temporal) validity is preserved under the formation of disjoint unions (see Theorem 3.14) no formula of tense logic defines trichotomy. Moreover, a little experimentation will convince the reader that canonical frames may have disjoint point generated subframes; such canonical frames are clearly not trichotomous. In short, to prove the desired completeness result we need to build a model with a property for which no modal formula is canonical. This is the problem we encounter time and time again when proving semantically driven results.

In the present case, a little lateral thinking leads to a solution. First, let us get rid of a possible preconception. Until now, we have always used the entire canonical model – but we do not need to do this. A point generated submodel suffices. More precisely, if $\mathfrak{M}^\Lambda, w \Vdash \Gamma$, then as modal satisfaction is preserved in generated submodels (see Proposition 2.6) $\mathfrak{S}, w \Vdash \Gamma$, where $\mathfrak{S}$ is the submodel of $\mathfrak{M}^\Lambda$ generated by w.

The observation is trivial, but its consequences are not. By restricting our attention to point-generated submodels, we increase the range of properties we can impose. In particular, we *can* impose trichotomy on point-generated submodels. We met the relevant axioms when working with the basic modal language. From our discussion of **S4.3** and **K4.3** (in particular, Exercise 4.3.3) we know that

$(.3_r) \quad (Fp \wedge Fq) \to F(p \wedge Fq) \vee F(p \wedge q) \vee F(q \wedge Fp)$

is canonical for no-branching-to-the-right. Analogously

$(.3_l) \quad (Pp \wedge Pq) \to P(p \wedge Pq) \vee P(p \wedge q) \vee P(q \wedge Pp).$

is canonical for no-branching-to-the-left. Call a frame with no branching to the left or right a *non-branching* frame.

Proposition 4.39 *Any trichotomous frame* (T, R) *is non-branching. Furthermore, if* R *is transitive and non-branching and* $t \in T$, *then the subframe of* (T, R) *generated by* t *is trichotomous.*

Proof. Trivial – though the reader should recall that when forming generated subframes for the basic temporal language, we generate on both the relation corresponding to F and that corresponding to P. That is, we generate both forwards and backwards along R. ⊣

In short, although no formula is canonical for trichotomy, there is a good 'approximation' to it (namely, the non-branching property) for which we do have a canonical formula (namely, the conjunction of $.3_l$ and $.3_r$). With this observed, the desired result is within reach.

Definition 4.40 Let $\mathbf{K}_t\mathbf{Q}$ be the smallest tense logic containing 4, D_l, D_r, *den*, $.3_l$ and $.3_r$. ⊣

Theorem 4.41 $\mathbf{K}_t\mathbf{Q}$ *is strongly complete with respect to the class of* DUWTO-*frames.*

Proof. If Γ is a $\mathbf{K}_t\mathbf{Q}$-consistent set of formulas, extend it to a $\mathbf{K}_t\mathbf{Q}$-MCS Γ^+. Let $\mathfrak{M}$ be the canonical model for $\mathbf{K}_t\mathbf{Q}$, and let $\mathfrak{S}$ be the submodel of $\mathfrak{M}$ generated by Γ^+. As we just noted, $\mathfrak{S}, \Gamma^+ \Vdash \Gamma$. Moreover, the frame underlying $\mathfrak{S}$ is a DUWTO-frame as required. First, as $\mathbf{K}_t\mathbf{Q}$ contains axioms that are canonical for transitivity, unboundedness, and density, $\mathfrak{M}$ has these properties; it is then not difficult to show that $\mathfrak{S}$ has them too. Moreover, as the conjunction of $.3_l$ and $.3_r$ is canonical for non-branching, $\mathfrak{M}$ is non-branching and $\mathfrak{S}$ trichotomous. ⊣

To conclude, two important remarks. First, the need to build models possessing properties for which no formula is canonical is the fundamental difficulty facing semantically driven results. In the present case, a simple idea enabled us to bypass the problem – but we will not always be so lucky and in the second part of the chapter we develop more sophisticated techniques for tackling the issue.

Second, the relationships between completeness, canonicity and correspondence are absolutely fundamental to the study of normal modal logics. These relationships are further discussed in the following section, and explored algebraically in Chapter 5, but let us immediately mention one of the most elegant positive results in the area: the *Sahlqvist Completeness Theorem*. In Chapter 3 we proved the Sahlqvist Correspondence Theorem: every Sahlqvist formula *defines* a first-order class of frames. Here is its completeness theoretic twin, which we will prove in Chapter 5:

Theorem 4.42 *Every Sahlqvist formula is canonical for the first-order property it defines. Hence, given a set of Sahlqvist axioms Σ, the logic* **KΣ** *is strongly complete with respect to the class of frames* F_Σ *(that is, the first-order class of frames defined by Σ).*

This is an extremely useful result. Most commonly encountered axioms in the basic modal language are Sahlqvist (the Löb and McKinsey formulas are the obvious exceptions) thus it provides an immediate answer to a host of completeness problems. Moreover, like the Sahlqvist Correspondence Theorem, the Sahlqvist Completeness Theorem applies to modal languages of *arbitrary* similarity type. Finally, the theorem generalizes to a number of extended modal logics, most notably D-logic (which we introduce in Chapter 7). Note that Kracht's Theorem (see Chapter 3) can be viewed as providing a sort of 'converse' to Sahlqvist's result, for it gives us a way of computing formulas that are canonical for certain first-order classes of frames.

Exercises for Section 4.3

4.3.1 Let 1.1 be the axiom $\Diamond p \to \Box p$. Show that **K1.1** is sound and strongly complete with respect to the class of all frames (W, R) such that R is a partial function.

4.3.2 Let Λ be a normal temporal logic containing the axioms $p \to GPp$ and $p \to HFp$. Show that if $R_P^\Lambda ts$ then $R_F^\Lambda st$, and if $R_F^\Lambda ts$ then $R_P^\Lambda st$.

4.3.3 Use canonical models to show that **K4.3** is strongly complete with respect to the class of frames that are transitive and have no branching to the right. Then, by proving suitable completeness results (and making use of the soundness results proved in Exercise 4.1.4), show that the normal logic axiomatized by 4 and $\Box(p \wedge \Box p \to q) \vee \Box(q \wedge \Box q \to p)$ is **K4.3**. Try proving the equivalence of these logics syntactically.

Formulate and prove similar results for **S4.3**.

4.3.4 Prove directly that $\Diamond\Box p \to \Box\Diamond p$ is canonical for the Church-Rosser property.

4.3.5 Let W5 be the formula $\Diamond\Box p \to (p \to \Box p)$, and let **S4W5** be the smallest normal logic extending **S4** that contains W5. Find a simple class of frames that characterizes this logic.

4.3.6 Show that **S5** is complete with respect to the the class of *globally related frames*, that is, those frames (W, R) in which R connects any two points.

4.3.7 Consider a similarity type τ with one binary operator $\triangle$. For each of the following Sahlqvist formulas, first compute the (global) first-order correspondent. Then, give a *direct* proof that the modal formula is canonical for the corresponding first-order property.

(a) $p \triangle q \to q \triangle p$,
(b) $(p \triangle q) \triangle r \to p \triangle (q \triangle r)$,
(c) $((q \triangle \neg (p \triangle q)) \wedge p) \to \bot$.

4.4 Limitative Results

Although completeness-via-canonicity is a powerful method, it is not infallible. For a start, not every normal modal logic is canonical. Moreover, not every normal logic is the logic of some class of frames. In this section we prove both claims and discuss their impact on modal completeness theory.

We first demonstrate the existence of non-canonical logics. We will show that **KL**, the normal modal logic generated by the Löb axiom $\Box(\Box p \rightarrow p) \rightarrow \Box p$, is not canonical. We prove this by showing that **KL** is not sound and strongly complete with respect to any class of frames. Now, every *canonical* logic is sound and strongly complete with respect to some class of frames. (For suppose Λ is a canonical logic and Γ is a Λ-consistent set of formulas. By the Truth Lemma, Γ is satisfiable on $\mathfrak{F}^\Lambda$; as Λ is canonical, $\mathfrak{F}^\Lambda$ is a frame for Λ.) Hence if **KL** is not sound and strongly complete with respect to any class of frames, it cannot be canonical either.

Theorem 4.43 **KL** *is not sound and strongly complete with respect to any class of frames, and hence it is not canonical.*

Proof. Let Γ be $\{\Diamond q_1\} \cup \{\Box(q_i \rightarrow \Diamond q_{i+1}) \mid 1 \leq i \in \omega\}$. We will show that Γ is **KL**-consistent, and that no model based on a **KL**-frame can satisfy all formulas in Γ at a single point. The theorem follows immediately.

To show that Γ is consistent, it suffices to show that every finite subset Ψ of Γ is consistent. Given any such Ψ, for some natural number n there is a finite set Φ of the form $\{\Diamond q_1\} \cup \{\Box(q_i \rightarrow \Diamond q_{i+1}) \mid 1 \leq i < n\}$ such that $\Psi \subseteq \Phi \subset \Gamma$. We show that Φ, and hence Ψ, is consistent.

Let $\widehat{\Phi}$ be the conjunction of all the formulas in Φ. To show that $\widehat{\Phi}$ is **KL**-consistent, it suffices to show that it can be satisfied in a model based on a frame for **KL**, for this shows that $\neg\widehat{\Phi}$ is *not* valid on all frames for **KL**, and hence is *not* one of its theorems. Let $\mathfrak{F}$ be the frame consisting of $\{0, \ldots, n\}$ in their usual order; as this is a transitive, converse well-founded frame, by Example 3.9 it is a frame for **KL**. Let $\mathfrak{M}$ be any model based on $\mathfrak{F}$ such that for all $1 \leq i \leq n$, $V(q_i) = \{i\}$. Then $\mathfrak{M}, 0 \Vdash \widehat{\Phi}$ and $\widehat{\Phi}$ is **KL** consistent.

Next, suppose for the sake of a contradiction that **KL** is sound and strongly complete with respect to some class of frames F; note that as **KL** is not the inconsistent logic, F must be non-empty. Thus any **KL**-consistent set of formulas can be satisfied at some point in a model based on a frame in F. In particular, there is a model $\mathfrak{M}$ based on a frame in F and a point w in $\mathfrak{M}$ such that $\mathfrak{M}, w \Vdash \Gamma$. But this is impossible: because $\mathfrak{M}, w \Vdash \Gamma$, we can inductively define an infinite path through $\mathfrak{M}$ starting at w; however as $\mathfrak{M}$ is based on a frame for **KL** it cannot contain such infinite paths. Hence **KL** is not sound and strongly complete with respect to any class of frames, and so cannot be canonical. $\dashv$

Remark 4.44 A normal logic Λ is said to be *compact* when any Λ-consistent set Σ can be satisfied in a frame for Λ at a single point. So the above proof shows that **KL** is not compact. Note that a non-compact logic cannot be canonical, and cannot be sound and strongly complete with respect to any class of frames. We will see a similar compactness failure when we examine PDL in Section 4.8.   ⊣

What are we to make of this result? The reader should *not* jump to the conclusion that it is impossible to characterize **KL** as the logic of some class of frames. Although no *strong* frame completeness result is possible, as we noted in Table 4.1 there is an elegant *weak* frame completeness result for **KL**, namely:

Theorem 4.45 **KL** *is weakly complete with respect to the class of all finite transitive trees.*

Proof. The proof uses the finitary methods studied later in the chapter. The reader is asked to prove the theorem in Exercises 4.8.7 and 4.8.8.   ⊣

Thus **KL** is the logic of all finite transitive trees – and there exist non-canonical but (weakly) complete normal logics. We conclude that, powerful though it is, the completeness-via-canonicity method cannot handle all interesting frame completeness results.

Let us turn to the second conjecture: are all normal logics weakly complete with respect to some class of frames? No: *incomplete* normal logics exist.

Definition 4.46 Let Λ be a normal modal logic. Λ is *(frame) complete* if there is a class of frames **F** such that $\Lambda = \Lambda_\mathsf{F}$, and *(frame) incomplete* otherwise.   ⊣

We now demonstrate the existence of incomplete logics in the basic temporal language. The demonstration has three main steps. First, we introduce a tense logic called $\mathbf{K}_t\mathbf{Tho}$ and show that it is consistent. Second, we show that no frame for $\mathbf{K}_t\mathbf{Tho}$ can validate the McKinsey axiom (which in tense logical notation is $GF\phi \rightarrow FG\phi$). It is tempting to conclude that $\mathbf{K}_t\mathbf{ThoM}$, the smallest tense logic containing both $\mathbf{K}_t\mathbf{Tho}$ and the McKinsey axiom, is the inconsistent logic. Surprisingly, this is *not* the case. $\mathbf{K}_t\mathbf{ThoM}$ is consistent – and hence is not the tense logic of any class of frames at all. We prove this in the third step with the help of general frames.

$\mathbf{K}_t\mathbf{Tho}$ is the tense logic generated by the following axioms:

$(.3_r)$ $Fp \wedge Fq \rightarrow F(p \wedge Fq) \vee F(p \wedge q) \vee F(Fp \wedge q)$
(D_r) $Gp \rightarrow Fp$
(L_l) $H(Hp \rightarrow p) \rightarrow Hp$

As we have already seen, the first two axioms are canonical for simple first-order conditions (no branching to the right, and right-unboundedness, respectively). The third axiom is simply the Löb axiom written in terms of the backwards looking operator H; it is valid on precisely those frames that are transitive and contain no infinite descending paths. (Note that such frames cannot contain reflexive points.) Let $\mathbf{K}_t\mathbf{Tho}$ be the tense logic generated by these three axioms. As all three axioms are valid on the natural numbers, $\mathbf{K}_t\mathbf{Tho}$ is consistent. If (T, R) is a frame for $\mathbf{K}_t\mathbf{Tho}$ and $t \in T$, then $\{u \in T \mid Rtu\}$ is a right-unbounded strict total order.

Now for the second step. Let $\mathbf{K}_t\mathbf{ThoM}$ be the smallest tense logic containing $\mathbf{K}_t\mathbf{Tho}$ and the McKinsey axiom $GFp \to FGp$. What are the frames for this enriched logic? The answer is: none at all, or, to put it another way, $\mathbf{K}_t\mathbf{ThoM}$ defines the empty class of frames. To see this we need the concept of cofinality.

Definition 4.47 Let $(U, <)$ be a strict total order and $S \subseteq U$. S is *cofinal in U* if for every $u \in U$ there is an $s \in S$ such that $u < s$. ⊣

For example, both the even numbers and the odd numbers are cofinal in the natural numbers. Indeed, they are precisely the kind of cofinal subsets we will use in the work that follows: mutually complementary cofinal subsets.

Lemma 4.48 *Let $\mathfrak{T}$ be any frame for $\mathbf{K}_t\mathbf{Tho}$. Then $\mathfrak{T} \not\Vdash GFp \to FGp$.*

Proof. Let t be any point in $\mathfrak{T}$, let $U = \{u \in T \mid Rtu\}$, and let $<$ be the restriction of R to U. As $\mathfrak{T}$ validates all the $\mathbf{K}_t\mathbf{Tho}$ axioms, $(U, <)$ is a right-unbounded strict total order. Suppose we could show that there is a non-empty proper subset S of U such that both S and $U \backslash S$ are cofinal in U. Then the lemma would be proved, for we would merely need to define a valuation V on $\mathfrak{T}$ such that $V(p) = S$, and $(\mathfrak{T}, V), t \not\Vdash GFp \to FGp$.

Such subsets S of U exist by (3.18) in Example 3.57. For a more direct proof, take a cardinal κ that is larger than the size of U. By ordinal induction, we will define a sequence of pairs of sets $(R_\alpha, S_\alpha)_{\alpha \leq \kappa}$ such that $R_\kappa \cap S_\kappa = \varnothing$ and both R_κ and S_κ are cofinal. We can easily prove the lemma from this by taking $S = S_\kappa$. The definition is as follows:

(i) For $\alpha = 0$, take some points r_0 and s_0 in U such that $r_0 < s_0$ and define $R_0 = \{r_0\}$ and $S_0 = \{s_0\}$.

(ii) If α is a successor ordinal $\beta + 1$, then distinguish two cases:

    (a) if R_β or S_β is cofinal, then define $R_\alpha = R_\beta$ and $S_\alpha = S_\beta$,

    (b) if neither R_β nor S_β is cofinal, then take some upper bound r_β of S_β (that is, $r_\beta > s$ for all $s \in S_\beta$), take some s_β bigger than r_β and define $R_\alpha = R_\beta \cup \{r_\beta\}$ and $S_\alpha = S_\beta \cup \{s_\beta\}$.

(iii) If α is a limit ordinal, then define $R_\alpha = \bigcup_{\beta < \alpha} R_\beta$ and $S_\alpha = \bigcup_{\beta < \alpha} S_\beta$.

It is easy to prove that $R_\alpha \cap S_\alpha = \varnothing$ for every ordinal $\alpha \leq \kappa$, so it remains to be shown that both R_κ and S_κ are cofinal. The key to this proof is the observation that if R_κ and S_κ were not cofinal, then the map $f : \kappa \to U$ given by $f : \alpha \mapsto r_{\alpha+1}$ would be injective (further proof details are left to the reader). This would contradict the assumption that κ exceeds the size of U.   ⊣

We are ready for the final step. As $\mathbf{K}_t\mathbf{ThoM}$ defines the empty class of frames, it is tempting to conclude that it is also complete with respect to this class; that is, that $\mathbf{K}_t\mathbf{ThoM}$ is the inconsistent logic. However, this is not the case.

Theorem 4.49 $\mathbf{K}_t\mathbf{ThoM}$ *is consistent and incomplete.*

Proof. Let $(\mathbb{N}, <)$ be the natural numbers in their usual order. Let A be the collection of finite and co-finite subsets of $\mathbb{N}$; we leave it to the reader to show that A is closed under boolean combinations and modal operations. Thus $(\mathbb{N}, <, A)$ is a general frame; we claim that it validates all the $\mathbf{K}_t\mathbf{ThoM}$ axioms. Now, it certainly validates all the $\mathbf{K}_t\mathbf{Tho}$ axioms, for these are already valid on the underlying frame. But what about M? As we noted in Example 1.34, $GFp \to FGp$ cannot be falsified under assignments mapping p to either a finite or a co-finite set. Hence all the axioms are valid and $\mathbf{K}_t\mathbf{ThoM}$ must be consistent.

Now, by Lemma 4.48, $\mathbf{K}_t\mathbf{ThoM}$ is not the logic of any non-empty class of frames. But as $\mathbf{K}_t\mathbf{ThoM}$ is consistent, it is not the logic of the empty class of frames either. In short, it is not the logic of any class of frames whatsoever, and is incomplete.   ⊣

Frame incompleteness results are not some easily fixed anomaly. As normal logics are sets of formulas closed under three rules of proof, the reader may be tempted to think that these rules are simply too weak. Perhaps there are yet-to-be-discovered rules which would strengthen our deductive apparatus sufficiently to overcome incompleteness? (Indeed, later in the chapter we introduce an additional proof rule, and it will turn out to be very useful.)

Nonetheless, no such strengthening of our deductive apparatus can eliminate frame incompleteness. Why is this? Ultimately it boils down to something we learned in Chapter 3: frame consequence is essentially a second-order relation. Moreover, as we discussed in the Notes to Chapter 3, it is a very strong relation indeed: strong enough to simulate the standard second-order consequence relation. Frame incompleteness results reflect the fact that (over frames) modal logic is second order logic in disguise. Hence, it will come as no surprise that incompleteness hits every modal similarity type: in Exercise 4.4.2 we meet an example in the basic modal similarity type. However, examples (such as $\mathbf{K}_t\mathbf{ThoM}$) of consistent logics with an *empty* frame class cannot be found for the basic modal similarity type, as the reader is asked to prove in Exercise 4.4.3.

There are many incomplete logics. Indeed, if anything, incomplete logics are the norm. An analogy may be helpful. When differential calculus is first encountered, most students have rather naive ideas about functions and continuity; poly-nomials, and other simple functions familiar from basic physics, are taken to be typical of all real-valued functions. The awakening comes with the study of anal-ysis. Here the student encounters such specimens as everywhere-continuous but nowhere-differentiable functions – and comes to see that the familiar functions are actually abnormally well-behaved. The situation is much the same in modal logic. The logics of interest to philosophers – logics such as **T**, **S4** and **S5** – were the first to be semantically characterized using frames. It is tempting to believe that such logics are typical, but they are actually fairly docile creatures; the lattice of normal logics contains far wilder inhabitants.

The significance of the incompleteness results depends on one's goals. Logi-cians interested in applications are likely to focus on certain intended classes of models, and completeness results for these classes. Beyond providing a salutary warning about the folly of jumping to hasty generalizations, incompleteness results are usually of little direct significance here. On the other hand, for those whose pri-mary interest is syntactically driven completeness results, the results could hardly be more significant: they unambiguously show the inadequacy of frame-based clas-sifications. Unsurprisingly, this has had considerable impact on the study of modal logic. For a start, it lead to a rebirth of interest in alternative tools – and, in partic-ular, to the renaissance of *algebraic semantics*, which we will study in Chapter 5. Moreover, it has lead modal logicians to study new types of questions. Let us consider some of the research themes that have emerged.

One response has been to look for general syntactic constraints on axioms which guarantee canonicity. The most elegant such result is the Sahlqvist Completeness Theorem, which we have already discussed. A second response has been to investi-gate the interplay between completeness, canonicity, and correspondence. Typical of the questions that can be posed is the following: *If $A_1, \ldots, A_n$ are axioms that define an elementary class of frames, is* $\mathbf{K}A_1 \ldots A_n$ *frame complete?* (In fact, the answer here is *no* – as the reader is asked to show in Exercise 4.4.4.) The most significant positive result that has emerged from this line of enquiry is the following:

Theorem 4.50 *If* F *is a first-order definable class of frames, then* Λ_F *is canonical.*

Again, we prove this in Chapter 5 using algebraic tools (see Theorem 5.56). Tanta-lizingly, at the time of writing the status of the converse was unknown: If a normal modal logic Λ is canonical, then there is a first-order definable class of frames F such that $\Lambda = \Lambda_F$. This conjecture seems plausible, but neither proof nor coun-terexample has been found.

A third response has been to examine particular classes of normal modal log-ics more closely. The entire lattice may have undesirable properties – but many sub-regions are far better behaved. We will examine a particularly well-behaved sub-region (namely, the normal logics extending **S4.3**) in the final section of this chapter.

This concludes our survey of basic completeness theory. The next four sections (all of which are on the basic track) explore the following issue: how are we to prove completeness results when we need to build a model that has a property for which no formula is canonical? Some readers may prefer to skip this for now and go straight on to the following chapter. This discusses completeness, canonicity and correspondence from an *algebraic* perspective.

Exercises for Section 4.4

4.4.1 Recall that any normal modal logic that has the finite model property also has the finite frame property. What are the consequences of this for incomplete normal modal logics?

4.4.2 The logic **KvB** consists of all formulas valid on the general frame $\mathfrak{J}$. The domain J of $\mathfrak{J}$ is $\mathbb{N} \cup \{\omega, \omega+1\}$ (the set of natural numbers together with two further points), and R is defined by Rxy iff $x \neq \omega+1$ and $y < x$ or $x = \omega+1$ and $y = \omega$. (The frame (J, R) is shown in Figure 6.2 on page 351.) A, the collection of subsets of J admissible in $\mathfrak{J}$, consists of all $X \subseteq J$ such that either X is finite and $\omega \notin X$, or X is co-finite and $\omega \in X$.

(a) Show that $\Box\Diamond(\top) \to \Box(\Box(\Box p \to p) \to p)$ is valid on $\mathfrak{J}$.
(b) Show that on any *frame* on which the previous formula is valid, $\Box\Diamond(\top) \to \Box(\bot)$ is valid too.
(c) Show that $\Box\Diamond(\top) \to \Box(\bot)$ is *not* valid on $\mathfrak{J}$.
(d) Conclude that **KvB** is incomplete.

4.4.3 Prove that any consistent normal modal logic in the basic modal similarity type is either valid on the frame consisting of a single reflexive point or valid on the frame consisting of a single irreflexive point. (Hint: use the fact that either $\Box\bot$ is Λ-consistent or $\Diamond\top$ is a Λ-theorem.)

Conclude that no consistent normal modal logic in the basic similarity type defines the empty frame class.

4.4.4 Consider the formulas (T) $p \to \Diamond p$, (M) $\Box\Diamond p \to \Diamond\Box p$, (E) $\Diamond(\Diamond p \wedge \Box q) \to \Box(\Diamond p \vee \Box q)$ and (Q) $(\Diamond p \wedge \Box(p \to \Box p)) \to p$. Let Λ denote the normal modal logic axiomatized by these formulas.

(a) Prove that E corresponds to the following first-order formula: $\forall xy_1y_2 ((Rxy_1 \wedge Rxy_2) \to (\forall z (Ry_1z \to Ry_2z) \vee \forall z (Ry_2z \to Ry_1z)))$.
(b) Prove that within the class of frames validating both T and E, Q defines the frames satisfying the condition $R^{\smile} \subseteq R^*$ (that is, if Rst then there is a finite path back from t to s).
(c) Prove that the conjunction of the four axioms defines the class of frames with a trivial accessibility relation – that is, T $\wedge$ M $\wedge$ E $\wedge$ Q corresponds to $\forall xy (Rxy \leftrightarrow x = y)$. (Hint: consider the effect of the McKinsey formula on the frames satisfying the condition $R^{\smile} \subseteq R^*$.)

(d) Consider the so-called *veiled recession frame* $(\mathbb{N}, R, A)$, where $\mathbb{N}$ is the set of natural numbers, Rmn holds iff $m \leq n+1$ and A is the collection of finite and co-finite subsets of $\mathbb{N}$. Show that all four axioms are valid on this general frame, but that the formula $p \rightarrow \Box p$ can be refuted.

(e) Conclude that Λ is incomplete, although it defines an elementary class of frames.

(f) Does this contradict Theorem 4.50?

4.4.5 Given a class K of frames, let $\Theta(\mathsf{K}) = \Lambda_{\mathsf{K}}$ denote the set $\{\phi \mid \mathfrak{F} \Vdash \phi \text{ for all } \mathfrak{F} \text{ in K }\}$ and given a logic Λ, let $\mathsf{Fr}(\Lambda)$ denote the class of frames on which Λ is valid.

(a) Show that the operations Θ and Fr form a so-called *Galois connection*. That is, prove that for all classes K and logics Λ:

$$\Lambda \subseteq \Theta(\mathsf{K}) \text{ iff } \mathsf{K} \subseteq \mathsf{Fr}(\Lambda).$$

(b) What does it mean for a logic Λ if $\Lambda = \Theta(\mathsf{Fr}(\Lambda))$? (Give an example of a logic for which it does *not* hold.)

(c) What does it mean for a frame class K if $\mathsf{K} = \mathsf{Fr}(\Theta(\mathsf{K}))$? (Give an example of a frame class for which it does *not* hold.)

4.5 Transforming the Canonical Model

What is the modal logic of partial orders? And what is the tense logic of strict total orders? Such questions bring us face to face with the fundamental problem confronting semantically driven completeness results. Partial orders are *antisymmetric*, and strict total orders are *irreflexive*. No modal formula defines either property, and (as the reader probably suspects) no formula is canonical for them either. Thus, to answer either question, we need to build a model for which we lack a canonical formula – and hence we will need to expand our repertoire of model building techniques. This is the main goal of the present section and the three that follow.

In this section we explore a particularly natural strategy: transforming the canonical model. Although a canonical model may lack some desired properties, it does get a lot of things right. Perhaps it is possible to reshape it, transforming it into a model with all the desired properties? We have done this once already, though in a very simple way: in the completeness proof for $\mathbf{K_t Q}$ (see Theorem 4.41 and surrounding discussion) we formed a point-generated submodel of the canonical model to ensure trichotomy. Here we will study two more sophisticated transformations – *unraveling* and *bulldozing* – and use them to answer the questions with which this section began.

It seems plausible that **S4** is the modal logic of partial orders: Theorem 4.29 tells us that **S4** is complete with respect to the class of reflexive transitive frames (that is, *preorders*) and there do not seem to be any modal formulas we could add to **S4** to reflect antisymmetry. Furthermore, it seems reasonable to hope that we could prove this using some sort of model transformation: as every **S4**-consistent set of formulas can be satisfied on a preorder, and as we know that modal languages are blind to antisymmetry (at least as far as frame definability is concerned) maybe we

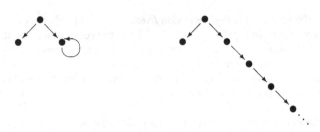

Fig. 4.1. A model and its unraveling

can find a way of transforming any satisfying preorder into a partial order without affecting satisfiability? (It is worth stressing that this informal line of argument is *not* a proof; it is intended solely to motivate the work that follows.)

A transformation called *unraveling* will enable us do this. Indeed, unraveling will let us prove the stronger result that **S4** is complete with respect to the class of *reflexive and transitive trees*. (This will be useful in Chapter 6 when we discuss decidability). We briefly discussed unraveling in Chapter 2, where we used it to show that modal logic has the tree property (see Proposition 2.15). Informally, given any model, unraveling builds a new model, whose points are *paths* of the original model. That is, transition sequences in the original model are explicitly represented as states in the unraveled model. More precisely:

Definition 4.51 (Unraveling) Let (W, R) be a frame generated by some point $w \in W$. The *unraveling* of (W, R) around w is the frame $(\vec{W}, \vec{R})$ where:

(i) $\vec{W}$ is the set of all finite sequences $(w, w_1, \ldots, w_n)$ such that $w_1, \ldots, w_n \in W$ and $Rww_1, \ldots, Rw_{n-1}w_n$, and

(ii) If $\vec{s}_1, \vec{s}_2 \in \vec{W}$, then $\vec{R}\vec{s}_1\vec{s}_2$ if there is some $v \in W$ such that $\vec{s}_1 + (v) = \vec{s}_2$, where $+$ denotes sequence concatenation.

If $\mathfrak{M} = (W, R, V)$ is a model and $(\vec{W}, \vec{R})$ is the unraveling of (W, R) around w, then we define the valuation $\vec{V}$ on $(\vec{W}, \vec{R})$ as follows:

$$\vec{V}(p) = \{(w, w_1, \ldots, w_n) \in \vec{W} \mid w_n \in V(p)\}$$

The model $\vec{\mathfrak{M}} = (\vec{W}, \vec{R}, \vec{V})$ is called the unraveling of $\mathfrak{M}$ around w.   ⊣

A simple example is given in Figure 4.1. As this example suggests (and as the reader should check) unraveling any frame around a generating point w yields an *irreflexive*, *intransitive*, and *asymmetric* frame. Indeed, note that unraveled frames are *trees*: the root node is the sequence (w), and the relation $\vec{R}$ is just the familiar (immediate) successor (or daughter-of) relation on trees.

Lemma 4.52 *Let* $\vec{\mathfrak{M}} = (\vec{W}, \vec{R}, \vec{V})$ *be the unraveling of* $\mathfrak{M} = (W, R, V)$ *around* *w. Then* (W, R) *is a bounded morphic image of* $(\vec{W}, \vec{R})$, *and* $\mathfrak{M}$ *is a bounded morphic image of* $\vec{\mathfrak{M}}$.

Proof. Let $f : \vec{W} \rightarrow W$ be defined by $f(w, w_1, \dots, w_n) = w_n$. It is easy to see that f is surjective, has the back and forth properties, and that for any $\vec{s} \in \vec{W}$, $\vec{s}$ and $f(\vec{s})$ satisfy the same propositional variables. ⊣

A simple corollary is that *any* satisfiable set of formulas is satisfiable on a (irreflexive, intransitive, and asymmetric) tree: for if a set of formulas is satisfiable, it is satisfiable on a point-generated model (take the submodel generated by the satisfying point), hence by unraveling we have the result. It follows that **K** is (strongly) complete with respect to this class of models.

But our real interest is **S4**. How do we use unraveling to make the *partially ordered* models we require for the completeness result? In the most obvious way possible: we simply take the reflexive transitive closures of unraveled models. More precisely, suppose we unravel $\mathfrak{M}$ around some generating point w to obtain $(\vec{W}, \vec{R}, \vec{V})$. Now consider the model $\mathfrak{M}^* = (\vec{W}, R^*, \vec{V})$ where R^* is the reflexive transitive closure of $\vec{R}$. Trivially, $\mathfrak{M}^*$ is an **S4** model. Moreover, as $(\vec{W}, \vec{R})$ is a tree, $(\vec{W}, R^*)$ is an *antisymmetric* frame. Indeed, it is a *reflexive and transitive tree*, for R^* is simply the familiar dominates (or ancestor-of) relation on trees. So only one question remains: is $\mathfrak{M}$ a bounded morphic image of $\mathfrak{M}^*$? In general, *no*. But if the model $\mathfrak{M}$ we started with was itself reflexive and transitive, *yes*:

Lemma 4.53 *Let* $\mathfrak{M} = (W, R, V)$ *be a reflexive transitive model generated by some* $w \in W$, *and let* $(\vec{W}, \vec{R}, \vec{V})$ *be the unraveling of* $\mathfrak{M}$ *around* w. *Let* R^* *be the reflexive transitive closure of* $\vec{R}$, *and define* $\mathfrak{M}^*$ *to be* $(\vec{W}, R^*, \vec{V})$. *Then* $\mathfrak{M}$ *is a bounded morphic image of* $\mathfrak{M}^*$.

Proof. It is easy to see that the function f defined in Lemma 4.52 remains the required bounded morphism; as far as surjectivity, the back property, and the distribution of proposition letters are concerned, nothing has changed. We only have to check that taking the reflexive transitive closure of $\vec{R}$ does not harm the forth property. But, as R is itself reflexive and transitive, the forth property survives. ⊣

Theorem 4.54 **S4** *is strongly complete with respect to the class of partially ordered reflexive and transitive trees.*

Proof. If Σ is an **S4**-consistent set of formulas, and Σ^+ is an **S4**-MCS extending Σ, then $\mathfrak{M}^{S4}, \Sigma^+ \Vdash \Sigma$. Moreover, as the **S4** axioms are canonical, $\mathfrak{M}^{S4}$ is a reflexive transitive model. We now transform this model into the required partial order in two steps.

Step 1. Let $\mathfrak{M}^S$ be the submodel of $\mathfrak{M}^{S4}$ generated by Σ^+. Clearly this is a reflexive, transitive, point-generated model such that $\mathfrak{M}^S, \Sigma^+ \Vdash \Sigma$.

Step 2. Let $\mathfrak{M}^* = (\vec{W}, R^*, \vec{V})$ be the reflexive transitive closure of the unraveling of $\mathfrak{M}^S$ around Σ^+.

By Lemma 4.53, $\mathfrak{M}^S$ is a bounded morphic image of $\mathfrak{M}^*$ under f, hence for all sequences $\vec{s} \in f^{-1}[\Sigma]$, we have $\mathfrak{M}^*, \vec{s} \Vdash \Sigma$, and by the surjectivity of f there is at least one such $\vec{s}$. Hence we have satisfied Σ on a reflexive and transitive tree.   ⊣

The previous proof could be summed up as follows: we found a way to use the information in a canonical model *indirectly*. The canonical model for **S4** did not have the structure we wanted – nonetheless, we successfully tapped into the information it contained via a short sequence of bisimulations ($\mathfrak{M}^*$ had $\mathfrak{M}^S$ as a bounded morphic image, and $\mathfrak{M}^S$ was a generated submodel of $\mathfrak{M}^{S4}$).

Unraveling is an intrinsically *global* transformation that can change a model's geometry drastically. This is in sharp contrast to the transformation we will now examine – *bulldozing* – which works locally, and (in spite of its name) rather more gently. We will use bulldozing to answer the second of the questions posed above. Recall that a *strict* total order (STO) is a relation that is transitive, trichotomous and *irreflexive*. The class of strict total orders contains such important structures as $(\mathbb{N}, <)$, $(\mathbb{Z}, <)$, $(\mathbb{Q}, <)$, and $(\mathbb{R}, <)$ (the natural numbers, the integers, the rationals and the reals in their usual order) and is widely used to model various temporal phenomena. What is its tense logic?

Once again, it is not hard to find a plausible candidate: $\mathbf{K}_t 4.3$, the tense logic generated by 4, $.3_l$ and $.3_r$, seems the only reasonable candidate. For a start, $\mathbf{K}_t 4.3$ is strongly complete with respect to the class of *weak* total orders. (To see this, observe that the axioms are canonical for transitivity and non-branching. Hence any point generated submodel $\mathfrak{M}^S$ of the canonical model is transitive and trichotomous, and the completeness result is immediate.) Moreover, there simply are no other plausible axioms – in particular, irreflexivity is not definable. Has this (somewhat dangerous) line of reasoning led to the right answer? Let us see.

If we could find a way of transforming weakly linear models into strictly linear models we would have the desired completeness result. Note that unraveling will not help – it would turn the weak total order into a tree, thus destroying trichotomy. If only we could find a method which replaced the undesirable parts of the model with some suitable STO, and left the good parts untouched: then trichotomy would not be affected, and we would have assembled the required strict total order. Bulldozing is a way of doing this. The first step is to pin down what the 'undesirable' parts of weak total orders are. The obvious response is 'reflexive points' – but while this is not exactly wrong, it misses the crucial insight. The entities we really need to think about are *clusters*, introduced in Chapter 2. We repeat the definition:

Definition 4.55 Let (T, R) be a transitive frame. A *cluster* on (T, R) is a subset C of T that is a maximal equivalence relation under R. That is, the restriction of

R to C is an equivalence relation, and this is *not* the case for any other subset D of T such that $C \subset D$. A cluster is *simple* if it consists of a single reflexive point, and *proper* if it contains more than one point. When we say that a model contains clusters, we mean that its underlying frame does. ⊣

The point is this: we should not think in terms of removing isolated reflexive points; rather, we should remove entire clusters at one stroke. (Intuitively, the information in a cluster is information that 'belongs together.') Any transitive trichotomous frame can be thought of as a strictly totally ordered collection of clusters (cf. Exercise 1.1.1). If we could remove each cluster as a single chunk, and replace it with something equivalent, we would have performed a local model transformation.

So the key question is: what should we replace clusters with? Clearly some sort of STO – but how can we do this in a truth preserving way? Note that any cluster C, even a simple one, introduces an infinity of information recurrence in both the forwards and backwards directions: we can follow paths within C, moving forwards and backwards, for as long as we please. Thus, when we replace a cluster C with a STO, we must ensure that the STO duplicates all the information in C infinitely often, in both directions. Bulldozing does precisely this in a straightforward way. We simply impose a strict total order on the cluster (that is, we pick some path through the cluster that visits each point once and only once) and then lay out infinitely many copies of this path in both the forwards and backwards direction. We then replace the cluster by the infinite repetition of the chosen path. We have squashed the clusters down into infinitely long STOs – hence the name 'bulldozing'.

Theorem 4.56 $\mathbf{K}_t\mathbf{4.3}$ *is strongly complete with respect to the class of strict total orders.*

Proof. Let Σ be a $\mathbf{K}_t\mathbf{4.3}$-consistent set of formulas; expand it to a $\mathbf{K}_t\mathbf{4.3}$-MCS Σ^+. Let $\mathfrak{M} = (T, R, V)$ be the canonical model for $\mathbf{K}_t\mathbf{4.3}$. By the canonicity of the axioms, $\mathfrak{M}$ is transitive and non-branching. Let $\mathfrak{M}^S = (S, R^S, V^S)$ be the submodel of $\mathfrak{M}$ generated by Σ^+; $\mathfrak{M}^S$ is a transitive and trichotomous model such that $\mathfrak{M}^S, \Sigma^+ \Vdash \Sigma$. But $\mathfrak{M}^S$ may contain clusters, which we will bulldoze away.

Step 1. Index the clusters in $\mathfrak{M}^S$ by some suitable set I.

Step 2. Define an arbitrary strict total order $<^i$ on each cluster C_i.

Step 3. Define $C_i^\flat$ to be $C_i \times \mathbb{Z}$. ($\mathbb{Z}$ is the set of integers.)

Step 4. Define B, the set underlying the bulldozed model, to be

$$S^- \cup \bigcup_{i \in I} C_i^\flat,$$

where S^- is the set $(S \setminus \bigcup_{i \in I} C_i)$ of points *not* belonging to any cluster.

Step 5. Define a mapping $\beta : B \to S$ by: $\beta(b) = b$, if $b \in S^-$; and $\beta(b) = s$, if $b = (s, z)$.

Step 6. Define an ordering $<^b$ on B by $b <^b b'$ iff

> **either** ($b \in S^-$ or $b' \in S^-$) and $\beta(b) R^S \beta(b')$;
> **or** $b = (s, z)$ and $b' = (s', z')$ and
>
>> **either** s and s' belong to distinct clusters and $\beta(b) R^S \beta(b')$;
>> **or** s and s' belong to the same cluster and $z <_{\mathbb{Z}} z'$ (where $<_{\mathbb{Z}}$ is the usual ordering on the integers);
>> **or** s and s' belong to the same cluster C_i and $z = z'$ and $s <^i s'$.

Step 7. Define a valuation V^b on $(B, <^b)$ by $b \in V(p)$ iff $\beta(b) \in V^S(p)$.
Step 8. Define $\mathfrak{M}^B$, *the bulldozed model*, to be $(B, <^b, V^b)$.

We now make the following claims:

Claim 1. *The mapping β is a surjective bounded morphism from $(B, <^b)$ to (S, R^S), and the model $\mathfrak{M}^S$ is a bounded morphic image of $\mathfrak{M}^B$ under β.*

Claim 2. $(B, <^b)$ *is a strict total order.*

Proving these claims is a matter of checking the definitions; we leave this to the reader as Exercise 4.5.5. With this done, the theorem is immediate. By Claim 1, for any $b \in \beta^{-1}(\Sigma^+)$ we have $\mathfrak{M}^B, b \Vdash \Sigma$, and since β is surjective, there is at least one such b. Thus $\mathfrak{B}$ is a model of Σ, and by Claim 2 it has the structure we want.   ⊣

Although it works more locally, like unraveling, bulldozing is a way of using the information in canonical models *indirectly*. Indeed, like unraveling, it accesses the information in the relevant canonical model via a sequence of bisimulations: the final model $\mathfrak{M}^B$ had $\mathfrak{M}^S$ as a bounded morphic image, and $\mathfrak{M}^S$ in turn was a generated submodel of $\mathfrak{M}$.

Bulldozing is a flexible method. For example, we are not forced to define C_i^b to be $C_i \times \mathbb{Z}$; any unbounded STO would do. Moreover, if we used a *reflexive* total order (for example $(\mathbb{Z}, \leq)$) instead, we could prove analogous completeness results for reflexive total orders; for example, the reader is asked to show in Exercise 4.5.6 that $S_t 4.3$ is the logic of this class of frames. Moreover, for modal languages, we only need to ensure infinite information repetition in the *forward* direction, so structures such as $(\mathbb{N}, <)$ and $(\mathbb{N}, \leq)$ suffice.

But there are more interesting variations. For example, instead of simply ordering the points in the cluster, one can *embed* the cluster in some suitable total order, and work with its embedded image instead. By embedding the clusters in a *dense* set, it is possible to build dense totally ordered models. And by combining such ideas with other transformations (notably filtrations) the method can be used to prove many classic completeness results of modal and tense logics.

Model manipulation methods, and completeness proofs making use of them, abound. Further examples are mentioned in the Notes, but it is not remotely possible to be encyclopedic: such methods trade on specific insights into the geometry of relational structures, and this gives rise to a wide variety of variants and combinations. The reader should certainly be familiar with such methods – they are often simple to adapt to specific problems – but it is just as important to appreciate the general point that has emerged from our discussion: even if the canonical model is not quite what we need, it can still be extremely useful. The following section further explores this theme.

Exercises for Section 4.5

4.5.1 **K** is complete with respect to the class of irreflexive frames. Unraveling shows this, but there is a much simpler transformation proof. (Hint: given a model $\mathfrak{M}$, tinker with the disjoint union of $\mathfrak{M}$ with itself.)

4.5.2 Formulate the unraveling method for modal languages containing two diamonds. Then formulate the method in such a way that bidirectional frames unravel into bidirectional frames.

4.5.3 Consider a similarity type τ with one binary operator $\triangle$. Call a τ-frame $\mathfrak{F} = (W, T)$ *acyclic* if the binary relation $R = \{(s, t) \in W^2 \mid Tstu \text{ or } Tsut \text{ for some } u \in W\}$ is acyclic (that is to say, R^+ is irreflexive). Prove that the basic modal logic $\mathbf{K}_\tau$ is strongly sound and complete with respect to the class of acyclic frames.

4.5.4 Show that the canonical model for $\mathbf{K}_t\mathbf{Q}$ contains proper clusters.

4.5.5 Prove Claims 1 and 2 of Theorem 4.56.

4.5.6 Let $\mathbf{K}_t\mathbf{QT}$ be the smallest normal temporal logic containing both $\mathbf{K}_t\mathbf{Q}$ and $p \to Fp$. Show, using a light bulldozing argument, that $\mathbf{K}_t\mathbf{QT}$ is strongly complete with respect to the class of all dense unbounded reflexive total orders. (In this context of reflexive orders, density refers to the property $\forall x \forall y \, ((Rxy \wedge x \neq y) \to \exists z \, (Rxz \wedge x \neq z \wedge Rzy \wedge z \neq y))$.)

4.6 Step by Step

Three main ideas underly the step-by-step method:

(i) Do not consider the entire canonical model to be the key ingredient of a completeness proof. Rather, think of *selections of* MCSs *from the canonical model* as the basic building blocks.

(ii) The standard way of proving completeness is by constructing a model for a consistent set of formulas. Take the term 'constructing' as literally as possible: break it down into a sequence of steps.

(iii) Putting the first two observations together, think of the construction of a model as the stepwise selection of the needed MCSs. More precisely, think of the model construction process as approaching a limit via a sequence of ever better approximations, using local configurations of the canonical model to make improvements at each step of the construction.

The method gives us enormous control over the models we build, and even at this stage it is easy to see why. First, we do not have to worry about unpleasant features of the canonical model (such as clusters) since we only work with selections of the information that canonical structures contain. Furthermore, as we select our information one step at a time, we obtain an iron grip on what ends up in the model.

To illustrate the method's potential, we use it to prove that the logic $\mathbf{K}_t\mathbf{Q}$ defined in Definition 4.40 is strongly complete with respect to $(\mathbb{Q}, <)$. In what follows, consistency means $\mathbf{K}_t\mathbf{Q}$-consistency, and $\mathfrak{M}^c = (T^c, R^c, V^c)$ is this logic's canonical model. Furthermore we fix a maximal consistent set Σ; the goal of our proof is to construct a model $\mathfrak{M} = (T, <, V)$ for Σ such that $(T, <)$ is an ordering which is isomorphic to $(\mathbb{Q}, <)$. At each step of the construction we will be dealing with an approximation of $\mathfrak{M}$ consisting of a strictly ordered finite set of points (that will ultimately end up) in T and for each of these, the set of all formulas that we want to be the point's modal type (that is, the set of formulas holding at the point).

Definition 4.57 A *network* is a triple $\mathcal{N} = (N, <, \nu)$ such that R is a binary relation on the set N, and ν is a labeling function mapping each point in N to a maximal consistent set. ⊣

We are not interested in networks that are blatantly faulty as approximations of our desired model. For example, we want $<$ to be a strict total ordering. Moreover, whenever a formula ψ is in the label set of a point s, then $F\psi$ should be in $\nu(t)$ for any t with $t < s$. Such requirements lead to the following definition.

Definition 4.58 A network $\mathcal{N} = (N, <, \nu)$ is *coherent* if it satisfies:

(C1) $<$ is a strict total ordering,
(C2) $\nu(s)R^c\nu(t)$ for all $s, t \in N$ such that $s < t$.

A *network for* Σ is a network such that Σ is the label set of some node.   ⊣

C1 and C2 are the minimal requirements for a network to be useful to us; note that both requirements are *universal*. (C2 is equivalent to the requirement that if $s < t$ then $F\phi \in \nu(s)$ for all $\phi \in \nu(t)$ and $P\phi \in \nu(t)$ for all $\phi \in \nu(s)$.) But if a network is to really resemble a model, it must also satisfy certain *existential* requirements.

Definition 4.59 A network $\mathcal{N} = (N, <, \nu)$ is *saturated* if it satisfies:

(S1) $<$ is unbounded to the left and to the right,
(S2) $<$ is dense,

(S3) $\mathcal{N}$ is modally saturated. That is, we demand that (F) if $F\psi \in \nu(s)$ for some $s \in N$, then there is some $t \in N$ such that $s < t$ and $\psi \in \nu(t)$, and (P) if $P\psi \in \nu(s)$ for some $s \in N$, then there is some $t \in N$ such that $t < s$ and $\psi \in \nu(t)$.

A network is *perfect* if it is both coherent and saturated. ⊣

We want networks to give rise to models. Let us now check that we have imposed sufficiently many criteria on networks to achieve this.

Definition 4.60 Let $\mathcal{N} = (N, <, \nu)$ be a network. The frame $\mathfrak{F}_{\mathcal{N}} = (N, <)$ is called the *underlying frame* of $\mathcal{N}$. The *induced valuation* $V_{\mathcal{N}}$ on $\mathfrak{F}$ is defined by $V_{\mathcal{N}}(p) = \{s \in N \mid p \in \nu(s)\}$. The structure $\mathfrak{I}_{\mathcal{N}} = (\mathfrak{F}_{\mathcal{N}}, V_{\mathcal{N}})$ is the *induced model*. ⊣

The following lemma shows that our definition of perfection is the right one.

Lemma 4.61 (Truth Lemma) *Let $\mathcal{N}$ be a countably infinite perfect network. Then for all formulas ψ, and all nodes s in N,*

$$\mathfrak{I}_{\mathcal{N}}, s \Vdash \psi \text{ iff } \psi \in \nu(s).$$

Moreover, $\mathfrak{F}_{\mathcal{N}}$ is isomorphic to the ordering of the rational numbers.

Proof. The first part of the proof is by induction on the degree of ψ. The base case is clear from the definition of the induced valuation, and the steps for the booleans are straightforward. As for the modal operators, the coherency of $\mathcal{N}$ drives the left to right implication through, and saturation takes care of the other direction.

Finally, the underlying frame of a perfect network must be a dense, unbounded, strict total ordering. Hence, if it is countably infinite, it must be isomorphic to $(\mathbb{Q}, <)$ by Cantor's Theorem. (Readers unfamiliar with this theorem should try to prove this classic result from first principles. The standard proof builds up the isomorphism using a step-by-step argument!) ⊣

It follows from Lemma 4.61 that we have reduced the task of finding a model for our MCS Σ to the quest for a countable, perfect network for Σ. And now we arrive at the heart of the step-by-step method: the crucial idea is that each witness to the imperfection of a coherent network can be removed, one step at a time. Such witnesses will be called *defects*. There are three kinds of defect: each corresponds to a violation of a saturation condition.

Definition 4.62 Let $\mathcal{N} = (N, <, \nu)$ be a network. An *S1-defect* of $\mathcal{N}$ consists of a node $s \in N$ that has no successor, or no predecessor; an *S2-defect* is a pair (s, t) of nodes for which there is no intermediate point. An *S3-defect* consists of (F) a node s and a formula $F\psi \in \nu(s)$ for which there is no t in N such that $s < t$ and

$\psi \in \nu(t)$, or (P) a node s and a formula $P\psi \in \nu(s)$ for which there is no t in N such that $t < s$ and $\psi \in \nu(t)$. $\dashv$

Now we need to say more about what it means to repair a defect. To do so, we need the notion of one network *extending* another.

Definition 4.63 Let $\mathcal{N}_0 = (N_0, <_0, \nu_0)$ and $\mathcal{N}_1 = (N_1, <_1, \nu_1)$ be two networks. We say that $\mathcal{N}_1$ *extends* $\mathcal{N}_0$ (notation: $\mathcal{N}_1 \rhd \mathcal{N}_0$) if $\mathfrak{F}_{\mathcal{N}_0}$ is a subframe of $\mathfrak{F}_{\mathcal{N}_1}$ and ν_0 agrees with ν_1 on N_0. $\dashv$

The key lemma of this (or for that matter, any) step-by-step proof states that any defect of a finite coherent network can be repaired. More precisely:

Lemma 4.64 (Repair Lemma) *For any defect of a finite, coherent network $\mathcal{N}$ there is a finite, coherent $\mathcal{N}' \rhd \mathcal{N}$ lacking this defect.*

Proof. Let $\mathcal{N} = (N, <, \nu)$ be a finite, coherent network and assume that $\mathcal{N}$ has some defect. We prove the Lemma by showing that all three types of defect can be removed.

S1-defects.
These are left as an exercise to the reader.

S2-defects.
Assume that there are nodes s and t in N for which there is no intermediate point.

How should we repair this defect? The basic idea is simple: just throw in a new point between s and t, and find an appropriate label for it. This can be done easily, since it follows by coherence of $\mathcal{N}$ that $\nu(s)R^c\nu(t)$, and by canonicity of the density axiom that there is some MCS Γ such that $\nu(s)R^c\Gamma R^c\nu(t)$. Hence, take some *new* node u (new in the sense that $u \notin N$) and define $\mathcal{N}' = (N', <', \nu')$ by

$$
\begin{aligned}
N' &:= N \cup \{u\}, \\
<' &:= < \cup \{(x, u) \mid x \leq s\} \cup \{(u, x) \mid t \leq x\}, \\
\nu' &:= \nu \cup \{(u, \Gamma)\}.
\end{aligned}
$$

It is clear that $\mathcal{N}'$ is a network that does not suffer from the old defect. But is $\mathcal{N}'$ coherent? Condition C1 is almost immediate by the definition, so we concentrate on C2. Let x and y be two arbitrary nodes in $\mathcal{N}'$ such that $x <' y$; we have to check that $\nu'(x)R^c\nu'(y)$. Now, as $<'$ is irreflexive, x and y are distinct. Moreover, there can only be a problem if one of the nodes is the new point u; assume that $y = u$ (the other case is similar). If $x = s$ then we have $\nu'(x)R^c\nu'(u)$ by our assumption on Γ, so suppose that $x \neq s$. By definition of $<'$ and the fact that there are no old nodes between s and t, this means that $x < s$, so by the coherency of $\mathcal{N}$ we have that $\nu(x)R^c\nu(s)$. Hence, it follows by the transitivity of R^c that $\nu(x)R^c\Gamma$; but then it is immediate by the definition of ν' that $\nu'(x)R^c\nu'(u)$.

S3-defects.

We only treat the P-defects; the case for F-defects follows by symmetry. Assume that there is a node s in N and a formula $P\psi$ in $\nu(s)$ for which there is no t in N such that $t < s$ and $\psi \in \nu(t)$.

Again, the basic strategy is simple: we insert a new point s' into the network (before s!) and choose an adequate label for it; this has to be a maximal consistent set containing ψ and preceding $\nu(s)$ in the preorder R^c. But *where* should s' be inserted? If we are not careful we will destroy the coherency of $\mathcal{N}$. The following maneuver (which takes advantage of the fact that $\mathfrak{F}_N$ is a *finite* STO) overcomes the difficulty.

Let m be the unique point in N such that (i) $(m, P\psi)$ is an S3-defect in $\mathcal{N}$, and (ii) for all $w < m$, $(w, P\psi)$ is *not* a defect. Such an m must exist (it is either s itself, or one of the finitely many points preceding s) and, as we will see, we can repair $(m, P\psi)$ without problems by simply inserting the new point s' immediately before m. Repairing this minimal defect automatically repairs the defect $(s, P\psi)$.

Choose some new point s' (that is, $s' \notin S$) and let Ψ be an MCS containing ψ such that $\Psi R^c\nu(m)$; such a Ψ exists by the Existence Lemma for normal logics. Define $\mathcal{N}' = (N', <', \nu')$ as follows:

$$
\begin{aligned}
N' &:= N \cup \{s'\}, \\
<' &:= < \cup \{(x, s') \mid x < m\} \cup \{(s', x) \mid m \le x\}, \\
\nu' &:= f \cup \{(s', \Psi)\}.
\end{aligned}
$$

Observe that $\mathfrak{F}_{\mathcal{N}'}$ is a strict total order, and that $\mathcal{N}'$ does *not* contain the defect $(s, P\psi)$. It only remains to ensure that $\mathcal{N}'$ satisfies the second coherency condition.

Consider two nodes $x, y \in N'$ such that $x <' y$. Again, the only cases worth checking are when either x or y is the new point s'. If we have $x = s'$ we are in a similar situation as in the case of S2-defects, so we do not go into details here.

Hence, assume that $y = s'$. By construction $\nu(s') = \Psi R^c\nu(m)$, and by the coherency of $\mathcal{N}$, $\nu(x)R^c\nu(m)$. But R^c is the canonical relation for $\mathbf{K}_t\mathbf{Q}$ – a relation with no branching to the left – hence either $\Psi R^c\nu(x)$, $\Psi = \nu(x)$ or $\nu(x)R^c\Psi$. We claim that the first two options are impossible. For, if $\Psi R^c\nu(x)$ then $\psi \in \Psi$ would imply that $P\psi \in \nu(x)$ and this contradicts the minimality of m; and if $\Psi = \nu(x)$, then $\psi \in \nu(x)$ would mean that $(s, P\psi)$ was not a defect in the first place! We conclude that $\nu(x)R^c\Psi$, which establishes coherence. $\dashv$

With both the Truth Lemma for induced models and the Repair Lemma at our disposal, we can prove the desired strong completeness result. The idea is straightforward. We start with a singleton network and extend it step by step to larger (but finite) networks by repeated use of the Repair Lemma. We obtain the required perfect network by taking the union of our sequence of networks.

Theorem 4.65 $K_t Q$ *is strongly complete with respect to* $(\mathbb{Q}, <)$.

Proof. Choose some set $S = \{s_i \mid i \in \omega\}$ (we will use its elements to build the required frame) and enumerate the set of potential defects (that is, the union of the sets S, $S \times S$ and $S \times \{F, P\} \times \textit{Form}$). Given a consistent set of formulas Σ, expand it to an MCS Σ_0. Let $\mathcal{N}_0$ be the network $(\{s_0\}, \varnothing, (s_0, \Sigma_0))$. Trivially, $\mathcal{N}_0$ is a finite, coherent network for Σ_0.

Let $n \geq 0$ and suppose $\mathcal{N}_n$ is a finite, coherent network. Let D be the defect of $\mathcal{N}_n$ that is minimal in our enumeration. Such a D exists, since any finite network must at least have S1- and S2-defects. Form $\mathcal{N}_{n+1}$ by repairing the defect D as described in the proof of the Repair Lemma. Observe that D will not be a defect of any network extending $\mathcal{N}_n$.

Let $\mathcal{N} = (N, <, \nu)$ be given by

$$N = \bigcup_{n \in \omega} N_n, \quad < = \bigcup_{n \in \omega} <_n, \quad \text{and} \quad \nu = \bigcup_{n \in \omega} \nu_n.$$

It is easy to see that $\mathfrak{F}_\mathcal{N}$ is a strict total order. Moreover, as we chose the points in $\mathcal{N}$ from a countably infinite set, $\mathcal{N}$ is countable.

It should be intuitively clear that $\mathcal{N}$ is perfect, but the actual proof has to take care of a subtlety. Suppose that $\mathcal{N}$ is not perfect; let D be the minimal (according to our enumeration) defect of $\mathcal{N}$, say $D = D_k$. By our construction, there must be an approximation $\mathcal{N}_i$ of $\mathcal{N}$ of which D is also a defect. Note that D need *not* be the minimal defect of $\mathcal{N}_i$ – this is the subtlety. Fortunately, there can be at most k defects that are more urgent, so D will be repaired before stage $k + i$ of the construction.

Finally, by the perfection of $\mathcal{N}$ it follows from Lemma 4.61 that the induced model $\mathfrak{I}_\mathcal{N}$ satisfies Σ at s_0. $\dashv$

The step-by-step method is one of the most versatile tools at the modal logician's disposal: a wide variety of results in modal and tense logic have been using this method, it is the tool of choice for many stronger modal systems such as arrow logic and since/until logic, and we will make use of step-by-step arguments when we discuss rules for the undefinable in the following section. We urge the reader to experiment with it. A good starting point is Exercise 4.6.1.

Exercises for Section 4.6

4.6.1 Consider a modal language with three diamonds $\Diamond_1$, $\Diamond_2$ and $\Diamond_3$. Give a complete axiomatization for the class of frames $\mathfrak{F} = (W, R_1, R_2, R_3)$ satisfying $R_3 = R_1 \cap R_2$.

4.6.2 Consider, for a modal language with two diamonds $\Diamond_0$ and $\Diamond_1$, the normal modal logic $(\mathbf{S5})^2$ axiomatized by **S5** axioms for both diamonds, and the commutativity axiom $\Diamond_0\Diamond_1 p \leftrightarrow \Diamond_1\Diamond_0 p$. Prove that this logic is complete for the class of square frames. A square frame for this language is of the form $\mathfrak{F} = (W, R_0, R_1)$ where for some set U we have

$$W = U^2,$$
$$R_i st \quad \text{iff} \quad s_i = t_i.$$

(Hint: take as approximations networks of the form (N, ν) where ν is a labeling mapping *pairs* over N to maximal consistent sets.)

4.6.3 Consider a similarity type τ with one binary operator $\circ$, as in arrow logic. Call a τ-frame $\mathfrak{F} = (W, T)$ a *relativized square* if W is some collection of pairs over a base set U, and $T \subseteq W^3$ satisfies $Tstu$ iff $s_0 = t_0$, $t_1 = u_0$ and $s_1 = u_1$.

(a) Prove that the basic modal logic $\mathbf{K}_\tau$ is strongly sound and complete with respect to the class of relativized squares.
(b) Try to axiomatize the logic of the class of frames (W, R) in which W is as above, but T satisfies $Tstu$ iff $s_0 = t_1$, $t_0 = u$ and $u_0 = s_1$.

4.7 Rules for the Undefinable

In the previous two sections we proved semantically driven completeness results by using standard canonical models indirectly. The present section takes a rather different approach: we enrich the deductive system with a special proof rule, and consider a special (not necessarily generated) submodel of the canonical model for this new logic. The submodel that we study contains only special *distinguishing* (or *witnessing*) MCSs. The completeness proof shows that this new canonical model has all the good properties of the original, and that, in addition, it is already in the right shape. We will make use of ideas introduced in our discussion of the step-by-step method in the previous section (in particular, the concept of a defect).

The running example in this section will (again) be the tense logic of dense unbounded strict total orderings. Recall that the difficulty when working with this logic is that there is no axiom ensuring the irreflexivity of the canonical frame – we have all the other required properties: point generated submodels of the candidate logic $\mathbf{K}_t\mathbf{Q}$ are transitive, trichotomous, dense, and unbounded. Now, in previous sections we achieved irreflexivity indirectly: either we bulldozed away clusters, or we used the canonical model for $\mathbf{K}_t\mathbf{Q}$ to induce a model on a carefully constructed irreflexive frame. In this section we will construct a canonical frame that is transitive, non-branching, dense *and irreflexive* right from the start. Indeed, if we work with a countably infinite language, every point generated subframe of this canonical model will be countable, and hence (by Cantor's Theorem) isomorphic to $(\mathbb{Q}, <)$.

The starting point of the enterprise is that irreflexivity, although not definable in basic modal languages, *can* be characterized in an alternative sense:

If a temporal formula ψ is satisfiable on an irreflexive frame, then for any proposition letter p not occurring in ψ, the conjunction $(\neg Pp \wedge p \wedge \neg Fp) \wedge \psi$ is also satisfiable on that frame.

For, if $\mathfrak{F}, V, s \Vdash \psi$, then $\mathfrak{F}, V', s \Vdash (\neg Pp \wedge p \wedge \neg Fp) \wedge \psi$, where V' is just like V except that it assigns the singleton $\{s\}$ to p. The condition that p does not occur in ψ is crucial here: it ensures that changing the set assigned to p does not affect the satisfaction of ψ.

Now, by taking the contrapositive of the above statement, we turn it into a proof rule:

(IRR)   if $\vdash (\neg Pp \wedge p \wedge \neg Fp) \rightarrow \phi$ then $\vdash \phi$, provided p does not occur in ϕ.

We have just seen that this rule is sound on the class of irreflexive frames. Moreover, note that on the class of strict total orders the formula $(\neg P\phi \wedge \phi \wedge \neg F\phi)$ is true at some state s iff s is the *only* state where ϕ holds (we need trichotomy and transitivity to guarantee this). That is, the formula $\neg P\phi \wedge \phi \wedge \neg F\phi$ acts as a sort of 'name' for the satisfying point. Call this formula $name(\phi)$. Bearing these remarks in mind, let us now see how adding this rule is of any help in proving the desired completeness result.

Definition 4.66 The logic $\mathbf{K}_t\mathbf{Q}^+$ is obtained by adding to $\mathbf{K}_t\mathbf{Q}$ the irreflexivity rule IRR. In what follows, consistency means $\mathbf{K}_t\mathbf{Q}^+$-consistency, $\vdash \phi$ means that ϕ is provable in $\mathbf{K}_t\mathbf{Q}^+$, and so on. The canonical model for $\mathbf{K}_t\mathbf{Q}^+$ is denoted by $\mathfrak{M}^c$, the canonical relation by R^c.   ⊣

The remainder of this section is devoted to proving completeness of the proof system $\mathbf{K}_t\mathbf{Q}^+$ with respect to $(\mathbb{Q}, <)$. Of course the *result* is not surprising: we have already seen that plain old $\mathbf{K}_t\mathbf{Q}$ is strongly complete with respect to $(\mathbb{Q}, <)$. It is the *method* that is important: rules such as IRR give us a way of forming more cleanly structured canonical models.

Our goal is to construct an irreflexive version of the canonical model for $\mathbf{K}_t\mathbf{Q}^+$. The basic idea is to work only with special *witnessing* MCSs:

Definition 4.67 A maximal consistent set is called *witnessing* if it contains a formula of the form $name(\phi)$.   ⊣

Why are these witnessing MCSs so interesting? Well, suppose that we are dealing with a collection W of witnessing maximal consistent sets. This collection induces a model in the obvious way: the relation is just the canonical accessibility relation restricted to W and likewise for the valuation. Now suppose that we can prove a Truth Lemma for this model; that is, suppose we can show that 'truth and membership coincide' for formulas and MCSs. It is then immediate that the underlying relation of the model is irreflexive: $name(\phi) \in \Gamma$ implies $\phi \in \Gamma$ and $F\phi \notin \Gamma$.

This is all very well, but it is obvious that we cannot just throw away non-witnessing MCSs from the canonical model without paying a price. How can we be sure that we did not throw away too many MCSs? An examination of the standard canonical completeness proof reveals that there are two spots where claims are made concerning the existence of certain MCSs.

(i) There is the Existence Lemma, which is needed to prove the Truth Lemma. In our case, whenever the formula $F\phi$ is an element of one of our witnessing MCSs (Γ, say) then there must be a *witnessing* Δ such that $\Gamma R^c \Delta$ and $\phi \in \Delta$. But if Δ is witnessing, then there is some δ with $name(\delta) \in \Delta$; it follows from the definition of the canonical accessibility relation that $F(\phi \wedge name(\delta)) \in \Gamma$. This shows that it will not do to just take the witnessing MCSs: the Existence Lemma requires stronger saturation conditions on MCSs, namely that whenever $F\phi \in \Gamma$, then there is some δ such that $F(\phi \wedge name(\delta)) \in \Gamma$ too.

(ii) If there are axioms in the logic that are canonical for some property with existential import, how can we make sure that the trimmed down version of the canonical model still validates these properties? Examples are the formulas $\Diamond\Box p \rightarrow \Box\Diamond p$, or, in the present case, the density axiom. The point is that from the density of the standard canonical frame we may not infer that its subframe formed by witnessing MCSs is dense as well: why should there be a *witnessing* MCS between two witnessing MCSs?

These two kinds of problems will be taken care of in two different ways. We first deal with the Existence Lemma. To start with, let us see how sets of MCSs give rise to models – the alternative versions of the canonical model that we already mentioned.

Definition 4.68 Let W be a set of maximal consistent sets of formulas. Define $\mathfrak{M}^c|_W$ to be the submodel of the canonical model induced by W; that is, $\mathfrak{M}|_W = (W, R, V)$ where R is the relation R^c restricted to W, and V is the canonical relation restricted to W. ⊣

Obviously, we are only interested in such models for which we can prove a Truth Lemma. The following definition gives a sufficient condition for that.

Definition 4.69 A set W of maximal consistent sets is called *diamond saturated* if it satisfies the requirement that for each $\Sigma \in W$ and each formula $F\psi \in \Sigma$ there is a set $\Psi \in W$ such that $\Sigma R^c \Psi$ and $\psi \in \Psi$, and the analogous condition holds for past formulas. ⊣

Lemma 4.70 (Truth Lemma) *Let W be a diamond saturated set of maximal consistent sets of formulas. Then for any $\Gamma \in W$ and any formula ϕ:*

$$\mathfrak{M}^c|_W, \Gamma \Vdash \phi \text{ iff } \phi \in \Gamma.$$

Proof. Straightforward by induction on ϕ.   ⊣

Our goal is now to prove the existence of diamond saturated collections of witnessing MCSs.

Proposition 4.71 *Let ξ be some consistent formula. Then there is a countable, diamond saturated collection W of witnessing MCSs such that $\xi \in \Xi$ for some $\Xi \in W$.*

Proof. The basic idea of the proof is to define W step by step, in a sort of parallel Lindenbaum construction on graphs. During the construction we are dealing with finite approximations of W. At each stage, one of the shortcomings of the current approximation is taken care of; this can be done in such a way that the limit of the construction has no shortcomings at all. A finite approximation of W will consist of a finite graph together with a labeling which assigns a finite set of formulas to each node of the graph. We associate a formula with each of these finite labeled graphs, and require that this corresponding formula be consistent for each of the approximations. The first graph has no edges, and just one point whose label set is the singleton $\{\xi\}$. The construction is such that the graph is growing in two senses: edges may be added to the graph, and formulas may be added to the label sets. (Some readers may find it helpful to think of this process as a rather abstract tableaux construction.) All this is done to ensure that in the limit we are dealing with a (possibly infinite) labeled graph meeting the requirements that (i) the label set of each point is an MCS, (ii) each label set contains a witness and (iii) if a formula of the form $F\phi$ ($P\phi$) belongs to the label set of some node, then there is an edge connecting this node to another one containing ϕ in its label set. Finally, W is defined as the range of this infinite labeling function – note that the label function will not be required to be injective.

Now for the technical details. Approximations to W will be called *networks*: a network is a quadruple $\mathcal{N} = (N, E, d, \Lambda)$ such that (N, E) is a finite, undirected, connected and acyclic graph; d is a direction function mapping each edge (s, t) of the graph to either R or its converse $\breve{R}$; and Λ is a label function mapping each node of N to a finite set of formulas.

As in our earlier example of a step-by-step construction, we first want to formulate coherence conditions on networks and define the notion of a defect of network with respect to its ideal, W. We start with a formulation of the coherence of a network. Since we are working in the basic temporal similarity type – that is, we have diamonds both for looking along R and along $\breve{R}$ – there is an obvious way of describing the network, from each of its nodes. Let $\mathcal{N} = (N, E, d, \Lambda)$ be some network, and let s and t be two adjacent nodes of $\mathcal{N}$. We use the following notational conventions:

$$\langle st \rangle := \begin{cases} F & \text{if } d(s,t) = R, \\ P & \text{if } d(t,s) = \breve{R}. \end{cases}$$

Moreover, let $E(s)$ denote the set of nodes adjacent to s. Finally, we let $\lambda(s)$ denote the conjunction $\bigwedge \Lambda(s)$. Define

$$\begin{aligned} \Delta(\mathcal{N}, s) &:= \lambda(s) \wedge \bigwedge_{v \in E(s)} \langle sv \rangle \theta(\mathcal{N}, v, s), \\ \theta(\mathcal{N}, t, s) &:= \lambda(t) \wedge \bigwedge_{s \neq v \in E(t)} \langle tv \rangle \theta(\mathcal{N}, v, t). \end{aligned}$$

In words, $\Delta(\mathcal{N}, s)$ starts with a local description $\lambda(s)$ of s and then proceeds to its neighbors. For each neighbor v, $\Delta(\mathcal{N}, s)$ writes a future operator if $d(s, v) = R$ (and a past operator if $d(s, v) = \breve{R}$) and then starts to describe the network after v by calling θ. $\theta(\mathcal{N}, v, s)$ first gives a local description $\lambda(v)$ of v, and then recursively proceeds to the neighbors of v – except for s. The omission of s, together with the finiteness and acyclicity of the graph, ensures that we end up with a finite formula.

The following claim shows that it does not really matter from which perspective we describe $\mathcal{N}$.

Lemma 4.72 *For any network $\mathcal{N}$ and any two nodes s, t in $\mathcal{N}$, $\Delta(\mathcal{N}, s)$ is consistent iff $\Delta(\mathcal{N}, t)$ is consistent.*

Proof. By the connectedness of $\mathcal{N}$ it is sufficient to prove the Lemma for adjacent s and t; the general case can be proved by a simple induction on the length of the path connecting the two nodes.

So suppose that s and t are adjacent; without loss of generality assume that $d(s, t) = R$. Since $\mathcal{N}$ is fixed it will not lead to confusion if we abbreviate $\Delta(\mathcal{N}, x)$ by $\Delta(x)$ and $\theta(\mathcal{N}, x, y)$ by $\theta(x, y)$. Then by definition, $\Delta(s)$ is given by

$$\begin{aligned} \Delta(s) &= \lambda(s) \wedge \bigwedge_{u \in E(s)} \langle su \rangle \theta(u, s) \\ &= \lambda(s) \wedge F\theta(t, s) \wedge \bigwedge_{t \neq u \in E(s)} \langle su \rangle \theta(u, s) \\ &= F\theta(t, s) \wedge \theta(s, t). \end{aligned}$$

Likewise, we can show that

$$\Delta(t) = \theta(t, s) \wedge P\theta(s, t).$$

But it is a general property of any logic extending $\mathbf{K}_t$ that for any two formulas α and β, $F\alpha \wedge \beta$ is consistent iff $\alpha \wedge P\beta$ is consistent. From this, the Lemma is immediate. $\dashv$

The upshot of Lemma 4.72 is a good definition of the coherence of a network: we will call a network $\mathcal{N}$ *coherent* if $\Delta(\mathcal{N}, s)$ is consistent for each of (equivalently: some of) its nodes s. However, being finite, our networks will never be perfect. What kinds of defects can they have?

A *defect* of a network is either (D1) a pair (s, ϕ) such that neither ϕ nor $\neg\phi$ belongs to $\Lambda(s)$; (D2) a pair $(s, F\phi)$ such that $F\phi \in \Lambda(s)$ while there is no witness for this (in the sense that $\phi \in \Lambda(t)$ for some node t with Est and $d(s, t) = R$); (D3) a similar pair $(s, P\phi)$; or (D4) a node s without a name; that is, $name(\phi) \in \Lambda(s)$ for no formula ϕ.

We will show that each kind of defect of a network can be repaired. For this we need some terminology. A network $\mathcal{N}'$ *extends* a network $\mathcal{N}$ (notation: $\mathcal{N}' \rhd \mathcal{N}$), if $N \subseteq N'$, while $E = E' \cap N \times N$, $d = d'|_N$ and $\Lambda(s) \subseteq \Lambda'(s)$ for each node s of $\mathcal{N}$.

Lemma 4.73 *For any defect of a finite, coherent network $\mathcal{N}$ there is a finite, coherent $\mathcal{N}' \rhd \mathcal{N}$ lacking this defect.*

Proof. Let $\mathcal{N} = (N, E, d, \Lambda)$ be a coherent network and assume that $\mathcal{N}$ has some defect. We will prove the Lemma by showing how to remove the various types of defect.

D1-defects.
Assume that there is a node s and a formula ϕ such that neither ϕ nor $\neg\phi$ belongs to Λ. Since the formula $\Delta(\mathcal{N}, s)$ is consistent, it follows that either $\Delta(\mathcal{N}, s) \wedge \phi$ or $\Delta(\mathcal{N}, s) \wedge \neg\phi$ is consistent; let $\pm\phi$ denote the formula such that $\Delta(\mathcal{N}, s) \wedge \pm\phi$ is consistent. Now define $\mathcal{N}'$ by $N' := N$, $E' := E$, $d' := d$, while Λ' is given by $\Lambda'(t) = \Lambda(t)$ for $t \neq s$ and

$$\Lambda(s) := \Lambda(s) \cup \{\pm\phi\}.$$

Clearly, $\mathcal{N}'$ is a finite network lacking the defect (s, ϕ). It is also obvious that $\Delta(\mathcal{N}', s)$ is the formula $\Delta(\mathcal{N}, s) \wedge \pm\phi$, so $\Delta(\mathcal{N}', s)$ is consistent, and hence, $\mathcal{N}'$ is coherent.

D2-defects.
Assume that there is a node s and a formula ϕ such that $F\phi \in \Lambda(s)$ while there is no witness for this. Take a *new* node t (that is, t does not belong to N) and define $\mathcal{N}'$ as follows:

$$
\begin{aligned}
N' &:= N \cup \{t\}, \\
E' &:= E \cup \{(s, t)\}, \\
d' &:= d \cup \{((s, t), R)\}, \\
\Lambda' &:= \Lambda \cup \{(t, \{\phi\})\}.
\end{aligned}
$$

It is obvious that $\mathcal{N}'$ extends $\mathcal{N}$ and that the defect has been repaired. Finally, it is clear by the definitions that $\Delta(\mathcal{N}', s) = \Delta(\mathcal{N}, s)$: the only information that the new node adds to the description is a conjunct $F\phi$ and by assumption this was already a member of $\Lambda(s)$, and thus a conjunct of $\lambda(s)$. Hence, the coherence of $\mathcal{N}'$ is an immediate consequence of the coherence of $\mathcal{N}$.

D3-defects.
Repaired analogously to D2-defects.

D4-defects.
These are repaired in the same way as D1-defects, using the fact that if $\Delta(\mathcal{N}, s)$ is consistent, then there is a propositional variable p that does not occur in any of the label sets. And here – at last – we use the IRR-rule to show that the formula $\Delta(\mathcal{N}, s) \wedge name(p)$ is consistent. This completes the proof of Lemma 4.73. ⊣

Finally, we return to the proof of Proposition 4.71. Assume that ξ is a consistent formula.

By a standard step-by-step construction we can define a sequence $(\mathcal{N}_i)_{i \in \mathbb{N}}$ of networks such that

(i) $\mathcal{N}_0$ is a one-node network with label set $\{\xi\}$,
(ii) $\mathcal{N}_j$ extends $\mathcal{N}_i$ whenever $i < j$, and
(iii) for every defect of any network $\mathcal{N}_i$ there is a network $\mathcal{N}_j$ with $j > i$ lacking this defect.

Let N be the set $\bigcup_{i \in \mathbb{N}} N_i$; and for $s \in N$, define $\Lambda(s) = \bigcup_{i \in \mathbb{N}} \Lambda_i(s)$. We claim that for every $s \in N$, $\Lambda(s)$ is a witnessing MCS. We first show that for all formulas ϕ, either ϕ or $\neg\phi$ belongs to $\Lambda(s)$. Let $i \in \mathbb{N}$ be such that s is already in existence in $\mathcal{N}_i$; if neither ϕ nor $\neg\phi$ belongs to $\Lambda_i(s)$, this constitutes a defect of $\mathcal{N}_i$. Hence, by the construction there is some $j > i$ such that either ϕ or $\neg\phi$ belongs to $\Lambda_j(s)$. But then the same formula belongs to $\Lambda(s)$. In the same manner we can prove that every set $\Lambda(s)$ contains a name. Now assume that $\Lambda(s)$ is not consistent; then there are formulas $\phi_1, \ldots, \phi_n$ in $\Lambda(s)$ such that $\phi_1 \wedge \cdots \wedge \phi_n$ is inconsistent. By construction, there must be a $k \in \mathbb{N}$ such that each ϕ_i belongs already to $\Lambda_k(s)$. But this contradicts the consistency of $\Delta(\mathcal{N}_k, s)$ and hence, the coherency of $\mathcal{N}_k$.

Finally, define W as the range of Λ. The preceding paragraphs show that W is a collection of witnessing MCSs. By our definition of $\mathcal{N}_0$, it follows that ξ belongs to some MCS in W.

Now let $F\phi$ be some formula in $\Gamma \in W$. By definition, there is some $s \in N$ such that $\Gamma = \Lambda(s)$, and thus, some $i \in \mathbb{N}$ such that $F\phi \in \Lambda_i(s)$. By our construction there is some $j \geq i$ and some $t \in N_j$ such that $E_j st$ and $\phi \in \Lambda_j(t)$. It follows that $\phi \in \Lambda(t)$, so it remains to prove that $\Lambda(s) R^c \Lambda(t)$. In order to reach a contradiction, suppose otherwise. Then there is a formula $\psi \in \Lambda(t)$ such that $F\psi \notin \Lambda(s)$. Since

$\Lambda(s)$ is an MCS, this implies that $\neg F\psi \in \Lambda(s)$. Now let $k \in \mathbb{N}$ be large enough that $\psi \in \Lambda_k(t)$ and $\neg F\psi \in \Lambda_k(s)$. From this it is immediate that $\Delta(\mathcal{N}_k, s)$ is inconsistent; this contradicts the coherency of $\mathcal{N}_k$. This proves that W is diamond saturated.

But then we have proved that W meets all requirements phrased in Proposition 4.71, and this completes its proof.   ⊣

This shows that we have more or less solved the first problem concerned with working in a trimmed down version of the canonical model: we have established that every consistent formula ξ can be satisfied in an *irreflexive* canonical-like model. Let us now think about the second kind of problem. Concretely, how can we prove that we have not destroyed the nice properties of the canonical frame by moving to a subframe? In particular, how can we ascertain *density*? We will see that here we will make good use of the special naming property of the formulas $name(\phi)$, namely that they can be used as identifiers of MCSs.

Lemma 4.74 *Let W be a diamond saturated collection of witnessing maximal consistent sets of formulas, and let $<$ denote the canonical relation R restricted to W. Then the frame $(W, <)$ is a non-branching, unbounded, dense, strict ordering.*

Proof. Let W and $<$ be as in the statement of the lemma. Clearly, $(W, <)$ is a subframe of the canonical frame; hence, it inherits every *universal* property of $\mathfrak{T}$, such as transitivity or non-branching. Irreflexivity follows from the fact that $\Gamma R \Gamma$ for no witnessing Γ. This shows that $<$ is a non-branching, strict ordering of W.

Unboundedness is not a universal condition, but nevertheless follows rather easily: simply use the fact that the formulas $F\top$ and $P\top$ are theorems of the logic and, hence, belong to every maximal consistent set. Unboundedness then follows by the diamond saturation of W.

The case of density is more difficult, and here is where names are genuinely useful. Assume that Γ and Δ are two MCSs such that $\Gamma < \Delta$. We have to find an MCS Θ *in W* that lies between Γ and Δ. Let δ be the formula such that $name(\delta) \in \Delta$. It follows from $\Gamma < \Delta$ that $F name(\delta) \in \Gamma$, so using the density axiom, we find that $FF name(\delta) \in \Gamma$. From this we may infer the existence of an MCS $\Theta \in W$ with $\Gamma < \Theta$ and $F name(\delta) \in \Theta$.

But is $\Theta < \Delta$? Note that since $<$ is non-branching to the right, we already know that $\Theta < \Delta$ or $\Theta = \Delta$ or $\Delta < \Theta$. But it clearly cannot be the case that $\Theta = \Delta$, since $F\delta \in \Theta$ and $\neg F\delta \in \Delta$. Neither is it possible that $\Delta < \Theta$, for suppose otherwise. It would follows from $F\delta \in \Theta$ that $FF\delta \in \Delta$, so by the transitivity axiom, $F\delta \in \Delta$; but this would contradict the fact that $\neg F\delta \in \Delta$.   ⊣

We now have all the ingredients for the main theorem of this section:

Theorem 4.75 $\mathbf{K}_t\mathbf{Q}^+$ *is complete with respect to* $(\mathbb{Q}, <)$.

Proof. Given any consistent formula ξ, construct a countable, diamond saturated set W of witnessing MCSs for ξ, as in the proof of Proposition 4.71. By the Truth Lemma 4.70, ξ is satisfiable at some MCS Ξ in the model $\mathfrak{M}^c|_W$ induced by W; and by Lemma 4.74, this model is based on a non-branching, unbounded, dense, strict ordering. But then the subframe generated by Ξ is based on a countable, dense, unbounded, strict total order and, hence, is isomorphic to the ordering of the rationals. $\dashv$

How widely applicable are these ideas? Roughly speaking, the situation is as follows. The basic idea is widely applicable; various rules for the undefinable have been employed in many different modal languages, and for many different classes of models (we will see further examples in Chapter 7). Moreover, the use of such rules can be fruitfully combined with other techniques, notably the step-by-step method (this combination sometimes succeeds when all else fails). Rules for the undefinable are fast becoming a standard item in the modal logicians' toolkit.

Nonetheless the method has its limitations, at least in the kinds of modal languages we have been considering so far. These limitations are centered on the problem of working with submodels of the original canonical model.

As we saw, the first problem – retaining sufficiently many MCSs for proving the Truth Lemma – has a fairly satisfactory solution. Two remarks are in order here.

(i) The method only works well when we are working in tense logic. In the proof of the 'multiple Lindenbaum Lemma', we crucially needed operators for looking in *both* directions in order to show that it does not matter from which perspective we describe a graph. If we have no access to the information of nodes lying 'behind,' we are forced to add a countably infinite *family* of more and more complex rules, instead of one single irreflexivity rule.

But there are no problems in generalizing the proof of Proposition 4.71 to similarity types with more than one tense diamond and/or versatile polyadic operators. For example, in Exercise 4.7.3 the reader is asked to use the method to prove completeness for the language of PDL with converse programs.

(ii) Observe that we only proved *weak* completeness for $\mathbf{K}_t\mathbf{Q}^+$. This is because our proof of Proposition 4.71 only works with finite networks. In the presence of names, however, it is possible to prove a stronger version of Proposition 4.71; the basic idea is that when an MCS Γ contains a name, other MCSs may have complete access to the information in Γ through the finite 'channel' of Γ's name. For details we refer to Exercise 4.7.2.

There is a second problem which seems to be more serious. Which properties of the canonical frame can we guarantee to hold in a trimmed down version? In general, very few. Obviously, universal properties of the canonical model hold in each of its submodels, and first-order properties that are the standard translation of closed modal formulas (such as $\forall x \exists y\, Rxy$) are valid in each subframe for which a Truth Lemma holds, but that is about it.

This is the point where the names come in very handy. In fact, in order to prove the inheritance of universal-existential properties like density, the names seem to be really indispensable. *If*, on the other hand, we have names at our disposal, we can prove completeness results for a wide range of logics. Roughly speaking, in case the logic is a tense logic, we can show that every Sahlqvist formula is 'distinguishing-canonical'. The crucial observation is that the witnessing submodel of the canonical model is a *named* model.

Definition 4.76 Let τ be some modal similarity type. A τ-model $\mathfrak{M}$ is called *named* if for every state s in $\mathfrak{M}$ there is a formula ϕ such that s is the only point in $\mathfrak{M}$ satisfying ϕ.   ⊣

Theorem 4.77 *Let τ be some modal similarity type, and suppose that $\mathfrak{M} = (\mathfrak{F}, V)$ is a named τ-model. Then for every very simple Sahlqvist formula σ:*

$$\mathfrak{M} \Vdash \sigma \;\text{iff}\; \mathfrak{F} \Vdash \sigma. \tag{4.1}$$

In the particular case of the basic temporal similarity type, if $\mathfrak{M}$ is in addition a bidirectional model, then (4.1) holds for every Sahlqvist formula.

Proof. Let $\mathfrak{M}$ be a named model. It was the aim of Exercise 1.4.7 to let the reader show that the collection

$$A := \{ V(\phi) \mid \phi \text{ a formula} \}$$

is closed under the boolean and modal operations. Hence, the structure $\mathfrak{g} = (\mathfrak{F}, A)$ is a general frame. Since $\mathfrak{M}$ is named, A contains all singletons. The result then follows from Theorem 5.90 in Chapter 5 – for the second part of the theorem, Exercise 5.6.1 is needed as well.   ⊣

The use of rules for the undefinable really comes into its own in some of the extended modal languages studied for Chapter 7. Two main paths have been explored, and we will discuss both. In the first, the *difference operator* is added to an orthodox modal language. It is then easy to state a rule for the undefinable (even if the underlying modal language does not contain converse operators) and (by extending the remarks just made) to prove a D-Sahlqvist theorem. In the second approach, atomic formulas called *nominals* and operators called *satisfaction operators* are added to an orthodox modal language. These additions make it straightforward to define simple rules for the undefinable (even if the underlying modal language does not contain converse operators) and to prove a general completeness result without making use of step-by-step arguments.

Exercises for Section 4.7

4.7.1 We are working in the basic modal similarity type. First, prove that a frame is intransitive ($\forall xyz \, (Rxy \wedge Ryz \rightarrow \neg Rxz)$) iff we can falsify the formula $\Box p \rightarrow \Diamond \Diamond p$ at every state of the frame.

Second, let $\mathbf{KB'}$ be the logic $\mathbf{K}$, extended with the symmetry axiom $p \rightarrow \Box \Diamond p$ and the rule

(ITR)   if $\vdash (\Box p \wedge \Box \Box \neg p) \rightarrow \phi$ then $\vdash \phi$, provided p does not occur in ϕ.

Show that $\mathbf{KB'}$ is sound and complete with respect to the class of symmetric, intransitive frames.

4.7.2 Assume that we are working with the logic $\mathbf{K}_t \mathbf{Q}^+$. Show that for each consistent set Σ there is a diamond saturated set of MCSs W such that $\Sigma \subseteq \Xi$ for some $\Xi \in W$.

(Hint: use a construction analogous to the one employed in the proof of Proposition 4.71. Add an infinite set of *new* variables to the language and first prove that $\Sigma \cup \{name(p)\}$ is consistent for any new variable p. A network is now allowed to have one special node with an *infinite* label set, which should contain $\Sigma \cup \{name(p)\}$. A description of a network is now an infinite set of formulas.)

4.7.3 Assume that we extend the language of PDL with a *reverse* program constructor:

- if π is a program then so is π^{-1}.

The intended accessibility relation of π^{-1} is the converse relation of R_π. Let $\mathbf{PDL}_\omega$ be the axiom system of PDL (see Section 4.8), modulo the following changes:

  (i) Add the converse axiom schemas $p \rightarrow [\pi]\langle \pi^{-1} \rangle p$ and $p \rightarrow [\pi^{-1}]\langle \pi \rangle p$,
  (ii) Replace the Segerberg induction axiom with the following infinitary rule:

  $(\omega{-}*)$ If $\vdash \phi \rightarrow [\pi^n]\psi$ for all $n \in \omega$, then $\vdash \phi \rightarrow [\pi^*]\psi$.

Prove that this logic is sound and complete with respect to the standard models.

4.8 Finitary Methods I

In this section we introduce finite canonical models. We use such models to prove weak completeness results for non-compact logics. We examine one of the best known examples – propositional dynamic logic – in detail. More precisely, we will axiomatize the validities of regular (test free) propositional dynamic logic. Recall from Chapter 1 that this has a set of diamonds $\langle \pi \rangle$ indexed by a collection of programs Π. Π consists of a collection of basic programs, and the programs generated from them using the constructors $\cup$, ; and $*$. A frame for this language is a transition system $\mathfrak{F} = (W, R_\pi)_{\pi \in \Pi}$, but we are only interested in *regular frames*, that is, frames such that for all programs π, π_1 and π_2:

$$R_{\pi_1 \cup \pi_2} = R_{\pi_1} \cup R_{\pi_2},$$
$$R_{\pi_1 ; \pi_2} = R_{\pi_1} ; R_{\pi_2},$$
$$R_{\pi^*} = (R_\pi)^*.$$

We say that a formula ϕ is a PDL-validity (written $\Vdash \phi$) if it is valid on all regular frames.

The collection of PDL-validities is not compact: consider the set

$$\Sigma = \{\langle a^*\rangle p, \neg p, \neg\langle a\rangle p, \neg\langle a\rangle\langle a\rangle p, \neg\langle a\rangle\langle a\rangle\langle a\rangle p, \ldots\}.$$

Any finite subset of Σ is satisfiable on a regular frame at a single point, but Σ itself is not. This compactness failure indicates that a *strong* completeness result will be out of reach (recall Remark 4.44) so our goal (as with **KL**) should be to prove a weak completeness result. It is not too hard to come up with a candidate axiomatization. For a start, the first two regularity conditions given above can be axiomatized by Sahlqvist axioms. The last condition is more difficult, but even here we have something plausible: recall that in Example 3.10 we saw that this last condition is *defined* by the formula set

$$\Delta = \{(p \wedge [\pi^*](p \to [\pi]p)) \to [\pi^*]p, \langle \pi^*\rangle p \leftrightarrow (p \vee \langle \pi\rangle\langle\pi^*\rangle p) \mid \pi \in \Pi\}.$$

This suggests the following axiomatization.

Definition 4.78 A logic Λ in the language of propositional dynamic logic is a *normal propositional dynamic logic* if it contains every instance of the following axiom schemas:

   (i) $[\pi](p \to q) \to ([\pi]p \to [\pi]q)$,
   (ii) $\langle\pi\rangle p \leftrightarrow \neg[\pi]\neg p$,
   (iii) $\langle\pi_1; \pi_2\rangle p \leftrightarrow \langle\pi_1\rangle\langle\pi_2\rangle p$,
   (iv) $\langle\pi_1 \cup \pi_2\rangle p \leftrightarrow \langle\pi_1\rangle p \vee \langle\pi_2\rangle p$,
   (v) $\langle\pi^*\rangle p \leftrightarrow (p \vee \langle\pi\rangle\langle\pi^*\rangle p)$,
   (vi) $[\pi^*](p \to [\pi]p) \to (p \to [\pi^*]p)$,

and is closed under modus ponens, generalization ($\vdash_\Lambda \phi$ implies $\vdash_\Lambda [\pi]\phi$, for all programs π) and uniform substitution. We call the smallest normal propositional dynamic logic **PDL**. In this section, $\vdash \phi$ means that ϕ is a theorem of **PDL**, consistency means **PDL**-consistency, and so on.   ⊣

As we have already remarked, axioms (iii) and (iv) are (conjunctions of) Sahlqvist axioms; they are canonical for the first two regularity conditions, respectively. Further, observe that axiom (v) is a Sahlqvist formula as well; it is canonical for the condition $R_{\pi^*} = Id \cup R_\pi; R_{\pi^*}$. Thus we have isolated the difficult part: axiom (vi), which we will call the *induction* axiom for obvious reasons, is the formula we need to think about if we are to understand how to cope with the canonicity failure. It is probably a good idea for the reader to attempt Exercise 4.8.1 right away.

    Proving the soundness of **PDL** is straightforward (though the reader should (re-)check that the induction axiom really is valid on all regular frames). We will prove completeness with the help of *finite* canonical models. Our work falls into two parts. First we develop the needed background material: finitary versions of MCSs, Lindenbaum's Lemma, canonical models, and so on. Following this, we turn to the completeness proof proper.

Recall that a set of formulas Σ is closed under subformulas if for all $\phi \in \Sigma$, if ψ is a subformula of ϕ then $\psi \in \Sigma$.

Definition 4.79 (Fischer-Ladner Closure) Let X be a set of formulas. Then X is *Fischer-Ladner closed* if it is closed under subformulas and satisfies the following additional constraints:

(i) If $\langle \pi_1; \pi_2 \rangle \phi \in X$ then $\langle \pi_1 \rangle \langle \pi_2 \rangle \phi \in X$.
(ii) If $\langle \pi_1 \cup \pi_2 \rangle \phi \in X$ then $\langle \pi_1 \rangle \phi \vee \langle \pi_2 \rangle \phi \in X$.
(iii) If $\langle \pi^* \rangle \phi \in X$ then $\langle \pi \rangle \langle \pi^* \rangle \phi \in X$.

If Σ is any set of formulas then $\mathrm{FL}(\Sigma)$ (the *Fischer-Ladner closure* of Σ) is the smallest set of formulas containing Σ that is Fischer-Ladner closed.

Given a formula ϕ, we define $\sim\!\phi$ as the following formula:

$$\sim\!\phi = \begin{cases} \psi & \text{if } \phi \text{ is of the form } \neg\psi, \\ \neg\phi & \text{otherwise.} \end{cases}$$

A set of formulas X is *closed under single negations* if $\sim\!\phi$ belongs to X whenever $\phi \in X$.

We define $\neg\mathrm{FL}(\Sigma)$, the *closure of* Σ, as the smallest set containing Σ which is Fischer-Ladner closed and closed under single negations. ⊣

It is convenient to talk as if $\sim\!\phi$ really is the negation of ϕ, and we often do so in what follows. The motivation of closing a set under *single* negations is simply to have a 'connective' that is just as good as negation, while keeping the set finite. (If we naively closed under ordinary negation, then any set would have an infinite closure.)

It is crucial to note that if Σ is finite, then so is its closure. Some reflection on the closure conditions will convince the reader of this fact, but it is not entirely trivial to give a precise proof. We leave this little combinatorial puzzle to the reader as Exercise 4.8.2.

We are now ready to define the generalization of the notion of a maximal consistent set that we will use in this section.

Definition 4.80 (Atoms) Let Σ be a set of formulas. A set of formulas A is an *atom* over Σ if it is a maximal consistent subset of $\neg\mathrm{FL}(\Sigma)$. That is, A is an atom over Σ if $A \subseteq \neg\mathrm{FL}(\Sigma)$, if A is consistent, and if $A \subset B \subseteq \neg\mathrm{FL}(\Sigma)$ then B is inconsistent. $At(\Sigma)$ is the set of all atoms over Σ. ⊣

Lemma 4.81 *Let Σ be any set of formulas, and A any element of $At(\Sigma)$. Then:*

   (i) *For all $\phi \in \neg FL(\Sigma)$: exactly one of ϕ and $\sim\phi$ is in A.*
  (ii) *For all $\phi \vee \psi \in \neg FL(\Sigma)$: $\phi \vee \psi \in A$ iff $\phi \in A$ or $\psi \in A$.*
 (iii) *For all $\langle \pi_1; \pi_2 \rangle \phi \in \neg FL(\Sigma)$: $\langle \pi_1; \pi_2 \rangle \phi \in A$ iff $\langle \pi_1 \rangle \langle \pi_2 \rangle \phi \in A$.*
 (iv) *For all $\langle \pi_1 \cup \pi_2 \rangle \phi \in \neg FL(\Sigma)$: $\langle \pi_1 \cup \pi_2 \rangle \phi \in A$ iff $\langle \pi_1 \rangle \phi \in A$ or $\langle \pi_2 \rangle \phi \in A$.*
  (v) *For all $\langle \pi^* \rangle \phi \in \neg FL(\Sigma)$: $\langle \pi^* \rangle \phi \in A$ iff $\phi \in A$ or $\langle \pi \rangle \langle \pi^* \rangle \phi \in A$.*

Proof. With the possible exception of the last item, obvious.   ⊣

Atoms are a straightforward generalization of MCSs. Note, for example, that if we choose Σ to be the set of all formulas, then $At(\Sigma)$ is just the set of all MCSs. More generally, the following holds:

Lemma 4.82 *Let $\mathcal{M}$ be the set of all MCSs, and Σ any set of formulas. Then*

$$At(\Sigma) = \{\Gamma \cap \neg FL(\Sigma) \mid \Gamma \in \mathcal{M}\}.$$

Proof. Exercise 4.8.3.   ⊣

Unsurprisingly, an analog of Lindenbaum's Lemma holds:

Lemma 4.83 *If $\phi \in \neg FL(\Sigma)$ and ϕ is consistent, then there is an $A \in At(\Sigma)$ such that $\phi \in A$.*

Proof. If Σ is infinite, the result is exactly Lindenbaum's Lemma, so let us turn to the more interesting finite case. There are two ways to prove this. We could simply apply Lindenbaum's Lemma: as ϕ is consistent, there is an MCS Γ that contains ϕ. Thus, by the previous lemma, $\Gamma \cap \neg FL(\Sigma)$ is an atom containing ϕ.

    But this is heavy handed: let us look for a finitary proof instead. Note that the information in an atom A can be represented by the single formula $\bigwedge_{\phi \in A} \phi$. We will write such conjunctions of atoms as $\widehat{A}$. Obviously $\widehat{A} \notin A$.

    Using this notation, we construct the desired atom as follows. Enumerate the elements of $\neg FL(\Sigma)$ as $\sigma_1, \ldots, \sigma_m$. Let A_0 be $\{\phi\}$. Suppose that A_n has been defined, where $n < m$. We have that

$$\vdash \widehat{A}_n \leftrightarrow (\widehat{A}_n \wedge \sigma_{n+1}) \vee (\widehat{A}_n \wedge \sim\sigma_{n+1}),$$

as this is a propositional tautology, thus either $A_n \cup \{\sigma_{n+1}\}$ or $A_n \cup \{\sim\sigma_{n+1}\}$ is consistent. Let A_{n+1} be the consistent extension, and let A be A_m. Then A is an atom containing ϕ.   ⊣

Note the technique: we forced a finite sequence of choices between σ and $\sim\sigma$. Actually, we did much the same thing in the proof of Lemma 4.26, the Existence Lemma for modal languages of arbitrary similarity type, and we will soon have other occasions to use the idea.

    Now that we have Lemma 4.83, it is time to define finite canonical models:

Definition 4.84 (Canonical Model over Σ) Let Σ be a finite set of formulas. The *canonical model over* Σ is the triple $(At(\Sigma), \{S_\pi^\Sigma\}_{\pi \in \Pi}, V^\Sigma)$ where for all propositional variables p, $V^\Sigma(p) = \{A \in At(\Sigma) \mid p \in A\}$, and for all atoms $A, B \in At(\Sigma)$ and all programs π,

$$AS_\pi^\Sigma B \text{ if } \widehat{A} \wedge \langle \pi \rangle \widehat{B} \text{ is consistent.}$$

V^Σ is called the *canonical valuation*, and the S_π are called the *canonical relations*. We generally drop the Σ superscripts. $\dashv$

Although we have defined it purely finitarily, the canonical model over Σ is actually something very familiar: a filtration. Which filtration? Exercise 4.8.4 asks the reader to find out. Further, note that although some of the above discussion is specific to propositional dynamic logic (for example, the use of the Fischer-Ladner closure) the basic ideas are applicable to any modal language. In Exercise 4.8.7 we ask the reader to apply such techniques to the logic **KL**.

But of course, the big question is: does this finite canonical model *work*? Given a consistent formula ϕ, we need to satisfy ϕ in a regular model. This gives two natural requirements on the canonical model: first, we need to prove some kind of Truth Lemma, and second, we want the model to be regular. The good news is that we can easily prove a Truth Lemma; the bad news is that we are unable to show regularity. This means that we cannot use the canonical model itself; rather, we will work with the canonical relations S_π for the atomic relations only, and define relations R_π for the other programs in a way that *forces* the model to be regular.

Definition 4.85 (Regular PDL-model over Σ) Let Σ be a finite set of formulas. For all basic programs a, define R_a^Σ to be S_a^Σ. For the complex programs, inductively define the PDL-relations R_π^Σ in the usual way using unions, compositions, and reflexive transitive closures. Finally, define $\mathfrak{R}$, the *regular* PDL-*model over* Σ, to be $(At(\Sigma), \{R_\pi^\Sigma\}_{\pi \in \Pi}, V^\Sigma)$, where V^Σ is the canonical valuation. Again, we generally drop the Σ superscripts. $\dashv$

But of course, *now* the main question is, will we be able to prove a Truth Lemma? Fortunately, we can prove the key element of this lemma, namely, an Existence Lemma (cf. Lemma 4.89 below). First the easy part. As the canonical relations S_a are identical to the **PDL**-relations R_a for all basic programs a, we have:

Lemma 4.86 (Existence Lemma for Basic Programs) *Let A be an atom, and let a be a basic program. Then for all formulas $\langle a \rangle \psi$ in $\neg FL(\Sigma)$, $\langle a \rangle \psi \in A$ iff there is a $B \in At(\Sigma)$ such that $AR_a B$ and $\psi \in B$.*

Proof. This can be proved by appealing to the standard Existence Lemma and then taking intersections (as in Lemma 4.83) – but it is more interesting to prove it finitarily. For the right to left direction, suppose there is a $B \in At(\Sigma)$ such that AR_aB and $\psi \in B$. As R_a and S_a are identical for basic programs, we have that AS_aB, thus $\widehat{A} \wedge \langle a\rangle\widehat{B}$ is consistent. As ψ is one of the conjuncts in $\widehat{B}$, $\widehat{A} \wedge \langle a\rangle\psi$ is consistent. As $\langle a\rangle\psi$ is in $\neg FL(\Sigma)$ it must also be in A, for A is an atom and hence *maximal* consistent in $\neg FL(\Sigma)$.

For the left to right direction, suppose $\langle a\rangle\psi \in A$. We construct an appropriate atom B by forcing choices. Enumerate the formulas in $\neg FL(\Sigma)$ as $\sigma_1, \ldots, \sigma_m$. Define B_0 to be $\{\psi\}$. Suppose as an inductive hypothesis that B_n is defined such that $\widehat{A} \wedge \langle a\rangle\widehat{B_n}$ is consistent (where $0 \le n < m$). We have

$$\vdash \langle a\rangle\widehat{B_n} \leftrightarrow \langle a\rangle((\widehat{B_n} \wedge \sigma_{n+1}) \vee (\widehat{B_n} \wedge \sim\sigma_{n+1}))$$

thus

$$\vdash \langle a\rangle\widehat{B_n} \leftrightarrow (\langle a\rangle(\widehat{B_n} \wedge \sigma_{n+1}) \vee \langle a\rangle(\widehat{B_n} \wedge \sim\sigma_{n+1})).$$

Therefore either for $B' = B_n \cup \{\sigma_{n+1}\}$ or for $B' = B_n \cup \{\sim\sigma_{n+1}\}$ we have that $\widehat{A} \wedge \langle a\rangle\widehat{B'}$ is consistent. Choose B_{n+1} to be this consistent expansion, and let B be B_m. B is the atom we seek.   ⊣

Now for the hard part. Axioms (v) and (vi) from Definition 4.78 cannot enforce the desired identity between S_π and R_π. But good news is at hand. These axioms are very strong and manage to 'approximate' the desired behavior fairly well. In particular, they are strong enough to ensure that $S_\pi \subseteq R_\pi$ for arbitrary programs π. This inclusion will enable us to squeeze out a proof of the desired Existence Lemma. The following lemma is the crucial one:

Lemma 4.87 *For all programs π, $S_{\pi^*} \subseteq (S_\pi)^*$.*

Proof. We need to show that for all programs π, if $AS_{\pi^*}B$ then there is a finite sequence of atoms $C_0, \ldots, C_n$ such that $A = C_0S_\pi C_1, \ldots, C_{n-1}S_\pi C_n = B$. Let $\mathcal{D}$ be the set of all atoms reachable from A by such a sequence. We will show that $B \in \mathcal{D}$.

Define δ to be $\bigvee_{D \in \mathcal{D}} \widehat{D}$. Note that $\delta \wedge \langle\pi\rangle\neg\delta$ is *inconsistent*, for suppose otherwise. Then $\delta \wedge \langle\pi\rangle\widehat{E}$ would be consistent for at least one atom E *not* in $\mathcal{D}$, which would mean that $\widehat{D} \wedge \langle\pi\rangle\widehat{E}$ was consistent for at least one $D \in \mathcal{D}$. But then by $DS_\pi E$, E could be reached from A in finitely many S_π steps, which would imply that $E \in \mathcal{D}$ – which it is not.

As $\delta \wedge \langle\pi\rangle\neg\delta$ is inconsistent, $\vdash \delta \to [\pi]\delta$, hence by generalization $\vdash [\pi^*](\delta \to [\pi]\delta)$. By axiom (vi), $\vdash \delta \to [\pi^*]\delta$. Now, as $A(S_\pi)^*A$, $\widehat{A}$ is one of the disjuncts in δ, thus $\vdash \widehat{A} \to \delta$ and hence $\vdash \widehat{A} \to [\pi^*]\delta$. As our initial assumption was that $\widehat{A} \wedge \langle\pi^*\rangle\widehat{B}$ is consistent, it follows that $\widehat{A} \wedge \langle\pi^*\rangle(\widehat{B} \wedge \delta)$ is consistent too. But this means that for one of the disjuncts $\widehat{D}$ of δ, $\widehat{B} \wedge \widehat{D}$ is consistent. As B and D are atoms, $B = D$ and hence $B \in \mathcal{D}$.   ⊣

With the help of this lemma, it is straightforward to prove the desired inclusion:

Lemma 4.88 *For all programs* π, $S_\pi \subseteq R_\pi$.

Proof. Induction on the structure of π. The base case is immediate, for we defined R_a to be S_a for all basic programs a. So suppose $AS_{\pi_1;\pi_2}B$, that is, $\widehat{A} \wedge \langle\pi_1;\pi_2\rangle\widehat{B}$ is consistent. By axiom (iii) of Definition 4.78, $\widehat{A} \wedge \langle\pi_1\rangle\langle\pi_2\rangle\widehat{B}$ is consistent as well. Using a 'forcing choices' argument we can construct an atom C such that $\widehat{A}\wedge\langle\pi_1\rangle\widehat{C}$ and $\widehat{C}\wedge\langle\pi_2\rangle\widehat{B}$ are both consistent. But then, by the inductive hypothesis, $AR_{\pi_1}C$ and $CR_{\pi_2}B$. It follows that $AR_{\pi_1;\pi_2}B$, as required. A similar argument using axiom (iv) from Definition 4.78 shows that $S_{\pi_1\cup\pi_2} \subseteq R_{\pi_1\cup\pi_2}$.

The case for reflexive transitive closures follows from the previous lemma and the observation that $S_\pi \subseteq R_\pi$ implies $(S_\pi)^* \subseteq (R_\pi)^*$. ⊣

We can now prove an Existence Lemma for *arbitrary* programs.

Lemma 4.89 (Existence Lemma) *Let A be an atom and let $\langle\pi\rangle\psi$ be a formula in* $\neg\mathrm{FL}(\Sigma)$. *Then $\langle\pi\rangle\psi \in A$ iff there is a B such that $AR_\pi B$ and $\psi \in B$.*

Proof. The left to right direction puts the crucial inclusion to work. Suppose $\langle\pi\rangle\psi \in A$. We can build an atom B such that $\psi \in B$ and $AS_\pi B$ by 'forcing choices' in the now familiar manner. But we have just proved that $S_\pi \subseteq R_\pi$, thus $AR_\pi B$ as well.

For the right to left direction we proceed by induction on the structure of π. The base case is just the Existence Lemma for basic programs, so suppose π has the form $\pi_1;\pi_2$, and further suppose that $AR_{\pi_1;\pi_2}B$ and $\psi \in B$. Thus there is an atom C such that $AR_{\pi_1}C$ and $CR_{\pi_2}B$ and $\psi \in B$. By the Fischer-Ladner closure conditions, $\langle\pi_2\rangle\psi$ belongs to $\neg\mathrm{FL}(\Sigma)$, hence by the inductive hypothesis, $\langle\pi_2\rangle\psi \in C$. Similarly, as $\langle\pi_1\rangle\langle\pi_2\rangle\psi$ is in $\neg\mathrm{FL}(\Sigma)$, $\langle\pi_1\rangle\langle\pi_2\rangle\psi \in A$. Hence by Lemma 4.81, $\langle\pi_1;\pi_2\rangle\psi \in A$, as required.

We leave the case $\pi = \pi_1 \cup \pi_2$ to the reader and turn to the reflexive transitive closure: suppose π is of the form ρ^*. Assume that $AR_{\rho^*}B$ and $\psi \in B$. This means there is a finite sequence of atoms $C_0, \ldots, C_n$ such that $A = C_0R_\rho C_1, \ldots,$ $C_{n-1}R_\rho C_n = B$. By a subinduction on n we prove that $\langle\rho^*\rangle\psi \in C_i$ for all i; the required result for $A = C_0$ is then immediate.

Base case: $n = 0$. This means $A = B$. From axiom (v) in Definition 4.78 we have that $\vdash \langle\rho^*\rangle\psi \leftrightarrow \psi \vee \langle\rho\rangle\langle\rho^*\rangle\psi$, and hence that $\vdash \psi \rightarrow \langle\rho^*\rangle\psi$. Thus $\langle\rho^*\rangle\psi \in A$.

Inductive step. Suppose the result holds for $n \leq k$, and that

$$A = C_0R_\rho C_1, \ldots, C_kR_\rho C_{k+1} = B.$$

By the inductive hypothesis, $\langle\rho^*\rangle\psi \in C_1$. Hence $\langle\rho\rangle\langle\rho^*\rangle\psi \in A$, for $\langle\rho\rangle\langle\rho^*\rangle\psi \in$ $\neg\mathrm{FL}(\Sigma)$. But $\vdash \langle\rho^*\rangle\psi \leftrightarrow \psi \vee \langle\rho\rangle\langle\rho^*\rangle\psi$. Hence $\langle\rho^*\rangle\psi \in A$.

This completes the subinduction, and establishes the required result for $\langle \beta^* \rangle$. It also completes the main induction and thus the proof of the lemma.   ⊣

Lemma 4.90 (Truth Lemma) *Let $\mathfrak{R}$ be the regular* **PDL**-*model over Σ. For all atoms A and all $\psi \in \neg\text{FL}(\Sigma)$, $\mathfrak{R}, A \Vdash \psi$ iff $\psi \in A$.*

Proof. Induction on the number of connectives. The base case follows from the definition of the canonical valuation over Σ. The boolean case follows from Lemma 4.81 on the properties of atoms. Finally, the Existence Lemma pushes through the step for the modalities in the usual way.   ⊣

The weak completeness result for propositional dynamic logic follows.

Theorem 4.91 PDL *is weakly complete with respect to the class of all regular frames.*

Exercises for Section 4.8

4.8.1 Show that the induction axiom is not canonical.

4.8.2 Prove that for a finite set Σ, its closure set $\neg\text{FL}(\Sigma)$ is finite as well.

4.8.3 Prove Lemma 4.82. That is, show that $At(\Sigma) = \{\Gamma \cap \neg\text{FL}(\Sigma) \mid \Gamma \in \mathcal{M}\}$, where $\mathcal{M}$ is the set of all MCSs, and Σ is any set of formulas.

4.8.4 Show that the finite models defined in the **PDL** completeness proofs are isomorphic to certain filtrations.

4.8.5 Show that for any collection of formulas Σ, $\vdash \bigvee_{A \in At(\Sigma)} \widehat{A}$.

4.8.6 Extend the completeness proof in the text to PDL with tests. Once you have found an appropriate axiom governing tests, the main line of the argument follows that given in the text. However, because tests build modalities from formulas you will need to think carefully about how to state and prove analogs of the key lemmas (such as Lemmas 4.87 and 4.88).

4.8.7 Use finite canonical models to show that **KL** is weakly complete with respect to the class of finite strict partial orders (that is, the class of finite irreflexive transitive frames). (Hint: given a formula ϕ, let Φ be the set of all ϕ's subformulas closed under single negations. Let the points in the finite canonical model be all the maximal **KL**-consistent subsets of Φ. For the relation R, define Rww' iff (1) for all $\Box\phi \in w$, both $\Box\phi$ and ϕ belong to w' and (2) there is some $\Box\phi \in w'$ such that $\Box\phi \notin w$. Use the natural valuation. You will need to make use of the fact that $\vdash_{KL} \Box\phi \to \Box\Box\phi$; bonus points if you can figure out how to prove this yourself!)

4.8.8 Building on the previous result, show that **KL** is weakly complete for the class of finite transitive trees. (Hint: unravel.)

4.9 Finitary Methods II

As we remarked at the end of Section 4.4, although the incompleteness results show that frame-theoretic tools are incapable of analyzing the entire lattice of normal modal logics, they are capable of yielding a lot of information about some of its subregions. The normal logics extending **S4.3** are particularly well-behaved, and in this section we prove three results about them. First, we prove Bull's Theorem: all such logics have the *finite frame property*. Next, we show that they are all *finitely axiomatizable*. Finally, we show that each of these logics has a *negative characterization in terms of finite sets of finite frames*, which will be important when we analyze their computational complexity in Chapter 6.

The logics extending **S4.3** are logics of frames that are rooted, transitive, and connected ($\forall xy \, (Rxy \lor Ryx)$)). To see this, recall that **S4.3** has as axioms 4, T, and .3. These formulas are canonical for transitivity, reflexivity, and no branching to the right, respectively. Hence any point-generated submodel of the canonical model for these logics inherits all three properties, and will in addition be rooted and connected. Now, any connected model is reflexive. Thus *rootedness*, *transitivity*, and *connectedness* are the fundamental properties, and we will call any frame that has them an **S4.3** *frame*. Note that any **S4.3** frame can be viewed as a chain of *clusters* (see Definition 2.43), a perspective which will frequently be useful in what follows.

Bull's Theorem

Our first goal is to prove Bull's Theorem: all extensions of **S4.3** have the finite frame property. In Definition 3.23 we defined the finite frame property as follows: Λ has the finite frame property with respect to a class of finite frames F if and only if F $\Vdash \Lambda$, and for every formula ϕ such that $\phi \notin \Lambda$ there is some $\mathfrak{F} \in$ F such that ϕ is falsifiable on $\mathfrak{F}$. Using the terminology introduced in this chapter, we can reformulate this more concisely as follows: Λ has the finite frame property if and only if there is a class of finite frames F such that $\Lambda = \Lambda_F$. So, to prove Bull's Theorem, we need to show that if Λ extends **S4.3**, then any Λ-consistent formula ϕ is satisfiable in a finite model (W, R, V) such that $(W, R) \Vdash \Lambda$. In short, Bull's Theorem is essentially a general weak completeness result covering all logics extending **S4.3**.

But how are we to build the required models? By transforming the canonical model. Suppose ϕ is Λ-consistent. Let w be any Λ-MCS containing ϕ, and let $\mathfrak{M}^w = (W^w, R^w, V^w)$ be the submodel of $\mathfrak{M}^\Lambda$ generated by w. Then $\mathfrak{M}^w, w \Vdash \phi$, and (as just discussed) $\mathfrak{M}^w$ is based on an **S4.3** frame. We are going to transform $\mathfrak{M}^w$ into a finite model $\mathfrak{M}^s$ that satisfies ϕ and is based on an **S4.3** frame that validates Λ.

Figure 4.2 shows what is involved. We are going to transform $\mathfrak{M}^w$ in two distinct

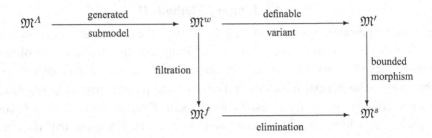

Fig. 4.2. The models we will construct, and their relationships

ways. One involves taking a filtration and eliminating certain points; this is the technical heart of the proof. The other involves defining a bounded morphism on a definable variant $\mathfrak{M}'$ of $\mathfrak{M}^w$; this part uses the results on definable variants and distinguishing models proved in Section 3.4. These transformations offer us two perspectives on the properties of $\mathfrak{M}^s$, and together yield enough information to prove the result.

And so to work. We first discuss the filtration/elimination transformation. Let Φ be the (finite) set consisting of all subformulas of $\Diamond\phi$, and let $\mathfrak{M}^f = (W^f, R^f, V^f)$ be the result of transitively filtrating $\mathfrak{M}^w$ through Φ. Recall that the relation R^f used in transitive filtrations is defined by $R^f|u||v|$ iff $\Diamond\psi \in v$ implies $\Diamond\psi \in u$, for all $\Diamond\psi \in \Phi$, and all $u, v \in W^w$; see Lemma 2.42. As Φ is finite, so is W^f. By the Filtration Theorem (Theorem 2.39), $\mathfrak{M}^f, |u| \Vdash \psi$ iff $\mathfrak{M}^w, u \Vdash \psi$, for all $\psi \in \Phi$, and all $u \in W^w$. Moreover, R^f is transitive, reflexive, and connected, and $|w|$ is a root of the filtration, thus $\mathfrak{M}^f$ is based on an **S4.3** frame. Hence the frame underlying $\mathfrak{M}^f$ is a finite chain of finite clusters.

Now for the key elimination step. We want to build a finite model based on a frame for Λ. Now, we do not know whether $\mathfrak{M}^w$ is based on such a frame, but we *do* know that $\mathfrak{M}^w \Vdash \Lambda$. If we could transfer the truth of Λ in $\mathfrak{M}^w$ to a finite *distinguishing* model, then by item (iii) of Lemma 3.27 we would immediately have Bull's Theorem. Unfortunately, while $\mathfrak{M}^f$ is finite, and also (being a filtration) distinguishing, we have no guarantee that $\mathfrak{M}^f \Vdash \Lambda$. This reflects something discussed in Section 2.3: the natural map associated with a filtration need not be a bounded morphism. It also brings us to the central idea of the proof: *eliminate all points in $\mathfrak{M}^f$ which prevent the natural map from being a bounded morphism.* Obviously, any model built from $\mathfrak{M}^f$ by eliminating points will be finite and distinguishing. So the crucial questions facing us are: which points should be eliminated, and how do we know that they can be thrown away without affecting the satisfiability of formulas in Φ?

Recall that the natural map associated with a filtration sends each point u in the original model to the equivalence class $|u|$ in the filtration. So if the natural map from the frame underlying $\mathfrak{M}^w$ to the frame underlying $\mathfrak{M}^f$ is *not* a bounded morphism, this means that for some $\beta, \alpha \in W^f$ we have that $R^f \beta \alpha$ but

$$\neg \forall v \in \beta \, \exists z \, (R^w vz \wedge z \in \alpha),$$

or equivalently, that $R^f \beta \alpha$ but

$$\exists v \in \beta \, \forall z \, (z \in \alpha \rightarrow \neg R^w vz).$$

This motivates the following definition:

Definition 4.92 Suppose $\beta, \alpha \in W^f$. We say that α is *subordinate* to β (α *sub* β) if there is a $v \in \beta$ such that for all $z \in \alpha$, it is *not* the case that $R^w vz$.   ⊣

So: if $\mathfrak{M}^f$ is *not* a bounded morphic image of $\mathfrak{M}^w$ under the natural map, then there is some $\alpha \in W^f$ such that for some $\beta \in W^f$, $R^f \beta \alpha$ and α *sub* β. We must get rid of all such αs; we will call them *eliminable* points. But to show that we can safely eliminate them, we need to understand the *sub* relation a little better.

Lemma 4.93   (i) *If α sub β, then there is a $v \in \beta$ such that for all $z \in \alpha$, $R^w zv$.*

(ii) *If α sub β then $R^f \alpha \beta$.*

(iii) *The sub relation is transitive and asymmetric.*

(iv) *Suppose $\alpha, \beta, \gamma \in W^f$ such that α sub γ and not α sub β. Then β sub γ.*

Proof. For item (i), note that by definition there is a $v \in \beta$ such that for all $z \in \alpha$, it is *not* the case that $R^w vz$. But R^w is a connected relation, hence for every $z \in \alpha$, $R^w zv$.

For item (ii), suppose α *sub* β. By item (i), this means that there is some element v of β, such that every element of α R^w-precedes v. Now if $\Diamond \psi \in \beta$, then $\mathfrak{M}^w, v \Vdash \Diamond \psi$. Hence (by the transitivity of R^w) for all $z \in \alpha$, $\mathfrak{M}^w, z \Vdash \Diamond \psi$ too. This means that $\Diamond \psi \in \alpha$, that is, $R^f \alpha \beta$. (It follows that if the natural map fails to be bounded morphism because of its behavior on the points β and α, then the eliminable point α belongs to the *same* cluster as β.)

Items (iii) and (iv) are left for the reader as Exercise 4.9.1.   ⊣

We are now ready for the key result: we can safely get rid of all the eliminable points; there are enough *non*-eliminable points left to prove an Existence Lemma:

Lemma 4.94 (Existence Lemma) *Let $u \in W^w$ and suppose $\Diamond \psi \in u \cap \Phi$. Then there is a $|v| \in W^f$ such that $R^f |u||v|$, $\psi \in |v|$, and $|v|$ is not eliminable.*

Proof. Construct a maximal sequence $\alpha_0, \alpha_1, \ldots$ through W^f with the following properties:

(i) $\alpha_0 = |u|$.

(ii) If $i > 0$ and odd, then α_i is some $|v|$ such that $\psi \in v$, $R^f \alpha_{i-1} |v|$, and not $|v|$ *sub* α_{i-1}.

(iii) If $i > 0$ and even, then α_i is some $|v|$ such that $R^f |v| \alpha_{i-1}$ and α_{i-1} *sub* $|v|$.

Here is the basic idea. Think of this sequence as a series of moves through the model. We are given $\Diamond \psi$, and our goal is to find an R^f-related ψ-containing point that is not eliminable. So, on our first move (an *odd* move) we select an R^f-related ψ-containing point (we are guaranteed to find one, pretty much as in any Existence Lemma). If the point is *not*-eliminable we have found what we need and are finished. Unfortunately, the point may well be eliminable. If so, we make a second move (an *even* move) to another point *in the same cluster* – namely a point to which the first point we found is subordinate. We iterate the process, and eventually we will find what we are looking for. We now make this (extremely sketchy) outline precise.

Claim 1. *For every item $\alpha_i = |v|$ in the sequence, $\Diamond \psi \in v$.*

If $i = 0$, $\alpha_i = |u|$ and by assumption $\Diamond \psi \in u$. If $i > 0$ and odd, then $\psi \in |v|$ by construction, hence $\psi \in v$. As v is a Λ-MCS it contains $\psi \rightarrow \Diamond \psi$, thus $\Diamond \psi \in v$ also. Finally, if $i > 0$ and even, then as we have just seen, $\Diamond \psi \in \alpha_{i-1}$. By construction, $R^f |v| \alpha_{i-1}$ hence $\Diamond \psi \in |v|$ and hence $\Diamond \psi \in v$. This proves Claim 1.

Claim 2. *The sequence terminates.*

Suppose i is even. By property (iii), α_{i+1} *sub* α_{i+2} and by property (ii), it is not the case that α_{i+1} *sub* α_i. Hence by item 3 of Lemma 4.93, α_i *sub* α_{i+2}. By item (ii) of Lemma 4.93, *sub* is a transitive and asymmetric relation, thus all α_i with even i are different. As there are only finitely many elements in W^f, the sequence must terminate. This proves Claim 2.

Claim 3. *The sequence does not terminate on even i.*

Suppose i is even. We need to show that there is an $\alpha_{i+1} \in W^f$ such that $R^f \alpha_i \alpha_{i+1}$ and not α_{i+1} *sub* α_i. Let $\{\beta_1, \ldots, \beta_m\}$ be $\{\beta \in W^f \mid \beta \ \text{*sub*} \ \alpha_i\}$. Then for each k $(1 \leq k \leq m)$ there is a $v_k \in \alpha_i$ such that not $R^w v_k z$, for all $z \in \beta_k$. Let v be one of these points v_k such that for all k, $R^w v_k v$, for $1 \leq k \leq m$. (It is always possible to choose such a v as R^w is connected.) As $\alpha_i = |v|$, by Claim 1 $\Diamond \psi \in v$. By the Existence Lemma for normal logics (Lemma 4.20), there is an $x \in W$ such that $\psi \in x$ and $R^w v x$. Moreover, not $|x|$ *sub* $|v|$. For suppose for the sake of a contradiction that $|x|$ *sub* $|v|$. Then $|x| = \beta_k$, for some $1 \leq k \leq m$, and hence not $R^w v_k x$. But $R^w v_k v$ and $R^w v x$, hence (by transitivity) $R^w v_k x$ – contradiction. We conclude that not $|x|$ *sub* $|v|$, hence (recalling that $|v| = \alpha_i$) we can always choose α_{i+1} to be $|x|$. This proves Claim 3.

We can now prove the result. By Claims 2 and 3, the sequence terminates on $\alpha_m = |v|$, for some odd number m. By construction, $\psi \in v$, hence $\psi \in |v|$.

Since α_{m+1} does not exist, α_m is not eliminable. By construction, for all even i, $R^f \alpha_i \alpha_{i+1}$. By item (ii) of Lemma 4.93, for all odd i, $R^f \alpha_i \alpha_{i+1}$. Hence by the transitivity of R^f, $R^f |u||v|$, and we are through. ⊣

We now define the model $\mathfrak{M}^s$. Let W^s be the set of non-eliminable points in W^f. (Note that by the previous lemma there must be at least one such point, for $\Diamond\phi \in w \cap \Phi$.) Then $\mathfrak{M}^s = (W^s, R^s, V^s)$ is $\mathfrak{M}^f$ restricted to W^s. Hence $\mathfrak{M}^s$ is a finite distinguishing model, and (W^s, R^s) is an **S4.3** frame.

Lemma 4.95 *$\mathfrak{M}^s$ satisfies ϕ.*

Proof. First, we show by induction on the structure of ψ that for all $\psi \in \Phi$, and all $|u| \in W^s$, $\mathfrak{M}^s, |u| \Vdash \psi$ iff $\psi \in u$. The only interesting case concerns the modalities. So suppose $\Diamond\psi \in u$. By the previous lemma, there is some $|v|$ such that $R^f|u||v|$, $\psi \in |v|$, and $|v|$ is not eliminable. As $\psi \in |v|$, $\psi \in v$, hence by the inductive hypothesis, $\mathfrak{M}^s, |v| \Vdash \psi$, hence $\mathfrak{M}^s, |u| \Vdash \Diamond\psi$ as desired. The converse is straightforward; we leave it to the reader.

It follows that ϕ is satisfied somewhere in $\mathfrak{M}^s$. For, as $\Diamond\phi \in w \cap \Phi$, by Lemma 4.94 there is a non-eliminable $|u|$ such that $R^f|w||u|$ and $\phi \in |u|$. Hence $\phi \in u$, and $\mathfrak{M}^s, |u| \Vdash \phi$. ⊣

We are almost there. If we can show that $\mathfrak{M}^s \Vdash \Lambda$, then as $\mathfrak{M}^s$ is a finite distinguishing model, its frame validates Λ and we are through. Showing that $\mathfrak{M}^s \Vdash \Lambda$, will take us along the other path from $\mathfrak{M}^w$ to $\mathfrak{M}^s$ shown in Figure 4.2. That is, we will show that $\mathfrak{M}^s$ is a bounded morphic image of a definable variant $\mathfrak{M}'$ of $\mathfrak{M}^w$.

The required bounded morphism f is easy to describe: it agrees with the natural map on all *non*-eliminable points, and where the natural map sent a point w to a point that has been eliminated, $f(w)$ will be a point 'as close as possible' to the eliminated point. Let us make this precise. Enumerate the elements of W^s. Define $f : W^w \to W^s$ by

$$
f(w) = \begin{cases} |w|, \text{ if } |w| \in W^s, \\ \text{the first element in the enumeration which is an } R^s\text{-minimal} \\ \text{element of } \{\alpha \in W^s \mid R^s|w|\alpha\}, \text{ otherwise.} \end{cases}
$$

As W^s is finite, the minimality requirement (which captures the 'as close as possible' idea) is well defined.

As we will show, f is a bounded morphism from (W^w, R^w) into (W^s, R^s). But we have no guarantee that f is a bounded morphism from the *model* $\mathfrak{M}^w$ to $\mathfrak{M}^s$, for while the underlying frame morphism is fine, we need to ensure that the valuations agree on propositional symbols. We fix this as follows. For any propositional symbol p, define $V'(p)$ to be $\{u \in W^w \mid f(u) \in V^s(p)\}$, and let $\mathfrak{M}'$ be (W^w, R^w, V'). That is, $\mathfrak{M}'$ is simply a variant of $\mathfrak{M}^w$ that agrees with $\mathfrak{M}^s$ under the mapping f. But it is not just any variant: as we will now see, it is a *definable* variant. It is time to pull all the threads together and prove the main result.

Theorem 4.96 (Bull's Theorem) *Every normal modal logic extending* **S4.3** *has the finite frame property.*

Proof. First we will show that $\mathfrak{M}'$ is a definable variant of $\mathfrak{M}^w$. If β is any of the equivalence classes that make up the filtration $\mathfrak{M}^f$, then $\beta \subseteq W^w$. Moreover, $\mathfrak{M}^w$ can define any such β: the defining formula $\hat{\beta}$ is simply a conjunction of all the formulas in some subset of Φ, the set we filtrated through. (Incidentally, we take the conjunction of the empty set to be $\bot$.) It follows that $\mathfrak{M}^w$ can define $V'(p)$ for any propositional symbol p. To see this, note that $V^s(p)$ is either the empty set or some finite collection of equivalence classes $\{\beta_1, \ldots, \beta_n\}$. In the former case, define δ_p to be $\bot$. In the latter case, define δ_p to be $\bigvee_{i \in n} \hat{\beta}_i$. Either way, δ_p defines $V'(p)$ in $\mathfrak{M}^w$, for $V'(p)$ is $\{u \in W^w \mid f(u) \in V^s(p)\}$. Thus $\mathfrak{M}'$ is a definable variant of $\mathfrak{M}^w$. (Note that this argument makes use of facts about all four models constructed in the course of the proof.)

Next we claim that f is indeed a surjective bounded morphism from $\mathfrak{M}^s$ onto $\mathfrak{M}'$; we show here that it satisfies the back condition and leave the rest to the reader. Suppose $R^s f(u) f(v)$. As $f(v) \in W^s$, it is not eliminable, hence not $f(v) \; sub \; f(u)$. But this means that every element in $f(u) \; R^w$-precedes an element in $f(v)$, as required.

But now Bull's Theorem follows. If Λ is a normal modal logic extending **S4.3** and ϕ is a Λ-consistent formula, build $\mathfrak{M}^s$ as described above. By Lemma 4.95, $\mathfrak{M}^s$ satisfies ϕ. Moreover $\mathfrak{M}^s \Vdash \Lambda$. To see this, simply follow the upper left to right path through Figure 4.2. $\mathfrak{M}^\Lambda \Vdash \Lambda$, hence so does $\mathfrak{M}^w$, for it is a generated submodel of $\mathfrak{M}^\Lambda$. As $\mathfrak{M}'$ is a definable variant of $\mathfrak{M}^w$, by Lemma 3.25 item (iii), $\mathfrak{M}' \Vdash \Lambda$. Hence, as $\mathfrak{M}^s$ is a bounded morphic image of $\mathfrak{M}'$, it too validates Λ as required. But $\mathfrak{M}^s$ is a finite distinguishing model, hence, by Lemma 3.27 item (iii), its frame validates Λ and we are through.   ⊣

Finite axiomatizability

We now show that every normal logic extending **S4.3** is *finitely axiomatizable*. (A logic Λ is finitely axiomatizable if there is a *finite* set of formulas Γ such that Λ is the logic generated by Γ.) The proof makes use of a special representation for finite **S4.3** frames.

Because every finite **S4.3** frame is a finite chain of finite clusters, any such frame can be represented as a list of positive integers: each positive integer in the list records the cardinality of the corresponding cluster. For example, the list $[3, 1, 2]$ represents the following frame:

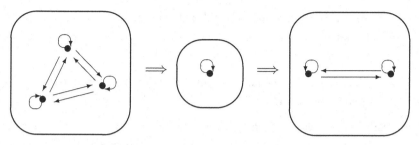

Such representations will allow us to reduce the combinatorial heart of the following proofs to a standard result about lists. The following definition pins down the relationship between lists that will be important.

Definition 4.97 A *list* is a finite non-empty list of positive integers. A list **t** *contains* a list **s** if **t** has a sublist of the same length as **s**, each item of which is greater than or equal to the corresponding item of **s**. A list **t** *covers* a list **s** if **t** contains **s** and the last item of **t** is greater than or equal to the last item of **s**. ⊣

For example, the list $[9, 40, 1, 9, 3]$ contains the list $[8, 2, 9]$, for it has $[9, 40, 9]$ as a sublist, but it does not cover this list. But $[9, 40, 1, 9, 10]$ covers $[8, 2, 9]$.

The modal relevance of list covering stems from the following lemma:

Lemma 4.98 *Let $\mathfrak{F}$ and $\mathfrak{G}$ be finite* **S4.3** *frames, and let* **f** *and* **g** *be their associated lists. Then* **f** *covers* **g** *iff there is a bounded morphism from $\mathfrak{F}$ onto $\mathfrak{G}$.*

Proof. Exercise 4.9.2. ⊣

In view of this result, the following well-known result can be viewed as asserting the existence of infinite sequences of bounded morphisms:

Theorem 4.99 (Kruskal's Theorem) *Every countably infinite sequence of lists* **t** *contains an infinite subsequence* **s** *such that for all lists s_i and s_j in* **s**, $i > j$ *implies s_i covers s_j.*

Proof. Let us call a (finite or infinite) subsequence $(t_i)_{i \in I}$ of a sequence of lists $\mathbf{t} = (t_i)_{i \in \omega}$ a *chain in* **t** if for all $i, j \in I$, t_j covers t_i whenever $j > i$. We assume familiarity with the notions of the head, the tail and the sum of a list. For instance, the head of $[8, 2, 9]$ is 8, its tail is $[2, 9]$ and its sum is 19. Call **s** *smaller* than **t** if the sum of **s** is smaller than that of **t**.

In order to prove the theorem, we will show the following holds:

> every countably infinite sequence of lists **t** contains a chain of length 2. (4.2)

Assume that (4.2) does not hold; that is, there are countably infinite sequences without chains of length 2 as subsequences.

Without loss of generality we may assume that **t** does not contain infinitely many lists of length 1. For otherwise, consider its subsequence $(t_i)_{i \in I}$ of these one-item lists. This subsequence may be identified with a sequence of natural numbers $(n_i)_{i \in \omega}$. But

> any sequence $(n_i)_{i \in \omega}$ of natural numbers contains a subsequence $(n_i)_{i \in I}$ such that for all $i, j \in I$, $i < j$ implies $n_i \leq n_j$, $\qquad$ (4.3)

as can easily be proved. But if $n_i \leq n_j$ then clearly t_j covers t_i. But then we may also assume that t does not contain one-item lists at all: simply consider the sequence found by eliminating all one-item lists.

Let t be a *minimal* such sequence. That is, t is a sequence of more-item lists, t has no 2-chains, and for all n, there are no more-item lists t'_n, t'_{n+1}, ... such that t'_n is smaller than t_n, while the sequence $t_0, t_1, \ldots, t_{n-1}, t'_n, t'_{n+1}, \ldots$ has no 2-chains.

Now we arrive at the heart of the argument. Define $(n_i)_{i \in \omega}$ and $(u_i)_{i \in \omega}$ as the sequences of the heads and the tails of t; that is, for each i, n_i is the head of t_i and u_i is the tail of t_i. By (4.3), there is a subsequence $(n_i)_{i \in I}$ such that $i < j$ implies $n_i \leq n_j$, whenever $i, j \in I$. Now consider the corresponding subsequence $(u_i)_{i \in I}$ of u. We need the following result:

> any subsequence $(v_i)_{i \in \omega}$ of tails of t contains a 2-chain. $\qquad$ (4.4)

By the same argument as before, we may assume that v contains only more-item lists. Let k be the natural number such that v_0 is the tail of t_k, and consider the sequence $t_0, t_1, \ldots, t_{k-1}, v_0, v_1, \ldots$. Since v_0 is the tail of t_k and, hence, *smaller* than t_k, it follows by the minimality of t_k that the mentioned sequence contains a 2-chain. But obviously this 2-chain can only occur in the v-part of the sequence. This proves (4.4).

But if u contains a 2-chain, this means that there are two numbers i and j in I with $i < j$ and u_j covers u_i. Also, by definition of I, $n_i \leq n_j$. But then $t_i = [m_i] * u_i$ is covered by $t_j = [m_j] * u_j$. This proves (4.2).

Finally, it remains to prove the theorem from (4.2). Let t be an arbitrary countably infinite sequence of lists. By successive applications of (4.2), it follows that t contains infinitely many chains. We claim that one of these chains is infinite. For if we suppose that there are only finite chains, we may consider the sequence z of last items of right-maximal finite chains in t (a chain is right-maximal if it can not be extended to the right). There must be infinitely many such right-maximal chains, so z is an infinite sequence. Hence, by yet another application of (4.2), z contains a chain of length 2. But then some chain was not right-maximal after all. $\qquad \dashv$

We now extract the consequences for logics extending **S4.3**:

Corollary 4.100 *There is no infinite sequence Λ_0, Λ_1, ... of normal logics containing* **S4.3** *such that for all i, $\Lambda_i \subset \Lambda_{i+1}$.*

Proof. Suppose otherwise. Then for some infinite sequence of logics Λ_0, Λ_1, ... extending **S4.3**, and for all natural numbers i, there is a formula ϕ_i such that $\phi_i \notin \Lambda_i$ and $\phi_i \in \Lambda_{i+1}$. So, by Bull's Theorem, for all natural numbers i there is a finite **S4.3** frame $\mathfrak{F}_i$ that validates Λ_i and does not satisfy ϕ_i. Let $\mathbf{t}$ be the infinite sequence of lists t_i associated with the frames $\mathfrak{F}_i$. By Kruskal's Theorem, there exist natural numbers k and l, such that $k > l$ and t_k covers t_l. Hence by Lemma 4.98 there is a bounded morphism from $\mathfrak{F}_k$ onto $\mathfrak{F}_l$. It follows that $\mathfrak{F}_l \Vdash \phi_l$ and we have a contradiction. ⊣

Theorem 4.101 *Every normal modal logic extending* **S4.3** *is finitely axiomatizable.*

Proof. To arrive at a contradiction, we will assume that there does exist an extension Λ of **S4.3** that is not finitely axiomatizable. We will construct an infinite sequence $\Lambda_0 \subset \Lambda_1 \subset \cdots$ of extensions of **S4.3**, thus contradicting Corollary 4.100.

As Λ is not finitely axiomatizable, it must be a proper extension of **S4.3**. Let ϕ_0 be an arbitrary formula in $\Lambda \setminus$ **S4.3**, and define Λ_0 to be the logic generated by **S4.3** $\cup \{\phi_0\}$. Then **S4.3** $\subset \Lambda_0 \subset \Lambda$. The latter inclusion is strict because Λ is not finitely axiomatizable. Hence, there exists $\phi_1 \in \Lambda \setminus \Lambda_0$. Let Λ_1 be the logic generated by $\Lambda_0 \cup \{\phi_1\}$. Continuing in this fashion we find the required infinite sequence $\Lambda_0 \subset \Lambda_1 \subset \cdots$ of extensions of **S4.3**. ⊣

A negative characterization

We turn to the final task: showing that every normal logic extending **S4.3** has a negative characterization in terms of finite sets of finite frames. Once again, the proof makes use of the representation of **S4.3** frames as lists of positive integers.

First some terminology. A set of lists X is *flat* if for every two distinct lists in X, neither covers the other. In view of Lemma 4.98, the modal relevance of flatness is this: if two frames are associated with distinct lists belonging to a flat set, then neither frame is a bounded morphic image of the other.

Lemma 4.102 *All flat sets are finite. Furthermore, for any set of lists Y there is a maximal set X such that $X \subseteq Y$ and X is flat.*

Proof. Easy consequences of Kruskal's Theorem. ⊣

If X is a flat set of lists, then $C(X)$ is the set of lists covered by some list in X. Note that $C(X)$ is finite and that $X \subseteq C(X)$. If X is a set of lists, then $B(X)$ is the class of all finite **S4.3** frames $\mathfrak{F}$ such that there is a bounded morphism from $\mathfrak{F}$ onto some frame whose list is in X.

Theorem 4.103 *For every normal modal logic Λ extending* **S4.3** *there is a finite set*

N *of finite* **S4.3** *frames with the following property: for any finite frame* $\mathfrak{F}$, $\mathfrak{F} \Vdash \Lambda$ *iff* $\mathfrak{F}$ *is an* **S4.3** *frame and there does not exist a bounded morphism from* $\mathfrak{F}$ *onto any frame in* N.

Proof. Let $\Lambda \supseteq$ **S4.3**, and let L' be the set of lists associated with finite **S4.3** frames which do not validate Λ. Let L be a maximal flat set such that $L \subseteq L'$. Note that $C(L) \subseteq L'$.

We claim that for any finite **S4.3** frame $\mathfrak{F}$, $\mathfrak{F} \Vdash \Lambda$ iff $\mathfrak{F} \notin B(C(L))$. The left to right implication is clear, for as no frame whose list belongs to $C(L)$ validates Λ, there cannot be a bounded morphism from $\mathfrak{F}$ onto any such frame. For the other direction, we show the contrapositive. Suppose that $\mathfrak{F} \not\Vdash \Lambda$. Let $\mathfrak{F}$'s list be **f**. Then $\mathbf{f} \in L'$. Now either $\mathbf{f} \in C(L)$ or $\mathbf{f} \in L' \setminus C(L)$. If $\mathbf{f} \in C(L)$, then the identity morphism on $\mathfrak{F}$ guarantees that $\mathfrak{F} \in B(C(L))$ as required. So suppose instead that $\mathbf{f} \in L' \setminus C(L)$. This means that $\mathbf{f} \notin L$ (as $L \subseteq C(L)$), hence as L is a *maximal* flat subset of L', **f** must cover some list **g** in L. Thus by Lemma 4.98, any **S4.3** frame $\mathfrak{G}$ whose list is **g** is a bounded morphic image of $\mathfrak{F}$, hence $\mathfrak{F} \in B(C(L))$ as required. This completes the proof of the claim.

We can now define the desired finite set N: for each $\mathbf{g} \in C(L)$, choose a frame whose list is **g**, and let N be the set of all our choices. $\dashv$

Exercises for Section 4.9

4.9.1 Show that the *sub* relation is transitive and asymmetric. Furthermore, show that if α *sub* γ and not α *sub* β, then β *sub* γ.

4.9.2 Prove Lemma 4.98. That is, let $\mathfrak{F}$ and $\mathfrak{G}$ be finite **S4.3** frames, and let **f** and **g** be their associated lists. Then show that **f** covers **g** iff there is a bounded morphism from $\mathfrak{F}$ onto $\mathfrak{G}$. (First hint: look at how we defined the bounded morphism used in the proof of Bull's Theorem. Second hint: look at the statement (but not the proof!) of Lemma 6.39.)

4.9.3 Give a complete characterization of all the normal logics extending **S5**. Your answer should include axiomatizations for all such logics.

4.9.4 Let $\mathbf{K}_t\mathbf{4.3}$ be the smallest tense logic containing 4, T, $.3_l$ and $.3_r$. Show that there are tense logics extending $\mathbf{K}_t\mathbf{4.3}$ that do not have the finite frame property. (Hint: look at the tense logic obtained by adding the Grzegorczyk axiom in the operator F. Is the Grzegorczyk axiom in P satisfiable in a model for this logic? Is the Grzegorczyk axiom in P satisfiable in a *finite* model for this logic?)

4.10 Summary of Chapter 4

▶ *Completeness*: A logic Λ is *weakly* complete with respect to a class of structures S if every formula valid on S is a Λ-theorem. It is *strongly* complete with respect to S if whenever a set of premises entails a conclusion over S, then the conclusion is Λ-deducible from the premises.

▶ *Canonical Models and Frames*: Completeness theorems are essentially model existence theorems. The most important model building technique is the canonical model construction. The points of the underlying canonical frames are maximal consistent sets of formulas, and the relations and valuation are defined in terms of membership of formulas in such sets.

▶ *Canonicity*: Many formulas are canonical for a property P. That is, they are valid on any frame with property P, and moreover, when used as axioms, they guarantee that the canonical frame has property P. When working with such formulas, it is possible to prove strong completeness results relatively straightforwardly.

▶ *Sahlqvist's Completeness Theorem*: Sahlqvist formulas not only define first-order properties of frames, each Sahlqvist formula is also canonical for the first-order property it defines. As a consequence, strong completeness is automatic for any logic that is axiomatized by axioms in Sahlqvist form.

▶ *Limitative Results*: The canonical model method is not universal: there are weakly complete logics whose axioms are not valid on any canonical frame. Indeed, no method is universal, for there are logics that are not sound and weakly complete with respect to any class of frames at all.

▶ *Unraveling and Bulldozing*: Often we need to build models with properties for which no modal formula is canonical. Sometimes this can be done by transforming the logic's canonical model so that it has the relevant properties. Unraveling and bulldozing are two useful transformation methods.

▶ *Step by Step*: Instead of modifying canonical models directly, the step-by-step method builds models by selecting MCSs. Because it builds these selections inductively, it offers a great deal of control over the properties of the resulting model.

▶ *Rules for the Undefinable*: By enriching our deductive machinery with special proof rules, it is sometimes possible to construct canonical models that have the desired properties right from the start, thus avoiding the need to massage the (standard) canonical model into some desired shape.

▶ *Finitary Methods*: The canonical model method establishes *strong* completeness. Only *weak* completeness results are possible for non-compact logics such as propositional dynamic logic, and finite canonical models (essentially filtrations of standard canonical models) are a natural tool for proving such results.

▶ *Logics extending* **S4.3**: Although the incompleteness results show that a frame based analysis of all normal logics is impossible, many subregions of the lattice of normal modal logics are better behaved. For example, the logics extending **S4.3** all have the finite frame property, are finitely axiomatizable, and have negative characterizations in terms of finite frames.

Notes

Modal completeness results can be proved using a variety of methods. Kripke's original modal proof systems (see [283, 284]) were tableaux systems, and completeness proofs for tableaux typically do not make use of MCSs (Fitting [137] is a good introduction to modal tableaux methods). Completeness via normal form arguments have also proved useful. For example, Fine [131] uses normal forms to prove the completeness of the normal logic generated by the McKinsey axiom; this logic is not canonical (see Goldblatt [187]).

Nonetheless, most modal completeness theory revolves, directly or indirectly, around canonical models; pioneering papers include Makinson [307] (who uses a method close to the step-by-step construction to pick out generated subframes of canonical models) and Cresswell [99]. But the full power of canonical models and completeness-via-canonicity arguments did not emerge clearly till the work of Lemmon and Scott [296]. Their monograph stated and proved the Canonical Model Theorem and used completeness-via-canonicity arguments to establish frame completeness results; these included a general canonicity result for axioms of the form $\Diamond^k \Box^j p \rightarrow \Box^m \Diamond^n p$, where $k, j, m, n \geq 0$. While less general than Sahlqvist's [396] later result (Theorem 4.42), this covered most of the better-known systems, and was testimony to the generality of the canonical model method.

That **KL** is weakly complete with respect to the class of finite transitive trees is proved in Segerberg [404]. (Strictly speaking, Segerberg proved that **KL4** is complete with respect to the transitive trees, as it was not then known that 4 was derivable in **KL**; derivations of 4 were independently found by de Jongh, Kripke, and Sambin: see Boolos [69, page 11] and Hughes and Cresswell [235, page 150].) Segerberg first proves weak completeness with respect to the class of finite strict partial orders (the result we asked the reader to prove in Exercise 4.8.7), however he does so by filtrating the canonical model for **KL**, whereas we asked the reader to use a finite canonical model argument. Of course, the two arguments are intimately related, but the finite canonical model argument (which we have taken from Hughes and Cresswell [235, Theorem 8.4]) is rather more direct. Segerberg then proves weak completeness with respect to finite trees by unraveling the resulting model (just as we asked the reader to do in Exercise 4.8.8).

The incomplete tense logic $\mathbf{K}_t\mathbf{ThoM}$ discussed in the text was the first known frame incomplete logic, and it is still one of the most elegant and natural examples. It can be found in Thomason [434], and the text follows Thomason's original incompleteness proof. Shortly afterward, both Fine [129] and Thomason [434] exhibited (rather complex) examples of incomplete logics in the the basic modal language. The (much simpler) incomplete logic **KvB** examined in Exercise 4.4.2 is due to van Benthem [39]; **KvB** is further examined in Cresswell [100]. In Exercise 4.4.4 we listed three formulas which jointly define a first-order class of frames.

but which when used as axioms give rise to an incomplete normal logic; this example is due to van Benthem [37, 43]. The logic of the veiled recession frame was first axiomatized by Blok [64]. It was also Blok [65, 66] who showed that incompleteness is the rule rather than the exception among modal logics. The result that every consistent normal modal logic in the basic modal similarity type defines a non-empty frame class is due to Makinson [309].

Although filtration and unraveling had been used earlier to prove completeness results, the systematic use of transformation methods stems from the work of Segerberg [404]. Segerberg refined the filtration method, developed the bulldozing technique, and used them (together with other transformations) to prove many important completeness results, including characterizations of the tense logics of $(\mathbb{N}, <)$, $(\mathbb{Z}, <)$, $(\mathbb{Q}, <)$, $(\mathbb{R}, <)$ and their reflexive counterparts.

We do not know who first developed the modal step-by-step method. Certainly the idea of building models inductively is a natural one, and has long been used in both algebraic logic (see [226]) and set-theory (see [417]). One influential source for the method is the work of Burgess: for example, in [78] he uses it to prove completeness results in since/until logic (see also Xu [466] for some instructive step-by-step proofs for this language). Moreover, in [79], his survey article on tense logic, Burgess proves a number of completeness results for the basic modal language using the method. A set of lecture notes by de Jongh and Veltman [250] is the source of the popularity among Amsterdam logicians. Recent work on Arrow Logic uses the method (and the related mosaic method) heavily, often combined with the use of rules for the undefinable (see, for example, [320]). Step-by-step arguments are now widely used in a variety of guises.

Gabbay [149] is one of the earliest papers on rules for the undefinable, and one of the most influential (an interesting precursor is Burgess [77], in which these rules are used in the setting of branching time logic). Gabbay and Hodkinson [155] is an important paper which shows that such rules can take a particularly simple form in the basic temporal language. For rules in modal languages equipped with the D-operator, see de Rijke [378] and Venema [444]. For rules in modal languages with nominals, see Passy and Tinchev [358], Gargov and Goranko [164], Blackburn and Tzakova [63], and Blackburn [56].

The axiomatization of **PDL** that we gave in the text is from [407], Segerberg's 1977 abstract. But there was a gap in Segerberg's completeness proof, and by the time he had published a full corrected version (see [409]) very different proofs by Parikh [353] and Kozen and Parikh [273], had appeared. It seems that several other unpublished completeness proofs were also in circulation at this time: see Harel's survey of dynamic logic [209] for details. The proof in the text is based on lecture notes by van Benthem and Meyer Viol [50].

Bull's Theorem was the first general result about the fine structure of the lattice of normal modal logics. Bull's original proof (in [74]) was algebraic; the model-

theoretic proof given in the text is due to Fine [128]. A discussion of the relationship between the two proofs may be found in Bull and Segerberg [75]. Moreover,
Goldblatt [177] presents Fine's proof from a rather different perspective, emphasizing a concept he calls 'clusters within clusters'; the reader will find it instructive
to compare Goldblatt's presentation with ours. Fine's [128] also contains the finite
axiomatizability result for logics extending **S4.3** (Theorem 4.101) and the (negative) characterization in terms of finite sets of finite frames (Theorem 4.103), and
the text follows Fine's original proofs here too.

The work of Bull and Fine initiated a (still flourishing) investigation into subregions of the lattice of normal modal logics. For example, the position of logics in
the lattice characterized by a single structure is investigated in Maksimova [311],
Esakia and Meskhi [124] and (using algebraic methods) Blok [66]. In [130, 133],
Fine adapts his methods to analyze the logics extending **K4.3** (the adaptation is
technically demanding as not all these logics have the f.f.p.). The Berlin school
has a long tradition in this area: see Rautenberg [372, 373, 374], Kracht [277, 278,
279], and Wolter [459]. More recently, the structure of the lattice of tense logics has received attention: see, for example, Kracht [275] and Wolter [457]. And
Wolter [458] investigates the transfer of properties when the converse operator P is
added to a logic (in the basic modal language) that extends **K4**, obtaining various
axiomatizability and decidability results.

Work by Zakharyaschev has brought new ideas to bear. As we pointed out in the
Notes to Chapter 3, in the 1960s (the early years following the introduction of relational semantics for modal logic) it was hoped that one could describe *any* modal
formula by imposing first-order conditions on its frames. But the incompleteness
results, and the discovery of modal formulas that do not correspond to any first-
order condition, destroyed this hope. In a series of papers Zakharyaschev [470,
471, 472, 473] has studied an alternative, purely frame-theoretic approach to the
classification of modal formulas. Given a modal (or intuitionistic) formula ϕ, one
can effectively construct finite rooted frames $\mathfrak{F}_1, \ldots, \mathfrak{F}_n$ such that a general frame
$\mathfrak{g}$ refutes ϕ iff there is a (not necessarily generated) subframe $\mathfrak{g}'$ of $\mathfrak{g}$ which satisfies
certain natural conditions and which can be mapped to one of the $\mathfrak{F}_i$ by a bounded
morphism. Conversely, with every finite rooted frame $\mathfrak{F}$ Zakharyaschev associates
a *canonical* formula which can be refuted on a frame iff that frame contains a subframe (satisfying certain natural conditions) that can be mapped to $\mathfrak{F}$ by a bounded
morphism. Like the search for first-order characterizations, the classification approach in terms of canonical formulas is not universal either. But its limitations are
of a different kind: it only characterizes *transitive* general frames – but for every
modal (and intuitionistic) formula. Zakharyaschev [467] is a very accessible survey of canonical formulas, with plenty of motivations, examples and definitions;
technical details and discussions of the algebraic and logical background of canonical formulas are provided by Chagrov and Zakharyaschev [88, Chapter 9].

5

Algebras and General Frames

In this chapter we develop an *algebraic* semantics for modal logic. The basic idea is to extend the algebraic treatment of classical propositional logic (which uses *boolean algebras*) to modal logic. The algebras employed to do this are called *boolean algebras with operators* (BAOs). The boolean part handles the underlying propositional logic, the additional operators handle the modalities.

But why algebraize modal logic? There are two main reasons. First, the algebraic perspective allows us to bring powerful new techniques to bear on modal-logical problems. Second, the algebraic semantics turns out to be better-behaved than frame-based semantics: we will be able to prove an algebraic completeness result for *every* normal modal logic. As our discussion of incompleteness in Section 4.4 makes clear, no analogous result holds for frames.

This chapter has three main parts. The first, consisting of the first three sections, introduces the algebraic approach: we survey the basic ideas in the setting of classical propositional logic, extend them to modal logic, and prove the Jónsson-Tarski Theorem. The second part, which consists of the fourth section, introduces *duality theory*, the study of correspondences between the universe of algebras and the universe of frames. The last part (the only part on the advanced track) is devoted to *general frames*. These turn out to be set-theoretic representations of boolean algebras with operators, and we examine their properties in detail, and use them to prove the Sahlqvist Completeness Theorem. Background information on universal algebra can be found in Appendix B.

Chapter guide

Section 5.1: Logic as Algebra (Basic track). What is algebraic logic? This section provides some preliminary answers by examining the relationship between propositional logic and boolean algebras.

Section 5.2: Algebraizing Modal Logic (Basic track). To algebraize modal logic, we introduce *boolean algebras with operators* (BAOs). We discuss BAOs

261

from a semantic perspective (introducing an important class of BAOs called *complex algebras*), and from a syntactic perspective (we use *Lindenbaum-Tarski algebras* to obtain abstract BAOs from normal modal logics).

Section 5.3: The Jónsson-Tarski Theorem (Basic track). Here we prove the theorem underlying algebraic approaches to modal completeness theory. First we learn how to construct a frame from an algebra by forming the *ultrafilter frame*. By turning this frame back into a complex algebra, we obtain the *canonical embedding algebra*. We then prove the Jónsson-Tarski Theorem: every boolean algebra with operators can be embedded in its canonical embedding algebra.

Section 5.4: Duality Theory (Basic track). Frames are inter-related by bounded morphisms, generated subframes, and disjoint union. Boolean algebras with operators are inter-related by homomorphisms, subalgebras, and direct products. Modal duality theory studies the relationship between these two mathematical universes. Two applications are given, one of which is an algebraic proof of the Goldblatt-Thomason Theorem.

Section 5.5: General Frames (Advanced track). We (re)introduce general frames and study them in detail, focusing on the relationship between general frames, frames, and boolean algebras with operators. We conclude with a brief discussion of some important topological aspects of general frames.

Section 5.6: Persistence (Advanced track). In this section we introduce a natural generalization of the notion of canonicity encountered in Chapter 4: *persistence*. We use it to prove the *Sahlqvist Completeness Theorem*.

5.1 Logic as Algebra

What do algebra and logic have in common? And why bring algebra into the study of logic? This section provides some preliminary answers: we show that algebra and logic share key ideas, and analyze classical propositional logic algebraically. Along the way we will meet a number of important concepts (notably formula algebras, the algebra of truth values, set algebras, abstract boolean algebras, and Lindenbaum-Tarski algebras) and results (notably the Stone Representation Theorem), but far more important is the overall picture. Algebraic logic offers a natural way of re-thinking many basic logical issues, but it is important not to miss the wood for the trees. The bird's eye view offered here should help guide the reader through the more detailed modal investigations that follow.

Algebra as logic

Most school children learn how to manipulate simple algebraic equations. Given the expression $(x + 3)(x + 1)$, they learn how to multiply these factors to form

$x^2 + 4x + 3$, and (somewhat later) study methods for doing the reverse (that is, for decomposing quadratics into factors).

Such algebraic manipulations are essentially logical. For a start, we have a well-defined syntax: we manipulate *equations* between *terms*. This syntax is rarely explicitly stated, but most students (building on the analogy with basic arithmetic) swiftly learn how to build legitimate terms using numerals, variables such as x, y and z, and $+$, $\cdot$ and $-$. Moreover, they learn the rules which govern this symbol manipulation process: replacing equals by equals, doing the same thing to both sides of an equation, appealing to commutativity, associativity and distributivity to simplify and rearrange expressions. High-school algebra is a form of proof theory.

But there is also a *semantic* perspective on basic algebra, though this usually only becomes clear later. As students learn more about mathematics, they realize that the familiar 'laws' do not hold for all mathematical objects: for example, matrix multiplication is not commutative. Gradually the student grasps that variables need not be viewed as standing for numbers: they can be viewed as standing for other objects as well. Eventually the semantic perspective comes into focus: there are various kinds of *algebras* (that is, sets equipped with collections of functions, or *operations*, which satisfy certain properties), and *terms denote elements in algebras*. Moreover, an equation such as $x \cdot y = y \cdot x$ is not a sacrosanct law: it is simply a property that holds for some algebras and not for others.

So algebra has a syntactic dimension (terms and equations) and a semantic dimension (sets equipped with a collection of operations). And in fact there is a tight connection between the proof theory algebra offers and its semantics. In Appendix B we give a standard derivation system for *equational* logic (that is, a standard set of rules for manipulating equations) and state a fundamental result due to Birkhoff: the system is strongly sound and complete with respect to the standard algebraic semantics. Algebra really can be viewed as logic.

But logic can also be viewed as algebra. We will now illustrate this by examining classical propositional logic algebraically. Our discussion is based around three main ideas: the algebraization of propositional semantics in the class of set algebras; the algebraization of propositional axiomatics in the class of abstract boolean algebras; and how the Stone Representation Theorem links these approaches.

Algebraizing propositional semantics

Consider any propositional formula, say $(p \lor q) \land (p \lor r)$. The most striking thing about propositional formulas (as opposed to first-order formulas) is their syntactic simplicity. In particular, there is no variable binding – all we have is a collection of atomic symbols (p, q, r, and so on) that are combined into more complex expressions using the symbols $\bot$, $\top$, $\neg$, $\lor$ and $\land$. Recall that we take $\bot$, $\neg$ and $\lor$ as the primitive symbols, treating the others as abbreviations.

Now, as the terminology 'proposition letters' suggests, we think of p, q, and r as symbols denoting entities called propositions, abstract bearers of information. So what do $\bot$, $\top$, $\neg$, $\vee$ and $\wedge$ denote? Fairly obviously: ways of combining propositions, or operations on propositions. More precisely, $\vee$ and $\wedge$ must denote binary operations on propositions (let us call these operations $+$ and $\cdot$ respectively), $\neg$ must denote a unary operation on propositions (let us call it $-$), while $\bot$ and $\top$ denote special nullary operations on propositions (that is, they are the names of two special propositions: let us call them 0 and 1 respectively). In short, we have worked our way towards the idea that *formulas can be seen as terms denoting propositions*.

But which kinds of algebras are relevant? Here is a first step towards an answer.

Definition 5.1 Let *Bool* be the algebraic similarity type having one constant (or nullary function symbol) $\bot$, one unary function symbol $\neg$, and one binary function symbol $\vee$. Given a set of propositional variables Φ, $Form(\Phi)$ is the set of *Bool*-terms in Φ; this set is identical to the collection of propositional formulas in Φ.

Algebras of type *Bool* are usually presented as 4-tuples $\mathfrak{A} = (A, +, -, 0)$. We make heavy use of the standard abbreviations $\cdot$ and 1. That is, $a \cdot b$ is shorthand for $-(-a + -b)$, and 1 is shorthand for -0.   ⊣

But this only takes us part of the way. There are many different algebras of this similarity type – and we are only interested in algebras which can plausibly be viewed as algebras of propositions. So let us design such an algebra. Propositional logic is about truth and falsehood, so let us take the set $2 = \{0, 1\}$ as the set A underlying the algebra; we think of '0' as the truth value *false*, and '1' as the value *true*. But we also need to define suitable operations over these truth values, and we want these operations to provide a natural interpretation for the logical connectives. Which operations are appropriate?

Well, the terms we are working with are just propositional formulas. So how would we go about evaluating a formula χ in the truth value algebra? Obviously we would have to know whether the proposition letters in χ are true or false, but let us suppose that this has been taken care of by a function $\theta : \Phi \to 2$ mapping the set Φ of proposition letters to the set 2 of truth values. Given such a θ (logicians will call θ a valuation, algebraists will call it an assignment) it is clear what we have to do: compute $\tilde{\theta}(\phi)$ according to the following rules:

$$\begin{aligned}
\tilde{\theta}(p) &= \theta(p), \text{ for all } p \in \Phi, \\
\tilde{\theta}(\bot) &= 0, \\
\tilde{\theta}(\neg\phi) &= 1 - \tilde{\theta}(\phi), \\
\tilde{\theta}(\psi \vee \psi) &= \max(\tilde{\theta}(\phi), \tilde{\theta}(\psi)).
\end{aligned} \tag{5.1}$$

Clearly the operations used here are the relevant ones; they simply restate the familiar truth table definitions. This motivates the following definition:

Definition 5.2 The *algebra of truth values* is $\mathbf{2} = (\{0, 1\}, +, -, 0))$, where $-$ and $+$ are defined by $-a = 1 - a$ and $a + b = \max(a, b)$, respectively. ⊣

Let us sum up our discussion so far. The crucial observations are that formulas can be viewed as terms, that valuations can be identified with algebraic assignments in the algebra $\mathbf{2}$, and that evaluating the truth of a formula under such a valuation/assignment is exactly the same as determining the meaning of the term in the algebra $\mathbf{2}$ under the assignment/valuation.

So let us move on. We have viewed meaning as a map $\tilde{\theta}$ from the set $Form(\Phi)$ to the set $\{0, 1\}$ – but it is useful to consider this *meaning function* in more mathematical detail. Note the 'shape' of the conditions on $\tilde{\theta}$ in (5.1): the resemblance to the defining condition of a *homomorphism* is too blatant to miss. But since homomorphisms are the fundamental maps between algebras (see Appendix B) why not try and impose algebraic structure on the *domain* of such meaning functions (that is, on the set of formulas/terms) so that meaning functions really are homomorphisms? This is exactly what we are about to do. We first define the needed algebraic structure on the set of formulas.

Definition 5.3 Let Φ be a set of proposition letters. The propositional *formula algebra* over Φ is the algebra

$$\mathfrak{Form}(\Phi) = (Form(\Phi), +, -, \perp),$$

where $Form(\Phi)$ is the collection of propositional formulas over Φ, and $-$ and $+$ are the operations defined by $-\phi := \neg\phi$ and $\phi + \psi := \phi \vee \psi$, respectively. ⊣

In other words, the carrier of this algebra is the collection of propositional formulas over the set of proposition letters Φ, and the operations $-$ and $+$ give us a simple mathematical picture of the dynamics of formula construction.

Proposition 5.4 *Let Φ be some set of proposition letters. Given any assignment $\theta : \Phi \to \mathbf{2}$, the function $\tilde{\theta} : Form(\Phi) \to \mathbf{2}$ assigning to each formula its meaning under this valuation, is a homomorphism from $\mathfrak{Form}(\Phi)$ to $\mathbf{2}$.*

Proof. A precise definition of homomorphism is given in Appendix B. Essentially, homomorphisms between algebras map elements in the source algebra to elements in the target algebra in an operation preserving way – and this is precisely what the conditions on $\tilde{\theta}$ in (5.1) express. ⊣

The idea of viewing formulas as terms, and meaning as a homomorphism, is fundamental to algebraic logic.

Another point is worth stressing. As the reader will have noticed, sometimes we call a sequence of symbols like $p \vee q$ a formula, and sometimes we call it a term. This is intentional. Any propositional formula can be viewed as – simply *is* – an algebraic term. The one-to-one correspondence involved is so obvious that it is not worth talking about 'translating' formulas to terms or vice-versa; they are simply two ways of looking at the same thing. We simply choose whichever terminology seems most appropriate to the issue under discussion.

But let us move on. As is clear from high-school algebra, algebraic reasoning is essentially *equational*. So a genuinely *algebraic* logic of propositions should give us a way of determining when two propositions are equal. For example, such a logic should be capable of determining that the formulas $p \vee (q \wedge p)$ and p denote the same proposition. How does the algebraic approach to propositional semantics handle this? As follows: an equation $s \approx t$ is valid in an algebra $\mathfrak{A}$ if for every assignment to the variables occurring in the terms, s and t have the same meaning in $\mathfrak{A}$ (see Appendix B for further details). Hence, an algebraic way of saying that a formula ϕ is a classical tautology (notation: $\models_C \phi$) is to say that the equation $\phi \approx \top$ is valid in the algebra of truth values.

Now, an attractive feature of propositional logic (a feature which extends to modal logic) is that not only terms, but *equations* correspond to formulas. There is nothing mysterious about this: we can define the bi-implication connective $\leftrightarrow$ in classical propositional logic, and viewed as an operation on propositions, $\leftrightarrow$ asserts that both terms have the same meaning:

$$\tilde{\theta}(\phi \leftrightarrow \psi) \;=\; \begin{cases} 1 & \text{if } \tilde{\theta}(\phi) = \tilde{\theta}(\psi), \\ 0 & \text{otherwise.} \end{cases}$$

So to speak, propositional logic is intrinsically equational.

Theorem 5.5 neatly summarizes our discussion so far: it shows how easily we can move from a logical to an algebraic perspective and back again.

Theorem 5.5 (2 Algebraizes Classical Validity) *Let ϕ and ψ be propositional formulas/terms. Then*

$$\models_C \phi \quad \textit{iff} \quad \mathbf{2} \models \phi \approx \top. \tag{5.2}$$
$$\mathbf{2} \models \phi \approx \psi \quad \textit{iff} \quad \models_C \phi \leftrightarrow \psi. \tag{5.3}$$
$$\models_C \phi \leftrightarrow (\phi \leftrightarrow \top). \tag{5.4}$$

Proof. Immediate from the definitions.   ⊣

Remark 5.6 The reader may wonder about the presence of (5.3) and in particular, of (5.4) in the Theorem. The point is that for a proper, 'full', algebraization of a logic, one has to establish not only that the membership of some formula ϕ in the logic can be rendered algebraically as the validity of some equation $\phi^\approx$ in some (class of) algebra(s). One also has to show that conversely, there is a translation of equations to formulas such that the equation holds in the class of algebras if and only if its translation belongs to the logic. And finally, one has to prove that translating a formula ϕ to an equation $\phi^\approx$, and then translating this equation back to a formula, one obtains a formula ϕ' that is *equivalent* to the original formula ϕ. The fact that our particular translations satisfy these requirements is stated by (5.3) and (5.4), respectively.

Since we will not go far enough into the theory of algebraic logic to use these 'full' algebraizations, in the sequel we will only mention the first kind of equivalence when we algebraize a logic. Nevertheless, in all the cases that we consider, the second and third requirements are met as well. ⊣

Set algebras

Propositional formulas/terms and equations may be interpreted in any algebra of type *Bool*. Most algebras of this type are uninteresting as far as the semantics of propositional logic is concerned – but other algebras besides **2** are relevant. A particularly important example is the class of *set algebras*. As we will now see, set algebras provide us with a second algebraic perspective on the semantics of propositional logic. And as we will see in the following section, the perspective they provide extends neatly to modal logic.

Definition 5.7 (Set Algebras) Let A be a set. As usual, we denote the *power set* of A (the set of all subsets of A) by $\mathcal{P}(A)$. The *power set algebra* $\mathfrak{P}(A)$ is the structure

$$\mathfrak{P}(A) = (\mathcal{P}(A), \cup, -, \varnothing),$$

where $\varnothing$ denotes the *empty* set, $-$ is the operation of taking the *complement* of a set relative to A, and $\cup$ that of taking the *union* of two sets. From these basic operations we define in the standard way the operation $\cap$ of taking the *intersection* of two sets, and the special element A, the *top set* of the algebra.

A *set algebra* or *field of sets* is a subalgebra of a power set algebra. That is, a set algebra (on A) is a collection of subsets of A that contains $\varnothing$ and is closed under $\cup$ and $-$ (so any set algebra contains A and is closed under $\cap$ as well). The class of all set algebras is called Set. ⊣

Set algebras provide us with a simple concrete picture of propositions and the way they are combined – moreover, it is a picture that even at this stage contains a number of traditional modal ideas. Think of A as a set of worlds (or situations, or states) and think of a proposition as a subset of A. And think of a proposition as a set of worlds – the worlds that make it true. So viewed, $\varnothing$ is a very special proposition: it is the proposition that is false in every situation, which is clearly a good way of thinking about the meaning of $\bot$. Similarly, A is the proposition true in all situations, which is a suitable meaning for $\top$. It should also be clear that $\cup$ is a way of combining propositions that mirrors the role of $\vee$. After all, in what worlds is $p \vee q$ true? In precisely those worlds that make p true or q true. Finally, complementation mirrors negation, for $\neg p$ is true in precisely those worlds where p is not true.

As we will now show, set algebras and the algebra **2** make precisely the same equations true. We will prove this algebraically by showing that the class of set algebras coincides (modulo isomorphism) to the class of subalgebras of powers of **2**. The crucial result needed is the following:

Proposition 5.8 *Every power set algebra is isomorphic to a power of* **2**, *and conversely.*

Proof. Let A be an arbitrary set, and consider the following function χ mapping elements of $\mathcal{P}(A)$ to 2-valued maps on A:

$$\chi(X)(a) = \begin{cases} 1 & \text{if } a \in X, \\ 0 & \text{otherwise.} \end{cases}$$

In other words, $\chi(X)$ is the *characteristic function* of X. The reader should verify that χ is an isomorphism between $\mathfrak{P}(A)$ and $\mathbf{2}^A$, where the latter algebra is as defined in Definition B.6.

Conversely, to show that every power of **2** is isomorphic to some power set algebra, let $\mathbf{2}^I$ be some power of **2**. Consider the map $\alpha : 2^I \to \mathcal{P}(I)$ defined by

$$\alpha(f) = \{i \in I \mid f(i) = 1\}.$$

Again, we leave it for the reader to verify that α is the required isomorphism between $\mathbf{2}^I$ and $\mathfrak{P}(I)$.   ⊣

Theorem 5.9 (Set **Algebraizes Classical Validity**) *Let ϕ and ψ be propositional formulas/terms. Then*

$$\models_C \phi \quad \textit{iff} \quad \text{Set} \models \phi \approx \top. \tag{5.5}$$

Proof. It is not difficult to show from first principles that the validity of equations is preserved under taking direct products (and hence powers) and subalgebras. Thus, with the aid of Theorem 5.5 and Proposition 5.8, the result follows.   ⊣

Algebraizing propositional axiomatics

We now have two equational perspectives on the semantics of propositional logic: one via the algebra **2**, the other via set algebras. But what about the syntactic aspects of propositional logic? It is time to see how the equational perspective handles such notions as theoremhood and provable equivalence.

Assume we are working in some fixed (sound and complete) proof system for classical propositional logic. Let $\vdash_C \phi$ mean that ϕ is a theorem of this system, and call two propositional formulas ϕ and ψ *provably equivalent* (notation: $\phi \equiv_C \psi$) if the formula $\phi \leftrightarrow \psi$ is a theorem. Theorem 5.11 is a syntactic analog of Theorem 5.9: it is the fundamental result concerning the algebraization of propositional axiomatics. Its statement and proof makes use of *boolean algebras*, so let us define these important entities right away.

Definition 5.10 (Boolean Algebras) Let $\mathfrak{A} = (A, +, -, 0)$ be an algebra of the boolean similarity type. Then $\mathfrak{A}$ is called a *boolean algebra* iff it satisfies the following identities (recall that $x \cdot y$ and 1 are shorthand for $-(-x + -y)$ and -0, respectively):

$(B0)$	$x + y = y + x$	$x \cdot y = y \cdot x$
$(B1)$	$x + (y + z) = (x + y) + z$	$x \cdot (y \cdot z) = (x \cdot y) \cdot z$
$(B2)$	$x + 0 = x$	$x \cdot 1 = x$
$(B3)$	$x + (-x) = 1$	$x \cdot (-x) = 0$
$(B4)$	$x + (y \cdot z) = (x + y) \cdot (x + z)$	$x \cdot (y + z) = (x \cdot y) + (x \cdot z)$

The operations $+$ and $\cdot$ are called *join* and *meet*, respectively, and the elements 1 and 0 are referred to as the *top* and *bottom* elements. We *order* the elements of a boolean algebra by defining $a \leq b$ if $a + b = b$ (or equivalently, if $a \cdot b = a$). Given a boolean algebra $\mathfrak{A} = (A, +, -, 0)$, the set A is called its *carrier set*. We call the class of boolean algebras BA. ⊣

By a famous result of Birkhoff (discussed in Appendix B) a class of algebras defined by a collection of equations can be structurally characterized as a *variety*. Thus in what follows we sometimes speak of the variety of boolean algebras, rather than the class of boolean algebras.

If you have not encountered boolean algebras before, you should check that the algebra **2** and the set algebras defined earlier are both examples of boolean algebras (that is, check that these algebras satisfy the listed identities). In fact, set algebras are what are known as *concrete* boolean algebras. As we will see when we discuss the Stone Representation Theorem, the relationship between abstract boolean algebras (that is, any algebraic structure satisfying the previous definition) and set algebras lies at the heart of the algebraic perspective on propositional soundness and completeness.

But this is jumping ahead: our immediate task is to state the syntactic analog of Theorem 5.9 promised above.

Theorem 5.11 (BA Algebraizes Classical Theoremhood) *Let ϕ and ψ be propositional formulas/terms. Then*

$$\vdash_C \phi \quad iff \quad \text{BA} \models \phi \approx \top. \tag{5.6}$$

Proof. Soundness (the direction from left to right in (5.6)) can be proved by a straightforward inductive argument on the length of propositional proofs. Completeness will follow from the Propositions 5.14 and 5.15 below.　　　⊣

How are we to prove this completeness result? Obviously we have to show that every non-theorem of classical propositional logic can be falsified on some boolean algebra (falsified in the sense that there is some assignment under which the formula does not evaluate to the top element of the algebra). So the key question is: how do we build falsifying algebras? Our earlier work on relational completeness suggests an answer. In Chapter 4 we made use of canonical models: that is, we manufactured models out of syntactical ingredients (sets of formulas) taking care to hardwire in all the crucial facts about the logic. So the obvious question is: can we construct algebras from (sets of) formulas in a way that builds in all the propositional logic we require? Yes, we can. Such algebras are called Lindenbaum-Tarski algebras. In essence, they are 'canonical algebras.'

First, some preliminary work. The observation underpinning what follows is that the relation of provable equivalence is a *congruence* on the formula algebra. A congruence on an algebra is essentially an equivalence relation on the algebra that respects the operations (a precise definition is given in Appendix B) and it is not hard to see that provable equivalence is such a relation.

Proposition 5.12 *The relation $\equiv_C$ is a congruence on the propositional formula algebra.*

Proof. We have to prove that $\equiv_C$ is an equivalence relation satisfying

$$\phi \equiv_C \psi \text{ only if } \neg\phi \equiv_C \neg\psi \tag{5.7}$$

and

$$\phi_0 \equiv_C \psi_0 \text{ and } \phi_1 \equiv_C \psi_1 \text{ only if } (\phi_0 \vee \phi_1) \equiv_C (\psi_0 \vee \psi_1). \tag{5.8}$$

In order to prove that $\equiv_C$ is reflexive, we have to show that for any formula ϕ, the formula $\phi \leftrightarrow \phi$ is a theorem of the proof system. The reader is invited to prove this in his or her favorite proof system for proposition calculus. The properties of symmetry and transitivity are also left to the reader.

But we want to prove that $\equiv_C$ is not merely an equivalence relation but a congruence. We deal with the case for negation, leaving (5.8) to the reader. Suppose that $\phi \equiv_C \psi$, that is, $\vdash_C \phi \leftrightarrow \psi$. Again, given that we are working with a sound and complete proof system for propositional calculus, this implies that $\vdash_C \neg\phi \leftrightarrow \neg\psi$. Given this, (5.7) is immediate.　　　⊣

The equivalence classes under $\equiv_C$ are the building blocks for what follows. As any such class is a maximal set of mutually equivalent formulas, we can think of such classes as propositions. And as $\equiv_C$ is a congruence, we can define a natural algebraic structure on these propositions. Doing so gives rise to Lindenbaum-Tarski algebras.

Definition 5.13 (Lindenbaum-Tarski Algebra) Given a set of proposition letters Φ, let $Form(\Phi)/\equiv_C$ be the set of equivalence classes that $\equiv_C$ induces on the set of formulas, and for any formula ϕ let $[\phi]$ denote the equivalence class containing ϕ. Then the *Lindenbaum-Tarski algebra* (for this language) is the structure

$$\mathfrak{L}_C(\Phi) := (Form(\Phi)/\equiv_C, +, -, 0),$$

where $+, -$ and 0 are defined by: $[\phi] + [\psi] := [\phi \vee \psi]$, $-[\phi] := [\neg \phi]$ and $0 := [\bot]$. Strictly speaking, we should write $[\phi]_\Phi$ instead of $[\phi]$, for ϕ's congruence class depends on the set Φ of proposition letters. But unless there is potential for confusion, we usually will not bother to do so. $\dashv$

Lindenbaum-Tarski algebras are easy to work with. For instance, it is easy to see that the meet operation in such an algebra is given by $[\phi] \cdot [\psi] = [\phi \wedge \psi]$, while the top element 1 is $[\top]$. As another example, we show that $a + (-a) = 1$ for all elements a of $\mathfrak{L}_C(\Phi)$. The first observation is that a, just like *any* element of $\mathfrak{L}_C(\Phi)$, is of the form $[\phi]$ for some formula ϕ. But then we have

$$a + (-a) = [\phi] + (-[\phi]) = [\phi] + [\neg \phi] = [\phi \vee (\neg \phi)] = [\top] = 1, \qquad (5.9)$$

where the fourth equality holds because $\vdash_C (\phi \vee \neg \phi) \leftrightarrow \top$.

It is fairly obvious that the structure of a Lindenbaum-Tarski algebra only depends on the cardinality of the set Φ of proposition letters; the reader is asked to prove this in Exercise 5.1.4.

We need two results concerning Lindenbaum-Tarski algebras. First, we have to show that they are indeed an 'algebraic canonical model' – that is, that they give us a counterexample for *every* non-theorem of propositional logic. Second, we have to show that they are counterexamples of the right kind: that is, we need to prove that any Lindenbaum-Tarski algebra is a boolean algebra.

Proposition 5.14 *Let ϕ be some propositional formula, and Φ a set of proposition letters of size not smaller than the number of proposition letters occurring in ϕ. Then*

$$\vdash_C \phi \text{ iff } \mathfrak{L}_C(\Phi) \models \phi \approx \top. \qquad (5.10)$$

Proof. We may and will assume that Φ actually contains all variables occurring in ϕ, cf. Exercise 5.1.4. We first prove the easy direction from right to left. Assume that ϕ is *not* a theorem of classical propositional logic. This implies that ϕ and $\top$ are not provably equivalent, whence we have $[\phi] \neq [\top]$. We have to find an assignment on $\mathfrak{L}_C(\Phi)$ that forms a counterexample to the validity of ϕ. There is one obvious candidate, namely the assignment ι given by $\iota(p) = [p]$. It can easily be verified (by a straightforward formula induction) that with this definition we obtain $\tilde{\iota}(\psi) = [\psi]$ for *all* formulas ψ that use variables from the set Φ. But then by our assumption on ϕ we find that

$$\tilde{\iota}(\phi) = [\phi] \neq [\top] = 1,$$

as required.

For the other direction we have to work a bit harder. If $\vdash_C \phi$ then it is obvious that $\tilde{\iota}(\phi) = [\phi] = [\top] = 1$, but only looking at ι is not sufficient now. We have to show that $\tilde{\theta}(\phi) = [\top]$ for *all* assignments θ.

So let θ be an arbitrary assignment. That is, θ assigns an equivalence class (under $\equiv_C$) to each proposition letter. For each variable p, take a representing formula $\rho(p)$ in the equivalence class $\theta(p)$; that is, we have $\theta(p) = [\rho(p)]$. We may view ρ as a *function* mapping proposition letters to formulas; in other words, ρ is a *substitution*. Let $\rho(\psi)$ denote the effect of performing this substitution on the formula ψ. It can be proved by an easy formula induction that, for any formula ψ, we have

$$\tilde{\theta}(\psi) = [\rho(\psi)]. \tag{5.11}$$

Now, the collection of propositional theorems is closed under uniform substitution (depending on the formulation of your favorite sound and complete proof system, this is either something that is hardwired in or can be shown to hold). This closure property implies that the formula $\rho(\phi)$ is a theorem, and hence that $\rho(\phi) \equiv_C \top$, or equivalently, $[\rho(\phi)] = [\top]$. But then it follows from (5.11) that

$$\tilde{\theta}(\phi) = [\top],$$

which is precisely what we need to show that $\mathfrak{L}_C(\Phi) \models \phi$. $\dashv$

Thus it only remains to check that $\mathfrak{L}_C(\Phi)$ is the right kind of algebra.

Proposition 5.15 *For any set Φ of proposition letters, $\mathfrak{L}_C(\Phi)$ is a boolean algebra.*

Proof. Fix a set Φ. The proof of this Proposition boils down to proving that all the identities B0–4 hold in $\mathfrak{L}_C(\Phi)$. In (5.9) we proved that the first part of B3 holds; we leave the reader to verify that the other identities hold as well. $\dashv$

Summarizing, we have seen that the axiomatics of propositional logic can be algebraized in a class of algebras, namely the variety of boolean algebras. We have also seen that Lindenbaum-Tarski algebras act as canonical representatives of the class of boolean algebras. (For readers with some background in universal algebra, we remark that Lindenbaum-Tarski algebras are in fact the *free* boolean algebras.)

Weak completeness via Stone

It is time to put our findings together, and to take one final step. This step is more important than any taken so far.

Theorem 5.9 captured tautologies as equations valid in set algebras:

$$\models_C \phi \text{ iff Set} \models \phi \approx \top.$$

On the other hand, in Theorem 5.11 we found an algebraic semantics for the notion of classical theoremhood:

$$\vdash_C \phi \text{ iff BA} \models \phi \approx \top.$$

But there is a fundamental *logical* connection between $\models_C$ and $\vdash_C$: the soundness and completeness theorem for propositional logic tells us that they are identical. Does this crucial connection show up algebraically? That is, is there an algebraic analog of the soundness and completeness result for classical propositional logic? There is: it is called the Stone Representation Theorem.

Theorem 5.16 (Stone Representation Theorem) *Any boolean algebra is isomorphic to a set algebra.*

Proof. We will make a more detailed statement of this result, and prove it, in Section 5.3. ⊣

(Incidentally, this immediately tells us that any boolean algebra is isomorphic to a subalgebra of a power of **2** – for Proposition 5.8 tells us that any power set algebra is isomorphic to a power of **2**.) But what really interests us here is the *logical* content of Stone's Theorem. In essence, it is the key to the weak completeness of classical propositional logic.

Corollary 5.17 (Soundness and Weak Completeness) *For any formula ϕ, ϕ is valid iff it is a theorem.*

Proof. Immediate from the equations above, since by the Stone Representation Theorem, the equations valid in Set must coincide with those valid in BA. ⊣

The relation between Theorem 5.11 and Corollary 5.17 is the key to much of our later work. Note that from a logical perspective, Corollary 5.17 is the interesting result: it establishes the soundness and completeness of classical propositional logic with respect to the standard semantics. So why is Theorem 5.11 important? After all, as it proves completeness with respect to an abstractly defined class of boolean algebras, it does not have the same independent logical interest. This is true, but given that the abstract algebraic counterexamples it provides can be represented as standard counterexamples – and this is precisely what Stone's Theorem guarantees – it enables us to prove the standard completeness result for propositional logic.

To put it another way, the algebraic approach to completeness factors the algebra building process into two steps. We first prove completeness with respect to an abstract algebraic semantics by building an abstract algebraic model. It is easy to do this – we just use Lindenbaum-Tarski algebras. We then try and *represent* the abstract algebras in the concrete form required by the standard semantics; that is, in terms of set algebras or of the algebra **2**.

In the next two sections we extend this approach to modal logic. Algebraizing modal logic is more demanding than algebraizing propositional logic. For a start, there is not just one logic to deal with – we want to be able to handle any normal modal logic whatsoever. Moreover, the standard semantics for modal logic is given in terms of frame-based models – so we are going to need a representation result that tells us how to represent algebras as relational structures.

But all this can be done. In the following section we will generalize boolean algebras to boolean algebras with operators; these are the *abstract* algebras we will be dealing with throughout the chapter. We also generalize set algebras to complex algebras; these are the *concrete* algebras which model the idea of set-based algebras of propositions for modal languages. We then define the Lindenbaum-Tarski algebras we need – and every normal modal logic will give rise to its own Lindenbaum-Tarski algebra. This is all a fairly straightforward extension of ideas we have just discussed. We then turn, in Section 5.3, to the crucial representation result: the Jónsson-Tarski Theorem. This is an extension of Stone's Representation Theorem that tells us how to represent a boolean algebra with operators as an ordinary modal model. It is an elegant result in its own right, but for our purposes its importance is the bridge it provides between completeness in the universe of algebras and completeness in the universe of relational structures.

Exercises for Section 5.1

5.1.1 Let A and B be two sets, and $f : A \to B$ some map. Show that $f^{-1} : \mathcal{P}(B) \to \mathcal{P}(A)$ given by $f^{-1}(Y) = \{a \in A \mid f(a) \in Y\}$ is a *homomorphism* from the power set algebra of B to that of A.

5.1.2 Prove that every power set algebra is isomorphic to a power of the algebra **2**, and that conversely, every power of **2** is isomorphic to a power set algebra. That is, fill in the details of the proof of Theorem 5.8.

5.1.3 Here is a standard set of axioms for propositional calculus: $p \to (q \to p)$, $(p \to (q \to r)) \to ((p \to q) \to (p \to r))$, and $(\neg p \to \neg q) \to (q \to p)$. Show that all three axioms are valid on any set algebra. That is, show that whatever subset is used to interpret the proposition letters, these formulas are true in all worlds. Furthermore, show that modus ponens and uniform substitution preserve validity.

5.1.4 Let Φ and Ψ be two sets of proposition letters.

(a) Prove that $\mathfrak{Form}(\Phi)$ is a subalgebra of $\mathfrak{Form}(\Psi)$ iff $\Phi \subseteq \Psi$.

(b) Prove that $\mathfrak{L}_C(\Phi)$ can be *embedded* in $\mathfrak{L}_C(\Psi)$ iff $|\Phi| \leq |\Psi|$.
(c) Prove that $\mathfrak{L}_C(\Phi)$ and $\mathfrak{L}_C(\Psi)$ are isomorphic iff $|\Phi| = |\Psi|$.
(d) Does $\Phi \subseteq \Psi$ imply that $\mathfrak{L}_C(\Phi)$ is a subalgebra of $\mathfrak{L}_C(\Psi)$?

5.2 Algebraizing Modal Logic

Let us adapt the ideas introduced in the previous section to modal logic. The most basic principle of algebraic logic is that formulas of a logical language can be viewed as terms of an algebraic language, so let us first get clear about the algebraic languages we will use in the remainder of this chapter:

Definition 5.18 Let τ be a modal similarity type. The *corresponding algebraic similarity type* $\mathcal{F}_\tau$ contains as function symbols all modal operators, together with the boolean symbols $\vee$ (binary), $\neg$ (unary), and $\bot$ (constant). For a set Φ of variables, we let $Ter_\tau(\Phi)$ denote the collection of $\mathcal{F}_\tau$-terms over Φ. $\dashv$

The algebraic similarity type $\mathcal{F}_\tau$ can be seen as the union of the modal similarity type τ and the boolean type *Bool*. In practice we often identify τ and $\mathcal{F}_\tau$, speaking of τ-terms instead of $\mathcal{F}_\tau$-terms. The previous definition takes the formulas-as-terms paradigm quite literally: by our definitions

$$Form(\tau, \Phi) = Ter_\tau(\Phi).$$

Just as boolean algebras were the key to the algebraization of classical propositional logic, in modal logic we are interested in *boolean algebras with operators* or BAOs. Let us first define BAOs abstractly; we will discuss concrete BAOs shortly.

Definition 5.19 (Boolean Algebras with Operators) Let $\tau = (O, \rho)$ be a modal similarity type. A *boolean algebra with τ-operators* is an algebra

$$\mathfrak{A} = (A, +, -, 0, f_\triangle)_{\triangle \in \tau}$$

such that $(A, +, -, 0)$ is a boolean algebra and every $f_\triangle$ is an *operator* of arity $\rho(\triangle)$; that is, $f_\triangle$ is an operation satisfying

(normality) $f_\triangle(a_1, \ldots, a_{\rho(\triangle)}) = 0$ whenever $a_i = 0$ for some i $(0 < i \leq \rho(\triangle))$.
(additivity) for all i (such that $0 < i \leq \rho(\triangle)$),

$$f_\triangle(a_1, \ldots, a_i + a_i', \ldots, a_{\rho(\triangle)}) =$$
$$f_\triangle(a_1, \ldots, a_i, \ldots, a_{\rho(\triangle)}) + f_\triangle(a_1, \ldots, a_i', \ldots, a_{\rho(\triangle)}).$$

If we abstract from the particular modal similarity type τ, or if τ is known from context, we simply speak of *boolean algebras with operators*, or BAOs. $\dashv$

Now, the boolean structure is obviously there to handle the propositional connectives, but what is the meaning of the normality and additivity conditions on the $f_\triangle$? Consider a unary operator f. In this case these conditions boil down to:

$$f(0) = 0,$$
$$f(x+y) = fx + fy.$$

But these equations correspond to the following modal formulas:

$$\Diamond\bot \leftrightarrow \bot,$$
$$\Diamond(p \vee q) \leftrightarrow \Diamond p \vee \Diamond q,$$

both of which formulas are modal validities. Indeed (as we noted in Remark 4.7) they can be even be used to axiomatize the minimal normal logic **K**. Thus, even at this stage, it should be clear that our algebraic *operators* are well named: their defining properties are modally crucial.

Furthermore, note that all operators have the property of *monotonicity*. An operation g on a boolean algebra is *monotonic* if $a \leq b$ implies $ga \leq gb$. (Here $\leq$ refers to the ordering on boolean algebra given in Definition 5.10: $a \leq b$ iff $a \cdot b = a$ iff $a + b = b$.) Operators are monotonic, because if $a \leq b$, then $a + b = b$, so $fa + fb = f(a + b) = fb$, and so $fa \leq fb$. Once again there is an obvious modal analog, namely the rule of proof mentioned in Remark 4.7: if $\vdash_\Lambda p \rightarrow q$ then $\vdash_\Lambda \Diamond p \rightarrow \Diamond q$.

Example 5.20 Consider the collection of binary relations over a given set U. This collection forms a set algebra on which we can define the operations $\mid$ (composition), $(\cdot)^{-1}$ (inverse) and Id (the identity relation); these are binary, unary and nullary operations respectively. It is easy to verify that these operations are actually operators; to give a taste of the kind of argumentation required, we show that composition is additive in its second argument:

$(x, y) \in R \mid (S \cup T)$
  iff   there is a z with $(x, z) \in R$ and $(z, y) \in S \cup T$
  iff   there is a z with $(x, z) \in R$ and $(z, y) \in S$ or $(z, y) \in T$
  iff   there is a z with $(x, z) \in R$ and $(z, y) \in S$,
         or there is a z with $(x, z) \in R$ and $(z, y) \in T$
  iff $(x, y) \in R \mid S$ or $(x, y) \in R \mid T$
  iff $(x, y) \in R \mid S \cup R \mid T$.

The reader should check the remaining cases.

Algebraizing modal semantics

However it is the next type of BAO that is destined to play the leading role: complex algebras. These structures make crucial use of the operations m_R that we met in Definition 1.30 and Definition 2.55.

Definition 5.21 (Complex Algebras) Let τ be a modal similarity type, and $\mathfrak{F} = (W, R_\triangle)_{\triangle \in \tau}$ a τ-frame. The *(full) complex algebra of* $\mathfrak{F}$ (notation: $\mathfrak{F}^+$), is the expansion of the power set algebra $\mathfrak{P}(W)$ with operations $m_{R_\triangle}$ for every operator $\triangle$ in τ. A *complex algebra* is a subalgebra of a full complex algebra. If K is a class of frames, then we denote the class of full complex algebras of frames in K by $\mathbf{Cm}$K.   ⊣

It is important that you fully understand this definition. For a start, note that complex algebras are set algebras (that is, concrete propositional algebras) to which m_R operations have been added. Recall that for a *binary* relation R, the unary operation m_R yields the set of all states which 'see' a state in a given subset X of the universe:

$$m_R(X) = \{y \in W \mid \text{ there is an } x \in X \text{ such that } Ryx\}.$$

For a relation of arity $n + 1$, the n-ary operation m_R maps an n-tuple of subsets of the universe to the set of all points which 'see' an n-tuple of states each of which belongs to the corresponding subset. It easily follows that if we have some model in mind and denote with $\tilde{V}(\phi)$ the set of states where ϕ is true, then

$$\tilde{V}(\triangle(\phi_1, \ldots, \phi_n)) = m_{R_\triangle}(\tilde{V}(\phi_1), \ldots, \tilde{V}(\phi_n)).$$

Thus it should be clear that complex algebras are intrinsically modal. In the previous section we said that set algebras model propositions as sets of possible worlds. By adding the m_R operations, we have modeled the idea that one world may be able to access the information in another. In short, we have defined a class of concrete algebras which capture the modal notion of access between states in a natural way.

How are complex algebras connected with abstract BAOs? One link is obvious:

Proposition 5.22 *Let τ be a modal similarity type, and $\mathfrak{F} = (W, R_\triangle)_{\triangle \in \tau}$ a τ-frame. Then $\mathfrak{F}^+$ is a boolean algebra with τ-operators.*

Proof. We have to show that operations of the form m_R are normal and additive. This rather easy proof is left to the reader; see Exercise 5.2.2.   ⊣

The other link is deeper. As we will learn in the following section (Theorem 5.43), complex algebras are to BAOs what set algebras are to boolean algebras:

every abstract boolean algebra with operators has a concrete set theoretic representation, for every boolean algebra with operators is isomorphic to a complex algebra.

But we have a lot to do before we are ready to prove this – let us continue our algebraization of the semantics of modal logic. We will now define the interpretation of τ-terms and equations in arbitrary boolean algebras with τ-operators. As we saw for propositional logic, the basic idea is very simple: given an assignment that tells us what the variables stand for, we can inductively define the meaning of any term.

Definition 5.23 Assume that τ is a modal similarity type and that Φ is a set of variables. Assume further that $\mathfrak{A} = (A, +, -, 0, f_\triangle)_{\triangle \in \tau}$ is a boolean algebra with τ-operators. An *assignment* for Φ is a function $\theta : \Phi \to A$. We can extend θ uniquely to a meaning function $\tilde{\theta} : Ter_\tau(\Phi) \to A$ satisfying:

$$\begin{aligned}
\tilde{\theta}(p) &= \theta(p), \text{ for all } p \in \Phi, \\
\tilde{\theta}(\bot) &= 0, \\
\tilde{\theta}(\neg s) &= -\tilde{\theta}(s), \\
\tilde{\theta}(s \vee t) &= \tilde{\theta}(s) + \tilde{\theta}(t), \\
\tilde{\theta}(\triangle(s_1, \ldots, s_n)) &= f_\triangle(\tilde{\theta}(s_1), \ldots, \tilde{\theta}(s_n)).
\end{aligned}$$

Now let $s \approx t$ be a τ-equation. We say that $s \approx t$ is *true* in $\mathfrak{A}$ (notation: $\mathfrak{A} \models s \approx t$) if for every assignment $\theta \colon \tilde{\theta}(s) = \tilde{\theta}(t)$.   ⊣

But now consider what happens when $\mathfrak{A}$ is a *complex* algebra $\mathfrak{F}^+$. Since elements of $\mathfrak{F}^+$ are *subsets* of the power set $\mathcal{P}(W)$ of the universe W of $\mathfrak{F}$, assignments θ are simply ordinary modal valuations! The ramifications of this observation are listed in the following proposition:

Proposition 5.24 *Let τ be a modal similarity type, ϕ a τ-formula, $\mathfrak{F}$ a τ-frame, θ an assignment (or valuation) and w a point in $\mathfrak{F}$. Then*

$$(\mathfrak{F}, \theta), w \Vdash \phi \quad \textit{iff} \quad w \in \tilde{\theta}(\phi), \tag{5.12}$$

$$\mathfrak{F} \Vdash \phi \quad \textit{iff} \quad \mathfrak{F}^+ \models \phi \approx \top, \tag{5.13}$$

$$\mathfrak{F}^+ \models \phi \approx \psi \quad \textit{iff} \quad \mathfrak{F} \Vdash \phi \leftrightarrow \psi. \tag{5.14}$$

Proof. We will only prove the first part of the proposition (for the basic modal similarity type); the second and third part follow immediately from this and the definitions.

Let ϕ, $\mathfrak{F}$ and θ be as in the statement of the theorem. We will prove (5.12) (for all w) by induction on the complexity of ϕ. The only interesting part is the modal case of the inductive step. Assume that ϕ is of the form $\Diamond\psi$. The key observation is that

$$\tilde{\theta}(\Diamond\psi) = m_{R_\Diamond}(\tilde{\theta}(\psi)). \tag{5.15}$$

We now have:

$$
\begin{array}{lll}
(\mathfrak{F},\theta), w \Vdash \Diamond\psi & \text{iff} & \text{there is a } v \text{ such that } R_\Diamond wv \text{ and } (\mathfrak{F},\theta), v \Vdash \psi \\
& \text{iff} & \text{there is a } v \text{ such that } R_\Diamond wv \text{ and } v \in \tilde{\theta}(\psi) \\
& \text{iff} & w \in m_{R_\Diamond}(\tilde{\theta}(\psi)) \\
& \text{iff} & w \in \tilde{\theta}(\Diamond\psi).
\end{array}
$$

Here the second equivalence is by the inductive hypothesis, and the last one by (5.15). This proves (5.12). ⊣

The previous proposition is easily lifted to the level of classes of frames and complex algebras. The resulting theorem is a fundamental one: it tells us that classes of complex algebras algebraize modal semantics. It is the modal analog of Theorem 5.9.

Theorem 5.25 *Let τ be a modal similarity type, ϕ and ψ τ-formulas, and K a class of τ-frames. Then*

$$\mathsf{K} \Vdash \phi \quad \textit{iff} \quad \mathbf{Cm}\mathsf{K} \models \phi \approx \top, \tag{5.16}$$

$$\mathbf{Cm}\mathsf{K} \models \phi \approx \psi \quad \textit{iff} \quad \mathsf{K} \Vdash \phi \leftrightarrow \psi. \tag{5.17}$$

Proof. Immediate by Proposition 5.24. ⊣

This proposition allows us to identify the *modal logic* Λ_K of a class of frames K (that is, the set of formulas that are valid in each $\mathfrak{F} \in \mathsf{K}$) with the *equational theory* of the class $\mathbf{Cm}\mathsf{K}$ of complex algebras of frames in K (that is, the set of equations $\{s \approx t \mid \mathfrak{F}^+ \models s \approx t, \text{ for all } \mathfrak{F} \in \mathsf{K}\}$).

Let us summarize what we have learned so far. We have developed an algebraic approach to the semantics of modal logic in terms of complex algebras. These complex algebras, *concrete* boolean algebras with operators, generalize to modal languages the idea of algebras of propositions provided by set algebras. And most important of all, we have learned that complex algebras embody *all* the information about normal modal logics that frames do. Thus, mathematically speaking, we can dispense with frames and instead work with complex algebras.

Algebraizing modal axiomatics

Turning to the algebraization of modal axiomatics, we encounter a situation similar to that of the previous section. Once again, we will see that the algebraic counterpart of a logic is an equational class of algebras. To give a precise formulation we need the following definition.

Definition 5.26 Given a formula ϕ, let $\phi^{\approx}$ be the equation $\phi \approx \top$. Now let τ be a modal similarity type. For a set Σ of τ-formulas, we define V_Σ to be the class of those boolean algebras with τ-operators in which the set $\Sigma^{\approx} = \{\sigma^{\approx} \mid \sigma \in \Sigma\}$ is valid. $\dashv$

We now state the algebraic completeness theorem for modal logic. It is the obvious analog of Theorem 5.11.

Theorem 5.27 (Algebraic Completeness) *Let τ be a modal similarity type, and Σ a set of τ-formulas. Then $\mathbf{K}_\tau \Sigma$ (the normal modal τ-logic axiomatized by Σ) is sound and complete with respect to V_Σ. That is, for all formulas ϕ we have*

$$\vdash_{\mathbf{K}_\tau \Sigma} \phi \ \text{ iff } \ V_\Sigma \models \phi^{\approx}.$$

Proof. We leave the soundness direction as an exercise to the reader. Completeness is an immediate corollary of Theorems 5.32 and 5.33. $\dashv$

As a corollary to the soundness direction of Theorem 5.27, we have that $V_{\mathbf{K}_\tau \Sigma} = V_\Sigma$, for any set Σ of formulas. In the sequel this will allow us to forget about axiom sets and work with logics instead.

   To prove the completeness direction of Theorem 5.27, we need a modal version of the basic tool used to prove algebraic completeness results: Lindenbaum-Tarski algebras. As in the the case of propositional languages, we will build an algebra on top of the set of formulas in such a way that the relation of provable equivalence between two formulas is a congruence relation. The key difference is that we do not have just one relation of provable equivalence, but many: we want to define the notion of Lindenbaum-Tarski algebras for arbitrary normal modal logics.

Definition 5.28 Let τ be an algebraic similarity type, and Φ a set of proposition letters. The *formula algebra* of τ over Φ is the algebra $\mathfrak{Form}(\tau, \Phi) = (Form(\tau, \Phi), +, -, \bot, f_\triangle)_{\triangle \in \tau}$ where $+, -$ and $\bot$ are given as in Definition 5.3, while for each modal operator $\triangle$, the operation $f_\triangle$ is given by

$$f_\triangle(t_1, \dots, t_n) = \triangle(t_1, \dots, t_n).$$ $\dashv$

Notice the double role of $\triangle$ in this definition: on the right-hand side of the equation, $\triangle$ is a 'static' part of the *term* $\triangle(t_1, \dots, t_n)$, whereas in the-left hand side we have a more 'dynamic' perspective on the *interpretation* $f_\triangle$ of the operation symbol $\triangle$.

Definition 5.29 Let τ be a modal similarity type, Φ a set of propositional variables, and Λ a normal modal τ-logic. We define $\equiv_\Lambda$ as a binary relation between τ-formulas (in Φ) by

$$\phi \equiv_\Lambda \psi \ \text{ iff } \ \vdash_\Lambda \phi \leftrightarrow \psi.$$

If $\phi \equiv_\Lambda \psi$, we say that ϕ and ψ are *equivalent modulo Λ*. $\dashv$

Proposition 5.30 *Let τ be a modal similarity type, Φ a set of proposition letters and Λ a normal modal τ-logic. Then $\equiv_\Lambda$ is a congruence relation on $\mathfrak{Form}(\tau, \Phi)$.*

Proof. We confine ourselves to proving the proposition for the basic modal similarity type. First, we have to show that $\equiv_\Lambda$ is an equivalence relation; this is easy, and we leave the details to the reader. Next, we must show that $\equiv_\Lambda$ is a *congruence* relation on the formula algebra; that is, we have to demonstrate that $\equiv_\Lambda$ has the following properties:

$$\begin{aligned}
\phi_0 \equiv_\Lambda \psi_0 \text{ and } \phi_1 \equiv_\Lambda \psi_1 &\quad \text{imply} \quad \phi_0 \vee \phi_1 \equiv_\Lambda \phi_0 \vee \psi_1, \\
\phi \equiv_\Lambda \psi &\quad \text{implies} \quad \neg\phi \equiv_\Lambda \neg\psi, \\
\phi \equiv_\Lambda \psi &\quad \text{implies} \quad \Diamond\phi \equiv_\Lambda \Diamond\psi.
\end{aligned} \tag{5.18}$$

The first two properties are easy exercises in propositional logic. The third is an immediate corollary of Lemma 4.6. $\dashv$

Proposition 5.30 tells us that the following are correct definitions of functions on the set $Form(\tau, \Phi)/\equiv_\Lambda$ of equivalence classes under $\equiv_\Lambda$:

$$\begin{aligned}
[\phi] + [\psi] &:= [\phi \vee \psi], \\
-[\phi] &:= [\neg\phi], \\
f_\Delta([\phi_1], \ldots, [\phi_n]) &:= [\Delta(\phi_1, \ldots, \phi_n)].
\end{aligned} \tag{5.19}$$

For unary diamonds, the last clause boils down to: $f_\Diamond[\phi] := [\Diamond\phi]$.

Given Proposition 5.30, the way is open to define the Lindenbaum-Tarski algebra for any normal modal logic Λ: we simply define it to be the *quotient algebra* of the formula algebra over the congruence relation $\equiv_\Lambda$.

Definition 5.31 (Lindenbaum-Tarski Algebras) Let τ be a modal similarity type, Φ a set of proposition letters, and Λ a normal modal τ-logic in this language. The *Lindenbaum-Tarski algebra of Λ over the set of generators Φ* is the structure

$$\mathfrak{L}_\Lambda(\Phi) := (Form(\tau, \Phi)/\equiv_\Lambda, +, -, f_\Delta),$$

where the operations $+, -$ and f_Δ are defined as in (5.19). $\dashv$

As with propositional logic, we need two results about Lindenbaum-Tarski algebras. First, we must show that modal Lindenbaum-Tarski algebras are boolean algebras with operators; indeed, we need to show that the Lindenbaum-Tarski algebra of any normal modal logic Λ belongs to V_Λ. Second, we need to prove that Lindenbaum-Tarski algebras provide canonical counterexamples to the validity of non-theorems of Λ in V_Λ. The second point is easily dealt with:

Theorem 5.32 *Let τ be a modal similarity type, and Λ a normal modal τ-logic.*

Let ϕ be some propositional formula, and Φ a set of proposition letters of size not smaller than the number of proposition letters occurring in ϕ. Then

$$\vdash_\Lambda \phi \ \text{iff} \ \mathfrak{L}_\Lambda(\Phi) \models \phi^\approx. \qquad (5.20)$$

Proof. This proof is completely analogous to that of Proposition 5.14 and is left to the reader.   ⊣

So let us verify that Lindenbaum-Tarski algebras are canonical algebraic models of the right kind:

Theorem 5.33 *Let τ be a modal similarity type, and Λ be a normal modal τ-logic. Then for any set Φ of proposition letters, $\mathfrak{L}_\Lambda(\Phi)$ belongs to V_Λ.*

Proof. Once we have shown that $\mathfrak{L}_\Lambda(\Phi)$ is a boolean algebra with τ-operators, the theorem immediately follows from Theorem 5.32. Now, that $\mathfrak{L}_\Lambda(\Phi)$ is a boolean algebra is clear, so the only thing that remains to be done is to show that the modalities really give rise to τ-operators.

As an example, assume that τ contains a diamond $\Diamond$; let us prove additivity of $f_\Diamond$. We have to show that

$$f_\Diamond(a + b) = f_\Diamond a + f_\Diamond b,$$

for arbitrary elements a and b of $\mathfrak{L}_\Lambda(\Phi)$. Let a and b be such elements; by definition there are formulas ϕ and ψ such that $a = [\phi]$ and $b = [\psi]$. Then

$$f_\Diamond(a + b) = f_\Diamond([\phi] + [\psi]) = f_\Diamond([\phi \vee \psi]) = [\Diamond(\phi \vee \psi)]$$

while

$$f_\Diamond a + f_\Diamond b = f_\Diamond([\phi]) + f_\Diamond([\psi]) = [\Diamond \phi] + [\Diamond \psi] = [\Diamond \phi \vee \Diamond \psi].$$

It is easy to check that

$$\vdash_\Lambda \Diamond(\phi \vee \psi) \leftrightarrow (\Diamond \phi \vee \Diamond \psi),$$

whence it follows that $[\Diamond(\phi \vee \psi)] = [\Diamond \phi \vee \Diamond \psi]$. We leave it for the reader to fill in the remaining details of this proof as Exercise 5.2.4.   ⊣

As an immediate corollary we have the following result: modal logics are always complete with respect to the variety of boolean algebras with operators where their axioms are valid. This is in sharp contrast to the situation in relational semantics, where (as we saw in Chapter 4) modal logics need *not* be complete with respect to the class of frames that they define.

This is an interesting result, but it is not what we really want, for it proves completeness with respect to abstract BAOs rather than complex algebras. Not only are complex algebras concrete algebras of propositions, we also know (recall Proposition 5.24) that complex algebras embody all the information of relevance to frame validity – so we really should be aiming for completeness results with respect to classes of complex algebras.

And that is why the long-promised Jónsson-Tarski Theorem, which we state and prove in the following section, is so important. This tells us that *every* boolean algebra with operators is isomorphic to a complex algebra, and thus guarantees that we can represent the Lindenbaum-Tarski algebras of any normal modal logics Λ as a complex algebra. In effect, it will convert Theorem 5.32 into a completeness result with respect to complex algebras. Moreover, because of the link between complex algebras and relational semantics, it will open the door to exploring frame completeness algebraically.

Exercises for Section 5.2

5.2.1 Let $\mathfrak{A}$ be a boolean algebra. Prove that $\cdot$ is an operator. How about $+$?

5.2.2 Show that every complex algebra is a boolean algebra with operators (that is, prove Proposition 5.22).

5.2.3 Let A be the collection of finite and co-finite subsets of $\mathbb{N}$. Define $f : A \to A$ by

$$f(X) = \begin{cases} \{y \in \mathbb{N} \mid y + 1 \in X\} & \text{if } X \text{ is finite,} \\ \mathbb{N} & \text{if } X \text{ is co-finite.} \end{cases}$$

Prove that $(A, \cup, -, \varnothing, f)$ is a boolean algebra with operators.

5.2.4 Let Λ be a normal modal logic. Prove that the Lindenbaum-Tarski algebra $\mathfrak{L}_\Lambda$ is a boolean algebra with τ-operators (that is, fill in the missing proof details in Theorem 5.33).

5.2.5 Let Σ be a set of τ-formulas. Prove that for any formula ϕ, $\vdash_{\mathbf{K}_\tau \Sigma} \phi$ implies $\mathsf{V}_\Sigma \models \phi^{\approx}$. That is, prove the soundness direction of Theorem 5.27.

5.2.6 Call a variety V of BAOs *complete* if it is generated by a class of full complex algebras, i.e., if $\mathsf{V} = \mathbf{HSPCm}\mathsf{K}$ for some frame class K. Prove that a logic Λ is complete iff the variety V_Λ is complete.

5.2.7 Let $\mathfrak{A}$ be a boolean algebra. In this exercise we assume familiarity with the notion of an infinite sum (supremum). An operation $f : A \to A$ is called *completely additive* if it distributes over infinite sums (in each of its arguments).

(a) Show that every operation of the form m_R is completely additive.
(b) Give an example of an operation that is additive, but not completely additive. (Hint: as the boolean algebra, take the set of finite and co-finite subsets of some frame.)

5.3 The Jónsson-Tarski Theorem

We already know how to construct a BAO from a frame: simply form the frame's complex algebra. We will now learn how to construct a frame from a BAO by forming the *ultrafilter frame* of the algebra. As we will see, this operation generalizes two constructions that we have met before: taking the ultrafilter extension of a model, and forming the canonical frame associated with a normal modal logic.

Our new construction will lead us to the desired representation theorem: by taking the complex algebra of the ultrafilter frame of a BAO, we obtain the *canonical embedding algebra* of the original BAO. The fundamental result of this section (and, indeed, of the entire chapter) is that *every boolean algebra with operators can be isomorphically embedded in its canonical embedding algebra*. We will prove this result and along the way discuss a number of other important issues, such as the algebraic status of canonical models and ultrafilter extensions, and the importance of canonical varieties of BAOs for modal completeness theory.

Let us consider the problem of (isomorphically) embedding an arbitrary BAO $\mathfrak{A}$ in a complex algebra. Obviously, the first question to ask is: what should be the underlying *frame* of the complex algebra? To keep our notation simple, let us assume for the moment that we are working in a similarity type with just one unary modality, and that $\mathfrak{A} = (A, +, -, 0, f)$ is a boolean algebra with one unary operator f. Thus we have to find a universe W and a binary relation R on W such that $\mathfrak{A}$ can be embedded in the complex algebra of the frame (W, R). Stone's Representation Theorem 5.16 gives us half the answer, for it tells us how to embed the boolean part of $\mathfrak{A}$ in the power set algebra of the set $Uf\mathfrak{A}$ of *ultrafilters* of $\mathfrak{A}$. Let us take a closer look at this fundamental result.

Stone's Representation Theorem

We have already met filters and ultrafilters in Chapter 2, when we defined the ultrafilter extension of a model. Now we generalize these notions to the context of *abstract* boolean algebras.

Definition 5.34 A *filter* of a boolean algebra $\mathfrak{A} = (A, +, -, 0)$ is a subset $F \subseteq A$ satisfying

(F1)　$1 \in F$,
(F2)　F is closed under taking meets; that is, if $a, b \in F$ then $a \cdot b \in F$,
(F3)　F is upward closed; that is, if $a \in F$ and $a \leq b$ then $b \in F$.

A filter is *proper* if it does not contain the smallest element 0, or, equivalently, if $F \neq A$. An *ultrafilter* is a proper filter satisfying

(F4)　For every $a \in A$, either a or $-a$ belongs to F.

The collection of ultrafilters of $\mathfrak{A}$ is called $Uf\mathfrak{A}$.　　　　　　　　　　⊣

Note the difference in terminology: an (ultra)filter *over* the set W is an (ultra)filter *of* the power set algebra $\mathfrak{P}(W)$.

Example 5.35 For any element a of a boolean algebra $\mathfrak{A}$, the set $a{\uparrow} = \{b \in A \mid a \leq b\}$ is a filter. In the field of finite and co-finite subsets of a countable set W, the collection of co-finite subsets of W forms an ultrafilter. ⊣

Example 5.36 Since the collection of filters of a boolean algebra is closed under taking intersections, we may speak of the *smallest* filter F_D containing a given set $D \subseteq A$. This filter can also be defined as the following set:

$$\{a \in A \mid \text{ there are } d_0, \ldots, d_n \in D \text{ such that } d_0 \cdot \ldots \cdot d_n \leq a\} \qquad (5.21)$$

which explains why we will also refer to F_D as the filter *generated by* D. This filter is proper if D has the so-called *finite meet property*; that is, if there is no finite subset $\{d_0, \ldots, d_n\}$ of D such that $d_0 \cdot \ldots \cdot d_n = 0$. ⊣

For future reference, we gather some properties of ultrafilters; the proof of the next proposition is left to the reader.

Proposition 5.37 *Let $\mathfrak{A} = (A, +, -, 0)$ be a boolean algebra. Then*

(i) *For any ultrafilter u of $\mathfrak{A}$ and for every pair of elements $a, b \in A$ we have that $a + b \in u$ iff $a \in u$ or $b \in u$.*

(ii) *$Uf\mathfrak{A}$ coincides with the set of maximal proper filters on $\mathfrak{A}$ ('maximal' is understood with respect to set inclusion).*

The main result that we need in the proof of Stone's Theorem is the Ultrafilter Theorem: this guarantees that there are enough ultrafilters for our purposes.

Proposition 5.38 (Ultrafilter Theorem) *Let $\mathfrak{A}$ be a boolean algebra, a an element of A, and F a proper filter of $\mathfrak{A}$ that does not contain a. Then there is an ultrafilter extending F that does not contain a.*

Proof. We first prove that every proper filter can be extended to an ultrafilter. Let G be a proper filter of $\mathfrak{A}$, and consider the set X of all proper filters H extending G. Suppose that Y is a *chain* in X; that is, Y is a nonempty subset of X of which the elements are pairwise ordered by set inclusion. We leave it to the reader to verify that $\bigcup Y$ is a proper filter; obviously, $\bigcup Y$ extends G; so $\bigcup Y$ belongs to X itself. This shows that X is closed under taking unions of chains, whence it follows from Zorn's Lemma that X contains a maximal element u. We claim that u is an ultrafilter.

For suppose otherwise. Then there is a $b \in A$ such that neither b nor $-b$ belongs to u. Consider the filters H and H' generated by $u \cup \{b\}$ and $u \cup \{-b\}$, respectively. Since neither of these can belong to X, both must be improper; that is, $0 \in H$ and

$0 \in H'$. But then by definition there are elements $u_1, \ldots, u_n, u'_1, \ldots, u'_m$ in u such that

$$u_1 \cdot \ldots \cdot u_n \cdot b \leq 0 \text{ and } u'_1 \cdot \ldots \cdot u'_m \cdot -b \leq 0.$$

From this it easily follows that

$$u_1 \cdot \ldots \cdot u_n \cdot u'_1 \cdot \ldots \cdot u'_m = 0,$$

contradicting the fact that u is a proper filter.

Now suppose that a and F are as in the statement of the proposition. It is not hard to show that $F \cup \{-a\}$ is a set with the finite meet property. In Example 5.36 we saw that there is a proper filter G extending F and containing $-a$. Now we use the first part of the proof to find an ultrafilter u extending G. But if u extends G it also extends F, and if it contains $-a$ it cannot contain a.   ⊣

It follows from Proposition 5.38 and the facts mentioned in Example 5.36 that any subset of a boolean algebra can be extended to an ultrafilter provided that it has the finite meet property. We now have all the necessary material to prove Stone's Theorem.

Theorem 5.16 (Stone Representation Theorem) *Any boolean algebra is isomorphic to a field of sets, and hence, to a subalgebra of a power of* **2**. *As a consequence, the variety of boolean algebras is generated by the algebra* **2**:

$$\mathsf{BA} = \mathbb{V}(\{\mathbf{2}\}).$$

Proof. Fix a boolean algebra $\mathfrak{A} = (A, +, -, 0)$. We will embed $\mathfrak{A}$ in the power set of $Uf\mathfrak{A}$. Consider the map $\rho : A \to \mathcal{P}(Uf\mathfrak{A})$ defined as follows:

$$\rho(a) = \{u \in Uf\mathfrak{A} \mid a \in u\}.$$

We first show that ρ is a homomorphism. As an example we treat the join operation:

$$
\begin{aligned}
\rho(a+b) &= \{u \in Uf\mathfrak{A} \mid a+b \in u\} \\
&= \{u \in Uf\mathfrak{A} \mid a \in u \text{ or } b \in u\} \\
&= \{u \in Uf\mathfrak{A} \mid a \in u\} \cup \{u \in Uf\mathfrak{A} \mid b \in u\} \\
&= \rho(a) \cup \rho(b).
\end{aligned}
$$

Note that the crucial second equality follows from Proposition 5.37.

It remains to prove that ρ is injective. Suppose that a and b are distinct elements of A. We may derive from this that either $a \not\leq b$ or $b \not\leq a$. Without loss of generality we may assume the second. But if $b \not\leq a$ then a does not belong to the filter $b{\uparrow}$ generated by $\{b\}$, so by Proposition 5.38 there is some ultrafilter u such that $b{\uparrow} \subseteq u$ and $a \notin u$. Obviously, $b{\uparrow} \subseteq u$ implies that $b \in u$. But then we have that $u \in \rho(b)$ and $u \notin \rho(a)$.

This shows that $\mathfrak{A}$ is isomorphic to a field of sets; it then follows by Proposition 5.8 that $\mathfrak{A}$ is isomorphic to a subalgebra of a power of **2**. From this it is immediate that BA is the variety generated by the algebra **2**. ⊣

Remark 5.39 That every boolean algebra is isomorphic to a subalgebra of a power of the algebra **2** can be proved more directly by observing that there is a one-to-one correspondence between ultrafilters of $\mathfrak{A}$ and homomorphisms from $\mathfrak{A}$ onto **2**. Given an ultrafilter u of $\mathfrak{A}$, define $\alpha_u : \mathfrak{A} \to \mathbf{2}$ by

$$\alpha_u(a) = \begin{cases} 1 & \text{if } a \in u, \\ 0 & \text{otherwise.} \end{cases}$$

And conversely, given a homomorphism $\alpha : \mathfrak{A} \to \mathbf{2}$, define the ultrafilter u_α by

$$u_\alpha = \alpha^{-1}(1) \ (= \{a \in \mathfrak{A} \mid \alpha(a) = 1\}).$$

We leave further details to the reader. ⊣

Ultrafilter frames

Now that we have a candidate for the universe of the ultrafilter frame of a given BAO $\mathfrak{A}$, let us see how to define a relation R on ultrafilters such that we can embed $\mathfrak{A}$ in the algebra $(Uf\mathfrak{A}, R)^+$. To motivate the definition of R, we will view the elements of the algebra as *propositions*, and imagine that $r(a)$ (the representation map r applied to proposition a) yields the set of states where a is *true* according to some valuation. Hence, reading fa as $\Diamond a$, it seems natural that a state u should be in $r(fa)$ if and only if there is a v with Ruv and $v \in r(a)$. So, in order to decide whether Ruv should hold for two arbitrary states (ultrafilters) u and v, we should look at all the propositions a holding at v (that is, all elements $a \in v$) and check whether fa holds at u (that is, whether $fa \in u$). Putting it more formally, the natural, 'canonical' choice for R seems to be the relation Q_f given by

$$Q_f uv \text{ iff } fa \in u \text{ for all } a \in v.$$

The reader should compare this definition with the definition of the canonical relation given in Definition 4.18. Although one is couched in terms of ultrafilters, and the other in terms of maximal consistent sets (MCSs), both clearly trade on the same idea. As we will shortly learn (and as the above identification of 'ultrafilters' and 'maximal sets of propositions' already suggests), this is no accident.

In the general case, we use the following definition (an obvious analog of Definition 4.24).

Definition 5.40 Given an n-ary operator f on a boolean algebra $(A, +, -, 0)$, we define the $(n + 1)$-ary relation Q_f on the set of ultrafilters of the algebra by

$$Q_f uu_1 \ldots u_n \text{ iff } f(a_1, \ldots, a_n) \in u \ \text{ for all } a_1 \in u_1, \ldots, a_n \in u_n.$$

Let $\mathfrak{A} = (A, +, -, 0, f_\Delta)_{\Delta \in \tau}$ be a boolean algebra with operators. The *ultrafilter frame* of $\mathfrak{A}$, notation: $\mathfrak{A}_+$, is the structure $(Uf\mathfrak{A}, Q_{f_\Delta})_{\Delta \in \tau}$. The complex algebra $(\mathfrak{A}_+)^+$ is called the *(canonical) embedding algebra* of $\mathfrak{A}$ (notation: $\mathfrak{Em}\mathfrak{A}$).   ⊣

We leave it to the reader to verify that the *ultrafilter extension* $\mathfrak{ue}\,\mathfrak{F}$ of a frame $\mathfrak{F}$ is nothing but the ultrafilter frame of the complex algebra of $\mathfrak{F}$, in symbols: $\mathfrak{ue}\,\mathfrak{F} = (\mathfrak{F}^+)_+$.

For later reference, we state the following proposition (an obvious analog of Lemma 4.25) which shows that we could have given an alternative but equivalent definition of the relation Q_f.

Proposition 5.41 *Let f be an n-ary operator on the boolean algebra $\mathfrak{A}$, and u, $u_1, \ldots, u_n$ an $(n+1)$-tuple of ultrafilters of $\mathfrak{A}$. Then*

$$Q_f u u_1 \ldots u_n \text{ iff } -f(-a_1, \ldots, -a_n) \in u \text{ implies that for some } i, a_i \in u_i.$$

Proof. We only prove the direction from left to right. Suppose that $Q_f u u_1 \ldots u_n$, and that $-f(-a_1, \ldots, -a_n) \in u$. To arrive at a contradiction, suppose that there is no i such that $a_i \in u_i$. But as $Q_f u u_1 \ldots u_n$, it follows that $f(-a_1, \ldots, -a_n) \in u$. But this contradicts the fact that $-f(-a_1, \ldots, -a_n) \in u$.   ⊣

As the above sequence of analogous definitions and results suggest, we have already encountered a kind of frame which is very much like an ultrafilter frame, namely the *canonical frame* of a normal modal logic (see Definition 4.18). The basic idea should be clear now: the states of the canonical frame are the MCSs of the logic, and an ultrafilter is nothing but an abstract version of an MCS. But this is no mere analogy: the canonical frame of a logic is actually *isomorphic* to the ultrafilter frame of its Lindenbaum-Tarski algebra, and the mapping involved is simple and intuitive. When making this connection, the reader should keep in mind that when we defined 'the' canonical frame in Chapter 4, we always had a fixed, countable set Φ of proposition letters in mind.

Theorem 5.42 *Let τ be a modal similarity type, Λ a normal modal τ-logic, and Φ the set of proposition letters used to define the canonical frame $\mathfrak{F}^\Lambda$. Then*

$$\mathfrak{F}^\Lambda \cong (\mathfrak{L}_\Lambda(\Phi))_+.$$

Proof. We leave it to the reader to show that the function θ defined by

$$\theta(\Gamma) = \{[\phi] \mid \phi \subset \Gamma\},$$

mapping a maximal Λ-consistent set Γ to the set of equivalence classes of its members, is the required isomorphism between $\mathfrak{F}^\Lambda$ and $(\mathfrak{L}_\Lambda(\Phi))_+$.   ⊣

The Jónsson-Tarski Theorem

We are ready to prove the Jónsson-Tarski Theorem: every boolean algebra with operators is embeddable in the full complex algebra of its ultrafilter frame.

Theorem 5.43 (Jónsson-Tarski Theorem) *Let τ be a modal similarity type, and $\mathfrak{A} = (A, +, -, 0, f_\Delta)_{\Delta \in \tau}$ be a boolean algebra with τ-operators. Then the representation function $r : A \to \mathcal{P}(Uf\mathfrak{A})$ given by*

$$r(a) = \{u \in Uf\mathfrak{A} \mid a \in u\}$$

is an embedding of $\mathfrak{A}$ into $\mathfrak{Em}\mathfrak{A}$.

Proof. To simplify our notation a bit, we work in a similarity type with a single n-ary modal operator, assuming that $\mathfrak{A} = (A, +, -, 0, f)$ is a boolean algebra with a single n-ary operator f. By Stone's Representation Theorem, the map $r : A \to \mathcal{P}(Uf\mathfrak{A})$ given by

$$r(x) = \{u \in Uf\mathfrak{A} \mid x \in u\}$$

is a boolean embedding. So, it suffices to show that r is also a *modal* homomorphism; that is, that

$$r(f(a_1, \ldots, a_n)) = m_{Q_f}(r(a_1), \ldots, r(a_n)). \tag{5.22}$$

We will first prove (5.22) for unary f. In other words, we have to prove that

$$r(fa) = m_{Q_f}(r(a)).$$

We start with the inclusion from right to left: assume $u \in m_{Q_f}(r(a))$. Then by definition of m_{Q_f}, there is an ultrafilter u_1 with $u_1 \in r(a)$ (that is, $a \in u_1$) and $Q_f u u_1$. By definition of Q_f this implies $fa \in u$, or $u \in r(fa)$.

For the other inclusion, let u be an ultrafilter in $r(fa)$, that is, $fa \in u$. To prove that $u \in m_{Q_f}(r(a))$, it suffices to find an ultrafilter u_1 such that $Q_f u u_1$ and $u_1 \in r(a)$, or $a \in u_1$. The basic idea of the proof is that we first pick out those elements of A (other than a) that we cannot avoid putting in u_1. These elements are given by the condition $Q_f u u_1$. By Proposition 5.41 we have that for every element of the form $-f(-y)$ in u, y has to be in u_1; therefore, we define

$$F := \{y \in A \mid -f(-y) \in u\}.$$

We will now show that there is an ultrafilter $u_1 \supseteq F$ containing a. First, an easy proof (using the additivity of f), shows that F is closed under taking meets. Second, we prove that

$$F' := \{a \cdot y \mid y \in F\}$$

has the finite meet property. As F is closed under taking meets, it is sufficient to show that $a \cdot y \neq 0$ whenever $y \in F$. To arrive at a contradiction, suppose that

$a \cdot y = 0$. Then $a \leq -y$, so by the monotonicity of f, $fa \leq f(-y)$; therefore, $f(-y) \in u$, contradicting $y \in F$.

By Proposition 5.38 there is an ultrafilter $u_1 \supseteq F'$. Note that $a \in u_1$, as $1 \in F$. Finally, $Q_f u u_1$ holds by definition of F: if $-f(-y) \in u$ then $y \in F \subseteq u_1$.

We now prove (5.22) for arbitrary $n \geq 1$ by induction on the arity n of f. We have just proved the base case. So, assume that the induction hypothesis holds for n. We only treat the direction from left to right, since the other direction can be proved as in the base case. Let f be a normal and additive function of rank $n + 1$, and suppose that $a_1, \ldots, a_{n+1}$ are elements of $\mathfrak{A}$ such that $f(a_1, \ldots, a_{n+1}) \in u$. We have to find ultrafilters $u_1, \ldots, u_{n+1}$ of $\mathfrak{A}$ such that (i) $a_i \in u_i$ for all i with $1 \leq i \leq n + 1$, and (ii) $Q_f u u_1 \ldots u_{n+1}$. Our strategy will be to let the induction hypothesis take care of $u_1, \ldots, u_n$ and then to search for u_{n+1}.

Let $f' : A^n \to A$ be the function given by

$$f'(x_1, \ldots, x_n) = f(x_1, \ldots, x_n, a_{n+1}).$$

That is, for the time being we fix a_{n+1}. It is easy to see that f' is normal and additive, so we may apply the induction hypothesis. Since $f'(a_1, \ldots, a_n) \in u$, this yields ultrafilters $u_1, \ldots, u_n$ such that $a_i \in u_i$ for all i with $1 \leq i \leq n$, and

$$f(x_1, \ldots, x_n, a_{n+1}) \in u, \text{ whenever } x_i \in u_i \ (1 \leq i \leq n). \qquad (5.23)$$

Now we will define an ultrafilter u_{n+1} such that $a_{n+1} \in u_{n+1}$ and $Q_f u u_1 \ldots u_{n+1}$. This second condition can be rewritten as follows (we abbreviate '$x_1 \in u_1, \ldots,$ $x_n \in u_n$' by '$\vec{x} \in \vec{u}$'):

$Q_f u u_1 \ldots u_{n+1}$
- iff   for all $\vec{x}, y$: if $\vec{x} \in \vec{u}$, then $y \in u_{n+1}$ implies $f(\vec{x}, y) \in u$
- iff   for all $\vec{x}, y$: if $\vec{x} \in \vec{u}$, then $f(\vec{x}, y) \notin u$ implies $y \notin u_{n+1}$
- iff   for all $\vec{x}, y$: if $\vec{x} \in \vec{u}$, then $-f(\vec{x}, y) \in u$ implies $-y \in u_{n+1}$
- iff   for all $\vec{x}, z$: if $\vec{x} \in \vec{u}$, then $-f(\vec{x}, -z) \in u$ implies $z \in u_{n+1}$.

This provides us with a minimal set of elements that u_{n+1} should contain; put

$$F := \{z \in A \mid \exists \vec{x} \in \vec{u} \, (-f(\vec{x}, -z) \in u)\}.$$

If $-f(\vec{x}, -z) \in u$, we say that $\vec{x}$ *drives* z *into* F. We now take the first condition into account as well, defining $F' := \{a_{n+1}\} \cup F$.

Our aim is to prove the existence of an ultrafilter u_{n+1} containing F'. It will be clear that this is sufficient to prove the theorem (note that $a_{n+1} \in F'$ as $1 \in F$). To be able to apply the Ultrafilter Theorem 5.38, we will show that F' has the finite meet property. We first need the following fact:

$$F \text{ is closed under taking meets.} \qquad (5.24)$$

Let z', z'' be in F; assume that z' and z'' are driven into F by $\vec{x}'$ and $\vec{x}''$, respectively. We will now see that $\vec{x} := (x_1' \cdot x_1'', \ldots, x_n' \cdot x_n'')$ drives $z := z' \cdot z''$ into F, that is, that $-f(\vec{x}, -z) \in u$.

Since f is monotonic, we have $f(\vec{x}, -z') \leq f(\vec{x}', -z')$, and hence we find that $-f(\vec{x}', -z') \leq -f(\vec{x}, -z')$. As u is upward closed and $-f(\vec{x}', -z') \in u$ by our 'driving assumption', this gives $-f(\vec{x}, -z') \in u$. In the same way we find $-f(\vec{x}, -z'') \in u$. Now

$$f(\vec{x}, -z) = f(\vec{x}, -(z' \cdot z'')) = f(\vec{x}, (-z') + (-z'')) = f(\vec{x}, -z') + f(\vec{x}, -z''),$$

whence

$$-f(\vec{x}, -z) = [-f(\vec{x}, -z')] \cdot [-f(\vec{x}, -z'')].$$

Therefore, $-f(\vec{x}, -z) \in u$, since u is closed under taking meets. This proves (5.24).

We can now finish the proof and show that indeed

$$F' \text{ has the finite meet property.} \tag{5.25}$$

By (5.24) it suffices to show that $a_{n+1} \cdot z \neq 0$ for all $z \in F$. To prove this, we reason by contraposition: suppose that $z \in F$ and $a_{n+1} \cdot z = 0$. Let $\vec{x} \in \vec{u}$ be a sequence that drives z into F, that is, $-f(\vec{x}, -z) \in u$. From $a_{n+1} \cdot z = 0$ it follows that $a_{n+1} \leq -z$, so by monotonicity of f we get $-f(\vec{x}, -z) \leq -f(\vec{x}, a_{n+1})$. But then $-f(\vec{x}, a_{n+1}) \in u$, which contradicts (5.23). This proves that indeed $a_{n+1} \cdot z \neq 0$ and hence we have shown (5.25) and thus, Theorem 5.43. ⊣

Canonicity: the algebraic perspective

To conclude this section, let us discuss the significance of this result. Clearly the Jónsson-Tarski Theorem guarantees that we can represent the Lindenbaum-Tarski algebras of normal modal logics as complex algebras, so it immediately converts Theorem 5.32 into a completeness result with respect to complex algebras.

But we want more: because of the link between complex algebras and relational semantics, it seems to offer a plausible algebraic handle on frame completeness. And in fact it does – but we need to be careful. As should be clear from our work in Chapter 4, even with the Jónsson-Tarski Theorem at our disposal, one more hurdle remains to be cleared. In Exercise 5.2.6 we defined the notion of a *complete* variety of BAOs: a variety V is complete if there is a frame class K that generates V in the sense that V = **HSPCmK**. The exercise asked the reader to show that any logic Λ is complete if and only if V_Λ is a complete variety. Now does the Jónsson-Tarski Theorem establish such a thing? Not really – it *does* show that every algebra $\mathfrak{A}$ is a complex algebra over *some* frame, thus proving that for any logic Λ we have that $V_\Lambda \subseteq \textbf{SCmK}$ for some frame class K. So, this certainly gives $V_\Lambda \subseteq \textbf{HSPCmK}$.

However, in order to prove completeness, we have to establish an equality instead of an inclusion. One way to prove this is to show that the complex algebras that we have found form a subclass of V_Λ. By Proposition 5.24 it would suffice to show that for any algebra $\mathfrak{A}$ in the variety V_Λ, the frame $\mathfrak{A}_+$ is a frame for the logic Λ. This requirement gives us an algebraic handle on the notion of canonicity.

Let us examine a concrete example. Recall that **K4** is the normal logic generated by the 4 axiom, $\Diamond\Diamond p \to \Diamond p$. We know from Theorem 4.27 that **K4** is complete with respect to the class of transitive frames. How can we prove this result algebraically?

A little thought reveals that the following is required: we have to show that the Lindenbaum-Tarski algebras for **K4** are embeddable in full complex algebras of *transitive* frames. Recall from Section 3.1 that the 4 axiom *characterizes* the transitive frames, thus in our proposed completeness proof, we would have to show that 4 is valid in the ultrafilter frame $(\mathfrak{L}_{\mathbf{K4}}(\Phi))_+$ of $\mathfrak{L}_{\mathbf{K4}}(\Phi)$, or equivalently, that $((\mathfrak{L}_{\mathbf{K4}}(\Phi))_+)^+$ belongs to the variety V_4. Note that by Theorem 5.32 we already know that $\mathfrak{L}_{\mathbf{K4}}(\Phi)$ belongs to V_4.

As this example suggests, proving frame completeness results for extensions of **K** algebraically leads directly to the following question: *which varieties of* BAOs *are closed under taking canonical embedding algebras?* In fact, this is the required algebraic handle on canonicity and motivates the following definition.

Definition 5.44 Let τ be a modal similarity type, and C a class of boolean algebras with τ-operators. C is *canonical* if it is closed under taking canonical embedding algebras; that is, if for all algebras $\mathfrak{A}$, $\mathfrak{Em}\mathfrak{A}$ is in C whenever $\mathfrak{A}$ is in C. Likewise, an equation is *canonical* if its validity is preserved when moving from a BAO to its canonical embedding algebra.   ⊣

Thus we now have two notions of canonicity, namely the logical one of Definition 4.30 and the algebraic one just defined. Using Theorem 5.32, we show that these two concepts are closely related.

Proposition 5.45 *Let τ be a modal similarity type, and Σ a set of τ-formulas. If V_Σ is a canonical variety, then Σ is canonical.*

Proof. Assume that the variety V_Σ is canonical, and let Φ be the fixed countable set of proposition letters that we use to define canonical frames. By Theorem 5.32, the Lindenbaum-Tarski algebra $\mathfrak{L}_{\mathbf{K\Sigma}}(\Phi)$ is in V_Σ; then, by assumption, its canonical embedding algebra $\mathfrak{Em}\mathfrak{L}_{\mathbf{K\Sigma}}$ is in V_Σ. However, from Theorem 5.42 it follows that this algebra is isomorphic to the complex algebra of the canonical frame of $\mathbf{K\Sigma}$:

$$\mathfrak{Em}\mathfrak{L}_{\mathbf{K\Sigma}}(\Phi) = ((\mathfrak{L}_{\mathbf{K\Sigma}}(\Phi))_+)^+ \cong (\mathfrak{F}^{\mathbf{K\Sigma}})^+.$$

Now the fact that $(\mathfrak{F}^{K\Sigma})^+$ is in V_Σ means that $\mathfrak{F}^{K\Sigma} \Vdash \Sigma$ by Proposition 5.24. But this implies that Σ is canonical. ⊣

An obvious question is whether the converse of Proposition 5.45 holds as well; that is, whether a variety V_Σ is canonical if Σ is a canonical set of modal formulas. However, note that canonicity of Σ only implies that one *particular* boolean algebra with operators has its embedding algebra in V_Σ, namely the Lindenbaum-Tarski algebra over a *countably* infinite number of generators. This is because throughout the completeness chapter we were working in a fixed, *countable* set of proposition letters. In fact, we are facing an open problem here:

Open Problem 1 *Let τ be a modal similarity type, and Σ a canonical set of τ-formulas. Is V_Σ a canonical variety?*

Equivalently, suppose that E is a set of equations such that for all countable *boolean algebras with τ-operators we have the following implication*

$$\text{if } \mathfrak{A} \models E \text{ then } \mathfrak{Em}\mathfrak{A} \models E. \tag{5.26}$$

Is V_E a canonical variety? In other words, does (5.26) hold for all *boolean algebras with τ-operators?*

This is an interesting problem. However, arguably the restriction of the notion of canonicity to countable languages that we adopted in Chapter 4 was not mathematically natural. Thus, let us redefine the logical notion of canonicity so that it refers to languages of arbitrary size. The definition of canonical frames and models can easily be parametrized by a set of proposition letters: the maximal consistent sets are supposed to be maximal within the induced set of formulas. We now simply define a logic Λ to be *canonical* if it is valid on *each* of its canonical frames. With this definition we can indeed establish equivalence between the logical and algebraic notions of canonicity. (An alternative, and mathematically quite interesting alternative, would be to introduce, both logically and algebraically, a *hierarchy* of canonicity notions, parametrized by cardinal numbers. Such an approach has indeed been studied in the literature, but this option will not be pursued here.)

Regardless of our approach towards this issue, the algebraic notion of canonicity can do a lot of work for us. The important point is that it offers a genuinely new perspective on what canonicity is, a perspective that will allow us to use algebraic arguments. This will be demonstrated in Section 5.6 when we introduce *persistence*, a generalization of the notion of canonicity, and prove the Sahlqvist Completeness Theorem.

Exercises for Section 5.3

5.3.1 Prove that for any frame $\mathfrak{F}$, we $\mathfrak{F} = (\mathfrak{F}^+)_+$.

5.3.2 Let Λ be a normal modal logic. Give a detailed proof that the canonical frame $\mathfrak{F}^{\Lambda}$ is isomorphic to the ultrafilter frame of $\mathfrak{L}_{\Lambda}$ (over a countable set of proposition letters).

5.3.3 Let A denote the collection of sets X of integers satisfying one of the following four conditions: (i) X is finite, (ii) X is co-finite, (iii) $X \oplus E$ is finite, (iv) $X \oplus E$ is co-finite. Here E denotes the set of all even integers, and $\oplus$ denotes symmetric difference: $X \oplus E = (X \setminus E) \cup (E \setminus X)$. Consider the following algebra $\mathfrak{A} = (A, \cup, -, \varnothing, f)$ where the operation f is given by

$$ f(X) = \begin{cases} \{x - 1 \mid x \in X\} & \text{if } X \text{ is of type (i) or (iii),} \\ \mathbb{Z} & \text{if } X \text{ is of type (ii) or (iv).} \end{cases} $$

(a) Show that $\mathfrak{A}$ is a boolean algebra with operators.
(b) Describe $\mathfrak{A}_+$.

5.3.4 Let W be the set $\mathbb{Z} \cup \{-\infty, \infty\}$ and let S be the successor relation on $\mathbb{Z}$, that is, $S = \{(z, z + 1) \mid z \in \mathbb{Z}\}$.

(a) Give a BAO whose ultrafilter frame is isomorphic to the frame $\mathfrak{F} = (W, R)$ with $R = S \cup \{(-\infty, -\infty), (\infty, \infty)\}$.
(b) Give a BAO whose ultrafilter frame is isomorphic to the frame $\mathfrak{F} = (W, R)$ with $R = S \cup (W \times \{-\infty, \infty\})$.
(c) Give a BAO whose ultrafilter frame is isomorphic to the frame $\mathfrak{F} = (W, R)$ with $R = S \cup \{(-\infty, -\infty)\} \cup (W \times \{\infty\})$.

5.3.5 An operation on a boolean algebra is called *2-additive* if it satisfies

$$ f(x + y + z) = f(x + y) + f(x + z) + f(y + z). $$

Prove an analog of the Jónsson-Tarski Theorem for boolean algebras augmented with 2-additive operations.

5.4 Duality Theory

We now know how to build frames from algebras and algebras from frames in ways that preserve crucial logical properties. But something is missing. Modal logicians rarely study frames in isolation: rather, they are interested in how to construct new frames from old using bounded morphisms, generated subframes, and disjoint unions. And algebraists adopt an analogous perspective: they are interested in relating algebras via such constructions as homomorphisms, subalgebras, and direct products. Thus modal logicians work in one mathematical universe, and algebraists in another, and it is natural to ask whether these universes are systematically related. They are, and duality theory studies these links.

In this section we will do two things. First, we will introduce the basic dualities that exist between the modal and algebraic universes. Second, we will demonstrate that these dualities are useful by proving two major theorems of modal logic. We assume that by this stage the reader has picked up the basic definitions and results concerning the algebraic universe (and in particular, what homomorphisms, subalgebras, and direct products are). If not, check out Appendix B.

Basic duality results

Theorems 5.47 and 5.48 below give a concise formulation of the basic links between the algebraic and frame-theoretic universes. They are stated using the following notation.

Definition 5.46 Let τ be a modal similarity type, $\mathfrak{F}$ and $\mathfrak{G}$ two τ-frames, and $\mathfrak{A}$ and $\mathfrak{B}$ two boolean algebras with τ-operators. We recall (define, respectively) the following notation for relations between these structures:

- $\mathfrak{F} \rightarrowtail \mathfrak{G}$ for $\mathfrak{F}$ is isomorphic to a generated subframe of $\mathfrak{G}$,
- $\mathfrak{F} \twoheadrightarrow \mathfrak{G}$ for $\mathfrak{G}$ is a bounded morphic image of $\mathfrak{F}$,
- $\mathfrak{A} \rightarrowtail \mathfrak{B}$ for $\mathfrak{A}$ is isomorphic to a subalgebra of $\mathfrak{B}$,
- $\mathfrak{A} \twoheadrightarrow \mathfrak{B}$ for $\mathfrak{B}$ is a homomorphic image of $\mathfrak{A}$. ⊣

Theorem 5.47 *Let τ be a modal similarity type, $\mathfrak{F}$ and $\mathfrak{G}$ two τ-frames, and $\mathfrak{A}$ and $\mathfrak{B}$ two boolean algebras with τ-operators.*

   (i) *If $\mathfrak{F} \rightarrowtail \mathfrak{G}$, then $\mathfrak{G}^{+} \twoheadrightarrow \mathfrak{F}^{+}$.*
   (ii) *If $\mathfrak{F} \twoheadrightarrow \mathfrak{G}$, then $\mathfrak{G}^{+} \rightarrowtail \mathfrak{F}^{+}$.*
   (iii) *If $\mathfrak{A} \rightarrowtail \mathfrak{B}$, then $\mathfrak{B}_{+} \twoheadrightarrow \mathfrak{A}_{+}$.*
   (iv) *If $\mathfrak{A} \twoheadrightarrow \mathfrak{B}$, then $\mathfrak{B}_{+} \rightarrowtail \mathfrak{A}_{+}$.*

Proof. This follows immediately from Propositions 5.51 and 5.52 below. ⊣

Theorem 5.48 *Let τ be a modal similarity type, and $\mathfrak{F}_i, i \in I$, a family of τ-frames. Then*

$$\left(\biguplus_{i \in I} \mathfrak{F}_i \right)^{+} \cong \prod_{i \in I} \mathfrak{F}_i^{+}.$$

Proof. We define a map η from the power set of the disjoint union $\biguplus_{i \in I} W_i$ to the carrier $\prod_{i \in I} \mathcal{P}(W_i)$ of the product of the family of complex algebras $(\mathfrak{F}_i^{+})_{i \in I}$.

Let X be a subset of $\biguplus_{i \in I} W_i$. Clearly, $\eta(X)$ has to be an element of the set $\prod_{i \in I} \mathcal{P}(W_i)$. And elements of the set $\prod_{i \in I} \mathcal{P}(W_i)$ are sequences σ such that $\sigma(i) \in \mathcal{P}(W_i)$. So it suffices to say what the i-th element of the sequence $\eta(X)$ is:

$$\eta(X)(i) = X \cap W_i.$$

We leave it as an exercise to show that η is an isomorphism; see Exercise 5.4.6. ⊣

Note that Theorem 5.48 (in contrast to Theorem 5.47) only states a connection in the direction from frames to algebras. This is because in general

$$\left(\prod_{i \in I} \mathfrak{A}_i \right)_{+} \not\cong \biguplus_{i \in I} (\mathfrak{A}_i)_{+}.$$

The reader is asked to give an example to this effect in Exercise 5.4.1.

In order to prove Theorem 5.47, the reader is advised to recall the definitions of the back and forth properties of bounded morphisms between frames (Definition 3.13). We also need some terminology for morphisms between boolean algebras with operators.

Definition 5.49 Let $\mathfrak{A}$ and $\mathfrak{A}'$ be two BAOs of the same similarity type, and let $\eta : A \to A'$ be a function. We say that η is a *boolean homomorphism* if η is a homomorphism from $(A, +, -, 0)$ to $(A', +', -', 0')$. We call η a *modal homomorphism* if η satisfies, for all modal operators $\triangle$:

$$\eta(f_\triangle(a_1, \ldots, a_{\rho(\triangle)})) = f'_\triangle(\eta a_1, \ldots, \eta a_{\rho(\triangle)}).$$

(Here ηa_i means $\eta(a_i)$; we will sometimes use this shorthand to keep the notation uncluttered.) Finally, η is a *(BAO-)homomorphism* if it is both a boolean and a modal homomorphism. ⊣

In the following definition, the construction of *dual* or *lifted* morphisms is given (here the word 'dual' is *not* used in the sense of $\diamond$ being the dual of $\square$).

Definition 5.50 Suppose θ is a map from W to W'; then its *dual*, $\theta^+ : \mathcal{P}(W') \to \mathcal{P}(W)$ is defined as:

$$\theta^+(X') = \{u \in W \mid \theta(u) \in X'\}.$$

In the other direction, let $\mathfrak{A}$ and $\mathfrak{A}'$ be two BAOs, and $\eta : \mathfrak{A} \to \mathfrak{A}'$ be a map from A to A'; then its *dual* is given as the following map from ultrafilters of $\mathfrak{A}'$ to subsets of A:

$$\eta_+(u') = \{a \in A \mid \eta(a) \in u'\}.$$ ⊣

The following propositions assert that the duals of bounded morphisms are nothing but BAO-homomorphisms:

Proposition 5.51 *Let $\mathfrak{F}$, $\mathfrak{F}'$ be frames, and $\theta : W \to W'$ a map.*

(i) θ^+ *is a boolean homomorphism.*

(ii) $m_R(\theta^+(Y_1'), \ldots, \theta^+(Y_n')) \subseteq \theta^+(m_{R'}(Y_1', \ldots, Y_n'))$, *if θ has the forth property.*

(iii) $m_R(\theta^+(Y_1'), \ldots, \theta^+(Y_n')) \supseteq \theta^+(m_{R'}(Y_1', \ldots, Y_n'))$, *if θ has the back property.*

(iv) θ^+ *is a BAO-homomorphism from $\mathfrak{F}'^+$ to $\mathfrak{F}^+$, if θ is a bounded morphism.*

(v) θ^+ *is surjective, if θ is injective.*

(vi) θ^+ *is injective, if θ is surjective.*

Proof. For notational convenience, we assume that τ has only one modal operator, so that we can write $\mathfrak{F} = (W, R)$.

(i) (Note that this was Exercise 5.1.1.) As an example, we treat complementation:

$$x \in \theta^+(-X') \text{ iff } \theta(x) \in -X' \text{ iff } \theta(x) \notin X' \text{ iff } x \notin \theta^+(X').$$

From this it follows immediately that $\theta^+(-X') = -\theta^+(X')$.

(ii) Assume that θ has the forth property. Then we have

$$x \in m_R(\theta^+(Y_1'), \ldots, \theta^+(Y_n'))$$
$$\implies \exists\, y_1, \ldots, y_n \text{ such that } \theta(y_i) \in Y_i' \text{ and } Rxy_1 \ldots y_n$$
$$\implies \exists\, y_1, \ldots, y_n \text{ such that } \theta(y_i) \in Y_i' \text{ and } R'\theta(x)\theta(y_1)\ldots\theta(y_n)$$
$$\implies \theta(x) \in m_{R'}(Y_1', \ldots, Y_n')$$
$$\implies x \in \theta^+(m_{R'}(Y_1', \ldots, Y_n')).$$

(iii) Now suppose $x \in \theta^+(m_{R'}(Y_1', \ldots, Y_n'))$. Then $\theta(x) \in m_{R'}(Y_1', \ldots, Y_n')$. So there are $y_1', \ldots, y_n'$ in W' with $y_i' \in Y_i'$ and $R'\theta(x)y_1' \ldots y_n'$. As θ has the back property, there are $y_1, \ldots, y_n \in W$ with $\theta(y_i) = y_i'$ for all i, and $Rxy_1 \ldots y_n$. But then $y_i \in \theta^+(Y_i')$ for every i, so $x \in m_R(\theta^+(Y_1'), \ldots, \theta^+(Y_n'))$.

(iv) This follows immediately from items (i), (ii) and (iii).

(v) Assume that θ is injective, and let X be a subset of W. We have to find a subset X' of W' such that $\theta^+(X') = X$. Define

$$\theta[X] := \{\theta(x) \in W' \mid x \in X\}.$$

We claim that this set has the desired properties. Clearly $X \subseteq \theta^+(\theta[X])$. For the other direction, let x be an element of $\theta^+(\theta[X])$. Then by definition, $\theta(x) \in \theta[X]$, so there is a $y \in X$ such that $\theta(x) = \theta(y)$. By the injectivity of θ, $x = y$. So $x \in X$.

(vi) Assume that θ is surjective, and let X' and Y' be distinct subsets of W'. Without loss of generality we may assume that there is an x' such that $x' \in X'$ and $x' \notin Y'$. As θ is surjective, there is an x in W such that $\theta(x) = x'$. So $x \in \theta^+(X')$, but $x \notin \theta^+(Y')$. So $\theta(X') \neq \theta(Y')$, whence θ^+ is injective. $\dashv$

Going in the opposite direction, that is, from algebras to relational structures, we find that the duals of BAO-homomorphisms are bounded morphisms:

Proposition 5.52 *Let $\mathfrak{A}, \mathfrak{A}'$ be boolean algebras with operators, and η a map from A to A'.*

(i) *If η is a boolean homomorphism, then η_+ maps ultrafilters to ultrafilters.*
(ii) *If $f'(\eta(a_1), \ldots, \eta(a_n)) \leq \eta(f(a_1, \ldots, a_n))$, then η_+ has the forth property.*

(iii) *If* $f'(\eta(a_1), \dots, \eta(a_n)) \geq \eta(f(a_1, \dots, a_n))$ *and* η *is a boolean homomorphism, then* η_+ *has the back property.*

(iv) *If* η *is a* BAO-*homomorphism, then* η_+ *is a bounded morphism from* $\mathfrak{A}'_+$ *to* $\mathfrak{A}_+$.

(v) *If* η *is an injective boolean homomorphism, then* $\eta_+ : Uf\mathfrak{A}' \rightarrow Uf\mathfrak{A}$ *is surjective.*

(vi) *If* η *is a surjective boolean homomorphism, then* $\eta_+ : Uf\mathfrak{A}' \rightarrow Uf\mathfrak{A}$ *is injective.*

Proof. Again, without loss of generality we assume that τ has only one modal operator, so that we can write $\mathfrak{A} = (A, +, -, 0, f)$.

(i) This item is left as Exercise 5.4.2.

(ii) Suppose that $Q_{f'} u' u'_1 \dots u'_n$ holds between some ultrafilters $u', u'_1, \dots, u'_n$ of $\mathfrak{A}'$. To show that $\mathfrak{A}_+ \models Q_f \eta_+ u' \eta_+ u'_1 \dots \eta_+ u'_n$, let $a_1, \dots, a_n$ be arbitrary elements of $\eta_+ u'_1, \dots, \eta_+ u'_n$ respectively. Then, by definition of η_+, $\eta a_i \in u'_i$, so $Q'_f u' u'_1 \dots u'_n$ gives $f'(\eta a_1, \dots, \eta a_n) \in u'$. Now the assumption yields $\eta f(a_1, \dots, a_n) \in u'$, as ultrafilters are upward closed. But then $f(a_1, \dots, a_n) \in \eta_+ u'$, which is what we wanted.

(iii) This item is left as Exercise 5.4.2.

(iv) This follows immediately from items (i), (ii) and (iii).

(v) Assume that η is injective, and let u be an ultrafilter of $\mathfrak{A}$. We want to follow the same strategy as in Proposition 5.51(v), and define

$$\eta[u] := \{\eta(a) \mid a \in u\}.$$

The difference with the proof of Proposition 5.51(v) is that here, $\eta_+(\eta[u])$ may not be defined. The reason for this is that, in general, $\eta[u]$ will not be upwards closed and hence, not an (ultra)filter, while η_+ is defined only for ultrafilters. Therefore, we define

$$F' := \{a' \mid \eta(a) \leq a' \text{ for some } a \in u \}.$$

Clearly, $\eta[u] \subseteq F'$. We will first show that F' is a *proper filter* of $\mathfrak{A}'$ (note that the clauses $(F1)$–$(F3)$ which define filters are given in Definition 5.34). For $(F1)$, observe that $1 \in u$, so $\eta(1) = 1 \in \eta[u] \subseteq F'$. For $(F2)$, assume $a', b' \in F'$. Then there are a, b in u such that $\eta a \leq a'$ and $\eta b \leq b'$. It follows that $\eta(a \cdot b) = \eta a \cdot \eta b \leq a' \cdot b' \in \eta[u]$; hence, $a' \cdot b' \in F'$ since $a \cdot b \in u$. This shows that F' is closed under taking meets. It is trivial to prove $(F3)$, that is, that F' is upwards closed. Finally, in order to show that F' is proper, suppose that $0' \in F'$. Then $0' = \eta a$ for some $a \in u$; as $0' = \eta(0)$, injectivity of η gives that $0 = a$, and hence, $0 \in u$. But then u is not an ultrafilter.

By the Ultrafilter Theorem 5.38, F' can be extended to an ultrafilter u'. We claim that $u = \eta_+(u')$. First let a be in u, then $\eta a \in \eta[u] \subseteq u'$, so $a \in \eta_+(u')$. This shows that $u \subseteq \eta_+(u')$. For the other inclusion, it suffices to show that $a \notin \eta_+(u')$ if $a \notin u$; we reason as follows:

$$
\begin{aligned}
a \notin u &\implies -a \in u \\
&\implies -\eta a = \eta(-a) \in \eta[u] \\
&\implies -\eta(a) \in u' \\
&\implies \eta a \notin u' \\
&\implies a \notin \eta_+(u').
\end{aligned}
$$

(vi) Similar to Proposition 5.51, item (vi); see Exercise 5.4.2. ⊣

Readers familiar with category theory will have noticed that the operation $(\cdot)^+$ is a functor from the category of τ-frames with bounded morphisms to the category of boolean algebras with τ-operators, and vice versa for $(\cdot)_+$. This categorial perspective is implicit in what follows, but seldom comes to the surface. In the remainder of the section we will see how our algebraic perspective on modal logic that we have developed can be applied.

Applications

In this subsection we tie a number of threads together and show how to use the duality between frames and algebras to give very short proofs of some major theorems of modal logic.

Our first example shows that all the results given in Theorem 3.14 on the preservation of *modal* validity under the fundamental frame operations fall out as simple consequences of well-known preservation results of universal algebra, namely that *equational* validity is preserved under the formation of subalgebras, homomorphic images and products of algebras.

Proposition 5.53 *Let τ be a modal similarity type, ϕ a τ-formula and $\mathfrak{F}$ a τ-frame. Then*

  (i) *If $\mathfrak{G}$ is a bounded morphic image of $\mathfrak{F}$, then $\mathfrak{G} \Vdash \phi$ if $\mathfrak{F} \Vdash \phi$.*

  (ii) *If $\mathfrak{G}$ is a generated subframe of $\mathfrak{F}$, then $\mathfrak{G} \Vdash \phi$ if $\mathfrak{F} \Vdash \phi$.*

  (iii) *If $\mathfrak{F}$ is the disjoint union of a family $\{\mathfrak{F}_i \mid i \in I\}$, then $\mathfrak{F} \Vdash \phi$ if for every $i \in I$, $\mathfrak{F}_i \Vdash \phi$.*

  (iv) *If $\mathfrak{ue}\,\mathfrak{F} \Vdash \phi$, then $\mathfrak{F} \Vdash \phi$.*

Proof. We only prove the first part of the proposition, leaving the other parts as exercises for the reader.

Assume that $\mathfrak{F} \twoheadrightarrow \mathfrak{G}$, and $\mathfrak{F} \Vdash \phi$. By Proposition 5.24, we have $\mathfrak{F}^+ \models \phi \approx \top$, and by Theorem 5.47, $\mathfrak{G}^+$ is a *subalgebra* of $\mathfrak{F}^+$. So by the fact that equational validity is preserved under taking subalgebras, we obtain that $\phi \approx \top$ holds in $\mathfrak{G}^+$. But then Proposition 5.24 implies that $\mathfrak{G} \Vdash \phi$. ⊣

Our second example is a simple proof of the Goldblatt-Thomason Theorem, which gives a precise structural characterization of the first-order definable classes of frames which are modally definable. We discussed this result in Chapter 3, and gave a proof which drew on the tools of first-order model theory (see Theorem 3.19 in Section 3.8). As we will now see, there is also an algebraic way of viewing the theorem: it is a more or less immediate corollary of Birkhoff's Theorem (see Appendix B) identifying equational classes and varieties. The version we prove here is slightly stronger than Theorem 3.19, since it applies to any class of frames that is closed under taking ultrapowers.

Theorem 5.54 (Goldblatt-Thomason Theorem) *Let τ be a modal similarity type, and let K be a class of τ-frames that is closed under taking ultrapowers. Then K is modally definable if and only if it is closed under the formation of bounded morphic images, generated subframes, and disjoint unions, and reflects ultrafilter extensions.*

Proof. The left to right direction is an immediate corollary of the previous proposition. For the right to left direction, let K be any class of frames satisfying the closure conditions given in the theorem. It suffices to show that any frame $\mathfrak{F}$ validating the modal theory of K is itself a member of K.

Let $\mathfrak{F}$ be such a frame. It is not difficult to show that Proposition 5.24 implies that $\mathfrak{F}^+$ is a model for the equational theory of the class $\mathbf{Cm}$K. It follows by Birkhoff's Theorem (identifying varieties and equational classes) that $\mathfrak{F}^+$ is in the variety generated by $\mathbf{Cm}$K, so $\mathfrak{F}^+$ is in $\mathbf{HSPCm}$K. In other words, there is a family $(\mathfrak{G}_i)_{i \in I}$ of frames in K, and there are boolean algebras with operators $\mathfrak{A}$ and $\mathfrak{B}$ such that

(i) $\mathfrak{B}$ is the product $\prod_{i \in I} \mathfrak{G}_i^+$ of the complex algebras of the $\mathfrak{G}_i$,
(ii) $\mathfrak{A}$ is a subalgebra of $\mathfrak{B}$, and
(iii) $\mathfrak{F}^+$ is a homomorphic image of $\mathfrak{A}$.

By Theorem 5.48, $\mathfrak{B}$ is isomorphic to the complex algebra of the disjoint union $\mathfrak{G}$ of the family $(\mathfrak{G}_i)_{i \in I}$:

$$\mathfrak{B} \cong \mathfrak{G}^+ = \left(\biguplus_{i \in I} \mathfrak{G}_i \right)^+ .$$

As K is closed under taking disjoint unions, $\mathfrak{G}$ is in K.

Now we have the following picture: $\mathfrak{F}^+ \twoheadleftarrow \mathfrak{A} \rightarrowtail \mathfrak{G}^+$. By Theorem 5.47 it follows that

$$(\mathfrak{F}^+)_+ \rightarrowtail \mathfrak{A}_+ \twoheadleftarrow (\mathfrak{G}^+)_+.$$

Since K is closed under ultrapowers, Theorem 3.17 implies that $(\mathfrak{G}^+)_+ = \mathfrak{ue}\, \mathfrak{G}$ is in K. As K is closed under the formation of bounded morphic images and generated subframes, it follows that $\mathfrak{A}_+$ and $\mathfrak{ue}\,\mathfrak{F} = (\mathfrak{F}^+)_+$ (in that order) are in K. But then $\mathfrak{F}$ itself is also a member of K, since K reflects ultrafilter extensions. ⊣

For our third example, we return to the concept of canonicity. We will prove an important result and mention an intriguing open problem, both having to do with the relation between canonical varieties and first-order definable classes of frames. Both the result and the open problem were mentioned in Chapter 4 (see Theorem 4.50 and the surrounding discussion), albeit in a slightly weaker form. To link the earlier statements with the versions discussed here, simply observe that any elementary class of frames is closed under the formation of ultraproducts.

First we need the following definition.

Definition 5.55 Let τ be modal similarity type, and K be a class of τ-frames. The variety generated by K (notation: $\mathbf{V_K}$) is the class **HSPCmK**. ⊣

Theorem 5.56 *Let τ be modal similarity type, and K be a class of τ-frames which is closed under ultraproducts. Then the variety $\mathbf{V_K}$ is canonical.*

Proof. Assume that the class K of τ-frames is closed under taking ultraproducts. We will first prove that the class **HSCmK** is canonical. Let $\mathfrak{A}$ be an element of this class; that is, assume that there is a frame $\mathfrak{F}$ in K and an algebra $\mathfrak{B}$ such that

$$\mathfrak{A} \twoheadleftarrow \mathfrak{B} \rightarrowtail \mathfrak{F}^+.$$

It follows from Theorem 5.47 that

$$\mathfrak{Em}\mathfrak{A} \twoheadleftarrow \mathfrak{Em}\mathfrak{B} \rightarrowtail \mathfrak{Em}\mathfrak{F}^+ = (\mathfrak{ue}\,\mathfrak{F})^+. \tag{5.27}$$

From Theorem 3.17 we know that $\mathfrak{ue}\,\mathfrak{F}$ is the bounded morphic image of some ultrapower $\mathfrak{G}$ of $\mathfrak{F}$. Note that $\mathfrak{G}$ is in K, by assumption. Now Theorem 5.47 gives

$$(\mathfrak{ue}\,\mathfrak{F})^+ \rightarrowtail \mathfrak{G}^+. \tag{5.28}$$

Since $\mathfrak{G}^+$ is in CmK, (5.27) and (5.28) together imply that $\mathfrak{Em}\mathfrak{A}$ is in **HSCmK**. Hence this class is canonical.

To prove that the *variety* generated by K is canonical, we need an additional fact. Recall that according to Proposition 3.63, the ultrapower of a disjoint union can be obtained as a bounded morphic image of a disjoint union of ultraproducts.

Now assume that $\mathfrak{A}$ is in $\mathbf{V_K} = \mathbf{HSPCmK}$. In other words, assume there is a family $\{\mathfrak{F}_i \mid i \in I\}$ of frames in K and an algebra $\mathfrak{B}$ such that

$$\mathfrak{A} \twoheadleftarrow \mathfrak{B} \rightarrowtail \prod_{i \in I} \mathfrak{F}_i^+.$$

To prove that $\mathfrak{Em}\mathfrak{A}$ is in $\mathbf{V_K}$, it suffices to show that $\mathfrak{Em}(\prod_{i \in I} \mathfrak{F}_i^+)$ is in **SPCmK**

– the remainder of the proof is as before. Let $\mathfrak{F}$ be the frame $\biguplus_{i \in I} \mathfrak{F}_i$, then by Theorem 5.48, $\mathfrak{F}^+ \cong \prod_{i \in I} \mathfrak{F}_i^+$. Hence, by Theorem 5.47:

$$\mathfrak{Em}\left(\prod_{i \in I} \mathfrak{F}_i^+\right) \cong ((\mathfrak{F}^+)_+)^+ = (\mathfrak{ue}\,\mathfrak{F})^+. \qquad (5.29)$$

By Theorem 3.17, there is an ultrapower $\mathfrak{G}$ of $\mathfrak{F}$ such that $\mathfrak{G} \twoheadrightarrow \mathfrak{ue}\,\mathfrak{F}$. Now we apply Proposition 3.63, yielding a frame $\mathfrak{H}$ such that (i) $\mathfrak{H}$ is a disjoint union of ultraproducts of frames in K and (ii) $\mathfrak{H} \twoheadrightarrow \mathfrak{G}$. Putting these observations together we have $\mathfrak{ue}\,\mathfrak{F} \twoheadleftarrow \mathfrak{G} \twoheadleftarrow \mathfrak{H}$. Hence, by Theorem 5.47:

$$(\mathfrak{ue}\,\mathfrak{F})^+ \rightarrowtail \mathfrak{G}^+ \rightarrowtail \mathfrak{H}^+. \qquad (5.30)$$

Note that $\mathfrak{H}$ is a disjoint union of frames in K, since K is closed under taking ultraproducts. This implies that $\mathfrak{H}^+$ is in **PCmK**. But then it follows from (5.29) and (5.30) that $\mathfrak{Em}(\prod_{i \in I} \mathfrak{F}_i^+)$ is in **SPCmK**, which is what we needed.   ⊣

Example 5.57 Consider the modal similarity type $\{\circ, \otimes, 1'\}$ of arrow logic, where $\circ$ is binary, $\otimes$ is unary and $1'$ is a constant. The standard interpretation of this language is given in terms of the *squares* (cf. Example 1.24). Recall that the square $\mathfrak{S}_U = (W, C, R, I)$ is defined as follows:

$$
\begin{aligned}
W &= U \times U, \\
C((u,v),(w,x),(y,z)) \quad &\text{iff} \quad u = w \text{ and } v = z \text{ and } x = y, \\
R((u,v),(w,x)) \quad &\text{iff} \quad u = x \text{ and } v = w, \\
I(u,v) \quad &\text{iff} \quad u = v.
\end{aligned}
$$

It may be shown that the class SQ of (isomorphic copies of) squares is first-order definable in the frame language with predicates C, R and I. Therefore, Theorem 5.56 implies that the variety generated by SQ is canonical. This variety is well known in the literature on algebraic logic as the variety RRA of Representable Relation Algebras. See Exercise 5.4.5.   ⊣

Rephrased in terminology from modal logic, Theorem 5.56 boils down to the following result.

Corollary 5.58 *Let τ be a modal similarity type, and K be a class of τ-frames which is closed under ultraproducts. Then the modal theory of K is a canonical logic.*

We conclude the section with the foremost open problem in this area: does the converse of Theorem 5.56 holds as well?

Open Problem 2 *Let τ be modal similarity type, and V a canonical variety of boolean algebras with τ-operators. Is there a class K of τ-frames, closed under taking ultraproducts, such that V is generated by K?*

Exercises for Section 5.4

5.4.1 Consider a countably infinite collection $(\mathfrak{A}_i)_{i \in I}$ of finite algebras that are non-trivial, that is, of size at least 2.

(a) Show that the product $\prod_{i \in I} \mathfrak{A}_i$ has uncountably many ultrafilters.
(b) Show that the ultrafilter frame of a finite algebra is finite, and that hence, the disjoint union $\biguplus_{i \in I} (\mathfrak{A}_i)_+$ is countable.
(c) Conclude that $\left(\prod_{i \in I} \mathfrak{A}_i \right)_+ \not\cong \biguplus_{i \in I} (\mathfrak{A}_i)_+$.

5.4.2 Prove Proposition 5.52(i), (iii) and (vi). Prove (iii) first for unary operators; for the general case, see the proof of the Jónsson-Tarski Theorem for inspiration.

5.4.3 Prove or disprove the following propositions:

(a) For any two boolean algebras with operators $\mathfrak{A}$ and $\mathfrak{B}$: $\mathfrak{A}_+ \cong \mathfrak{B}_+$ only if $\mathfrak{A} \cong \mathfrak{B}$. (Hint: first consider the question for plain boolean algebras, thinking of specimens like the ones occurring in Exercise 5.2.3 and Exercise 5.3.3.)
(b) For any two frames $\mathfrak{F}$ and $\mathfrak{G}$: $\mathfrak{F}^+ \cong \mathfrak{G}^+$ only if $\mathfrak{F} \cong \mathfrak{G}$.

5.4.4 Consider the frames $\mathfrak{F} = (X, R)$ and $\mathfrak{G} = (Y, S)$ given by

$$X = \mathbb{N} \qquad\qquad Y = \mathbb{N} \cup \{\infty\}$$
$$R = \{(x, y) \in X \times X \mid x \neq y\} \qquad S = \{(x, y) \in Y \times Y \mid x \neq y\} \cup \{(\infty, \infty)\}.$$

(a) Show that $\mathfrak{F}$ is not a bounded morphic image of $\mathfrak{G}$.
(b) Show that on the other hand, $\mathfrak{F}^+$ can be embedded in $\mathfrak{G}^+$ (that is, define an injective homomorphism $\eta \colon \mathfrak{F}^+ \rightarrowtail \mathfrak{G}^+$).

5.4.5 Show that the class SQ of (isomorphic copies of) square arrow frames is first-order definable. See Example 5.57.

5.4.6 Show that the embedding η used in the proof of Theorem 5.48 is an isomorphism.

5.5 General Frames

Although the algebraic semantics for modal logic has the nice property that there is a fundamental completeness result (Theorem 5.27), many modal logicians still prefer frame-based semantics, either because they find it more intuitive, or because frames are the structures in which they take an (application-driven) interest. In this (and the following) section we will discuss an intermediate semantics which in a sense unifies relational and algebraic semantics. As we will see below, a general frame is an ordinary frame and a boolean algebra with operators rolled into one. The nice thing about general frames is that we can prove a fundamental completeness theorem for modal logic and general frames – and modal semantics based on general frames is almost as intuitive as the familiar relational semantics.

Although we have already met general frames (we briefly introduced them in Section 1.4, and used them when we studied frame incompleteness in Section 4.4), we have not yet discussed them systematically. We will now put that right. We will reintroduce them, discuss some important classes of general frames, pin down the relationship between general frames, frames and boolean algebras with operators, and briefly discuss them from a topological perspective.

Here is how general frames are defined for an arbitrary similarity type τ:

Definition 5.59 (General Frame) Let τ be a modal similarity type. A *general τ-frame* is a pair $\mathfrak{g} = (\mathfrak{F}, A)$ such that $\mathfrak{F} = (W, R_\triangle)_{\triangle \in \tau}$ is a τ-frame, and A is (the carrier of) a complex algebra over $\mathfrak{F}$. That is, A is a non-empty collection of subsets of W which is closed under the boolean operations and under the modal operation $m_{R_\triangle}(X_1, \ldots, X_n)$ for each $\triangle \in \tau$.

A valuation V on $\mathfrak{F}$ is called *admissible* for $\mathfrak{g}$ if for each proposition letter p, $V(p)$ is an *admissible* subset of W, that is, an element of A. A *model based on a general frame* is a triple $(\mathfrak{F}, A, V)$ where $(\mathfrak{F}, A)$ is a general frame and V is an admissible valuation for $(\mathfrak{F}, A)$. Truth in such a model is defined in the obvious way, that is, as if we were talking about the model $(\mathfrak{F}, V)$.   ⊣

Convention 5.60 To avoid confusion, in this section we will use the term 'Kripke frame' when talking about (ordinary) frames.   ⊣

It is easy to verify that the closure conditions mentioned in Definition 5.59 ensure that if V is an admissible valuation on a general frame, then the set $V(\phi)$ is admissible for every formula ϕ. Conversely, every model gives rise to a general frame.

Example 5.61 Given a model $\mathfrak{M} = (\mathfrak{F}, V)$, it is obvious that the collection

$$A_\mathfrak{M} = \{V(\phi) \mid \phi \text{ a modal formula } \}$$

is closed under the boolean operations and under each $m_{R_\triangle}$. Hence the structure $(\mathfrak{F}, A_\mathfrak{M})$ is a general frame.

Note that we can apply this technique to the canonical model $\mathfrak{M}_\Lambda$ of any normal modal logic Λ. It follows from the Truth Lemma (Lemma 4.21) that the resulting structure $(\mathfrak{F}_\Lambda^c, A_{\mathfrak{M}_\Lambda^c})$ is isomorphic to the *canonical general frame* $\mathfrak{f}_\Lambda$ which is defined as

$$\mathfrak{f}_\Lambda^c = (\mathfrak{F}_\Lambda^c, \{\widehat{\phi} \mid \phi \text{ a formula } \}), \tag{5.31}$$

where $\widehat{\phi}$ is the set of maximal consistent sets Γ such that $\phi \in \Gamma$.

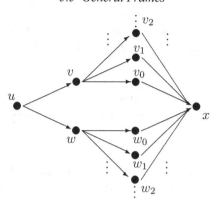

Fig. 5.1.

The notions of validity and semantic consequence of classes of general frames are defined in the expected way:

Definition 5.62 Let $\mathfrak{g}$ be a general frame. A formula ϕ is *valid* on $\mathfrak{g}$ if ϕ holds in every state of $\mathfrak{g}$ under every admissible valuation V; a similar definition holds for sets of formulas and classes of general frames.

For a set of formulas, in particular for a normal modal logic Λ, a general frame is called a Λ-*frame* if Λ is valid on the frame.

Now, let K be a class of general frames, Σ a set of formulas and ϕ a formula. We say that ϕ is a *semantic consequence* of Σ over K if for every general frame $\mathfrak{g}$ in K, every admissible valuation V on $\mathfrak{g}$ and every state s in $\mathfrak{g}$, we have that $(\mathfrak{g}, V), s \Vdash \Sigma$ implies $(\mathfrak{g}, V), s \Vdash \phi$. ⊣

The following example shows that a formula may be valid on a general frame, while it is not valid on the underlying Kripke frame:

Example 5.63 Consider the following Kripke frame $\mathfrak{C} = (C, R)$. Its set C of states is given as $C = \{u, v, w, x\} \cup \{v_n, w_n \mid n \in \omega\}$, while the accessibility relation R is defined as follows: Ruv, Ruw, Rvv_n and Rww_n (for all n), Rv_nx and Rw_nx (for all n), and Rxx; see Figure 5.1.

We leave it as an exercise to verify that the pair $\mathfrak{c} = (\mathfrak{C}, F)$ is a general frame, where F is the collection of all finite and co-finite subsets of C. We claim that $\mathfrak{c} \Vdash \Diamond\Box p \to \Box\Diamond p$, while $\mathfrak{C} \nVdash \Diamond\Box p \to \Box\Diamond p$. Let us first prove the latter: consider the valuation V given by $V(p) = \{v_n \mid n \in \mathbb{N}\}$. Now, from $(\mathfrak{C}, V), v \Vdash \Box p$ we derive $(\mathfrak{C}, V), u \Vdash \Diamond\Box p$, but $(\mathfrak{C}, V), u \nVdash \Box\Diamond p$, as $(\mathfrak{C}, V), w \nVdash \Diamond p$.

To prove that $\mathfrak{c} \Vdash \Diamond\Box p \to \Box\Diamond p$, it suffices to look at u (why?). Suppose that for some admissible valuation V, $\Diamond\Box p$ holds at u. Without loss of generality we may assume that $(\mathfrak{c}, V), v \Vdash \Box p$, so p holds at all v_i. Then $V(p)$ is not finite and since

$V(p)$ must be admissible, it follows from the definition of F that $V(p)$ is co-finite. Hence there are (co-finitely many) w_i with $(\mathfrak{c}, V)$, $w_i \Vdash p$. But then $\Diamond p$ holds at w and thus $\Box\Diamond p$ at u.   ⊣

And now for the promised general completeness result: general frames share with boolean algebras with operators the property of providing an adequate semantics for all normal modal logics.

Theorem 5.64 *Let Λ be a normal modal logic. Then Λ is sound and strongly complete with respect to the class of general Λ-frames.*

Proof. As usual, the soundness proof is rather trivial and is left to the reader. For completeness, consider the canonical frame $\mathfrak{f}_\Lambda$ as defined in (5.31) in Example 5.61. It is easy to verify that $\mathfrak{f}_\Lambda \Vdash \Lambda$ (see Exercise 5.5.4). Now assume that $\Sigma \not\vdash_\Lambda \phi$. It follows that the set $\Sigma \cup \{\neg\phi\}$ is Λ-consistent. By definition of the canonical frame and the Truth Lemma, there is a state s of $\mathfrak{f}_\Lambda$ such that $(\mathfrak{f}_\Lambda^c, V^c), s \Vdash \Sigma \cup \{\neg\phi\}$. Since the canonical valuation V^c is admissible, this means that we have falsified the claim that ϕ is a semantical consequence of Σ on the class of general Λ-frames.   ⊣

Properties of general frames

The following definition singles out some important properties and classes of general frames.

Definition 5.65 (Properties of General Frames) Let τ be a modal similarity type, and assume that $\mathfrak{g} = (\mathfrak{F}, A)$ is a general τ-frame, with $\mathfrak{F}$ being the Kripke τ-frame $(W, R_\Delta)_{\Delta \in \tau}$. Then $\mathfrak{g}$ is called

differentiated if for all s, t in W:

$$s = t \text{ iff } \forall a \in A\, (s \in a \iff t \in a),$$

tight if for all $\Delta \in \tau$ (assume that $\rho(\Delta) = n$) and for all $s, s_1, \ldots, s_n$ in W:

$$R_\Delta s s_1 \ldots s_n \text{ iff } \forall a_1, \ldots, a_n \in A\left((\textstyle\bigwedge_i s_i \in a_i) \Rightarrow s \in m_{R_\Delta}(a_1, \ldots, a_n)\right),$$

compact if $\bigcap A_0 \neq \varnothing$ for every subset A_0 of A which has the finite intersection property,

refined if $\mathfrak{g}$ is differentiated and tight,

descriptive if $\mathfrak{g}$ is refined and compact,

full if every subset of W is admissible, that is, if $A = \mathcal{P}(W)$, and

discrete if every singleton is admissible, that is, if $\{s\} \in A$ for every $s \in W$.   ⊣

Convention 5.66 When discussing general frames we will usually drop the adjective 'general' when it is clear that we are talking about general frames and not about Kripke frames. For instance, 'a descriptive frame' is short for: 'a descriptive general frame.' ⊣

Let us try to develop some intuitions concerning the notions defined above by considering a number of examples.

The best way to understand the concept of differentiation is by observing that a general frame is differentiated if and only if for every distinct pair of states s and t there is an admissible set a *witnessing* this difference in the sense that $s \in a$, $t \notin a$. Likewise (now we confine ourselves to the basic similarity type), a general frame is tight if and only if for every state s and every state t which is not accessible from s, there is an admissible set a witnessing this in the sense that $t \in a$, $s \notin m_{R_\Diamond}(a)$. Thus both differentiation and tightness are an indication that there are many admissible sets. On the other hand, compactness means that there are a multitude of *states*, the basic intuition being as follows. Consider a collection A_0 of admissible sets; if A_0 is not finitely contradictory (in the sense that there is a state in the intersection of any finite subcollection of A_0), then there is some state belonging to every set in A_0. The other definitions speak for themselves.

The notions of differentiation and tightness are independent, as is shown in the following example:

Example 5.67 Consider the structure $\mathfrak{h} = (\mathbb{N}, \equiv_2, A)$, where $\mathbb{N}$ is the set of natural numbers, $m \equiv_2 n$ iff $m - n$ is even, and A is the set $\{\varnothing, \mathbb{N}, E, O\}$, such that E and O are the sets of even and odd numbers, respectively. We leave it to the reader to check that this is indeed a general frame, and that $\mathfrak{h}$ is tight but not differentiated; see Exercise 5.5.6.

Conversely, let $\mathfrak{g}$ be the structure $(W, >, B)$ where W is the set $\mathbb{N} \cup \{\omega\}$ and B is given by

$$B = \{b \subseteq W \mid b \text{ is finite and } \omega \notin b\} \cup \{b \subseteq W \mid b \text{ is co-finite and } \omega \in b\}.$$

It is easy to see that $m_>(b)$ is co-finite and $\omega \in m_>(b)$ for every non-empty $b \subseteq W$. From this observation it is easy to deduce that $\mathfrak{g}$ is a general frame. It should also be obvious that $\mathfrak{g}$ is differentiated: for any two non-identical states s and t, at least one, say s, belongs to $\mathbb{N}$; but then the singleton $\{s\}$ is admissible, and $s \in \{s\}$, but $t \notin \{s\}$.

Now suppose that $\mathfrak{g}$ is tight. Consider an arbitrary co-finite set a such that $\omega \in a$. Then $a \neq \varnothing$, whence the set $m_>(a)$ is co-finite as well, and hence, $\omega \in m_>(a)$. But since a was arbitrary, by tightness we may infer that $\omega > \omega$, which is clearly not the case. It follows that $\mathfrak{g}$ is not tight. ⊣

An example of a refined frame is given by the structure $\mathfrak{c}$ of Example 5.63. In fact, its refinedness follows from the fact that $\mathfrak{c}$ is discrete.

Proposition 5.68 *Discrete frames are refined.*

Proof. We only treat tightness and confine ourselves to the basic similarity type. Let $\mathfrak{g} = (W, R, A)$ be a discrete frame. Assume that the state t is not a successor of the state s. Then $s \notin m_R(\{t\})$, while obviously $t \in \{t\}$. By definition of discreteness, $\{t\}$ is admissible.   ⊣

The converse of Proposition 5.68 does not hold. Examples of refined frames that are not discrete are provided by canonical general frames, cf. Proposition 5.69 below.

Of the classes mentioned in Definition 5.65, the class of descriptive frames is the most important one. One reason for this is that in a certain (category-theoretic) sense, descriptive general frames and boolean algebras with operators are the same mathematical objects. For instance, below we will see that every boolean algebra with operators can be represented as a descriptive general frame. Let us first consider a particular example of this.

Proposition 5.69 *Let τ be a modal similarity type, and Λ a normal modal τ-logic. Then $\mathfrak{f}_\Lambda^c$ is a descriptive general frame.*

Proof. To show that $\mathfrak{f}_\Lambda^c$ is differentiated, assume that Γ and Δ are distinct maximal Λ-consistent sets of formulas. In other words, there is a formula γ such that $\gamma \in \Gamma$, $\gamma \notin \Delta$. But then $\Gamma \in \widehat{\gamma}$, while $\Delta \notin \widehat{\gamma}$. It is almost as easy to prove tightness – simply use the definition of the canonical accessibility relation.

For compactness, assume that $S := \{\widehat{\sigma} \mid \sigma \in \Sigma\}$ is a collection of admissible sets and that it has the finite intersection property. It follows that Σ is consistent, for suppose otherwise; then there are $\sigma_1, \ldots, \sigma_n \in \Sigma$ such that $\vdash_\Lambda (\sigma_1 \wedge \cdots \wedge \sigma_n) \to \bot$. This implies that there is no maximal Λ-consistent Γ such that $\sigma_i \in \Gamma$ for all $i \le n$. But then $\widehat{\sigma}_1 \cap \cdots \cap \widehat{\sigma}_n = \varnothing$, which contradicts our assumption on S.

But if Σ is consistent, it can be extended to a maximal Λ-consistent set Σ^+. It is almost immediate that $\Sigma^+ \in \bigcap\{\widehat{\sigma} \mid \sigma \in \Sigma\}$; this obviously bears witness to the fact that S has a non-empty intersection.   ⊣

The following example shows that the condition of compactness really adds to the definition of a descriptive frame.

Example 5.70 In this example we will exhibit a general frame $\mathfrak{h} = (W, R, A)$ that is refined but not descriptive. To define W, let S and T denote the sets $S = \{s_n \mid n \in \mathbb{N}\}$ and $T = \{t_n \mid n \in \mathbb{N}\}$, respectively. Now, put

$$W = \{r, \infty\} \cup S \cup T,$$

$$R = \{(r, \infty)\} \cup \{(r, s_n), (s_n, s_n), (s_n, t_n) \mid n \in \mathbb{N}\}.$$

The set A is defined as follows: a subset a of W is admissible if either $a \cap T$ is finite and $\infty \notin a$, or $a \cap T$ is co-finite and $\infty \in a$.

We will not give a detailed proof that this $\mathfrak{h}$ is indeed a general frame. We will only observe here that for all $a \in A$, the set $m_R(a)$ satisfies $m_R(a) \cap T = \varnothing$ and $\infty \notin m_R(a)$, whence it is admissible as well.

To show that $\mathfrak{h}$ is refined, we first consider differentiation. Let v and w be different states of W. One of them, say v, must be different from ∞. But then the set $\{v\}$ is admissible, and $w \notin \{v\}$. For tightness, assume that Rvw does not hold, for some states v and w. We will define an admissible set a such that $w \in a$ while $v \notin m_R(a)$. Now, if $w \neq \infty$, then the set $a = \{w\}$ suffices, so assume that $w = \infty$. Then clearly, $v \neq r$. Again distinguish cases.

First assume that $v \in \{\infty\} \cup T$; in this case, take $a = \{\infty\} \cup T$, then $m_R(a) = \{r\} \cup S$, so indeed, $v \notin m_R(a)$. The only case that is left is where $w = \infty$ and $v = s_n$ for some natural number n. Now consider the set $a = (\{\infty\} \cup T) \setminus \{t_n\}$. Since t_n is the only successor of s_n, this gives $s_n \notin m_R(a)$.

Finally, we have to show that $\mathfrak{h}$ is not compact. But this is rather easy; consider for instance the set of co-finite subsets of S. This set has the finite intersection property, but there is no state in W that belongs to all co-finite subsets of S.   ⊣

Operations on general frames

In this subsection we look at the relation between ordinary frames, general frames and boolean algebras with operators. First, however, we adapt familiar notions from the theory of Kripke frames to the setting of general frames.

Definition 5.71 Let $\mathfrak{g} = (\mathfrak{F}, A)$ and $\mathfrak{g}' = (\mathfrak{F}', A')$ be two general frames. Assume that $\mathfrak{F} = (W, R_\Delta)_{\Delta \in T}$ and $\mathfrak{F}' = (W', R'_\Delta)_{\Delta \in T}$. A map $\theta : W \to W'$ is a *bounded morphism between* $\mathfrak{g}$ *and* $\mathfrak{g}'$ (notation: $\theta : \mathfrak{g} \to \mathfrak{g}'$) if θ is a bounded morphism between the frames $\mathfrak{F}$ and $\mathfrak{F}'$ such that

$$\theta^{-1}(a') \in A \text{ for all } a' \in A'. \tag{5.32}$$

Such a bounded morphism θ is called an *embedding* if it is injective and satisfies

$$\text{for all } a \in A \text{ there is an } a' \in A' \text{ such that } \theta[a] = \theta[W] \cap a'. \tag{5.33}$$

Here $\theta[a]$ is defined as the set $\{\theta(s) \mid s \in a\}$. We write $\mathfrak{g} \rightarrowtail \mathfrak{g}'$ to denote that $\mathfrak{g}$ can be embedded in $\mathfrak{g}'$.

The general frame $\mathfrak{g}'$ is called a *bounded morphic image* of $\mathfrak{g}$ (notation: $\mathfrak{g} \twoheadrightarrow \mathfrak{g}'$) if there is a surjective bounded morphism from $\mathfrak{g}$ to $\mathfrak{g}'$. Two general frames $\mathfrak{g}$ and $\mathfrak{g}'$ are *isomorphic* if there is a surjective embedding $\theta : \mathfrak{g} \to \mathfrak{g}'$.

Assume that for all $i \in I$, $\mathfrak{g}_i$ is the general frame $(\mathfrak{F}_i, A_i)$; we define the *disjoint union* $\biguplus_i \mathfrak{g}_i$ of this family $(\mathfrak{g}_i)_{i \in I}$ as the general frame $(\biguplus_i \mathfrak{F}_i, A)$ where A consists of those subsets $a \subseteq \bigcup_i W_i$ that satisfy $a \cap W_i \in A_i$ for all $i \in I$.   ⊣

As a special example of an embedding, consider two general frames $\mathfrak{g} = (\mathfrak{F}, A)$ and $\mathfrak{g}' = (\mathfrak{F}', A')$ such that $\mathfrak{F}$ is a generated subframe of $\mathfrak{F}'$. Then the identity inclusion $\iota : W \to W'$ is a bounded morphism from $\mathfrak{F}$ into $\mathfrak{F}'$, but ι is only a bounded morphism from $\mathfrak{g}$ to $\mathfrak{g}'$ if $a' \cap W \in A$ for all $a' \in A'$. For ι to be an *embedding*, condition (5.33) requires that for all $a \in A$, there is an $d \in A'$ such that $a = W \cap a'$; in other words, every admissible set $a \in A$ must be the 'W-part' of an admissible set $a' \in A'$. Finally, it is not difficult to see that an isomorphism between $\mathfrak{g} = (\mathfrak{F}, A)$ and $\mathfrak{g}' = (\mathfrak{F}', A')$ is just an isomorphism θ between $\mathfrak{F}$ and $\mathfrak{F}'$ such that θ^{-1} is an isomorphism between the complex algebras with carriers A and A'.

As with Kripke frames, the constructions defined in Definition 5.71 are truth-preserving. The proof of Proposition 5.72 below is straightforward and left to the reader.

Proposition 5.72 *Let τ be a modal similarity type, and ϕ a τ-formula.*

(i) *Let $\{\mathfrak{g}_i \mid i \in I\}$ be a family of general frames. Then $\biguplus \mathfrak{g}_i \Vdash \phi$ if $\mathfrak{g}_i \Vdash \phi$ for every i in I.*

(ii) *Assume that $\mathfrak{g}' \rightarrowtail \mathfrak{g}$. Then $\mathfrak{g}' \Vdash \phi$ if $\mathfrak{g} \Vdash \phi$.*

(iii) *Assume that $\mathfrak{g} \twoheadrightarrow \mathfrak{g}'$. Then $\mathfrak{g}' \Vdash \phi$ if $\mathfrak{g} \Vdash \phi$.*

Now we define a number of operations that enable the construction of general frames out of Kripke frames or out of BAOs, and conversely. The reader is advised to recall from earlier sections of this chapter the definition of the complex algebra $\mathfrak{F}^+$ of a frame $\mathfrak{F}$, and the ultrafilter frame $\mathfrak{A}_+$ of an algebra $\mathfrak{A}$.

Definition 5.73 Let τ be a modal similarity type, and let $\mathfrak{g} = (W, R_\Delta, A)_{\Delta \in \tau}$ be a general τ-frame. The *underlying τ-frame* of $\mathfrak{g}$ is given by $\mathfrak{g}_\sharp = (W, R_\Delta)_{\Delta \in \tau}$. It follows from the closure conditions on A that the structure

$$\mathfrak{g}^* = (A, \cup, -, \varnothing, m_{R_\Delta})_{\Delta \in \tau}$$

is a boolean algebra with τ-operators; this algebra is called the *underlying boolean algebra with τ-operators of $\mathfrak{g}$.*

Conversely, the *full τ-frame* of a τ-frame $\mathfrak{F} = (W, R_\Delta)_{\Delta \in \tau}$ is given by

$$\mathfrak{F}^\sharp = (\mathfrak{F}, \mathcal{P}(W)).$$

And finally, the *general ultrafilter frame* of a BAO $\mathfrak{A} = (A, +, -, 0, f_\Delta)_{\Delta \in \tau}$ is defined as

$$\mathfrak{A}_* = (\mathfrak{A}_+, \{\widehat{a} \mid a \in A\}),$$

where $\widehat{a} \subseteq Uf\mathfrak{A}$ is defined as the set of ultrafilters u such that $a \in u$:

$$\widehat{a} = \{u \in Uf\mathfrak{A} \mid a \in u\}. \qquad \dashv$$

There are many obvious connections between the operations $(\cdot)^+$, $(\cdot)_+$, $(\cdot)^\sharp$, $(\cdot)_\sharp$, $(\cdot)^*$ and $(\cdot)_*$. We list a few of these in the next proposition, and invite the reader to find more.

Proposition 5.74 *Let τ be a modal similarity type, and let $\mathfrak{F}$, $\mathfrak{g}$ and $\mathfrak{A}$ be a Kripke τ-frame, a general τ-frame and a boolean algebra with τ-operators, respectively. Then*

(i) $(\mathfrak{F}^\sharp)_\sharp = \mathfrak{F}$,

(ii) $(\mathfrak{F}^\sharp)^* = \mathfrak{F}^+$,

(iii) $(\mathfrak{A}_*)_\sharp = \mathfrak{A}_+$,

(iv) $(\mathfrak{g}_\sharp)^\sharp = \mathfrak{g}$ *if and only if $\mathfrak{g}$ is full.*

Next we devote our attention to the relation between BAOs and general frames. The following theorem is a counterpart of Proposition 5.24 for general frames.

Theorem 5.75 *Let τ be a modal similarity type, and let $\mathfrak{g}$ and $\mathfrak{A}$ be a general τ-frame and a boolean algebra with τ-operators, respectively. Then for every τ-formula ϕ:*

$$\mathfrak{g} \Vdash \phi \quad \text{iff} \quad \mathfrak{g}^* \models \phi \approx \top,$$
$$\mathfrak{A} \models \phi \approx \top \quad \text{iff} \quad \mathfrak{A}_* \Vdash \phi.$$

We omit the relatively easy proof of Theorem 5.75. Note however, that the theorem actually states something stronger than Proposition 5.24, namely that truth is also preserved when going from an algebra to its associated general frame. The point is that we may infer from $\mathfrak{A} \models \phi \approx \top$ that $\mathfrak{A}_* \Vdash \psi$ because on $\mathfrak{A}_*$ only *admissible* valuations over the Kripke frame $\mathfrak{A}_+$ are allowed. The implication 'if $\mathfrak{A} \models \phi \approx \top$, then $\mathfrak{A}_+ \Vdash \phi$' only holds for *canonical* formulas ϕ (we will come back to this point in the next section).

The following theorem substantiates our earlier claim that BAOs and descriptive general frames are really the same mathematical objects.

Theorem 5.76 *Let τ be a similarity type, and let $\mathfrak{g}$ and $\mathfrak{A}$ be a general τ-frame and a boolean algebra with τ-operators, respectively. Then*

(i) $\mathfrak{A}_*$ *is a descriptive general frame,*

(ii) $(\mathfrak{A}_*)^* \cong \mathfrak{A}$,

(iii) $(\mathfrak{g}^*)_* \cong \mathfrak{g}$ *if and only if $\mathfrak{g}$ is descriptive.*

Proof. For notational convenience, we assume that τ has only one modal operator $\triangle$, say, of arity n.

(i) This part of the proof is more or less the same as the proof that every canonical general frame is descriptive (see Proposition 5.69).

(ii) Let $\mathfrak{A} = (A, +, -, f)$ be a boolean algebra with operator. Then $\mathfrak{A}_*$ is a general frame whose underlying Kripke frame is of the form $(Uf\mathfrak{A}, Q_f)$, while the admissible sets are of the form $\widehat{a}$, where $a \in A$. Hence, the carrier of the algebra $(\mathfrak{A}_*)^*$ consists of all elements $\widehat{a}$, and its operator is of the form m_{Q_f}. (For the definition of Q and m we refer to Definitions 5.40 and 5.21, respectively). Clearly, it suffices to show that the map $r : A \to \mathcal{P}(Uf\mathfrak{A})$ given by

$$r : a \mapsto \widehat{a}$$

is a BAO-isomorphism. But it follows from the proof of Theorem 5.43 that r is an injective homomorphism, and surjectivity is immediate by the definitions.

(iii) The 'only if' direction follows immediately from item (i). For the other direction, assume that $\mathfrak{g} = (W, R, A)$ is a descriptive frame. For each state s in W, define U_s to be the set of admissible sets a such that $s \in a$. We claim that these sets constitute precisely the set of ultrafilters of $\mathfrak{g}^*$, that is,

$$\{U_s \mid s \in W\} = Uf\mathfrak{g}^*. \tag{5.34}$$

The inclusion '$\subseteq$' is obvious. For the other inclusion, let u be an arbitrary ultrafilter of the BAO $\mathfrak{g}^*$, or, to be more precise, of the boolean algebra $(A, \cup, -, \varnothing)$. Since ultrafilters are closed under intersection and never contain the empty set, it follows that u has the finite intersection property. By compactness, there is a state s such that $s \in \bigcap u$. This implies $u \subseteq U_s$; but since U_s is also an ultrafilter of $(A, \cup, -, \varnothing)$, we obtain that $u = U_s$. This proves (5.34).

Now consider the general frame $(\mathfrak{g}^*)_*$. Since its states are the ultrafilters of $\mathfrak{g}^*$, (5.34) implies that the map θ given by $\theta : s \mapsto U_s$ is a map from the universe W of $\mathfrak{g}$ onto the universe $Uf\mathfrak{g}^*$ of $(\mathfrak{g}^*)_*$. Injectivity follows from differentiation of $\mathfrak{g}$. Hence, the map $\theta : W \to Uf\mathfrak{g}^*$ is a bijection.

We will show that θ is an isomorphism between general frames. Let R' denote the accessibility relation of $\triangle$ in $(\mathfrak{g}^*)_*$. Unraveling the definition of R', we find that for ultrafilters $u, u_1, \ldots, u_n$ we have

$$R'uu_1 \ldots u_n \text{ iff } \forall a_1, \ldots, a_n \in A \left(\bigwedge a_i \in u_i \Rightarrow m_R(a_1, \ldots, a_n) \in u \right). \tag{5.35}$$

Now let $s, s_1, \ldots, s_n$ be arbitrary points of $\mathfrak{g}$. By tightness, we have

$$Rss_1 \ldots s_n \text{ iff}$$
$$\forall a_1, \ldots, a_n \in A \left(\bigwedge s_i \in a_i \Rightarrow s \in m_R(a_1, \ldots, a_n) \right). \tag{5.36}$$

But by definition of θ, we have $t \in a$ iff $a \in \theta(t)$, for all $t \in W$. Hence, from

(5.36) it follows that

$$Rss_1 \ldots s_n \text{ iff}$$

$$\forall a_1, \ldots, a_n \in A \left(\bigwedge a_i \in \theta(s_i) \Rightarrow m_R(a_1, \ldots, a_n) \in \theta(s) \right). \quad (5.37)$$

But then we may infer from (5.35) and (5.37) that

$$Rss_1 \ldots s_n \text{ iff } R'\theta(s)\theta(s_1) \ldots \theta(s_n).$$

In other words, θ is an isomorphism between the underlying Kripke frames of $\mathfrak{g}$ and $(\mathfrak{g}^*)_*$.

Finally, consider an arbitrary admissible set a of $\mathfrak{g}$. We will show that $\theta(a)$ is an admissible set of $(\mathfrak{g}^*)_*$. By definition of the operation $(\cdot)^*$, a is a member of the carrier of $\mathfrak{g}^*$; hence by definition of the operation $(\cdot)_*$, the set $\widehat{a} := \{u \in Uf(\mathfrak{g}^*) \mid a \in u\}$ is an admissible set of $(\mathfrak{g}^*)_*$. We leave it to the reader to verify that $\widehat{a} = \theta(a)$ and hence, that $a = \theta^{-1}[\widehat{a}]$. ⊣

As an immediate corollary of Theorems 5.75 and 5.76 we have that every general frame has an equivalent descriptive general frame.

Theorem 5.77 *Let $\mathfrak{g}$ be a general τ-frame for some similarity type τ. Then $(\mathfrak{g}^*)_*$ is a descriptive general frame equivalent to $\mathfrak{g}$; that is, for every τ-formula ϕ:*

$$\mathfrak{g} \Vdash \phi \text{ iff } (\mathfrak{g}^*)_* \Vdash \phi.$$

Just as in the case of the basic duality between Kripke frames and BAOs, we can extend the constructions of Definition 5.73 to *morphisms* between algebras or between general frames.

Definition 5.78 Let τ be a modal similarity type, and let $\mathfrak{g} = (W, R_\Delta, A)_{\Delta \in \tau}$ and $\mathfrak{g}' = (W', R'_\Delta, A')_{\Delta \in \tau}$ be two general τ-frames. Given a bounded morphism $\theta : W \twoheadrightarrow W'$, its *dual* $\theta^* : A' \to A$ is defined by

$$\theta^*(a') = \theta^{-1}[a'] \ (= \{s \in W \mid \theta s \in a'\}).$$

Now let $\mathfrak{A} = (A, +, -, 0, f_\Delta)_{\Delta \in \tau}$ and $\mathfrak{A}' = (A', +', -', 0', f'_\Delta)_{\Delta \in \tau}$ be two boolean algebras with τ-operators, and η a map from A to A'. Then we define the *dual* η_* of η to be the following map from $Uf\mathfrak{A}'$ to $\mathcal{P}(A)$:

$$\eta_*(u') = \eta^{-1}[u'] \ (= \{a \in A \mid \eta a \in u'\}). \quad ⊣$$

For the maps defined in Definition 5.78 we can prove results analogous to Propositions 5.51 and 5.52.

Proposition 5.79 *Let τ be a modal similarity type, and let $\mathfrak{g}$ and $\mathfrak{h}$ be two general τ-frames, and θ a map from the universe of $\mathfrak{g}$ to the universe of $\mathfrak{h}$. Then*

(i) *If θ is a bounded morphism, then θ^* is a BAO-homomorphism from $\mathfrak{h}^*$ to $\mathfrak{g}^*$.*

(ii) *If $\theta : \mathfrak{g} \rightarrowtail \mathfrak{h}$, then $\theta^* : \mathfrak{h}^* \twoheadrightarrow \mathfrak{g}^*$.*

(iii) *If $\theta : \mathfrak{g} \twoheadrightarrow \mathfrak{h}$, then $\theta^* : \mathfrak{h}^* \rightarrowtail \mathfrak{g}^*$.*

Proposition 5.80 *Let τ be a modal similarity type, and let $\mathfrak{A}$ and $\mathfrak{B}$ be two boolean algebras with τ-operators, and η a map from A to B. Then*

(i) *If η is a BAO-homomorphism, then η^* is a bounded morphism from $\mathfrak{B}_*$ to $\mathfrak{A}_*$.*

(ii) *If $\eta : \mathfrak{A} \rightarrowtail \mathfrak{B}$, then $\eta_* : \mathfrak{B}_* \twoheadrightarrow \mathfrak{A}_*$.*

(iii) *If $\eta : \mathfrak{A} \twoheadrightarrow \mathfrak{B}$, then $\eta_* : \mathfrak{B}_* \rightarrowtail \mathfrak{A}_*$.*

The proofs of Propositions 5.79 and 5.80 are similar to the proofs of Propositions 5.51 and 5.52. We leave it as an exercise to check where conditions (5.32) and (5.33) are needed in the proof of Proposition 5.79; see Exercise 5.5.8.

By combining Theorem 5.76 and Propositions 5.79 and 5.80 together, we add further support to our claim that boolean algebras with operators and descriptive general frames are really equivalent objects.

Theorem 5.81 *Let τ be a modal similarity type, and let $\mathfrak{g}$ and $\mathfrak{h}$ be two descriptive general frames, and $\mathfrak{A}$ and $\mathfrak{B}$ two boolean algebras with τ-operators. Then*

(i) *$\mathfrak{g} \rightarrowtail \mathfrak{h}$ iff $\mathfrak{h}^* \twoheadrightarrow \mathfrak{g}^*$,*

(ii) *$\mathfrak{g} \twoheadrightarrow \mathfrak{h}$ iff $\mathfrak{h}^* \rightarrowtail \mathfrak{g}^*$,*

(iii) *$\mathfrak{A} \rightarrowtail \mathfrak{B}$ iff $\mathfrak{B}_* \twoheadrightarrow \mathfrak{A}_*$,*

(iv) *$\mathfrak{A} \twoheadrightarrow \mathfrak{B}$ iff $\mathfrak{B}_* \rightarrowtail \mathfrak{A}_*$.*

For disjoint unions we cannot obtain an equivalence with a simple algebraic operation in the spirit of Theorem 5.81. The reason for this is that the disjoint union of an *infinite* family of frames can never be descriptive itself. This does not cast a shadow on the duality between descriptive general frames and BAOs, it simply indicates that the disjoint union of descriptive frames as we defined it in Definition 5.71, is not the proper categorical notion of a sum. For that one has to take the *descriptive union* $((\biguplus_i \mathfrak{g}_i)^*)_*$ of a family $(\mathfrak{g}_i)_{i \in I}$ of descriptive frames; see Exercise 5.5.10.

Topology

The reader familiar with topology will have realized that we are drawing on a number of topological concepts. Examples are the notions of compactness and of differentiation (the latter being very similar to topological separation axioms). Furthermore, given a general frame $\mathfrak{g} = (W, R, A)$ for the basic similarity type, one can consider A as a base for a topology $\mathcal{T}_A$. The general frame $\mathfrak{g}$ is descriptive iff $(W, \mathcal{T}_A)$ is a boolean space with A as the set of clopens, and R is a point-closed relation; that is, $R[s]$ (defined as $\{t \in W \mid Rst\}$) is closed for every s in W.

Indeed the whole theory of (descriptive) general frames is permeated with topological concepts, and an entire chapter could be devoted to the topic. We will restrict ourselves to a brief discussion of closed sets.

Definition 5.82 Let $\mathfrak{g} = (\mathfrak{F}, A)$ be a general frame. A subset c of the universe is called *closed* if it is the intersection of a (possibly infinite) collection of admissible sets, or equivalently, if

$$c = \bigcap \{a \in A \mid c \subseteq a\}. \qquad \dashv$$

By way of example, consider the canonical general frame $\mathfrak{f}_\Lambda$ of any normal modal logic Λ. Here, an admissible set a is of the form $\hat{\phi}$ for some formula ϕ. We could also say that a represents the formula ϕ, since by the Truth Lemma, a consists of those states s such that ϕ is true at s. Likewise, a closed set c, being the intersection of a family $\{\hat{\sigma} \mid \sigma \in \Sigma\}$, represents the *set of formulas* Σ in $\mathfrak{f}_\Lambda$. For, we have that (every formula from) Σ holds at a state s iff $s \in c$.

Proposition 5.83 *Let $\mathfrak{g}$ be a descriptive general frame. Then*

- (i) *Every singleton is closed.*
- (ii) *The collection of closed sets is closed under finite unions and arbitrary intersections.*
- (iii) *If c is a closed set, then so is the set $R_\Diamond[c]$ (defined as $R_\Diamond[c] = \{t \mid \exists s \in c \; R_\Diamond st\}$), for every diamond $\Diamond$.*
- (iv) *For every state s and every sequence β of diamonds, the set $R_\beta[s]$ is closed.*
- (v) *Let C be a family of closed sets with the finite intersection property. Then C has a non-empty intersection.*

Proof. Assume that $\mathfrak{g} = (\mathfrak{F}, A)$ is a descriptive frame, and that $\mathfrak{F}$ is the frame $(W, R_\Delta)_{\Delta \in \tau}$.

(i) We claim that for every point $s \in W$:

$$\{s\} = \bigcap \{a \in A \mid s \in a\}.$$

In fact, this claim is nothing but a reformulation of the differentiation condition.

(ii) We only sketch the proof for finite unions. Suppose that $c_i = \bigcap \{a \in A \mid c_i \subseteq a\}, i \in \{0, 1\}$. We leave it to the reader to verify that

$$c_0 \cup c_1 = \bigcap \{a_0 \cup a_1 \mid c_0 \subseteq a_0 \text{ and } c_1 \subseteq a_1\}.$$

(iii) Assume that c is a closed set, that is, that $c = \bigcap \{a \in A \mid c \subseteq a\}$. Let $\Diamond$ be a diamond of τ; we abbreviate $R_\Diamond$ by R. We will prove that

$$R[c] = \bigcap\{b \in A \mid c \subseteq l_R(b)\}. \tag{5.38}$$

The left to right inclusion in (5.38) is trivial. For the other inclusion, assume that $t \notin R[c]$ while $t \in b$ whenever $c \subseteq l_R(b)$. We will derive a contradiction from this. Define

$$A_0 = \{a \in A \mid c \subseteq a\} \cup \{m_R(b) \mid t \in b \text{ and } b \in A\}.$$

We first prove that

$$A_0 \text{ has the finite intersection property.} \tag{5.39}$$

For, suppose that there is a finite subcollection of A_0 with an empty intersection. This means that there are $a_1, \dots, a_m$ and $b_1, \dots, b_m$ in A such that $c \subseteq a_i$ for all $i \le m$, $t \in b_j$ for all $j \le n$ and

$$a_1 \cap \cdots \cap a_m \cap m_R(b_1) \cap \cdots \cap m_R(b_n) = \varnothing.$$

Define $a = a_1 \cap \cdots \cap a_m$ and $b = b_1 \cap \cdots \cap b_n$, then we have $a, b \in A$, $c \subseteq a$, $t \in b$ and

$$a \cap m_R(b) = \varnothing \tag{5.40}$$

since $m_R(b) \subseteq m_R(b_1) \cap \cdots \cap m_R(b_n)$. But (5.40) implies that $a \subseteq l_R(-b)$, whence $c \subseteq l_R(-b)$. It follows from the assumption that $t \in -b$. But $t \in -b$ contradicts $t \in b$. This proves (5.39).

Now since $\mathfrak{g}$ is a *compact* frame, (5.39) implies that A_0 has a non-empty intersection, so there is an s in W with

$$s \in \bigcap\{a \mid c \subseteq a\} \tag{5.41}$$

and

$$s \in \bigcap\{m_R(b) \mid t \in b\}. \tag{5.42}$$

It follows immediately from (5.41) that $s \in c$. But also, it follows from (5.42) that Rst. For suppose otherwise: by tightness of $\mathfrak{g}$, there would be a $b \in A$ witnessing $\neg Rst$; that is, $t \in b$ while $s \notin m_R(b)$. But this clearly contradicts (5.42).

(iv) The reader is asked to supply this proof in Exercise 5.5.11.

(v) Let $\{c_i \mid i \in I\}$ be a collection of closed sets with the finite intersection property. By definition of closedness, for every $i \in I$ there is a collection $\{a_{ij} \mid j \in J_i\}$ such that $c_i = \bigcap_{j \in J_i} a_{ij}$. It is rather easy to see that the collection $\{a_{ij} \mid i \in I, j \in J_i\}$ has the finite intersection property. Hence, by compactness,

$$\bigcap_{i \in I, j \in J_i} a_{ij} \ne \varnothing.$$

But then

$$\bigcap_{i \in I} c_i = \bigcap_{i \in I} \bigcap_{j \in J_i} a_{ij} = \bigcap_{i \in I, j \in J_i} a_{ij} \neq \varnothing. \qquad \dashv$$

The main result in the next section is the Sahlqvist Completeness Theorem, and closed sets will play a key role in our proof of it. Phrased in terms of general frames, canonicity is the following question. Given a descriptive general frame and a formula ϕ that is valid on it, can we infer that ϕ is valid on the underlying Kripke frame as well? That is, if we know that any admissible valuation makes ϕ true everywhere in the frame, does the same hold for an arbitrary valuation? The role of closed sets in the answer to this question is that we will need to consider closed valuations (that is, valuations that map variables to closed sets); the crucial observation then will be that such closed valuations behave almost as well as admissible ones.

Exercises for Section 5.5

5.5.1 Which properties of general frames (for example: refinedness, fullness,...) are preserved under taking generated subframes, bounded morphic images or disjoint unions?

5.5.2 Let $\mathfrak{f} = (\mathfrak{F}, A)$ be a refined general frame. Prove that the following are equivalent:

(a) $\mathfrak{f}$ is compact,
(b) every ultrafilter of A is of the form $\{a \in A \mid s \in a\}$ for some state s.

5.5.3 Prove that a full frame is descriptive if and only if it is finite.

5.5.4 Let $\mathfrak{f}^c_\Lambda$ be the canonical general frame for the normal modal logic Λ. Prove that $\mathfrak{f}^c_\Lambda \Vdash \Lambda$.

5.5.5 Consider the structure $\mathfrak{c} = (\mathfrak{C}, F)$ of Example 5.63.

(a) Prove that $\mathfrak{c}$ is a general frame if we view $\mathfrak{C}$ as a frame for the basic modal similarity type.
(b) What happens if we view $\mathfrak{C}$ as a standard frame for the basic *tense* similarity type?

5.5.6 Consider the structure $\mathfrak{h}$ defined in Example 5.67.

(a) Show that $\mathfrak{h}$ is a general frame.
(b) Show that $\mathfrak{h}$ is tight but not differentiated.

5.5.7 Let τ be a similarity type, and let $\mathfrak{F}$, $\mathfrak{f}$ and $\mathfrak{g}$ be a τ Kripke frame and two general τ-frames, respectively. Prove or disprove the following:

(a) $(\mathfrak{F}^\sharp)_\sharp = \mathfrak{F}$,
(b) $\mathfrak{g} \rightarrowtail (\mathfrak{g}_\sharp)^\sharp$,
(c) $\mathfrak{g}^* \cong \mathfrak{f}^*$ only if $\mathfrak{g} \cong \mathfrak{f}$.

5.5.8 Prove Propositions 5.79 and 5.80.

5.5.9 Given a frame $\mathfrak{F}$, let $\mathfrak{F}^\circ$ denote the subalgebra of $\mathfrak{F}^+$ that is generated by the atoms of $\mathfrak{F}^+$ (that is, by the singleton subsets of the frame). Prove the following:

(a) If ϕ is a canonical formula (as defined after Open Problem 1), then $\mathfrak{F} \Vdash \phi$ only if $(\mathfrak{F}^\circ)_+ \Vdash \phi$.

(b) Show that the converse does not always hold.

5.5.10 Show that the disjoint union of an infinite family of general frames can never be descriptive, even if those frames are descriptive. Develop and describe the notion of a descriptive union $((\biguplus_i \mathfrak{g}_i)^*)_*$ of a family $(\mathfrak{g}_i)_{i \in I}$ of descriptive frames.

5.5.11 Let $\mathfrak{g}$ be a descriptive general frame. Prove that for every state s and every sequence β of diamonds, the set $R_\beta[s]$ is closed. That is, prove item (iv) of Proposition 5.83.

5.6 Persistence

This section is devoted to the notion of persistence, a generalization of the concept of canonicity. With the help of this notion we will show that all Sahlqvist formulas are canonical.

Definition 5.84 Let τ be a similarity type. Let ϕ be a τ-formula, and let X denote a property (or class) of general frames (as defined in Definition 5.65). Then ϕ is called X-*persistent* (or *persistent with respect to* X), if, for every general τ-frame $\mathfrak{g}$ in X, $\mathfrak{g} \Vdash \phi$ only if $\mathfrak{g}_\sharp \Vdash \phi$. Persistence with respect to the classes of refined, descriptive and discrete frames is called *r-persistence*, *d-persistence* and *di-persistence*, respectively.   ⊣

The best way to understand persistence is as follows. Let $\mathfrak{g} = (W, R, A)$ be a general frame, and ϕ a modal formula (in the basic similarity type). First, note that the implication '$\mathfrak{g}_\sharp \Vdash \phi \Rightarrow \mathfrak{g} \Vdash \phi$' is immediate: if $V(\phi) = W$ for every valuation, then certainly $V(\phi) = W$ for every admissible one. The converse implication, which will obviously not hold in general, indicates that in order to determine whether ϕ holds in the underlying Kripke frame $\mathfrak{g}_\sharp = (W, R)$, it suffices to look at admissible valuations only. If this is always the case when we take the general frame from a given class X of general frames, then we call ϕ X-persistent.

There is also an algebraic interpretation of the notion of X-persistence. This stems from the observation that given a general frame $\mathfrak{g}$, its associated algebra $\mathfrak{g}^*$ is a *subalgebra* of the full complex algebra $(\mathfrak{g}_\sharp)^+$ of the underlying Kripke frame $\mathfrak{g}_\sharp$ of $\mathfrak{g}$. Preservation of validity under taking subalgebras means that $(\mathfrak{g}_\sharp)^+ \models s \approx t \Rightarrow \mathfrak{g}^* \models s \approx t$ for every equation $s \approx t$; again, the preservation of validity in the other direction will not always hold. The concept of persistence gives us a way of describing the special situation (in the sense that $\mathfrak{g}^*$ is a special kind of subalgebra of $(\mathfrak{g}_\sharp)^+$) where this 'upward preservation' does hold.

From the discussion of descriptive frames in Section 5.5 it is easy to see that d-persistence and canonicity are really the same notion (provided that we have a logical notion of canonicity as defined after Open Problem 1).

Proposition 5.85 *Let ϕ be a modal formula in a similarity type τ. Then ϕ is canonical if and only if ϕ is d-persistent.*

Example 5.86 As a first example, we will prove that $\Box p \to p$ is r-persistent. Let $\mathfrak{g} = (W, R, A)$ be a refined frame such that $\mathfrak{g} \Vdash \Box p \to p$. We have to show that $\mathfrak{g}_\sharp \Vdash \Box p \to p$. By Sahlqvist correspondence, this is equivalent to showing that $\mathfrak{g}_\sharp \models \forall x\, Rxx$.

Suppose, in order to arrive at a contradiction, that there is an *irreflexive* state s in $\mathfrak{g}$. It follows from *tightness* that there must exist a set $a \in A$ such that $s \in a$, but $s \notin m_R(a)$. Now consider the valuation V given by $V(p) = -a$. Clearly V is admissible, and since $s \in l_R(-a)$ and $s \notin -a$, we find that $\mathfrak{g}, V, s \nVdash \Box p \to p$. Hence, we have contradicted the assumption that $\mathfrak{g} \Vdash \Box p \to p$. ⊣

Example 5.87 The formula $\Diamond p \to \Diamond\Diamond p$ is di-persistent, but not r-persistent. For di-persistence, let $\mathfrak{g} = (W, R, A)$ be a discrete frame such that $\mathfrak{g} \Vdash \Diamond p \to \Diamond\Diamond p$. Denote the underlying Kripke frame of $\mathfrak{g}$ by $\mathfrak{F}$ – we have to prove that $\Diamond p \to \Diamond\Diamond p$ is valid in $\mathfrak{F}$. Again we use Sahlqvist correspondence, by which we may confine ourselves to proving that $\mathfrak{F} \models \forall x x_1 \, (Rxx_1 \to \exists z\, (Rxz \land Rzx_1))$. Let s and s_1 be such that Rss_1. Following Example 3.7, we define the *minimal* valuation V_m by

$$V_m(p) = \{s_1\}.$$

Our first observation is that V_m is admissible, since $\mathfrak{g}$ is discrete. This means that we can proceed as in Example 3.7. Our second observation is that

$$(\mathfrak{F}, V_m), s \Vdash \Diamond\Diamond p \text{ iff } \mathfrak{F} \models \exists z\, (Rxz \land Rzs_1)[s].$$

But $(\mathfrak{F}, V_m), s \Vdash \Diamond\Diamond p$ is immediate by the validity of $\Diamond p \to \Diamond\Diamond p$ in $\mathfrak{g}$.

To show that the formula $\Diamond p \to \Diamond\Diamond p$ is not r-persistent, we consider the refined frame $\mathfrak{h}$ of Example 5.70. It is rather easy to see that $\mathfrak{h}_\sharp \nVdash \Diamond p \to \Diamond\Diamond p$; take for instance the valuation V given by $V(p) = \{\infty\}$. Then $r \Vdash \Diamond p$, but r has no successors where $\Diamond p$ is true. Hence, $r \nVdash \Diamond\Diamond p$.

On the other hand, we will now show that $\mathfrak{h} \Vdash \Diamond p \to \Diamond\Diamond p$. Consider a valuation V and a state x of $\mathfrak{h}$ such that $(\mathfrak{h}, V), x \Vdash \Diamond p$. We have to show that $x \Vdash \Diamond\Diamond p$. From $x \Vdash \Diamond p$ it follows that there is a y with Rxy and $y \Vdash p$. The only tricky case is where $x = r$ and $y = \infty$ – all the other cases are easily solved by using the reflexivity of the states in S. But if $\infty \in V(p)$ and $V(p)$ is admissible, then $V(p) \cap T$ is a co-finite set. Hence, there is a state t_n such that $t_n \Vdash p$. But then $s_n \Vdash \Diamond p$ and $r \Vdash \Diamond\Diamond p$. ⊣

Convention 5.88 Fix a modal similarity type τ and a set of propositional variables Φ. In Section 2.4 we saw that models for this language can also be seen as structures for the corresponding first-order language $\mathcal{L}^1_\tau(\Phi)$. In this and the following section we will make good use of this fact; in fact, it will be very convenient to have a sort of 'hybrid' notation, something between modal and first-order logic.

Let $\mathfrak{F}$ be a τ-frame, V a valuation on $\mathfrak{F}$ and $\vec{s}$ a finite sequence of states in $\mathfrak{F}$. Further, consider a first-order formula $\rho(\vec{x})$ in $\mathcal{L}^1_\tau(\Phi)$. We will write $(\mathfrak{F}, V), \vec{s} \Vdash \rho$ to denote $(\mathfrak{F}, V) \models \rho[\vec{s}]$. Here we assume that $[\vec{s}]$ is an assignment sending x_0 to s_0, x_1 to s_1, and so on. ⊣

Example 5.89 Although the formula $\Diamond_1 \Box_2 p \to \Box_3 \Diamond_4 p$ is d-persistent, it is not di-persistent. A counterexample to di-persistence is provided by a simple modification of the discrete frame $\mathfrak{c}$ of Example 5.63.

For d-persistence, let $\mathfrak{g} = (W, R, A)$ be a descriptive frame; define $\mathfrak{F} = (W, R)$. Assume that the formula holds on $\mathfrak{g}$; we will prove that it is also valid on $\mathfrak{F}$. By Sahlqvist correspondence it suffices to show that

$$\mathfrak{F} \models \forall x x_1 (R_1 x x_1 \to \forall z_0 (R_3 x z_0 \to \exists z_1 (R_2 x_1 z_1 \wedge R_4 z_0 z_1))). \tag{5.43}$$

Let s and s_1 be states in W such that $R_1 s s_1$. Define the *minimal* valuation V_m by

$$V_m(p) = \{u \in W \mid R_2 s_1 u\}.$$

Now assume that V_m is admissible. Then it follows from $(\mathfrak{F}, V_m), s \Vdash \Diamond_1 \Box_2 p$ that $(\mathfrak{F}, V_m), s \Vdash \Box_3 \Diamond_4 p$. Hence, every R_3-successor t_0 of s has its R_4-successor z_1 such that $(\mathfrak{F}, V_m), t_1 \Vdash p$. But by definition of V_m, $t_1 \Vdash p$ iff $R_2 s_1 t_1$. This would prove (5.43).

However, the problem is that V_m need not be admissible. Nevertheless, a close inspection of the 'proof' in the preceding paragraph reveals that we only need that in $(\mathfrak{F}, V_m)$ at the pair s, s_1 the following formula holds:

$$(R_1 x x_1 \wedge \forall y (R_2 x_1 y \to Py)) \to \forall z_0 (R_3 x z_0 \to \exists z_1 (R_4 z_0 z_1 \wedge P z_1)). \tag{5.44}$$

We will not prove (5.44) here; it follows from a far more general claim in the proof of Theorem 5.91. The two important observations in the proof are that $V_m(p)$ is a closed set, and that the right-hand side of the formula in (5.44) is the standard translation of a positive formula.

We will give three theorems concerning persistence. The most important result is Theorem 5.91 stating that all Sahlqvist formulas are d-persistent and hence, canonical. As a warming-up exercise for its proof, we show that every very simple Sahlqvist formula is di-persistent (Theorem 5.90). Finally, Theorem 5.92 states that every r-persistent formula is elementary; that is, it has a first-order frame correspondent.

Theorem 5.90 *Every very simple Sahlqvist formula is di-persistent.*

Proof. Recall from Definition 3.41 that a very simple Sahlqvist formula is of the form $\phi \to \psi$, where

- ϕ is built up from atoms, using conjunctions and (existential) modal operators, and
- ψ is positive.

Now let $\chi \equiv \phi \to \psi$ be such a very simple Sahlqvist formula, and let $\mathfrak{g} = (\mathfrak{F}, A)$ be a discrete general frame such that $\mathfrak{g} \Vdash \chi$. We want to prove that $\mathfrak{F} \Vdash \chi$. From the discussion in the proof of Theorem 3.42 we know that this is equivalent to the following, cf. (3.10):

$$\mathfrak{F} \models \forall P_1 \dots \forall P_n \forall x \forall x_1 \dots \forall x_m \, (\text{REL} \wedge \text{AT} \to \text{POS}),$$

where

REL is a conjunction of first-order statements of the form $R_\triangle x_0 \dots x_n$, corresponding to occurrences of (existential) modal operators in ϕ,
AT is a conjunction of translations of atomic formulas (these correspond to the proposition letters and constants in ϕ), and
POS is the standard translation of the formula ψ.

Let $\vec{s}$ be a fixed but arbitrary sequence of states in $\mathfrak{F}$, and define valuation V_m by

$$V_m(p) = \{ s_i \mid Px_i \text{ is a conjunct of AT } \}.$$

It is easy to see that V_m is the *minimal* valuation U such that $(\mathfrak{F}, U), \vec{s} \Vdash \text{AT}$; in other words, $(\mathfrak{F}, U), \vec{s} \Vdash \text{AT}$ only if U is an extension of V_m. (V_m is in fact the semantical version of the minimal substitution that we used in the proof of the correspondence theoretic part of Sahlqvist's Theorem in Chapter 3.)

The key observations to the proof of this theorem are (5.45) and (5.46) below:

$$V_m \text{ is admissible} \tag{5.45}$$

and

$$(\mathfrak{F}, V_m), \vec{s} \Vdash (\text{REL} \wedge \text{AT}) \to \text{POS} \text{ iff}$$
$$(\mathfrak{F}, V), \vec{s} \Vdash (\text{REL} \wedge \text{AT}) \to \text{POS} \text{ for } every \text{ valuation } V. \tag{5.46}$$

The assumption that χ is true in every admissible model over $\mathfrak{F}$ means we must have that $(\mathfrak{F}, U), \vec{s} \Vdash (\text{REL} \wedge \text{AT}) \to \text{POS}$ for every admissible valuation U. Hence, the theorem follows immediately from (5.45) and (5.46).

The proof of (5.45) is rather easy: for every p, $V_m(p)$ is a finite union of finitely many singletons, and, since A contains all singletons and is closed under unions, each $V_m(p)$ is in A.

Concerning (5.46), the direction from right to left is of course trivial; for the other direction, assume that $(\mathfrak{F}, V_m), \vec{s} \Vdash (\text{REL} \wedge \text{AT}) \to \text{POS}$, and that $(\mathfrak{F}, V), \vec{s} \Vdash$ REL $\wedge$ AT. From $(\mathfrak{F}, V), \vec{s} \Vdash$ AT it follows that V is an extension of V_m. From $(\mathfrak{F}, V), \vec{s} \Vdash$ REL it follows that, regardless of the valuation U, $(\mathfrak{F}, U), \vec{s} \Vdash$ REL. In particular, we have $(\mathfrak{F}, V_m), \vec{s} \Vdash$ REL. But then $(\mathfrak{F}, V_m), \vec{s} \Vdash$ REL $\wedge$ AT and hence, $(\mathfrak{F}, V_m), \vec{s} \Vdash$ POS. Since V is an extension of V_m, the monotonicity of positive formulas implies that $(\mathfrak{F}, V), \vec{s} \Vdash$ POS.   ⊣

Theorem 5.91 *Every Sahlqvist formula is d-persistent and hence, canonical.*

Proof. We will prove the theorem for simple Sahlqvist formulas in a modal similarity type containing diamonds only. Recall that such formulas are of the form $\phi \to \psi$ where

- ϕ is built up from constants and boxed atoms (cf. Definition 3.45), using conjunctions and (existential) modal operators only, and
- ψ is positive.

Let $\chi \equiv \phi \to \psi$ be such a formula, and let $\mathfrak{g} = (\mathfrak{F}, A)$ be a descriptive general frame such that $\mathfrak{g} \Vdash \chi$. We have to show that χ is valid on the underlying Kripke frame $\mathfrak{F}$ of $\mathfrak{g}$. Again, by Sahlqvist *correspondence* this is equivalent to the following, cf. (3.10):

$$\mathfrak{F} \models \forall P_1 \ldots \forall P_n \forall x \forall x_1 \ldots \forall x_m \,(\text{REL} \wedge \text{BOX-AT} \to \text{POS}), \qquad (5.47)$$

where

REL is a conjunction of first-order statements of the form $R_\diamond x_i x_j$ (corresponding to occurrences of diamonds in ϕ),

BOX-AT is a conjunction of formulas of the form $\forall y \,(R_\beta x_i y \to Py)$, corresponding to the occurrences $\Box_\beta p$ of boxed atoms in ϕ (see Definition 3.45), and

POS is the standard translation of the formula ψ.

As in the proof of the previous theorem, statement (5.47) itself is equivalent to the following:

for every $\vec{s}$ in $\mathfrak{F}$: for every valuation V, $(\mathfrak{F}, V), \vec{s} \Vdash (\text{REL} \wedge \text{BOX-AT} \to \text{POS})$.

We fix a sequence $\vec{s}$ of states in $\mathfrak{F}$, and define a valuation V_m by

$$V_m(p) = \bigcup \{R_\beta[s_i] \mid \text{the formula } \forall y \,(R_\beta x_i y \to Py) \text{ occurs in BOX-AT}\}.$$

Again, V_m is the *minimal* valuation U such that $(\mathfrak{F}, U), \vec{s} \Vdash$ BOX-AT, and with equal ease as in the proof of Theorem 5.90 we can show that

$$\begin{aligned}(\mathfrak{F}, V_m), \vec{s} \Vdash (\text{REL} \wedge \text{BOX-AT}) &\to \text{POS iff} \\ (\mathfrak{F}, V), \vec{s} \Vdash (\text{REL} \wedge \text{BOX-AT}) &\to \text{POS for } every \text{ valuation } V,\end{aligned} \qquad (5.48)$$

but – and this is different from the previous case – V_m need not be admissible now. Nevertheless, it does hold that

$$(\mathfrak{F}, V_m), \vec{s} \Vdash (\text{REL} \wedge \text{BOX-AT}) \to \text{POS}, \qquad (5.49)$$

but we have to work much harder to prove it. The remainder of the proof is devoted to establishing (5.49) – note that the theorem follows immediately from (5.48) and (5.49).

We first define the following concepts. For valuations U and V, let $U \lessdot V$ mean that V is an admissible extension of U. (We say that V is an extension of U if $U(p) \subseteq V(p)$ for every proposition letter p.) It is easy to see that the set of valuations on a given frame is closed under taking argument-wise intersections and unions.

Recall that a set $c \subseteq W$ is *closed* if it is the intersection of a (possibly infinite) set of admissible sets. Now we call a valuation V *closed* if $V(p)$ is a closed set for every proposition letter p. It is then not very difficult to prove that a valuation U is closed iff $U = \bigcap_{U \lessdot V} V$.

Now (5.49) follows from (5.50) and (5.51) below:

$$V_m \text{ is closed}, \qquad (5.50)$$

and

$$\begin{array}{l} \text{if } U \text{ is a closed valuation and } \gamma \text{ a positive formula,} \\ \text{then } U(\gamma) = \bigcap_{U \lessdot V} V(\gamma). \end{array} \qquad (5.51)$$

For, suppose that we have proved (5.50) and (5.51), and assume that

$$(\mathfrak{F}, V_m), \vec{s} \Vdash \text{REL} \wedge \text{BOX-AT}.$$

Let V be an arbitrary admissible valuation such that $V_m \lessdot V$. Then $(\mathfrak{F}, V), \vec{s} \Vdash$ REL $\wedge$ BOX-AT, so by the assumption, $(\mathfrak{F}, V), s \Vdash$ POS. If we look up again what formula POS is, we see that this means $s \in V(\psi)$, where ψ was the consequent of the Sahlqvist formula. Since V was arbitrary, this gives $s \in \bigcap_{V_m \lessdot V} V(\psi)$. But then (5.50) and (5.51) imply that $s \in V_m(\psi)$. Hence, $(\mathfrak{F}, V_m), \vec{s} \Vdash$ POS. This proves (5.49).

It remains to prove (5.50) and (5.51). The first statement follows from Proposition 5.83, item (iv), and the fact that finite unions of closed sets are closed.

For (5.51), we assume (without loss of generality) that γ is built up from atomic formulas using $\wedge$, $\vee$, $\diamond$ and $\square$. This allows us to give the following inductive proof:

Atomic case. If γ is a variable p, then it follows immediately (by the closedness of U) that $U(p) = \bigcap_{U \lessdot V} V(p)$. If γ is a constant, then $U(\gamma) = V(\gamma)$ for every valuation V (admissible or not), and hence $U(\gamma) = \bigcap_{U \lessdot V} V(\gamma)$.

Conjunction. Assume γ is of the form $\gamma_1 \wedge \gamma_2$. Then $U(\gamma) = U(\gamma_1) \cap U(\gamma_2) =$

$$\bigcap_{U \lessdot V} V(\gamma_1) \cap \bigcap_{U \lessdot V} V(\gamma_2) = \bigcap_{U \lessdot V} (V(\gamma_1) \cap V(\gamma_2)) = \bigcap_{U \lessdot V} V(\gamma).$$

Disjunction. Assume γ is of the form $\gamma_1 \vee \gamma_2$. It is easy to see that $U(\gamma) \subseteq \bigcap_{U \lessdot V} V(\gamma)$. Then $U(\gamma) = U(\gamma_1) \cup U(\gamma_2) =$

$$\bigcap_{U \lessdot V} V(\gamma_1) \cup \bigcap_{U \lessdot V} V(\gamma_2) \subseteq \bigcap_{U \lessdot V} (V(\gamma_1) \cup V(\gamma_2)) = \bigcap_{U \lessdot V} V(\gamma).$$

For the other direction, assume that $u \notin U(\gamma)$. It follows immediately that $u \notin U(\gamma_1)$ and $u \notin U(\gamma_2)$. By the inductive hypothesis, this implies the existence of admissible valuations V_1 and V_2 such that $U \lessdot V_i$ and $u \notin V_i(\gamma_i)$ ($i \in \{1, 2\}$). Now let V_{12} be the intersection of the valuations V_1 and V_2, that is,

$$V_{12}(p) = V_1(p) \cap V_2(p)$$

for every p. It is easy to see that V_{12} is admissible and that $U \lessdot V_{12}$. However, since γ_1 and γ_2 are positive and hence monotone, we also have $V_{12}(\gamma_1) \subseteq V_1(\gamma_1)$ and $V_{12}(\gamma_2) \subseteq V_2(\gamma_2)$. So we find that $u \notin V_{12}(\gamma_1)$ and $u \notin V_{12}(\gamma_2)$, whence $u \notin V_{12}(\gamma_1 \vee \gamma_2) = V_{12}(\gamma)$. But then $U \lessdot V_{12}$ implies $u \notin \bigcap_{U \lessdot V} V(\gamma)$.

Box. Assume that γ is of the form $\Box \gamma'$. Then (writing R for $R_\diamond$):

$$U(\gamma) = l_R(U(\gamma')) = l_R\left(\bigcap_{U \lessdot V} V(\gamma') \right) = \bigcap_{U \lessdot V} l_R(V(\gamma')) = \bigcap_{U \lessdot V} V(\gamma).$$

Diamond. Finally, assume that γ is of the form $\Diamond \gamma'$. Again, the inclusion $U(\gamma) \subseteq \bigcap_{U \lessdot V} V(\gamma)$ is rather easy:

$$U(\gamma) = m_R(U(\gamma')) = m_R\left(\bigcap_{U \lessdot V} V(\gamma') \right) \subseteq \bigcap_{U \lessdot V} m_R(V(\gamma')) = \bigcap_{U \lessdot V} V(\gamma).$$

The other direction is the hard part of the proof. Assume that $u \in \bigcap_{U \lessdot V} V(\gamma)$. Thus for every admissible extension V of U, there is a t_V such that Rut_V and $t_V \in V(\gamma')$.

We want to prove that there is a t such that Rut and $t \in U(\gamma')$. By the inductive hypothesis, this is equivalent to showing that there is a *single* t such that Rut and $t \in V(\gamma')$ for every admissible V with $U \lessdot V$. In other words, it is sufficient to prove that

$$R[u] \cap \bigcap_{U \lessdot V} V(\gamma') \neq \varnothing. \tag{5.52}$$

We will first prove that the set

$$\mathcal{X} = \{R[u]\} \cup \{V(\gamma') \mid U \lessdot V\} \text{ has the finite intersection property.}$$

Consider an arbitrary finite subcollection of $\mathcal{X}$; without loss of generality, $R[u]$ is part of it, hence we may assume that we have taken the sets $R[u]$, $V_1(\gamma')$, ..., $V_n(\gamma')$. Let V_0 be the valuation given by

$$V_0(p) = V_1(p) \cap \cdots \cap V_n(p)$$

for every p. Then obviously we have $U \lessdot V_0$, and so by assumption there is a state t_0 in $\mathfrak{F}$ such that Rut_0 and $t_0 \in V_0(\gamma')$. But since γ' is positive and V_i is an extension of V_0 for all i, it follows that $V_0(\gamma') \subseteq V_i(\gamma')$ for all i. So $t_0 \in R[u] \cap V_1(\gamma') \cap \cdots \cap V_n(\gamma')$. This shows that $\mathcal{X}$ has the finite intersection property.

It follows from Proposition 5.83(iv) that $R[u]$ is closed; hence, item (v) of the same proposition implies (5.52).

This completes the proof of (5.51), and hence, of Theorem 5.91. ⊣

Theorem 5.92 *Every r-persistent formula is elementary.*

The reader is asked to supply the proof of Theorem 5.92 in Exercise 5.6.4.

Exercises for Section 5.6

5.6.1 Show that if we are working in the basic *temporal* similarity type, and we confine ourselves to bidirectional general frames (that is, those in which the accessibility relations of the diamonds are each other's converse), then every Sahlqvist formula is di-persistent.

5.6.2 Show that the formula $\Box M \wedge 4$ (see Examples 3.11 and 3.57) is not equivalent to a Sahlqvist formula. (Hint: use the previous exercise.)

5.6.3 In this exercise we define the notion of an ultraproduct of a general frame. Let $\{\mathfrak{g}_i \mid i \in I\}$ be a family of general frames, where each $\mathfrak{g}_i$ is given as the triple (W_i, R_i, A_i). Let U be an ultrafilter over I; for elements $s \in \prod_{i \in I} W_i$ and $a \in \prod_{i \in I} A_i$, define $[\![s \in a]\!]$ to be the set $\{i \in I \mid s_i \in a_i\}$. Now define, for an arbitrary $a_U \in \prod_U A_i$:

$$(a_U)^\circ := \{s_U \mid [\![s \in a]\!] \in U\}.$$

(a) Prove that this is a correct definition, that is, show that for arbitrary elements $s, t \in \prod_{i \in I} W_i$, $a, b \in \prod_{i \in I} A_i$:

    (i) if $s \sim_U t$, then $[\![s \in a]\!] \in U$ iff $[\![t \in a]\!] \in U$,

    (ii) if $a \sim_U b$, then $[\![s \in a]\!] \in U$ iff $[\![s \in b]\!] \in U$.

The ultraproduct $\prod_U \mathfrak{g}_i$ is defined as the structure $(\prod_U (W_i, R_i), A_U)$, where A_U is given as the set

$$A_U := \{(a_U)^\circ \mid a \in \prod_{i \in I} A_i\}.$$

(b) Prove that $\prod_U \mathfrak{g}_i$ is a general frame.

(c) Prove that for every modal formula ϕ:

$$\prod_U \mathfrak{g}_i \Vdash \phi \text{ iff } \{i \in I \mid \mathfrak{g}_i \Vdash \phi\} \in U.$$

(d) Prove that the ultraproduct of a family of refined general frames is again a refined general frame. How about the other properties of general frames?

5.6.4 In this exercise the reader is asked to supply the proof of Theorem 5.92. Let ϕ be an r-persistent formula.

(a) Let $\{\mathfrak{F}_i \mid i \in I\}$ be a family of frames such that $\mathfrak{F}_i \Vdash \phi$, for all $i \in I$, and let U be an ultrafilter over I. Prove that $\prod_U \mathfrak{F}_i \Vdash \phi$. (Hint: use Exercise 5.6.3.)
(b) Why is this sufficient to prove Theorem 5.92? (Hint: use Exercise 3.8.3.)

5.7 Summary of Chapter 5

▶ *The Algebra of Propositional Logic*: Both the algebra **2** of truth values and the class Set of set algebras algebraize classical validity. The class BA of boolean algebras algebraizes classical theoremhood.

▶ *Stone's Representation Theorem*: This classical result states that every boolean algebra can be embedded in the power set algebra of the collection of its ultrafilters. It is the key to the algebraic proof of the soundness and completeness theorem for classical propositional logic.

▶ *Modal Formulas as Terms*: Modal similarity types, extended with the boolean connectives, can be seen as algebraic similarity types. Modal formulas can be identified with algebraic terms.

▶ *Boolean Algebras with Operators* (BAOs): BAOs are boolean algebras augmented with a normal additive operator for each modal operator. They are the abstract algebras used to interpret modal logic.

▶ *The Semantic Approach to Algebraization*: Complex algebras are boolean algebras with operators based on the power set algebra of a frame. Complex algebras are the concrete BAOs that algebraize relational semantics.

▶ *The Axiomatic Approach to Algebraization*: Provable equivalence of two formulas in a normal modal logic Λ is a congruence relation on the formula algebra of the modal language. The Lindenbaum-Tarski Algebra of Λ is the induced quotient structure. Such algebras act as algebraic canonical models.

▶ *Completeness and Representation*: Modal completeness theorems correspond to algebraic representation theorems.

▶ *Ultrafilter Frames*: The ultrafilter frame of a boolean algebra with operators is a relational structure based on the collection of ultrafilters of the algebra.

▶ *Canonical Embedding Algebras*: The complex algebra of the ultrafilter frame of an algebra $\mathfrak{A}$ is called the canonical embedding algebra of $\mathfrak{A}$.

▶ *The Jónsson-Tarski Theorem:* This is the fundamental theorem underlying the algebraization of modal logic. It states that every boolean algebra with operators can be embedded in the complex algebra of its ultrafilter frame.

▶ *Algebraic Canonicity:* A class of algebras is canonical if it is closed under taking canonical embedding algebras; this concept is closely connected to the logical notion of canonicity.

▶ *Basic Duality:* Bounded morphisms between frames correspond to homomorphisms between their complex algebras, and homomorphisms between algebras give rise to bounded morphisms between their ultrafilter frames. This links generated subframes to homomorphic images, bounded morphic images to subalgebras and disjoint unions to direct products.

▶ *Goldblatt-Thomason Theorem:* This can be derived from Birkhoff's identification of varieties with equational classes using basic duality arguments.

▶ *Ultraproducts and Canonicity:* If K is a class of frames which is closed under ultraproducts, then the variety V_K is canonical.

▶ *General Frames:* A general frame combines a frame and a boolean algebra with operators in one structure. Like boolean algebras with operators, general frames provide an adequate semantics for normal modal logics. Important properties of general frames include refinedness, discreteness and descriptiveness.

▶ *Descriptive Frames and* BAOs: There is a full categorical duality between the categories of descriptive τ-frames with bounded morphisms and boolean algebras with τ-operators with homomorphic images.

▶ *Persistence:* A generalization of the notion of canonicity.

▶ *Sahlqvist's Completeness Theorem.* All Sahlqvist formulas are d-persistent, and hence canonical.

Notes

The main aim of algebraic logic is to gain a better understanding of logic by treating it in universal algebraic terms – in fact, the theory of universal algebra was developed in tandem with that of algebraic logic. Given a logic, algebraic logicians try to find a class of algebras that algebraizes it in a natural way. When a logic is algebraizable, natural properties of a logic will correspond to natural properties of the associated class of algebras, and the apparatus of universal algebra can be applied to solve logical problems. For instance, we have seen that representation theorems are the algebraic counterpart of completeness theorems in modal logic. The algebraic approach has had a profound influence on the development of logic, especially non-classical logic; readers interested in the general methodology of algebraic logic should consult Blok and Pigozzi's [67] or Andréka *et al.* [5].

The field has a long and strong tradition dating back to the nineteenth century. In fact, nineteenth century mathematical logic *was* algebraic logic: to use the terminology of Section 5.1, propositions were represented as algebraic terms, not logical formulas. Boole is generally taken as the founding father of both propositional logic and modern algebra – the latter because, in his work, terms for the first time refer to objects other than numbers, and operations very different from the arithmetical ones are considered. The work of Boole was taken up by de Morgan, Peirce, Schröder and others; their contributions to the theory of binary relations formed the basis of Tarski's development of relational algebra. In the Historical Overview in Chapter 1 we mentioned MacColl, the first logician in this tradition to treat modal logic. A discussion of the nineteenth century roots of algebraic logic is given by Anellis and Houser in [10].

However, when the quantificational approach to logic became firmly established in the early twentieth century, interest in algebraic logic waned, and it was only the influence of a relatively small number of researchers such as Birkhoff, Stone, Tarski, and Rasiowa and Sikorski [370, 371], that ensured that the tradition was passed on to the present day. The method of basing an algebra on a collection of formulas (or equivalence classes of formulas), due to Lindenbaum and Tarski, proved to be an essential research tool. This period also saw the distinction between logical languages and their semantics being sharpened; an algebraic semantics for non-classical logics was provided by Tarski's matrix algebras. But the great success story of algebraic logic was its treatment of classical propositional logic in the framework of boolean algebras, which we sketched in Section 5.1. Here, the work of Stone [425] was a milestone: not only did he prove the representation theorem for boolean algebras (our Theorem 5.16), he also recognized the importance of topological notions for the area (something we did not discuss in the text). This enabled him to prove a duality theorem permitting boolean algebras to be viewed as essentially the same objects as certain topologies (now called Stone spaces). Stone's work has influenced many fields of mathematics, as is witnessed by Johnstone [248].

McKinsey and Tarski [324] drew on Stone's work in order to prove a representation theorem for so-called closure algebras (that is, S4-algebras); this result significantly extended McKinsey's [322] which dealt with finite closure algebras. However, when it comes to the algebraization of modal logics, the reader is now in a position to appreciate the full significance of the work of Jónsson and Tarski [255]. Although modal logic is not mentioned in their paper, the authors simultaneously invented relational semantics, and showed (via their representation theorem) how this new relational world related to the algebraic one. Both Theorem 5.43 from the present chapter and important results on canonicity, overlapping with our Theorem 5.91, are proved here. It is also obvious from their terminology (for instance, the use of the words 'closed' and 'open' for certain elements of the canonical embedding algebra) that hidden beneath the surface of the paper lies a duality theory that extends Stone's result to cover operators on boolean algebras.

For many years after the publication of the Jónsson-Tarski paper, research in modal logic and in BAO theory pretty much took place in parallel universes. In algebraic circles, the work of Jónsson and Tarski was certainly not neglected. The paper fitted well with a line of work on *relation algebras*. These were introduced by Tarski [427] to be to binary relations what boolean algebras are to unary ones; the concrete, so-called representable relation algebras have, besides the boolean repertoire, operations for taking the converse of a relation and the composition of two relations, and as a distinguished element, the identity relation. (From the perspective of our book, the class RRA of representable relation algebras is nothing but the variety generated by the complex algebras of the two-dimensional arrow frames; see Example 5.57. In fact, one of the motivations behind the introduction of arrow logic was to give a modal account of the theory of relation algebras.) Much attention was devoted to finding an analog of Stone's result for boolean algebras: that is, a nice equational characterization of the representable relation algebras. But this nut turned out to be hard to crack. Lyndon [304] showed that the axioms that Tarski had proposed did not suffice, and Monk [334] proved that the variety does not even have a finite first-order axiomatization. Later work showed that equational axiomatizations will be very complex (Andréka [6]) and not in Sahlqvist form (Hodkinson [230] and Venema [446]), although Tarski proved that RRA is a canonical variety. As a positive result, a nice game-theoretical characterization was given by Hirsch and Hodkinson [225]. But these Notes can only provide a lop-sided account of one aspect of the theory of relation algebras; for more, the reader is referred to Jónsson [251, 252], Maddux [306] or Hirsch and Hodkinson [226]. One last remark on relation algebras: it is a *very* powerful theory. In fact, as Tarski and Givant show in [428], one can formalize all of set theory in it.

Tarski and his students developed other branches of algebraic logic as well: for example, the theory of cylindric algebras. The standard reference here is Henkin, Monk and Tarski [218]. Cylindric algebras (and also the polyadic algebras of Halmos [202]), are boolean algebras with operators that were studied as algebraic counterparts of first-order logic. For an introductory survey of these and other algebras of relations we refer the reader to Németi [341]; modal logic versions of these algebras are discussed in Section 7.5.

But in modal circles, the status of algebraic methods was very different. Indeed, with the advent of relational semantics for modal logic in the 1960s, it seemed that algebraic methods were to be swept away: model theoretic tools seemed to be *the* route to a brave new modal world. (Bull's work was probably the most important exception to this trend. For example, his theorem that all normal extensions of **S4.3** are characterized by classes of finite *models* was proved using *algebraic* arguments.) Indicative of the spirit of the times is the following remark made by Lemmon in his two part paper on algebraic semantics for modal logic. After thanking Dana Scott for ideas and stimulus, he remarks:

...I alone am responsible for the ugly algebraic form into which I have cast some of his elegant semantics. [295, page 191]

Such attitudes only seemed reasonable because Jónsson and Tarski's work had been overlooked by the leading modal logicians, and neither Jónsson nor Tarski had drawn attention to its modal significance. Only when the frame incompleteness results (which began to appear around 1972) showed that not all normal modal logics could be characterized in terms of frames were modal logicians forced to reappraise the utility of algebraic methods.

The work of Thomason and Goldblatt forms the next major milestone in the story: Thomason [433] not only contains the first incompleteness results and uses BAOs, it also introduced general frames (though similar, language dependent, structures had been used in earlier work by Makinson [308] and Fine [127]). Thomason showed that general frames can be regarded as simple set theoretic representations of BAOs, and notes the connection between general frames and Henkin models for second-order logic (we briefly noted this link in Chapter 3.2). In [435], Thomason developed a duality between the categories of frames with bounded morphisms and that of complete and atomic modal algebras with homomorphisms that preserve infinite meets. It was Goldblatt, however, who did the most influential work: in [184] the full duality between the categories of modal algebras with homomorphisms and descriptive general frames with bounded morphisms is proved, a result extending Stone's. Independently, Esakia [123] came up with such a duality for closure algebras. Goldblatt generalized his duality to arbitrary similarity types in [186]; a more explicitly topological version can be found in Sambin and Vaccaro [398]. Ever since their introduction in the seventies, general frames, a nice compromise between algebraic and the relational semantics, have occupied a central place in the theory of modal logic. Kracht [276, 279] developed an interesting calculus of internal descriptions which connects the algebraic and first-order side of general frames. Zakharyaschev gave an extensive analysis of transitive general frames in his work on canonical formulas – see the Notes to Chapter 4 for further information.

The first proof of the canonicity of Sahlqvist formulas, for the basic modal similarity type, was given by Sahlqvist [396], although many particular examples and less general classes of Sahlqvist axioms were known to be canonical. In particular, Jónsson and Tarski proved canonicity, not only for certain equations, but also for various boolean combinations of suitable equations. This result overlaps with Sahlqvist's in the sense that canonicity for simple Sahlqvist formulas follows from it, but on the other hand, Sahlqvist formulas allowing properly boxed atoms in the antecedent do not seem to fall under the scope of the results in Jónsson and Tarski [255]. Incidentally, Sahlqvist's original proof is well worth consulting: it is non-algebraic, and very different from the one given in the text. Our proof of the Sahlqvist Completeness Theorem is partly based on the one given in Sambin and Vaccaro [399].

Recent years have seen a revived interest in the notion of canonicity. De Rijke and Venema [387] defined the notion of a Sahlqvist *equation* and generalized the theory to arbitrary similarity types. Jónsson became active in the field again; in [254] he gave a proof of the canonicity of Sahlqvist equations by purely algebraic means, building on the techniques of his original paper with Tarski. Subsequent work of Gehrke, Jónsson and Harding (see for instance [169, 168]) generalized the notion even further by weakening the boolean base of the algebra to that of a (distributive) lattice. Ghilardi and Meloni [171] proved canonicity of a wide class of formulas using a different representation of the canonical extension of algebras with operators. Whereas the latter lines of research tend to separate the canonicity of formulas from correspondence theoretic issues, a reverse trend is visible in the work of Kracht (already mentioned) and Venema [449]. The latter work shows that for some non-Sahlqvist formulas there is still an algorithm generating a first-order formula which is now not equivalent to the modal one, but does define a property for which the modal formula is canonical.

The applications of universal algebraic techniques in modal logic go much further than we could indicate in this chapter. Various properties of modal logics have been successfully studied from an algebraic perspective; of the many examples we only mention the work of Maksimova [310] connecting interpolation with amalgamation properties. Also, most work by Blok, Kracht, Wolter and Zakharyaschev on mapping the lattice of modal logics makes essential use of algebraic concepts such as splitting algebras; a good starting point for information on this line of research would be Kracht [279].

Most of the results that we present in Sections 5.4 and 5.5 are simplified versions of results in Goldblatt [184]. The proof of the Goldblatt-Thomason theorem given in this chapter (as opposed to the model-theoretic one given in Section 3.8) treats it as a corollary of Birkhoff's Theorem [53] identifying varieties with equational classes; our proof is essentially the original proof of Goldblatt and Thomason [188]. The proof of Theorem 5.56 is a generalization and algebraization by Goldblatt [186] of results due to Fine [132] and van Benthem [41]. The Open Problems 1 and 2 seem to have been formulated first in Fine [132]; readers who would like to try and solve the second one should definitely consult Goldblatt [180]. The now standard terminology concerning properties of general frames – refined, descriptive, and so on – is due to Thomason [433], Goldblatt [184] and Fine [130]. An exception, as far as we know, is the notion of discreteness, which did not play a role until Venema [444], where our Theorem 5.90 was proved. The generalization of canonicity to the notion of persistence stems from Goldblatt [185], and the proof that r-persistent formulas are elementary (Theorem 5.92) was first given by Lachlan [291].

6

Computability and Complexity

In this chapter we investigate the computability and complexity of normal modal logics. In particular, we examine the computability of *satisfiability problems* (given a modal formula ϕ and a class of models M, is it computable whether ϕ is M-satisfiable?) and *validity problems* (given a modal formula ϕ and a class of models M, is it computable whether ϕ is valid on M?). When the answer is 'yes', we probe further: how complex is the problem – in particular, what resources of time (that is, computation steps) or space (that is, memory) are needed to carry out the required computations? When the answer is 'no', we pose a similar question: how uncomputable is the problem? There are vast differences in the complexities of modal satisfiability problems: some are no worse than the satisfiability problem for propositional calculus, while others are highly undecidable.

This chapter has two main parts. The first, consisting of the five sections on the basic track, introduces the basic ideas and discusses modal (un-)decidability. Three techniques for proving decidability are discussed (finite models, interpretations in monadic second-order theories of trees, and quasi-models and mosaics) and undecidability is approached via tiling problems. In the second part, consisting of the last three sections of the chapter, we examine the complexity of some key modal satisfiability problems. These sections are on the advanced track, but the initial part of each of them should be accessible to all readers.

Basic ideas about computability and complexity are reviewed in the first section, and further background information can be found in Appendix C. Throughout the chapter we assume we are working with countable languages.

Chapter guide

Section 6.1: Computing Satisfiability (Basic track). We discuss the key concepts assumed throughout the chapter: satisfiability and validity problems, and how to compute them on Turing machines.

Section 6.2: Decidability via Finite Models (Basic track). We discuss the use of finite models for proving decidability results. Three basic theorems are proved, and many of the logics discussed in Chapter 4 are shown to be decidable.

Section 6.3: Decidability via Interpretations (Basic track). Another way of proving modal decidability results is via interpretations in powerful decidable theories such as monadic second-order theories of trees. This technique is useful for showing the decidability of logics without the finite model property.

Section 6.4: Decidability via Quasi-models and Mosaics (Basic track). For logics lacking the finite model property it may also be possible to prove decidability results by computing with more abstract kinds of finite structure; quasi-models and mosaics are important examples of such structures.

Section 6.5: Undecidability via Tiling (Basic track). In this section we show just how easily undecidable – and even highly undecidable – modal logics can arise. We do so by introducing an important proof method: tiling arguments.

Section 6.6: NP (Advanced track). This section introduces the concept of NP algorithms, illustrates the modal content of this idea using some simple examples, and then proves Hemaspaandra's Theorem: every normal logic extending **S4.3** is NP-complete.

Section 6.7: PSPACE (Advanced track). The key complexity class for the basic modal language is PSPACE, the class of problems solvable in polynomial space. We give a PSPACE algorithm for the satisfiability problem for **K**, and prove Ladner's Theorem: every normal logic between **K** and **S4** is PSPACE-hard.

Section 6.8: EXPTIME (Advanced track). We show that the satisfiability problem for **PDL** is EXPTIME-complete. EXPTIME-hardness is shown by reduction from a tiling problem, and the EXPTIME algorithm introduces an important technique called elimination of Hintikka sets.

6.1 Computing Satisfiability

The work of this chapter revolves around satisfiability and validity problems. Here is an abstract formulation.

Definition 6.1 (Satisfiability and Validity Problems) Let τ be a modal similarity type, ϕ be a τ-formula and M a class of τ-models. The M-*satisfiability problem* is to determine whether or not ϕ is satisfiable in some model in M. The M-*validity problem* is to determine whether or not ϕ is true in all models in M; that is, whether or not M $\Vdash \phi$. (We call this the validity problem because we are mostly interested in cases where M is the class of all models over some class of frames.) The M-validity and M-satisfiability problem are each other's *duals*.   ⊣

In fact, as far as discussions of computability (or non-computability) are concerned, we are free to talk in terms of either satisfiability or validity problems.

Lemma 6.2 *Let* τ *be a modal similarity type, and suppose that* M *is a class of* τ-*models. Then there is an algorithm for solving the* M-*satisfiability problem iff there is an algorithm for solving the* M-*validity problem.*

Proof. As $\neg\phi$ is *not* satisfiable in M iff M ⊪ ϕ, given an algorithm for M-satisfiability, we can test for the validity of ϕ by giving it the input $\neg\phi$. In a similar fashion, an algorithm for M-validity can be used to test for M-satisfiability.   ⊣

This argument does not give us any interesting information about the relative *complexity* of dual satisfiability and validity problems; and indeed, they may well be different.

How do the themes of this chapter relate to the *normal modal logics* introduced in Section 1.6 and discussed in Chapters 4 and 5? Clearly we should investigate the following two problems.

Definition 6.3 Let τ be a modal similarity type, Λ be a normal modal logic in a language for τ, and ϕ a τ-formula. The problem of determining whether or not ϕ is Λ-consistent is called the Λ-*consistency problem*, and the problem of determining whether or not $\Lambda \vdash \phi$ is called the Λ-*provability problem*.   ⊣

Note that Λ-consistency and Λ-provability problems are satisfiability and validity problems in disguise. In particular, if Λ is a normal modal logic, and M is any class of models such that $\Lambda = \Lambda_M$, then the Λ-consistency problem is the M-satisfiability problem, and the Λ-provability problem is the M-validity problem. As every normal modal logic is determined by at least one class of models (namely, the singleton class containing its canonical model; see Theorem 4.22), we are free to think of consistency and provability problems in terms of satisfiability and validity problems. We do so in this chapter, and to emphasize this we usually call the Λ-consistency problem the Λ-*satisfiability problem*, and the Λ-provability problem the Λ-*validity problem*.

Our discussion so far has given an *abstract* account of the problems we will explore, and most of our results will be stated, proved, and discussed at this level. But what does it mean to have an algorithm for solving (say) a validity problem? And what does it mean to talk about the complexity of (say) a satisfiability problem? After all, computation is the finitary manipulation of finite structures – but both formulas and models are abstract set-theoretical objects. To show that our abstrac account really makes sense, we need to choose a well-understood method of computation and show that formulas and models can be *represented* in a way that is suited to our method.

We have chosen *Turing machines* (Appendix C) as our fundamental model of computation. The most relevant fact about Turing machines for our purposes is that they compute by manipulating finite strings of symbols; hence we need to represent models and formulas as symbol strings. As far as mere computability is concerned, the key demand is that these symbol string representations be *finite*. For complexity analyses more is required: representations must also be *efficient*. Let us discuss these requirements.

Clearly modal formulas can be represented as finite strings over a finite set of symbols: proposition letters can be represented by a single symbol (say, p) followed by (the representation of) a number. Thus, instead of working with an infinite collection of primitive symbols we could work with (say) $p1$, $p10$, $p11$, $p100$ and so on, where the numeric tail is represented in binary. Fine – but what about models? Models are set-theoretic entities of the form (W, R, V), and each component may be infinite. However, the difficulty is more apparent than real. For a start, when evaluating a formula ϕ in some model, the only relevant information in the valuation is the assignments made to proposition letters actually occurring in ϕ (see Exercise 1.3.1). Thus, instead of working with V, we can work with the finite valuation V' which is defined on the (finite) language consisting of exactly the proposition letters in ϕ, and which agrees with V on these letters. Secondly, much of our work will revolve around models based on *finite* frames (or more generally, the frames of *finite character* defined below).

We already know quite a lot about finite models and their logics. For a start, in Section 2.3 we introduced two techniques for building finite models (selection and filtration) and defined the finite model property for the basic modal language. In Section 3.4 we introduced the finite frame property (again, for the basic modal language) and proved Theorem 3.28: a normal modal logic has the finite frame property iff it has the finite model property. Since then we have learned what a normal modal logic in a language of arbitrary similarity type is (Definition 4.13), so let us now define the finite frame property and the finite model property for modal languages of arbitrary similarity type, and generalize Theorem 3.28.

Definition 6.4 Let τ be a modal similarity type. A frame of type τ has *finite character* if it contains finitely many states, and finitely many non-empty relations. If Λ is a normal modal logic in a language for τ, and F is a class of τ-frames of finite character, and $\Lambda = \Lambda_{\mathsf{F}}$, then Λ is said to have the *finite frame property (f.f.p.)* with respect to F. If $\Lambda = \Lambda_{\mathsf{F}}$ for some class of τ-frames F of finite character, then Λ has the finite frame property.

A class of τ-models M is *finitely based* if every model in M is based on a τ-

frame of finite character. If Λ is a normal modal logic in a language for τ, and M is a class of finitely based τ-models, and $\Lambda = \Lambda_M$, then Λ has the *finite model property (f.m.p.)* with respect to M. If $\Lambda = \Lambda_M$ for some class of of finitely based τ-models M, then Λ has the finite model property.   ⊣

A few remarks may be helpful. First, the concept of finite character is a natural way of coping with similarity types containing infinitely many relations. Second, note that the way the finite frame property is defined here (where we simply insist that $\Lambda = \Lambda_F$) is somewhat simpler than that used in Definition 3.23 (where we insisted that F ⊩ Λ, and for every formula ϕ such that $\phi \notin \Lambda$ there is some $\mathfrak{F} \in$ F such that ϕ is falsifiable on $\mathfrak{F}$). It is easy to see that these definitions are equivalent. Finally, a class of frames of finite character (or indeed, a class of finite frames) may well be a proper class. Nonetheless, up to isomorphism, there are only denumerably many frames in any such class; hence, if Λ has the finite frame property, it has the finite frame property with respect to a denumerably infinite *set* of frames, and we take this for granted without further comment throughout the chapter.

Given this definition, it is straightforward to generalize Theorem 3.28.

Theorem 6.5 *Let τ be a modal similarity type. Any normal modal logic in a language for τ has the finite model property iff it has the finite frame property.*

Proof. This is a matter of verifying that the proof of Theorem 3.28 extends to arbitrary similarity types; see Exercise 6.1.1.   ⊣

There are many ways to represent a frame of finite character, together with a valuation V' defined on finitely many proposition letters, as a finite symbol string. While any such finitization is sufficient for discussions of computability, we need to exercise more care when it comes to complexity. Complexity theory measures the difficulty of problems in terms of the resources required to solve them – and these are measured as a function of the size of the input. A highly inefficient representation of the input can render such resource measures vacuous, so we must be careful not to smuggle in sources of inefficiency. For the complexity classes we will be dealing with, this is pretty much a matter of common sense, but the following point should be made explicit: we must *not* represent the numeric subscripts on propositional variables and states in unary notation.

The point is this. Even binary representations (which are longer than the more familiar decimal representations) are exponentially more compact than unary ones. For example, the representation of the number 64 in unary is a string of 64 consecutive ones, whereas its representation in binary is 1000000. If we represent our subscripts in unary, we are using a *highly* inefficient representation of the problem.

For this reason we will regard modal formulas (for the basic modal language) as strings over the alphabet $\{p, 0, 1, (,), \wedge, \neg, \diamond\}$, and proposition letters will be represented by strings consisting of p followed by the *binary* representation of a number (without leading zeroes). Similarly, we will regard models as strings over the alphabet $\{w, p, 0, 1, ;, \langle, \rangle\}$. A state in a model will be represented by w followed by the binary representation of a number (without leading zeroes), and the representation of proposition letters (which we need to encode the valuation) will be as just described. A string representing a model will have the following form:

$$\langle\langle w_1; \ldots; w_n\rangle;$$
$$\langle\langle w_i; w_j\rangle; \ldots; \langle w_k; w_l\rangle\rangle;$$
$$\langle\langle p_x; \langle w_r; \ldots; w_s\rangle\rangle; \ldots; \langle p_y; \langle w_t; \ldots; w_u\rangle\rangle\rangle,$$

where $1 \leq i, j, k, l, r, s, t, u \leq n$. Such triples represent models in the obvious way: the first component gives the states, the second the relation, and the third the valuation. The subscripted ws and ps are metavariables over our representations of states and proposition letters, respectively. We assume that our representations of models contain no repetitions in any of the three components, and that they satisfy obvious well-formedness conditions (in particular, the third component represents a *function*, thus we cannot have the same representation p_y appearing as the first item in different tuples). Here is a simple example (though to keep things readable we have represented the numbers in decimal):

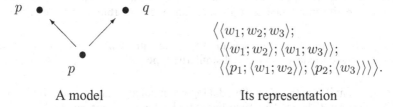

| A model | Its representation |

Such representations open the door to all the standard concepts of computability theory and computational complexity. For a start, it now makes sense to describe sets of formulas (including normal modal logics), sets of models, and sets of frames as being *recursively enumerable (r.e.)*, or as being *recursive*. Saying that a set is r.e. means that it is possible to write a Turing machine that will successively output all and only its elements. Saying that a set is recursive means that it is possible to write a Turing machine which, when given any input, will perform a *finite* number of computation steps, halt, and then correctly tell us whether the input represents a member of the set or not. (In short, recursive sets are those for which we can decide membership using a terminating computation.)

Furthermore, it is clearly possible to program a Turing machine so that when it is presented with (the representations of) a formula, a model, and a point, it will evaluate (the representation of) the formula in (the representation of) the model at (the representation of) the point. Admittedly it would be rather painful to write out such a Turing machine in detail – but it is straightforward to write a program to carry out this task in most high-level programming languages; hence, by Church's Thesis (see Appendix C), it is possible to write a Turing machine to do the job as well. Thus it makes perfectly good sense to talk about writing Turing machines which test for the satisfiability or validity of a formula on a class of finitely based models and to inquire about the complexity of such problems.

Apart from asking the reader to generalize the above representation schema to cover modal languages of arbitrary similarity type (see Exercise 6.1.2) we will not discuss the issue of representation further. In most of what follows we talk as if the abstract definition of satisfiability and validity problems given earlier was the focus of our computational investigations. For example, we will often call $|\phi|$ the size of the input formula; strictly speaking, it is the size of its representation. Nor do we mention Turing machines very often. The results of this chapter rest on the fact that there is an efficient representation which enables us to compute satisfiability and validity problems; for many purposes we can ignore the details.

Exercises for Section 6.1

6.1.1 Prove Theorem 6.5. That is, show that for any modal similarity type τ, any normal modal logic in a language for τ has the finite model property if and only if it has the finite frame property. This is simply a matter of verifying that the proof of Theorem 3.28 extends to arbitrary similarity types – but note that there will be a gap in your proof if you have not yet proved the Filtration Theorem for modal languages of arbitrary similarity type.

6.1.2 Modify the representation schema for models given above so that it can represent any finitely based model of any modal similarity type.

6.1.3 Show that if Λ is the normal modal logic generated by an r.e. set of formulas, then Λ itself is an r.e. set. (The reader unfamiliar with this type of proof may find it useful to look at the proof of Lemma 6.12 below.)

6.2 Decidability via Finite Models

Call a normal modal logic Λ *decidable* if the Λ-satisfiability (or equivalently: Λ-validity) problem is decidable, and *undecidable* if it is not. How should we establish decidability results? A lot depends on our 'access' to the logic. For example, we may know Λ purely semantically: it is given as the logic of some class of frames of interest. However, we may also have a syntactic handle on Λ; in particular, we may know that it is the logic generated by some set of axioms. Whether Λ is semantically or syntactically specified, establishing that it has the finite model property is a useful first step towards proving decidability, for if we can prove this two plausible strategies for establishing decidability suggest themselves, as we will now explain.

- **Decidability for semantically specified logics: informal argument.** Suppose we only have a semantic specification of Λ, but that we have been able to prove that Λ possesses a strong form of the finite model property: not only does Λ have the f.m.p. with respect to some set of models, but for any formula ϕ there is a computable function f such that $f(|\phi|)$ is an upper bound on the size of these models needed to satisfy ϕ. Write a Turing machine that takes ϕ as input, generates all the finite models belonging to this set up to size $f(|\phi|)$, and tests for the satisfiability of ϕ on these models. Because ϕ is Λ-satisfiable iff it is satisfied in a Λ-model of size at most $f(|\phi|)$, and because the machine systematically examines all these models, our machine decides Λ-satisfiability.

- **Decidability for syntactically specified logics: informal argument.** Suppose Λ is given axiomatically, and we have been able to show that Λ has the f.m.p. with respect to some set of models M. First, construct a Turing machine that makes use of the axiomatization to recursively enumerate the Λ-validities. Second, construct a Turing machine that recursively enumerates all the finite models in M. Given two such machines we can effectively test the Λ-validity of any formula ϕ: if ϕ is valid it will eventually be generated by the first machine; if it is not, we will eventually be able to falsify it on a model generated by the second. One of the machines must eventually settle ϕ's fate, and thus decide Λ-validity.

Such arguments underly most applications of the finite model property to decidability. We have deliberately phrased both arguments rather loosely; the fundamental goal of this section is to explore the underlying ideas more carefully, and formulate them rigorously. Our investigation will yield three main theorems. The first is a precise formulation of the argument for semantically specified logics. The second and third are distinct reformulations of the argument for syntactically specified logics. We will consider a number of applications of these theorems, and will put both of the methods introduced in Section 2.3 for constructing finite models (namely *filtration* and *selection*) to work.

Let us begin by scrutinizing the first of the above arguments. This revolves around a strong form of the finite model property.

Definition 6.6 (Strong Finite Model Property) Let Λ be a normal modal logic, M a set of finitely based models such that $\Lambda = \Lambda_M$, and f a function mapping natural numbers to natural numbers. Λ has the $f(n)$-*size model property* with respect to M if every Λ-consistent formula ϕ is satisfiable in a model in M containing at most $f(|\phi|)$ states.

Λ has the *strong finite model property* with respect to M if there is a *computable* function f such that Λ has the $f(n)$-size model property with respect to M. Λ has the *polysize model property* with respect to M if there is a *polynomial* p such that Λ has the $p(n)$-size model property with respect to M.

Λ has the $f(n)$-size model property (respectively, strong finite model property, polysize model property) if there is a set of finitely based models M such that $\Lambda = \Lambda_M$ and Λ has the $f(n)$-size model property (respectively, strong finite model property, polysize model property) with respect to M.   ⊣

If a logic Λ has the polysize model property, any Λ-satisfiable formula is satisfiable not just on a finite model, but a genuinely *small* model. Even this very strong form of the f.m.p does *not* guarantee decidability: as the reader is asked to prove in Exercise 6.2.4, there are uncountably many normal modal logics which possess the polysize model property but have *undecidable* satisfiability problems.

In view of this result, the first informal argument sketch is clearly inadequate – but where does its deficiency lie? It makes the following (false) assumption: that for any set of models, and any natural number n, it is possible to generate all and only the models in M of size at most n. This assumption is warranted only if M is a recursive set (that is, only if a Turing machine can decide exactly which finite models belong to M). But this is the only shortcoming of the informal argument.

Theorem 6.7 *If Λ is a normal modal logic that has the strong finite model property with respect to a recursive set of models M, then Λ is decidable.*

Proof. First, observe that for any natural number n it is possible to generate all distinct (representations of) models in M that have size at most n: we simply need to write a machine that generates *all* distinct (representations of) models that have size at most n, tests each model (representation) as it is generated to see whether it belongs to M (this is the key point: we can effectively test for membership in M precisely because M is a recursive set) and then outputs exactly those models (representations) which do belong to M. (From now on we drop all mention of representations, and will speak simply of 'generating all models' or 'generating all models up to size n', and so on.)

So, given ϕ, we use this machine to generate all models of the appropriate set up to size $f(|\phi|)$, and test whether ϕ is satisfiable on any of the models it produces. If ϕ is satisfiable on at least one of them, it is Λ-satisfiable; if not, it is not Λ-satisfiable, for Λ has the strong f.m.p. with respect to M.   ⊣

Theorem 6.7 is an important result. If we are to apply it, how do we establish that a logic has the strong finite model property? Unfortunately, no fully general answer to this question is known – nonetheless, both filtration and selection can be useful. We start by illustrating the utility of filtrations.

Corollary 6.8 K, T, KB, K4, S4, S5, K_t, $K_t4.3$ *and* K_tQ *are decidable.*

Proof. First, all these logics have the f.m.p. with respect to the expected sets of models; for example, K4 has the f.m.p. with respect to the set of finite transitive models, and $\mathbf{K}_t\mathbf{Q}$ has the f.m.p with respect to the finite dense unbounded weak total orders (that is, the finite DUWTO frames; see Theorem 4.41). The easiest way to prove this is to use filtrations. In Section 2.3 we defined filtrations for both the basic modal language and the basic temporal language. Given a model $\mathfrak{M}$ that satisfies a formula ϕ at some state, by filtrating $\mathfrak{M}$ through the set of all ϕ's subformulas we obtain a *finite* model $\mathfrak{M}^f$ that satisfies ϕ. Of course, we need to be careful that $\mathfrak{M}^f$ has all the right properties; for example, if $\mathfrak{M}$ was a **K4**-model, we want $\mathfrak{M}^f$ to be a **K4**-model as well. By and large this is straightforward, though the reader will need to think a little about how to handle density; see Exercise 6.2.1.

Such filtration arguments actually establish the *strong* f.m.p. for these logics. If we form $\mathfrak{M}^f$ by filtrating $\mathfrak{M}$ through the subformulas of ϕ, then $\mathfrak{M}^f$ has at most $2^{|\phi|}$ nodes, thus we have a computable (though, unfortunately, exponential) upper bound on the size of satisfying models for all these logics; see Section 2.3.

It remains to check that the relevant sets of finite models are recursive. Checking for membership in these sets boils down to checking that the models possess (various combinations of) such properties as reflexivity, transitivity, trichotomy, and so on. It is clearly possible to devise algorithms to test for the relevant properties, hence (by Church's thesis) we can program a Turing machine to do so. Thus Theorem 6.7 applies, and all these logics are decidable. ⊣

Filtration is a widely used technique for showing that logics have the strong finite model property, but it has limitations. Suppose we are working with a modal language containing n unary modal operators ($n > 0$) and no others. Let F_1^n be the set of frames for this language such that for each $\mathfrak{F} \in \mathsf{F}_1^n$, the relation corresponding to each modality is a partial function, let M_1^n be the set of models built over F_1^n, and let $\mathbf{K}_n\mathbf{Alt}_1$ be its logic. Now, $\mathbf{K}_n\mathbf{Alt}_1$ has the strong finite model property, but there is no obvious way of using filtrations to show this; see Exercise 6.2.3.

However – at least in the present case – it is straightforward to use *selection*, the other method of building finite models discussed in Section 2.3, to establish the strong finite model property.

Corollary 6.9 $\mathbf{K}_n\mathbf{Alt}_1$ *is decidable.*

Proof. We argue as follows. Suppose $\mathfrak{M}$ is in M_1^n and $\mathfrak{M}, w \Vdash \phi$. Let $\mathfrak{M}'$ be the model that is identical to $\mathfrak{M}$ save possibly that any relations in $\mathfrak{M}$ *not* corresponding to modal operators in ϕ are empty. Clearly $\mathfrak{M}'$ is also in M_1^n and $\mathfrak{M}', w \Vdash \phi$. Let m be the degree of ϕ (that is, the maximal depth of nested modalities; see Definition 2.28). Let $\mathfrak{M}''$ be the submodel of $\mathfrak{M}'$ formed by selecting all and only those nodes reachable from w in m or fewer steps. Clearly $\mathfrak{M}''$ is in M_1^n and $\mathfrak{M}'', w \Vdash \phi$. Moreover, because each relation is a partial function, $\mathfrak{M}''$ has only finitely many nodes: indeed, it can contain at most $t^m + 1$ nodes, where t is the number of distinct types of modality that occur in ϕ. Hence $\mathbf{K}_n\mathbf{Alt}_1$ has the strong finite model property with respect to M_1^n.

It is clear that the set of finitely based M_1^n models is recursive, for testing whether a finite model $\mathfrak{M}$ belongs to it essentially boils down to checking that each of $\mathfrak{M}$'s (finitely many non-empty) transition relations is a partial function. Decidability follows by Theorem 6.7.   ⊣

Selection is not as general a method as filtration – but it can be useful, especially when working with non-transitive models. As we will see when we discuss NP-completeness, selection is a natural way of turning a finite model (perhaps produced via a filtration) into a truly small (that is, polysize) model.

Theorem 6.7, together with such methods as filtration and selection, can be a useful tool for establishing modal decidability results, for it does not require us to have an axiomatization. Very often we do have an axiomatization at our disposal, and it is natural to ask whether (and how) we can make use of it to help establish decidability. This is what the second informal argument attempts to do. The key idea it embodies is the following: if a logic is both axiomatizable and has the finite model property with respect to some (recursively enumerable) set of models M, then we should be able to prove decidability. This is an important idea that can be developed in two different ways, depending on the kind of axiomatization we have, and what we know about the computational properties of M.

When we discussed completeness in Chapter 4, we viewed axiomatizations very abstractly: we simply said that if Λ was a normal modal logic, Σ a set of modal formulas, and $\mathbf{K}\Sigma$ (the smallest normal logic generated by Σ) equaled Λ, then Σ was an axiomatization of Λ. To give computational content to the phrase 'generated by' we need to impose restrictions on Σ, for under the definition just given every normal logic Λ generates itself. This is too abstract to be useful here, so we will introduce various notions of *axiomatizability* that offer more computational leverage.

Definition 6.10 A logic Λ is *finitely axiomatizable* if it has a *finite* axiomatization Σ; it is *recursively axiomatizable* if it has a *recursive* axiomatization Σ; and it is *axiomatizable* if it has a *recursively enumerable* axiomatization Σ.   ⊣

Although it will not play a major role in what follows, there is a neat result called Craig's Lemma that readers should know: *every axiomatizable logic is recursively axiomatizable*. So the following lemma is essentially Craig's Lemma for modal logic:

Lemma 6.11 *If Λ is axiomatizable, then Λ is recursively axiomatizable.*

So, given a computationally reasonable notion of axiomatizability, the idea of using axiomatizations to generate validities is correct. But how do we use this fact to turn the informal argument into a theorem? Here is the most obvious way: demand that M be an r.e. set. As the following lemma shows, this ensures that we can recursively enumerate the formulas that are *not* valid on M.

Lemma 6.12 *If* M *is a recursively enumerable set of finite models, then the set of formulas falsifiable in* M *is recursively enumerable.*

Proof. As M is an r.e. set, we can construct a machine *M1* to generate all its elements, and clearly we can construct a machine *M2* that generates all the formulas. So, construct a machine *M3* that operates as follows: it calls on *M1* to generate a model, and on *M2* to generate a formula, and then stores both the model and the formula. It then tests all stored formulas on all stored models (*M3* is not going to win any prizes for efficiency) and outputs any of the stored formulas it can falsify on some stored model. *At any stage there are only finitely many stored formulas and models, hence this testing process terminates.* When the testing process is finished, *M3* calls on *M1* and *M2* once more to generate another model and formula, stores them, performs another round of testing, and so on *ad infinitum*.

Suppose ϕ is falsifiable on some model $\mathfrak{M}$ in M. At some finite stage both ϕ and $\mathfrak{M}$ will be stored by *M3*, hence ϕ will eventually be tested on $\mathfrak{M}$, falsified, and returned as output. This means that the set of formulas falsifiable on $\mathfrak{M}$ is recursively enumerable. ⊣

Theorem 6.13 *If* Λ *is an axiomatizable normal modal logic that has the finite model property with respect to an r.e. set of models* M, *then* Λ *is decidable.*

Proof. It is not difficult to see that Λ, being axiomatizable, is recursively enumerable. But the set of formulas *not* in Λ is also r.e. for $\Lambda = \Lambda_M$ and the set of formulas that are *not* M-valid is r.e. by the previous lemma. Any formula ϕ must eventually turn up on one of these enumerations, hence Λ is decidable. ⊣

As an application, we will show that the minimal propositional dynamic logic is decidable.

Corollary 6.14 PDL *is decidable.*

Proof. By Theorem 4.91, **PDL** is complete with respect to the set of all regular PDL-models. The axioms of **PDL** clearly form a *recursive* set, so trivially they form a recursively enumerable set, thus to be able to apply the Theorem 6.13 it only remains to show that **PDL** has the finite model property with respect to an r.e. set of models.

This follows easily from our completeness proof for **PDL**. Recall that we proved completeness by constructing, for any consistent formula ϕ, a finite model $\mathfrak{P}$ that satisfied ϕ. This gives us what we want, modulo the following glitch: although

$\mathfrak{P}$ contains only finitely many nodes, it may contain infinitely many non-empty relations, thus it may not be of finite character and thus (strictly speaking) our completeness proof does not establish that **PDL** has the finite model property. This is a triviality: for any formula ϕ, only finitely many of the relations on $\mathfrak{P}$ are relevant to the satisfiability of ϕ, namely those that actually occur in ϕ. Let $\mathcal{R}_\phi$ be the smallest set that contains all the relations in $\mathfrak{P}$ corresponding to modalities in ϕ and is downward closed under the usual relation constructors (that is, if $R_{\pi;\pi'} \in \mathcal{R}_\phi$ then so are R_π and $R_{\pi'}$, and analogously for relations defined by union and transitive closure). Note that $\mathcal{R}_\phi$ is finite. Let $\mathfrak{P}'$ be the model that is identical to $\mathfrak{P}$ save that all the relations *not* in $\mathcal{R}_\phi$ are empty; we call $\mathfrak{P}'$ a *reduced model*. Clearly $\mathfrak{P}'$ is a finitely based model that satisfies ϕ. This shows that **PDL** has the finite model property.

The set of reduced models is a recursive set, since checking that a finite model is a reduced model boils down to showing that the relations corresponding to non-basic modalities really are generated out of simpler relations via composition, union, or transitive closure, and this is obviously something we can write a program to do. Hence, the relevant models are recursively enumerable, thus the conditions of Theorem 6.13 are satisfied, and **PDL** is decidable.	⊣

We can also show that **PDL** is decidable by appealing to Theorem 6.7. As we have just seen, our completeness proof for **PDL** gives us the finite model property for **PDL** – but in fact it even gives us the *strong* finite model property. To see this, recall that for any consistent ϕ, we constructed $\mathfrak{P}$ out of *atoms*, that is, maximal consistent subsets of the Fischer-Ladner closure of $\{\phi\}$. As there are at most $2^{c|\phi|}$ such atoms for some constant c, we have a computable upper bound on the size of the models needed to satisfy ϕ. We noted in the proof of Corollary 6.14 that the relevant finite models (the reduced models) form a recursive set, hence we have established everything we need to apply Theorem 6.7.

Theorem 6.13 is a fundamental one and is useful in practice. It does not make use of axiomatizations in a particularly interesting way: it uses them merely to enumerate validities. To apply the theorem we need to know that the set of relevant finite models is recursively enumerable. We often have much stronger syntactic information at our disposal: we may know that a logic is *finitely* axiomatizable. Our next theorem is based on the following observation: if a logic with the f.m.p. is *finitely* axiomatizable, we can use the axiomatization not only to recursively enumerate the validities, but to help us enumerate the non-validities as well.

Theorem 6.15 *If Λ is a finitely axiomatizable normal modal logic with the finite model property, then Λ is decidable.*

Proof. As in the proof of Theorem 6.13, we can use the axiomatization to recursively enumerate Λ, so if we can show that the set of formulas *not* in Λ is also r.e. we will have proved the theorem.

By Theorem 6.5, if Λ has the finite *model* property it also has the finite *frame* property, thus there is some set of finite frames F such that $\Lambda = \Lambda_F$. Hence, if $\phi \notin \Lambda$, ϕ is falsifiable in some model based on a frame in F. Obviously all such frames must validate every axiom of Λ, hence if $\phi \notin \Lambda$, ϕ is falsifiable in some model based on a frame that validates the Λ axioms. Now for the crucial observation: we can write a machine M which decides whether or not a finite frame validates the Λ axioms, for as Λ has only finitely many axioms, each frame can be checked in finitely many steps. With the help of M, we can recursively enumerate the formulas falsifiable in some F-based model, but these are just the formulas which do not belong to Λ. It follows that Λ is decidable. $\dashv$

Can Theorem 6.15 be strengthened by replacing its demand for a finite axiomatization with a demand for a *recursive* axiomatization? No – in Exercise 6.2.5 we give an example of an *undecidable* recursively axiomatizable logic $\mathbf{KU}_X$ with the finite model property; the result hinges on Craig's Lemma.

Theorem 6.15 has many applications, for many common modal and tense logics have the f.m.p. and are finitely axiomatizable. For example, Theorem 6.15 yields another proof that $\mathbf{K}$, $\mathbf{T}$, $\mathbf{KB}$, $\mathbf{K4}$, $\mathbf{S4}$, $\mathbf{S5}$, $\mathbf{K}_t$, $\mathbf{K}_t 4.3$, and $\mathbf{K}_t \mathbf{Q}$ are decidable, for all these logics were shown to be finitely axiomatizable in Chapter 4, and we saw above that they all have the (strong) finite model property. However, a more interesting application follows from our work on logics extending $\mathbf{S4.3}$ in Section 4.9.

Corollary 6.16 *Every normal logic extending* $\mathbf{S4.3}$ *is decidable.*

Proof. By Bull's Theorem (Theorem 4.96) every normal logic extending $\mathbf{S4.3}$ has the finite model property, and by Theorem 4.101 every normal logic extending $\mathbf{S4.3}$ is finitely axiomatizable. Hence the result is an immediate corollary of Theorem 6.15. $\dashv$

Corollary 6.16 completes the main discussion of the section. To summarize what we have learned so far, in Theorems 6.7, 6.13, and 6.15 we have results that pin down three important situations in which the finite model property implies decidability – and indeed, most modal decidability results make use of one of these three theorems.

Exercises for Section 6.2

6.2.1 Provide full proof details for Corollary 6.8. Pay particular attention to showing that $\mathbf{K}_t \mathbf{Q}$ has the f.m.p. with respect to the finite DUWTO-frames (see Theorem 4.41). Filtrations generally do not preserve density, so how do we know that this filtration is dense? Hint: trichotomy.)

6.2.2 Show that if Λ is a finitely axiomatizable normal modal logic with the finite model property, then Λ has the finite frame property with respect to a recursive set of frames.

6.2.3 In this exercise we ask you to show that there is no method of filtrating a partial function that guarantees that the resulting relation is again a partial function.

Consider the model $\mathfrak{M} = (\mathbb{N}, S, V)$ where S is the successor relation on the set $\mathbb{N}$ of natural numbers, and V makes the proposition letter p true at precisely the even numbers. Let Σ be the set $\{\Diamond\neg p, \Diamond p, \neg p, p\}$. Prove that no filtration of $\mathfrak{M}$ through Σ is based on a frame in which S^f is a partial function.

6.2.4 In this exercise we ask the reader to prove that there are uncountably many undecidable normal modal logics with the polysize model property.

Let F_{suc} be the set of all finite frames (W, R) such that $W = \{0, \ldots, k\}$ (for some $k \in \omega$) and for all distinct m and n in W, Rnm iff $m = n + 1$. (Note that this definition permits reflexive points. Indeed, any frame in this set is uniquely determined by its size and which points, if any, are reflexive.) Then, for each $j \in \omega$ define F_j to be the set containing: (1) all the irreflexive frames in F_{suc}; (2) all the frames in F_{suc} whose last point is reflexive; and (3) the (unique) F_{suc} frame containing $j + 1$ nodes such that 0 is the only reflexive point; call this frame $\mathfrak{F}_j$. Now define, for any non-empty $I \subseteq \omega$, F_I as the set $\bigcup_{i \in I} \mathsf{F}_i$; let Λ_I be its logic.

Define ϕ_j to be the formula $p \wedge \Diamond p \wedge \Diamond(\neg p \wedge \Diamond^{j-1}\Box \bot)$.

(a) Prove that ϕ_j is satisfiable in F_i iff $i = j$.
(b) Prove that if ϕ is F_i-satisfiable, then it is satisfiable on a frame in F_i that contains at most $m + 2$ points, where m is the number of modalities in ϕ.
(c) Prove that if I and J are distinct (non-empty) subsets of ω then there is a formula that is satisfiable in F_I but not in F_J.
(d) Prove that each Λ_I has the polysize model property.
(e) Prove that there can only be countably many decidable logics. (This step is actually the easiest one: after all, how many distinct Turing machines can there be?)
(f) Conclude that there are uncountably many undecidable normal modal logics with the polysize model property.

6.2.5 Let X be an r.e. subset of the natural numbers that is *not* recursive; assume that $0 \in X$ but $1 \notin X$. Then $\mathbf{KU}_X$ is the smallest normal modal logic containing the following formulas:

(U1) $\Diamond(\Diamond p \wedge \Diamond q) \rightarrow \Diamond\Diamond(p \wedge q)$,
(U2) $\Diamond(p \wedge \Box \bot) \wedge \Diamond(q \wedge \Box \bot) \rightarrow \Diamond(p \wedge q)$,
(U3) $\Diamond(p \wedge \Diamond\top) \wedge \Diamond(q \wedge \Diamond\top) \rightarrow \Diamond(p \wedge q)$,
(U4)$_k$ $(\Diamond\Box \bot \wedge \Diamond\Diamond\top) \rightarrow \Box^k\Diamond\top$, where $k \in X$.

Note that by Craig's Lemma $\mathbf{KU}_X$ has a *recursive* axiomatization.

(a) Use Sahlqvist's Correspondence and Completeness Theorems to find a first order definable class U of frames for which $\mathbf{KU}_X$ is sound and complete.
(b) Prove that $\mathbf{KU}_X$ has the finite model property.
(c) Show that $\mathbf{KU}_X$ is undecidable.
(Hint: prove that any formula U4$_j$ with j *not* in X, is not satisfiable in U.)

6.3 Decidability via Interpretations

For all its usefulness, decidability via finite models has a number of limitations. One is absolute: as we will shortly see, there are decidable logics that lack the finite model property. Another is practical: it may be difficult to establish the finite model property, for although filtration or selection work in many cases, no universal approach is known. Thus we need to become familiar with other techniques for establishing decidability, and in this section we introduce an important one: *decidability via interpretations*, and in particular, *interpretations in* SnS.

A general strategy for proving a problem decidable is to effectively reduce it to a problem already known to be decidable. But there are many decidable problems; which of them can help us prove modal decidability results? We would like to find a decidable problem, or class of problems, to which modal satisfiability problems can be reduced in a natural manner. Moreover, we would like the approach to be as general as possible: not only should a large number of modal satisfiability problems be so reducible, but the required reductions should be reasonably uniform.

A suitable group of problems is the satisfiability problem for SnS (where $n \in \omega$ or $n = \omega$), the monadic second-order theory of trees of infinite depth, where each node has n successors. Because these problems are themselves satisfiability problems – and indeed, satisfiability problems for monadic second-order languages, the kinds of language used in correspondence theory – it can be relatively straightforward to reduce modal satisfiability to SnS satisfiability. Moreover, the various reductions share certain core ideas; for example, analogs of the standard translation play a useful role. The method can also be used for strong modal languages, such as languages containing the until operator U; see Exercise 2.2.4.

In this section we introduce the reader to such reductions (or better, for reasons which will become clear, *interpretations*). We first introduce the theories SnS, note some examples of their expressivity, and state the crucial decidability results on which subsequent work depends. We then illustrate the method of interpretations with two examples. First, we prove that **KvB**, a logic lacking the finite model property, is decidable. As **KvB** is characterized by a *single* structure (namely, a certain general frame) this example gives us a relatively straightforward introduction to the method. We then show how the decidability of **S4** can be proved via interpretation. The result itself is rather unexciting – we already know that **S4** has the finite model property and is decidable (see Corollary 6.8) – but the proof is important and instructive. **S4** is most naturally characterized as the logic of transitive and reflexive frames, but this is a characterization in terms of an uncountable class of structures. How can this characterization be 'interpreted' in SnS? In fact, it can be done rather naturally, and the ideas involved open the doors to a wide range of further decidability results.

Let us set about defining SnS. If $\mathcal{A}$ is some fixed set (our alphabet), then $\mathcal{A}^*$ is the set of all finite sequences of elements of $\mathcal{A}$, including the null-sequence λ. We introduce the following apparatus:

(i) Define an ordering $\leq$ on $\mathcal{A}^*$ by $x \leq y$ if $y = xz$ for some $z \in \mathcal{A}^*$. Clearly this 'initial-segment-of' relation is a partial order. If $x \leq y$ and $x \neq y$ we write $x < y$.

(ii) Suppose $\mathcal{A}$ is totally ordered by a relation $<_\mathcal{A}$. Then we define $\preceq$ to be the *lexicographic ordering of* $\mathcal{A}^*$ *induced by* $<_\mathcal{A}$. That is, $x \preceq y$ if and only if $x \leq y$, or $x = zau$ and $y = zbv$ where $a, b \in \mathcal{A}$ and $a <_\mathcal{A} b$. Note that $\preceq$ totally orders $\mathcal{A}^*$.

(iii) For any $a \in \mathcal{A}$ we define $r_a : \mathcal{A}^* \to \mathcal{A}^*$, the *a-th successor function*, by $r_a(x) = xa$.

Definition 6.17 (SnS) For any n such that n is a natural number, or $n = \omega$, let T_n be $\{i \in \omega \mid i < n\}^*$. The structure $\mathfrak{N}_n$ is $(T_n, r_i, \leq, \preceq)_{i<n}$, where $\preceq$ is the lexicographic ordering induced by $<_\omega$, the usual ordering of the natural numbers. $\mathfrak{N}_n$ is called the *structure of n successor functions*. (Note that all these structures are countably infinite.)

The *monadic second-order theory of n successor functions* is the monadic second-order theory of $\mathfrak{N}_n$ in the monadic second-order language of appropriate signature (we spell out the details of this language below); this theory is usually referred to as SnS.   ⊣

Let us spell out the intuitions underlying this machinery. First, note that each structure $\mathfrak{N}_n$ really is an infinite tree where each node has n immediate successors (or *daughters*, in standard tree terminology). For example, consider $\mathfrak{N}_1$; that is $(\{0\}^*, r_0, \leq, \preceq)$. This is the infinite tree in which each node has exactly one daughter; that is, it is simply an isomorphic copy of the natural numbers in their usual order. Next, consider $\mathfrak{N}_2$, that is $(\{0,1\}^*, r_0, r_1, \leq, \preceq)$. This is the full binary tree (that is, the infinite tree in which every node has exactly two daughters). An initial segment of $\mathfrak{N}_2$ is shown in Figure 6.1. Note that λ is the *root node* of the tree depicted in Figure 6.1, and that r_0 and r_1 are the *first daughter* and *second daughter* relations, respectively. Further, note that $\leq$ has a natural tree-geometric interpretation: it is simply the *dominates* relation. That is, $x \leq y$ iff it is possible to reach x by moving upwards in the tree from y. Similarly, $\preceq$ is the *dominates-or-to-the-left-of* relation. The tree-like nature of these models plays an important role in the work that follows, and must be properly understood. In particular, the reader should check that $\mathfrak{N}_\omega$ really is an infinite tree in which every node has ω daughters.

So much for the structures – what about the theories? Each of the theories SnS is a monadic *second-order* theory in the *appropriate* language. For example, the monadic second-order language appropriate for talking about $\mathfrak{N}_2$ contains two function symbols for talking about r_0 and r_1 (we will be economical with our

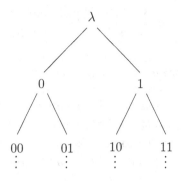

Fig. 6.1. An initial segment of $\mathfrak{N}_2$

notation and use r_0 and r_1 for these symbols) and two binary predicate symbols for talking about $\leq$ and $\preceq$ (we use $\leq$ and $\preceq$ for this purpose). In addition, the language contains a denumerably infinite set of individual variables $x, y, z, \ldots$, a denumerably infinite set of predicate (or set) variables $P, Q, S, \ldots$, that range over subsets of the domain, and the usual quantifiers and boolean operators. The syntax and semantics of the language is standard; see Appendix A for further discussion of monadic second-order logic.

Using these languages, we can say many useful things about $\mathfrak{N}_n$. First, note that although we did not include a primitive equality predicate, an equality predicate is definable over $\mathfrak{N}_n$:

$$x = y \text{ iff } x \preceq y \wedge y \preceq x.$$

Next, note that we can define a unary predicate symbol ROOT that is true only of the root node λ:

$$\text{ROOT}(x) \text{ iff } \neg \exists y \, (y < x). \tag{6.1}$$

We can define the unary higher-order predicate 'P is a finite set.' Recall that a total ordering R on a set S is a *well-ordering* if every non-empty subset of S has an R-least element; it is a standard observation that S is well-ordered by R iff S contains no infinitely descending R-chains. It follows that a subset P of T_n is finite iff it is well-ordered by both $\preceq$ and its converse, for such a set contains no infinitely descending $\preceq$-chains and no infinitely ascending $\preceq$-chains.

$$\text{FINITE}(P) \text{ iff} \tag{6.2}$$
$$\forall Q \, ((\exists x \, Qx \wedge \forall y \, (Qy \rightarrow Py)) \rightarrow \exists u \, (Qu \wedge \forall w \, (Qw \rightarrow u \preceq w))$$
$$\wedge \, \exists v \, (Qv \wedge \forall w \, (Qw \rightarrow w \preceq v))).$$

That is, P is finite if every non-empty subset of P has a $\preceq$-first and a $\preceq$-last element.) In short, monadic second-order logic is an extremely powerful language for talking about trees – which makes the following result all the more remarkable.

Theorem 6.18 (Rabin) *For any natural number* n, *or* $n = \omega$, $\mathrm{S}n\mathrm{S}$ *is decidable.*

That is, for any n, it is possible to write a Turing machine which, when given a monadic second-order formula (in the language of appropriate signature), correctly decides whether or not the formula is satisfiable in $\mathfrak{N}_n$. The proof of this beautiful result is beyond the scope of this book; we refer the reader to the Notes for discussion and references.

Given a modal logic Λ, how can we use the fact of $\mathrm{S}n\mathrm{S}$-decidability to establish Λ-decidability? Suppose $\Lambda = \Lambda_\mathsf{M}$ for some class of *countable* models M. The essence of the interpretation method is to attempt to construct, for any modal formula ϕ, a monadic second-order formula *Sat-$\Lambda(\phi)$* that does three things.

- It must encode the information in ϕ; this is usually achieved by using some variant of the standard translation.
- It must define a set of substructures of $\mathfrak{N}_n$ (for some choice of n) which are isomorphic copies of the models in M.
- It must bring the two previous steps together. That is, *Sat-$\Lambda(\phi)$* must be constructed so that it is satisfiable in $\mathfrak{N}_n$ iff (the translation of) ϕ is satisfiable in (a definable substructure of $\mathfrak{N}_n$ that is isomorphic to) a model in M – that is, iff ϕ is Λ-satisfiable.

If such a formula *Sat-$\Lambda(\phi)$* can be constructed, the ramifications for modal decidability are clear: as $\mathrm{S}n\mathrm{S}$ is decidable, we can decide whether or not *Sat-$\Lambda(\phi)$* is satisfiable on $\mathfrak{N}_n$. As this is equivalent to deciding the Λ-satisfiability of ϕ, we will have established that Λ is decidable.

As our first example of the method in action, we will prove the decidability of **KvB**. We met this logic briefly in Exercise 4.4.2; it is the logic of a certain general frame $\mathfrak{J}$. The domain J of $\mathfrak{J}$ consists of $\mathbb{N} \cup \{\omega, \omega + 1\}$ (that is, the set of natural numbers together with two further points), and the relation R is defined by Rxy iff $x \neq \omega + 1$ and $y < x$ or $x = \omega + 1$ and $y = \omega$. The frame (J, R) is shown in Figure 6.2. A, the collection of subsets of J admissible in $\mathfrak{J}$, consists of all $X \subseteq J$ such that either X is finite and $\omega \notin X$, or X is co-finite and $\omega \in X$.

As the reader was asked to show in Exercise 4.4.2, **KvB** is incomplete; that is, there is no class of *frames* F such that $\mathbf{KvB} = \Lambda_\mathsf{F}$. By Theorem 6.5 it follows that **KvB** lacks the finite model property. Even though it lacks the finite model property, **KvB** *is* decidable, and we will demonstrate this via an interpretation in S2S.

Theorem 6.19 KvB *is decidable.*

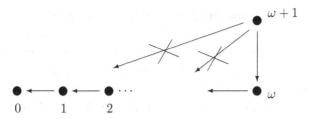

Fig. 6.2. The frame underlying $\mathfrak{J}$. Note that $\omega + 1$ is related only to ω

Proof. Let us make two initial assumptions; we will shortly show that both assumptions are correct. First, let us suppose that $\mathfrak{J} = (J, R, A)$ can be isomorphically embedded in $\mathfrak{N}_2$. We will refer to this isomorphic copy as $\mathfrak{J}$; no confusion should arise because of this double usage. Furthermore, let us suppose that this isomorphic image is *definable* in the monadic second-order language for $\mathfrak{N}_2$. That is, suppose that there are formulas $\widehat{J}(x)$, $\widehat{R}(x, y)$ and $\widehat{A}(P)$ (containing 1 free individual variable x, 2 free individual variables x and y, and 1 free predicate variable P, respectively) such that

$$
\begin{aligned}
J &= \{t \in T_2 \mid \mathfrak{N}_2 \models \widehat{J}(x)[t]\}, \\
R &= \{(t, t') \in T_2 \times T_2 \mid \mathfrak{N}_2 \models \widehat{R}(x, y)[t, t']\}, \\
A &= \{U \subseteq T_2 \mid \mathfrak{N}_2 \models \widehat{A}(P)[U]\}.
\end{aligned}
$$

Given these assumptions, it is easy to reduce the satisfiability problem for the general frame to the satisfiability problem for S2S. First, with the help of the formula $\widehat{R}$, we can define a translation T from the modal language into the second-order language:

$$
\begin{aligned}
T_x(p) &= Px, \\
T_x(\neg\phi) &= \neg T_x(\phi), \\
T_x(\phi \wedge \psi) &= T_x(\phi) \wedge T_x(\psi), \\
T_x(\Diamond\phi) &= \exists y\, (\widehat{R}xy \wedge T_y(\phi)).
\end{aligned}
$$

Note that T_x is just the standard translation with the formula $\widehat{R}$ replacing the use of a fixed relation symbol. We leave it as an exercise to show that for any modal formula ϕ (built out of proposition letters $p_1, \ldots, p_n$):

$$((J, R, A), V), w \Vdash \phi \quad \text{iff} \quad \mathfrak{N}_2 \models T_x(\phi)[w, V(p_1), \ldots, V(p_n)]. \tag{6.3}$$

(See Exercise 6.3.1; the notation $[w, V(p_1), \ldots, V(p_n)]$ means assign the state w to the free variable x, and assign the subset $V(p_i)$ to the predicate variable P_i.)

For any modal formula ϕ, let *Sat-KvB*(ϕ) be the following monadic second-order sentence:

$$\exists P_1 \ldots \exists P_n \exists x\, (\widehat{A}(P_1) \wedge \cdots \wedge \widehat{A}(P_n) \wedge \widehat{J}(x) \wedge T_x(\phi)).$$

It follows that ϕ is satisfiable in (J, R, A) iff $\mathfrak{N}_2 \models Sat\text{-}KvB(\phi)$. Thus – *given our two initial assumptions* – we have effectively reduced the satisfiability problem for **KvB** to the S2S-satisfiability problem, for ϕ is satisfiable on $\mathfrak{J}$ iff $Sat\text{-}KvB(\phi)$ belongs to S2S, and by Rabin's result it is possible to decide the latter.

Hence, to complete the proof that **KvB** is decidable, it only remains to show that our assumptions were justified; that is, to show that $\mathfrak{J}$ really does have a definable isomorphic image in $\mathfrak{N}_2$. Given the expressive power at our disposal, this is actually rather easy to do. We will make use of the general predicates $=$, ROOT, and FINITE defined in (6.1) and (6.2). In addition, we will use

$$x <_1 y \text{ iff } r_1(x) \le y \land \neg\exists z (x \le z \land r_0(z) \le y).$$

Note that $x <_1 y$ means that x is a proper initial subsequence of y such that y extends x by a finite sequence of 1s – or, in terms of tree geometry, it is possible to move down from x to y by using only the 'second daughter' relation.

We will now define an isomorphic image of $\mathfrak{J}$ in $\mathfrak{N}_2$. First, we can define the numeric part of the underlying frame as follows:

$$N(x) \text{ iff } \text{ROOT}(x) \lor \exists y (\text{ROOT}(y) \land y <_1 x).$$

The isomorphism involved should be clear: the natural number zero is taken to be the empty sequence, and the positive integer n is taken to be the sequence of n 1s. Next, we will represent ω by 0, and $\omega + 1$ by 00. Defining these choices is easy:

$$\text{OMEGA}(x) \quad \text{iff} \quad \exists y (\text{ROOT}(y) \land x = r_0(y)),$$
$$\text{OMEGA+1}(x) \quad \text{iff} \quad \exists y (\text{ROOT}(y) \land x = r_0(r_0(y))).$$

Putting it all together, we define the required predicates $\widehat{J}$ and $\widehat{R}$ as follows:

$$\widehat{J}(x) \quad = \quad Nx \lor \text{OMEGA}(x) \lor \text{OMEGA+1}(x),$$
$$\widehat{R}(x, y) \quad = \quad (Ny \land (y <_1 x \lor \text{OMEGA}(x))) \lor (\text{OMEGA}(y) \land \text{OMEGA+1}(x)).$$

Clearly these two formulas define a subset of the tree domain isomorphic to (J, R). Thus it merely remains to define A, the class of allowable valuations. With the help of the FINITE predicate, this is straightforward.

$\widehat{A}(P)$ iff

$$\forall x (Px \to \widehat{J}(x)) \land ((\text{FINITE}(P) \land \forall z (\text{OMEGA}(z) \to \neg Pz)) \lor$$
$$\forall Q \forall x (Qx \leftrightarrow (\widehat{J}x \land \neg Px) \to (\text{FINITE}(Q) \land \forall z (\text{OMEGA}(z) \to \neg Qz)))).$$

In short, a definable isomorphic image of $\mathfrak{J}$ really does live inside $\mathfrak{N}_2$. We conclude that **KvB** is decidable.

While the above result is a nice introduction to decidability via interpretation, in one respect it is rather misleading. **KvB** is characterized by a single structure (and a rather simple one at that) thus we only had to define a single isomorphic image, and were able to do this fairly straightforwardly using S2S. However, as we saw in Chapter 4, it is usual to characterize logics in terms of a *class* of structures; for example, **S4** is usually characterized as the logic of the class of reflexive and transitive models. Do class-based characterizations mesh well with the idea of decidability via interpretations? Classes of models may contain uncountable structures – and only countable structures can be isomorphically embedded in $\mathfrak{N}_n$. And why should we expect to be able to isomorphically embed even countable models in infinite *trees*?

Two simple observations clear the way. First, in many important cases, only the *countable* structures in characterizing classes are required. Second, there is a standard method for converting a model into a tree-based model, namely the *unraveling* method studied in Chapters 2 and 4. Taken together, these observations enable us to view the classes of structures characterizing many important logics as a collection of definable substructures of $\mathfrak{N}_\omega$. We will illustrate the key ideas involved by proving the decidability of **S4** via interpretation in SωS.

As a first step, we claim that **S4** is sound and strongly complete with respect to the class of *countable* reflexive and transitive models. We could prove this directly (for example, using the step-by-step method discussed in Section 4.6) but it also follows from the following general observation. (Recall that for the duration of this chapter, we are only working with countable languages.)

Theorem 6.20 *If Λ is a normal logic that is sound and strongly complete with respect to a first-order definable class of models* M, *then Λ is also sound and strongly complete with respect to the class of all* countable *models in* M.

Proof. Left as Exercise 6.3.3. ⊣

Lemma 6.21 **S4** *is sound and strongly complete with respect to the class of countable (reflexive and transitive) trees.*

Proof. By Theorems 4.29 and 6.20, **S4** is sound and strongly complete with respect to the class of *countable* reflexive, transitive models; that is, every **S4**-consistent set of sentences Σ is satisfiable on such a model $\mathfrak{M} = (W, R, V)$ at some point w. Now, (as in the proof of Theorem 4.54) let $\vec{\mathfrak{M}} = (\vec{W}, \vec{R}, \vec{V})$ be the unraveling of $\mathfrak{M}$ around w, and let $\mathfrak{M}^*$ be $(\vec{W}, R^*, \vec{V})$, where R^* is the reflexive transitive closure of $\vec{R}$; this model is a reflexive transitive tree that verifies Σ at its root. Moreover, it is a *countable* model, for its nodes are all the finite sequences of states in $\mathfrak{M}$ that start at w, and as $\mathfrak{M}$ is countable, there are only countably many such sequences. The result follows. ⊣

Corollary 6.22 **S4** *is decidable, and its decidability can be proved via interpretations.*

Proof. Let us call a subset of $\mathfrak{N}_\omega$ an *initial subtree* if it contains λ and is closed under the inverse of $\leq$ (that is, if y belongs to the subset, and $x \leq y$, then x belongs to the subset). If S is such a subtree, then $\leq_S$ denotes the restriction of $\leq$ to S. Now for the key observation. Let $(\vec{W}, \vec{R})$ be the unraveling of some *countable* **S4**-frame (W, R) around a point w, and let R^* be the reflexive transitive closure of $\vec{R}$. Then $(\vec{W}, R^*)$ is isomorphic to a pair $(S, \leq_S)$ for some initial subtree S. To see this, note that we can inductively construct an isomorphism f from $(\vec{W}, R^*)$ to some initial subtree as follows. First, we stipulate that f maps the root of $(\vec{W}, R^*)$ to λ. Next, suppose that for some $\vec{u} \in \vec{W}$, $f(\vec{u})$ has been defined to be m. Now, $\{\vec{s} \in \vec{W} \mid \vec{u}R\vec{s}\}$ is a countable set as $\vec{W}$ is countable, so we can enumerate its elements. Then, if $\vec{s}$ is the i-th element in this enumeration, we stipulate that $f(\vec{s}) = r_i(m)$. (That is, the successor of $\vec{u}$ that is i-th in our enumeration is mapped to the i-th successor of m.) In short, $\mathfrak{N}_\omega$ is 'wide enough' to accommodate a copy of every branch through a tree-like **S4** model in a very obvious way. In fact, it is precisely because the required isomorphisms are so simple that we have elected to work with $\mathfrak{N}_\omega$.

With this observed, the interpretation is easy to define. First, we define a predicate $\textsc{Isubtree}(S)$, which picks out the initial subtrees of $\mathfrak{N}_\omega$:

$$\textsc{Isubtree}(S) \text{ iff } \exists y\,(\textsc{Root}(y) \wedge Sy) \wedge \forall z \forall u\,((Sz \wedge u \leq z) \to Su).$$

Second, we define a predicate $\leq_S$ that defines the restriction of $\leq$ to a subset S of $\mathfrak{N}_\omega$ by

$$x \leq_S y \text{ iff } Sx \wedge Sy \wedge x \leq y.$$

Third, we define a translation T^ω from the basic modal language to the monadic second-order language for $\mathfrak{N}_\omega$. Like the translation T we used when proving the decidability of **KvB** this translation is a simple variant of the standard translation. In fact, it is identical to T save in the clause for modalities, which is given by:

$$T^\omega_{x,S}(\Diamond\phi) = \exists y\,(x \leq_S y \wedge T^\omega_{y,S}(\phi)).$$

Note that as well as containing the free individual variable x, the translation of $\Diamond\phi$ contains a free set variable S; when written in full the above expression becomes:

$$T^\omega_{x,S}(\Diamond\phi) = \exists y\,(Sx \wedge Sy \wedge x \leq y \wedge T^\omega_{y,S}(\phi)).$$

We need the free variable here because we are not working with one fixed isomorphic image (as we were when proving the decidability of **KvB**). Rather, we have a separate relation for each initial subtree, and the presence of the free variable allows all our definitions to be relativized in the appropriate way.

It simply remains to put it all together. Suppose ϕ is a modal formula constructed out of the proposition letters $p_1, \ldots, p_n$. Define *Sat-S4*(ϕ) to be the following sentence:

$$\exists S \exists P_1 \ldots \exists P_n \exists x \left(\text{ISUBTREE}(S) \wedge \right.$$
$$\left. \forall z \left(P_1 z \to Sz \right) \wedge \cdots \wedge \forall z \left(P_n z \to Sz \right) \wedge Sx \wedge T^\omega_{x,S}(\phi) \right).$$

Recall that $T^\omega(\phi)$ contains free occurrences of S and x; these become bound in this sentence. Bearing this in mind, it is clear that this sentence asserts the existence of an initial subtree S of $\mathfrak{N}_\omega$, a collection of n subsets P_i of this subtree, and a state x in the subtree, that satisfy the translation of ϕ. That is, it asserts the existence of a tree-like **S4** model for the (translation of) ϕ, and we have reduced the S4-satisfiability problem to the SωS-satisfiability problem. ⊣

This completes our discussion of interpretations in SnS – though we should immediately admit that we have barely scratched the surface of the method's potential: Rabin's theorem is very strong, the ideas underlying it make contact with many branches of mathematics, and it has become a fundamental tool in many branches of logic and theoretical computer science. Nonetheless, our discussion has unearthed themes relevant to modal logic: the importance of establishing completeness results with respect to classes of *countable* structures, the use of *unraveling* to produce tree-like models, and the particular utility of $\mathfrak{N}_\omega$ in allowing reasonably straightforward isomorphic embeddings. These three ideas enable a wide range of modal decidability results to be proved via interpretations.

One final remark: while SnS is important, it is certainly not the only logical system in which modal logics can be interpreted. Many fragments of classical logic, or theories in classical logics, are known to be decidable, and offer opportunities for proving modal decidability results. Indeed we have already met a (very simple) example. We pointed out in Section 2.4 that the basic modal language translates into the 2 variable fragment of classical logic, (see Proposition 2.49), from which it immediately follows that **K** (and some simple extensions such as **T**) are decidable. Moreover, on occasions it can be useful to interpret a modal logic in another modal logic already known to be decidable. See the Notes for further discussion.

Exercises for Section 6.3

6.3.1 We claimed that the general frame for **KvB** is isomorphically embedded in the tree domain, and that $\widehat{R}$ defines the accessibility relation of this isomorphic image. Check this claim, and show that

$$((W, R, A), V), w \Vdash \phi \text{ iff } \mathfrak{N}_2 \models T_x(\phi)[w, V(p_1), \ldots, V(p_n)],$$

for any modal formula ϕ (see (6.3)).

6.3.2 Show by interpretation in S2S that both the tense logic of the natural numbers, and the tense logic of the integers, are decidable. Now add the until operator U to your language (this operator was defined in Chapter 2 in Exercise 2.2.4). Are the logics of the natural numbers and the integers in this richer language still decidable?

6.3.3 Prove Theorem 6.20. That is, show that if Λ is a normal logic that is sound and strongly complete with respect to a first-order definable class of models M, then Λ is also sound and strongly complete with respect to the class of all *countable* models in M. (Hint: use the standard translation and the Downward Löwenheim-Skolem Theorem.)

6.4 Decidability via Quasi-models and Mosaics

In this section we will show that such familiar techniques as filtration can be employed to prove decidability, even for logics lacking the finite model property. The key move is simply to think more abstractly: instead of trying to work with finite models themselves, we will work with finite structures which encode information *about* models.

Quasi-models for KvB

For our first example we will re-examine the logic **KvB**, which we proved decidable in the previous section via interpretation in S2S. Recall that **KvB** is the logic of a single general frame $\mathfrak{J}$ whose universe J is $\mathbb{N} \cup \{\omega, \omega + 1\}$, and whose accessibility relation is R. Also recall that **KvB** is an *incomplete* logic, which implies that it does *not* have the finite model property. Nonetheless, we can establish the decidability of **KvB** using a filtration argument. We cannot use filtration to build a finite **KvB** *model* (no such model exists), but we can use it to build a finite *quasi-model*.

Consider a model $\mathfrak{M} = (J, R, V)$, where V is an admissible valuation for $\mathfrak{J}$. What kind of filtration seems natural for this structure? If it were not for the point $\omega + 1$, it is obvious that we would go for the transitive filtration. Very well then – let us adopt the following procedure: first delete the point $\omega + 1$, then take the transitive filtration of the remainder of the frame, and finally glue a copy of the point $\omega + 1$ back on to the resulting finite structure. Of course, we know that this will not result in a finite **KvB** model; but hopefully it will yield something from which we can *construct* a **KvB** model.

First we need the notion of a *closure* of a set of sentences. We will not filtrate through arbitrary subformula-closed sets of sentences; rather, we will insist on working with sets of sentences that are closed under single negations as well.

Definition 6.23 (Closed Sets and Closures) A set of formulas Σ is said to be *closed* if it is closed under subformulas and single negations. That is, if $\sigma \in \Sigma$ and θ is a subformula of σ, then $\theta \in \Sigma$; and moreover if $\sigma \in \Sigma$, and σ is not of the form $\neg\theta$, then $\neg\sigma \in \Sigma$.

If Γ is a set of formulas, then $\mathrm{Cl}(\Gamma)$, the *closure* of Γ, is the smallest closed set of formulas containing Γ. Note that if Γ is finite then so is $\mathrm{Cl}(\Gamma)$. If $\Gamma = \{\phi\}$

where ϕ is any modal formula, then we usually write $\mathrm{Cl}(\phi)$ for $\mathrm{Cl}(\{\phi\})$ and call this set the closure of ϕ. ⊣

Advanced track readers should note that they have already met a more elaborate version of this idea when we proved the completeness of **PDL**: any Fischer-Ladner closed set (see Definition 4.79) is closed in the sense of the previous definition.

Now for quasi-models. Let ϕ be some basic modal formula. A **KvB** *quasi-model for* ϕ is a pair $\mathfrak{Q} = (\mathfrak{F}, \lambda)$ where:

(i) $\mathfrak{F} = (Q, S)$ is a finite frame, containing two distinct distinguished points called c and ∞, that satisfies conditions F1–F5 below; and

(ii) λ is a function mapping states of $\mathfrak{F}$ to subsets of $\mathrm{Cl}(\phi)$ that satisfies the conditions L0–L3 below. We call λ a *labeling*.

Let us first consider the conditions F1–F5. These are very simple, and should be checked against Figure 6.2. If you read c as 'co-finite' and view this element as the quasi-model's analog of ω, and view ∞ as the analog of $\omega + 1$, the resemblance between finite frames fulfilling these conditions and the frame (J, R) should be clear.

(F1) On $Q \setminus \{\infty\}$, S is trichotomous and transitive,
(F2) Scw iff $w \neq \infty$,
(F3) Swc iff $w = c$ or $w = \infty$,
(F4) $S\infty w$ iff $w = c$, and
(F5) $Sw\infty$ for no w in Q.

Note that c is *reflexive*. Intuitively, the filtration process described above squashes ω down into a cluster.

There are also conditions on the labeling. One of these conditions is that every label should be a *Hintikka set*. This is an important concept, and one we will use again later in this chapter.

Definition 6.24 (Hintikka Sets) Let Σ be a subformula closed set of formulas. A *Hintikka set* H over Σ is a maximal subset of Σ that satisfies the following conditions:

(i) $\bot \notin H$.
(ii) If $\neg\phi \in \Sigma$, then $\neg\phi \in H$ iff $\phi \notin H$.
(iii) If $\phi \wedge \psi \in \Sigma$, then $\phi \wedge \psi \in H$ iff $\phi \in H$ and $\psi \in H$. ⊣

It is important to realize that Hintikka sets also satisfy conditions such as the following: if $\phi \vee \psi \in \Sigma$, then $\phi \vee \psi \in H$ iff $\phi \in H$ or $\psi \in H$. This is because in this book we define $\vee$ (and also $\rightarrow$, $\leftrightarrow$, and $\top$) in terms of $\bot$, $\neg$, and $\wedge$ (see Definition 1.12). Hintikka sets need *not* be satisfiable (the reader is asked to construct a non-satisfiable Hintikka set in Exercise 6.4.2) but items (i) and (ii) above guarantee that they contain no blatant propositional contradictions. If a Hintikka set is satisfiable we call it an *atom*. Note that both the MCSs used to build canonical models, and the special atoms used to prove the completeness of **PDL** are examples of (consistent) Hintikka sets.

We are now ready for the quasi-model labeling conditions:

(L0) $\phi \in \lambda(w)$ for some $w \in Q$,
(L1) $\lambda(w)$ is a Hintikka set, for each $w \in Q$,
(L2) for all $\Diamond\psi \in Cl(\phi)$, $\Diamond\psi \in \lambda(w)$ iff $\psi \in \lambda(v)$ for some v with Swv,
(L3) if $\Diamond\psi \in \lambda(w)$, then $\psi \in \lambda(v)$ for some v with Swv *and not* Svw.

We take the *size* of a quasi-model (Q, S, λ) to be the size of its universe Q.

Lemma 6.25 *Let ϕ be a formula in the basic modal language. Then ϕ is satisfiable in $\mathfrak{J}$ if and only if there is a quasi-model for ϕ, of size at most $2^{|\phi|}$.*

Proof. We leave it to the reader to prove the left to right direction; this is simply a matter filling in the details of the 'delete $\omega + 1$, filtrate, glue $\omega + 1$ back on' strategy sketched above (the filtration must be made through $Cl(\phi)$) and the upper bound on the size of the quasi-model follows as in any filtration argument. So let us look at the right to left direction.

Let $\mathfrak{Q} = (Q, S, \lambda)$ be a quasi-model for ϕ, let c and ∞ be the distinguished points of the quasi-model, and let Q_0 denote the set $Q \setminus \{\infty\}$. We now define an equivalence relation $\sim_0$ on Q_0 by

$$w \sim_0 v \text{ iff } w = v \text{ or } (Swv \text{ and } Svw).$$

This really is an equivalence relation, and a more-or-less familiar one at that: the equivalence class $\bar{w}$ containing a *reflexive* point w is simply the *cluster* that w belongs to (see Definition 4.55), while the equivalence class $\bar{w}$ containing an *irreflexive* point w is simply $\{w\}$. The equivalence classes on Q_0 are naturally ordered by the relation $\succ$ defined as follows:

$$\bar{w} \succ \bar{v} \text{ iff } Swv \text{ and not } Svw.$$

It follows from F1 that $\succ$ is a strict total ordering. Now consider an enumeration $q_0, q_1, \ldots, q_N$ of the elements of Q_0, such that q first enumerates all elements of the leftmost equivalence class, then all elements of its rightmost neighbor, and so on. We may extend this enumeration to a map $f : J \to Q$ by putting

$$f(w) = \begin{cases} q_w & \text{if } w \leq N, \\ c & \text{if } w > N \text{ or } w = \omega, \\ \infty & \text{if } w = \omega + 1. \end{cases}$$

It is straightforward to check that for all w, v in J, Rwv implies $Sf(w)f(v)$

Consider, for instance, the case where $w = \omega$; Rwv implies that $v = n$ for some natural number n. But then $f(w) = c$ and $f(v) \in Q_0$, so $Sf(w)f(v)$ follows from F2. The other cases are left to the reader. What we have shown is that

$$f \text{ is a homomorphism mapping } (J, R) \text{ onto } (Q, S). \qquad (6.4)$$

Now consider the following valuation V on (J, R):

$$V(p) = \{w \in J \mid p \in \lambda(f(w))\}.$$

It is easy to see that V is admissible in the general frame $\mathfrak{J}$: if $\omega \in V(p)$ then by definition of V, $n \in V(p)$ for all $n > N$, so $V(p)$ is co-finite.

Hence, in order to prove the lemma, it is sufficient to show that ϕ holds somewhere in the model $(\mathfrak{J}, V)$; but this follows from L0 and the following claim:

$$\text{for all } \psi \in \text{Cl}(\phi), \text{ and all } w \in J: \quad \mathfrak{J}, V, w \Vdash \psi \text{ iff } \psi \in \lambda(f(w)). \qquad (6.5)$$

We will prove this claim by induction on the complexity of ψ. The base case, where ψ is a propositional variable, holds by definition of V, and the induction step for the boolean connectives is trivial since λ labels with *Hintikka sets* only. Hence, the only interesting case is where ψ is of the form $\Diamond\chi$. Note that the inductive hypothesis applies to χ and that $\chi \in \text{Cl}(\phi)$ since the set is closed under taking subformulas.

First assume that $\mathfrak{J}, V, w \Vdash \Diamond\chi$. There is a state v with Rwv and $v \Vdash \chi$. By the fact that f is a homomorphism it is immediate that $Sf(w)f(v)$, while the inductive hypothesis implies that $\chi \in \lambda(f(v))$. From this and L2 it follows that $\Diamond\chi \in \lambda(f(w))$.

Now suppose, in order to prove the other direction of (6.5), that $\Diamond\chi \in \lambda(f(w))$. We have to show that $\mathfrak{J}, V, w \Vdash \Diamond\chi$. Distinguish the following cases:

(i) $w = \omega + 1$. From this it follows that $f(w) = \infty$, so from L2 and the fact (F4) that c is the *only* successor of ∞, it follows that $\chi \in \lambda(c)$. Hence from $c = f(\omega)$ and the inductive hypothesis it follows that $\omega \Vdash \chi$. But then it is immediate that $\omega + 1 \Vdash \Diamond\chi$.

(ii) $w \neq \omega + 1$. By L3 we may assume the existence of an element $q \in Q$ satisfying $Sf(w)q$, not $Sqf(w)$ and $\chi \in \lambda(q)$. It is obvious from $Sf(w)q$ and F5 that $q \neq \infty$. Let v be a pre-image of q; from $q \neq \infty$ it follows that $v \neq \omega + 1$. Since R is trichotomous on $J \setminus \{\omega + 1\}$, we have Rvw or $w = v$ or Rwv. The first two options are impossible: Rvw would imply $Sf(v)f(w)$, while $w = v$ is incompatible with the fact that $Sf(v)f(w)$ but not $Sf(w)f(v)$. Hence, we find that Rwv; but the induction hypothesis gives $v \Vdash \chi$, so indeed we have $w \Vdash \Diamond\chi$.

This finishes the proof of (6.5) and hence, of the lemma. ⊣

Theorem 6.26 *The logic* **KvB** *is decidable.*

Proof. By Lemma 6.25 it suffices to show that it is decidable whether there is a quasi-model for ϕ of size not exceeding $2^{|\phi|}$. But this is easy to see: we first make a finite list of all triples (Q, S, λ) such that $|Q| \leq 2^{|\phi|}$, $S \subseteq Q \times Q$ and $\lambda : Q \to \mathcal{P}(\mathrm{Cl}(\phi))$; we then check for each member of this list whether it is a quasi-model for ϕ. And clearly it is possible to write a terminating program which does this.   ⊣

The important lesson is that in order to prove decidability of a logic, not only finite *models* are useful: rather, any finite structure that *encodes* a model is potentially valuable. Now, the finite structure employed in the previous example was still very much *like* a model – as our name 'quasi-model' indicates – but in the general case one can push the idea much further. The satisfiability of ϕ does not need to be witnessed by a finite model for ϕ, or indeed by anything that looks very much like a model; all we need is a finite *toolkit* which contains the instructions needed to construct a model for ϕ. The concept of a *mosaic* develops this line of thought.

Mosaics for the tense logic of the naturals

Consider the frame $\mathfrak{N} = (\mathbb{N}, <)$ with $<$ the standard ordering of the natural numbers, and let $\mathbf{K}_t\mathbf{N}$ be its tense logic. $\mathbf{K}_t\mathbf{N}$ does *not* have the finite model property (see Exercise 6.4.1), but it *is* decidable, as we will now show using mosaics.

We use the following terminology and notation. For a given formula ϕ in the basic temporal similarity type, let $\mathrm{Cl}(\phi)$ denote the smallest set containing ϕ, $P\top$ and $F\top$ which is closed under taking subformulas and single negations (cf. 4.79).

Definition 6.27 (Bricks) A *brick* is a pair $b = (\Phi, \Lambda)$ such that Φ and Λ are Hintikka sets satisfying

(B0) if $F\psi$ or ψ belongs to Λ, then $F\psi \in \Phi$,
(B1) if $P\psi$ or ψ belongs to Φ, then $P\psi \in \Lambda$.
(B2) $F\top \in \Lambda$.

A brick is called *small* if it satisfies, in addition:

(B3) if $F\psi \in \Phi$, then either ψ or $F\psi$ is in Λ,
(B4) if $P\psi \in \Lambda$, then either ψ or $P\psi$ is in Φ.

What we are really interested in are *sets* of bricks satisfying certain saturation conditions. A brick set B is a *saturated set of bricks for* ϕ (in short: a ϕ-SSB) if it satisfies

(S0) for some $(\Phi, \Lambda) \in B$, $\neg P\top \in \Phi$ and $\phi \in \Lambda \cup \Phi$,
(S1) for all $(\Phi, \Lambda) \in B$, if $F\psi \in \Lambda$ then there is a $(\Lambda, \Gamma) \in B$ with $\psi \in \Gamma$,

(S2) for all $(\Phi, \Lambda) \in B$ there is a path of small bricks leading from Φ to Λ.

Here we say that a *path* of (small) bricks from Φ to Λ is a sequence $(\Phi_0, \Lambda_0), \ldots,$ (Φ_n, Λ_n) $(n \geq 0)$ of (small) bricks such that $\Phi = \Phi_0$, $\Lambda = \Lambda_n$ and $\Lambda_i = \Phi_{i+1}$ for all $i < n$. Finally, we simply define the *size* of an SSB B to be the number of bricks in B. ⊣

The best way of grasping the intuitive meaning of these notions is by reading the proof of the next lemma.

Lemma 6.28 *If ϕ is satisfiable in $\mathfrak{N}$, then there is a ϕ-SSB of size at most $2^{|\phi|}$.*

Proof. Assume that we have a valuation V on $\mathfrak{N}$ such that ϕ is satisfied in the model $(\mathfrak{N}, V)$. For any number n, let Γ_n denote the truth set of n:

$$\Gamma_n = \{\psi \in \mathrm{Cl}(\phi) \mid \mathfrak{N}, V, n \Vdash \psi\}.$$

Define B as the set

$$B = \{(\Gamma_n, \Gamma_m) \mid n < m\},$$

and call a brick *sequential* if it is of the form (Γ_n, Γ_{n+1}) for some number n. We claim that B is a saturated set of bricks for ϕ.

We first show that all elements of B are indeed bricks. Let (Γ_n, Γ_m) with $n < m$ be an arbitrary element of B. For B0, assume that $F\psi$ or ψ belongs to Γ_m for some $m > n$; we have to prove that $F\psi \in \Gamma_n$. But from the assumption it easily follows that $\mathfrak{N}, V, k \Vdash \psi$ for some $k \geq m$, and transitivity of the ordering gives $k > n$. So it holds that $\mathfrak{N}, V, n \Vdash \psi$, immediately yielding, as required, that $F\psi \in \Gamma_n$. B1 is proved in a similar way, and B2 is trivial to show.

It is likewise straightforward to prove the saturation conditions. For example, in order to prove S0, let k be the point where ϕ is true, and consider the brick (Γ_0, Γ_{n_0}) if $n_0 > 0$, or the brick (Γ_0, Γ_1) in case $n_0 = 0$. It is immediate from the definitions and assumptions that $\neg P\top \in \Gamma_0$ and $\phi \in \Gamma_0 \cup \Gamma_1$. Proving S2 is straightforward from the observation that sequential bricks are small, and to see why that is the case, take an arbitrary sequential brick (Γ_n, Γ_{n+1}). We only discuss B3: assume that $F\psi \in \Gamma_n$. By definition, $\mathfrak{N}, V, n \Vdash F\psi$, so there must be some $m > n$ with $m \Vdash \psi$. Note that either $m = n + 1$ or $m > n + 1$; in the first case, we obtain $\psi \in \Gamma_{n+1}$, in the second, $F\psi \in \Gamma_{n+1}$. In either case we have B3.

Finally, the collection $\{\Gamma_n \mid n \in \mathbb{N}\}$ is a subset of the power set of $\mathrm{Cl}(\phi)$, whence its cardinality does not exceed $2^{|\phi|}$; but then the size of B can be at most $(2^{|\phi|})^2 = 2^{2|\phi|}$. ⊣

We now show that we have recorded enough information in the definition of a saturated set of bricks for ϕ to construct an $\mathfrak{N}$-based model for ϕ.

Lemma 6.29 *If there is a ϕ-SSB, then ϕ is satisfiable in $\mathfrak{N}$.*

Proof. Assume that B is a saturated set of bricks for ϕ. We will use these bricks to build, step by step, the required model for ϕ. As usual, in each finite stage of the construction we are dealing with a finite approximation of this model: a *history* is a pair (L, λ) such that L is a natural number and λ is a function on the set $\{0, \ldots, L\}$ to the set of Hintikka sets. Such a history (L, λ) is supposed to satisfy the following constraints:

(H0) $\neg P\top \in \lambda(0)$,
(H1)   for all m with $m < L$, $(\lambda(m), \lambda(m+1))$ is a small brick.

We leave it to the reader to verify that any history (L, λ) has the following properties:

(H2)   if $F\psi \in \lambda(n)$ for some $n < L$, then there is some m with $n < m \leq L$ and $\psi \in \lambda(m)$, or otherwise $F\psi \in \lambda(L)$,
(H3)   if $P\psi \in \lambda(n)$ for some $n \leq L$, then there is some m with $m < n$ and $\psi \in \lambda(m)$.

The importance of the properties H2 and H3 is that they show that the only essential shortcomings of a history (regarded as a finite approximation of a model) are of the form '$F\psi \in \lambda(L)$, and there is no witness for this fact; that is, no $m > L$ such that $\psi \in \lambda(m)$.'

Of course, we are not going to use histories in isolation; we say that one history (L', λ') is an *extension* of another history (L, λ), notation: $(L, \lambda) \lhd (L', \lambda')$, if $L < L'$, while λ and λ' agree on the domain of λ. The crucial extension lemma of the step-by-step construction is given in the following claim:

$$\text{Every history } (L, \lambda) \text{ with } F\psi \in \lambda(L) \tag{6.6}$$
$$\text{has an extension } (L', \lambda') \text{ such that } \psi \in \lambda'(L').$$

To prove (6.6), let (L, λ) be a history and ψ a formula such that $F\psi \in \lambda(L)$. It follows from H1 that $(\lambda(L-1), \lambda(L))$ is a brick, so by S1 there is a brick (Φ, Λ) in B such that $\Phi = \lambda(L)$ and $\psi \in \Lambda$. We now use S2 to find a path of small bricks $(\Sigma_0, \Sigma_1), (\Sigma_1, \Sigma_2), \ldots, (\Sigma_{k-1}, \Sigma_k)$, such that $\Sigma_0 = \Phi$ and $\Sigma_k = \Lambda$. Obviously we are going to 'glue' this path to the old history, thus creating a new history (L', λ'). To be precise, L' is defined as $L' = L + k$, while λ' is given by

$$\lambda'(n) = \begin{cases} \lambda(n) & \text{if } n \leq L, \\ \Sigma_i & \text{if } n = L + i. \end{cases}$$

With this definition (L', λ') satisfies the condition of (6.6).

Using (6.6), by a standard step-by-step construction one can define a sequence $(L_0, \lambda_0) \lhd (L_1, \lambda_1) \lhd \ldots$ of histories such that $\neg P\top \in \lambda_0(0)$ and $\phi \in \lambda_0(0) \cup \lambda(L_0)$, while for each i and each formula $F\psi \in \lambda_i(L_i)$ there is a $j > i$ and a number $L_i < m \leq L_j$ such that $\psi \in \lambda_j(m)$. This sequence of nested histories wil be our guideline for the definition of a valuation on $\mathfrak{N}$. Note that for all formulas ψ and all i and n, we have that

$$\text{if } n \leq L_i, \text{ then } \psi \in \lambda_i(n) \text{ iff } \psi \in \lambda_j(n) \text{ for all } j \geq i.$$

In other words, the histories always agree where they are defined; this fact will be used below without explicit comment.

Now consider the following valuation V on $\mathfrak{N}$:

$$V(p) = \{n \in \mathbb{N} \mid p \in \lambda_i(n) \text{ for some } i\}.$$

We are now ready to prove the crucial claim of this lemma.

$$\text{For all } \psi \in \text{Cl}(\phi) \text{ and all } n: \ \mathfrak{N}, V, n \Vdash \psi \text{ iff } \psi \in \lambda_i(n) \text{ for some } i. \quad (6.7)$$

Obviously, (6.7) will be proved by induction on ψ. The base step and the boolean cases of the induction step are straightforward and we leave them to the reader; we concentrate on the modal cases.

First assume that ψ is of the form $F\chi$. For the direction from left to right, assume that $\mathfrak{N}, V, n \Vdash F\chi$. There must be a number $m > n$ with $m \Vdash \chi$; so by the inductive hypothesis, there is an i with $\chi \in \lambda_i(m)$. It is easy to show (by backward induction and H1) that this implies $F\chi \in \lambda_i(k)$ for all k with $n \leq k < m$.

For the other direction, assume that $F\chi \in \lambda_i(n)$ for some i. It follows from H2 that there is either a number m with $n < m \leq L_i$ and $\chi \in \lambda_i(m)$, or otherwise $F\chi \in \lambda_i(L_i)$. In the first case we use the inductive hypothesis to establish that $m \Vdash \chi$ and hence, $n \Vdash F\chi$. Hence, assume that we are in the other case: $F\chi \in \lambda_i(L_i)$. Now our sequence of histories is such that this implies the existence of a history (L_j, λ_j) with $j > i$ and such that $\chi \in \lambda_j(m)$ for some m with $L_i, m \leq L_j$. It follows from the inductive hypothesis that $m \Vdash \chi$; thus the truth definition gives us that $n \Vdash F\chi$.

Now assume that ψ is of the form $P\chi$. The direction from left to right is as in the previous case. For the other direction, assume that $P\chi \in \lambda_i(n)$ for some i; it follows by H3 that there is an $m < n$ with $\chi \in \lambda_i(m)$. The inductive hypothesis yields that $\mathfrak{N}, V, m \Vdash \chi$, so by the truth definition we get $n \Vdash F\chi$. $\dashv$

Theorem 6.30 $\mathbf{K_t N}$ *is decidable.*

Proof. Immediate by Lemmas 6.28 and 6.29, and the obvious fact that it is decidable whether there is a ϕ-SSB of size at most $2^{2|\phi|}$. $\dashv$

A wide range of modal satisfiability problems can be studied in using quasi-models and mosaics. Indeed, such methods are not only useful for establishing decidability results, they can be used to obtain complexity results as well; see the Notes for further references.

Exercises for Section 6.4

6.4.1 Prove that $\mathbf{K}_t\mathbf{N}$ does not have the finite model property.

6.4.2 Give an example of an unsatisfiable Hintikka set. (Hint: work with the closure of $\{\Box(p \wedge q), \neg\Box p, \neg\Box q\}$.)

6.4.3 Extend our proof of the decidability of the tense logic of the natural numbers to a similarity type including the *next time* operator X. The semantics of this operator is given by

$$(\mathfrak{N}, V), n \Vdash X\phi \text{ iff } (\mathfrak{N}, V), n + 1 \Vdash \phi.$$

6.4.4 Let F_2 be the class of frames for the basic modal similarity type in which every point has exactly two successors. Use a mosaic argument to prove that this class has a decidable satisfiability problem.

6.4.5 In this exercise we consider a version of deterministic PDL in which *every* program is interpreted as a partial function – at least, in the intended semantics. The syntax of this language is given by

$$\phi \quad ::= \quad p \mid \bot \mid \neg\phi \mid \phi_1 \wedge \phi_2 \mid \langle\pi\rangle\phi \quad (p \text{ a proposition letter}),$$
$$\pi \quad ::= \quad a \mid \pi_1 ; \pi_2 \mid \text{if}(\phi, \pi_1, \pi_2) \mid \text{repeat}(\pi, \phi) \quad (a \text{ an atomic program}).$$

In a regular model $\mathfrak{M}$ for this language, each relation R_a is a partial function, and the composed programs obtain the obvious interpretation. That is, $R_{\pi_1;\pi_2}$ is the relational composition of R_{π_1} and R_{π_1}; $R_{\text{if}(\phi,\pi_1,\pi_2)}st$ holds if either $\mathfrak{M}, s \Vdash \phi$ and $R_{\pi_1}st$ or else $\mathfrak{M}, s \not\Vdash \phi$ and $R_{\pi_2}st$. Finally, we have $R_{\text{repeat}(\pi,\phi)}st$ if there is a path $sR_\pi t_1 R_\pi t_2 \ldots R_\pi t_n = t$ of length $n \geq 1$ from s to t such that $t = t_n$ is the *first* t_i where ϕ holds.

(a) Prove that in a regular model, each program π is interpreted as a partial function.
(b) Prove that the class of regular models has a decidable theory over this language. (Hint: use bricks of the form (Φ, Λ, Π) where Φ and Λ are Hintikka sets and Π is a set of programs closed under some natural conditions associating a unique 'first' atomic program with the brick. Also, make sure that any formula $\langle\pi\rangle\phi \in \Phi$ induces a unique formula of the form $\langle a\rangle\phi' \in \Phi$.)

6.5 Undecidability via Tiling

There are lots of undecidable modal logics; indeed, even uncountably many with the polysize model property (see Exercise 6.2.4). Moreover, there are undecidable modal logics which in many other ways are rather well-behaved (we saw an example in Exercise 6.2.5). Nice as they are, these examples do not really make clear just how easily undecidable modal logics can arise, nor how serious the undecidability can be. This is especially relevant if we are working with the richer modal languages (such as PDL) typically used in computer science and other applications, and the first goal of this section is to show that natural (and on the face of it straightforward) ideas can transform simple decidable logics into undecidable (or even *highly* undecidable) systems. While the examples are interesting in their own right, this section has a second goal: to introduce the concept of *tiling problems*.

Given a modal satisfiability problem S, to prove that S is undecidable we must reduce some known undecidable problem U to S. But which problems are the interesting candidates for reduction? Unsurprisingly, there is no single best answer to this question. As with decidability proofs, proving undecidability is something of an art: it can be very difficult, and there is no substitute for genuine insight into the satisfiability problem. Certain problems lend themselves rather naturally to modal logic, and tiling problems are a particularly nice example.

What is a tiling problem? In essence, a jigsaw puzzle. A *tile T* is simply a 1×1 square, fixed in orientation, each side of which has a *color*. We refer to these four colors as $right(T)$, $left(T)$, $up(T)$, and $down(T)$. Figure 6.3 depicts an example. (We have used different types of shading to represent the different colors.)

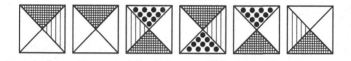

Fig. 6.3. Six distinct tile types

Six tiles are shown in Figure 6.3. Note that if we rotated the third tile 180 degrees clockwise, it would look just like the fourth tile, and that if we rotated the first tile 180 degrees clockwise it would look just like the sixth tile. We ignore such similarities. (This is what we meant when we said that tiles are 'fixed in orientation.') That is, the diagram shows six distinct types of tile.

Now for a simple tiling problem:

Is it possible to arrange tiles of the type just shown on a 2×4 grid in such a way that adjacent tiles have the same color on the common side?

A little experimentation shows that this *is* possible. A solution is given in Figure 6.4.

Fig. 6.4. A 2×4 tiling

This simple idea of pattern-matching underlying tiling problems gives rise to a family of problems which can be used to analyze computational complexity and demonstrate undecidability. This is the general form that tiling problems take:

Given a finite set of tile types $\mathcal{T}$, can we cover a certain part of $\mathbb{Z} \times \mathbb{Z}$ in such a way that adjacent tiles have the same color on the common edge? (Below, covering a grid with tiles so that adjacent colors match will be called 'tiling.')

Some tiling problems impose additional constraints on what counts as a successful tiling (we will shortly see an example) and some are formulated as games to be played between two players (we will see an example at the end of this chapter).

To spell this out somewhat, we might describe our previous example as an instance of the 2×4 tiling problem. That is, we were given a finite set of tile types (six, to be precise), asked to tile a 2×4 grid, and no further constraints were imposed. In the remainder of this section, we are going to make use of two much harder tiling problems. The first is the:

$\mathbb{N} \times \mathbb{N}$ **tiling problem**. *Given a finite set of tile types $\mathcal{T}$, can $\mathcal{T}$ tile $\mathbb{N} \times \mathbb{N}$?*

Here is a simple instance of this problem: can we tile $\mathbb{N} \times \mathbb{N}$ using the six tile types shown? Of course! We need simply 'slot-together' copies of our solution to the 2×4 problem.

In general, however, the $\mathbb{N} \times \mathbb{N}$ tiling problem is hard, and in fact it is known to be *undecidable*. Indeed, this problem is Π_1^0-complete; that is, it is a paradigmatic example of 'ordinary undecidability.' (See Appendix C for further discussion of degrees of undecidability.) We will not prove this result here – see the Notes for references – but it is really quite straightforward: think of each row of tiles as encoding Turing machine tapes and states, and the matching process as governing the state transitions.

The second problem we will use is the:

$\mathbb{N} \times \mathbb{N}$ **recurrent tiling problem**. *Given a finite set of tile types $\mathcal{T}$, which includes some distinguished tile type T_1, can $\mathcal{T}$ tile $\mathbb{N} \times \mathbb{N}$ in such a way that T_1 occurs infinitely often in the first row?*

As an easy example, note that our previous six tile types recurrently tile $\mathbb{N} \times \mathbb{N}$ when either the first, the third, the fourth, or the sixth tile type is distinguished.

Now our new problem is just the $\mathbb{N} \times \mathbb{N}$ tiling problem with an additional constraint imposed – but what a difference this constraint makes! Not only is this problem undecidable, it is Σ_1^1-complete (again, see Appendix C).

We will prove two modal undecidability results with the aid of these problems. Both examples are based around a natural variant of Deterministic Propositional Dynamic Logic, with intersection replacing choice and iteration as program constructors; we call this variant KR. We obtain our undecidability results as follows. First we enrich KR with the *global modality*. As we will show, the combination of the intersection construct with the global modality is a powerful one: it is possible to give an extremely straightforward reduction of the $\mathbb{N} \times \mathbb{N}$ tiling problem. We then enrich KR with a modality called the *master modality*. This is also a natural operator – indeed, perhaps more natural than the global modality. As a very easy reduction from the $\mathbb{N} \times \mathbb{N}$ recurrent tiling problem reveals, the resulting system is highly undecidable.

Intersection and the global modality

Our first example vividly illustrates how easily undecidability can arise. We are going to mix two simple ingredients together, both of which are decidable, and show that the result has an undecidable satisfiability problem.

The first ingredient is a variant of DPDL, with intersection replacing choice and iteration as program constructors. Recall from Example 1.15 that DPDL is simply PDL interpreted over deterministic PDL structures (that is, PDL structures in which the relations R_a corresponding to atomic programs a are partial functions). Further, recall from Example 1.26 that modalities built with the intersection constructor (that is, modalities of the form $\langle \pi_1 \cap \pi_2 \rangle$) are interpreted by the relation $R_{\pi_1} \cap R_{\pi_2}$, where R_{π_1} is the relation corresponding to $\langle \pi_1 \rangle$ and R_{π_2} the relation corresponding to $\langle \pi_2 \rangle$.

In what follows we will not use the entire language; instead we will work with a fragment (called KR) which consists of all formulas without occurrences of $*$ and $\cup$. That is, KR contains precisely the following formulas ϕ:

$$\phi \quad ::= \quad p \mid \bot \mid \neg \phi \mid \phi_1 \wedge \phi_2 \mid \langle \pi \rangle \phi \quad (p \text{ a proposition letter}),$$

$$\pi \quad ::= \quad a \mid \pi_1 ; \pi_2 \mid \pi_1 \cap \pi_2 \quad (a \text{ an atomic program}).$$

The KR language is rather simple: essentially it allows us to state whether or not different sequences of (deterministic) programs terminate in the same state when executed in parallel. Note that (over deterministic PDL structures) a selection argument immediately shows that KR has a decidable satisfiability problem. Over deterministic structures, *every* modal operator in KR is interpreted by a partial function. (This is because all atomic programs are modeled by partial functions, and the only program constructors we have at our disposal are composition and intersection.) It follows that if a sentence ϕ from KR is satisfiable in a deterministic model, then it is satisfiable in a *finite* deterministic model; the proof is essentially the same as that of Corollary 6.9.

The second ingredient is even simpler. We are going to add the *global modality* A to our fragment. This is an interesting operator that we are going to discuss in detail in Section 7.1; for present purposes we only need to know two things about it. First, it is interpreted as follows: $\mathfrak{M}, w \Vdash A\phi$ if for all v in $\mathfrak{M}$ we have $\mathfrak{M}, v \Vdash \phi$. Thus, as its name suggests, the global modality is a modal operator which allows us to express global facts. Second, A has a decidable satisfiability problem. (To see this, simply observe that A is an **S5** operator, and we know that **S5** is decidable.) Thus, on its own, A is pretty harmless.

But what happens when we add A to KR? The resulting language called KRA can talk about computations in a very natural (and very powerful) way. For example,

$$A(\langle a\rangle\top \rightarrow \psi)$$

expresses that in every state of a computation, ψ is a precondition for the program a to have a terminating execution. As we will now show, KRA has crossed the border into undecidability.

Theorem 6.31 *Assume that the language has at least two atomic programs. Then the satisfiability problem for* KRA *is undecidable. To be precise, it is* Π_1^0-*hard.*

Proof. We show this by reducing the $\mathbb{N} \times \mathbb{N}$ tiling problem to the KRA satisfiability problem; the undecidability (and Π_1^0-hardness) of the satisfiability problem will follow from the known undecidability (Π_1^0-hardness) of the $\mathbb{N} \times \mathbb{N}$ tiling problem.

Recall that the $\mathbb{N} \times \mathbb{N}$ tiling problem asks: given a finite set of tile types $\mathcal{T}$, can $\mathcal{T}$ tile $\mathbb{N} \times \mathbb{N}$? Putting this more formally: does there exist a function $t : \mathbb{N} \times \mathbb{N} \rightarrow \mathcal{T}$ such that

$$\begin{aligned} right(t(n,m)) &= left(t(n+1,m)), \\ up(t(n,m)) &= down(t(n,m+1))? \end{aligned}$$

We will reduce $\mathbb{N} \times \mathbb{N}$ tiling to the satisfiability problem as follows. Let $\mathcal{T} = \{T_1, \ldots, T_k\}$ be the given set of tile types. We will construct a formula $\phi_{\mathcal{T}}$ such that

$$\mathcal{T} \text{ tiles } \mathbb{N} \times \mathbb{N} \text{ iff } \phi_{\mathcal{T}} \text{ is satisfiable.} \tag{6.8}$$

If we succeed in constructing such a formula it follows that the KRA-satisfiability problem is undecidable. (For suppose it was decidable. Then we could solve the $\mathbb{N} \times \mathbb{N}$ tiling problem as follows: given $\mathcal{T}$, form $\phi_{\mathcal{T}}$, and use the putative KRA-satisfiability algorithm to check for satisfiability. By (6.8) this would solve the tiling problem – which is impossible.)

The construction of $\phi_{\mathcal{T}}$ proceeds in three steps. First, we show how to use KRA to demand 'gridlike' models. Second, we show how to use KRA to demand that a tiling exists on this 'grid.' Finally we prove (6.8).

Step 1. Forcing the grid. The basic idea is to let the nodes in $\mathfrak{M}$ mimic the nodes in $\mathbb{N} \times \mathbb{N}$, and to use two relations R_r and R_u to mimic the 'to-the-right' and the 'up' functions of $\mathbb{N} \times \mathbb{N}$. To get the gridlike model we want, we simply demand that R_r and R_u commute:

$$\phi_{grid} := A\langle(r\,;u) \cap (u\,;r)\rangle\top.$$

This says that everywhere in the model it is possible to make a 'to-the-right transition followed by an up transition' and an 'up transition followed by a to-the-right transition,' and both these transition sequences lead to the same point. (Note that this is all we need to say, since by assumption R_r and R_u are partial functions.)

Step 2. Tiling the model. We will 'tile the model' by making use of proposition letters $t_1, \ldots, t_k$ which correspond to the tile types in $\mathcal{T}$. The basic idea is simple: we want t_i to be true at a node w iff a tile of type T_i is placed on w. Of course, not any placement of tiles will do: we want a genuine tiling. But the following three demands ensure this:

(i) Exactly one tile is placed at each node:

$$\phi_1 := A \left(\bigvee_{i=1}^{k} t_i \wedge \bigwedge_{1 \leq i < j \leq k} \neg(t_i \wedge t_j) \right).$$

(ii) Colors match going right:

$$\phi_2 := A \left(\bigvee_{right(T_i)=left(T_j)} (t_i \wedge \langle r \rangle t_j) \right).$$

(iii) Colors match going up:

$$\phi_3 := A \left(\bigvee_{up(T_i)=down(T_j)} (t_i \wedge \langle u \rangle t_j) \right).$$

Putting this together, we define $\phi_{\mathcal{T}} := \phi_{grid} \wedge \phi_1 \wedge \phi_2 \wedge \phi_3$.

Step 3. Proving the equivalence. We now show that (6.8) holds. Assume first that $t : \mathbb{N} \times \mathbb{N} \to \mathcal{T}$ is a tiling of $\mathbb{N} \times \mathbb{N}$. Construct a satisfying model for $\phi_{\mathcal{T}}$ as follows:

$$
\begin{aligned}
W &= \{w_{n,m} \mid n, m \in \mathbb{N}\}, \\
R_r &= \{(w_{n,m}, w_{n+1,m}) \mid n, m \in \mathbb{N}\}, \\
R_u &= \{(w_{n,m}, w_{n,m+1}) \mid n, m \in \mathbb{N}\}, \\
V(t_i) &= \{w_{n,m} \mid n, m \in \mathbb{N} \text{ and } t(n, m) = T_i\}.
\end{aligned}
$$

Clearly, $\phi_{\mathcal{T}}$ holds at any state w of $\mathfrak{M}$.

For the converse, let $\mathfrak{M}$ be a model such that $\mathfrak{M}, w_0 \Vdash \phi_{\mathcal{T}}$. It follows from $\mathfrak{M}, w_0 \Vdash \phi_{grid}$ that there exists a function $f : \mathbb{N} \times \mathbb{N} \to W$ such that $f(0,0) = w_0$, $R_r f(n, m) f(n+1, m)$ and $R_u f(n, m) f(n, m+1)$. Define the tiling $t : \mathbb{N} \times \mathbb{N} \to \mathcal{T}$ by

$$t(n, m) = T_i \text{ iff } \mathfrak{M}, f(n, m) \Vdash t_i.$$

By ϕ_1, t is well-defined and total. Moreover, if $t(n, m) = T_i$ and $t(n+1, m) = T_j$, then $R_r f(n,m) f(n+1,m)$, and both $\mathfrak{M}, f(n,m) \Vdash t_i$ and $\mathfrak{M}, f(n+1,m) \Vdash t_j$. Given that w_0 satisfies ϕ_2, we conclude that $right(T_i) = left(T_j)$. Similarly, because of ϕ_3, if $t(n,m) = T_i$ and $t(n,m+1) = T_j$, then $up(T_i) = down(T_j)$. Thus, $\mathcal{T}$ tiles $\mathbb{N} \times \mathbb{N}$.   ⊣

The above proof clearly depends on having two deterministic atomic programs at our disposal. But what happens if we only have one? It should be clear that then $\cap$ cannot do any interesting work for us, and in fact the language has a *decidable* satisfiability problem; see Exercise 6.5.1.

We now know that KRA-satisfiability is undecidable (given more than one atomic program) but *how undecidable* is it? In particular can we also prove a Π_1^0 upper bound to match the Π_1^0-hardness result? (That is, can we show that we are dealing with a case of 'ordinary undecidability'?) To prove this, it suffices to show that the validities of KRA form an r.e. set. Now we could do this by devising a recursive axiomatization of the KRA-validities, but by making use of a general lemma from correspondence theory we can establish the result more straightforwardly.

Lemma 6.32 *If* K *is a class of frames defined by a first-order formula, then its modal logic is recursively enumerable.*

Proof. Assume that the first-order formula α defines K, where α is built using only relation symbols of arity 2 or higher, and identity. Then, a modal formula ϕ is valid on K iff it is valid on all frames in K iff

$$\alpha \models \forall x \forall P_1 \ldots \forall P_n \, ST_x(\phi), \tag{6.9}$$

where $P_1, \ldots, P_n$ are unary predicate symbols corresponding to the proposition letters in ϕ. As the predicate variables $P_1, \ldots, P_n$ do not occur in α, (6.9) is equivalent to $\alpha \models \forall x \, ST_x(\phi)$. But this is an ordinary first-order implication, which is an r.e. notion. Hence, modal validity on K is an r.e. notion as well.   ⊣

Theorem 6.33 *Assume that our language has at least two, but at most finitely many atomic programs. Then the satisfiability problem for* KRA *is* Π_1^0*-complete.*

Proof. The Π_1^0 lower bound is given by the encoding of the $\mathbb{N} \times \mathbb{N}$ tiling problem in the proof of Theorem 6.31. For the Π_1^0 upper bound we show that the validity problem for KRA is r.e. The standard translations for the constructors ; and $\cap$ are given in Section 2.4; both are first-order. (Recall that the $*$ constructor is the only part of PDL that takes us out of first-order logic.) The standard translation for A is obvious (and clearly first-order):

$$ST_x(A\phi) = \forall y \, ST_y(\phi).$$

Thus – assuming we are working with a language of KRA that contains at most finitely many atomic programs – the required class of frames is defined by

$$\bigwedge_{\alpha \text{ atomic}} \forall xyz \, (R_\alpha xy \wedge R_\alpha xz \to y = z).$$

Hence, by Lemma 6.32, the modal logic of the class of frames for KRA is r.e. as required. ⊣

Intersection and the master modality

Our next example illustrates how easily *high* undecidability can arise. Once again, we will enrich the KR language, but this time with the *master modality*. As we will see, the resulting language KR[*] has a Σ_1^1-complete satisfiability problem.

Like the global modality, the master modality [*] is a tool for expressing general constraints in the object language, but it works rather differently. A formula of the form [*]ϕ is true at a node w iff ϕ is true at all nodes reachable by any finite sequence of atomic transitions from w. Formally,

$$w \Vdash [*]\phi \text{ iff } v \Vdash \phi \text{ for all } v \text{ such that } (w, v) \in \left(\bigcup_{a \text{ atomic}} R_a \right)^*.$$

That is, [*] explores the reflexive transitive closure of the union of all the relations used to interpret the atomic programs. If we only have finitely many atomic programs $a_1, \ldots, a_n$, the master modality is simply shorthand for the PDL modality $[(a_1 \cup \cdots \cup a_n)^*]$. From a computational perspective, this modality is arguably even more natural than the global modality: it is a way of looking at what must happen throughout the space of possible computations. (It has other natural interpretations as well. For example, if we interpret our basic modalities as in multi-agent epistemic logic – that is, $[a]\phi$ means 'agent a knows that ϕ' – then [*] is the 'common knowledge' operator.)

But, for all its naturalness, the master modality can be extremely dangerous. Let us see what happens when we add it to KR. First, observe that KR[*] must be undecidable. (There is nothing new to prove here; simply observe that if we systematically replace every occurrence of A in the proof of Theorem 6.31 by [*], the argument still goes through.) But can we prove a matching Π_1^0 upper bound? We certainly cannot appeal to Lemma 6.32; while the global modality was essentially first-order, the master modality is not. (As with the * constructor of PDL, its natural correspondence language is infinitary; see Section 2.4.) And indeed, any attempt to recursively enumerate the validities of KR[*] is bound to fail.

Theorem 6.34 *The satisfiability problem for* KR[*] *is highly undecidable. To be precise, it is* Σ_1^1-*hard.*

Proof. We show this by reducing the recurrent tiling problem to the KR[∗]-satisfiability problem; the Σ_1^1-hardness of the satisfiability problem will follow from the known Σ_1^1-hardness of the recurrent tiling problem.

Recall that the recurrent tiling problem asks: given a finite set of tile types $\mathcal{T}$, which includes some distinguished tile type T_1, can $\mathcal{T}$ tile $\mathbb{N} \times \mathbb{N}$ in such a way that T_1 occurs infinitely often in the first row? Putting this more formally: does there exist a function $t : \mathbb{N} \times \mathbb{N} \to \mathcal{T}$ such that

$$
\begin{aligned}
right(t(n,m)) &= left(t(n+1,m)), \\
up(t(n,m)) &= down(t(n,m+1)), \\
\{n \mid t(n,0) = T_1\} \quad &\text{is infinite?}
\end{aligned}
$$

We reduce $\mathbb{N} \times \mathbb{N}$ recurrent tiling to KR[∗]-satisfiability as follows. Let $\mathcal{T} = \{T_1, \ldots, T_k\}$ be the set of tile types. We will define a formula $\phi_{\mathcal{T},T_1}$ such that

$$\mathcal{T} \text{ and } T_1 \text{ recurrently tile } \mathbb{N} \times \mathbb{N} \text{ iff } \phi_{\mathcal{T},T_1} \text{ is satisfiable.} \qquad (6.10)$$

Most of the real work was done in the proof of Theorem 6.31. Let us simply take the earlier encoding $\phi_{\mathcal{T}}$ and replace every occurrence of A with [∗]. Call the result $\phi_{\mathcal{T}}^*$. This formula reduces the $\mathbb{N} \times \mathbb{N}$ tiling problem to the KR[∗]-satisfiability problem.

To reduce the *recurrent* tiling problem, it remains to ensure that our distinguished tile T_1 occurs infinitely often on the first row. As t_1 is the proposition letter corresponding to T_1, this means we want to force t_1 to be true at nodes of the form $t(n,0)$ for infinitely many n. To do this, we will introduce a new proposition letter *first-row* and then define:

$$\phi_{rec} := \textit{first-row} \land [∗][u]\neg\textit{first-row} \land [∗](\textit{first-row} \to \langle r \rangle \langle ∗ \rangle (\textit{first-row} \land t_1)).$$

Suppose that ϕ_{rec} is satisfied at some point w_0 of a grid-like model. It follows that *first-row* is satisfied at w_0; that *first-row* can only be satisfied at points reachable by a finite number of R_r transitions from w_0; and that for infinitely many distinct natural numbers n, $w_0 \Vdash \langle r \rangle^n (\textit{first-row} \land t_1)$.

So, let $\phi_{\mathcal{T},T_1}$ be the conjunction of $\phi_{\mathcal{T}}^*$ and ϕ_{rec}. Then (6.10) holds. $\dashv$

To conclude this section, two general remarks. First, the examples in this section were clearly chosen to make the undecidability proofs run as smoothly as possible. In particular, our examples hinged on the use of $\cap$ to force the existence of the grid. What happens if we are working in languages without this constructor? That is, how widely applicable is this method for proving undecidability?

Suppose we are working with an arbitrary modal language, and we want to establish the undecidability of its satisfiability problem. If we abstract from the proof of Theorem 6.31, we see that there is one ingredient that will always be needed to make similar arguments go through: sufficient 'global' expressive power. Thi power may arise directly through the presence of additional operators, or it may arise indirectly through special features of the class of models under consideration, but one way or another we will need it. On the other hand, we do not need the ∩ constructor; Exercise 6.5.2 is a nice example.

Second, we have discussed tiling problems as if they were useful only for establishing different grades of undecidability. In fact, they can also be used to analyze the complexity of *decidable* problems: for example, there are NP-hard, PSPACE-hard, and EXPTIME-hard tiling problems (see the Notes for further references). At the end of this chapter we will use a 2-player tiling problem to show that the satisfiability problem for **PDL** is EXPTIME-hard.

Exercises for Section 6.5

6.5.1 Show that KRA-satisfiability is decidable if we have only one atomic program at our disposal. (This result can be proved via a finite model property argument.)

6.5.2 (i) Show that the satisfiability problem of the following 'tiling' logic **Tile**$_1$ is undecidable. **Tile**$_1$ is a normal modal logic with three diamonds $\langle u \rangle$, $\langle r \rangle$ and $\Diamond$, defined by the following (Sahlqvist) axioms:

$$\langle u \rangle p \to [u]p \text{ and } \langle r \rangle p \to [r]p, \qquad (6.11)$$
$$\langle r \rangle \langle u \rangle p \to [u]\langle r \rangle p,$$
$$\Diamond \Diamond p \to \Diamond p,$$
$$\langle u \rangle p \to \Diamond p \text{ and } \langle r \rangle p \to \Diamond p.$$

(ii) Now use this logic plus the standard translation to conclude that the three variable fragment of first-order logic (without function symbols, but possibly with equality) is undecidable.

(iii) Let **Tile**$_2$ be obtained from **Tile**$_1$ by omitting axiom (6.11). Show that **Tile**$_2$ is still undecidable. (Hint: Reduce the satisfiability problem of **Tile**$_1$ to that of **Tile**$_2$.)

(iv) Conclude that first-order logic with three variables, but without equality is undecidable.

(v) Use a similar tiling logic to show that first-order logic with one variable, two unary function symbols, and only unary predicate symbols is undecidable. (Hint: adjust the standard translation so that it exploits the unary function symbols directly.)

6.6 NP

The interpretation method (and in particular, interpretations in SnS) is a powerful and widely applicable way of proving decidability. Nevertheless, it has disadvantages. Reducing the satisfiability problems of what are often rather simple modal logics to SnS is using a sledgehammer to crack a nut. The decision problem for SnS is non-elementary. This means that the time required to decide whether an arbitrary formula ϕ is decidable cannot be bounded by any finite tower of exponentials of the form

$$2^{2^{\cdot^{\cdot^{\cdot 2^{|\phi|}}}}}.$$

The use of filtrations to establish decidability is open to similar objections. A filtration is typically $2^{|\phi|}$ in the size of the input formula. But it is not feasible to enumerate all the models up to this size even for quite small values of $|\phi|$. And even a nondeterministic Turing machine, which could 'guess' a filtration in one move (see Appendix C and the discussion below), would still be faced with the immensely costly task of checking that ϕ was true on this huge structure (to use the terminology discussed in Appendix C, filtrations typically offer us NEXPTIME algorithms). Indeed, of the three decidability techniques discussed so far, only the mosaic method (which 'deconstructs' models locally) respects what is special about modal logic; and as we will learn in Section 7.4, the mosaic method can be used to give essentially optimal satisfaction algorithms.

But this is jumping ahead. In this section and the three that follow, we will use concepts drawn from *computational complexity theory* to present a more fine-grained analysis of modal satisfiability. This analysis is interesting for two reasons. First, by making use of only three central complexity classes (NP, PSPACE and EXPTIME), we will be able to present a classification of modal satisfiability that covers many important logics. Secondly, in many cases the techniques involved have a distinctly modal flavor: essentially, the work boils down to a refined analysis of the finite model property.

We begin our analysis with the class NP, the class of problems solvable using nondeterministic polynomial time algorithms. We first review the central ideas underlying this complexity class and their import for modal satisfiability problems. Then, using examples from multi-modal and tense logic, we show how simple selection arguments can be used to prove NP-completeness results. Finally, we apply the same method to prove a more general result: every normal modal logic extending **S4.3** has an NP-complete satisfiability problem.

When a problem $\mathcal{P}$ is said to be complete with respect to a complexity class C, two things are being claimed. The first is that $\mathcal{P}$ belongs to C; that is, there is an algorithm using only the resources permitted by C that solves $\mathcal{P}$. For example, if C = NP this means that there exists a non-deterministic polynomial time algorithm for solving $\mathcal{P}$. The second claim is that $\mathcal{P}$ is C-hard; that is, any other problem in C is polynomial time reducible to $\mathcal{P}$.

Now, as far as the satisfiability problem for normal modal logics is concerned, NP-hardness is a triviality: all (consistent) normal modal logics have NP-hard satisfiability problems. The point is this. The classic NP-hard problem is the satisfiability problem for propositional logic. But as every normal modal logic is a extension of propositional logic, every (consistent) normal modal logic has a satis-fiability problem at least as hard as that for propositional logic. Thus – for the class NP – our work is somewhat simplified: we are simply looking for normal modal logics whose satisfiability problem belongs to NP.

What sort of problems belong to NP? Many problems decompose naturally into the following two steps: a *search for a solution* followed by a *verification of the solution*. In general, search is expensive, but by thinking in terms of non-deterministic algorithms we can abstract away from this expense: if a solution exists, such an algorithm will find it in one non-deterministic step. (If necessary, consult Appendix C for further discussion.) This abstraction leaves us free to con-centrate on the verification step, and leads us to isolate the class NP: a problem be-longs to NP iff it has the above general profile (that is, a non-deterministic choice of a solution followed by a verification) *and moreover* the verification step is tractable (that is, solvable in polynomial time).

How do such ideas bear on modal satisfiability? The key idea we need is em-bodied in the following lemma.

Lemma 6.35 *Let τ be a finite similarity type. Let Λ be a consistent normal modal logic over τ with the polysize model property with respect to some class of models* M. *If the problem of deciding whether $\mathfrak{M} \in$ M is computable in time polynomial in $|\mathfrak{M}|$, then Λ has an NP-complete satisfiability problem.*

Proof. As noted above, the NP-hardness of the problem is immediate, so it remains to prove the existence of an algorithm in NP that solves Λ-satisfiability. Given ϕ, non-deterministically choose a model $\mathfrak{M}$ whose size is polynomial in the size of ϕ. Because $\mathfrak{M}$ is polysize in $|\phi|$, we can check in time polynomial in $|\phi|$ whether $\mathfrak{M}$ verifies ϕ. For the special case of the basic modal language, this may be seen as follows.

Let $||\mathfrak{M}||$ denote the sum of the number of states in $\mathfrak{M}$ and the number of pairs in $\mathfrak{M}$'s binary relation $R^{\mathfrak{M}}$. Let $\psi_1, \ldots, \psi_k$ be an enumeration of the subformulas of ϕ, in increasing length. So $\psi_k = \phi$ and if ψ_i is a subformula of ψ_j, then $i < j$. Notice that $k \leq |\phi|$. One can show by induction on m that we can mark each state w in $\mathfrak{M}$ with ψ_j or $\neg\psi_j$, for $j = 1, \ldots, m$, depending on whether or not ψ_j is true at w in time $\mathcal{O}(m \cdot ||\mathfrak{M}||)$. The only non-trivial case is if $\psi_{m+1} = \Diamond\psi_j$, for some $j < m + 1$. But in that case we mark w with $\Diamond\psi_j$ if some v with Rwv is marked with ψ_j. By our induction hypothesis, every state is already marked with ψ_j or $\neg\psi_j$, this step can be carried out in time $\mathcal{O}(||\mathfrak{M}||)$. Since $\mathfrak{M}$ is polysize in $|\phi|$, so is $||\mathfrak{M}||$. Hence, checking whether $\mathfrak{M}$ satisfies ϕ can indeed be done in time polynomial in $|\phi|$.

Finally, then, because membership in M is decidable in time polynomial in $|\mathfrak{M}|$, and $|\mathfrak{M}|$ is polynomial in $|\phi|$, we can check in time polynomial in $|\phi|$ that $\mathfrak{M}$ is in M. ⊣

Where did we use the assumption that τ is a finite similarity type in the proof of Lemma 6.35? Essentially, it allows us to check whether $\mathfrak{M}$ verifies ϕ in time polynomial in $|\phi|$ and in $|\mathfrak{M}|$. The key point is this: when working with a fixed finite similarity type, we are actually working within a finite-variable fragment, say with l variables. This allows us to restrict our attention to only finitely many relations of arity at most l in $\mathfrak{M}$. While the total number of tuples in all relations in $\mathfrak{M}$ may be huge, it is nonetheless independent of ϕ; see Exercise 6.6.2 for further elaborations.

Note that the second demand – that M-membership be polynomial time decidable – is vital. As the reader was asked to show in Exercise 6.2.4, the polysize model property alone is insufficient to ensure decidability, let alone the existence of a solution in NP. However, for many important logics this property can be established by appealing to the following standard result.

Lemma 6.36 *If* F *is a class of frames definable by a first-order sentence, then the problem of deciding whether $\mathfrak{F}$ belongs to* F *is decidable in time polynomial in the size of $\mathfrak{F}$.*

Proof. Left as Exercise 6.6.1. ⊣

We will show that many normal modal logics are NP-complete. The proofs revolve around one central idea: the construction of polysize models by the selection of polynomially many points from some given satisfying model.

For our first example, we return to the multi-modal language containing n unary modal operators discussed earlier (see Corollary 6.9). Recall that F_1^n is the class of frames for this language in which each relation is a partial function, M_1^n is the class of models built over F_1^n, and $\mathbf{K}_n\mathbf{Alt}_1$ is its logic.

Theorem 6.37 $\mathbf{K}_n\mathbf{Alt}_1$ *has an NP-complete satisfiability problem.*

Proof. We already showed that this logic has the strong f.m.p., but the selection argument we used generated models exponential in size of the input formula. A simple refinement of the method shows that $\mathbf{K}_n\mathbf{Alt}_1$ actually has the *polysize* model property.

Given a formula ϕ of this language and a model $\mathfrak{M} = (W, R, V)$ we define a selection function s as follows:

$$
\begin{aligned}
s(p, w) &= \{w\}, \\
s(\neg\phi, w) &= s(\phi, w), \\
s(\phi \wedge \psi, w) &= s(\phi, w) \cup s(\psi, w),
\end{aligned}
$$

$$s(\langle a\rangle\psi, w) = \{w\} \cup \bigcup_{\{w'|R_aww'\}} s(\psi, w').$$

Intuitively, $s(\phi, w)$ selects the nodes actually needed when evaluating ϕ in $\mathfrak{M}$ at w – and indeed, it follows by induction on the structure of ϕ that for all nodes w of $\mathfrak{M}$, and all formulas ϕ

$$\mathfrak{M}, w \Vdash \phi \text{ iff } \mathfrak{M} \restriction s(\phi, w), w \Vdash \phi.$$

It is clear that $\mathfrak{M} \restriction s(\phi, w) \in \mathsf{M}_1^n$. So let us look at the size of the new model. If $\mathfrak{M} \in \mathsf{M}_1^n$, we claim that $|s(\phi, w)| \leq |\phi|+1$. To see this, note that only occurrences of modalities in ϕ cause new nodes to be adjoined to $s(\phi, w)$. This adjunction of points is carried out in the fourth clause of the inductive definition for s, which tells us to adjoin every state w' such that R_aww'. Because $\mathfrak{M} \in \mathsf{M}_1^n$, every relation R_a is a partial function; hence if such a w' exists, it is unique. In short, $\mathbf{K}_n\mathbf{Alt}_1$ has the polysize model property: simply counting the number of occurrences of modal operators in ϕ and adding one gives us an upper bound on the size of the domain of the required satisfying model.

By Lemma 6.36, membership in $\mathbf{K}_n\mathbf{Alt}_1$ is decidable in polynomial time, for this is a class of frames definable by a first-order sentence – namely the conjunction of sentences that say that each of the n relations is a partial function.

The result follows by Lemma 6.35. $\dashv$

The argument for $\mathbf{K}_n\mathbf{Alt}_1$ shows the selection method in its simplest form: given *any* model for ϕ we build a new polysize model for ϕ by making a suitable selection of polynomially many points. This simple form of argumentation is applicable to a number of logics, a particularly noteworthy example being **S5**. Given any **S5** model for ϕ, it is possible to select $m + 1$ points from this model (where m is the number of modality occurrences in ϕ) which suffice to construct a new **S5** model for ϕ, and the NP-completeness of **S5** follows straightforwardly. We leave the details as Exercise 6.6.4 and turn our attention to a modification of the point selection method frequently needed in practice: a *detour via finite models*.

Both $\mathbf{K}_n\mathbf{Alt}_1$ and **S5** are very simple logics; in neither case is it difficult to determine which points should be selected. In other cases, we may not be so fortunate. Suppose we are trying to show that a logic Λ has the polysize model property, and we already know that Λ has the f.m.p. Then, instead of trying to select points from an arbitrary model, we are free to select points from a finite model, or even a point-generated submodel of a finite model. This often gives us an easy way of zooming in on the crucial points. In particular, when we are working with models based on finite orderings it makes sense to talk of choosing points that are maximal (or minimal) in the frame ordering that satisfy some subformula; such extremal points are often the vital ones. As an example of such an argument, let us consider $\mathbf{K}_t\mathbf{4.3}$, the temporal logic of linear frames (in the basic temporal language).

Theorem 6.38 $\mathbf{K}_t\mathbf{4.3}$ *has an NP-complete satisfiability problem.*

Proof. We will first show that $\mathbf{K}_t\mathbf{4.3}$ has the polysize model property. Let ϕ be a formula of the basic temporal language that is satisfiable on a $\mathbf{K}_t\mathbf{4.3}$ model. As $\mathbf{K}_t\mathbf{4.3}$ has the f.m.p. with respect to the class of weak total orders (see Definition 4.37 and Corollary 6.8), there is a finite weakly totally ordered model $\mathfrak{M} = (T, \leq, V)$ containing a node t such that $\mathfrak{M}, t \Vdash \phi$. We now build a polysize model for ϕ by selecting points from $\mathfrak{M}$.

Let $F\psi_1, \ldots, F\psi_k$ and $P\theta_1, \ldots, P\theta_l$ be all subformulas of ϕ of the form $F\psi$ and $P\theta$, respectively, that are satisfied in $\mathfrak{M}$. For each formula $F\psi_i$ choose a point u_i such that $\mathfrak{M}, u_i \Vdash \psi_i$ and u_i is a *maximal* point in the $\leq$-ordering with this property. Similarly, for each formula $P\theta_j$ choose a point v_j satisfying θ_j that is *minimal* in the $\leq$-ordering with respect to this property. Let $\mathfrak{M}'$ $(= (T', \leq', V'))$ be $\mathfrak{M} \restriction \{t, u_1, \ldots, u_k, v_1, \ldots, v_l\}$. As $\leq$ is a weak total ordering of T, $\leq'$ is a weak total ordering of T'. Furthermore, the number of nodes in $\mathfrak{M}'$ does not exceed $m + 1$, where m is the number of modalities in ϕ, thus $\mathfrak{M}'$ is a polysize model in the correct class. It remains to show that $\mathfrak{M}', t \Vdash \phi$, but this follows straightforwardly by induction on the structure of ϕ.

As the class of weak total orders is definable using a first-order sentence, the NP-completeness of $\mathbf{K}_t\mathbf{4.3}$ follows from Lemma 6.36 and the polysize model property that we have just established. ⊣

We are ready to prove a general complexity result for the basic modal language: all normal logics extending **S4.3** have an NP-complete satisfiability problem. Recall from our discussion of Bull's Theorem in Section 4.9 that an **S4.3** frame is a frame that is rooted, transitive, and connected ($\forall xy\,(Rxy \lor Ryx)$); note that all such frames are reflexive. Bull's Theorem tells us that all normal modal logics extending **S4.3** have the finite frame property with respect to a class of **S4.3** frames. By making a suitable selection from models based on such frames, we can prove that every such logic has the polysize model property. Then, by using the fact that every normal logic extending **S4.3** has a negative characterization in terms of finite sets of finite frames (Theorem 4.103), we will be able to prove that all these satisfiability problems are NP-complete.

First we need the following lemma; it is really just Lemma 4.98, which linked bounded morphisms and covering lists, stated in purely modal terms.

Lemma 6.39 *Let $\mathfrak{F}$ and $\mathfrak{G}$ be two finite **S4.3** frames. Then the following two statements are equivalent:*

(i) *There exists a surjective bounded morphism from $\mathfrak{F}$ to $\mathfrak{G}$.*

(ii) *$\mathfrak{G}$ is isomorphic to a subframe of $\mathfrak{F}$ that contains a maximal point of $\mathfrak{F}$.*

Proof. First suppose that f is a surjective bounded morphism from $\mathfrak{F}$ to $\mathfrak{G}$. Le

w_{max} be a maximal point in $\mathfrak{F}$, and let $\widehat{W}$ consist of w_{max} together with exactly one maximal world in $f^{-1}[v]$ for every point v of $\mathfrak{G}$ such that $v \neq f(w_{max})$. Then $\widehat{\mathfrak{F}} = \mathfrak{F} \upharpoonright \widehat{W}$ is the subframe we want.

Conversely, suppose that $\widehat{W}$ is a subset of the points in $\mathfrak{F}$, such that $\widehat{W}$ contains a maximal point w_{max}, and $\mathfrak{F} \upharpoonright \widehat{W}$ is isomorphic to $\mathfrak{G}$. We claim that the following defines a bounded morphism from $\mathfrak{F}$ onto $\mathfrak{F} \upharpoonright \widehat{W}$: $f(w) = w$, for $w \in \widehat{W}$; and if $w \notin \widehat{W}$, then $f(w)$ is a minimal world $\widehat{w} \in \widehat{W}$ such that $Rw\widehat{w}$ (that is, for any w', if Rww' then $R\widehat{w}w'$). Note that such a minimal world must always exist, since $w_{max} \in \widehat{W}$, thus f is well-defined. (In short, f maps 'missing points' to successors that are as close as possible. We used the same idea to define the bounded morphism in the proof of Bull's Theorem.) Clearly f is surjective. So suppose Rww'. Since $Rw'f(w')$ and R is transitive, we have $Rwf(w')$. By definition, $f(w)$ is a minimal element in $\widehat{W}$ such that $Rwf(w)$, thus $Rf(w)f(w')$ and f satisfies the forth condition on bounded morphisms. Finally, suppose $Rf(w)f(w')$. As $Rwf(w)$, by the transitivity of R we have $Rwf(w')$. Since $f(f(w')) = f(w')$, the back condition for bounded morphisms is also satisfied and we have shown that $\mathfrak{F} \upharpoonright \widehat{W}$ is a bounded morphic image of $\mathfrak{F}$. As $\mathfrak{F} \upharpoonright \widehat{W}$ is isomorphic to $\mathfrak{G}$, $\mathfrak{G}$ is a bounded morphic image of $\mathfrak{F}$ as well.   ⊣

We now show that any normal modal logic extending **S4.3** has the polysize model property.

Lemma 6.40 *Let Λ be a normal modal logic such that $\mathbf{S4.3} \subseteq \Lambda$. Any formula ϕ that is satisfiable on a frame for Λ is satisfiable on a frame for Λ that contains at most $m + 2$ states, where m is the number of occurrences of modal operators in ϕ.*

Proof. Suppose ϕ is satisfiable on a frame for Λ. By Bull's Theorem, Λ has the finite frame property, thus there is a finite model based on a Λ-frame that satisfies ϕ at some point w_0. Let $\mathfrak{M}$ be the submodel of this model that is generated by w_0. Clearly $\mathfrak{M}, w_0 \Vdash \phi$, and as formation of generated submodels preserves modal validity, $\mathfrak{M}$ is based on a frame for Λ.

Now we select points. Let $\Diamond\psi_1, \ldots, \Diamond\psi_k$ be all the $\Diamond$-subformulas of ϕ that are satisfied at w_0. For each $1 \leq i \leq k$, select a point w_i that is maximal with respect to the property of satisfying ψ_i. These are the points needed to ensure that ϕ is satisfied in the polysize model at w_0, but if we select only w_0 and these points, we have no guarantee that we have constructed a Λ-frame. However, as we will now see, we *can* guarantee this if we glue on a maximal point. So, let w_{k+1} be such a point and define

$$\widehat{\mathfrak{M}} := \mathfrak{M} \upharpoonright \{w_0, w_1, \ldots, w_k, w_{k+1}\}.$$

$\widehat{\mathfrak{M}}$ contains at most $m + 2$ points, where m is the number of modal operators in ϕ. Moreover, it *is* based on a Λ-frame. To see this, note that the frame underlying $\widehat{\mathfrak{M}}$

is a subframe of the frame underlying $\mathfrak{M}$ that satisfies the requirements of item (ii) of Lemma 6.39; hence there is a surjective bounded morphism from $\mathfrak{M}$ to $\widehat{\mathfrak{M}}$. Such morphisms preserve modal validity, thus as $\mathfrak{M}$ is a Λ-model, so is $\widehat{\mathfrak{M}}$.

It remains to ensure that $\widehat{\mathfrak{M}}, w_0 \Vdash \phi$. We prove by induction for all subformulas ψ of ϕ, and all i such that $0 \le i \le k$, that

$$\mathfrak{M}, w_i \Vdash \psi \text{ iff } \widehat{\mathfrak{M}}, w_i \Vdash \psi.$$

The only interesting step is for formulas of the form $\Diamond\psi$. Suppose that $\mathfrak{M}, w_i \Vdash \Diamond\psi$ (thus $\psi = \psi_j$ for some $1 \le j \le k$). Since $\mathfrak{M}$ is point-generated by w_0 and transitive, it follows that Rw_0w_i, hence $\mathfrak{M}, w_0 \Vdash \Diamond\psi$. We chose w_j to be a world maximal with respect to the property of satisfying ψ_j, hence Rw_iw_j. By the induction hypothesis, $\widehat{\mathfrak{M}}, w_j \Vdash \psi_j$. Hence $\widehat{\mathfrak{M}}, w_i \Vdash \Diamond\psi$. The converse implication is left to the reader. ⊣

Theorem 6.41 (Hemaspaandra's Theorem) *Every normal modal logic extending* **S4.3** *has an NP-complete satisfiability problem.*

Proof. Lemma 6.40 established the polysize model property for Λ, so it remains to check that membership for Λ-frames can be decided in polynomial time. How can we show this? Recall Theorem 4.103:

> For every normal modal logic Λ extending **S4.3** there is a finite set N of finite **S4.3** frames with the following property: for any finite frame $\mathfrak{F}$, $\mathfrak{F} \Vdash \Lambda$ iff $\mathfrak{F}$ is an **S4.3** frame and there does not exist a bounded morphism from $\mathfrak{F}$ onto any frame in N.

This gives us a possible strategy: given any frame $\mathfrak{F}$, check whether it is an **S4.3** frame, and whether there is a surjective bounded morphism onto any frame in N. Now, as **S4.3** frames are first-order definable, by Lemma 6.36 the first part can be performed in polynomial time. But what about the second? First, note that because N is a *fixed* finite set, we need only ensure that the task of checking whether there is a bounded morphism from $\mathfrak{F}$ to a *fixed* frame $\mathfrak{G}$ can be performed in polynomial time. But the naive strategy of examining all the functions from $\mathfrak{F}$ to $\mathfrak{G}$ is completely unsuitable: the number of such functions is $|\mathfrak{G}|^{|\mathfrak{F}|}$, which is exponential in the size of $\mathfrak{F}$. However, applying Lemma 6.39, we see that the task can be simplified: we only need to check whether there is a set $\widehat{W}$ of worlds in $\mathfrak{F}$ such that $\mathfrak{F} \upharpoonright \widehat{W}$ is isomorphic to $\mathfrak{G}$ and $\widehat{W}$ contains a maximal world. Thus we need to check less than $|\mathfrak{F}|^{|\mathfrak{G}|}$ embeddings. But this number is *polynomial* in the size of $\mathfrak{F}$ for $\mathfrak{G}$ is fixed. By Lemma 6.35, NP-completeness follows. ⊣

The results of this section tell us something about the complexity of validity problems. The complement of NP is called co-NP. As a formula ϕ is not Λ-satisfiable iff $\neg\phi$ is Λ-valid, it follows that an NP-completeness result for Λ-satisfiability tells us that Λ-validity is co-NP complete (see Appendix C for further discussion). It is standardly conjectured that NP $\neq$ co-NP, thus the validity and satisfiability problems for these logics probably have different complexities.

Exercises for Section 6.6

6.6.1 Prove Lemma 6.36. That is, show that if F is a class of frames definable by a first-order sentence, then the problem of deciding whether $\mathfrak{F}$ belongs to F is decidable in time polynomial in the size of $\mathfrak{F}$.

6.6.2 Explain why the argument given in the proof of Lemma 6.35 may break down when we lift the restriction to finite similarity types. In particular, examine the situation when the similarity type contains modal operators of arbitrarily high arities.

6.6.3 Extend the proof of Theorem 6.38 to show that $\mathbf{K}_t\mathbf{Q}$ has the polysize model property, and is NP-complete.

6.6.4 Use a selection of points argument to show that **S5** has the polysize model property, and is NP-complete.

6.6.5 Show that if we restrict attention to a fixed finite set of proposition letters Φ, then the satisfiability problem for **S5** is decidable in linear time. (Hint: if Φ is finite, the number of models we have to check to determine whether a given formula ϕ is satisfied in them, is independent of ϕ.)

6.7 PSPACE

PSPACE, the class of problems solvable by a deterministic Turing machine using only polynomial space, is the complexity class of most relevance to the basic modal language. As we will see, some important modal satisfiability problems belong to PSPACE, and many modal logics have PSPACE-hard satisfiability problems. This suggests that modal satisfiability problems are typically tougher than the satisfiability problem for propositional calculus, for it is standardly conjectured that PSPACE-hard problems are not solvable in NP.

The work of this section revolves around *trees*. We first show that **K** lacks the polysize model property by forcing the existence of binary-tree-based models using short formulas. We then take a closer look at **K**-satisfiability and show that it is in PSPACE. The proof also shows that every **K**-satisfiable formula is satisfiable on a tree-based model of polynomial *depth*. We then put all this work together to prove Ladner's Theorem: every normal logic between **K** and **S4** has a PSPACE-hard satisfiability problem.

Forcing binary trees

The NP-completeness results of the previous section were proved using polysize model property arguments. So, before going any further, we will show that **K** does *not* have the polysize model property. We do so by showing that **K** can force the existence of binary trees. Many of the ideas introduced here will be reused in the proof of Ladner's Theorem.

For any natural number m, we are going to devise a satisfiable formula $\phi^B(m)$ with the following properties:

(i) the size of $\phi^B(m)$ is polynomial (indeed, quadratic) in m, but
(ii) when ϕ^B is satisfied in any model $\mathfrak{M}$ at a node w_0, then the submodel of $\mathfrak{M}$ generated by w_0 contains an isomorphic copy of the binary tree of depth m.

As the binary branching tree of depth m contains 2^n nodes, the size of the smallest satisfying model of $\phi^B(m)$ is exponential in $|\phi^B(m)|$. Thus we will have shown that small formulas can force the existence of large models.

We will define these formulas by mimicking truth tables. For any natural number m, $\phi^B(m)$ will be constructed out of the following variables: $q_1, \ldots, q_m$, and $p_1, \ldots, p_m$. The q_is play a supporting role. They will be used to mark the *level* (or *depth*) in the model; that is, they will mark the number of upward steps that need to be taken to reach the satisfying node. But any satisfying model for $\phi^B(m)$ will give rise to a full truth table for $p_1, \ldots, p_m$: every possible combination of truth values for $p_1, \ldots, p_m$ will be realized at some node, and hence any model for $\phi^B(m)$ must contain at least 2^m nodes.

That is the basic idea. To carry it out, we first define two macros: B_i, and $S(p_i, \neg p_i)$. For $i = 0, \ldots, m - 1$, B_i is defined as follows:

$$B_i \quad := \quad q_i \rightarrow (\Diamond(q_{i+1} \land p_{i+1}) \land \Diamond(q_{i+1} \land \neg p_{i+1})). \tag{6.12}$$

Given that we are going to use the q_is to mark the levels, the effect of B_i should be clear: it will force a *branching* to occur at level i, set the value of p_{i+1} to true at one successor at level $i + 1$, and set p_{i+1} to false at another.

Our other macro is closely related. For $i = 0, \ldots, m - 1$, $S(p_i, \neg p_i)$ is defined as follows:

$$S(p_i, \neg p_i) \quad := \quad (p_i \rightarrow \Box p_i) \land (\neg p_i \rightarrow \Box \neg p_i). \tag{6.13}$$

This formula *sends* the truth values assigned to p_i and its negation one level down. The idea is that once B_i has forced a branching in the model by creating a p_{i+1} and a $\neg p_{i+1}$ successor, $S(p_{i+1}, \neg p_{i+1})$ ensures that these newly set truth values are sent further down the tree; ultimately we want them to reach the leaves.

We are ready to define $\phi^B(m)$. It is the conjunction of the formulas listed in Figure 6.5. Note that $\phi^B(m)$ has the required effect. The first conjunct, q_0, ensures

(i) q_0

(ii) $\square^{(m)}(q_i \rightarrow \bigwedge_{i \neq j} \neg q_j)$ $\quad (0 \leq i \leq m)$

(iii) $B_0 \wedge \square B_1 \qquad \wedge \square^2 B_2 \qquad \wedge \square^3 B_3 \qquad \wedge \cdots \wedge \square^{m-1} B_{m-1}$

(iv) $\quad \square S(p_1, \neg p_1) \wedge \square^2 S(p_1, \neg p_1) \wedge \square^3 S(p_1, \neg p_1) \wedge \cdots \wedge \square^{m-1} S(p_1, \neg p_1)$

$\qquad \wedge \square^2 S(p_2, \neg p_2) \wedge \square^3 S(p_2, \neg p_2) \wedge \cdots \wedge \square^{m-1} S(p_2, \neg p_2)$

$\qquad \qquad \wedge \square^3 S(p_3, \neg p_3) \wedge \cdots \wedge \square^{m-1} S(p_3, \neg p_3)$

$$\vdots$$

$$\wedge \square^{m-1} S(p_{m-1}, \neg p_{m-1})$$

Fig. 6.5. The formula $\phi^{\mathcal{B}}(m)$

that any node that satisfies $\phi^{\mathcal{B}}(m)$ is marked as having level 0. The effect of (ii) is to ensure that no two distinct level marking atoms q_i and q_j can be true at the same node (at least, this will be the case all the way out to level m, which is all we care about). To see this, recall that $\square^{(m)}\phi$ is shorthand for $\phi \wedge \square \phi \wedge \square^2 \phi \wedge \cdots \wedge \square^m \phi$. Thus our level markers are beginning to work as promised.

But the real work is carried out by (iii) and (iv). Because of the prefixed blocks of $\square$ modalities, the B_i macros in (iii) force m successive levels of branching; and each such branching 'splits' the truth value of one of the ps. Then, again because of the prefixed $\square$ modalities, (iv) uses the $S(p_i, \neg p_i)$ macro to send each of these newly split truth values all the way down to the m-th level. In short, (iii) creates branching, and (iv) preserves it. It is worthwhile sitting down with a pencil and paper to check the details. If you do, it will become clear that $\phi^{\mathcal{B}}(m)$ is satisfiable, and that any satisfying model for $\phi^{\mathcal{B}}(m)$ must contain a submodel that is isomorphic to the binary branching tree of depth m. It follows that any model of $\phi^{\mathcal{B}}(m)$ must contain at least 2^m nodes, as we claimed.

In spite of its appearance, $\phi^{\mathcal{B}}(m)$ is indeed a *small* formula. To see this, consider what happens when we increment m by 1. The answer is: not much. For example (iii) simply gains an extra conjunct, becoming

$$\square B_0 \wedge \square^2 B_1 \wedge \square^3 B_2 \wedge \cdots \wedge \square^{m-1} B_{m-1} \wedge \square^m B_m.$$

Similarly, each row in (iv) gains an extra conjunct (as does the next empty row) thus we gain a new column containing m formulas. The biggest change occurs in (ii). If you write (ii) out in full, you will see that it gains an extra row, and an extra column, and an extra atomic symbol in each embedded conjunct, and this means that $|\phi^{\mathcal{B}}(m)|$ increases by $\mathcal{O}(m^2 \log m)$ (that is, slightly faster than quadratically). This is negligible compared with the explosion in the size of the smallest satisfying model: this doubles in size every time we increase m by one.

Theorem 6.42 **K** *lacks the polysize model property.*

That is, **K** lacks a property enjoyed by all the NP-complete logics examined in the previous section, and there is no obvious way of using NP guess-and-check algorithms to solve **K**-satisfiability. What sort of algorithms will work?

A PSPACE algorithm for K

We will now define a PSPACE-algorithm called *Witness* whose successful termination guarantees the **K**-satisfiability of the input. It may seem surprising that we can do this. After all, we have just seen that there are satisfiable formulas $\phi^B(m)$ whose smallest satisfying model contains 2^m nodes. What happens if we give $\phi^B(m)$ as input to *Witness*? Will it be forced to use an exponential amount of space to determine the satisfiability of $\phi^B(m)$? The answer is: *no*. *Witness* will take an exponential amount of *time* to terminate on difficult input, but it uses *space* efficiently. As we will see, if a formula ϕ is satisfiable in some model, it is satisfiable in a tree-based model of polynomial depth. While some formulas require models with exponentially many nodes, we can always find a shallow satisfying model: the length of each branch is polynomial in $|\phi|$. *Witness* tests for the existence of shallow models, and does so one branch at a time. It does not need to keep track of the entire model, and hence can be made to run in PSPACE.

Witness is essentially an abstract tableaux system for **K**: it explores spaces of *Hintikka sets* (see Definition 6.24). Recall that Hintikka sets need not be satisfiable, and that we call satisfiable Hintikka sets *atoms*. *Witness* will take two finite sets of formulas H and Σ as input, and determine whether or not H is an atom over Σ. It does so by looking at the demands that H makes and recursively calculating whether all these demands can be met. The following definition makes the idea of a demand precise (compare Definition 4.62, and recall our convention concerning single negations from Definition 4.79):

Definition 6.43 Suppose H is a Hintikka set over Σ, and $\Diamond\psi \in H$. Then the *demand* that $\Diamond\psi$ creates in H (notation: $\text{Dem}(H, \Diamond\psi)$) is

$$\{\psi\} \cup \{\sim\theta \mid \neg\Diamond\theta \in H\}.$$

We use $H_{\Diamond\psi}$ to denote the set of Hintikka sets over $\text{Cl}(\text{Dem}(H, \Diamond\psi))$ that contain $\text{Dem}(H, \Diamond\psi)$. (Recall that for any set of sentences Σ, $\text{Cl}(\Sigma)$ denotes the closure of Σ; see Definition 6.23.)

Remark 6.44 Suppose that A is an atom over Σ, and that $\Diamond\psi \in A$. As A is satisfiable, so is $\text{Dem}(A, \Diamond\psi)$. From this it follows that there is at least one atom in $A_{\Diamond\psi}$ that contains $\text{Dem}(A, \Diamond\psi)$. For suppose $\mathfrak{M}, w \Vdash \text{Dem}(A, \Diamond\psi)$. Let Ψ be the set of all formulas satisfied in $\mathfrak{M}$ at w. Then $\Psi \cap \mathrm{Cl}(\mathrm{Dem}(A, \Diamond\psi))$ is an atom over $\mathrm{Cl}(\mathrm{Dem}(A, \Diamond\psi))$ that contains $\mathrm{Dem}(A, \Diamond\psi)$.

Furthermore, as the reader can easily ascertain, for any formula ϕ, ϕ is satisfiable iff there is an atom A over $\mathrm{Cl}((\,)\phi)$ that contains ϕ.   ⊣

Definition 6.45 Suppose H and Σ are finite sets of formulas such that H is a Hintikka set over Σ. Then $\mathcal{H} \subseteq \mathcal{P}(\Sigma)$ is a *witness set generated by H on Σ* if $H \in \mathcal{H}$ and

(i) if $I \in \mathcal{H}$, then for each $\Diamond\psi \in I$, there is a $J \in I_{\Diamond\psi}$ such that $J \in \mathcal{H}$,

(ii) if $J \in \mathcal{H}$ and $J \neq H$ then for some $n > 0$ there are $I^0, \ldots, I^n \in \mathcal{H}$ such that $H = I^0$, $J = I^n$, and for each $0 \leq i < n$ there is some formula $\Diamond\psi \in I^i$ such that $I^{i+1} \in I^i_{\Diamond\psi}$.

The *degree* of a finite set of formulas Σ is simply the maximum of the degrees of the formulas contained in Σ; that is, $\deg(\Sigma) = \max\{\deg(\phi) \mid \phi \in \Sigma\}$.   ⊣

For all choices of H and Σ, any witness set $\mathcal{H}$ generated by H on Σ must be finite, for $\mathcal{H} \subseteq \mathcal{P}(\Sigma)$, which is a finite set. Further, observe that if $I, J \in \mathcal{H}$ and $J \in I_{\Diamond\psi}$ then the degree of J is strictly less than that of I. Moreover, observe that item (ii) of the previous definition is essentially a 'no junk' condition: if J belongs to $\mathcal{H}$, it is there because it is generated by some other elements of $\mathcal{H}$, and ultimately by H itself.

Lemma 6.46 *Suppose that H and Σ are finite sets of formulas such that H is a Hintikka set over Σ. Then H is an atom iff there is a witness set generated by H on Σ.*

Proof. For the left to right direction we proceed by induction on the degree of Σ. Let $\deg(\Sigma) = 0$, and suppose H is an atom. Trivially, $\mathcal{H} = \{H\}$ is a witness set generated by H. For the inductive step, suppose the required result holds for all pairs H' and Σ' such that H' is an atom of Σ' and $\deg(\Sigma') < n$. Let H be an atom of Σ such that $\deg(\Sigma) = n$. Then, as we noted in Remark 6.44, for all $\Diamond\psi \in H$ there exists at least one atom I^ψ in $H_{\Diamond\psi}$. As the degree of $\mathrm{Cl}(\mathrm{Dem}(H, \Diamond\psi)) < n$, for all $\Diamond\psi \in H$, the inductive hypothesis applies and every such atom I^ψ generates a witness set $\mathcal{I}^\psi$ on $\mathrm{Cl}(\mathrm{Dem}(H, \Diamond\psi))$. Define

$$\mathcal{H} = \{H\} \cup \bigcup_{\Diamond\psi \in H} \mathcal{I}^\psi.$$

Clearly $\mathcal{H}$ is a witness set generated by H on Σ.

For the right to left direction, we will show that if $\mathcal{H}$ is a witness set on Σ generated by H, then H can be satisfied in a model $(\mathfrak{F}, V)$ where $\mathfrak{F}$ is a finite tree of depth at most $\deg(H)$. This is stronger than the stated result, and later it will help us understand why **K**-satisfiability is solvable in PSPACE. Assume we have a countably infinite set of new entities $W = \{w_0, w_1, w_2, w_3, \ldots\}$ at our disposal. We will use (finitely many) elements of W to build a model for H, using a finitary version of the step-by-step method discussed in Section 4.6. This model will be a tree, thus showing once again that **K** has the tree model property.

Define $W_0 = \{w_0\}$, $R_0 = \varnothing$, $f_0(w_0) = H$. Suppose W_n, R_n and f_n have been defined. If for all $w \in W_n$ such that $\Diamond \psi \in f_n(w)$ there exists a $w' \in W_n$ such that (i) $\psi \in f_n(w')$ and (ii) $f_n(w') \in f_n(w)_{\Diamond \psi}$, then halt the step-by-step construction. Otherwise, if there is a $w \in W_n$ such that $\Diamond \psi \in f_n(w)$, while for no $w' \in W_n$ are these two conditions satisfied, then carry on to stage $n + 1$ and define:

$$
\begin{aligned}
W_{n+1} &= W_n \cup \{w_{n+1}\}, \\
R_{n+1} &= R_n \cup \{(w, w_{n+1})\}, \\
f_{n+1} &= f_n \cup \{(w_{n+1}, I)\},
\end{aligned}
$$

where $I \in \mathcal{H}$ is such that $I \in f_n(w)_{\Diamond \psi}$. Note that because $\mathcal{H}$ is a witness set it will always be possible to find such an I.

This step-by-step procedure halts after finitely many steps since each $I \in \mathcal{H}$ contains only finitely many formulas of the form $\Diamond \psi$ (thus ensuring that the tree we are constructing is finitely branching), and whenever $R_n w w'$, then

$$
\deg(f_n(w')) < \deg(f_n(w))
$$

(thus ensuring that the tree is not only finite, but shallow: it has depth at most $\deg(H)$). Let m be the stage at which it halts, and define $\mathfrak{F}$ to be (W_m, R_m). To construct the desired model for H, it only remains to define a suitable valuation V, and we do this as follows: choose V to be any function from Σ to $\mathcal{P}(W_m)$ satisfying $w \in V(p)$ iff $p \in f_m(w)$, for all $p \in \Sigma$. Let $\mathfrak{M} = (\mathfrak{F}, V)$. Exercise 6.7.1 asks the reader to show that $\mathfrak{M}, w_0 \Vdash H$; an immediate consequence is that H is an atom. $\dashv$

Two remarks. The above proof shows that every atom is satisfiable in a shallow tree-based model – a fact which will prove to be important below. Second, we now have a syntactic criterion – namely the existence or non-existence of witness set – for determining whether a Hintikka set is **K**-satisfiable. (In short, we have just proved a completeness result.) Moreover, the criterion is intuitively computable witness sets are simple finite structures, thus it seems reasonable to expect that w can algorithmically test for their existence. And indeed we can.

We now define the *Witness* algorithm. This takes as input two finite sets c formulas H and Σ and returns the value **true** if and only if there is a witness s generated by H on Σ.

*function *Witness*(H, Σ) returns boolean*

begin
  if H is a Hintikka set over Σ
  and for each subformula $\Diamond\psi \in H$ there is a set of formulas
    $I \in H_{\Diamond\psi}$ such that $Witness(I, \mathrm{Cl}(\mathrm{Dem}(H, \Diamond\psi)))$
  then return **true**
  else return **false**
end

Note that $Witness$ is an intuitively acceptable algorithm – and hence (by Church's thesis) implementable on a Turing machine. Checking that H is a Hintikka set over Σ involves ascertaining that Σ is subformula closed, and that H satisfies the properties demanded of Hintikka sets; these tasks involve only simple syntactic checking. Moreover, both the '**and** for each subformula ... there is' clause and the recursive call to $Witness$ are clearly computable: the first involves search through a finite space, while the recursive call performs the same tasks on input of lower degree. Thus $Witness$ is indeed an algorithm. Moreover, it is *correct*: if H and Σ are finite sets of formulas, then $Witness(H, \Sigma)$ returns **true** iff H is a Hintikka set over Σ that generates a witness set in Σ. This follows by induction on the degree of Σ. The right to left direction is easy, while the left to right direction is similar to the proof of Lemma 6.46; see Exercise 6.7.2.

We are now ready for the main result.

Theorem 6.47 K-*satisfiability is in PSPACE.*

Proof. It follows from Lemma 6.46 and the correctness of $Witness$ that for any formula ϕ, ϕ is satisfiable iff there is an $H \subseteq \mathrm{Cl}((\,)\phi)$ such that $\phi \in H$ and $Witness(H, \mathrm{Cl}((\,)\phi))$ returns the value **true**. Thus, if we can show that $Witness$ can be given a PSPACE implementation, we will have the desired result. We will implement $Witness$ on a non-deterministic Turing machine. Given any formula ϕ, this machine will non-deterministically pick a Hintikka set H in $\mathrm{Cl}((\,)\phi)$ that contains ϕ, and run $Witness(H, \mathrm{Cl}((\,)\phi))$. It will be easy to show that this machine runs in non-deterministic PSPACE (that is, NPSPACE). But then it follows by an appeal to Savitch's Theorem (PSPACE = NPSPACE; see Appendix C) that the required PSPACE implementation exists.

So how do we implement $Witness$ on a non-deterministic Turing machine? The key points are the following:

(i) All sets of formulas used in the execution of the program are subsets of $\mathrm{Cl}((\,)\phi)$, and we can represent any such subset by using pointers to the connectives and proposition letters in ϕ's representation: a pointer to a proposition letter will mean that the letter belongs to the subset, and a pointer to a connective means that the subformula built using that connective belongs to it. Thus encoding a subset of $\text{Cl}(()\phi)$ requires only space $\mathcal{O}(|\phi|)$ (that is, space of the order of the size of ϕ).

(ii) The '**and** for each subformula $\Diamond\psi \in H$' part can be handled by treating each subformula in turn. As any subformula can be represented using a pointer to ϕ's representation, we can cycle through all possible subformulas, using only polynomial space, by cycling through these pointers. Moreover, as we are using a non-deterministic Turing machine, the 'there is a set of formulas ...' clause can be implemented by making non-deterministic choices. Note that, even though $H_{\Diamond\psi}$ is a *set of* sets of formulas, it is a rather trivial task to verify whether I belongs to $H_{\Diamond\psi}$, given the definition of $H_{\Diamond\psi}$.

(iii) To enable recursive calls to be made, we implement a stack on our Turing machine. To perform the recursion, we copy the formula ϕ onto the stack and point to proposition letters and connectives to indicate the subsets of interest.

So, suppose we run *Witness* on input H and Σ. The crucial point that must be investigated is whether the recursive calls to *Witness* cause a blow-up in space requirements. From items (i), (ii) and (iii) it is clear that at each level of recursion we use space $\mathcal{O}(|\phi|)$. How long does it take for the recursion to bottom out? Note that after $\deg(\phi)$ recursive calls, $\Sigma = \varnothing$. That is, the depth of recursion is bounded by $\deg(\phi)$ and hence by $|\phi|$. Thus, when we implement *Witness* on a non-deterministic Turing machine the total amount of space required is $\mathcal{O}(|\phi|^2)$, hence the algorithm runs in NPSPACE. Thus, by Savitch's Theorem, we conclude that **K**-satisfiability is in PSPACE.   ⊣

The appeal to Savitch's Theorem in the above proof can be avoided: *Witness* can be implemented on a deterministic Turing machine. This involves replacing the non-deterministic choice used in item (iii) by a brute force search through subsets of $\text{Cl}(()\phi)$ that uses only polynomial space, and the reader is asked to do this in Exercise 6.7.4. But the above proof illustrates why Savitch's Theorem is so useful in practice: by freeing us to think in terms of non-deterministic computations, it reduces the required bookkeeping to a minimum.

Let us try and pin down the key intuition underlying Theorem 6.47. **K** lacks the polysize model property, but in spite of this the **K**-satisfiability problems can be determined in PSPACE. Why? The key lies in the proof of Lemma 6.46 which showed that every atom is satisfiable in a *shallow* finite tree-based model. Such models make it easy to visualize the explorations that *Witness* makes as it tests the satisfiability of ϕ: it just works out what each branch of such a model must contain. While the size of the entire model may be exponential in $|\phi|$ it is not necessary to keep track of all this information. The locally relevant information is simply the information on each branch – and we know that the tree has depth at most $\deg(\phi) + 1$. In short, *Witness* exploits the fact that only shallow tree-based models are needed to determine **K**-satisfiability.

PSPACE algorithms have been devised for a number of well-known logics including **T**, **K4** and **S4**, the temporal counterparts of **K**, **T**, **K4** and **S4**, and multi-modal **K**, **T**, **K4**, **S4** and **S5**. While proofs of these results are essentially refinements of the proof of Theorem 6.47, some are rather tricky. The reader who does Exercise 6.7.3, which asks for a PSPACE algorithm for **K4**, will find out why. In some cases alternative methods are preferable; see the Notes for pointers.

Ladner's Theorem

We are ready to prove the major result of the section: every normal modal logic between **K** and **S4** is PSPACE-hard, and hence (assuming PSPACE $\neq$ NP) the satisfiability problems for all these logics are tougher than the satisfiability problem for propositional logic. We prove this by giving a polynomial time reduction of the validity problem for prenex quantified boolean formulas to all these modal satisfiability problems. The reduction boils down to forcing the existence of certain tree-based models, and we will be able to reuse much of our previous work.

Definition 6.48 The set of *quantified boolean formulas* is the smallest set X containing all formulas of propositional calculus such that if $\beta \in X$ and p is a proposition letter, then both $\forall p\, \beta$ and $\exists p\, \beta \in$ S. The quantifiers range over the truth values 1 (true) and 0 (false), and a quantified boolean formula without free variables is *valid* if and only if it evaluates to 1.

A quantified boolean formula is said to be in *prenex* form if it is of the form

$$Q_1 p_1 \cdots Q_m p_m\, \theta(p_1, \ldots, p_m);$$

here Q is either $\forall$ or $\exists$, and $\theta(p_1, \ldots, p_m)$ is a formula of propositional logic. We will refer to such prenex formulas as QBFs. $\dashv$

The problem of deciding whether a QBF containing no free variables is valid is called the *QBF-validity problem*, and it is known to be PSPACE-complete.

We are going to define a polynomial time translation f_L from QBFs to modal formulas, and prove that it has the following two properties:

(i) If β is a QBF-validity, then $f_L(\beta)$ is **S4**-satisfiable.
(ii) If $f_L(\beta)$ is **K**-satisfiable, then β is a QBF-validity.

These two properties – together with the known PSPACE-hardness of the QBF-validity problem – will lead directly to the desired theorem.

Let us think about what is involved in evaluating a QBF. We start by peeling off the outermost quantifier. If it is of the form $\exists p$ we choose one of the truth values or 0 and substitute for the newly freed occurrences of p. On the other hand, if it is of the form $\forall p$ we must substitute both 1 and 0 for the newly freed occurrences of p. In this fashion, we work our way successively through the prefixed list of quantifiers until we reach the matrix, a formula of propositional logic.

We are essentially generating a tree. This tree consists of the root node, and then – working inwards along the quantifier string – each existential quantifier extends it by adding a single branch, and each universal quantifier extends it by adding two branches. Indeed, we are even generating an annotated tree: we can label each node with the substitution it records. For example, corresponding to the QBF $\forall p \exists q\, (p \leftrightarrow \neg q)$ we have the following annotated tree:

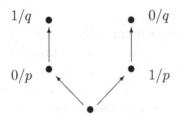

The information in such annotated trees – we will call them *quantifier trees* – will play a crucial role. For a start, QBF-validity is witnessed by certain quantifier trees: β is a QBF-validity if and only if there is a quantifier tree for β such that the substitutions it records ensure that the matrix evaluates to 1. Moreover, quantifier trees give us a bridge between the QBF world and the modal world: $f_L(\beta)$ will be a modal formula that describes the structure of a quantifier tree evaluating β.

We define the translation f_L by modifying the way we forced the existence of binary trees in the proof of Theorem 6.42, and we will reuse the macros B_i and $S(p_i, \neg p_i)$ defined on page 382 in (6.12) and (6.13), respectively.

Definition 6.49 Given any QBF $\beta = Q_1 p_1 \cdots Q_m p_m\, \theta(p_1, \ldots, p_m)$, choose new propositional variables $q_0, \ldots, q_m$. Then $f_L(\beta)$ is the conjunction of the formulas displayed in Figure 6.6.   ⊣

The idea underlying f_L is this: for any QBF β, $f_L(\beta)$ describes the peel-off-quantifiers-and-substitute evaluation process for β. (That is, it describes how we generate a quantifier tree for β.) Moreover, it does so using ideas we have met already: note that (i), (ii) and (iv) are exactly the same formulas we used when forcing the existence of binary trees.

In fact, the major difference between these formulas and our earlier work lies in the word *binary*. Here we *do not* always want binary branching: we only want it when we encounter the quantifier $\forall$. Thus, instead of the earlier (iii) which forced branching all the way down to level m, we have the pair of formulas (iiia) and (iiib). (iiia) guarantees that if q_i is true and $i < m$ then there is a next level q_{i+1} which simply amounts to saying that if $i < m$ then we have not yet peeled of

(i) q_0

(ii) $\Box^{(m)}(q_i \rightarrow \bigwedge_{i \neq j} \neg q_j)$ $(0 \leq i \leq m)$

(iiia) $\Box^{(m)}(q_i \rightarrow \Diamond q_{i+1})$ $(0 \leq i < m)$

(iiib) $\bigwedge_{\{i | Q_i = \forall\}} \Box^i B_i$

(iv) $\Box S(p_1, \neg p_1) \wedge \Box^2 S(p_1, \neg p_1) \wedge \Box^3 S(p_1, \neg p_1) \wedge \cdots \wedge \Box^{m-1} S(p_1, \neg p_1)$

$$\wedge \Box^2 S(p_2, \neg p_2) \wedge \Box^3 S(p_2, \neg p_2) \wedge \cdots \wedge \Box^{m-1} S(p_2, \neg p_2)$$

$$\wedge \Box^3 S(p_3, \neg p_3) \wedge \cdots \wedge \Box^{m-1} S(p_3, \neg p_3)$$

$$\vdots$$

$$\wedge \Box^{m-1} S(p_{m-1}, \neg p_{m-1})$$

(v) $\Box^m(q_m \rightarrow \theta)$

Fig. 6.6. The formula $f_L(\beta)$

all the quantifiers and a new level will be necessary. But it does *not* force binary branching. The task of forcing binary branching, when necessary, is left to (iiib). Note that this formula is simply a selection of conjuncts from our earlier (iii). There is only one other difference: (v) insists that after m quantifiers have been peeled off, the propositional matrix θ must be true.

Clearly, $f_L(\beta)$ is polysize in $|\beta|$, thus this translation causes no blowup in space requirements.

Theorem 6.50 (Ladner's Theorem) *If Λ is a normal modal logic such that $\mathbf{K} \subseteq \Lambda \subseteq \mathbf{S4}$, then Λ has a PSPACE-hard satisfiability problem. Moreover, Λ has a PSPACE-hard validity problem.*

Proof. Fix a modal logic Λ with $\mathbf{K} \subseteq \Lambda \subseteq \mathbf{S4}$. We are going to prove that f_L is a (polynomial time) reduction from the QBF-validity problem to the Λ-satisfiability problem. The crucial step in this proof is summarized in the following two statements:

$$\text{if } \beta \text{ is a QBF-validity, then } f_L(\beta) \text{ is satisfiable on a frame for } \mathbf{S4}, \quad (6.14)$$

and

$$\text{if } f_L(\beta) \text{ is satisfied in a } \mathbf{K}\text{-model then } \beta \text{ is a QBF-validity}. \quad (6.15)$$

From these two statements the desired result follows immediately. For suppose β is a QBF-validity. Then by (6.14) $f_L(\beta)$ is $\mathbf{S4}$-satisfiable and hence Λ-satisfiable. Conversely, if $f_L(\beta)$ is Λ-satisfiable then it is also $\mathbf{K}$-satisfiable, and by (6.15) β is a QBF-validity. Thus Λ-satisfiability is PSPACE-hard. That the Λ-validity problem is also PSPACE-hard follows immediately from the fact that PSPACE = co-PSPACE.

It remains to prove (6.14) and (6.15). For (6.14), assume that β is a QBF-validity Generate a quantifier tree witnessing the validity of β; if β is valid, such a tree must exist. This tree gives rise to an **S4**-model for $f_L(\beta)$ as follows. First, take the transitive and reflexive closure of the 'daughter-of' relation of the tree; this gives us the **S4**-frame we require. Then make the variable q_i true precisely at the nodes of level i; p_i is to be made true at a node of level $j \geq i$ iff the substitution connected to that node, or its predecessor at level i returns the value 1 for p_i. (For nodes at level $j < i$ it does not matter what truth value we choose for p_i.) It is straightforward to check that the formula $f_L(\beta)$ is true in this model at the root of the tree; see Exercise 6.7.5.

For (6.15), suppose that β is a QBF of quantifier depth m, and that $f_L(\beta)$ is **K**-satisfiable. Note that $\deg(f_L(\beta)) = m$, hence from the proof of Lemma 6.46 we know that $f_L(\beta)$ holds at the root r of a tree-based model $\mathfrak{M} = (T, R, V)$ of depth at most m. Using clauses (iiia) and (iiib) of the definition of $f_L(\beta)$, it is easily verified that we may cut off branches from this tree such that in the resulting tree, a node at level $i < m$ has either one or two successors. This number is one iff $Q_{i+1} = \exists$. And if $Q_{i+1} = \forall$, then one of the successors satisfies p_{i+1} and the other one, $\neg p_{i+1}$. But then this reduced tree model is a quantifier tree witnessing the validity of β.   -

Among other things, Ladner's Theorem tells us that **K**, **T**, **K4** and **S4** all have PSPACE-hard satisfiability problems. It follows that the temporal counterpart of **K**, **T**, **K4** and **S4**, and multi-modal **K**, **T**, **K4**, and **S4**, are PSPACE-hard too for they contain the unimodal satisfiability problems as a special case. Hence, as PSPACE algorithms are known for these logics, they all have PSPACE-complete satisfiability problems. As PSPACE = co-PSPACE, these logics have PSPACE complete validity problems too.

Exercises for Section 6.7

6.7.1 Show that in the model $\mathfrak{M}$ constructed in the proof of Lemma 6.46, $\mathfrak{M}, w_0 \Vdash H$.

6.7.2 We claimed that *Witness* is a correct algorithm. That is, if H and Σ are finite sets of formulas, then *Witness*(H, Σ) returns **true** iff H is a Hintikka set over Σ that generates a witness set in Σ. Prove this.

6.7.3 Adapt the *Witness* algorithm so that it decides **K4** satisfiability correctly. (Hint: since you cannot consider smaller and smaller Hintikka sets (why not?) make use of lists of Hintikka sets, rather than the single Hintikka sets used in the proof for **K**, and show that the length of such lists can always be kept polynomial.)

6.7.4 Show how to avoid the use of Savitch's Theorem in the proof of Theorem 6.47. That s, show that the *Witness* function can be implemented on a *deterministic* Turing machine. Hint: implement the '**and** for each subformula . . . there is' clause by cycling through all possible subsets of $\mathrm{Cl}(()\phi)$. This cycling process has a simple implementation using only space $\mathcal{O}(|\phi|)$: generate all binary strings of length $|\phi|$, and decide of each whether or not t encodes a subset of $\mathrm{Cl}(\phi)$.)

6.7.5 Supply the missing details in the proof of Ladner's Theorem.

6.7.6 Show that the satisfiability problem for bimodal **S5** is PSPACE-hard.

6.7.7 In this exercise we examine the effects of bounding the number of proposition letters and of restricting the degree of formulas.

(a) Show that for any fixed k, the satisfiability problem for **K** with respect to a language consisting of all formulas whose degree is at most k, is NP-complete.
(b) Show that, in contrast, the satisfiability problem for **S4** remains PSPACE-complete for languages consisting of all formulas of degree at most k ($k \geq 2$).
(c) Now suppose that Φ, the set of proposition letters, is finite. Show that for any fixed k, the satisfiability problems for **K** and **S4** with respect to a language consisting of all formulas whose degree is at most k, is decidable in linear time.

6.8 EXPTIME

EXPTIME, the class of problems deterministically solvable in exponential time, is an important complexity class for many modal languages. In particular, when a modal language has operators $[a]$ and $[a^*]$ which explore a relation R_a and its reflexive transitive closure $(R_a)^*$, its satisfiability problem is likely to be EXPTIME-hard, which means that the worst cases are computationally intractable. As such operator pairs are important in many applications, we need to understand the complexity theoretic issues they give rise to. In this section we examine the satisfiability problem for **PDL**; our discussion illustrates some key themes and introduces some useful techniques.

Forcing exponentially deep models

By Corollary 6.14 we know that **PDL** has a decidable satisfiability problem – but just how difficult is it? Clearly it is PSPACE-hard, for each basic modality $[a]$ is a **K** operator, and we saw in the previous section (Theorem 6.50) that **K** has a PSPACE-hard satisfiability problem. But can we prove a matching PSPACE upper bound?

We used a tableaux-like algorithm called *Witness* to show that **K**-satisfiability was solvable in PSPACE. *Witness* traded on the following insight: while a **K**-consistent formula ϕ may require a satisfying model of size $2^{|\phi|}$, it is always possible to build a satisfying tree model of this size in which each branch has less than $|\phi|$ nodes. *Witness* tests for **K**-satisfiability by building such trees one branch

at a time; as each branch is polynomial in the size of the input, *Witness* runs i
PSPACE. However, as we will now show, even small fragments of PDL are stron
enough to force the existence of exponentially *deep* models.

Proposition 6.51 *For every natural number n there is a satisfiable PDL formula κ_n
of size $\mathcal{O}(n^2)$ such that every model which satisfies κ_n contains an R_a-path containing 2^n distinct nodes. Moreover, κ_n contains occurrences of only two modali
ties $[a]$ and $[a^*]$, where a is an atomic program.*

Proof. We will show how to count using this PDL-fragment. Given a natural num
ber n, we select n distinct proposition letters $q_1, \ldots, q_n$. Using 1 for true, and (
for false, the list of truth values $[V(q_n, w), \ldots, V(q_i, w), \ldots, V(q_1, w)]$ is the n-bi
binary encoding of a natural number. We take $V(q_1, w)$ to be the least significan
digit, and $V(q_n, w)$ to be the most significant.

We now construct a formula κ_n which, when satisfied at some state w_0, force:
the (n-bit representation of) zero to hold at w_0, and forces the existence of a patl
of distinct successors of w_0 which correctly count from 0 to 2^{n-1} in binary. Fo
example, if $n = 2$, the model will contain a path of length 4 from w_0 to w_3, and a:
we move along this path we will successively encounter the following truth valu
lists: $[0, 0], [0, 1], [1, 0], [1, 1]$.

To do the encoding, we need to know what happens when we add 1 to a binar
number m. First suppose that the least significant bit of m is 0; for example
suppose that m is 010100. When we add 1 we obtain 010101; that is, we flip th
least significant digit to 1 and leave everything else unchanged. We can force thi
kind of incrementation in PDL as follows:

$$INC_0 := \neg q_1 \to \left([a]q_1 \wedge \bigwedge_{j>1} ((q_j \to [a]q_j) \wedge (\neg q_j \to [a]\neg q_j)) \right).$$

This guarantees that the value of q_1 changes to 1 at any successor state, while th
truth values of all the other q_js remain unchanged.

Now suppose that the least significant digit of m is 1. For example, suppose tha
m is 01011. When we add 1 we obtain 01100. We can describe this incrementatio
as follows. First, we locate the longest unbroken block of 1s containing the lea:
significant digit and flip all these 1s to 0s. Second, we flip the following digit from
to 1 (we have to 'carry one'). Finally, we leave all remaining digits unchanged. Th
following formula forces this kind of incrementation when the longest unbroke
block of 1s containing the least significant digit has length i, where $0 < i < n$:

$$INC_1(i) :=$$
$$\left(\neg q_{i+1} \wedge \bigwedge_{j=1}^{i} q_j \right) \to$$

$$\left([a](q_{i+1} \wedge \bigwedge_{j=1}^{i} \neg q_j \wedge \bigwedge_{k>i+1} ((q_k \to [a]q_k) \wedge (\neg q_k \to [a]\neg q_k)) \right).$$

We can now define the required formula κ_n:

$$(\neg q_n \wedge \cdots \wedge \neg q_1) \wedge [a^*]\langle a \rangle \top \wedge [a^*] \left(INC_0 \wedge \bigwedge_{i=1}^{n-1} INC_1(i) \right).$$

The first conjunct of κ_n initializes the counting at 0, the second guarantees that there will always be successor states, while the third guarantees that incrementation is carried out correctly. Clearly κ_n is of size $\mathcal{O}(n^2)$ and uses only the allowed modalities. ⊣

Proposition 6.51 is suggestive. It does not *prove* that no PSPACE algorithm is possible, but it does tend to confirm our suspicions that **PDL**-satisfiability is computationally difficult. And indeed it is. The remainder of the chapter is devoted to proving the following result: the **PDL**-satisfiability problem is EXPTIME-complete. The proof methods we use are important in their own right and well worth mastering: we will prove EXPTIME-hardness by reduction from the two person corridor tiling game, and demonstrate the existence of an EXPTIME algorithm using elimination of Hintikka sets.

EXPTIME-hardness via tiling

In Section 6.5 we used tiling problems to prove two undecidability results. We remarked that tiling problems were also useful for proving complexity results, and in this section we give an example. We will describe the *two person corridor tiling game* and use it to prove the EXPTIME-hardness of **PDL**-satisfiability; we make use of notation and ideas introduced in our discussion of undecidability.

As with our earlier tiling problems, the two person corridor tiling game involves placing tiles on a grid so that colors match, but there are some extra ingredients. There are two players, and we assume that there is a third person present – the referee – who starts the game correctly and keeps it flowing smoothly. The referee will give the players a finite set $\{T_1, \ldots, T_s\}$ of tile types; the players will use tiles of these types to attempt to tile a grid so that colors match. In addition, the referee will set aside two special tile types: T_0 and T_{s+1}. T_0 is there solely to mark the boundaries of the corridor (we think of the boundaries as having some distinctive color, say white), while T_{s+1} is a special winning tile, whose role will be described later.

At the start of play, the referee places n initial tiles $I_1, \ldots, I_n$ in a row. To the left of I_1 and to the right of I_n he places copies of the white tile T_0. That is, the following sequence of tiles is the initial position:

This is the first row of the corridor. The white tiles in column 0 and column $n +$ mark the boundaries of the corridor. Columns 1 through n are the corridor prope Actually, we may as well stipulate that the referee immediately fills in columns and $n + 1$ with the special boundary-marking white tile. That is, the players ar going to be playing into the grid inside the following n-column corridor:

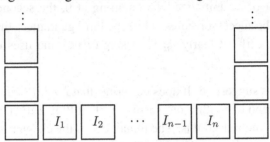

Now the players are ready to start. There are two players, Eloise and Abelard. Th players take turn placing tiles in the corridor, and it is always Eloise who move first. The rules for tile placement are strict: the corridor has to be filled in fron the bottom, from left to right. For example, after Eloise has placed her first tile th corridor will look like this:

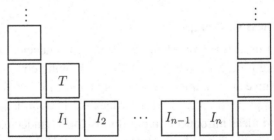

When Abelard replies, he must place his tile immediately to the right of tile T When the players have completed tiling a row, they start tiling the next one, startir at column 1. In short, the players have no choice about *where* to place a til only about which type of tile they will place there. The player's choice of tiles subject to the usual color-matching rules of tiling, and any tile placed in colum 0 or column n has to match the white of the corridor tile. (For example, in th previous diagram, it must be the case that $left(T) = white$.)

When do the players win or lose? As follows. If after finitely many rounds tiling is constructed in which the special winning tile T_{s+1} is placed in column Eloise wins. Otherwise (that is, if one of the players cannot make a legal move ar T_{s+1} is not in column 1, or if the game goes on infinitely long) Abelard wins.

Now for the EXPTIME-complete problem: *given a game, does Eloise have* winning strategy in that game? That is, can she win the game no matter what Abelard does? It is useful to think of winning strategies in terms of *game trees*. For any game, a game tree for that game records all possible responses Abelard can make to Eloise's moves. (Note that we do not insist that game trees encode all of Eloise's options; but it is *vital* that game trees record all of Abelard's options.) Note that Abelard has only finitely many possible responses, for there are only finitely many tile types. Clearly, if Eloise has a winning strategy in a game, then there is a game tree that describes that strategy: such a tree spells out exactly what she has to do, and takes all Abelard's possible responses into account.

Now that we know about game trees, let us think about winning strategies for Eloise. In fact, we can recursively characterize this concept. We first define the notion of a *winning position for Eloise* in a game tree:

(i) Whenever the winning tile T_{s+1} is placed in column 1, that position is a winning position for Eloise.

(ii) In case Eloise is to move in position x, then x is a winning position for Eloise if there exists a move to a winning position for Eloise.

(iii) In case Abelard is to move in position x, then x is a winning position for Eloise if Abelard can make a move and all his moves lead to a winning position for Eloise.

We now say that Eloise has a *winning strategy* iff there is a game tree such that the root of the game tree is a winning position for her. The problem of determining whether Eloise has a winning strategy is called the *two person corridor tiling problem*, which is known to be EXPTIME-complete (see the Notes for references).

Theorem 6.52 *The satisfiability problem for* **PDL** *is EXPTIME-hard.*

Proof. We show this by reducing the two person corridor tiling problem to the **PDL** satisfiability problem. We will view a game tree as a rooted regular PDL model with one atomic transition R_m which codes one move of the game. Given an instance $T = (n, \{T_0, \ldots, T_{s+1}\})$ of the two person corridor tiling game (here n is the width of the corridor, and the T_i are the tile types), we will show how to create a formula ϕ_T such that

(i) If Eloise has a winning strategy, ϕ_T is satisfiable at the root of some game tree for T (viewed as a regular PDL model).

(ii) If ϕ_T is satisfiable, then Eloise has a winning strategy in the game T; in fact, she will be able to read off her winning strategy by following a path through the satisfying model (starting at the point that satisfies ϕ_T).

(iii) The formula ϕ_T can be computed in time polynomial in n and s.

The formula ϕ_T contains two kinds of information: it fully describes the structure of the game tree, and states necessary and sufficient conditions for Eloise to win. The first part boils down to using PDL to describe the initial configuration, that players move alternately, that colors match, and so on; this is a little tedious, but straightforward. Stating necessary and sufficient conditions for Eloise to win involves finding PDL formulas that capture the recursive characterization of winning strategies, and prevent the game from running for infinitely many moves; this is the interesting part of the proof.

We use the following proposition letters to construct ϕ_T:

(i) $t_0, t_1, \ldots, t_s$, and t_{s+1}. These will be used to represent the tiles. We will often write t_0 as *white*.

(ii) *eloise*. This will be used to indicate that Eloise has the next move. Its negation will indicate that Abelard has the next move.

(iii) $pos_1, \ldots, pos_n$. We use pos_i to indicate that in the current round, a tile is to be placed in column i.

(iv) $col_i(t)$, for all $0 \leq i \leq n+1$ and all $t \in \{t_0, t_1, \ldots, t_s, t_{s+1}\}$. These will be used to indicate that the tile previously placed in column i is of type t.

(v) *win*. This means that the current position is a winning position for Eloise.

In addition, we make use of the modalities $[m]$ and $\langle m \rangle$ ('after every possible move' and 'after some possible move' respectively) and $[m^*]$, which can be read as 'after every possible sequence of moves.'

So let us describe the structure of the game tree. The following formula records the situation at the start of play:

$$eloise \wedge pos_1 \wedge col_0(white) \wedge col_1(t_{I_1}) \wedge \cdots \wedge col_n(t_{I_n}) \wedge col_{n+1}(white).$$

The first conjunct says that Eloise has to make the first move, while the second says that she has to place her tile in column 1. The remaining conjuncts simply say that the tiles previously placed in all columns are those of the initial configuration. (Of course, these were not placed by the players but by the referee.) That is, they say that columns 1 through n contain the initial tiles $I_1, \ldots, I_n$, and that there is a white corridor tile on each side.

We now write down a series of formulas which regulate the way that further play takes place. (Note that all these conditions are preceded by the $[m^*]$ modality, thus ensuring that they continue to hold after any finite sequence of moves.) We start by giving the desired meaning to pos_i and $col_i(t)$.

- Tiles always have to be placed in one of columns 1 through n:

$$[m^*](pos_1 \vee \cdots \vee pos_n),$$

and indeed, in *exactly* one of these columns:

$$[m^*](pos_i \rightarrow \neg pos_j) \qquad (1 \leq i \neq j \leq n).$$

- In every column i, at least one tile type was previously placed:

$$[m^*](col_i(t_0) \vee \cdots \vee col_i(t_{s+1})) \qquad (0 \le i \le n+1).$$

- In every column i, at most one tile type was previously placed:

$$[m^*](col_i(t_u) \rightarrow \neg col_i(t_v)) \qquad (0 \le i \le n+1 \text{ and } 0 \le u \ne v \le s+1).$$

- Moreover, the referee has already placed white tiles in columns 0 and $n+1$:

$$[m^*](col_0(white) \wedge col_{n+1}(white)).$$

- In the course of play, tiles are placed left to right (flipping back to column 1 when a row has been completed):

$$[m^*]((pos_1 \rightarrow [m]pos_2) \wedge (pos_2 \rightarrow [m]pos_3) \wedge \cdots \wedge (pos_n \rightarrow [m]pos_1)).$$

- In columns where no tile is placed, nothing changes when a move is made:

$$[m^*](\neg pos_i \rightarrow ((col_i(t_u) \rightarrow [m]col_i(t_u)) \wedge (\neg col_i(t_u) \rightarrow [m]\neg col_i(t_u)).$$

(Here $0 \le i \le n+1$ and $0 \le u \le s+1$.)

With these preliminaries behind us, we can now describe the structure of the game tree.

First of all, players alternate:

$$[m^*]((eloise \rightarrow [m]\neg eloise) \wedge (\neg eloise \rightarrow [m]eloise)).$$

Next, both players make legal moves; that is, they only place tiles which correctly match adjacent tiles. It will be helpful to define the following ternary relation of 'compatibility' between propositional variables:

$$C(t', t, t'') \text{ iff } right(T') = left(T) \text{ and } down(T) = up(T''),$$

where T, T' and T are the tiles that correspond to the propositional variables t, t' and t'' respectively. That is, $C(t', t, t'')$ holds iff the tile T can be placed to the right of tile T' and above tile T''. With the aid of this relation we can formulate the first constraint on tile placement as follows:

$$[m^*]\left(pos_i \wedge col_{i-1}(t') \wedge col_i(t'') \rightarrow [m]\bigvee\{col_i(t) \mid C(t', t, t'')\}\right).$$

(Here $0 \le i \le n$, and, by convention, $\bigvee \varnothing = \bot$.)

However this constraint is not quite enough; it only ensures matching to the left and downwards. We also need to ensure that tiles placed in column n match the white corridor tile to their right, and we can do this as follows:

$$[m^*]\left(pos_n \rightarrow [m]\bigvee\{col_n(t) \mid right(T) = white\}\right).$$

(Here t is the proposition letter corresponding to tile T.)

- Next, we need to ensure that all of Abelard's possible responses are encoded i‎
  the model:

$$[m^*] \left(\neg eloise \wedge pos_i \wedge col_i(t'') \wedge col_{i-1}(t') \rightarrow \bigwedge \{\langle m \rangle col_i(t) \mid C(t', t, t'')\} \right)$$

(Here $1 \leq i < n$, and, by convention, $\bigwedge \varnothing = \top$.)

That completes our description of the game tree. So let us turn to our other task‎
ensuring that Eloise indeed has a winning strategy. We will do this with the help c‎
our recursive characterization of winning strategies, thus the first step is easy; w‎
simply state that the initial position is a winning position for Eloise:

$$win.$$

Next, we spell out the recursive conditions:

$$[m^*] \left(win \rightarrow (col_1(t_{s+1}) \vee (\neg eloise \wedge \langle m \rangle \top \wedge [m]win) \vee (eloise \wedge \langle m \rangle win)) \right).$$

We are almost there – but we do not have quite enough. If a game does not te‎
minate, Abelard wins, so we need to rule out this possibility. Now, any infinit‎
branch must involve repetition of rows. Indeed, if $N = n^{s+2}$, then if a game run‎
N moves, repetition must have occurred.

Repetitions do not help Eloise: if she can win, she can do so in fewer tha‎
N moves. So we are simply going to insist that games run fewer than N move‎
– and we can do this with the help of the PDL counter defined in the proof c‎
Proposition 6.51. To use the notation of that proof, we make use of propositiona‎
variables $q_1, \ldots, q_n$, all initially set to zero, and increment the counter by 1 at ever‎
move. If the counter reaches N (that is, if all these propositional variables are tru‎
in some successor state) then the game has gone on too long and Abelard win‎
The following formula encodes this observation:

$$[m^*]((counter = N) \rightarrow [m]\neg win).$$

Let ϕ^T be the conjunction of all these formulas. We must now verify the thre‎
claims made about ϕ^T at the start of the proof.

First we need to show that if Eloise has a winning strategy, then there is a gam‎
tree such that ϕ_T is satisfiable at the root of the game tree viewed as a PDL model. ‎
Eloise has a winning strategy, then she can win in at most N moves. If $\mathfrak{M}$ is the PD‎
model corresponding to this at-most-N move strategy, then it is straightforward ‎
check that ϕ^T is satisfied at the root of $\mathfrak{M}$.

The second claim is more interesting: we need to show that if $\mathfrak{M}, w \Vdash \phi_T$, the‎
Eloise has a winning strategy in the game T – and that her winning strategy ‎
encoded in $\mathfrak{M}$. So suppose there is such a model. Imagine Eloise facing Abelar‎
across the playing board and consulting this model to choose her moves. As ϕ^T i‎
satisfied at w, *win* is satisfied at w (remember that the initial position is marked a‎

winning), hence *eloise* $\land$ $\langle m \rangle win$, the third disjunct of our recursive characterization of winning strategy, is true at w too. Eloise simply needs to pick a successor state in the model marked as winning to see which tile to place. In short, she plays the move described in the model, and continues doing so in subsequent rounds.

Though this guarantees that Eloise can keep moving to winning positions, can she actually win the game after finitely many moves? Yes! In fact she can win in at most N moves. For suppose the N-th move has just been played (that is, *counter* $= N$ has just become true). As Eloise has always been moving to winning positions, the N-th position is also winning, which means that one of the following formulas is satisfied there:

- $col_1(t_{s+1})$, or
- $\neg eloise \land \langle m \rangle \top \land [m]win$, or
- $eloise \land \langle m \rangle win$.

As the counter has reached N, $[m]\neg win$ is satisfied too. This means there are no further winning positions, and so the second and third disjuncts are false. Hence $col_1(t_{s+1})$ is satisfied: the winning tile was placed in the first column in the previous round. Thus Eloise has already won.

It remains to check that ϕ^T is polynomial in n and s. The only point that requires comment is that we can encode N. Encoding any natural number $m \geq 2$ in binary requires at most $\lg(m) + 1$ bits ($\lg$ denotes the logarithm to base 2). So encoding N takes at most $\lg(n^{s+2}) = (s+2)\lg(n) \leq (s+2)n$ bits, which is polynomial in n and s. Thus we have reduced the two person corridor tiling problem to the satisfiability problem for **PDL**, hence the latter is EXPTIME-hard. $\dashv$

As the previous proof makes clear, the EXPTIME-hardness of **PDL** largely stems from the fact that it contains a pair of modalities, one for working with a relation R_m and the other for reflexive transitive closure $(R_m)^*$; this is what enabled us to force exponentially deep models, and to code the corridor tiling problem. Now, this is not the entire story, for there *are* logics containing such modality pairs whose satisfiability problem is in PSPACE: one example is the modal logic of the frame $(\mathbb{N}, S, <)$, in a language with two diamonds $\langle s \rangle$ and $\langle < \rangle$. Here $\mathbb{N}$ is the set of natural numbers, S is the successor-of relation, and $<$ is the usual ordering of $\mathbb{N}$. And an even more expressive language – one involving the until operator U – has a PSPACE-complete satisfiability problem over $(\mathbb{N}, S, <)$; see the Notes for references.

Despite this, the following is a reliable rule of thumb: when working with a modal language containing a pair of modalities for working with a relation and its transitive closure, suspect EXPTIME-hardness. Do not begin your investigations by looking for a PSPACE-algorithm, unless you are working with a class of frames that allows little or no branching. And as this section has demonstrated, an elegant way of proving EXPTIME-hardness is via the two person corridor tiling game.

Elimination of Hintikka sets

By Theorem 6.52 there are instances of the **PDL**-satisfiability problem which will require exponentially many steps to solve. As yet we have no matching upper bound. In fact, so far the best solution to **PDL**-satisfiability we have is the following *non-deterministic* algorithm: given a formula ϕ, let Σ be the set of all ϕ' subformulas, form the collection of all Hintikka sets in Σ, non-deterministically choose a model of size at most $2^{c|\phi|}$, and check ϕ on this model. By the decidability result for **PDL** (Corollary 6.14), if ϕ is satisfiable, it is satisfiable in a model of at most this size, hence **PDL**-satisfiability is solvable in NEXPTIME.

As we will now show, the EXPTIME-hardness result of the previous section can be matched by an EXPTIME algorithm. Like the PSPACE *Witness* algorithm developed in the previous section, the EXPTIME algorithm for **PDL** is based around the idea of *Hintikka sets*. Here is how we define this notion for **PDL**.

Definition 6.53 (Hintikka set for PDL) Let Σ be a set of PDL formulas and $\neg FL(\Sigma)$ the closure under single negations of its Fischer-Ladner closure (see Definition 4.79). A *Hintikka set* over Σ is any maximal subset of $\neg FL(\Sigma)$ that satisfies the following conditions:

(i) If $\neg\phi \in \neg FL(\Sigma)$, then $\neg\phi \in H$ iff $\phi \notin H$.
(ii) If $\phi \wedge \psi \in \neg FL(\Sigma)$, then $\phi \wedge \psi \in H$ iff $\phi \in H$ and $\psi \in H$.
(iii) If $\langle\pi_1;\pi_2\rangle\phi \in \neg FL(\Sigma)$, then $\langle\pi_1;\pi_2\rangle\phi \in H$ iff $\langle\pi_1\rangle\langle\pi_2\rangle\phi \in H$.
(iv) If $\langle\pi_1 \cup \pi_2\rangle\phi \in \neg FL(\Sigma)$, then $\langle\pi_1 \cup \pi_2\rangle\phi \in H$ iff $\langle\pi_1\rangle\phi$ or $\langle\pi_2\rangle\phi \in H$.
(v) If $\langle\pi^*\rangle\phi \in \neg FL(\Sigma)$, then $\langle\pi^*\rangle\phi \in H$ iff $\phi \in H$ or $\langle\pi\rangle\langle\pi^*\rangle\phi \in H$.

We denote the set of all Hintikka sets over Σ by $Hin(\Sigma)$.

The first clause of Definition 6.53 ensures the maximality of Hintikka sets: if $H \in Hin(\Sigma)$ then there is no $H' \in Hin(\Sigma)$ such that $H \subset H'$. So, when the effect of clause (ii) is taken into account, we see that Hintikka sets are maximal subsets of $\neg FL(\Sigma)$ that contain no blatant propositional inconsistencies. Hintikka sets for PDL are a generalization of something we met in Chapter 4, namely *atoms* (see Definition 4.80). Clearly $At(\Sigma) \subseteq Hin(\Sigma)$; indeed, $At(\Sigma)$ contains precisely the **PDL**-consistent Hintikka sets.

We use Hintikka sets as follows. We define a model $\mathfrak{M}^0$ that is built out of $Hin(\Sigma)$. We then iteratively *eliminate Hintikka sets* from this model, thus forming a sequence of ever smaller models. This process is deterministic, and terminates after at most exponentially many steps yielding a model $\mathfrak{M}$. We will then show that a PDL formula ϕ is satisfiable iff it is satisfiable in $\mathfrak{M}$.

Elimination of Hintikka sets:

Base case. Let Σ be a finite set of PDL formulas, and let Π be the set of programs that occur in Σ. Define W^0 to be $Hin(\Sigma)$. For all basic programs a, and all $H, H' \in W^0$, define a binary relation Q_a^0 by $H Q_a^0 H'$ iff for every $\phi \in H'$, if $\langle a \rangle \phi \in \neg FL(\Sigma)$ then $\langle a \rangle \phi \in H$. For all other programs $\pi \in \Pi$, define Q_π^0 to be the usual inductively defined PDL relations, and let $\mathfrak{F}^0$ be $(W^0, Q_\pi^0)_{\pi \in \Pi}$. Define V^0 by $V^0(p) = \{H \in W^0 \mid p \in H\}$, for all propositional variables p. Finally, let $\mathfrak{M}^0$ be $(\mathfrak{F}^0, V^0)$.

Inductive step. Suppose that $n \geq 0$ and that $\mathfrak{F}^n = (W^n, Q_\pi^n)_{\pi \in \Pi}$ and $\mathfrak{M}^n = (\mathfrak{F}^n, V^n)$ are defined. Say that $H \in W^n$ is *demand-satisfied* iff for all $\pi \in \Pi$, and all formulas ψ, if $\langle \pi \rangle \psi \in H$ then there is an $H' \in W^n$ such that $H Q_\pi^n H'$ and $\psi \in H'$. Then define:

   (i) $W^{n+1} = \{H \in W^n \mid H \text{ is demand-satisfied}\}$.
   (ii) Q_π^{n+1} is $Q_\pi^n \cap (W^{n+1} \times W^{n+1})$, and $\mathfrak{F}^{n+1}$ is $(W^{n+1}, Q_\pi^{n+1})_{\pi \in \Pi}$.
   (iii) V^{n+1} is $V^n \restriction W^{n+1}$, and $\mathfrak{M}^{n+1}$ is $(\mathfrak{F}^{n+1}, V^{n+1})$.

As $Hin(\Sigma)$ is finite and $W^{n+1} \subseteq W^n$, then for some $m \geq 0$ this inductive process stops creating new structures. (That is, for all $j \geq m$, $\mathfrak{M}^j = \mathfrak{M}^m$.) Define $\mathfrak{F}\ (= (W, Q_\pi)_{\pi \in \Pi})$ to be $\mathfrak{F}^m$ and define $\mathfrak{M}\ (= (\mathfrak{F}, V))$ to be $\mathfrak{M}^m$.

The reader should contrast this use of Hintikka sets with the way we used them in our discussion of PSPACE. The *Witness* algorithm carefully builds sequences of ever smaller Hintikka sets using only PSPACE resources. In sharp contrast to this, the first step of *Elimination of Hintikka sets* forms all possible Hintikka sets (and there are exponentially many), and subsequent steps filter out the useless ones.

Theorem 6.54 *The satisfiability problem for **PDL** is solvable in deterministic exponential time.*

Proof. Given a PDL formula ψ, we will test for its satisfiability as follows. Letting Σ be the set of all ψ's subformulas, we form $Hin(\Sigma)$ and perform elimination of Hintikka sets. This process terminates yielding a model $\mathfrak{M} = (W, Q_\pi, V)_{\pi \in \Pi}$. We will shortly prove the following claim, for all formulas $\phi \in \Sigma$:

$$\phi \text{ is satisfiable iff } \phi \in H \text{ for some } H \in W. \tag{6.16}$$

If we can prove this claim, the theorem follows. To see this, note that the number of Hintikka sets over Σ is exponential in the size of ψ, and the process of constructing $\mathfrak{M}^{n+1}$ out of $\mathfrak{M}^n$ is a deterministic process that can be performed in time polynomial in the size of the model, and hence elimination of Hintikka sets is an EXPTIME algorithm.

So it remains to establish (6.16). For the right to left direction, we will show that if $\phi \in H$ for some $H \in W$, then $\mathfrak{M}$ itself satisfies ϕ at H. Indeed, we will show that for all $\phi \in \neg FL(\Sigma)$ and all $H \in W$, $\mathfrak{M}, H \Vdash \phi$ iff $\phi \in H$. This proof is by induction. The clause for propositional symbols is clear, and the step for boolean combinations follows using clauses (i) and (ii) in the definition of Hintikka sets. For the step involving the modal operators we need the following subclaim:

$$\text{for all } \langle \pi \rangle \chi \in \neg FL(\Sigma), \langle \pi \rangle \chi \in H \text{ iff} \atop \text{for some } H' \in W \text{ we have } Q_\pi H H' \text{ and } \chi \in H'. \tag{6.17}$$

The left to right direction of (6.17) is immediate from the construction of $\mathfrak{M}$, for at the end of the elimination process only the demand-satisfied Hintikka sets remain. The right to left direction follows by induction on the structure of π; we demonstrate the base case and the step for modalities constructed using $*$. Suppose that for some basic program a there are Hintikka sets H and H' such that $H Q_a H'$ and $\chi \in H'$. As we built the relation Q_a by a sequence of eliminations and restrictions, it follows that if $H Q_a H'$ then $H Q_a^0 H'$ – and hence it follows by definition that $\langle a \rangle \chi \in H'$. Next, suppose that for some program π^* there are Hintikka sets H and H' such that $H Q_{\pi^*} H'$ and $\chi \in H'$. But this means there is a finite sequence

$$H = H_0 Q_\pi H_1 \ldots H_{n-1} Q_\pi H_n = H'.$$

As $\chi \in H'$ it follows by the definition of Hintikka sets that $\langle \pi^* \rangle \chi \in H' = H_n$, so inductively we find that $\langle \pi \rangle \langle \pi^* \rangle \chi \in H_{n-1}$; hence, by the definition of Hintikka sets, $\langle \pi^* \rangle \chi \in H_{n-1}$. Again, by induction on π it follows that $\langle \pi \rangle \langle \pi^* \rangle \chi \in H_{n-2}$, whence $\langle \pi^* \rangle \chi \in H_{n-2}$ since this set if Fischer-Ladner closed. By repeating this argument we obtain that $\langle \pi^* \rangle \chi \in H_0 = H$. This establishes the inductive proof of (6.17), which in turn completes the inductive proof of the right to left direction of (6.16).

The fastest way to prove the left to right direction of (6.16) is to make use of ideas developed when proving the completeness of **PDL** in Chapter 4. Recall that we defined $\mathfrak{P}$, the **PDL** model over Σ, to be $(At(\Sigma), \{R_\pi^\Sigma\}_{\pi \in \Pi}, V^\Sigma)$. Here $At(\Sigma)$ is the set of all atoms over Σ, V^Σ is the natural valuation, and R_π^Σ is defined as follows: for any two atoms A and B, and any basic program a, $A R_a^\Sigma B$ holds iff $\widehat{A} \wedge \langle a \rangle \widehat{B}$ is consistent. We defined R_π for arbitrary programs by closing these basic relations under composition, union, and reflexive transitive closure in the usual way.

Now, we first claim that for all programs π, $R_\pi^\Sigma \subseteq Q_\pi^0$. To see this, first observe that as $At(\Sigma) \subseteq Hin(\Sigma)$, all atoms A and B are in W^0. So suppose $A R_a^\Sigma B$. Then, as $\widehat{A} \wedge \langle a \rangle \widehat{B}$ is consistent, by the maximality of Hintikka sets we have that for all $\phi \in B$, if $\langle a \rangle \phi \in \neg FL(\Sigma)$ then $\langle a \rangle \phi \in A$, that is, $A Q_\pi^0 B$. Thus for all atomic programs, the desired inclusion holds. But the relations R_π and Q_π^C corresponding to arbitrary programs π are generated out of R_a and Q_a^0 in the usual way, hence the inclusion follows for all programs.

The importance of this observation is the following consequence: atoms ca never be discarded in the process of elimination of Hintikka sets. This follows from the Existence Lemma for **PDL** (Lemma 4.89), which states that for all atoms A, and all formulas $\langle \pi \rangle \psi \in \neg FL(\Sigma)$, if $\langle \pi \rangle \psi \in A$, there is an atom B such that $A R_\pi^\Sigma B$ and $\psi \in B$. As all atoms belong to W^0, and as $R_\pi^\Sigma \subseteq Q_\pi^0$, it follows that every atom in W^0 is demand-satisfied. Moreover, this demand satisfiability depends only on the presence of other atoms. It follows that Hintikka elimination cannot get rid of atoms; that is, $At(\Sigma) \subseteq W$.

But now the left to right direction of (6.16) follows easily. Suppose that ϕ is satisfiable. Then ϕ is **PDL**-consistent, which means it belongs to at least one atom in Σ. This atom will survive the elimination process, and we have the result. $\dashv$

This establishes the result we wanted: an EXPTIME algorithm for deciding the satisfiability problem for PDL. One question may be bothering some readers: what is the relationship between the models $\mathfrak{M}$ and $\mathfrak{P}$ in the proof of Theorem 6.54? Let us consider the matter. In the proof, we observed that all atoms survive the Hintikka elimination process. In fact, only atoms can survive. (To see this, simply observe that if some inconsistent Hintikka set H survived the Hintikka process, then by (6.16), every formula in H would be satisfied in $\mathfrak{M}$ at H. But as $\mathfrak{M}$ is a regular model, this is impossible.) Hence $\mathfrak{M}$, like $\mathfrak{P}$, is a model built over the set of atoms. Moreover, we showed in the course of proving the previous theorem that every relation in $\mathfrak{P}$ is a subrelation of the corresponding relation in $\mathfrak{M}$. It follows that $\mathfrak{P}$ is a submodel of $\mathfrak{M}$.

Actually, we can say a little more. Recall from Exercise 4.8.4 that $\mathfrak{P}$ is isomorphic to a certain filtration. In fact, $\mathfrak{M}$ is isomorphic to a filtration over the same set of sentences. Which filtration? We leave this as an exercise for the reader; see Exercise 6.8.4.

Exercises for Section 6.8

6.8.1 Enrich the basic modal language with the global modality A. (This was defined in Section 6.5.) Show that the satisfiability problem for the enriched language over the class of all frames is EXPTIME-hard.

6.8.2 As in the previous exercise, enrich the basic modal language with the global modality A. Use elimination of Hintikka sets to show that the satisfiability problem for the enriched language over the class of all frames is solvable in EXPTIME.

6.8.3 In this exercise we investigate the complexity of deterministic PDL.

(a) Change the PDL-hardness proof so that it works for deterministic PDL. How many programs do you need? Are two programs sufficient?

(b) Encode with just one functional program that a model has an exponential deep path. Use this to describe n-corridor tiling. What can you conclude?

(c) So by now we might have a suspicion that with only one program, the satisfiability problem for deterministic PDL might be in PSPACE. But how to prove that? The best way is to find a proof in the literature which can be used almost immediately. What are the crucial features of functional PDL with one program? Think of a temporal logic which has precisely these same features. Can you interpret functional PDL into that temporal logic, using some kind of translation function? If so, what is the complexity of that function? What can you conclude?

6.8.4 Determine the exact relationship between the models $\mathfrak{P}$ and $\mathfrak{M}$ discussed following the proof of Theorem 6.54.

6.8.5 PDL has an EXPTIME-complete satisfiability problem. Suppose we add the global modality to the language. What is the complexity of the resulting satisfiability problem?

6.9 Summary of Chapter 6

▶ *Decidability and Undecidability*: A logic is called decidable if its satisfiability problem (or equivalently, its validity problem) is decidable. Otherwise it is called undecidable.

▶ *Decidability via the Finite Model Property*: While possession of the finite model property does not guarantee decidability, finite models can be used to prove decidability given some extra information about the models or the logic. The decidability of many of the more important modal logics, including **PDL**, can be established using such arguments.

▶ *Decidability via Interpretations*: Another important technique for establishing decidability is via interpretation in decidable logical theories, most notably the monadic second-order theories of countable finitely- or ω-branching trees. If a modal logic is complete with respect to a class of models that can be viewed as monadic second-order definable substructures of such a tree, its decidability follows.

▶ *Quasi-Models and Mosaics*: Even when a modal logic lacks the finite model property, it is sometimes possible to prove decidability using finite *representations* of the information contained in satisfying models. Quasi-models and mosaics are such representations.

▶ *Undecidability*: Undecidability arises easily in modal logic. Moreover, not all undecidable modal logics have the simplest degree of undecidability; many are highly undecidable.

▶ *Tiling Problems*: Tiling problems can be used to classify the difficulty of both decidable and undecidable problems. The simple geometric ideas underlying them make them a useful tool for investigating modal satisfiability problems.

▶ *The Modal Significance of NP*: Only modal logics with the polysize model property with respect to particularly simple classes of structures can be expected to have satisfiability problems in NP. Some important logics, such as the normal logics extending **S4.3**, fall into this category.

▶ *The Modal Significance of PSPACE*: Assuming that PSPACE $\neq$ NP, most modal satisfiability problems are *not* solvable in NP, but are at least PSPACE-hard. For example, every normal logic between **K** and **S4** has a PSPACE-hard satisfiability problem. Explicit PSPACE algorithms are known for some of these logics.

▶ *The Modal Significance of EXPTIME*: Modal languages containing a modality $\langle r \rangle$ and a matching reflexive transitive closure modality $\langle r^* \rangle$ often have EXPTIME-hard satisfiability problems. The two person corridor tiling game is an attractive tool for proving modal EXPTIME-hardness results, and elimination of Hintikka sets is a standard way of defining EXPTIME algorithms.

Notes

'inite models have long been used to establish decidability, both in modal logic and lsewhere. Arguments based on *finite* axiomatizability together with the f.m.p. are ⱳidely used (Theorem 6.15); this approach traces back to Harrop [213]. Also popⲗar is the use of the *strong* finite model property; our formulation (Theorem 6.7) is ⲁsed on Goldblatt's [177]. The fact that a *recursive* axiomatization together with ⲏe f.m.p. with respect to a *recursively enumerable* class of models guarantees deⲥidability (Theorem 6.13) seems to have first been made explicit in Urquhart [440]. ⲏe main point of Urquhart's article is to prove the result we presented as Exerⲥise 6.2.5: there is a normal modal logic which is recursively axiomatizable, and ⲁs the f.m.p., but is undecidable. This shows that the use of finite axiomatizations ⲛ the statement of Theorem 6.15 cannot be replaced by recursive axiomatizations, ⲛd Urquhart states Theorem 6.13 as the correct generalization. Exercise 6.2.4 ⲓ due to Hemaspaandra (*née* Spaan); see Spaan [419]. For Craig's Lemma, see ⲥraig [98].

The original proof of Rabin's Tree Theorem may be found in Rabin [368]. Rabin ⲏows that the decidability of SnS for $n > 2$ or $n - \omega$ is reducible to the decidabilⲓy of S2S, and the bulk of his paper is devoted to proving that S2S is decidable. ⲁbin's paper is demanding, and simpler proofs have subsequently been found; ⲟr an up to date survey of Rabin's Theorem and related material, see Gecseg and ⲥteinby [167] and Thomas [432]. Rabin's Theorem was applied in modal logic alⲙost immediately: Fine [127] used it to prove decidability results in second-order ⲙodal logic (that is, modal logic in which it is possible to bind propositional variⲟles), and Gabbay [145, 146, 147] applied it to a wide range of modal logics in ⲙany different languages. Gabbay, Hodkinson, and Reynolds [156] is a valuable ⲟurce on the subject.

Two kinds of variations on Rabin's Tree Theorem are relevant to our readers. ⲓrst, the *weak* monadic second-order theory of n successor functions (WSnS) ⲟnstrains the set variables to range over finite sets only. The decidability of WSnS

– which is due to Thatcher and Wright [429] and Doner [113] – is based on a clos
correspondence between formulas in WSnS and finite automata; any relation
definable in WS2S can also be defined by a tree automaton A_ϕ that encodes th
satisfying assignments to the formula in the labels on the nodes of the tree that
accepts. The MONA system [220] implements this decision procedure. Despit
the non-elementary worst-case complexity of WS2S, MONA works well in prac
tice on a large range of problems; Basin and Klarlund [29] offer empirical evidenc
and an analysis of why this is the case. At the time of writing there are no expe
imental results evaluating the performance of tools such as MONA on logics suc
as propositional dynamic logic. Muller et $al.$ [338] use reductions to WS2S t
explain why many temporal and dynamic logics are decidable in EXPTIME.

A second variation is important when working with expressive modal language
(for example, those containing the until operator U) over highly restricted classe
of models (for example, models isomorphic to the real numbers in their usual o
der). It may be necessary to appeal to stronger results about specific classes c
structures; Burgess and Gurevich [80] and Gurevich and Shelah [201] are essenti
reading here.

Prenex normal form fragments of first-order logic are defined using strings ov
$\{\exists, \exists^*, \forall, \forall^*\}$; for instance, $\exists\forall^*$ represents the class of first-order formulas i
prenex normal form where the quantifier prefix starts with an existential quant
fier and is followed by a (possibly empty) sequence of universal quantifiers. Th
decidability of prenex normal form fragments seems to have been studied at lea
since the early 1920s, which is when Skolem showed that $\forall^*\exists^*$ is undecidable. I
1928, Bernays and Schönfinkel gave a decision procedure for the satisfiability c
$\exists^*\forall\exists^*$ sentences. Gödel, Kalmár and Schütte, independently in 1931, 1933 an
1934 respectively, discovered decision procedures for the satisfiability of $\exists^*\forall^2\exists$
sentences. In 1933, Gödel showed that $\forall^3\exists^*$ sentences form a reduction class fc
satisfiability (that is, arbitrary first-order satisfiability problems can be reduced 1
such satisfiability problems). More recently, Kahr in 1962 proved the undecidabi
ity of $\forall\exists\forall$. Consult Börger et $al.$ [71] for references and an encyclopedic account c
prenex normal form fragments. For recent work on the relevance of such fragmen
to modal logic, see Hustadt [237].

That the two-variable fragment of any first-order language is decidable is re
evant to a number of modal decidability problems. The first decidability resu
for this fragment (without equality) was obtained by Scott [401]; Mortimer [33
established decidability of the two-variable fragment with equality. In contras
for $k \geq 3$, the k-variable fragment is undecidable. Consult Grädel, Kolaitis, ar
Vardi [195] for complexity results, and Grädel, Otto, and Rosen [196] for relate
results.

But perhaps the most natural way to reduce a modal logic is – to another mod
logic! Such reductions are far likelier to yield not only decidability results, b

information about complexity as well. Embeddings of temporal logic into the basic modal language were first studied by Thomason in the mid 1970s (see, for example, [437]). The approach has gained a new lease of life recently – important results on the approach can be found in Kracht and Wolter [282] and Kracht [279].

Our use of quasi-models and mosaics has its roots in the work of Zakharyaschev and others. In particular, Zakharyaschev and Alekseev [468] use such arguments to show that all finitely axiomatizable normal logics extending **K4.3** are decidable, and Wolter [457] uses them to show that all finitely axiomatizable tense logics extending $K_t 4.3$ are decidable too.

The mosaic method for proving decidability of a logic stems from Németi [340] who proved that various classes of relativized cylindric algebras have a decidable equational theory. It has since been used for a wide range of logics, often with a multi-dimensional flavor; see for instance Marx and Venema [320], Mikulás [329], Reynolds [377], Wolter and Zakharyaschev [462], Wolter [460], or the references in our Notes on the guarded fragment in Chapter 7. With hindsight, even Gödel's proof of the decidability of the satisfiability problem for the $\forall^2 \exists^*$ prenex sentences can be called a mosaic style proof as well; see the very clear exposition in the monograph [71]. Mosaics can also be used to investigate modal complexity theory; see Marx [316] for further details.

Constructing specific examples of undecidable modal logics is not trivial, and Thomason [434] contains the earliest explicit example of an undecidable normal logic in the basic modal language that we know of. Undecidable logics can be constructed in a variety of ways. Urquhart's [440] definition of Λ_U (see Exercise 6.2.5) is neat, if abstract. For undecidable logics in the basic modal language constructed by detailed simulation of a concrete model of computation (namely, Minsky machines), see Chagrov and Zakharyaschev [88, Chapter 16].

We have chosen to focus on tiling problems (or domino problems, as they are sometimes called). These were introduced in Wang [453] and have since been used in a variety of forms to prove undecidability and complexity results. Proofs that the $\mathbb{N} \times \mathbb{N}$ tiling problem is undecidable can be found in Berger [52], Robinson [389], and Lewis and Papadimitriou [301]. Two important papers on tiling are Harel [210, 211]: these demonstrate the flexibility of the method as a tool for measuring the complexity of logics. Harel uses tiling to give an intuitive account of highly undecidable (and in particular, Σ_1^1-complete) problems, and these two papers are probably the best starting point for readers interested in learning more. The logic KR used in the text to illustrate the tiling method is a notational variant of Kasper Rounds logic, which is used in computational linguistics to analyze the notion of feature structure unification. Decidability and complexity results for (various versions of) Kasper Rounds logic can be found in Kasper and Rounds [266] and Blackburn and Spaan [61]; the latter is the source for Theorems 6.31 and 6.34. A wide range of related results can be found in the literature (see for example

Harel [209], Halpern and Vardi [206], and Passy and Tinchev [358]). Even in quite modest languages, asserting something about all paths through a model can lead to extremely high complexity; for a deeper understanding of why this is so, we refer the reader to Harel [210, 211], and to Harel, Kozen and Tiuryn [212].

As to complexity-theoretic classifications of modal satisfiability and validity problems, Ladner [292] is one of the earliest analyses; this classic paper is the source of Ladner's Theorem and much else besides – it is required reading! Halpern and Moses [205] is an excellent introduction to the decidability and complexity of multi-modal languages. We strongly recommend this article to our readers, especially those who are encountering complexity theoretic ideas for the first time.

But to return to the results in this chapter, the NP-completeness of **S5** was proved in Ladner [292]. Ono and Nakamura [348] is the source of Theorem 6.38; in that paper it is also shown that the complexity of the satisfiability problems in the language with F and P with respect to the following flows of time are all NP-complete: linear transitive flows of time without endpoints, and dense linear transitive flows of time without endpoints (see Exercise 6.6.3). Hemaspaandra's Theorem, that all normal modal logics extending **S4.3** are NP-complete, may be found in Spaan [419] and Hemaspaandra [215]. As an aside, the satisfiability problem for the flow of time $(\mathbb{N}, \leq)$ in the language with just F was shown to be NP-complete by Sistla and Clarke [415]; the satisfiability problem is also shown to be NP-complete for formulas using F and the so-called nexttime operator. NP-complete modal-like logics were also investigated in the area of description logic: see below for references. Many NP-completeness results make use of Lemma 6.36, that frame membership is decidable in polynomial time for first-order definable frame classes (see in Exercise 6.6.1). This is a standard result in finite model theory, and you can find a proof in Ebbinghaus and Flum [118].

The key results on PSPACE come from Ladner [292]. Ladner first establishes the existence of PSPACE algorithms for **K**, **T**, and **S4**. His proof of the PSPACE completeness of **K** is like that given in the text, save that Ladner uses 'concrete tableaux' (that is, his algorithm specifies how to construct the required atoms) rather than 'abstract tableaux' (which factor out the required boolean reasoning). Concrete tableaux are also used by Halpern and Moses [205] to construct PSPACE algorithms for multi-modal versions of **K**, **S4** – and indeed **S5**; as they show, logics containing two **S5** modalities are PSPACE-hard. This paper gives a very clear exposition of how to use tableaux systems to establish decidability and complexity results. The abstract tableaux systems used in this chapter are based on the work of Hemaspaandra [420, 419, 215]. In the description logic community, tableaux systems are often called *constraint systems* [115]; *description* logics (also known as *concept* languages or *terminological* logics) are essentially multi-modal languages equipped with additional operators to facilitate the representation of knowledge with global constraints (the so-called TBox), or with means to reason about individuals and properties (the so-called ABox). Unlike the modal logic community, in the description logic community considerable attention has been paid to reasoning tasks other than satisfiability or validity checking, such as subsumption checking, instance checking, and reasoning in the presence of a background theory [114].

In the text (page 401) we also mentioned the fact that, over the natural numbers (with $<$ and the successor function S), the temporal logic with the until operator has a PSPACE-complete satisfiability problem; this result is due to Sistla and Clarke [415]. In the same paper, the authors also show that the satisfiability problem for $(\mathbb{N}, <)$ is PSPACE-complete for each of the following systems: F and X; U (until); U, S (since), X; and the extended temporal logic ETL due to Wolper [456].

The effect of bounding the number of proposition letters and the degree of modal formulas has been studied by Halpern [203]. In addition to the results mentioned in Exercises 6.6.5 and 6.7.7, he shows that the PSPACE-completeness results of Ladner and Halpern and Moses hold for multi-modal versions of **K**, **T**, **S4**, **S5**, even with a single proposition letter in the language. If we restrict to a finite degree, the satisfiability problem is NP-complete for all the logics considered, but **S4**; if we impose both restrictions, the complexity goes down to linear time in all cases.

The EXPTIME-hardness of **PDL** (Theorem 6.52) is due to Fischer and Ladner [135], who explicitly construct a PDL formula which simulates the actions of a linear space bounded alternating Turing machine. The (simpler) proof given in the text stems from Chlebus [93], which establishes the EXPTIME hardness of the two person corridor tiling game (via a reduction from alternating Turing machines) and uses it to provide a new proof of EXPTIME hardness for **PDL**. Another proof of this via two person corridor tiling can be found in van Emde Boas [120]; the recursive formulation of the game halting condition is due to Marx.

The existence of an EXPTIME algorithm for **PDL**, and the method of eliminating Hintikka sets, comes from Pratt [363]. Other applications of the method can be found in multi-modal logics of knowledge equipped with a common knowledge operator (see Halpern and Vardi [206], or Fagin *et al.* [125]); in computational tree logic (CTL; see Emerson [121]); in expressive description logics (see Donini *et al.* [115]); and in work on the global modality (see Marx [316] or Spaan [419]).

One important approach to the analysis of modal complexity has not been discussed in this chapter: the use of finite automata. The theory of automata has been a subject of research since the 1960s (Büchi [72], Thatcher and Wright [429], Rabin [368]). Especially relevant to temporal and dynamic logics has been a resurgence of interest in finite automata on infinite objects in the 1980s and 1990s; see Gecseg and Steinby [167], Hayashi [214], and Thomas [431, 432]. A wide variety of automata have been studied, and complexity results for their acceptance problems are known. It is often possible to analyze the complexity of modal satisfiability problems by reducing them to acceptance problems for automata. For example, general automata-theoretic techniques for reasoning about relatively simple logics using Büchi tree automata have been described by Vardi and Wolper [442].

We conclude on a more general note. In this chapter we have focused mainly on satisfiability and validity problems – what about the decidability and complexity of other reasoning tasks? For a start, the *global* satisfiability problem (whether there is a model which satisfies a formula at *all* its points) is important in many applications and quite different from the (local) satisfiability problem discussed here. The discussion of the global modality in Section 6.5 and Exercise 6.8.1 has given the reader some of the flavor of such problems; for more, see Marx [316]. Other reasoning tasks that are closely related to the *global* satisfiability problem, are often studied in the area of description logic mentioned before; see De Giacomo [106] or Areces and de Rijke [15].

Furthermore, there is a great deal of interest in building practical systems that evaluate formulas (not necessarily modal ones) in models; this field is known as *model checking*. Many interesting problems can be usefully viewed as model checking problems, and representations which enable evaluation to be performed efficiently – even when the models contain a very large number of states – have been developed. For an intuitive, modally oriented, introduction to the basic ideas, see Halpern and Vardi [207]. For further pointers to the model checking literature, see [326, 95, 96, 239].

Third, it is interesting to inquire into the decidability or otherwise of a wide range of metalogical properties of logics. One such result was mentioned in Section 3.7: Chagrova's Theorem tells us that it is undecidable whether a first-order property of frames can be defined by a modal formula. And many other questions along these lines can be raised (for example: is it decidable whether a new proof rule is admissible in a given logic?). The best sources for further information on such topics are Chagrov and Zakharyaschev [88, Chapters 16 and 17] and Kracht [279]. Another line of results that we should mention here is work on the following question: given two (finite) models $\mathfrak{M}$ and $\mathfrak{N}$, how hard is it to decide whether they are bisimilar? Ponse *et al.* [360] contains a number of valuable starting points for such questions.

7

Extended Modal Logic

As promised in the preface, this chapter is the party at the end of the book. We have chosen six of our favorite topics in extended modal logic, and we are going to tell you a little about them. There is no point in offering detailed advice here: simply read these introductory remarks and the following Chapter Guide and turn to whatever catches your fancy.

Roughly speaking, the chapter works its way from fairly concrete to more abstract. A recurrent theme is the interplay between modal and first-order ideas. We start by introducing a number of important *logical modalities* (and learn that we have actually been using logical modalities all through the book). We then examine languages containing the *since* and *until* operators, and show that first-order expressive completeness can be used to show modal deductive completeness. We then explore two contrasting strategies, namely the strategy underlying *hybrid logic* (import first-order ideas into modal logic, notably the ability to refer to worlds) and the strategy that leads to the *guarded fragment* of first-order logic (export the modal locality intuition to classical logic). Following this we discuss *multi-dimensional modal logic* (in which evaluation is performed at a sequence of states), and see that first-order logic itself can be viewed as modal logic. We conclude by proving a *Lindström Theorem* for modal logic.

Chapter guide

Section 7.1: Logical Modalities (Basic track). Logical modalities have a fixed interpretation in every model. We introduce two of the most important (the *global modality*, and the *difference operator*) and briefly discuss *Boolean Modal Logic* (a system which contains an entire *algebra of diamonds*).

Section 7.2: Since and Until (Basic track). We introduce the since and until operators (and their stronger cousins, the *Stavi connectives*), discuss the expressive completeness results they give rise to, and use expressive completeness to prove deductive completeness.

Section 7.3: Hybrid Logic (Basic track). Hybrid languages are modal languages which can refer to worlds. They do so using atomic formulas called *nominals* which are true at exactly one world in any model. We introduce the basic hybrid language and discuss its completeness theory.

Section 7.4: The Guarded Fragment (Advanced track). As is clear from the standard translation, modal operators perform a 'guarded' form of quantification across states. What happens when this idea is exported to first-order logic and generalized? This section provides some answers.

Section 7.5: Multi-Dimensional Modal Logic (Advanced track). By viewing assignments as possible worlds and quantifiers as diamonds, one can treat first-order logic itself as a modal formalism. In fact, orthodox Tarskian semantics for first-order logic provides a prime example of multi-dimensional modal logic: formulas are evaluated at a sequence of points.

Section 7.6: A Lindström Theorem for Modal Logic (Advanced track). As a famous theorem due to Lindström tells us, any logic satisfying completeness, compactness, and Löwenheim-Skolem is essentially first-order logic. Is there an analogous abstract characterization of modal logic?

7.1 Logical Modalities

Pure first-order logic has a significant expressive weakness: it is not strong enough to express the concept of equality in arbitrary structures. But because equality is such an important relation, logicians introduce a special binary relation symbol (namely $=$) and *stipulate* that it denotes the equality relation. As the interpretation of $=$ is fixed, and as the relation it denotes is so fundamental, the equality symbol is called a *logical predicate*.

Logical modalities trade on the same idea. Are there important relations which ordinary modal languages cannot express? Very well then: let us add new modalities and stipulate that they be interpreted by the relation in question. In this section we will discuss two of the most important logical modalities: the *global modality* (which is interpreted by the relation $W \times W$) and the *difference operator* (which is interpreted by $\neq$, the inequality relation). We will also make a few remarks about *Boolean Modal Logic* (BML), a system containing an entire family of logical modalities.

But before going any further, let us get one thing absolutely clear: *we have been using logical modalities all through the book.* Here is the simplest example. Suppose we are working with the basic modal language. Now, for many purpose we may be happy simply using $\Diamond$ to talk about the relation R – but sometime we may want to talk about $\breve{R}$, the converse of R, as well. Now, we know (se Exercise 2.1.2) that this cannot be done in the basic modal language, so we hav to add a new backward-looking modality as a primitive; doing so, of course, give us the basic temporal language. But note: we *do not* have to bring in the concept of time to justify this extension. If a binary relation R is important, its converse is likely to be too – so it is simply common sense to consider adding a diamond for $R^{\smile}$. In short, the 'temporal operator' P is really a logical modality.

The other important example is PDL. To motivate PDL we told a story about programs and transition systems – but a more abstract motivation is not only possible, it is more satisfying. The point is this. As soon as we fix a collection of relations R_α, regular algebra is staring us in the face: we can combine these relations using union and composition, and form transitive closures. Any model containing the initial R_α relations implicitly contains many other interesting relations as well – so it is natural to add extra modalities to deal with them explicitly, and doing so yields PDL. As this example shows, we can go way beyond the idea of adding a single new logical modality: we can add an entire *algebra of diamonds*. We will see another example of this when we discuss BML.

The global modality

Throughout the book we have emphasized the locality of modal logic, and for many purposes local languages are ideal. For example, suppose we are working with a modal language for talking about computer networks, and in this language ϕ means Server 1 is active and ψ means Server 2 is active. Then we can check whether the network makes it possible for Server 1 to be active by checking whether ϕ is satisfiable, and we can check whether it is possible for Server 2 to be inactive by testing for the satisfiability of $\neg\psi$.

But suppose we want to know if *whenever* Server 1 is active, then so is Server 2. There is no obvious way to test this. Testing for the satisfiability of $\phi \rightarrow \psi$ does *not* answer this question: if $\phi \rightarrow \psi$ is satisfiable, this only means that there is a state where either ϕ is false or ψ is true. We want to know whether *every* state that makes ϕ true is also a state that makes ψ true. This is clearly a global query. What are we to do?

Here is an elegant answer: enrich the language with the *global modality*. To keeps things simple, suppose we are working in the basic modal language over some fixed choice of proposition letters; to simplify our notation, let us call this language $ML(\Diamond)$. We will now add a second diamond, written E, and call the resulting language $ML(\Diamond, E)$. The interpretation of E is *fixed*: in any model $\mathfrak{M} = W, R, V$), E must be interpreted using the relation $W \times W$. That is:

$$\mathfrak{M}, w \Vdash E\phi \text{ iff there is a } u \in W \text{ such that } \mathfrak{M}, u \Vdash \phi.$$

Thus E scans the entire model for a state that satisfies ϕ. Its dual $A\phi := \neg E\neg\phi$ has the following interpretation:

$$\mathfrak{M}, w \Vdash A\phi \text{ iff } \mathfrak{M}, u \Vdash \phi, \text{ for all } u \in W.$$

That is, $A\phi$ asserts that ϕ holds at *all* points in the model. In effect, A brings the metatheoretic notion of global truth in a model down into the object language: for any model $\mathfrak{M}$, and any formula ϕ, we have that $\mathfrak{M} \Vdash \phi$ iff $A\phi$ is satisfiable in $\mathfrak{M}$. We will call E the *global diamond*, and A the *global box*. When it is irrelevant whether we mean E or its dual, we will simply say *global modality*.

It should now be clear how to handle the computer network problem: to test whether `Server 2` is active whenever `Server 1` is, we test the satisfiability not of $\phi \to \psi$, but of $A(\phi \to \psi)$. This query has exactly the global force required.

Well – this looks appealing. But what are the properties of this (obviously richer) new language? Maybe introducing the global modality destroys the properties that make model logic attractive in the first place! We have made an important change, and we need to take a closer look at the consequences.

Now, we could begin by discussing the sublanguage $ML(\text{E})$ – but this is not very interesting (it is easy to see that E is just an **S5** modality). Anyway (as our server example shows) the main reason for adding logical modalities is to have them available as *additional* tools. So the real question is: what does $ML(\Diamond, \text{E})$ offer that $ML(\Diamond)$ does not? The most obvious answer is *expressivity*. Let us first consider expressivity at the level of frames:

$$(R = W^2) \qquad \text{E}p \to \Diamond p,$$
$$(R \neq \varnothing) \qquad \text{E}\Diamond\top,$$
$$(\exists x \forall y \,\neg Rxy) \qquad \text{E}\Box \perp,$$
$$(\forall x \exists y \, Ryx) \qquad p \to \text{E}\Diamond p,$$
$$(|W| = 1) \qquad \text{E}p \to p,$$
$$(|W| \leq n) \qquad \bigwedge_{i=1}^{n+1} \text{E}p_i \to \bigvee_{i \neq j} \text{E}(p_i \wedge p_j),$$
$$(R \text{ is trichotomous}) \quad (p \wedge \Box q) \to A(q \vee p \vee \Diamond p),$$
$$(R^{\smile} \text{ is well-founded}) \quad A(\Box p \to p) \to p.$$

None of the frame classes listed is definable in $ML(\Diamond)$, but (as we ask the reader to check in Exercise 7.1.1) the $ML(\Diamond, \text{E})$ formulas to their right do define the corresponding property.

Where does this extra frame expressivity come from? From trivializing the notion of generated submodel (generating on $W \times W$ always yields $W \times W$) and rendering inapplicable the notion of disjoint union (for any disjoint frames $(W, R$ and (W', R'), $(W \times W) \uplus (W' \times W') \neq (W \uplus W') \times (W \uplus W')$). By insisting that E be interpreted using $W \times W$, we have trashed two of the classic modal preservation results and thereby bought ourselves more expressivity. How much more? For first-order definable frame classes, the answer is elegant:

Theorem 7.1 *A first-order definable class of frames is definable in $ML(\Diamond, \text{E})$ iff is closed under taking bounded morphic images, and reflects ultrafilter extension* his is *exactly* the Goldblatt-Thomason Theorem – minus closure under disjoint ιnions and generated subframes.

There is also a gain of expressivity at the level of models (the server example ιakes this clear, and we already know from Section 2.1 that the global modality not definable in the basic modal language). Moreover, we can measure the gain ϵing our old friends: bisimulations. It is an easy exercise to adapt the definition of ιsimulation for the basic modal language to $ML(\Diamond, E)$, and a rather more demand-ιg one to prove a bisimulation-based characterization result for the language. The ϶ader is asked to attend to these matters in Exercises 7.1.3 and 7.1.4.

What about completeness? The set of valid $ML(\Diamond, E)$ formulas can be axioma-zed as follows. Take the minimal normal logic in $\Diamond$ and E (that is, apply Defini-ɔn 4.13 to this two-diamond similarity type), and add the following axioms:

eflexivity)	$p \to Ep$,
ymmetry)	$p \to AEp$,
ransitivity)	$EEp \to Ep$,
nclusion)	$\Diamond p \to Ep$.

ote that the first three axioms are the familiar T, B, and 4 axioms (written in E ιd A rather than $\Diamond$ and $\Box$). We discussed *inclusion* in Example 1.29(iv). We will ιll this logic $\mathbf{K}_g$.

heorem 7.2 $\mathbf{K}_g$ *is strongly complete with respect to the class of all frames.*

his theorem says that to lift the minimal logic **K** (for the basic modal language) $ML(\Diamond, E)$, we need merely treat the global modality as a normal operator that ιtisfies four further axioms. In fact, we can lift *any* canonical $ML(\Diamond)$ logic in is way. If $\mathbf{K}\Gamma$ is a normal modal logic in $ML(\Diamond)$, let $\mathbf{K}_g\Gamma$ be the normal modal gic in $ML(\Diamond, E)$ obtained by treating E as a normal operator and adding the four ɔioms listed above. Then:

heorem 7.3 *Let Γ be a set of $ML(\Diamond)$ formulas, and let* F *be the class of frames* ɑt Γ *defines. If* $\mathbf{K}\Gamma$ *is canonical, then* $\mathbf{K}_g\Gamma$ *is strongly complete with respect to*

ʳoof. Let $\mathfrak{M} = (W, R_\Diamond, R_E, V)$ be the canonical model for $\mathbf{K}_g\Gamma$. Note that as $\Gamma \subseteq \mathbf{K}_g\Gamma$, we have that $(W, R_\Diamond)$ belongs to F, for $\mathbf{K}\Gamma$ is canonical. Indeed, ιy generated subframe of $(W, R_\Diamond)$ belongs to F, for validity in the basic modal ιnguage is closed under generated subframes.

Given a $\mathbf{K}_g\Gamma$-consistent set of sentences Σ, use Lindenbaum's Lemma to ex-ιnd it to a $\mathbf{K}_g$-MCS Σ^+. By the Canonical Model Theorem, $\mathfrak{M}, \Sigma^+ \Vdash \Sigma$. Now, *eflexivity*), (*symmetry*), and (*transitivity*) are canonical formulas, thus R_E is an ιuivalence relation. And although there is no guarantee that R_E is $W \times W$, this is easy to correct: let $\mathfrak{M}' = (W', R'_\diamond, R'_E, V)$ be the submodel of $\mathfrak{M}$ generated by Σ^+ using the R_E-relation. Then $R'_E = W' \times W'$, so we have the global relation we need. Furthermore, because of *inclusion*, $R_\diamond \subseteq R_E$, thus $\mathfrak{M}'$ is also a generated submodel of $\mathfrak{M}$ with respect to $R_\diamond$, hence $\mathfrak{M}', \Sigma^+ \Vdash \Sigma$. It only remains to observe that (by our initial remarks) $(W', R'_\diamond)$ is in F, hence the result follows (Theorem 7.2 is the special case in which $\Gamma = \varnothing$.)

Example 7.4 Suppose we are working with $ML(\diamond)$ over transitive frames (so the relevant logic is **K4**, which is canonical). Now, we may want to state global constraints on models, or insist that certain information holds somewhere or other, and of course we can do this if we add the global modality. But how do we obtain complete logic for transitive frames in the enriched language?

Simply enrich **K4** by treating the global modality as a normal operator and adding the (*reflexivity*), (*transitivity*), (*symmetry*), and (*inclusion*) axioms. Doing so yields K_g4, and by the theorem just proved this logic is strongly complete with respect to the class of transitive frames.

What about decidability and complexity? We briefly met the global modality in Section 6.5, and we saw that its global reach makes it possible to force the existence of gridlike models. This led to undecidability results for languages containing several diamonds, and it is not difficult to adapt these arguments to find frame classes with decidable $ML(\diamond)$ logics and undecidable $ML(\diamond, E)$ logics (we give such an example in Exercise 7.1.5). Moreover, although undecidability does not strike over the class of all frames, K_g is probably more complex than **K**, for K_g has an EXPTIME-complete satisfiability problem (the reader was asked to prove this in Exercises 6.8.1 and 6.8.2) while **K** is PSPACE-complete (see Section 6.7). On the other hand, there is a rather nice transfer result concerning the filtration method: if we can prove the decidability of an $ML(\diamond)$ logic by using filtrations to establish the strong finite frame property, then we can also do so after adding the global modality. For example, it follows that the logic K_g4 (see Example 7.4) is decidable. We will state and prove a stronger version of this result when we discuss the difference operator.

All in all, the global modality is a strikingly natural extension of modal logic and at first glance this seems surprising. How can something so obviously global blend so well with the locality of modal logic? Basically, because the enriched language still takes an *internal* perspective on relational structure. Although we now have a global operator at our disposal, we still place formulas *inside* models and evaluate them at a particular state. To put it another way, the intuition that a modal formula is an automaton scanning accessible states is remarkably robust; even if we add a special automaton programmed to regard *all* states as accessible, we retain much of the characteristic flavor of ordinary modal logic.

A lot more could be said about the global modality. For a start, it is natural when viewed from an algebraic perspective (it gives rise to *discriminator varieties*). Moreover, the global modality can be added to many richer modal systems, including **PDL** and the hybrid and multi-dimensional logics discussed later in the chapter, often without raising the computational complexity (for example **PDL** is EXPTIME-complete, and adding E does not change this). But for more information the reader will have to consult the Notes and Exercises, for it is time to discuss an even more powerful logical modality.

The difference operator

At the bottom of every toolbox lies a heavy cast-iron hammer. It is not the sort of tool we use every day – for delicate jobs it is inappropriate, and we may feel slightly embarrassed about using it at all. Still, there will always come a time when something simply will not budge, and then we find ourselves reaching for it. Think of the difference operator as that hammer.

Once again, we will start with $ML(\Diamond)$. We will add a second diamond D, the *difference operator*, and call the resulting language $ML(\Diamond, D)$. The interpretation of D is *fixed*: in any model $\mathfrak{M} = (W, R, V)$, D must be interpreted using the inequality relation $\neq$. That is:

$$\mathfrak{M}, w \Vdash D\phi \text{ iff there is a } u \neq w \text{ such that } \mathfrak{M}, u \Vdash \phi.$$

Thus the difference operator scans the entire model looking for a *different* state that satisfies ϕ. Its dual $\overline{D} := \neg D\neg\phi$ has the following interpretation

$$\mathfrak{M}, w \Vdash \overline{D}\phi \text{ iff } \mathfrak{M}, u \Vdash \phi \text{ for all } u \neq w.$$

In what follows we discuss $ML(\Diamond, D)$, but the sublanguage $ML(D)$ is quite interesting in its own right, and we ask the reader is asked to explore it in Exercise 7.1.6.

Using the difference operator, we can define the global modality: $E\phi := \phi \vee D\phi$. Thus all our earlier examples of frame classes definable in $ML(\Diamond, E)$ are definable in $ML(\Diamond, D)$ too. But $ML(\Diamond, D)$ can define even more:

irreflexivity)	$\Diamond p \to Dp,$
antisymmetry)	$(p \wedge \neg Dp) \to \Box(\Diamond p \to p),$
$\exists xy\,(x \neq y))$	$D\top,$
$|W| > n)$	$A(\bigvee_{1 \leq i \leq n} p_i) \to E\bigvee_{1 \leq i \leq n}(p_i \wedge Dp_i).$

None of these frame classes is closed under bounded morphic images hence (by Theorem 7.1) none of them is definable in $ML(\Diamond, E)$; but it is easy to see that the listed $ML(\Diamond, D)$ formulas successfully capture them. Incidentally, we have already seen that $ML(\Diamond, E)$ can define $|W| \leq n$, thus as $ML(\Diamond, D)$ can define $|W| > n$, the difference operator can count states, at least as far as *frames* are concerned; in

Exercise 7.1.7 we ask the reader to investigate whether it can count over *models* ‹ well. Furthermore, note the $p \wedge \neg Dp$ antecedent in the definition of antisymmetr This is only true when p is true at exactly one state in the model: in effect we a‹ using the power of D to force p to act as 'name' for a state; we will put this pow‹ to good use shortly.

What about completeness? The set of valid $ML(\Diamond, D)$ formulas can be axioma tized as follows. Take the minimal normal logic in $\Diamond$ and D, and add the followin axioms:

(*symmetry*) $p \to \overline{D}Dp,$

(*pseudo-transitivity*) $DDp \to (p \vee Dp),$

(*D-inclusion*) $\Diamond p \to p \vee Dp.$

We will call this logic $\mathbf{K}_d$. Now, it is not particularly difficult to prove the com pleteness of $\mathbf{K}_d$ (we ask the reader to do so in Exercise 7.1.8) – but it is hard‹ than with $\mathbf{K}_g$ (we have to do more than simply take a generated submodel) and th result does not extend to stronger logics so easily (there is no obvious analog ‹ Theorem 7.3). It is also easy to find frame incompleteness results, indeed we ca even find them in the sublanguage $ML(D)$! Things are not looking too good ....

Enter the hammer. When we discussed rules for the undefinable (Section 4.7) w learned that proof rules which rely on 'names' can lead to general frame complet‹ ness results. And as we noted above, the difference operator is powerful enoug to simulate state names, thus we can formulate the following rule of proof (th D-rule):

$$\frac{\vdash (p \wedge \neg Dp) \to \theta}{\vdash \theta} .$$

(Here p is a proposition letter that does not occur in θ. The intuitions underlyir this rule are analogous to those underlying the IRR rule discussed in Section 4. and we will leave it to the reader to verify that it preserves validity.) And now f‹ a remarkable result. The D-rule neatly meshes with our earlier work on Sahlqvi formulas to yield one of the most general completeness results known in mod logic, the D-*Sahlqvist theorem*.

Here we only formulate a version in the basic temporal language. Consider th language with operators F, P and D; let, for a set Σ of axioms in this logic, $\mathbf{K}_{td}\Sigma$ be the normal modal logic generated by the axioms of basic temporal logic, th D-axioms and D-rule given above, and the formulas in Σ.

Theorem 7.5 *Let Σ be a collection of Sahlqvist formulas in the basic tempor‹ language. Then $\mathbf{K}_{td}\Sigma$ is strongly sound and complete with respect to the class ‹ bidirectional frames defined by (the first-order frame correspondents of) the axiom in Σ.*

roof. We will prove weak completeness only. The first step of the proof is to
·ove the existence of a collection W of maximal consistent sets such that

(i) each Γ in W contains a name, that is, a formula of the form $\phi \wedge \neg D\phi$,
(ii) for each Γ in W and each formula $F\psi \in \Gamma$, there is a Δ in W such that Γ
   and Δ are in the canonical accessibility relation R_F^c for F; and, likewise,
   for the operators P and D.
(iii) for each pair of distinct points Γ and Δ in W we have $R_D^c \Gamma \Delta$.

ll of this can be proved in the style of Proposition 4.71.

It easily follows from (i) and (iii) above that R_D^c is the inequality relation on W.
ut then the model on W given by $V(p) = \{\Gamma \in W \mid p \in \Gamma\}$ is *named*; that
, for every point in the model there is a formula which is true *only* at this point,
·e Definition 4.76. However, condition (ii) allows us to prove a *Truth Lemma*
hich implies that all axioms of the logic are true throughout the model. But then
follows from Theorem 4.77 that the Sahlqvist axioms are valid on the underlying
ame as well.   ⊣

ιe pinch of Theorem 7.5 lies in the fact that the first-order frame correspondents
mentions use *inequality* for the 'relation symbol' referring to the accessibility
lation of D. This means that we can automatically axiomatize frame properties
.e irreflexivity or antisymmetry. The reader may doubt the usefulness of this:
.rely, the logic of the class of irreflexive frames is identical to the logic of the class
˙all frames? True, but this may change when we consider irreflexivity in *addition*
ith other properties. Conditions like irreflexivity, undefinable in themselves, may
·vertheless have 'side effects' so to speak. What we mean is that there are frame
asses K such that the logic of K *differs* from the logic of the irreflexive frames in
. In such cases the above theorem can be of tremendous help.

In a surprisingly large number of cases we find ourselves in the situation that
·er a certain class of frames, the difference operator is *definable* in the underlying
odal language. For example, over the class of strict linear orders, the temporal
rmula $Fp \vee Pp$ holds at a point if and only if p holds at a *different* point. In
·neral, we say that a formula $\delta(p)$ *acts as* D on a frame $\mathfrak{F}$ if $\mathfrak{F} \Vdash \delta(p) \leftrightarrow Dp$; if
$p)$ acts as the difference operator on every frame in a class K then we say that δ
·fines D *over* K.

Definability of the difference operator is of great use for axiomatizability, as the
llowing result shows. For a formula $\delta(p)$, let $\mathbf{K}_{t\delta}\Sigma$ be the 'δ'-version of $\mathbf{K}_{td}$,
at is, the logic in the language without the D-operator obtained by replacing, in
ll axioms and derivation rules of $\mathbf{K}_{td}$, every formula $D\phi$ with $\delta(\phi)$.

ιeorem 7.6 *Let Σ be a collection of Sahlqvist formulas. Then $\mathbf{K}_{t\delta}\Sigma$ is strongly*
und and complete with respect to the class of those bidirectional frames on which
* is valid and on which δ acts as the difference operator.*

In the section on multi-dimensional modal logic we will see an application of this theorem; for a proof, we refer the reader to Exercise 7.1.9. We will examine another name-driven proof rule (called PASTE) in detail when we discuss hybrid logic. First we turn to decidability issues concerning the difference operator.

$ML(\Diamond, D)$ is a strong language. As it can define the global modality, $\mathbf{K}_d$ must have an EXPTIME-hard satisfiability problem (in fact, the problem is EXPTIME complete; see Exercise 7.1.10) and it is even easier to find undecidable logics than in $ML(\Diamond, E)$. Nonetheless, decidability is often retained. In particular, if the $ML(\Diamond)$ logic of a class of frames can be proved decidable by using a filtration argument to establish the strong finite frame property, then the $ML(\Diamond, D)$ logic of that same frame class can be proved decidable in the same way. Let us prove this.

Definition 7.7 Let Λ be a logic, and let F be a class of frames for Λ. We say that Λ *admits filtrations on* F if for any model $\mathfrak{M}$ which is based on a frame in F, and for any finite subformula closed set Σ of $ML(\Diamond)$ formulas, there is a filtration $\mathfrak{M}^f$ of $\mathfrak{M}$ through Σ which is based on a frame in F.

Theorem 7.8 *Suppose that* F *is a class of frames, and that* Λ_F *(the set of all* $ML(\Diamond)$*-formulas valid on* F*) admits filtrations on* F*. Then the logic* Λ_F^d *(the set of all* $ML(\Diamond, D)$*-formulas valid on* F*) has the strong finite frame property with respect to* F*.*

Proof. Let ξ be an $ML(\Diamond, D)$-formula satisfiable in a model $\mathfrak{M} = (W, R, V)$ of which the underlying frame (W, R) is in F. We want to show that ξ is satisfiable in an F-frame of bounded size.

Let Σ be the set of subformulas of ξ. First consider the relation $\leftrightsquigarrow_\Sigma$ which holds between two points if they satisfy the same formulas in Σ. As the points of our finite model we would like to take the equivalence classes of this relation but this would not work out well (it is instructive to see how the proof of the filtration lemma fails in the inductive step of the difference operator). The key idea of the proof of the theorem is to solve this problem by splitting each equivalence class in two parts – unless the original class is a singleton. To achieve this we add a new proposition letter d to the language and we make d true at exactly one point of each equivalence class. We would then like to filtrate the new model according to the equivalence relation $\leftrightsquigarrow_{\Sigma \cup \{d\}}$.

There is still a problem, however: we can only guarantee that the underlying frame of the filtrated model is in F if we filtrate through a set of $ML(\Diamond)$ formulas. But Σ may contain formulas with occurrences of D. In order to get rid of these, we employ a little technical trick. For every formula of the form $D\psi$ in Σ, choose a distinct propositional variable q_ψ that does not occur in any formula in Σ. Let V' be the valuation that differs from V, if at all, only in that $V'(q_\psi) = \{w \mid \mathfrak{M}, w \Vdash D\psi\}$ and that $V'(d)$ is as indicated above. Let $\mathfrak{M}'$ be the model (W', R', V').

Now define the set Σ' as follows. It is not difficult to see that for every $\phi \in \Sigma$ there is a *unique $ML(\Diamond)$* formula ϕ' such that ϕ can be obtained from ϕ' by replacing in ϕ' every proposition letter q_ψ by $D\psi$. Put

$$\Sigma' = \{\phi' \mid \phi \in \Sigma\} \cup \{d, q_\psi \mid D\psi \in \Sigma\}.$$

Observe that the formulas in Σ' are D-free and that Σ' is subformula closed. The model $\mathfrak{M}'$ is (or can be seen as) an $ML(\Diamond)$-model satisfying

$$\mathfrak{M}, s \Vdash \phi \text{ iff } \mathfrak{M}', s \Vdash \phi' \tag{7.1}$$

for all formulas ϕ in Σ. Let $\leftrightsquigarrow_{\Sigma'}$ hold between two points iff they satisfy the same formulas in Σ'; it is easy to see that every $\leftrightsquigarrow_\Sigma$-equivalence class $|s|$ splits into either one or two $\leftrightsquigarrow_{\Sigma'}$-equivalence classes, depending on whether $|s|$ has one or more elements.

In any case, it follows from the assumption in the theorem that there is a filtration $\mathfrak{M}^f$ through Σ' which is based on a frame in F. Note that by definition, the points of $\mathfrak{M}^f$ are the $\leftrightsquigarrow_{\Sigma'}$-equivalence classes. We claim that this model $\mathfrak{M}^f$ satisfies the following property for all $ML(\Diamond, D)$-formulas ϕ in Σ and all states s in $\mathfrak{M}$:

$$\mathfrak{M}, s \Vdash \phi \text{ iff } \mathfrak{M}^f, |s| \Vdash \phi. \tag{7.2}$$

From this, the theorem is almost immediate.

The proof of (7.2) proceeds by a formula induction of which we omit the standard inductive steps concerning the boolean operators; the clauses for $\Diamond$ are fairly easy as well – but note that for one direction, one needs (7.1). For the case that ϕ is of the form $D\psi$ we also omit the easy right to left direction of (7.2). For the other direction, suppose that $\mathfrak{M}, s \Vdash D\psi$. Then there is a point $s' \neq s$ such that $\mathfrak{M}, s' \Vdash \psi$. If $|s|$ and $|s'|$ are distinct then we are finished, so suppose otherwise. But from $s \leftrightsquigarrow_{\Sigma'} s'$ it follows on the one hand that $\mathfrak{M}, s \Vdash d$ iff $\mathfrak{M}, s' \Vdash d$, and on the other hand, that s and s' belong to the same $\leftrightsquigarrow_\Sigma$-equivalence class. Since we chose *exactly* one point in each $\leftrightsquigarrow_\Sigma$-class to satisfy d, this means that neither s nor s' can be this special point. Hence, there must be *another* point s'' in this $\leftrightsquigarrow_\Sigma$-equivalence class which does make d true. From $s' \leftrightsquigarrow_\Sigma s''$ it follows that $\mathfrak{M}, s'' \Vdash \psi$, so by the inductive hypothesis we have that $\mathfrak{M}^f, |s''| \Vdash \psi$. But $|s''|$ is distinct from $|s|$ since d holds at s'' and not at s. This gives that $\mathfrak{M}^f, |s| \Vdash D\psi$, as required. ⊣

How does decidability follow? Any logic Λ that admits filtrations on F has the strong finite frame property with respect to F – so if F is recursive we can apply theorem 6.7 and conclude that Λ_F is decidable. But then by the result just proved, we know that Λ_F^d also has the strong finite frame property with respect to F, so we can apply the model enumeration idea underlying the proof of Theorem 6.7 to formulas of the richer languages. As D is always interpreted by the inequality relation, and as this relation is obviously computable on finite structures, the decidability of Λ_F^d follows.

A great deal more could be said about the difference operator (in particular, bisimulations are easily adapted to cope with D, and a bisimulation-based characterization result is forthcoming; see Exercises 6.8.1 and 6.8.2) but it is time to take a brief look at a system containing a whole family of logical modalities.

Boolean Modal Logic

As we have remarked, as soon as we fix a collection of relations R_α, we can form the regular algebra over this base; building an algebra of diamonds corresponding to these leads to PDL. But an even more obvious algebra demands attention: we can also form the *boolean algebra* over base relations R_α. Why not define an algebra of diamonds corresponding to 1, $-$, $\cap$, and $\cup$? Doing so leads to Boolean Modal Logic (BML).

We define the language of BML as follows. As with PDL, we fix a set of primitive relation symbols a, b, c, ..., and in addition a distinguished relation symbol 1. From these we build complex relations using the relation constructors $-$, $\cap$ and $\cup$: that is, if α and β are relation symbols, then so are $\neg\alpha$, $\alpha \cap \beta$, and $\alpha \cup \beta$. BML is the modal language containing a diamond $\langle\alpha\rangle$ for each relation symbol α. In principle we can interpret BML on any model of appropriate similarity type – that is triples $\mathfrak{M} = (W, \{R_\alpha \mid \alpha \text{ is a relation symbol}\}, V)$ – but most such models are inappropriate. We are only interested in *boolean models*, the models in which $R_1 = W \times W$, and such that, for all relation symbols α and β, $R_{-\alpha} = \overline{R_\alpha}$ (that is, $(W \times W) \setminus R_\alpha$), $R_{\alpha\cap\beta} = R_\alpha \cap R_\beta$, and $R_{\alpha\cup\beta} = R_\alpha \cup R_\beta$.

BML is an expressive language – for a start, it contains the global modality – and it may seem that we have bitten off more than we can chew. While the $\cup$ constructor is well-behaved (in particular $\mathfrak{F} \Vdash \langle\alpha \cup \beta\rangle p \leftrightarrow \langle\alpha\rangle\phi \vee \langle\beta\rangle p$ iff $R_{\alpha\cup\beta} = R_\alpha \cup R_\beta$), the $\cap$ constructor is difficult to work with. However, as we will now see, with the help of the $-$ constructor we can get an exact grip on the relations of interest.

First we define the following operator (often called *window*): for any relation symbol α:

$$\|\alpha\|\phi := [-\alpha]\neg\phi.$$

That is:

$$\mathfrak{M}, w \Vdash \|\alpha\|\phi \text{ iff } \forall u \, (\mathfrak{M}, u \Vdash \phi \Rightarrow R_\alpha wu).$$

Window is an extremely natural operator – once you have seen it, you wonder how you ever managed without it (see the Notes for various interpretations). But what concerns us here is the following result: window allows very smooth definitions of the relations we are interested in.

Proposition 7.9 *Let $\mathfrak{F}$ be a frame* $(W, \{R_\alpha \mid \alpha$ *is a relation symbol* $\})$. *Then:*

(i) $\mathfrak{F} \Vdash [-\alpha]p \leftrightarrow \|\alpha\| \neg p$ *iff* $R_{\overline{\alpha}} \subseteq \overline{R_\alpha}$,

(ii) $\mathfrak{F} \Vdash [\alpha]\neg p \leftrightarrow \| - \alpha\|p$ *iff* $\overline{R_\alpha} \subseteq R_{\overline{\alpha}}$,

(iii) $\mathfrak{F} \Vdash \|\alpha \cap \beta\|p \leftrightarrow \|\alpha\|p \wedge \|\beta\|p$ *iff* $R_{\alpha \cap \beta} = R_\alpha \cap R_\beta$.

Proof. We prove the third claim. The right to left direction is trivial. For the left to right direction, assume that $\mathfrak{F} \Vdash \|\alpha \cap \beta\|p \leftrightarrow \|\alpha\|p \wedge \|\beta\|p$. We need to show that $R_{\alpha \cap \beta} = R_\alpha \cap R_\beta$. To see that $R_{\alpha \cap \beta} \subseteq R_\alpha \cap R_\beta$, suppose that $R_{\alpha \cap \beta}wu$, and let V be any valuation on $\mathfrak{F}$ such that $V(p) = \{u\}$. Then $(\mathfrak{F}, V), w \Vdash \|\alpha \cap \beta\|p$. As $\mathfrak{F} \Vdash \|\alpha \cap \beta\|p \leftrightarrow \|\alpha\|p \wedge \|\beta\|p$ we have $(\mathfrak{F}, V), w \Vdash \|\alpha\|p \wedge \|\beta\|p$. But u is the only point satisfying p, hence $R_\alpha wu$ and $R_\beta wu$. A similar argument shows that $R_\alpha \cap R_\beta \subseteq R_{\alpha \cap \beta}$. ⊣

In a sense, the relations are divided into two kingdoms: the ordinary $[\alpha]$ modalities govern relations built with $\cup$, the window modalities $\|\alpha\|$ govern the relations built with $\cap$, and the $-$ constructor acts as a bridge between the two realms. Moreover the bridging function of $-$ also finds expression in a new rule of proof, BR. Unlike the other additional rules discussed in this book, BR is not name-driven:

$$\frac{\vdash [\alpha]p \to ([\beta]p \to [\gamma]p)}{\vdash [\alpha]p \to (\|\gamma\|\neg p \to \|\beta\|\neg p)} \tag{BR}$$

While it is possible to prove a completeness result for BML without using BR, its use leads to an elegant axiomatization, for it enables us to thread negations through the structured modalities.

A final surprise is in store. In Theorem 6.31 we showed that the fragment containing the $\cap$ constructor and the global modality was undecidable over deterministic frames. Nonetheless, the minimal logic in BML actually turns out to be *decidable*. All in all, BML is a fascinating system. For more information, see the Notes.

Exercises for Section 7.1

7.1.1 We listed numerous frame conditions definable in $ML(\diamond, \text{E})$ and $ML(\diamond, \text{D})$ which were not definable in $ML(\diamond)$. Show that these definability claims are correct.

7.1.2 Show that $ML(\diamond, \text{E})$ validity is preserved under bounded morphisms and reflects ultrafilter extensions. (That is, show the easy direction of the Goldblatt-Thomason style result for $ML(\diamond, \text{E})$ stated in Theorem 7.1.) Can you prove the (far more demanding) converse?

7.1.3 Extend the standard translation to the global modality and the difference operator. Extend the notion of bisimulation for the basic modal language to $ML(\diamond, \text{E})$ and $ML(\diamond, \text{D})$, and show that your definition leads to an invariance result.

7.1.4 Building on the previous exercise, characterize the expressivity of $ML(\Diamond, E)$ and $ML(\Diamond, D)$ over models.

7.1.5 Let **2-3** be the class of frames (W, R) such that every state has 2 R-successors, and 3 R-successors of R-successors. First show that the satisfiability problem in $ML(\Diamond)$ over **2-3** is *decidable* (note: this *can not* be proved using a filtration argument). Then show that the satisfiability problem in $ML(\Diamond, E)$ over **2-3** is undecidable. (It may be helpful to note that this exercise is related to Exercise 6.5.2.)

7.1.6 Show that a class of frames is definable in $ML(D)$ if and only if it is definable in the first-order language over $=$ (that is, the first-order language of equality). What is the complexity of the satisfiability problem for $ML(D)$?

7.1.7 Clearly we can define in $ML(\Diamond, D)$ an operator Q with the following satisfaction definition: for any model $\mathfrak{M}$, any state w in $\mathfrak{M}$, and any formula ϕ, $\mathfrak{M}, w \models Q\phi$ iff there is exactly one state u in $\mathfrak{M}$ such that $\mathfrak{M}, u \models Q\phi$. But it is also possible to define modalities $Q_2\phi$, $Q_3\phi$, $Q_4\phi$, and so on, that are satisfied when ϕ holds at precisely Q_n states ($n \geq 2$) in the model?

7.1.8 Show that $\mathbf{K}_d$ is complete with respect to the class of all frames. (No need to try anything fancy here – just fiddle with the canonical model.)

7.1.9 Prove Theorem 7.6. That is, let Σ be a collection of Sahlqvist formulas in the basic modal language. Show that $\mathbf{K}_{t\delta}\Sigma$ is strongly sound and complete with respect to the class of those frames on which Σ is valid and on which δ acts as the difference operator. (Hint: use an auxiliary logic $\mathbf{K}_{t\delta}\Sigma^+$ in the temporal language expanded with the difference operator. Simply define this logic as having *both* the D *and* the δ versions of the D-axioms and rules. Now first use Theorem 7.5 to prove that this logic is sound and strongly complete with respect to the class of Σ-frames on which δ acts as the difference operator. Then, prove that $\mathbf{K}_{t\delta}\Sigma^+$ is conservative over $\mathbf{K}_{t\delta}\Sigma$; that is, show that for every purely temporal formula ϕ, we have that ϕ belongs to $\mathbf{K}_{t\delta}\Sigma$ iff it belongs to $\mathbf{K}_{t\delta}\Sigma^+$.)

7.1.10 Use an elimination of Hintikka sets argument to show that the $\mathbf{K}_d$ satisfiability problem is solvable in EXPTIME.

7.2 Since and Until

The modal operators considered in previous chapters all have satisfaction defini tions involving only existential or only universal quantifiers. In this section w look at a popular temporal logic whose operators are based on modalities wit more complex satisfaction definitions: S (since) and U (until). The main rea son for considering these modalities is, again, to achieve an increase in expressiv power. We will first give some examples demonstrating why the increased expres sivity is useful. We will then learn that (over Dedekind complete frames) we hav actually achieved expressive *completeness*: any expression in the first-order corr spondence language (in one free variable) has an equivalent in the modal languag in S and U. Finally, we will show that this (first-order) *expressive* completene leads to (modal) *deductive* completeness.

asic definitions

he basic operators needed for temporal reasoning seem to be F and P. These
low us to say things like 'Something good will happen' and 'Something bad has
appened.'

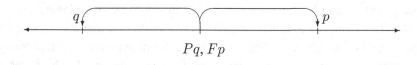

$$Pq, Fp$$

ut in several application areas this is not enough. For example, in the semantics of
oncurrent programs one often needs to be able to express properties of executions
f programs that have the general format 'Something good is going to happen, *and
ntil that time nothing bad will happen.*' Or, more concretely: p will be the case,
nd until that time q will hold:

$$U(p, q)$$

uch properties are sometimes called *guarantee properties* in the computational
terature. To state them, the binary *until* operator U can be used; its satisfaction
efinition reads:

$t \Vdash U(\phi, \psi)$ iff

there is a $v > t$ such that $v \Vdash \phi$ and for all s with $t < s < v$: $s \Vdash \psi$.

he mirror image of U is the *since* operator S:

$t \Vdash S(\phi, \psi)$ iff

there is a $v < t$ such that $v \Vdash \phi$ and for all s with $v < s < t$: $s \Vdash \psi$.

nat is the basic idea – but before going further, let us make our discussion a
tle more precise. The set of S, U-*formulas* is built up from a collection Φ of
oposition letters, the usual boolean connectives, and the *binary* operators S and
. The *mirror image* of a formula ϕ is obtained by simultaneously substituting S
r U and U for S in ϕ.
S, U-formulas are interpreted on frames of the form $\mathfrak{F} = (T, <)$, where T is a
t of time points and $<$ is a binary relation on T. U looks forwards along $<$, and
looks backwards. We use the notation $(T, <)$ for frames (rather than our usual
, R)) because here we are primarily interested in the temporal interpretation of
and U. In fact, will be working with frames $(T, <)$ such that $<$ is a Dedekind
mplete order – more on this below. To emphasize our interest in the temporal

interpretation, we will often refer to frames as *flows of time*. As usual, a valuation is a function assigning subsets of T to the proposition letters in the language.

How does the language in S and U relate to the basic temporal language? First, observe that F and P are definable in the language with S and U: we can define $F\phi := U(\phi, \top)$, $P\phi := S(\phi, \top)$, $G\phi := \neg F\neg\phi$ and $H\phi := \neg P\neg\phi$. Thus the language with S and U is at least as strong as the basic temporal language. In fact, it is strictly stronger. For a start, we saw in Exercise 2.2.4 that the basic temporal language could not define U. Moreover, as the following proposition shows, even if we restrict attention to models based on the real numbers, the basic temporal language still is not strong enough to define U.

Proposition 7.10 U *is not definable over* $(\mathbb{R}, <)$ *using F and P.*

Proof. We will give two models that agree on all formulas in the language with F and P only, but that can be distinguished using the until operator. Consider the following model $\mathfrak{M}_1$ based on the reals:

$$0 \Vdash U(p, q)$$

So, $V_1(p) = \{r \mid r \in \mathbb{Z}\}$, and $V_1(q) = \{0\} \cup \{r \mid \exists n \in \mathbb{N} (-2n - 1 < r < -2n)\} \cup \{r \mid \exists n \in \mathbb{N} (2n < r < 2n + 1)\}$.

Next, consider the model $\mathfrak{M}_2$ given by the following picture:

$$0 \nVdash U(p, q)$$

We leave it to the reader to show that the models $\mathfrak{M}_1$ and $\mathfrak{M}_2$ agree on all formulas in F and P, but that $\mathfrak{M}_1, 0 \Vdash U(p, q)$, whereas $\mathfrak{M}_2, 0 \nVdash U(p, q)$ (see Exercise 7.2.1).　　　　　　　　　　　　　　　　　　　　　　　　　　　　 ⊣

So the temporal language in S and U is expressive – but just how expressive i it? To answer such questions we need a correspondence language and a standar translation of S and U into the correspondence language. Let Φ be a collection c proposition letters, and let $\mathcal{L}^1_<(\Phi)$, or simply $\mathcal{L}^1_<$, be the first-order language wit unary predicate symbols corresponding to the proposition letters in Φ, and with and $<$ as binary relation symbols. We use $\mathcal{L}^1_<(x)$ to denote the set of $\mathcal{L}^1_<$ formula having one free variable x. Note: this is the familiar correspondence language f the basic temporal language, except that we are using $<$ rather than R as the bina relation symbol.

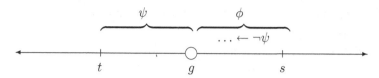

Fig. 7.1. The Stavi connectives

The *standard translation* ST_x for the until operator U is

$$ST_x(U(\phi, \psi)) \;=\; \exists z\,(x < z \wedge ST_z(\phi) \wedge \forall y\,(x < y < z \rightarrow ST_y(\psi))).$$

The standard translation of S is the mirror image of that of U. Observe that we need 3 variables to specify the translation of since and until! We only needed 2 variables to specify the translation of the basic modal operators (see Proposition 2.49).

Let K be a class of models, ML a modal or temporal language, and $\mathcal{L}$ a classical language. Then ML is *expressively complete over* K, if every $\mathcal{L}^1_<(x)$-formula has an equivalent (over K) in the modal language ML. The study of expressive completeness is an important theme in temporal logics with since and until because of the following remarkable result: the language with S and U is expressively complete over the class of all Dedekind complete flows of time (we will define this class shortly). Moreover, below we will define an even richer temporal language that is expressively complete for the class of *all* linear flows of time. In the remainder of this section we will briefly explain these expressive completeness results, and use them to obtain a deductive completeness result for since and until over well-ordered flows of time.

Further preliminaries

A flow of time is called *Dedekind complete* if every subset with an upper bound has a least upper bound. The standard examples are the reals $(\mathbb{R}, <)$ and the natural numbers $(\mathbb{N}, <)$. A flow of time is *well-ordered* if every non-empty subset has a smallest element; the canonical example here is $(\mathbb{N}, <)$.

To arrive at our goal of axiomatizing the well-ordered flows of time, we make a detour through a still richer temporal language built using the *Stavi connectives*.

Definition 7.11 (The Stavi Connectives) To introduce the Stavi connectives we need the notion of a gap. A *gap* of a frame $\mathfrak{F} = (T, <)$ is a proper subset $g \subset T$ which is downward closed (that is, $t \in g$ and $s < t$ implies $s \in g$), and which does not have a supremum. One can think of a gap as a hole in a Dedekind-complete flow of time; see Figure 7.1 Now, $U'(\phi, \psi)$ holds at a point t if the situation depicted in the above figure holds; that is, if

(i)   there are a point s and a gap g such that $t \in g$ and $s \notin g$;

(ii) ψ holds between t and g;

(iii) ϕ holds between s and g; and

(iv) $\neg\psi$ is true arbitrarily soon after g.

$S'(\phi, \psi)$ is the mirror image of $U'(\phi, \psi)$.

The above informal second-order definition (we quantify over gaps, and hence over sets) can be replaced by a first-order definition; see Exercise 7.2.2.   ⊣

Theorem 7.12 (Expressive Completeness)

(i)   *U, S are complete over Dedekind complete flows of time.*

(ii) *U, S, U', S' are complete over all linear flows of time.*

Next, we need a complete axiom system for the class of linear flows of time:

Definition 7.13 Consider the following collection of axioms:

(A1a)	$G(p \rightarrow q) \rightarrow (U(p,r) \rightarrow U(q,r))$
(A2a)	$G(p \rightarrow q) \rightarrow (U(r,p) \rightarrow U(r,q))$
(A3a)	$p \wedge U(q,r) \rightarrow U(q \wedge S(p,r), r)$
(A4a)	$U(p,q) \wedge \neg U(p,r) \rightarrow U(q \wedge \neg r, q)$
(A5a)	$U(p,q) \rightarrow U(p, q \wedge U(p,q))$
(A6a)	$U(q \wedge U(p,q), q) \rightarrow U(p,q)$
(A7a)	$U(p,q) \wedge U(r,s) \rightarrow$
	$\qquad U(p \wedge r, q \wedge s) \vee U(p \wedge s, q \wedge s) \vee U(q \wedge r, q \wedge s)$
(Aib)	the mirror images of (A1a)–(A7a)
(D)	$(F\top \rightarrow U(\top, \bot)) \wedge (P\top \rightarrow S(\top, \bot))$
(L)	$H\bot \vee PH\bot$
(W)	$Fp \rightarrow U(p, \neg p)$
(N)	$D \wedge L \wedge F\top$

⊣

Axioms (D), (L), (W), and (N) are discussed in Lemma 7.14 and Exercise 7.2.3 below. As to the other axioms, (A1a) and (A2a) can be viewed as counterparts of the familiar distribution or K axiom $\Box(p \rightarrow q) \rightarrow (\Box p \rightarrow \Box q)$. (A3a) captures the fact that U and S explore relations that are each other's converse. (A4a) and (A5a) connect the current and the future point (at which something good is going to happen) on the one hand with the points in between on the other hand. (A6a) expresses transitivity of the flow of time, and, finally, (A7a) forces the flow of time to be linearly ordered.

Lemma 7.14 *Let $\mathfrak{F}$ be a linear flow of time. Then*

(i)   *$\mathfrak{F} \models D$ iff $\mathfrak{F}$ is a discrete ordering.*

(ii) $\mathfrak{F} \models W \wedge L$ *iff $\mathfrak{F}$ is a well-ordering.*

(iii) $\mathfrak{F} \models W \wedge N$ *iff $\mathfrak{F} \cong (\mathbb{N}, <)$.*

he proof of Lemma 7.14 is left as Exercise 7.2.3.

Next, we define three axiom systems: **B**, **BW**, and **BN**. The set of axioms of **B** onsists of all classical tautologies, (A1a)–(A7a), and (A1b)–(A7b). **BW** extends with W, and **BN** extends **BW** with N. All three derivation systems have modus onens, temporal generalization, and uniform substitution as derivation rules:

(MP) If $\vdash \phi$ and $\vdash \phi \rightarrow \psi$, then $\vdash \psi$.

(TG) If $\vdash \phi$, then $\vdash G\phi$ and $\vdash H\phi$.

(UB) If $\vdash \phi$, then $\vdash [\psi/p]\phi$.

model $\mathfrak{M}$ is called an **X**-*model* if it has $\mathfrak{M} \models \phi$ for all **X**-theorems ϕ, where $\in \{B, BW, BN\}$.

For future use we state the following axiomatic completeness result:

heorem 7.15 *For all sets of S, U-formulas Σ and formulas ϕ: $\Sigma \vdash_B \phi$ iff $\Sigma \models_B$*

'e need one more preliminary result, on definable properties. By Exercise 7.2.4, ell-foundedness is a condition on linear frames which cannot be expressed in first-der logic: it involves an essential second-order quantification over all subsets of e universe. However, to arrive at our expressive completeness result we can get ' with less, namely the condition that every *first-order definable* non-empty subset ust have a smallest element; one can show that definably well-ordered models are fficiently similar to genuine well-ordered models.

The following definition and lemma capture what we need.

efinition 7.16 Let α be a first-order formula in $\mathcal{L}_<^1(x)$, $\mathfrak{M} = (T, <, V)$ a model r $\mathcal{L}_<^1$. Define X_α to be the set defined by α, that is, $X_\alpha := \{t \in T \mid \mathfrak{M} \models \alpha[t]\}$. nen, $\mathfrak{M}$ is called *definably well-ordered* if for all $\alpha(x) \in \mathcal{L}_<^1$, the set X_α has a nallest element.

Two $\mathcal{L}_<^1$-models $\mathfrak{M}_1$ and $\mathfrak{M}_2$ are called *n-equivalent*, notation $\mathfrak{M}_1 \equiv_{FOL}^n \mathfrak{M}_2$, for all first-order sentences $\alpha \in \mathcal{L}_<^1$ of quantifier depth at most n, $\mathfrak{M}_1 \models \alpha$ iff $_2 \models \alpha$. ⊣

'oviso. For the remainder of this section we will assume that our collection of oposition letters Φ is finite. This is not an essential restriction, but it simplifies me of the arguments below (see Exercise 7.2.5 for a way of circumventing the sumption).

emma 7.17 *Let $n \in \mathbb{N}$. Then every definably well-ordered linear model is n-uivalent to a fully well-ordered model.*

Proof. Let $\mathfrak{M} = (T, <, V)$ be a definably well-ordered linear model. For $a, b \in$? such that $b < a$, define $[b, a) = \{t \in T \mid b \leq t < a\}$, and $(\infty, a) = \{t \in T \mid t \cdot a\}$. Obviously, we can view such sets – with the ordering and valuation induced b $\mathfrak{M}$ – as linear $\mathcal{L}^1_<$-models in their own right. Define

$$Z := \{a \in T \mid \forall b < a \, ([b, a) \text{ has a well-ordered } n\text{-equivalent})\}.$$

By Exercise 7.2.6 there are only finitely many first-order formulas $\alpha(x, y)$ of quar tifier depth at most n, say $\alpha_1(x, y), \ldots, \alpha_m(x, y)$. Let $\beta_1(x, y), \ldots, \beta_k(x, y)$ $\{\alpha_1(x, y), \ldots, \alpha_m(x, y)\}$ be such that if $\mathfrak{M} \models \beta_i(x, y)[ab]$ then $[b, a)$ has a wel ordered n-equivalent. Then Z is defined by the formula

$$\alpha(x) := \forall y \left(y \leq x \rightarrow \bigvee_{i \leq k} \beta(x, y) \right).$$

As a consequence, $T \setminus Z$ (the complement of Z in $\mathfrak{M}$) is definable as well. W will now show that $T \setminus Z$ is empty. For, suppose otherwise. Then Z must hav a smallest element a (as $\mathfrak{M}$ is definably well-ordered). Distinguish the followir cases:

  (i) a is the first element of T,
 (ii) a has an immediate successor, and
(iii) there exists an ascending sequence $(b_\xi)_{\xi < \lambda}$, which is cofinal in $[b, a)$ an such that $b_0 = b$. (That is, $b_0 = b$, $b_i < b_j$ whenever $i < j$, and for a $c \in [b, a)$ there exists a $b_i > c$.)

It is easy to see that the first two cases lead to contradictions. As to the thi case, since a is the minimal element of $T \setminus Z$, all b_ξ are in Z. So, by definitio every interval $[b_\xi, b_{\xi+1})$ has a well-ordered n-equivalent $\mathfrak{M}_\xi$. By Exercise 7.2. the lexicographic sum $\sum_{\xi < \lambda} \mathfrak{M}_\xi$ is well-ordered and an n-equivalent to $[b, a)$. B then $a \in Z$ – a contradiction.

Therefore $T \setminus Z = \varnothing$, and hence $Z = T$, so every interval $[b, a)$ of T has a n-equivalent well-ordered model. By using Exercise 7.2.7 again, we see that Σ must have a well-ordered n-equivalent, as required.

Completeness via completeness

With the above preliminaries out of the way, we are now in a position to use tl expressive completeness result recorded in Theorem 7.12 to arrive at an axiomat completeness result for **BW** over well-ordered flows of time.

We need the following lemma:

Lemma 7.18 *Every linear **BW**-model is definably well-ordered.*

Proof. Let $\mathfrak{M}$ be a linear model satisfying all instances of the **BW**-theorems. We will prove that every non-empty $\mathcal{L}^1_<$-definable subset of T has a smallest element via detour using the Stavi connectives S' and U'.

Let X be a non-empty $\mathcal{L}^1_<$-definable subset of T. By Theorem 7.12(ii) it follows that X has a defining formula ϕ in the language with S, U, S', U'. If we can show that ϕ does in fact belong to the sublanguage with S and U, then we are done, because then we can use the validity of the axioms W and L to show that there must be a *minimal* element in X.

It suffices to show that *every* formula in the language with S, U, S', U' is equivalent to an S, U-formula over $\mathfrak{M}$. To this end we argue by induction of formulas in the richer language. The only non-trivial case is for formulas of the form $U'(\phi, \psi)$ (and their mirror images), where ϕ and ψ are already assumed to be equivalent to S, U formulas by the induction hypothesis. So assume $\mathfrak{M}, t \Vdash U'(\phi, \psi)$. Then there is a gap g after t such that (i) ψ holds everywhere between t and g, and (ii) ψ is false arbitrarily soon after g. Now (i) implies that $\mathfrak{M}, t \Vdash F\psi$, so by the validity of the W axiom in $\mathfrak{M}$ it follows that $\mathfrak{M}, t \Vdash U(\neg\psi, \psi)$, which contradicts (ii). $\dashv$

Theorem 7.19 **BW** *is (weakly) complete for the class of all well-ordered flows of time.*

Proof. Let ϕ be a **BW**-consistent formula. Construct a maximal **BW**-consistent set Δ with $\phi \in \Delta$. As **BW** extends **B**, Δ must also be **B**-consistent. By Theorem 7.15 there exists a linear model $\mathfrak{M} = (T, <, V)$ in which Δ is satisfiable. Clearly, for every S, U-formula ψ, the formula $HW(\psi) \wedge W(\psi) \wedge GW(\psi)$ is in Δ, where $W(\psi)$ is the W axiom instantiated for ψ. Thus $\mathfrak{M}$ is a **BW**-model, and hence, by Lemma 7.18 it is definably well-ordered.

Now, for the final step, let n be the quantifier rank of $ST(\phi)$. By Lemma 7.17 there is a well-ordered model $\mathfrak{M}'$ that is $(n + 1)$-equivalent to $\mathfrak{M}$. Therefore, $\mathfrak{M}' \models \exists x\, ST(\phi)(x)$, and we are done. $\dashv$

Using Theorem 7.19 it is easy to obtain a further completeness result, for the temporal logic of the natural numbers.

Theorem 7.20 **BN** *is weakly complete for* $(\omega, <)$*, the natural numbers with their standard ordering.*

The proof of Theorem 7.20 is left as Exercise 7.2.8.

Exercises for Section 7.2

2.1 Supply the missing details for the proof of Proposition 7.10.

2.2 Give a first-order definition for the Stavi connectives introduced in Definition 7.11 – you may assume that we are working on linear flows of time.

7.2.3 Prove Lemma 7.14. That is, show that D defines discrete orderings, that W ∧ L defines well-orderings, and that W ∧ N picks out the natural numbers in their usual ordering up to isomorphism.

7.2.4 Show that well-foundedness is a condition on linear frames which cannot be expressed in first-order logic.

7.2.5 Throughout this section we assumed that the collection of proposition symbols that we are working with is finite. Show that this assumption can be lifted.

7.2.6 Show that, over a finite vocabulary, there are only finitely many non-equivalent first-order formulas $\alpha(x, y)$ of quantifier depth at most n.

7.2.7 Show that the lexicographic sum of a collection of structures that are well-ordered and n-equivalent to a given structure $\mathfrak{M}$, is again well-ordered and n-equivalent to $\mathfrak{M}$.

7.2.8 Prove Theorem 7.20: show that **BN** is weakly complete for $(\omega, <)$, the natural numbers with their standard ordering.

7.3 Hybrid Logic

An oddity lurks at the heart of modal logic: although states are the cornerstone of modal semantics, they are not directly reflected in modal syntax. We evaluate formulas inside models, at some state, and use the modalities to scan accessible states. But modal syntax offers no grip on the states themselves: it does not let us name them, and it does not let us reason about state equality. Modal syntax and semantics dance to different tunes.

For many applications, this is a drawback. As we mentioned in Example 1.1, both feature and description logics can be viewed as modal logics – or at least they can up to a point. Real feature logics contain mechanisms for asserting that two sequences of transitions lead to the same state, and description logics allow us to name and reason about individuals. Such capabilities (which are crucial) take us beyond the kinds of modal language we have considered so far. Similarly, it is often important to reason about what is going on at particular times, and the temporal formalisms used in artificial intelligence usually provide expressions such as $Holds(i, \phi)$, asserting that the information ϕ holds at the time named by i, to make this possible. The modal logics considered so far contain no analogs of these important tools.

In their simplest form, *hybrid languages* are modal languages which put this right. Hybrid languages treat states as first class citizens, and they do so in a particularly simple way. The key idea is simply to *sort* the atomic formulas, and use one sort of atom – the *nominals* – to refer to states. Because this mechanism is so simple, many of the attractive properties of modal logic (such as robust decidability) are unaffected. Indeed, in certain respects hybrid logics are arguably better-behaved than their ordinary modal counterparts: their completeness theory is particularly straightforward, and they are proof theoretically natural.

In this section we examine one of the simplest hybrid languages: a two-sorted system with names for states. To build such a language, take a basic modal language (built over propositional variables p, q, r, and so on) and add a second sort of atomic formula. These new atoms are called *nominals*, and are typically written i, j and k. Both types of atom can be freely combined to form more complex formulas in the usual way. For example,

$$\Diamond(i \wedge p) \wedge \Diamond(i \wedge q) \rightarrow \Diamond(p \wedge q)$$

is a well-formed formula. And now for the key idea: insist that *each nominal be true at exactly one state in any model*. Thus a nominal names a state by being true there and nowhere else. This simple idea gives rise to richer logics. Note, for example, that the previous formula is valid: if the antecedent is satisfied at a state w, then the unique state named by i must be accessible from w, and both p and q must be true there. And note that the use of the nominal i is crucial: if we substituted the ordinary propositional variable r for i, the resulting formula could be falsified.

Actually, what we call the *basic hybrid language* offers more than this: it also enables us to build formulas of the form $@_i\phi$, where i is a nominal. The composite symbol $@_i$ is called a *satisfaction operator*, and it has the following interpretation: $@_i\phi$ is true at any state in a model if and only if ϕ is satisfied at the (unique) state named by i (so $@_i\phi$ is analogous to *Holds*(i, ϕ)). Satisfaction operators play an important role in hybrid proof theory.

Our discussion of basic hybrid logic is largely confined to a single topic: the link between frame definability and completeness. We will show that when *pure* formulas are used as axioms they always yield systems which are complete with respect to the class of frames they define. Now, a pure formula is simply a formula whose only atoms are nominals, so in effect this result tells us that frame completeness is automatic for axioms constructed solely out of names. Our discussion will center on a proof rule called PASTE which is related to the IRR rule discussed in Section 4.7 and the D-rule of Section 7.1.

The basic hybrid language

Given a basic modal language built over propositional variables $\Phi = \{p, q, r, \ldots\}$, let $\Omega = \{i, j, k, \ldots\}$ be a nonempty set disjoint from Φ. The elements of Ω are called *nominals*; they are a second sort of atomic formula which will be used to name states. We call $\Phi \cup \Omega$ the set of *atoms* and define *basic hybrid language* (over $\Phi \cup \Omega$) as follows:

$$\phi ::= i \mid p \mid \bot \mid \neg\phi \mid \phi \wedge \psi \mid \Diamond\phi \mid @_i\phi.$$

For any nominal i, the symbol $@_i$ is called a satisfaction operator. Note that, syntactically speaking, the basic hybrid language is simply a multimodal language (the modalities being $\Diamond$ and all the $@_i$), whose atomic symbols are subdivided into two sorts. If a formula contains no propositional variables (that is, if its only atoms are nominals) we call it a *pure* formula. In what follows we assume that we are working with a fixed basic hybrid language $\mathcal{L}$ in which both Φ and Ω are countably infinite.

The basic hybrid language is interpreted on models. As usual, a model $\mathfrak{M}$ is a triple (W, R, V), where (W, R) is a frame, and V is a valuation. But although the definition of a frame is unchanged, we want nominals to act as names, so we will insist that a valuation V on a frame (W, R) is a function with domain $\Phi \cup \Omega$ and range $\mathcal{P}(W)$ such that for all $i \in \Omega$, $V(i)$ is a *singleton* subset of W. That is, as usual, we place no restrictions on the interpretation of ordinary propositional variables, but we insist that a valuation makes each nominal true at a *unique* state. We call the unique state w that belongs to $V(i)$ the *denotation* of i under V. We interpret the basic hybrid language by adding the following two clauses to the satisfaction definition for the basic modal language:

$$\mathfrak{M}, w \Vdash i \quad \text{iff} \quad w \in V(i),$$
$$\mathfrak{M}, w \Vdash @_i\phi \quad \text{iff} \quad \mathfrak{M}, d \Vdash \phi \text{ where } d \text{ is the denotation of } i \text{ under } V.$$

As usual, $\mathfrak{M} \Vdash \phi$ means that ϕ is true at all states in $\mathfrak{M}$, $\mathfrak{F} \Vdash \phi$ means that ϕ is valid on the frame $\mathfrak{F}$, and $\Vdash \phi$ means that ϕ is valid on all frames.

Note that a formula of the form $@_i j$ expresses the *identity* of the states named by i and j. Further, note that a formula of the form $@_i \Diamond j$ says that the state named by i has as an R-successor the state named by j.

Although it allows us to refer to states, and talk about state equality, the basic hybrid language is very much a modal language. Nominals name, but they are simply a second sort of atomic formula. Moreover, satisfaction operators are *normal* modal operators: note that for every nominal i, $\Vdash @_i(\phi \to \psi) \to (@_i\phi \to @_i\psi)$, is valid; and if $\Vdash \phi$, then $\Vdash @_i\phi$.

Moreover, the basic hybrid language is quite a simple modal language. For example, its satisfiability problem is known to be no more complex than the satisfiability problem for the basic modal language:

Theorem 7.21 *The satisfiability problem for the basic hybrid logic is PSPACE complete.*

But in spite of its simplicity the basic hybrid language is surprisingly strong when it comes to frame definability. For a start, many properties definable in the basic modal language can be defined using pure formulas:

(*reflexivity*) $i \to \Diamond i,$

(symmetry)	$i \to \Box \Diamond i,$
(transitivity)	$\Diamond \Diamond i \to \Diamond i,$
(density)	$\Diamond i \to \Diamond \Diamond i,$
(determinism)	$\Diamond i \to \Box i.$

Moreover, pure formulas also enable us to define many properties *not* definable in the basic modal language, as the reader can easily verify:

(irreflexivity)	$i \to \neg \Diamond i,$
(asymmetry)	$i \to \neg \Diamond \Diamond i,$
(antisymmetry)	$i \to \Box (\Diamond i \to i),$
(intransitivity)	$\Diamond \Diamond i \to \neg \Diamond i,$
(universality)	$\Diamond i,$
(trichotomy)	$@_j \Diamond i \vee @_j i \vee @_i \Diamond j,$
(at most 2 states)	$@_i (\neg j \wedge \neg k) \to @_j k.$

All the frame properties defined above are first-order. This is no coincidence: all pure formulas define first-order frame conditions. This is easy to prove: there is a natural way of extending the standard translation to cover nominals and satisfaction operators which explains why (see Exercise 7.3.1).

But not only do pure formulas define first-order properties, when used as axioms they are automatically *complete* with respect to the class of frames they define. More precisely, there is a proof system called $\mathbf{K}_h + \text{RULES}$ such that for any set of pure formulas Π:

> If **P** is the normal hybrid logic (which we will shortly define) obtained by adding the formulas in Π as axioms to $\mathbf{K}_h + \text{RULES}$, then **P** is complete with respect to the class of frames defined by P.

The rest of the section is devoted to proving this, but before diving into the technicalities it is worth noting that the result hinges on a rather simple observation. Let us say that a model (W, R, V) is *named* if every state in the model is the denotation of some nominal (that is, for all $w \in W$ there is some nominal i such that $V(i) = \{w\}$). Furthermore, if ϕ is a pure formula, we say that ψ is a *pure instance* of ϕ if ψ is obtained from ϕ by uniformly substituting nominals for nominals. Then we have:

Lemma 7.22 *Let $\mathfrak{M} = (\mathfrak{F}, V)$ be a named model and ϕ a pure formula. Suppose that for all pure instances ψ of ϕ, $\mathfrak{M} \Vdash \psi$. Then $\mathfrak{F} \Vdash \phi$.*

Proof. Exercise 7.3.3. ⊣

That is, for named models and pure formulas the gap between truth in a model and validity in a frame is non-existent. So if we had a way of building named models, we would not need to appeal to relatively complex syntactic criteria (such as being a Sahlqvist formula) to obtain general completeness results: *any* pure formula would give rise to a strongly complete logic for the class of frames it defined. In essence, the work that follows can be summed as follows: we are going to isolate the logic $\mathbf{K}_h +$ RULES and show that we can build named models from its MCSs and prove an Existence Lemma. Once this is done, a wide range of frame completeness results will be immediate by an appeal to Lemma 7.22.

Pure extensions of $\mathbf{K}_h +$ RULES

Let us first say what a normal hybrid logic is:

Definition 7.23 A set of formulas Λ in the basic hybrid language is a *normal hybrid logic* if it contains all tautologies, $\Box(p \to q) \to (\Box p \to \Box q)$, $\Diamond p \leftrightarrow \neg\Box\neg p$, the axioms listed below, and it is closed under the following rules of proof: modus ponens, generalization, $@_i$-*generalization* (if ϕ is provable then so is $@_i\phi$, for any nominal i) and *sorted substitution* (if $\phi \in \Lambda$, and θ results from ϕ by uniformly replacing *proposition letters by arbitrary formulas*, and *nominals by nominals*, then $\theta \in \Lambda$). We call the smallest normal hybrid logic $\mathbf{K}_h$. ⊣

The motivation for the sorted substitution rule should be clear: while propositional variables are placeholder for arbitrary information, nominals are names, and substitution must respect the distinction.

The axioms needed to complete our definition of $\mathbf{K}_h$ fall into three groups. The first identifies the basic logic of satisfaction operators:

(K$_@$) $@_i(p \to q) \to (@_i p \to @_i q)$,
(*self-dual*) $@_i p \leftrightarrow \neg@_i\neg p$,
(*introduction*) $i \wedge p \to @_i p$.

As satisfaction operators are normal modal operators, the inclusion of $K_@$ should come as no surprise. As for *self-dual*, note that self-dual modalities are those whose transition relation is a *function*: given the jump-to-the-named-state interpretation of satisfaction operators, this is exactly the axiom we would expect. *Introduction* tells us how to place information under the scope of satisfaction operators. Actually it also tells us how to get hold of such information, for if we replace p by $\neg p$, contrapose, and make use of *self-dual*, we obtain $(i \wedge @_i p) \to p$; we call this the *elimination* formula.

The second group is a modal theory of naming (or to put it another way, a modal theory of state equality):

(*ref*) $@_i i$,
(*sym*) $@_i j \leftrightarrow @_j i$,

(*nom*) $@_i j \land @_j p \to @_i p,$

(*agree*) $@_j @_i p \leftrightarrow @_i p.$

Note that the transitivity of naming follows from the *nom* axiom; for example, substituting the nominal k for the propositional variable p yields $@_i j \land @_j k \to @_i k$. The final axiom pins down the interaction between $@$ and $\Diamond$:

(*back*) $\Diamond @_i p \to @_i p.$

Note that $\Diamond i \land @_i p \to \Diamond p$ is another valid $@$-$\Diamond$ interaction principle; it is called *bridge* and we will use it when we prove the Existence Lemma. However *bridge* is provable in $\mathbf{K}_h$ as the reader is asked to show in Exercise 7.3.4.

The soundness of these axioms is clear – but what about completeness? Let us say that a $\mathbf{K}_h$-MCS is *named* if and only if it contains a nominal, and call any nominal belonging to a $\mathbf{K}_h$-MCS a *name* for that MCS. Now, $\mathbf{K}_h$ is strong enough to prove a lemma which is fundamental to our later work: hidden inside any $\mathbf{K}_h$-MCS are a collection of named MCSs with a number of desirable properties:

Lemma 7.24 *Let Γ be a $\mathbf{K}_h$-MCS. For every nominal i, let Δ_i be $\{\phi \mid @_i \phi \in \Gamma\}$. Then:*

  (i) *For every nominal i, Δ_i is a $\mathbf{K}_h$-MCS that contains i.*

  (ii) *For all nominals i and j, if $i \in \Delta_j$, then $\Delta_j = \Delta_i$.*

  (iii) *For all nominals i and j, $@_i \phi \in \Delta_j$ iff $@_i \phi \in \Gamma$.*

  (iv) *If k is a name for Γ, then $\Gamma = \Delta_k$.*

Proof. (i) First, for every nominal i we have the *ref* axiom $@_i i$, hence $i \in \Delta_i$. Next, Δ_i is consistent. For assume for the sake of a contradiction that it is not. Then there are $\delta_1, ..., \delta_n \in \Delta_i$ such that $\vdash \neg(\delta_1 \land \cdots \land \delta_n)$. By $@_i$-necessitation, $\vdash @_i \neg(\delta_1 \land \cdots \land \delta_n)$, hence $@_i \neg(\delta_1 \land \cdots \land \delta_n)$ is in Γ, and thus by *self-dual* $\neg @_i(\delta_1 \land \cdots \land \delta_n)$ is in Γ too. On the other hand, as $\delta_1, ..., \delta_n \subset \Delta_i$, we have $@_i \delta_1, ..., @_i \delta_n \in \Gamma$. As $@_i$ is a normal modality, $@_i(\delta_1 \land \cdots \land \delta_n) \in \Gamma$ as well, contradicting the consistency of Γ. So Δ_i is consistent.

Is Δ_i maximal? Assume it is not. Then there is a formula χ such that neither χ nor $\neg\chi$ is in Δ_i. But then both $\neg @_i \chi$ and $\neg @_i \neg\chi$ belong to Γ, and this is impossible: if $\neg @_i \chi \in \Gamma$, then by self-duality $@_i \neg\chi \in \Gamma$ as well. We conclude that Δ_i is a $\mathbf{K}_h$-MCS named by i.

(ii) Suppose $i \in \Delta_j$; we will show that $\Delta_j = \Delta_i$. As $i \in \Delta_j$, $@_j i \in \Gamma$. Hence, by *sym*, $@_i j \in \Gamma$ too. But now the result is more-or-less immediate. First, $\Delta_j \subseteq \Delta_i$. For if $\phi \in \Delta_j$, then $@_j \phi \in \Gamma$. Hence, as $@_i j \in \Gamma$, it follows by *nom* that $@_i \phi \in \Gamma$, and hence that $\phi \in \Delta_i$ as required. A similar *nom*-based argument shows that $\Delta_i \subseteq \Delta_j$.

(iii) By definition $@_i \phi \in \Delta_j$ iff $@_j @_i \phi \in \Gamma$. By *agree*, $@_j @_i \phi \in \Gamma$ iff

$@_i\phi \in \Gamma$. (We call this the $@$-*agreement property*; it plays an important role in the completeness proof.)

(iv) Suppose Γ is named by k. Let $\phi \in \Gamma$. Then as $k \in \Gamma$, by *introduction* $@_k\phi \in \Gamma$, and hence $\phi \in \Delta_k$. Conversely, if $\phi \in \Delta_k$, then $@_k\phi \in \Gamma$. Hence, as $k \in \Gamma$, by *elimination* we have $\phi \in \Gamma$.   ⊣

In what follows, if Γ is a $\mathbf{K}_h$-MCS and i is a nominal, then we will call $\{\phi \mid @_i\phi \in \Gamma\}$ a *named set yielded by* Γ.

We have reached an important crossroad. It is now reasonably straightforward to prove that $\mathbf{K}_h$ is the minimal hybrid logic. We would do so as follows. Given a $\mathbf{K}_h$-consistent set of sentences Σ, use the ordinary Lindenbaum's Lemma to expand it to a $\mathbf{K}_h$-MCS Σ^+, and build a model by taking the submodel of the ordinary canonical model generated by $\Sigma^+ \cup \{\Delta_i \mid \Delta_i$ is a named set yielded by $\Sigma^+\}$. The reader is asked to do this in Exercise 7.3.5.

But we have a more ambitious goal in mind: we do not want to build just any model, we want a named model. This will enable us to apply Lemma 7.22 and prove the completeness of pure axiomatic extensions. However we face two problems. The first is this. Given a $\mathbf{K}_h$-consistent set of formula, we can certainly expand it to an MCS using Lindenbaum's Lemma – but nothing guarantees that this MCS will be named. The second problem is much deeper. Suppose we overcame the first problem and learned how to expand any consistent set of sentences Σ to a named MCS Σ^+. Now, as we want to build a named model, this pretty much dictates that only the named MCSs yielded by Σ^+ should be used in the model construction. And now for the tough part: nothing we have seen so far guarantees that there are enough MCSs here to support an Existence Lemma. Incidentally, note that the completeness-via-generation method sketched in the previous paragraph does not face this problem: generation automatically gives us all successor MCSs, so we can make use of the ordinary modal Existence Lemma. Unfortunately, not all these successor MCSs need be named, so the generation method will not help with the stronger result we have in mind.

But these difficulties are similar to those we faced when discussing rules for the undefinable, and this suggests a solution. In Section 4.7 we simulated names using tense operators, and used the forwards-and-backwards interplay of F and P to create a coherent network of named MCSs which supported a suitable Existence Lemma. Moreover, simulated names were used to define the D-rule mentioned in Section 7.1. But nominals are genuine names, and satisfaction operators are an excellent way of enforcing coherence – surely it must be possible to define analogous proof rules for the basic hybrid language? Indeed it is:

$$(\text{NAME}) \quad \frac{\vdash j \to \theta}{\vdash \theta} \qquad\qquad (\text{PASTE}) \quad \frac{\vdash @_i\Diamond j \wedge @_j\phi \to \theta}{\vdash @_i\Diamond\phi \to \theta}$$

In both rules, j is a nominal distinct from i that does not occur in ϕ or θ. The NAME rule is going to solve our first problem, the PASTE rule our second. These rules are clearly close cousins of the IRR rule and the D-rule, but let us defer further discussion till the end of the section, and put them to work right away.

Let $\mathbf{K}_h$ + RULES be the logic obtained by adding the NAME and PASTE rules to $\mathbf{K}_h$. We say that a $\mathbf{K}_h$ + RULES-MCS Γ is *pasted* iff $@_i\Diamond\phi \in \Gamma$ implies that for some nominal j, $@_i\Diamond j \wedge @_j\phi \in \Gamma$. And now for the key observation: our new rules guarantee we can extend any $\mathbf{K}_h$ + RULES-consistent set of sentences to a *named and pasted* $\mathbf{K}_h$ + RULES-MCS, provided we enrich the language with new nominals:

Lemma 7.25 (Extended Lindenbaum Lemma) *Let Ω' be a (countably) infinite collection of nominals disjoint from Ω, and let $\mathcal{L}'$ be the language obtained by adding these new nominals to $\mathcal{L}$. Then every $\mathbf{K}_h$ + RULES-consistent set of formulas in language $\mathcal{L}$ can be extended to a named and pasted $\mathbf{K}_h$ + RULES-MCS in language $\mathcal{L}'$.*

Proof. Enumerate Ω'. Given a consistent set of $\mathcal{L}$-formulas Σ, define Σ_k to be $\Sigma \cup \{k\}$, where k is the first new nominal in our enumeration. Σ_k is consistent. For suppose not. Then for some conjunction of formulas θ from Σ, $\vdash k \rightarrow \neg\theta$. But as k is a new nominal, it does not occur in θ; hence, by the NAME rule, $\vdash \neg\theta$. But this contradicts the consistency of Σ, so Σ_k must be consistent after all.

We now paste. Enumerate all the formulas of $\mathcal{L}'$, define Σ^0 to be Σ_k, and suppose we have defined Σ^m, where $m \geq 0$. Let ϕ_{m+1} be the $(m + 1)$-th formula in our enumeration of $\mathcal{L}'$. We define Σ^{m+1} as follows. If $\Sigma^{m+1} \cup \{\phi_{m+1}\}$ is inconsistent, then $\Sigma^{m+1} = \Sigma^m$. Otherwise:

(i) $\Sigma^{m+1} = \Sigma^m \cup \{\phi_{m+1}\}$ if ϕ_{m+1} is not of the form $@_i\Diamond\phi$. (Here i can be any nominal.)

(ii) $\Sigma^{m+1} = \Sigma^m \cup \{\phi_{m+1}\} \cup \{@_i\Diamond j \wedge @_j\phi\}$, if ϕ_{m+1} is of the form $@_i\Diamond\phi$. (Here j is the first nominal in the new nominal enumeration that does not occur in Σ^m or $@_i\Diamond\phi$.)

Let $\Sigma^+ = \bigcup_{n\geq 0} \Sigma^n$. Clearly this set is named (by k), maximal, and pasted. Furthermore, it is consistent, for the only non-trivial aspects of the expansion is that defined by the second item, and the consistency of this step is precisely what the PASTE rule guarantees. Note the similarity of this argument to the standard completeness proof for first-order logic: in essence, PASTE gives us the deductive power required to use nominals as Henkin constants. ⊣

And now we can define the models we need. We are basically going to use the named sets examined in Lemma 7.24, but with one small but crucial change: instead of starting with an arbitrary $\mathbf{K}_h$-MCS, we will insist on using the named sets yielded by a *named and pasted* $\mathbf{K}_h$ + RULES-MCS.

Definition 7.26 Let Γ be a named and pasted $\mathbf{K}_h$ + RULES-MCS. The *named model yielded by* Γ, is $\mathfrak{M}^\Gamma = (W^\Gamma, R^\Gamma, V^\Gamma)$. Here W^Γ is the set of all named sets yielded by Γ, R is the restriction to W^Γ of the usual canonical relation between MCSs (so $R^\Gamma uv$ iff for all formulas ϕ, $\phi \in v$ implies $\Diamond\phi \in u$) and V^Γ is the usual canonical valuation (so for any atom a, $V^\Gamma(a) = \{w \in W^\Gamma \mid a \in w\}$).   ⊣

Note that $\mathfrak{M}^\Gamma$ really is a *model*: by items (i) and (ii) of Lemma 7.24, V^Γ assigns every nominal a *singleton* subset of W^Γ. And, because we insisted that Γ be named and pasted, we can prove the Existence Lemma we require:

Lemma 7.27 (Existence Lemma) *Let* Γ *be a named and pasted* $\mathbf{K}_h$ + RULES-MCS*, and let* $\mathfrak{M} = (W, R, V)$ *be the named model yielded by* Γ. *Suppose* $u \in W$ *and* $\Diamond\phi \in u$. *Then there is a* $v \in W$ *such that* Ruv *and* $\phi \in v$.

Proof. As $u \in W$, for some nominal i we have that $u = \Delta_i$. Hence as $\Diamond\phi \in u$, $@_i\Diamond\phi \in \Gamma$. But Γ is pasted so for some nominal j, $@_i\Diamond j \wedge @_j\phi \in \Gamma$, and so $\Diamond j \in \Delta_i$ and $\phi \in \Delta_j$. If we could show that $R\Delta_i\Delta_j$, then Δ_j would be a suitable choice of v. So suppose $\psi \in \Delta_j$. This means that $@_j\psi \in \Gamma$. By @-agreement (item (iii) of Lemma 7.24) $@_j\psi \in \Delta_i$. But $\Diamond j \in \Delta_i$. Hence, by *bridge*, $\Diamond\psi \in \Delta_i$ as required.   ⊣

In short, we have successfully blended the first-order idea of Henkin constants with the modal idea of canonical models, and it is plain sailing all the way to the desired completeness result.

Lemma 7.28 (Truth Lemma) *Let* $\mathfrak{M} = (W, R, V)$ *be the named model yielded by a named and pasted* $\mathbf{K}_h$+RULES-MCS Γ, *and let* $u \in W$. *Then, for all formulas* ϕ, $\phi \in u$ *iff* $\mathfrak{M}, u \Vdash \phi$.

Proof. Induction on the structure of ϕ. The atomic, boolean, and modal cases are obvious (we use the Existence Lemma just proved for the modalities). What about the satisfaction operators? Suppose $\mathfrak{M}, u \Vdash @_i\psi$. This happens iff $\mathfrak{M}, \Delta_i \Vdash \psi$ (for by items (i) and (ii) of Lemma 7.24, Δ_i is the only MCS containing i, and hence, by the the atomic case of the present lemma, the only state in $\mathfrak{M}$ where i is true) iff $\psi \in \Delta_i$ (inductive hypothesis) iff $@_i\psi \in \Delta_i$ (using the fact that $i \in \Delta_i$ together with *introduction* for the left to right direction and *elimination* for the right to left direction) iff $@_i\psi \in u$ (@-agreement).

Theorem 7.29 (Completeness) *Every* $\mathbf{K}_h$ + RULES-*consistent set of formulas i language* $\mathcal{L}$ *is satisfiable in a countable named model. Moreover, if* Π *is a s* *of pure formulas (in $\mathcal{L}$), and* **P** *is the normal hybrid logic obtained by adding all the formulas in Π as extra axioms to* $\mathbf{K}_h$ + RULES, *then every* **P**-*consistent set of sentences is satisfiable in a countable named model based on a frame which validates every formula in Π.*

Proof. For the first claim, given a $\mathbf{K}_h$ + RULES-consistent set of formulas Σ, use the Extended Lindenbaum Lemma to expand it to a named and pasted set Σ^+ in a countable language $\mathcal{L}'$. Let $\mathfrak{M} = (W, R, V)$ be the named model yielded by Σ^+. By item (iv) of Lemma 7.24, because Σ^+ is named, $\Sigma^+ \in W$. By the Truth Lemma, $\mathfrak{M}, \Sigma^+ \Vdash \Sigma$. The model is countable because each state is named by some $\mathcal{L}'$ nominal, and there are only countably many of these.

For the 'moreover' claim, given a **P**-consistent set of formulas Ξ, use the Extended Lindenbaum Lemma to expand it to a named pasted **P**-MCS Ξ^+. The named model $\mathfrak{M}^\Xi$ that Ξ^+ gives rise to will satisfy Ξ at Ξ^+; but in addition, as every formula in Π belongs to every **P**-MCS, we have that $\mathfrak{M}^\Xi \Vdash \Pi$. Hence, by Lemma 7.22, the frame underlying $\mathfrak{M}^\Xi$ validates Π. ⊣

Example 7.30 We know that $i \rightarrow \neg\Diamond i$ defines irreflexivity and $\Diamond\Diamond i \rightarrow \Diamond i$ defines transitivity, hence adding these formulas as axioms to $\mathbf{K}_h$ + RULES yields a logic (let us call it **I4**) which is complete with respect to the class of strict preorders. Hence $\Diamond\Diamond p \rightarrow \Diamond p$, the ordinary modal transitivity axiom, must be **I4**-provable. Furthermore, as $i \rightarrow \neg\Diamond\Diamond i$ is valid on any asymmetric frame, and $i \rightarrow \Box(\Diamond i \rightarrow i)$ is valid on any antisymmetric frame, these must be **I4**-provable too. The reader is asked to supply **I4**-proofs in Exercise 7.3.6. ⊣

The PASTE rule has played a pivotal role in our work; is there anything we can say about it apart from 'Hey, it works!'? There is. As we will now see, PASTE is actually a lightly-disguised sequent rule.

A *sequent* is an expression of the form $\Gamma \longrightarrow \Theta$, where Γ and Θ are multisets of formulas (that is, Γ and Θ may contain multiple occurrences of the same formula). Note that the sequent arrow $\longrightarrow$ is longer than the material implication arrow $\rightarrow$. Sequents can be read as follows: whenever all the formulas in Γ are true at some state in a model, at least one formula in Θ is true at that state too. A *sequent rule* takes a sequent as input, and returns another sequent as output.

Now, here is PASTE as we stated it above:

$$\frac{\vdash @_i\Diamond j \wedge @_j\phi \rightarrow \theta}{\vdash @_i\Diamond\phi \rightarrow \theta} \ .$$

Let us get rid of the $\vdash$ symbols and replace the implications by sequent arrows:

$$\frac{@_i\Diamond j \wedge @_j\phi \longrightarrow \theta}{@_i\Diamond\phi \longrightarrow \theta} \ .$$

Splitting the formula in the top line into two simpler formulas yields:

$$\frac{@_i \Diamond j, @_j \phi \longrightarrow \theta}{@_i \Diamond \phi \longrightarrow \theta}.$$

This rule works in arbitrary deductive contexts, so let us add a left-hand multiset Γ, and turn θ into a right-hand multiset Θ, thus obtaining:

$$\frac{@_i \Diamond j, @_j \phi, \Gamma \longrightarrow \Theta}{@_i \Diamond \phi, \Gamma \longrightarrow \Theta}.$$

But this is just a sequent rule, and a useful one at that. Let us read it from bottom to top: to prove Θ given the information $@_i \Diamond \phi$ and Γ (that is, the bottom line), introduce a brand new nominal j and try to prove Θ from $@_i \Diamond j$, $@_j \phi$ and Γ (that is, the top line). That is, we should search for a proof by decomposing the formula $@_i \Diamond \phi$ into a near-atomic formula $@_i \Diamond j$ and simpler formula $@_j \phi$. In fact, this decomposition is the key idea needed to define sequent calculi, tableaux, and natural deduction systems for hybrid logics, and several systems which work this way have been developed (see the Notes for details). In effect, such systems discard $\mathbf{K}_h$ from $\mathbf{K}_h$ + RULES (after all, why bother keeping the clumsy Hilbert-style part?) and strengthen the RULES component so it can assume full deductive responsibility.

To conclude, a general remark. As should now be clear (especially if you have already done Exercises 7.3.1, 7.3.2, and 7.3.3), the basic hybrid language is a genuine hybrid between first-order and modal logic: it makes available a number of key first-order capabilities (such as names for states and state-equality assertions) in a decidable (indeed, PSPACE-complete) propositional modal logic. But now that we are used to viewing names as formulas, it is easy to go even further. For example, instead of thinking of nominals as names, we could think of them as variables over states and bind them with quantifiers. For example, we could allow ourselves to form expressions such as

$$\exists x \, (x \wedge \Diamond \exists y \, (y \wedge \phi \wedge @_x \Box (\Diamond y \rightarrow \psi))).$$

This expression captures the effect of the until operator: it says $U(\phi, \psi)$. Note that in this example the $\exists$ quantifier is only used to bind nominals to the *current* state. This is such an important operation that a special notation, $\downarrow$, has been introduced for it. Using this notation the definition of $U(\phi, \psi)$ can be written as

$$\downarrow x \, (x \wedge \Diamond \downarrow y \, (y \wedge \phi \wedge @_x \Box (\Diamond y \rightarrow \psi))).$$

It turns out that when the basic hybrid language is enriched only with $\downarrow$ (that is not with the full power of $\exists$) then the resulting language picks out *exactly* the fragment of the first-order correspondence language that is invariant under generated submodels. See the Notes for more details.

Exercises for Section 7.3

7.3.1 Extend the standard translation to the basic hybrid language by adding clauses for nominals and satisfaction operators. Use your translation to show that all classes of frames defined by pure formulas are first-order definable. (Hint: translate nominals to free first-order variables.)

7.3.2 For any $n \geq 1$, let $R^n xy$ be the first-order formula $\exists z_1 \cdots \exists z_n \, (Rxz_1 \wedge Rz_1z_2 \wedge \cdots \wedge Rz_n y)$. Let ψ be a first-order formula that is a boolean combination of formulas of the form $R^n xy$, Rxy, and $x = y$. Show that the class of frames defined by the universal closure of ψ is definable in the basic hybrid language. (Hint: look at the way we defined trichotomy.)

7.3.3 Prove Lemma 7.22. That is, if $\mathfrak{M} = (\mathfrak{F}, V)$ is a named model and ϕ is a pure formula and for all pure instances ψ of ϕ we have that $\mathfrak{M} \Vdash \psi$, then $\mathfrak{F} \Vdash \phi$.

7.3.4 Show that $\Diamond i \wedge @_i p \to \Diamond p$, the *bridge* formula, is provable in $\mathbf{K}_h$. (Hint: prove the contraposed form $\Diamond i \wedge \Box p \to @_i p$ with the help of $\Diamond q \wedge \Box p \to \Diamond(q \wedge p)$, *introduction*, and *back*.)

7.3.5 Prove that $\mathbf{K}_h$ is the minimal hybrid logic by fleshing out the completeness-via-generation argument sketched in the text.

7.3.6 Find **I4**-proofs of $\Diamond\Diamond p \to \Diamond p$, $i \to \neg\Diamond\Diamond i$, and $i \to \Box(\Diamond i \to i)$. (The logic **I4** was introduced in Example 7.30.)

7.3.7 The PASTE rule makes crucial use of @-operators. Prove an analog of Theorem 7.29 for the @-free sublanguage of the basic hybrid language. (Hint: you need to simulate the satisfaction operators using the modalities. So for all $n, m \geq 0$, add the axiom $\Diamond^n(i \wedge p) \to \Box^m(i \to p)$. Furthermore, let $\Diamond_i\phi$ be shorthand for $\Diamond(i \wedge \phi)$, and add all rules of the form

$$\frac{\vdash \Diamond_k \cdots \Diamond_i \Diamond_j \phi \to \theta}{\vdash \Diamond_k \cdots \Diamond_i \Diamond \phi \to \theta} \; .$$

Here j is a nominal distinct from $k, \cdots, i$ that does not occur in ϕ or θ.)

7.3.8 Let **I4D** be the normal hybrid logic obtained by adding the axiom $\Diamond(i \vee \neg i)$ to **I4**. Clearly **I4D** lacks the finite frame property. Show that it possesses the finite *model* property (and hence that Theorem 3.28 fails for hybrid languages). Exploit this by proving the decidability of **I4D** using a filtration argument.

7.3.9 Add the global diamond E to the basic hybrid language. Use a filtration argument to show that the satisfiability problem for the resulting language is decidable. What is its complexity? (Note that $@_i\phi$ can be defined to be $E(i \wedge \phi)$, so you do not have to deal explicitly with the satisfaction operators.) Show that a class of frames is definable in this language if and only if it is definable in the basic modal language enriched with the D-operator. (Here 'definable' means definable by an arbitrary formula, not just a pure formula.)

7.4 The Guarded Fragment

In Chapter 2 we saw that modal languages can be viewed as fragments of first-order logic, and in Chapter 6 we discovered that these fragments have some nice computational properties. It thus seems natural to try and see how far we can generalize these properties to larger fragments of first-order logic. This will be the main aim of this section: we will define and discuss two extensions of the modal fragment with reasonably nice computational behavior.

In order to isolate such fragments, what properties of the modal fragment of first-order logic should we concentrate on? In particular, what makes modal logic decidable? If we confine ourselves to the basic modal language, is it perhaps the fact that the standard translation can be carried out entirely within the two variable fragment of first-order logic (which has a decidable satisfiability problem)? This argument immediately breaks down if we consider languages with modal operators of higher arity: while giving rise to decidable logics as well, these languages have standard translations that really need *more* than two variables. But as soon as we are considering n-variable fragments of first-order logic with $n > 2$, we face an *undecidable* satisfiability problem.

Rather, it seems to be the fact that the modal fragment of first-order logic allows quantification only in a very restricted form, as is obvious from the modal clause in the definition of the standard translation function:

$$ST_x(\Diamond\phi) = \exists y\,(Rxy \wedge ST_y(\phi)). \tag{7.3}$$

It is this restricted form of quantification which ensures that modal logic is the bisimulation invariant fragment of first-order logic, and bisimulation invariance of modal truth was critical in the first method of proving the finite model property for the basic modal language (see Section 2.3). Recall that the starting point of this method was the observation that modal logic has the *tree model property* (meaning that every satisfiable modal formula is satisfiable on a tree model), and that bisimulation invariance is pivotal in proving this result. In short, there seems to be a direct line from the restricted quantifier pattern in (7.3), via bisimulation invariance and the tree model property, to the finite model property and decidability.

This provides our first search direction: look for first-order fragments characterized by restricted quantification. It turns out that one can easily relax many constraints applying to the (basic) modal fragment. For example, we do not have to confine ourselves to formulas using two variables only, to formulas having precisely one free variable, or to formulas with predicates of arity at most two. Relaxing these constraints naturally leads to the so-called *guarded fragment* of first-order logic; the idea here is that quantifiers may appear only in the following form:

$$\exists \overline{y}\,(G(\overline{x},\overline{y}) \wedge \psi(\overline{x},\overline{y})) \tag{7.4}$$

in which $G(\overline{x},\overline{y})$ is an atomic formula that we will call the *guard* of the quanti cation (or, of the formula). The crucial ingredient that we *keep* from (7.3) is that all free variables of ψ are also free in the guard $G(\overline{x}, \overline{y})$. And indeed, it can be shown that the guarded fragment has various nice properties, such as a decidable satisfiability problem and the finite model property.

However, there are some very natural modal-like languages, or alternative but intuitive interpretations for standard modal languages, that correspond to a decidable fragment of first-order logic as well, but are not covered by this definition. For example, consider the language with the since and until operators: it is straightforward to turn the truth definitions for these operators into a standard translation to first-order logic. The interesting clauses are

$$ST_x(U(\phi, \psi)) = \exists y \, (Rxy \wedge ST_y(\phi) \wedge \forall z \, ((Rxz \wedge Rzy) \rightarrow ST_z(\psi))), \quad (*)$$

and a similar one for the since operator. We can prove that this kind of clause takes us outside the guarded fragment of first-order logic: the problem concerns the 'betweenness conjunct' $\forall z \, ((Rxz \wedge Rzy) \rightarrow ST_z(\psi))$ which has a 'composite' guard, $(Rxz \wedge Rzy)$. Nevertheless, the language with since and until has a decidable satisfiability problem; apparently, *some* composite guards are admissible as well.

Examples such as $(*)$ lead to extensions of the guarded fragment to fragments in which one is more liberal in the precise conditions imposed on the guard. One *can* be a bit more liberal here because in the 'direct line' mentioned earlier there are some steps that could be skipped on the way. In particular, if we are interested in *decidability* rather than the finite model property, we could just as well settle for fragments of first-order logic to which we may apply the *mosaic method* of Section 6.4. Recall that the mosaic method is a way of proving decidability by 'deconstructing' a model into a finite number of finite pieces, and then using such finite toolboxes for constructing models again, models that usually hang together quite loosely (in a sense to be made precise later). This provides the second direction in our quest: try to find fragments of first-order logic to which the mosaic method applies, leading to a *loose model property*. Implementing this idea one naturally finds quantifier restrictions of the form

$$\exists \overline{y} \, (\pi(\overline{x}, \overline{y}) \wedge \psi(\overline{x}, \overline{y})) \tag{7.5}$$

in which there are constraints on the presence of variables in certain subformulas of the guard π. For such fragments one may find a direct line from the restricted quantifier pattern in (7.5), via an appropriate notion of bisimulation invariance and the *loose* model property, to some finite *mosaic* property and decidability.

The particular extension that we discuss in this section is that of the *packed fragment*; it fits very nicely in the mosaic approach. On a first reading of the section the reader may choose to skip the parts referring to this packed fragment, and concentrate on the guarded fragment.

The guarded and the packed fragment

We need some preliminaries. The first-order language that we will be working in is purely relational, with equality; the language contains neither constants nor function symbols. For a sequence of variables $\bar{x} = x_1, \ldots, x_n$, we frequently write $\exists \bar{x}\, \phi$, which, as usual, has the same meaning as $\exists x_1 \ldots \exists x_n\, \phi$. However, in this section we view $\exists \bar{x}$ not as an abbreviation, but as a primitive operator. In particular this means that the subformulas of $\exists \bar{x}\, \phi$ are just $\exists \bar{x}\, \phi$ itself, together with the subformulas of ϕ. As usual, by writing $\phi(\bar{x})$ we indicate that the free variables of ϕ are among $x_1, \ldots, x_n$.

Definition 7.31 We say that a formula ϕ *packs* a set of variables $\{x_1, \ldots, x_k\}$ if (i) *Free*$(\phi) = \{x_1, \ldots, x_k\}$ and (ii) ϕ is a conjunction of formulas of the form $x_i = x_j$ or $R(x_{i_1}, \ldots, x_{i_n})$ or $\exists x_{j_1} \ldots \exists x_{j_m} R(x_{i_1}, \ldots, x_{i_n})$ such that (iii) for every $x_i \neq x_j$, there is a conjunct in ϕ in which x_i and x_j both occur free.

The *packed fragment* *PF* is defined as the smallest set of first order formulas which contains all atomic formulas and is closed under the boolean connectives and under *packed quantification*. That is, whenever ψ is a packed formula, π packs *Free*(π), and *Free*$(\psi) \subseteq$ *Free*(π), then $\exists \bar{x}\, (\pi \wedge \psi)$ is packed as well; π is called the *guard* of this formula. The *guarded fragment* *GF* is the subfragment of *PF* in which we only allow *guarded quantification* as displayed in (7.4); that is, packed quantification in which the guard π is an *atomic* formula.

PF$_n$ and *GF*$_n$ denote the restrictions to n variables and at most n-ary predicate symbols of *PF* and *GF*, respectively.   ⊣

Examples of guarded formulas are

   (i) the standard translation of any modal formula (in any language),
  (ii) the standard translation of any formula in the basic temporal language,
 (iii) formulas like $\forall xy\, (Rxy \rightarrow Ryx)$, $\exists xy\, (Rxy \wedge Ryx \wedge (Rxx \vee Ryy))$, ...

For an example of a packed formula which is not guarded, consider $\exists xyz\, ((Rxy \wedge Rxz \wedge Ryz) \wedge \neg Cxyz)$. For another example, first consider the standard translation $\exists y\, (Rxy \wedge Py \wedge \forall z\, ((Rxz \wedge Rzy) \rightarrow Qz))$ of the formula $U(p, q)$. This formula is not packed *itself*, because the guard of the subformula $\forall z\, ((Rxz \wedge Rzy) \rightarrow Qz)$ has no conjunct in which the variables x and y occur together. But of course, the formula is *equivalent* to

$$\exists x\, (Rxy \wedge Py \wedge \forall z\, ((Rxz \wedge Rzy \wedge Rxy) \rightarrow Qz))$$

which is packed. It is not hard to convert this example into a proof showing that *every* formula in the since and until language is equivalent to a packed formula.

Second, note that the notion of packedness only places meaningful restrictions on pairs of *distinct* variables: since the formula $x = x$ packs the set of variables

$\{x\}$, the formula $\exists x\,(x = x \wedge \psi(x))$, (that is, with a *single* quantification over the variable x) is a packed formula, at least, provided that $\psi(x)$ is packed. Since the given formula is equivalent to $\exists x\,\psi(x)$ this shows that packedness allows a fairly mild form of ordinary quantification, namely over formulas with one free variable only. A nice corollary of this is that we may perform the standard translation of the global diamond E within the two variable guarded fragment:

$$ST_x(E\phi) = ST_y(E\phi) = \exists x\,(ST_x(\phi)) \equiv \exists x\,(x = x \wedge ST_x(\phi)).$$

Finally, not *all* formulas are packed, or equivalent to a packed formula. For example, the *transitivity* formula $\forall yz\,((Rxy \wedge Ryz) \to Rxz)$ is not packed, and neither is the standard translation of the difference operator: $\exists y\,(x \neq y \wedge Py)$.

Nice properties

Having defined the packed and the guarded fragment of first-order logic, let us see now what we can *prove* about these fragments. To start with, for each of the two fragments we can find a suitable notion of bisimulation which characterizes the fragment in the same way as the ordinary bisimulation characterizes the modal fragment of first-order logic. Unfortunately we do not have the space to go into detail here. Nevertheless, we will show that both fragments have what we call the *loose model property*: in Theorem 7.33 we will show that every satisfiable packed formula can be satisfied on a loose model. What, then, is a loose model?

Definition 7.32 Let $\mathfrak{A} = (A, I)$ be a first-order structure. A tuple $(a_1, \ldots, a_n)$ of objects in A is called *live in* $\mathfrak{A}$ if either $a_1 = \cdots = a_n$ or $(a_1, \ldots, a_n) \in I(P)$ for some predicate symbol P. A subset X of A is called *guarded* if there is some live tuple $(a_1, \ldots, a_n)$ such that $X \subseteq \{a_1, \ldots, a_n\}$. In particular, singleton sets are always guarded; note also that guarded sets are always finite. X is *packed* or *pairwise guarded* if it is finite and each of its two-element subsets is guarded.

We say that $\mathfrak{A}$ is a *loose model of degree* $k \in \mathbb{N}$ if there is some acyclic connected graph $\mathfrak{G} = (G, E)$ and a function f mapping nodes of $\mathfrak{G}$ to subsets of A of size not exceeding k such that for every live tuple $\bar{s}$ from $\mathfrak{A}$, the set $L(\bar{s}) = \{g \in G \mid s_i \in f(g) \text{ for all } s_i\}$, is a non-empty and connected subset of $\mathfrak{G}$. ⊣

In words, we call a model $\mathfrak{A} = (A, I)$ loose if we can associate a connected graph $\mathfrak{G} = (G, E)$ with it in the following way. Each node t of the graph corresponds to a *small* subset $f(t)$ of the model; a good way of thinking about this is that t describes' $f(t)$. One then requires that the graph 'covers' the entire model in the sense that any $a \in A$ belongs to one of these sets (this follows from the fact that for any $a \in A$, the 'tuple' a is live). The fact that each set $L(\bar{a})$ is connected whenever is live, implies that various nodes of the graph will not give contradictory descriptions of the model. Finally, the *looseness* of the model stems from the acyclicity of

$\mathfrak{G}$ and the connectedness of the sets $L(\bar{a})$; for, this ensures that in walking through the graph we may describe different parts of the model, *but we never have to worry about returning to the same part once we have left it.* Summarizing, we may see the graph as a loose, coherent collection of descriptions of local submodels of the model. Loose models are the ones for which we can find such a graph.

The following result states that the packed fragment of first-order logic has the *loose model property.*

Theorem 7.33 *Every satisfiable packed formula can be satisfied on a loose model (of degree at most the number of $\exists \bar{x}$ subformulas of ξ).*

But the big question is of course whether following this looseness principle we have indeed arrived at a decidable fragment of first-order logic. The next theorem states that we have.

Theorem 7.34 *The satisfiability problems for the guarded and the packed fragment are decidable; both problems are DEXPTIME-complete (complete for doubly exponential time). However, for a fixed natural number n, the satisfiability problem for formulas in the packed fragment PF_n is decidable in EXPTIME.*

And finally, what about the finite model property? Will every satisfiable packed formula have a finite model? Here as well, the packed fragment displays very nice behavior. Unfortunately, we do not have the space for a proof of the finite model property for the packed fragment – suffice it to say that it involves some quite advanced techniques from finite model theory. For some further information the reader is referred to the Notes at the end of the section.

Mosaics

The remainder of the section is devoted to proving the Theorems 7.33 and 7.34. The main idea behind the proof is to use the *mosaic method* that we met in Chapter 6. Roughly speaking, this method is based on the idea of deconstructing model into a finite collection of finite submodels, and, conversely, of building up new 'loose' models from such parts. We will see that the packed fragment is in a sense tailored towards making this idea work.

The proof is structured as follows. We start by formally defining mosaics and some related concepts. After that we state the main result concerning the mosaic method, namely the *Mosaic Theorem* stating that a packed formula is satisfiable and only if there is a so-called *linked* set of mosaics for it, of bounded size. Th equivalence enables us to define our decision algorithm and establish the com plexity upper bounds mentioned in Theorem 7.34. We then continue to prove th

Mosaic Theorem. In doing so we obtain the loose model property for the packed fragment as a spin-off.

For a formal definition of the concept of a mosaic we first need some syntactic preliminaries. Given a first-order formula ξ, we let $Var(\xi)$ and $Free(\xi)$ denote the sets of variables and free variables occurring in ξ, respectively. Let V be a set of variables. A V-*substitution* is any partial map $\sigma : V \to V$. The result of performing the substitution σ on the formula ψ is denoted by ψ^σ. (We can and may assume that such substitutions can be carried out without increasing the total number of variables involved; more precisely, we assume that if $Var(\psi) \subseteq V$ then $Var(\psi^\sigma) \subseteq V$.)

As usual, we will employ a notion of closure to delineate a finite set of *relevant* formulas, that is formulas that for some reason critically influence the truth of a given formula ξ. Let the *single negation* $\sim\phi$ of a formula ϕ denote the formula ψ if ϕ is of the form $\neg\psi$; otherwise, $\sim\phi$ is the formula $\neg\phi$; we say that a set Σ of formulas is closed under single negations if $\sim\phi \in \Sigma$ whenever $\phi \in \Sigma$.

Definition 7.35 Let Σ be a set of packed formulas in the set V of variables. We call Σ V-*closed* if it is closed under subformulas, single negations and V-substitutions (that is, if ψ belongs to Σ, then so does ψ^σ for every V-substitution σ). With $\mathrm{Cl}_g(\xi)$ we denote the smallest $Var(\xi)$-closed set of formulas containing ξ. $\dashv$

For the remainder of this section, we fix a packed formula ξ – all definitions to come should be understood as being relativized to ξ. The number of variables occurring in ξ (free or bound) is denoted by k; that is, k is the size of $Var(\xi)$. It can easily be verified that the sets of guarded and packed formulas are both closed under taking subformulas; hence, the set $\mathrm{Cl}_g(\xi)$ consists of guarded (packed, respectively) formulas. An easy calculation shows that the cardinality of $\mathrm{Cl}_g(\xi)$ is bounded by $k^k \cdot (2|\xi|)$.

The following notion is the counterpart of the atoms that we have met in earlier decidability proofs (see Lemma 6.29, for instance). All three defining conditions are fairly obvious.

Definition 7.36 Let $X \subseteq Var(\xi)$ be a set of variables. An X-*type* is a set $\Gamma \subseteq \mathrm{Cl}_g(\xi)$ with free variables in X satisfying, for all formulas $\phi \wedge \psi$, $\sim\phi$, ϕ in $\mathrm{Cl}_g(\xi)$ with free variables in X, the conditions (i) $\phi \wedge \psi \in \Gamma$ iff $\phi \in \Gamma$ and $\psi \in \Gamma$, (ii) $\phi \notin \Gamma$ iff $\sim\phi \in \Gamma$ and (iii) if $\phi, x_i = x_j \in \Gamma$ then $\phi^\sigma \in \Gamma$ for any substitution σ mapping x_i to x_j and/or x_j to x_i, while leaving all other variables fixed. $\dashv$

The next definition introduces our key tool in proving the decidability of the packed fragment: mosaics and linked sets of them. Basically, a mosaic consists of a subset X of $Var(\xi)$ together with a set Γ encoding the relevant information on some small part of a model. Here 'small' means that its size is bounded by the number of objects that can be named using variables in X, and 'relevant' refers to all formulas in $Cl_g(\xi)$ whose free variables are in X. It turns out that a finite set of such mosaics contains sufficient information to construct a model for ξ provided that the set links the mosaics together in a nice way. Here is a more formal definition.

Definition 7.37 A *mosaic* is a pair (X, Γ) such that $X \subseteq Var(\xi)$ and $\Gamma \subseteq Cl_g(\xi)$. A mosaic is *coherent* if it satisfies the following conditions:

(C1) Γ is an X-type,
(C2) if $\psi(\overline{x}, \overline{z})$ and $\pi(\overline{x}, \overline{z})$ are in Γ, then so is $\exists \overline{y} \, (\pi(\overline{x}, \overline{y}) \wedge \psi(\overline{x}, \overline{y}))$,
        (provided that the latter formula belongs to $Cl_g(\xi)$).

A *link* between two mosaics (X, Γ) and (X', Γ') is a renaming (that is, an injective substitution) σ with $\mathrm{dom}\, \sigma \subseteq X$ and $\mathrm{range}\, \sigma \subseteq X'$ which satisfies, for all formulas $\phi \in Cl_g(\xi)$: $\phi \in \Gamma$ iff $\phi^\sigma \in \Gamma'$.

A *requirement* of a mosaic is a formula of the form $\exists \overline{y} \, (\pi(\overline{x}, \overline{y}) \wedge \psi(\overline{x}, \overline{y}))$ belonging to Γ. A mosaic (X', Γ') *fulfills* the *requirement* $\exists \overline{y} \, (\pi(\overline{x}, \overline{y}) \wedge \psi(\overline{x}, \overline{y}))$ of a mosaic (X, Γ) *via* the link σ if for some variables $\overline{u}$, $\overline{v}$ in X' we have that $\sigma(\overline{x}) = \overline{u}$ and $\pi(\overline{u}, \overline{v})$ and $\psi(\overline{u}, \overline{v})$ belong to Γ'. A set S of mosaics is *linked* if every requirement of a mosaic in S is fulfilled via a link to some mosaic in S. S is a linked set of mosaics for ξ if it is linked and $\xi \in \Gamma$ for some (X, Γ) in S.   ⊣

Note that a mosaic (X, Γ) may fulfill its own requirements, either via the identity map or via some other map from X to X.

The key result concerning mosaics is the following Mosaic Theorem:

Theorem 7.38 (Mosaic Theorem) *Let ξ be a packed formula. Then ξ is satisfiable if and only if there is a linked set of mosaics for ξ.*

Proof. The hard, right to left, direction of the theorem is treated in Lemma 7.39 below; here we only prove the other direction.

Suppose that ξ is satisfied in the model $\mathfrak{A} = (A, I)$. In a straightforward way we can 'cut out' from $\mathfrak{A}$ a linked set of mosaics for ξ. Consider the set of partial assignments of elements in A to variables in $Var(\xi)$. For each such α, let $(X_\alpha, \Gamma_\alpha)$ be the mosaic given by $X_\alpha = \mathrm{dom}\, \alpha$ and

$$\Gamma_\alpha = \{\phi \in Cl_g(\xi) \mid \mathfrak{A} \models \phi[\alpha]\}.$$

We leave it to the reader to verify that this collection forms a linked set of mosaics for ξ.   ⊣

When establishing the hard direction of this proposition we will in fact prove something stronger: starting from a linked set of mosaics for a formula ξ we will show via a step-by-step argument, that there is a *loose* or *tree-like* model for ξ. Firs however, we want to show that the Mosaic Theorem is the key towards proving the decidability of the packed fragment, and also for finding an upper bound for its complexity.

The decision algorithm and its complexity

The mosaic theorem tells us that any packed formula ξ is satisfiable if and only if there is a linked set of mosaics for ξ. Thus an algorithm answering the question whether a linked set of mosaics exists for ξ, also decides whether ξ is satisfiable. By providing such an algorithm we establish the upper complexity bound for the satisfiability problem of the packed fragment.

Recall that k denotes the number of variables occurring in ξ. The following observations are fairly straightforward consequences of our definitions:

(i) up to isomorphism there are at most $2^k \cdot 2^{2|\xi| \cdot k^k}$ mosaics. Using the $\mathcal{O}$ notation, this is at most $2^{\mathcal{O}(|\xi|) \cdot 2^{k \log k}}$,

(ii) given sets X, Γ with $|X| \leq k$ and $\Gamma \subseteq \mathrm{Cl}_g(\xi)$ it is decidable in time polynomial in k^k and $|\xi|$ whether (X, Γ) is a coherent mosaic,

(iii) given a set X of coherent mosaics and a requirement $\phi(\overline{x})$ it is decidable in time polynomial in $|X|$ and $|\phi(\overline{x})|$ whether X fulfills the requirement $\phi(\overline{x})$.

Using methods similar to the elimination of Hintikka sets that we saw in the decidability proof for propositional dynamic logic (see Section 6.8), we now give an algorithm which decides the existence of a linked set of mosaics for ξ. Let S_0 be the set of *all* coherent mosaics. By the observations above, S_0 contains at most $2^{\mathcal{O}(|\xi|) \cdot 2^{k \log k}}$ elements and can be constructed in time polynomial in $|S_0|$. We now inductively construct a sequence of sets of mosaics $S_0 \supseteq S_1 \supseteq S_2 \supseteq S_3 \cdots$. If every requirement of a mosaic μ in a set S_i is fulfilled we call μ *happy*. If every mosaic in S_i is happy then return 'there is a linked set of mosaics for ξ' if S_i contains a mosaic (X, Γ) with $\xi \in \Gamma$, and return 'there is no linked set of mosaics for ξ' otherwise. If, on the other hand, S_i contains *unhappy* mosaics, let S_{i+1} consist of all happy mosaics in S_i and continue the construction. Since our sets decrease in size at every step, the construction must halt after at most $|S_0|$ many stages. By the observations above, computing which states in S_i are happy can be done in time polynomial in ξ and $|S_i|$. Thus the entire computation can be performed in time polynomial in $|S_0|$. Clearly the algorithm is correct.

Hence, if we consider a formula ξ in a packed fragment with a *fixed number of variables*, $|S_0|$ is exponential in $|\xi|$. In general, however, the number of variables occurring in a formula depends on the formula's length and, hence, $|S_0|$ is doubly exponential in $|\xi|$. Thus, pending the correctness of Lemma 7.39 below, this establishes the complexity upper bounds in Theorem 7.34.

Loose models

Finally, we show the hard direction of the Mosaic Theorem; as a spin-off we establish the 'loose model property' mentioned in Theorem 7.33.

Lemma 7.39 *Let ξ be a packed formula. If there is a linked set of mosaics for ξ, then ξ is satisfiable in a loose model of degree $|Var(\xi)|$.*

Proof. Assume that S is a linked set of mosaics for ξ. Using a step-by-step construction, we will build a loose model for ξ, together with an acyclic graph associated with the model. At each stage of the construction we will be dealing with some kind of approximation of the final model and tree; these approximations will be called networks and are slightly involved structures.

A *network* is a quintuple $(\mathfrak{A}, \mathfrak{G}, \mu, \alpha, \sigma)$ such that $\mathfrak{A} = (A, I)$ is a model for the first-order language; $\mathfrak{G} = (G, E)$ is a connected, acyclic graph; $\mu : G \to S$ is a map associating a mosaic $\mu_t = (X_t, \Gamma_t)$ in S with each node t of the graph; α is a map associating an assignment $\alpha_t : X_t \to A$ with each node t of the graph; and finally, σ is a map associating with each edge (t, t') of the graph a link $\sigma_{tt'}$ from μ_t to $\mu_{t'}$ (we will usually simplify our notation by writing σ instead of $\sigma_{tt'}$).

The idea is that each mosaic μ_t is meant to give a complete description of the relevant requirements that we impose on a small part of the model-to-be. Which part? This is given by the assignment α_t. And the word 'relevant' refers to the fact that we are only interested in the formulas influencing the truth of ξ; that is, the formulas in $Cl_g(\xi)$. The links between neighboring mosaics are there to ensure that distinct mosaics agree on the part of the model that they both have access to.

Now obviously, if we want all of this to work properly we have to impose some conditions on networks. In order to formulate these, we need some auxiliary notation. For a subset $Q \subseteq A$, let $L(Q)$ denote the set of nodes in $\mathfrak{G}$ that have 'access' to Q; formally, we define $L(Q) = \{t \in G \mid Q \subseteq \text{range}(\alpha_t)\}$. For a tuple $\bar{a} = (a_1, \ldots, a_n)$ of elements in A we set $L(\bar{a}) = L(\{a_1, \ldots, a_n\})$. Now a network is called *coherent* if it satisfies the following conditions (all to be read universally quantified):

(C1) $P\overline{x} \in \Gamma_t$ iff $\mathfrak{A} \models P\overline{x}[\alpha_t]$,
(C2) $x_i = x_j \in \Gamma_t$ iff $\alpha_t(x_i) = \alpha_t(x_j)$,
(C3) $L(Q)$ is non-empty for every guarded set $Q \subseteq A$,
(C4) $L(Q)$ is connected for every guarded set $Q \subseteq A$,
(C5)   if Ett' then $\sigma_{tt'}(x) = x'$ iff $\alpha_t(x) = \alpha_{t'}(x')$.

A few words of explanation about these conditions: (C1) and (C2) ensure that every mosaic is a complete description of the atomic formulas holding in the part of the model it refers to. Condition (C3) states that no live tuple of the model remains unseen from the graph, while the conditions (C4) and (C5) are the crucial on making that remote parts of the graph cannot contain contradictory information about the model – how this works precisely will become clear further on. Note that condition (C5) has two directions: the left-to-right direction states that neighboring mosaics have common access to part of the model, while the other direction ensures that they agree on their requirements concerning this common part.

The motivation for using these networks is that in the end we want any formula $\phi(\overline{x}) \in \text{Cl}_g(\xi)$ to hold in $\mathfrak{A}$ under the assignment α_t if and only if $\phi(\overline{x})$ belongs to Γ_t. Coherence on its own is not sufficient to make this happen. A *defect* of a network consists of a formula $\exists \overline{y} \, (\pi(\overline{x}, \overline{y}) \wedge \psi(\overline{x}, \overline{y}))$ which is a requirement of the mosaic μ_t for some node t while there is no neighboring node t' such that $\mu_{t'}$ fulfills $\exists \overline{y} \, (\pi(\overline{x}, \overline{y}) \wedge \psi(\overline{x}, \overline{y}))$ via the link $\sigma_{tt'}$. A coherent network $\mathfrak{N}$ is *perfect* if it has no defects. We say that $\mathfrak{N}$ is a network *for* ξ if for some $t \in G$, $\mu_t = (X_t, \Gamma_t)$ is such that $\xi \in \Gamma_t$.

Claim 1 *If $\mathfrak{N} = (\mathfrak{A}, \mathfrak{G}, \mu, \alpha, \sigma)$ is a perfect network for ξ, then*

(i) *$\mathfrak{A}$ is a loose model of degree $|Var(\xi)|$, and*
(ii) *for all formulas $\phi(\overline{x}) \in \text{Cl}_g(\xi)$ and all nodes t of $\mathfrak{G}$: $\phi \in \Gamma_t$ iff $\mathfrak{A} \models \phi[\alpha_t]$.*

Proof of Claim. For part (i) of the claim, let $\mathfrak{N} = (\mathfrak{A}, \mathfrak{G}, \mu, \alpha, \sigma)$ be the perfect network for ξ. Let $\mathfrak{A} = (A, I)$. As the function f mapping nodes of $\mathfrak{G}$ to *subsets* of A, simply take the map that assigns the *range* of α_t to the node t. Since the domain of each map α_t is always a subset of $Var(\xi)$, it follows immediately that $f(t)$ will always be a set of size at most $|Var(\xi)|$. Now take an arbitrary live tuple $\overline{s}$ in $\mathfrak{A}$; it follows from (C3) and (C4) that $L(\overline{s})$ is a non-empty and connected part of the graph $\mathfrak{G}$. Thus $\mathfrak{A}$ is a loose model of degree $|Var(\xi)|$.

We prove part (ii) of the claim by induction on the complexity of ϕ. For atomic formulas the claim follows by conditions (C1) and (C2), and the boolean case of the induction step is straightforward (since Γ_t is an X–type) and left to the reader. We concentrate on the case that $\phi(\overline{x})$ is of the form $\exists \overline{y} \, (\pi(\overline{x}, \overline{y}) \wedge \psi(\overline{x}, \overline{y}))$.

First assume that $\phi(\overline{x}) \in \Gamma_t$. Since $\mathfrak{N}$ is perfect there is a node t' in G and variables $\overline{u}, \overline{v}$ in $X_{t'}$ such that Ett', $\pi(\overline{u}, \overline{v})$ and $\psi(\overline{u}, \overline{v})$ belong to $\Gamma_{t'}$, while the link σ from μ_t to $\mu_{t'}$ maps $\overline{x}$ to $\overline{u}$. By the induction hypothesis we find that

$$\mathfrak{A} \models \pi(\overline{u}, \overline{v}) \wedge \psi(\overline{u}, \overline{v})[\alpha_{t'}]. \tag{7.6}$$

But from condition (C5) it follows that $\alpha_{t'}(\overline{x}) = \alpha_t(\overline{u})$, whence (7.6) implies that

$$\mathfrak{A} \models \exists \overline{y} \, (\pi(\overline{x}, \overline{y}) \wedge \psi(\overline{x}, \overline{y}))[\alpha_t],$$

which is what we were after. Here and in the sequel, if $\overline{x} = (x_1, \ldots, x_n)$, then $\alpha(\overline{x})$ abbreviates $(\alpha(x_1), \ldots, \alpha(x_n))$.

Now suppose, in order to prove the converse direction, that $\mathfrak{A} \models \phi(\overline{x})[\alpha_t]$. Let $\overline{a}$

denote $\alpha_t(\overline{x})$, then there are $\overline{b}$ in A such that $\mathfrak{A} \models \pi(\overline{x}, \overline{y})[\overline{a}\overline{b}]$ and $\mathfrak{A} \models \psi(\overline{x}, \overline{y})[\overline{a}\overline{b}]$. Our first aims are to prove that

$$L(\overline{a}\overline{b}) \neq \varnothing, \tag{7.7}$$

and

$$L(Q) \text{ is connected for every } Q \subseteq \{\overline{a}, \overline{b}\}. \tag{7.8}$$

Note that if we are working in the guarded fragment, then $\pi(\overline{x}, \overline{y})$ is an atomic formula, whence it follows from $\mathfrak{A} \models \pi(\overline{x}, \overline{y})[\overline{a}\overline{b}]$ that $\overline{a}\overline{b}$ is live. Thus $\{\overline{a}, \overline{b}\}$ is guarded, and hence (7.7) is immediate by condition (C3). In fact, *every* $Q \subseteq \{\overline{a}, \overline{b}\}$ is guarded in this case, so (7.8) is immediate by condition (C4).

In the more general case of the packed fragment we have to work a bit harder. First, observe that it *does* follow from $\mathfrak{A} \models \pi(\overline{x}, \overline{y})[\overline{a}\overline{b}]$ and the conditions on $\pi(\overline{x}, \overline{y})$ in the definition of packed quantification, that $\{c, d\}$ is guarded, and thus, $L(c, d) \neq \varnothing$, for every *pair* (c, d) of points taken from $\overline{a}\overline{b}$. It follows from (C4) that $\{L(c) \mid c \text{ taken from } \overline{a}\overline{b}\}$ is a collection of non-empty, connected, pairwise overlapping subgraphs of the acyclic graph $\mathfrak{G}$. It is fairly straightforward to prove, for instance, by induction on the size of the graph $\mathfrak{G}$, that any such collection must have a non-empty intersection. From this, (7.7) and (7.8) are almost immediate.

Thus, we may assume the existence of a node t' in $\mathfrak{G}$ such that $\{\overline{a}, \overline{b}\} \subseteq \text{range } \alpha_{t'}$. Let $\overline{u}$ and $\overline{v}$ in $X_{t'}$ be the variables such that $\alpha_{t'}(\overline{u}) = \overline{a}$ and $\alpha_{t'}(\overline{v}) = \overline{b}$. The induction hypothesis implies that $\pi(\overline{u}, \overline{v})$ and $\psi(\overline{u}, \overline{v})$ belong to $\Gamma_{t'}$, whence $\phi(\overline{u}) \in \Gamma_{t'}$ by coherence of $\mu_{t'}$. Since both t and t' belong to $L(\overline{a})$, it follows from (7.8) that there is a path from t to t' *within* $L(\overline{a})$, say $t' = s_0 E s_1 E \ldots E s_n = t$. Let σ_i be the link between the mosaics of s_i and s_{i+1}, and define ρ to be the composition of these maps. It follows by an easy inductive argument on the length of the path that ρ is a link between $\mu_{t'}$ and μ_t such that $\rho(\overline{u}) = \overline{x}$. Hence, by definition of a link we have that $\phi(\overline{x}) \in \Gamma_{t'}$.

By Claim 1, in order to prove Lemma 7.39 it suffices to construct a perfect network for ξ. This construction uses a step-by-step argument; to start the construction we need *some* coherent network for ξ.

Claim 2 *There is a coherent network for ξ.*

Proof of Claim. By our assumption on ξ there is a coherent mosaic $\mu = (X, \Gamma)$ such that $\xi \in \Gamma$. Without loss of generality we may assume that X is the set $\{x_1, \ldots, x_n\}$ (otherwise, take an isomorphic copy of μ in which X does have this form). Let $a_1, \ldots, a_n$ be a list of objects such that for all i and j we have that $a_i = a_j$ if and only if the formula $x_i = x_j$ belongs to Γ. Define $A = \{a_1, \ldots, a_n\}$ and put the tuple $(a_{i_1}, \ldots, a_{i_k})$ in the interpretation $I(P)$ of the k-ary predicate symbol P precisely if $P x_{i_1} \ldots x_{i_n} \in \Gamma$. Let $\mathfrak{A}$ be the resulting model (A, I) and define $\mathfrak{G}$ as the trivial graph with one node 0 and no edges. Let $\mu(0)$ be the mosaic μ; $\alpha_0 : X \to A$ is given by $\alpha(x_i) = a_i$; and finally, σ_{00} is the identity map from X to X.

We leave it to the reader to verify that the quintuple $(\mathfrak{A}, \mathfrak{G}, \mu, \alpha, \sigma)$ is a coherent network for ξ. $\dashv$

The crucial step of this construction will be to show that any defect of a coherent network can be repaired.

Claim 3 *For any coherent network* $\mathfrak{N} = (\mathfrak{A}, \mathfrak{G}, \mu, \alpha, \sigma)$ *and any defect of* $\mathfrak{N}$ *there is a coherent network* $\mathfrak{N}^+$ *extending* $\mathfrak{N}$ *and lacking this defect.*

Proof of Claim. Suppose that $\phi(\overline{x})$ is a defect of $\mathfrak{N}$ because it is a requirement of the mosaic μ_t and not fulfilled by any neighboring mosaic $\mu_{t'}$. We will define an extension $\mathfrak{N}^+$ of $\mathfrak{N}$ in which this defect is repaired.

Since S is a linked set of mosaics and μ_t belongs to S, μ_t is linked to a mosaic $(X', \Gamma') \in S$ in which the requirement is fulfilled via some link ρ. Let Y be the set of variables in X' that do not belong to the range of ρ; suppose that $Y = \{y_1, \ldots, y_k\}$ (with all y_i being distinct). For the sake of a smooth presentation, assume that Γ' contains the formulas $\neg x' = y$ for all variables $x' \in X'$ and $y \in Y$ (this is not without loss of generality – we leave the general case as an exercise to the reader). Take a set $\{c_1, \ldots, c_k\}$ of fresh objects (that is, no c_i is an element of the domain A of $\mathfrak{A}$), and let γ be the assignment with domain X' defined as follows:

$$\gamma(x') = \begin{cases} \alpha_t(x) & \text{if } x' = \rho(x), \\ c_i & \text{if } x' = y_i, \end{cases}$$

and let t' be an object not belonging to G. Now define the network $\mathfrak{N}^+ = (\mathfrak{A}^+, \mathfrak{G}^+, \mu^+, \alpha^+, \sigma^+)$ as follows:

$$\begin{aligned} A^+ &= A \cup \{c_1, \ldots, c_k\}, \\ I^+(P) &= I(P) \cup \{\overline{d} \mid \text{ for some } \overline{x}, \overline{d} = \gamma(\overline{x}) \text{ and } P\overline{x} \in \Gamma'\}, \\ G^+ &= G \cup \{t'\}, \\ E^+ &= E \cup \{(t, t')\}, \end{aligned}$$

while μ^+, α^+ and σ^+ are given as the obvious extensions of μ, α and σ, namely by putting $\mu_{t'}^+ = (X', \Gamma')$, $\alpha_{t'}^+ = \gamma$ and $\sigma_{tt'} = \rho$.

Since the interpretation I^+ agrees with I on 'old' tuples it is a straightforward exercise to verify that the new network $\mathfrak{N}^+$ satisfies the conditions (C1)–(C3) and (C5).

In order to check that condition (C4) holds, take some guarded subset Q from 1^+; we will show that $L^+(Q)$ is a connected subgraph of $\mathfrak{G}^+$. It is rather easy to see that $L^+(Q)$ is identical to either $L(Q)$ or $L(Q) \cup \{t'\}$; hence by the connectedness of $L(Q)$ it suffices to prove, on the assumptions that $t \in L^+(Q)$ and $L(Q) \neq \varnothing$, that $t \in L(Q)$. Hence, suppose that $t' \in L^+(Q)$; that is, each $a \in Q$ is in the range of γ. But if $L(Q) \neq \varnothing$, each such point a must be old; hence, by definition of γ, each $a \in Q$ must belong to range α_t. This gives that $t \in L(Q)$, as required. $\dashv$

As in our earlier step-by-step proofs, the previous two claims show that using some standard combinatorics we can construct a chain of networks such that their *limit* is a perfect network. This completes the proof of Lemma 7.39. $\dashv$

Exercises for Section 7.4

7.4.1 In the *loosely guarded fragment* the following quantification patterns are allowed: $\exists \overline{x}\, (\pi(\overline{x}, \overline{y}) \wedge \psi(\overline{x}, \overline{y}))$ is a loosely guarded formula if $\psi(\overline{x}, \overline{y})$ is loosely guarded, $\pi(\overline{x}, \overline{y})$ is a conjunction as in the packed fragment, and any pair z, z' of distinct variables from $\overline{xy}$ occurs free in some conjunct of the guard π, unless z and z' are both from $\overline{y}$. For example, $\exists x\, ((Ryx \wedge Rxy') \wedge \neg Cxyy')$ is loosely guarded, but not packed since there is no conjunct having both y and y' free.

Show that for every loosely guarded *sentence* ξ there exists an equivalent packed sentence ξ' in the same language.

7.4.2 Define the *universal packed fragment* as the fragment of first-order logic that is generated from atoms, negated atoms, conjunction, disjunction, ordinary existential quantification, and packed universal quantification. (With the latter we mean that $\forall \overline{x}\, (\pi \rightarrow \psi)$ is in the fragment if ψ is universally packed, π packs its own free variables, and $Free(\psi) \subseteq Free(\pi)$.)

Show that *satisfiability* is decidable for the universal packed fragment.

7.4.3 Fix a natural number n, and suppose that we are working in an *n-bounded* first-order signature; that is, all predicate symbols have arity at most n. Prove that in such a signature, every guarded sentence is equivalent to a guarded sentence using at most n variables. Does this hold for packed sentences as well? What are the consequences for the complexity of the respective satisfiability problems?

7.4.4 Let ξ be a packed formula, and suppose that ξ is satisfiable. Prove that ξ is satisfiable in a loose model with an associated graph $\mathfrak{G}$ of which the *out-degree* is bounded by some recursive function on ξ. In particular, this out-degree should be *finite*. (The out-degree of a node k of a graph (G, E) is defined as the number of its neighbors, or, formally, as the size of the set $\{k' \in G \mid kEk'\}$; the out-degree of a graph is defined as the supremum of the out-degrees of the individual nodes.)

7.5 Multi-Dimensional Modal Logic

In Chapter 2 we backed up our claim that logical formalisms do not live in isola tion by developing the correspondence theory of modal logic: we studied mod languages as fragments of first-order languages. In this section we will turn th looking glass around and examine first-order logic as if it were a modal formalism. The basic observations enabling this perspective are that we may view *assignments* (the functions that give first-order variables their value in a first-order structure) as *states* of a modal model, and that this makes standard first-order *quantifiers* behave just like modal diamonds and boxes. First-order logic thus forms an example of a *multi-dimensional* modal system. Multi-dimensional modal logic is a branch of modal logic dealing with special relational structures in which the states, rather than being abstract entities, have some inner structure. More specifically, these states are tuples or sequences over some base set, in our case, the domain of the first-order structure. Furthermore, the accessibility relations between these states are (partly) determined by this inner structure of the states.

Reverse correspondence theory

To simplify our presentation, in this section we will not treat modal versions of first-order logic in general, but restrict our attention to certain finite variable fragments. A precise definition of these fragments will be given later on (see Definition 7.40). For the time being, we fix a natural number $n \geq 2$ and invite the reader to think of a first-order language with equality, but without constants or function symbols, in which all predicates are n-adic. Consider the basic declarative statement in first-order logic concerning the truth of a formula in a model under an assignment s:

$$\mathfrak{M} \models \phi \, [s]. \tag{7.9}$$

The basic observation underlying our approach, is that we can read (7.9) from a modal perspective as: 'the formula ϕ is true in $\mathfrak{M}$ *at state s*.' But since we have only n variables at our disposal, say $v_0, \ldots, v_{n-1}$, we can identify assignments with maps: $n \, (= \{0, \ldots, n-1\}) \to U$, or equivalently, with n-tuples over the domain U of the structure $\mathfrak{M}$ – we will denote the set of such n-tuples by U^n. But then we find ourselves in the setting of multi-dimensional modal logic: the universe of our modal models will be of the form U^n for some base set U. Now recall that the truth definition of the quantifiers reads as follows:

$$\mathfrak{M} \models \exists v_i \, \phi[s] \text{ iff there is an } u \in U \text{ such that } \mathfrak{M} \models \phi \, [s_u^i],$$

where s_u^i is the assignment defined by $s_u^i(k) = u$ if $k = i$ and $s_u^i(k) = s(k)$ otherwise. We can replace the above truth definition with the more 'modal' equivalent,

$$\mathfrak{M} \models \exists v_i \, \phi[s] \text{ iff there is an assignment } s' \text{ with } s \equiv_i s' \text{ and } \mathfrak{M} \models \phi \, [s'],$$

where $\equiv_i$ is given by

$$s \equiv_i s' \text{ iff for all } j \neq i, \, s_j = s_j'. \tag{7.10}$$

In other words: existential quantification behaves like a modal *diamond*, having $\equiv_i$ as its *accessibility relation*.

Since the semantics of the boolean connectives in the predicate calculus is the same as in modal logic, this shows that the inductive clauses in the truth definition of first-order logic neatly fit a modal mould. So let us now concentrate on the atomic formulas. To start with, we observe that *equality* formulas do not cause any problem: the formula $v_i = v_j$, with truth definition

$$\mathfrak{M} \models v_i = v_j[s] \quad \text{iff} \quad s \in Id_{ij},$$

can be seen as a modal *constant*. Here Id_{ij} is defined by

$$s \in Id_{ij} \quad \text{iff} \quad s_i = s_j. \tag{7.11}$$

The case of the other atomic formulas is more involved, however. Since we confined ourselves to the calculus of n-adic relations and do not have constants or function symbols, our atomic predicate formulas are of the form $Pv_{\sigma(0)} \cdots v_{\sigma(n-1)}$. Here σ is an n-*transformation*, that is, a map: $n \to n$. In the model theory of first-order logic the predicate symbol P will be interpreted as a subset of U^n; but this is precisely how modal valuations treat propositional variables in models where the universe is of the form U^n! Therefore, we can identify the set of propositional variables of the modal formalism with the set of predicate symbols of our first-order language. In this way, we obtain a modal reading of (7.9) for the case where ϕ is the atomic formula $Pv_0 \cdots v_{n-1}$: $\mathfrak{M} \models Pv_0 \cdots v_{n-1}[s]$ iff s belongs to the interpretation of P. However, as a consequence of this approach our set-up will not enjoy a one-to-one correspondence between atomic first-order formulas and atomic modal ones: the atomic formula $Pv_{\sigma(0)} \cdots v_{\sigma(n-1)}$ will correspond to the modal atom p only if σ is the identity function on n. For the cases where σ is not the identity map we still have to find some kind of solution. There are many options here.

Since we are working in a first-order language with equality, atomic formulas with *multiple* occurrences of a variable can be rewritten as formulas with only 'unproblematic' atomic subformulas, for instance

$$
\begin{aligned}
Pv_1v_0v_0 \quad &\leftrightarrow \quad \exists v_2\,(v_2 = v_0 \wedge Pv_1v_2v_2) \\
&\leftrightarrow \quad \exists v_2\,(v_2 = v_0 \wedge \exists v_0\,(v_0 = v_1 \wedge Pv_0v_2v_2)) \\
&\leftrightarrow \quad \exists v_2\,(v_2 = v_0 \wedge \exists v_0\,(v_0 = v_1 \wedge \exists v_1\,(v_1 = v_2 \wedge Pv_0v_1v_2))).
\end{aligned}
$$

This leaves the case of what to do with atoms of the form $Pv_{\sigma(0)} \cdots v_{\sigma(n-1)}$, where σ is a permutation of n, or in other words, atomic formulas where variables have been substituted *simultaneously*. The previous trick does not work here: for example, to write an equivalent of the formula $Pv_1v_0v_2$ one needs *extra* variables as buffers, for instance, when replacing $Pv_1v_0v_2$ by

$$\exists v_3 \exists v_4\,(v_3 = v_0 \wedge v_4 = v_1 \wedge \exists v_0 \exists v_1\,(v_0 = v_4 \wedge v_1 = v_3 \wedge Pv_0v_1v_2)).$$

One might consider a solution where a predicate P is translated into *various* modal propositional variables p_σ, one for every permutation σ of n, but this is not very elegant. One might also forget about simultaneous substitutions and confine oneself to a *fragment* of n-variable logic where all atomic predicate formulas are of the form $Pv_0 \ldots v_{n-1}$ – this fragment of *restricted* first-order logic is defined below. A third solution is to take substitution seriously, so to speak, by adding special 'substitution operators' to the language. The key observation is that for any transformation $\sigma \in n^n$, we have that

$$\mathfrak{M} \models Pv_{\sigma(0)} \ldots v_{\sigma(n-1)}[s] \text{ iff } \mathfrak{M} \models Pv_0 \ldots v_{n-1}\,[s \circ \sigma], \qquad (7.12)$$

where $s \circ \sigma$ is the composition of σ and s (recall that s is a map: $n \to U$). So, if we define the relation $\bowtie_\sigma \subseteq U^n \times U^n$ by

$$s \bowtie_\sigma t \text{ iff } t = s \circ \sigma, \qquad (7.13)$$

we have rephrased (7.12) in terms of an accessibility relation (in fact, a function): $\mathfrak{M} \models Pv_{\sigma(0)} \ldots v_{\sigma(n-1)}[s]$ iff, for some t with $s \bowtie_\sigma t$, $\mathfrak{M} \models Pv_0 \ldots v_{n-1}\,[t]$. So if we add an operator $\bigcirc_\sigma$ to the modal language for every n-transformation σ in n^n, with $\bowtie_\sigma$ as its intended accessibility relation, we have got the desired modal equivalent for any atomic formula $Pv_{\sigma(0)} \ldots v_{\sigma(n-1)}$ – in the form $\bigcirc_\sigma p$. (As a special case, for the formula $Pv_0 \ldots v_{n-1}$ one can take the identity map on n.)

Definition 7.40 Let n be an arbitrary but fixed natural number. The alphabet of L_n and of L_n^r consists of a set of variables $\{v_i \mid i < n\}$, a countable set of n-adic relation symbols ($P_0, P_1, \ldots$), equality ($=$), the boolean connectives $\neg, \vee$ and the quantifiers $\exists v_i$. The collection of formulas is defined as usual in first-order logic, with the restriction that the atomic formulas of L_n^r are of the form $v_i = v_j$ or $P_l(v_0 \ldots v_{n-1})$; for L_n we allow all atomic formulas (but note that all predicates are of arity n).

A *first-order structure for L_n (L_n^r)* is a pair $\mathfrak{M} = (U, V)$ such that U is a set called the domain of the structure and V is an interpretation function mapping every P to a subset of U^n. The notion of a formula ϕ being *true* in a first-order structure $\mathfrak{M}$ under an assignment s is defined as usual. For instance, given our notation we have, for any atomic formula:

$$\mathfrak{M} \models P(v_0 \ldots v_{n-1})\,[s] \text{ if } s \in V(P),$$
$$\mathfrak{M} \models P(v_{\sigma(0)} \ldots v_{\sigma(n-1)})\,[s] \text{ if } s \circ \sigma\,(= (s_{\sigma(0)} \ldots s_{\sigma(n-1)})) \in V(P).$$

An L_n-formula ϕ is *true in $\mathfrak{M}$* (notation: $\mathfrak{M} \models \phi$), if $\mathfrak{M} \models \phi\,[s]$ for all $s \in U^n$; it is *valid* (notation: $\models_{fo} \phi$), if it is true in every first-order structure of L_n. The same definition applies to L_n^r. ⊣

From now on, we will concentrate on the *modal* versions of L_n^r and L_n, which are given in the following definition:

Definition 7.41 Let n be an arbitrary but fixed natural number. MLR_n (short for: *modal language of relations*) is the modal similarity type having constants $\iota\delta_{ij}$ and diamonds $\Diamond_i, \bigcirc_\sigma$ (for all $i, j < n, \sigma \in n^n$). CML_n, the similarity type of *cylindric modal logic*, is the fragment of MLR_n-formulas in which no substitution operator $\bigcirc_\sigma$ occurs.

A first-order structure $\mathfrak{M} = (U, V)$ can be seen as a modal model based on the universe U^n, and formulas of these modal similarity types are interpreted in such a structure in the obvious way; for instance, we have

$$\mathfrak{M}, s \Vdash \iota\delta_{ij} \quad \text{iff} \quad s_i = s_j,$$
$$\mathfrak{M}, s \Vdash \bigcirc_\sigma\phi \quad \text{iff} \quad \mathfrak{M}, s \circ \sigma \Vdash \phi$$
$$\text{(iff} \quad \text{there is a } t \text{ with } s \bowtie_\sigma t \text{ and } \mathfrak{M}, t \Vdash \phi),$$
$$\mathfrak{M}, s \Vdash \Diamond_i\phi \quad \text{iff} \quad \text{there is a } t \text{ with } s \equiv_i t \text{ and } \mathfrak{M}, t \Vdash \phi.$$

If an MLR_n-formula ϕ holds throughout any first-order structure, we say that it is *first-order valid*, notation: $\mathsf{C}_n \Vdash \phi$ (this notation will be clarified further on). $\dashv$

The modal disguise of L_n in MLR_n and of L_n^r in CML is so thin, that we give the translations mapping first-order formulas to modal ones without further comments.

Definition 7.42 Let $(\cdot)^t$ be the following translation from L_n to MLR_n:

$$(Pv_{\sigma(0)} \cdots v_{\sigma(n-1)})^t = \bigcirc_\sigma p,$$
$$(v_i = v_j)^t = \iota\delta_{ij},$$
$$(\neg\phi)^t = \neg\phi^t,$$
$$(\phi \vee \psi)^t = \phi^t \vee \psi^t,$$
$$(\exists v_i \, \phi)^t = \Diamond_i\phi^t. \qquad \dashv$$

This translation allows us to see L_n^r and CML_n as syntactic variants: $(\cdot)^t$ is easily seen to be an *isomorphism* between the formula algebras of L_n^r and CML_n. Note that in the case of L_n versus MLR_n, we face a different situation: where in MLR the simultaneous substitution of two variables for each other is a *primitive* operator in first-order logic it can only be defined by induction. Nevertheless, we could easily define a translation mapping MLR_n-formulas to equivalent L_n^r-formulas. In any case, the following proposition shows that we really have developed a reverse correspondence theory; we leave the proof as an exercise to the reader.

Proposition 7.43 *Let ϕ be a formula in L_n, then*

(i) *for any first-order structure $\mathfrak{M}$, and any n-tuple/assignment s, we have th*
    $\mathfrak{M} \models \phi[s]$ if and only if $\mathfrak{M}, s \Vdash \phi^t$;

(ii) *as a corollary, we have that $\models_{fo} \phi$ iff $\mathsf{C}_n \Vdash \phi^t$.*

Let us now put the modal machinery to work and see whether we can find out something new about first-order logic.

Degrees of validity

Perhaps the most interesting aspect of this modal perspective on first-order logic is that it allows us to generalize the semantics of first-order logic, and thus offers a wider perspective on the standard Tarskian semantics. The basic idea is fairly obvious: now that we are talking about *modal* languages, it is clear that the first-order structures of Definition 7.41 are *very specific* modal models for these languages. We may abstract from the first-order background of these models, and consider modal models in which the universe is an *arbitrary* set and the accessibility relations are *arbitrary* relations (of the appropriate arity).

Definition 7.44 An MLR_n-*frame* is a tuple $(W, T_i, E_{ij}, F_\sigma)_{i,j<n, \sigma \in n^n}$ such that every E_{ij} is a subset of the universe W, and such that every T_i and every F_σ is a binary relation on W. An MLR_n-*model* is a pair $\mathfrak{M} = (\mathfrak{F}, V)$ with $\mathfrak{F}$ an MLR_n-frame and V a *valuation*, that is, a map assigning subsets of W to propositional variables. CML_n-models and frames are defined likewise. $\dashv$

For such models, *truth* of a formula at a state is defined via the usual modal induction, for instance:

$$\mathfrak{M}, w \Vdash \bigcirc_\sigma \phi \text{ iff there is a } v \text{ with } F_\sigma wv \text{ and } \mathfrak{M}, v \Vdash \phi.$$

In this very general semantics, states (that is, elements of the universe) are no longer real assignments, but, rather, abstractions thereof. First-order logic now really has become a poly-modal logic, with quantification and substitution diamonds. It is interesting and instructive to see how familiar laws of the predicate calculus behave in this new set-up. For example, the axiom schema $\phi \to \exists v_i \, \phi$ will be valid only in n-frames where T_i is a reflexive relation (this follows from the fact that the modal formula $p \to \Diamond_i p$ corresponds to the frame condition $\forall x \, T_i xx$). Likewise, the axiom schemes $\exists v_i \exists v_i \, \phi \to \exists v_i \, \phi$ and $\phi \to \forall v_i \exists v_i \, \phi$ will be valid only in frames where the relation T_i is transitive and symmetric, respectively.

Later on we will see more of such correspondences; the point to be made here s that the abstract perspective on the semantics of first-order logic imposes a certain 'degree of validity' on well-known theorems of the predicate calculus. Some heorems are valid in *all* abstract assignment frames, like distribution:

$$\forall v_i \, (\phi \to \psi) \to (\forall v_i \, \phi \to \forall v_i \, \psi),$$

which is nothing but the modal K-axiom. Other theorems of the predicate calculus, like the ones mentioned above, are only valid in *some* classes of frames.

Narrowing down the class of frames means increasing the set of valid formulas, and vice versa. In particular, we now have the option to look at classes of frames that are only slightly more general than the standard first-order structures, but have much nicer computational properties. This new perspective on first-order logic, which was inspired by the literature on algebraic logic, provides us with enormous freedom to play with the semantics for first-order logic. In particular, consider the fact that first-order structures can be seen as frames of the form $(U^n, \equiv_i, Id_{ij}, \bowtie_\sigma)_{i,j<n,\sigma \in n^n}$ where *all* assignments $s \in U^n$ are available. But why not study a semantics where states are still real assignments on the base set U, but *not all such assignments are available?*

There are at least two good reasons to make such a move. First, it turns out that the logic of such generalized assignment frames has much nicer meta-properties than the logic of the cubes such as decidability, see for instance Theorem 7.46. These logics will provide less laws than the usual predicate calculus, but their supply of theorems may be sufficient for particular applications. Note for instance, that the schemes $\phi \to \exists v_i\, \phi$, $\exists v_i \exists v_i\, \phi \to \exists v_i\, \phi$ and $\phi \to \forall v_i \exists v_i\, \phi$ are still valid in every generalized assignment frame, since $\equiv_i \lceil_W$ is always an equivalence relation.

In some situations it may even be *useful* not to have all familiar validities. Consider for instance the schema

$$\exists v_i \exists v_j\, \phi \to \exists v_j \exists v_i\, \phi. \tag{7.14}$$

It follows from correspondence theory that (7.14) is valid in a frame $\mathfrak{F}$ iff (7.15) below holds in $\mathfrak{F}$.

$$\forall xz\, (\exists y\, (T_i xy \wedge T_j yz) \to \exists u\, (T_j xu \wedge T_i uz)). \tag{7.15}$$

The point is this. The schema (7.14) prevents us from making the dependency of variables explicit in the language (that is, whether v_j is dependent of v_i or the other way around), while these dependencies play an important role in some proof theoretical approaches. So, the second motivation for generalizing the semantics of first-order logic is that it gives us a finer sieve on the notion of equivalence between first-order formulas. Note for instance that (7.14) is not valid in frames with assignment 'holes': take $n = 2$. In a square (that is, 2-cubic) frame we have $(a, b) \equiv_0 (a', b) \equiv_1 (a', b')$, but if (a', b') is not an available tuple, then there is no s such that $(a, b) \equiv_1 s \equiv_0 (a', b')$ – hence this frame will not satisfy (7.15). So the schema (7.14) will not be valid in this frame.

In this new paradigm, a whole landscape of frame classes and corresponding logics arises. In the most general approach, *any* subset of U^n may serve as the universe of a multi-dimensional frame, but it seems natural to impose restrictions on the set of available assignments. Unfortunately, for reasons of space limitations we cannot go into further detail here, confining ourselves to the following definition.

Definition 7.45 Let U be some set, and W a set of n-tuples over U, that is, $W \subseteq U^n$. The *cube over* U or *full assignment frame over* U is defined as the frame

$$\mathfrak{C}_n(U) = (U^n, \equiv_i, Id_{ij}, \bowtie_\sigma)_{i,j<n,\sigma \in n^n}.$$

The W-*relativized cube over* U or W-assignment frame on U is defined as the frame

$$\mathfrak{C}_n^W(U) = (W, \equiv_i \upharpoonright_W, Id_{ij} \cap W, \bowtie_\sigma \upharpoonright_W)_{i,j<n,\sigma \in n^n}.$$

C_n and R_n are the classes of cubes and relativized cubes, respectively. ⊣

Observe that this definition clarifies our earlier notation '$\mathfrak{C}_n \Vdash \phi$' for the fact that the modal formula ϕ is 'first-order valid'.

Decidability

As we already mentioned, one of the reasons for developing the abstract and generalized assignment semantics is to 'tame' first-order logic by looking for core versions with nicer computational behavior. This idea is substantiated by the following theorem.

Theorem 7.46 *It is decidable in exponential time whether a given MLR_n-formula is satisfiable in a given relativized cube. As a corollary, the problem whether a given first-order formula in L_n can be satisfied in a general assignment frame is also decidable in exponential time.*

Proof. This theorem can be proved directly by using the *mosaic method* that we encountered in Section 6.4 – in fact, the mosaic method was developed for this particular proof! However, space limitations prevent us from giving the mosaic argument here. Therefore, we prove the theorem by a reduction of the R_n satisfiability problem to the satisfiability problem of the n-variable *guarded fragment* of Section 7.4.

This reduction is quite interesting in itself: the key idea is that we find a syntactic counterpart to the semantic notion of restricting the set of *available* assignments. There is a very simple way of doing so, namely by introducing a special n-adic predicate G that will be interpreted as the collection of available assignments. One can then translate modal formulas (or L_n-formulas) into first-order ones, with the proviso that this translation is *syntactically relativized* to G. The formula $Gv_0 \ldots v_{n-1}$ acts as a *guard* of the translated formula, and, indeed, it will easily be seen that the range of this translation falls inside the guarded fragment.

Now for the technical details. Given a collection Φ of propositional variables, assume that with each $p \in \Phi$ we have an associated n-adic predicate symbol P. Also, fix a *new* n-adic predicate symbol G; let Φ^+ denote the expanded signature

$\{P \mid p \in \Phi\} \cup \{G\}$. Consider the following translation $(\cdot)^{\bullet}$ mapping MLR_n-formulas to first-order formulas:

$$
\begin{aligned}
p^{\bullet} &= Pv_0 \ldots v_{n-1}, \\
\iota\delta_{ij}^{\bullet} &= v_i = v_j, \\
(\neg\phi)^{\bullet} &= Gv_0 \ldots v_{n-1} \wedge \neg\phi^{\bullet}, \\
(\phi \wedge \psi)^{\bullet} &= \phi^{\bullet} \wedge \psi^{\bullet}, \\
(\bigcirc_{\sigma}\phi)^{\bullet} &= (Gv_0 \ldots v_{n-1} \wedge \phi^{\bullet})^{\sigma}, \\
(\Diamond_i\phi)^{\bullet} &= \exists v_i\, (Gv_0 \ldots v_{n-1} \wedge \phi^{\bullet}).
\end{aligned}
$$

Here, for a given transformation σ, $(\cdot)^{\sigma}$ denotes the corresponding syntactic substitution operation on first-order formulas.

Claim 1 *For any MLR_n-formula ϕ we have that $\mathsf{R}_n \Vdash \phi$ if and only if the formula $Gv_0 \ldots v_{n-1} \to \phi^{\bullet}$ is a first-order validity.*

Proof of Claim. In order to prove this claim, we need a correspondence between modal models and first-order models for the new language. Given a relativized assignment model $\mathfrak{M} = (\mathfrak{C}_n^W(U), V)$, define the corresponding first-order model $\mathfrak{M}^{\bullet}$ as the structure (U, I) where $I(P) = V(p)$ for every propositional variable p, and $I(G) = W$. Conversely, given a first-order structure $\mathfrak{A} = (A, I)$ for the expanded first-order signature Φ, let $\mathfrak{A}_{\bullet}$ be the relativized cube model $(\mathfrak{C}_n^{I(G)}(A), V)$, where the valuation V is given by $V(p) = I(P)$.

For any relativized assignment model $\mathfrak{M}$, and any available assignment s, we have

$$
\mathfrak{M}, s \Vdash \phi \text{ iff } \mathfrak{M}^{\bullet} \models \phi^{\bullet}[s]. \tag{7.16}
$$

This suffices to prove Claim 1, because of the following. First suppose that the modal formula ϕ is satisfiable in some relativized cube model $\mathfrak{M}$, say at state s. Since s is an available tuple, it follows from (7.16) that $\phi^{\bullet}$ is satisfiable in the first-order structure $\mathfrak{M}^{\bullet}$ under the assignment s; but also, since s is available we have $\mathfrak{M}^{\bullet} \models Gv_0 \ldots v_{n-1}[s]$. This shows that $\phi^{\bullet} \wedge Gv_0 \ldots v_{n-1}$ is satisfiable.

Conversely, if the latter formula is satisfiable, there is some first-order structure $\mathfrak{A}$ for the language Φ^+, and some assignment s such that $\mathfrak{A} \models \phi^{\bullet} \wedge Gv_0 \ldots v_{n-1}[s]$. It is not difficult to see that $(\mathfrak{A}_{\bullet})^{\bullet} = \mathfrak{A}$. Since $\mathfrak{A} \models Gv_0 \ldots v_{n-1}[s]$, it follows by definition that s is an available assignment of $\mathfrak{A}_{\bullet}$. But then we may apply (7.16) which yields that $\mathfrak{A}_{\bullet}, s \Vdash \phi$; in particular, ϕ is satisfiable in R_n. The proof of (7.16) proceeds by a standard induction, which we leave to the reader.

Finally, we leave it to the reader to verify that the range of $(\cdot)^{\bullet}$ indeed falls entirely inside the n-variable guarded fragment $\mathfrak{F}_n$. From Claim 1 and this observation the theorem is immediate.

Axiomatization

To finish off the section we will sketch how to prove completeness for the class of cube models. For simplicity we confine ourselves to the similarity type of cylindric modal logic – but observe that this completeness result will immediately transfer to the restricted n-variable fragment L_n^r.

Multi-dimensional modal logic is an area with a very interesting completeness theory. For instance, if one only admits the standard modal derivation rules (modus ponens, necessitation and uniform substitution), then *finite* axiomatizations are few and far between. For instance, concerning the CML_n-theory of the class C_n, it is known that if Σ is a set of CML_n-formulas axiomatizing C_n, then for each natural number m, Σ contains infinitely many formulas that contain all diamonds $\Diamond_i$, at least one diagonal constant $\iota\delta_{ij}$ and at least m propositional variables .... However, if we allow special derivation rules, in the style of Section 4.7, then a nice finite axiomatization can be obtained, as we will see now. A key role in our axiomatization and in our proof will be played by a defined operator $D_n p$ which acts as the *difference operator* on the class of cube frames, see Section 7.1. For its definition we need some auxiliary operators:

$$
\begin{aligned}
O_{ij}\phi &= \Diamond_i(\iota\delta_{ij} \wedge \phi) & (i \neq j), \\
E_i^n \phi &= \Diamond_0 \ldots \Diamond_{i-1}\Diamond_{i+1} \ldots \Diamond_{n-1}\phi, \\
D_n\phi &= \bigvee_{j\neq i} O_{ji}\Diamond_i(\neg\iota\delta_{ij} \wedge E_i^n \phi).
\end{aligned}
$$

The definition of D_n may look fairly complex, but it is directly based on the observation that two n-tuples s and t are *distinct* if and only for some coordinate i, s is distinct from t_i.

Proposition 7.47 D_n *acts as the difference operator on the class of cubes.*

Proof. Let $\mathfrak{M} = (\mathfrak{C}_n(U), V)$ be a cube model. We will show that

$$\mathfrak{M}, s \Vdash D_n p \text{ iff } \mathfrak{M}, t \models p \text{ for some } t \text{ such that } s \neq t. \qquad (7.17)$$

For the sake of a clear exposition we assume that $n = 3$, so that we may write $s = (s_0, s_1, s_2)$.

For the left to right direction of (7.17), suppose that $\mathfrak{M}, s \Vdash D_n p$. Without loss of generality we may assume that $s \Vdash O_{10}\Diamond_0(\neg\iota\delta_{01} \wedge E_0^n p)$. By definition of O_{10} it follows that $(s_0, s_0, s_2) \Vdash \Diamond_0(\neg\iota\delta_{01} \wedge E_0^n p)$. This in its turn implies that there is some s_0' such that $(s_0', s_0, s_2) \Vdash \neg\iota\delta_{01}$ and $(s_0', s_0, s_2) \Vdash E_0^n p$. It is easily seen that the meaning of E_0^n is given by

$$\mathfrak{M}, u \Vdash E_i^n \psi \text{ iff } \mathfrak{M}, v \models \psi \text{ for some } v \text{ such that } u_i = v_i,$$

$(s_0', s_0, s_2) \Vdash E_0^n p$ means that there is some n-tuple t such that $t \Vdash p$ and $s = t_0$. But it follows from $(s_0', s_0, s_2) \Vdash \neg\iota\delta_{01}$ that $s_0 \neq s_0'$, so that we find that

$t_0 \neq s_0$. But then, indeed, t is distinct from s. We leave it to the reader to prove the right to left direction of (7.17).   ⊣

However, the connection between D_n and the class of cubes is far tighter than this proposition suggests. In fact, the cubes are the *only* frames on which D_n acts as the difference operator, at least, against the right background of the class HCF_n of *hypercylindric frames*.

Definition 7.48 A CML_n-frame is called *hypercylindric* if the following formulas are valid on it:

$(CM1_i)$ $p \rightarrow \Diamond_i p$,

$(CM2_i)$ $p \rightarrow \Box_i \Diamond_i p$,

$(CM3_i)$ $\Diamond_i \Diamond_i p \rightarrow \Diamond_i p$,

$(CM4_{ij})$ $\Diamond_i \Diamond_j p \rightarrow \Diamond_j \Diamond_i p$,

$(CM5_i)$ $\iota\delta_{ii}$,

$(CM6_{ij})$ $\Diamond_i(\iota\delta_{ij} \wedge p) \rightarrow \Box_i(\iota\delta_{ij} \rightarrow p))$ $(i \neq j)$,

$(CM7_{ijk})$ $\iota\delta_{ij} \leftrightarrow \Diamond_k(\iota\delta_{ik} \wedge \iota\delta_{kj})$ $(k \notin \{i, j\})$,

$(CM8_{ij})$ $(\iota\delta_{ij} \wedge \Diamond_i(\neg p \wedge \Diamond_j p)) \rightarrow \Diamond_j(\neg\iota\delta_{ij} \wedge \Diamond_i p)$ $(i \neq j)$.   ⊣

All these axioms are Sahlqvist formulas and thus express first-order properties of frames. Clearly, the axioms $CM1$–3 together say that each T_i is an equivalence relation. $CM6_{ij}$ then means that in every T_i-equivalence class there is *at most one* element on the diagonal E_{ij} $(i \neq j)$. One can combine this fact with the (first-order translations of) $CM5_j$ and $CM7_{jji}$ to show that every T_i-equivalence class contains *exactly* one representative on the E_{ij}-diagonal. Apart from this effect, the contribution of $CM7$ is rather technical. Finally, the meaning of $CM4$ and $CM8$ is best made clear by Figure 7.2, where the straight lines represent the antecedent of the first-order correspondents, and the dotted lines, the relations holding of the 'old' states and the 'new' ones given by the consequent.

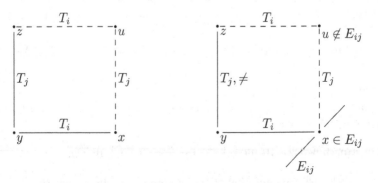

Fig. 7.2. The meaning of $CM4_{ij}$ (left) and $CM8_{ij}$ (right)

he key theorem in our completeness proof is the following.

Theorem 7.49 *For any frame $\mathfrak{F}$ in* $\mathsf{HCF_n}$*,* D_n *acts as the difference operator on* $\mathfrak{F}$ *if $\mathfrak{F}$ is a cube.*

Proof. We have already proved the left to right direction of this equivalence in Proposition 7.47. The proof of the other direction is technically rather involved and falls outside the scope of this book.   ⊣

In fact, with Theorem 7.49 we have all the material in our hands to prove the desired completeness result.

Definition 7.50 Consider the following modal derivation system Ω_n. Its axioms are (besides the ones of the minimal modal logic for the similarity type CML_n), the formulas $CM1$–8; as its derivation rules we take, besides the standard ones, also the D_n-rule:

$$\frac{\vdash (p \wedge \neg D_n p) \to \theta}{\vdash \theta}.$$

As usual, Ω_n will also denote the *logic* generated by this derivation system.   ⊣

Theorem 7.51 Ω_n *is sound and strongly complete with respect to the class* C_n.

Proof. It follows immediately from Theorem 7.6 and Theorem 7.49 that we obtain a complete axiomatization for C_n if we extend Ω_n with the D_n-versions of the axioms *symmetry*, *pseudo-transitivity* and D-*inclusion*. However, as its turns out, these axioms are valid on the class of hypercylindric frames, so they are already derivable in Ω_n (even without the use of the D_n-rule). From this, the theorem is immediate.   ⊣

Exercises for Section 7.5

5.1 Let n and m be natural numbers such that $n < m$, and consider a CML_n-formula ϕ. First, observe that ϕ is also a CML_m-formula. Prove that $\mathsf{C}_n \Vdash \phi$ iff $\mathsf{C}_m \Vdash \phi$. Conclude that our definition of an MLR_n-formula being *first-order valid*, is unambiguous.

5.2 Prove that the formula $\Diamond_0 \cdots \Diamond_{n-1} p$ acts as the global modality on the class of hypercylindric frames. That is, show that for any model $\mathfrak{M}$ based on such a frame we have that

$$\mathfrak{M}, s \Vdash \Diamond_0 \cdots \Diamond_{n-1} p \text{ iff } \mathfrak{M}, t \Vdash p \text{ for some } t \text{ in } \mathfrak{M}.$$

Which of the axioms $CM1$–$CM8$ are actually needed for this?

5.3 Let L_n^- denote the equality-free fragment of L_n^r; that is, all atomic formulas are of the form $P v_0 \ldots v_{n-1}$. In an obvious way we can define relativized assignment frames for this language. Prove that the satisfiability problem for L_n^- in this class of frames can be solved in PSPACE.

7.5.4 Prove that every hypercylindric CML_2-frame is the bounded morphic image of a square frame (that is, a 2-cube). Use this fact to find a complete axiomatization for the class C_2 that only uses the standard modal derivation rules.

7.5.5 Let CF_n be the class of cylindric frames, that is, those CML_n-frames that satisfy the axioms $CM1-CM7$. The class of n-dimensional cylindric algebras is defined as $CA_n = \mathbb{SPCm}CF_n$. The classes HCF_n and HCA_n are defined similarly, now using all axioms $CM1-CM8$.

(a) Prove that CA_n and HCA_n are canonical, that is, closed under taking canonical embedding algebras.
(b) Prove that CA_n and HCA_n are varieties.

7.5.6 A *full n-dimensional cylindric set algebra* is an algebra of the form

$$(\mathcal{P}(U^n), \cup, -, \varnothing, C_i, Id_{ij})_{i,j<n}.$$

Here the *i-th cylindrification* is defined as the map $C_i : \mathcal{P}(U^n) \to \mathcal{P}(U^n)$ given by

$$C_i(X) = \{s \in U^n \mid t \in X \text{ for some } t \text{ in } X \text{ with } s \equiv_i t\}.$$

If we close the class of these algebras under products and subalgebras, we arrive at the variety RCA_n of *representable n-dimensional cylindric algebras*.

(a) Prove that every representable n-dimensional cylindric algebra is a boolean algebra with operators.
(b) Prove that RCA_n is contained in the classes CA_n and HCA_n of the previous exercise.
(c) Prove that RCA_n is canonical. (Hint: use Theorem 7.49 to show that the class C_n of n-dimensional cubes is first-order definable in the frame language of CML_n.)

7.6 A Lindström Theorem for Modal Logic

Throughout this book we have seen many examples of modal languages, especially in the present chapter. To get a clear picture of the emerging spectrum, these languages may be classified according to their expressive power or their semantic properties. But what – if any – is the special status of the familiar modal languages defined in Chapter 1. If we focus on characteristic semantic properties, then clearly their invariance under bisimulations must be a key feature. But what else is needed to single out the (standard) modal languages?

The answer to this question is a modal analog of a classic result in first-order model theory: Lindström's Theorem. It states that, given a suitable explication of what 'classical logic' is, first-order logic is the strongest logic to possess the Compactness and Löwenheim-Skolem properties. To prove an analogous characterization result for modal logic we need to agree on a number of things:

- What will be the distinguishing property of the logic that we want to characterize (on top of its invariance for bisimulations)? To answer this question we will exploit the notion of degree introduced in Definition 2.28.

What is a suitable notion of an abstract modal logic? To answer this question we will introduce some bookkeeping properties from the formulation of the original Lindström Theorem for first-order logic, and add a further property having to do with invariance under bisimulations.

Our plan for this section is to discuss each of the above items, one after the other, and to conclude with a Lindström Theorem for modal logic.

Background material

Throughout this section models for modal languages are *pointed models* of the form $(\mathfrak{M}, w)$, where $\mathfrak{M}$ is a relational structure and w is an element of $\mathfrak{M}$ (its *distinguished point*) at which evaluation takes place. Our main reasons for adopting this convention are the following. First, the basic semantic unit in modal logic simply *is* a structure together with a distinguished node at which evaluation takes place. Second, some of the results below admit smoother formulations when we adopt the *local* perspective of pointed models.

Bisimulations between pointed models $(\mathfrak{M}, w)$ and $(\mathfrak{N}, v)$ are required to link the distinguished points w and v.

Definition 7.52 (In-degree) Let τ be a modal similarity type, and let $\mathfrak{M}$ be a τ-model. The *in-degree* of a state u in $\mathfrak{M}$ is the number of times u occurs as a non-first argument in a relation: $Rw \ldots u \ldots$. More formally, it is defined as

$$|\{\vec{w} \in \mathfrak{M}^{<\omega} \mid \text{for some } R \text{ and } i > 1, u = w_i \text{ and } R^{\mathfrak{M}} w_1 \ldots w_i \ldots w_n)\}|,$$

where $\mathfrak{M}^{\omega}$ is the collection of all finite sequences of elements in $\mathfrak{M}$. ⊣

In addition to the in-degree of an element of a model, we will also need to use the notion of *height* as defined in Definition 2.32.

Below we will need models with nice properties, such as a low in-degree or finite height for each of its elements. To get such models, we use the notion of forcing. Fix a similarity type τ. A property P of models is $\underline{\leftrightarrow}_\tau$-*enforceable*, or *enforceable*, iff for every pointed τ-model $(\mathfrak{M}, w)$, there is a pointed τ-model $(\mathfrak{N}, v)$ with $(\mathfrak{M}, w) \underline{\leftrightarrow}_\tau (\mathfrak{N}, v)$ and $(\mathfrak{N}, v)$ has P. For example, the property 'every element has finite height' is enforceable. To see this, let $(\mathfrak{M}, w)$ be a pointed τ-model; we may assume that $\mathfrak{M}$ is generated by w. Let $(\mathfrak{N}, w)$ be the submodel of $\mathfrak{M}$ whose domain consists of all elements of finite height. Then $(\mathfrak{M}, w) \underline{\leftrightarrow}_\tau (\mathfrak{N}, w)$.

Proposition 7.53 *The following properties of models are enforceable:*

(i) *tree-likeness, and*

(ii) *the conjunction of 'having a root with in-degree 0' and 'every element (except the root) has in-degree at most 1'.*

Proof. Item (ii) follows from item (i). A proof of item (i) for similarity type
only involving diamonds is given in Proposition 2.15; for the general case, consu
Exercise 2.1.7.

We will characterize modal logic (in the sense of Definitions 1.12 and 1.23) t
showing that it is the only modal logic satisfying a modal counterpart of the origin
Lindström conditions: having a notion of *finite degree* which gives a fixed upp
bound on the height of the elements that need to be considered to verify a formul
recall Definition 2.28 for the definition.

To wrap up our discussion of background material needed for our Lindströ
Theorem, let us briefly recall some basic facts related to degrees and height. He
is the first of these facts; recall that $((\mathfrak{M}, w) \upharpoonright n, w)$ denotes the submodel of $\mathfrak{L}$
that is generated from w and that only has states of height at most n.

Proposition 7.54 *Let ϕ be a modal formula with $\deg(\phi) \leq n$. Then $(\mathfrak{M}, w) \Vdash
iff $((\mathfrak{M}, w) \upharpoonright n, w) \Vdash \phi$.*

Next, recall from Proposition 2.29 that, up to logical equivalence, there are on
finitely many non-equivalent modal formulas with a fixed finite degree over a fini
similarity type.

We say that $(\mathfrak{M}, w)$ and $(\mathfrak{N}, v)$ are *n-equivalent* if w and v satisfy the san
modal formulas of degree at most n.

Proposition 7.55 *Let τ be a finite similarity type. Let $(\mathfrak{M}, w)$, $(\mathfrak{N}, v)$ be tw
rooted models such that the roots have in-degree 0, every element different fro
the root has in-degree at most 1, and all nodes have height at most n.*
If $(\mathfrak{M}, w)$ and $(\mathfrak{N}, v)$ are $(n+1)$-equivalent, then $(\mathfrak{M}, w) \leftrightarrow (\mathfrak{N}, v)$.

Proof. Define $Z \subseteq A \times B$ by xZy iff:

$\text{height}(x) = \text{height}(y) = m$ and $(\mathfrak{M}, x)$ and $(\mathfrak{N}, y)$ are $(n-m)$-equivalent.

We claim that $Z : (\mathfrak{M}, w) \leftrightarrow (\mathfrak{N}, v)$. To prove this, we only show the for
condition. Assume xZy and $R^{\mathfrak{M}}xx_1 \ldots x_k$, where $\text{height}(x) = \text{height}(y) = n$
Then $n - m \geq 1$. Let $\triangle$ be the modal operator whose semantics is based on R.

As τ is finite, there are only finitely many non-equivalent formulas of degree
most $n - m - 1$. Let ψ_i be the conjunction of all non-equivalent modal form
las of at most this degree that are satisfied at x_i ($1 \leq i \leq k$). Then $(\mathfrak{M}, x)$
$\triangle(\psi_1, \ldots, \psi_k)$, and $\triangle(\psi_1, \ldots, \psi_k)$ has degree $n - m$. Hence, as xZy, $(\mathfrak{N}, y)$
$\triangle(\psi_1, \ldots, \psi_k)$. So there are $y_1, \ldots, y_k$ in $\mathfrak{N}$ such that $R^{\mathfrak{N}}yy_1 \ldots y_k$ and $(\mathfrak{N}, y_i)$
ψ_i ($1 \leq i \leq k$).

Now, as all states have in-degree at most 1, $\text{height}(x_i) = \text{height}(y_i) = m +
and $(\mathfrak{M}, x_i)$ and $(\mathfrak{N}, y_i)$ ($1 \leq i \leq k$) are $(n - (m+1))$-equivalent. Henc
$(\mathfrak{M}, x_i) \leftrightarrow_\tau (\mathfrak{N}, y_i)$. This proves the forth condition.

Abstract modal logic

The original Lindström Theorem for first-order logic starts from a definition of an abstract classical logic as a pair $(\mathcal{L}, \models_{\mathcal{L}})$ consisting of a set of formulas $\mathcal{L}$ and a satisfaction relation $\models_{\mathcal{L}}$ between $\mathcal{L}$-structures and $\mathcal{L}$-formulas that satisfies three bookkeeping conditions, an Isomorphism property, and a Relativization property which allows one to consider definable submodels. Then, an abstract logic extending first-order logic coincides with first-order logic if and only if it satisfies the Compactness and Löwenheim-Skolem properties. We will now set up our modal analog of Lindström's Theorem along similar lines.

The definition runs along the same lines as the definition of an abstract classical logic. An abstract modal logic is characterized by three properties: two bookkeeping properties, and a Bisimilarity property to replace the Isomorphism property.

Definition 7.56 (Abstract Modal Logic) By an *abstract modal logic* we mean a pair $(\mathcal{L}, \Vdash_{\mathcal{L}})$ with the following properties (here $\mathcal{L}$ is the set of formulas, and $\Vdash_{\mathcal{L}}$ is its satisfaction relation, that is, a relation between (pointed) models and $\mathcal{L}$-formulas):

(i) *Occurrence property.* For each ϕ in $\mathcal{L}$ there is an associated finite language $\mathcal{L}(\tau_\phi)$. The relation $(\mathfrak{M}, w) \Vdash_{\mathcal{L}} \phi$ is a relation between $\mathcal{L}$-formulas ϕ and structures $(\mathfrak{M}, w)$ for languages $\mathcal{L}$ containing $\mathcal{L}(\tau_\phi)$. That is, if ϕ is in $\mathcal{L}$, and $\mathfrak{M}$ is an $\mathcal{L}$-model, then the statement $(\mathfrak{M}, w) \Vdash_{\mathcal{L}} \phi$ is either true or false if $\mathcal{L}$ contains $\mathcal{L}(\tau_\phi)$, and undefined otherwise.

(ii) *Expansion property.* The relation $(\mathfrak{M}, w) \Vdash_{\mathcal{L}} \phi$ depends only on the *reduct* of $\mathfrak{M}$ to $\mathcal{L}(\tau_\phi)$. That is, if $(\mathfrak{M}, w) \Vdash_{\mathcal{L}} \phi$ and $(\mathfrak{N}, w)$ is an expansion of $(\mathfrak{M}, w)$ to a larger language, then $(\mathfrak{N}, v) \Vdash_{\mathcal{L}} \phi$.

(iii) *Bisimilarity property.* The relation $(\mathfrak{M}, w) \Vdash_{\mathcal{L}} \phi$ is preserved under bisimulations: if $(\mathfrak{M}, w) \underline{\leftrightarrow}_\tau (\mathfrak{N}, v)$ and $(\mathfrak{M}, w) \Vdash_{\mathcal{L}} \phi$, then $(\mathfrak{N}, v) \Vdash_{\mathcal{L}} \phi$. ⊣

If we compare the above definition to the list of properties defining an abstract classical logic, we see that it is the Bisimilarity property that determines the *modal character* of an abstract modal logic.

Obviously, ordinary modal formulas provide an example of an abstract modal logic, but so does propositional dynamic logic. In contrast, the language of basic temporal logic provides an example of a logic that is *not* an abstract modal logic, as formulas from basic temporal logic are not preserved under bisimulations.

Next, we need to say what we mean by '$(\mathcal{L}, \Vdash_{\mathcal{L}})$ extends basic modal logic' and by closure under negation.

Definition 7.57 We say that $(\mathcal{L}, \Vdash_{\mathcal{L}})$ *extends* modal logic if for every basic modal formula there exists an equivalent $\mathcal{L}$-formula, that is, if for each basic modal formula ϕ there exists an $\mathcal{L}$-formula ψ such that for any model $(\mathfrak{M}, w)$ we have $(\mathfrak{M}, w) \Vdash \phi$ iff $(\mathfrak{M}, w) \Vdash_{\mathcal{L}} \psi$.

Also, $(\mathcal{L}, \Vdash_{\mathcal{L}})$ is *closed under negation* if for all $\mathcal{L}$-formulas ϕ there exists an $\mathcal{L}$-formula $\neg\phi$ such that for all models $(\mathfrak{M}, w)$, $(\mathfrak{M}, w) \Vdash \phi$ iff $(\mathfrak{M}, w) \not\Vdash \neg\phi$. $\dashv$

Of course, propositional dynamic logic is an example of an abstract modal logic that extends (basic) modal logic.

Logics in the sense of Definition 7.56 deal with the same class of pointed models as (basic) modal logic, and only the formulas and satisfaction relation may be different. This implies, for example, that intuitionistic logic or the hybrid logics considered in Section 7.3 are not abstract modal logics: their models need to satisfy special constraints. The original Lindström characterization of first-order logic suffers from similar limitations (by not allowing ω-logic as a logic, for example).

As a final step in our preparations, we need to say what the notion of degree means in the setting of an abstract modal logic.

Definition 7.58 (Notion of Finite Degree) An abstract modal logic has a *notion of finite degree* if there is a function $\deg_{\mathcal{L}} : \mathcal{L} \to \omega$ such that for all $(\mathfrak{M}, w)$, all ϕ in $\mathcal{L}$,

$$(\mathfrak{M}, w) \Vdash_{\mathcal{L}} \phi \quad \text{iff} \quad ((\mathfrak{M}, w) \upharpoonright \deg_{\mathcal{L}}(\phi)), w \Vdash_{\mathcal{L}} \phi.$$

If $\mathcal{L}$ extends (basic) modal logic, we assume that $\deg_{\mathcal{L}}$ behaves regularly with respect to standard modal operators and proposition letters. That is, if $\triangle$ is a modal operator (see Definition 1.12), then $\deg_{\mathcal{L}}(p) = 0$ and $\deg_{\mathcal{L}}(\triangle(\phi_1, \dots, \phi_n)) = 1 + \max\{\deg_{\mathcal{L}}(\phi_i) \mid 1 \leq i \leq n\}$.

Finally, two models $(\mathfrak{M}, w)$ and $(\mathfrak{N}, v)$ for the same language are *$\mathcal{L}$-equivalent* if for every ϕ in $\mathcal{L}$, $(\mathfrak{M}, w) \Vdash \phi$ iff $(\mathfrak{N}, v) \Vdash \phi$. $\dashv$

Having a finite degree is a very restrictive property, which is not implied by the finite model property (f.m.p.). To see this, recall that propositional dynamic logic has the f.m.p.: it has the property that every satisfiable formula ϕ is satisfiable on a model of size at most $|\phi|^3$, where ϕ is the length of ϕ. However, it does not have a notion of finite degree. To see this, consider the model (ω, R_a, V), where R_a is the successor relation and V is an arbitrary valuation, and let $\phi = [a^*]\langle a \rangle \top$; clearly $(\omega, R_a, V), 0 \Vdash \phi$. But for no $n \in \omega$ does the restriction $(\omega, R_a, V) \upharpoonright n$ satisfy ϕ at 0. It follows that PDL does not have a notion of finite degree.

Characterizing modal logic

We are almost ready now to prove our characterization result. The following lemma is instrumental.

Lemma 7.59 *Let* $(\mathcal{L}, \Vdash_{\mathcal{L}})$ *be an abstract modal logic which is closed under negation. Assume* $\mathcal{L}$ *has a notion of finite degree* $\deg_{\mathcal{L}}$. *Let* ϕ *be an* $\mathcal{L}$-*formula with* $\deg_{\mathcal{L}}(\phi) = n$. *Then, for any two models* $(\mathfrak{M}, w)$, $(\mathfrak{N}, v)$ *such that* $(\mathfrak{M}, w)$ *and* $(\mathfrak{N}, v)$ *are* n-*equivalent, we have that* $(\mathfrak{M}, w) \Vdash_{\mathcal{L}} \phi$ *implies* $(\mathfrak{N}, v) \Vdash_{\mathcal{L}} \phi$.

Proof. Assume that the conclusion of the lemma does not hold. Let $(\mathfrak{M}, w)$, $(\mathfrak{N}, v)$ be such that $(\mathfrak{M}, w)$ and $(\mathfrak{N}, v)$ are n-equivalent, but $(\mathfrak{M}, w) \Vdash_{\mathcal{L}} \phi$ and $(\mathfrak{N}, v) \Vdash_{\mathcal{L}} \neg\phi$.

By the Occurrence and Expansion properties we may assume that $\mathcal{L} = \mathcal{L}(\tau_\phi)$, where $\mathcal{L}(\tau_\phi)$ is the finite language in which ϕ lives.

By Proposition 7.53 we can assume that $(\mathfrak{M}, w)$ and $(\mathfrak{N}, v)$ are rooted such that the roots have in-degree 0, while all other nodes have in-degree at most 1. Then $((\mathfrak{M}, w) \restriction n, w)$ and $((\mathfrak{N}, v) \restriction n, v)$ are n-equivalent, and $((\mathfrak{M}, w) \restriction n, w) \Vdash_{\mathcal{L}} \phi$ but $((\mathfrak{N}, v) \restriction n, v) \Vdash_{\mathcal{L}} \neg\phi$. In addition $((\mathfrak{M}, w) \restriction n, w)$ and $((\mathfrak{N}, v) \restriction n, v)$ both have in-degree 1 and roots of in-degree 0. By Proposition 7.55 it follows that $((\mathfrak{M}, w) \restriction n, w)$ and $((\mathfrak{N}, v) \restriction n, v)$ are bisimilar – but now we have a contradiction with the Bisimilarity property as $((\mathfrak{M}, w) \restriction n, w)$ and $((\mathfrak{N}, v) \restriction n, v)$ are bisimilar but do not agree on ϕ. $\dashv$

Theorem 7.60 *Let* $(\mathcal{L}, \Vdash_{\mathcal{L}})$ *extend modal logic. If* $(\mathcal{L}, \Vdash_{\mathcal{L}})$ *has a notion of finite degree, then it is equivalent to the modal language as defined in Definition 1.12.*

Proof. We must show that every $\mathcal{L}$-formula ϕ is $\mathcal{L}$-equivalent to a basic modal formula ψ, that is, for all $(\mathfrak{M}, w)$, $(\mathfrak{M}, w) \Vdash_{\mathcal{L}} \phi$ iff $(\mathfrak{M}, w) \Vdash_{\mathcal{L}} \psi$. As before, by the Occurrence and Expansion properties we may restrict ourselves to a finite language. Moreover, ϕ has a basic modal equivalent iff it has such an equivalent with the same degree; so we have to locate the equivalent we are after among the basic modal formulas whose degree equals the $\mathcal{L}$-degree of ϕ.

Assume $n = \deg_{\mathcal{L}}(\phi)$. By Proposition 2.29 there are only finitely many (non-equivalent) basic modal formulas whose degree equals n; assume that they are all contained in Γ_n. It suffices to show the following:

> if $(\mathfrak{M}, w)$ and $(\mathfrak{N}, v)$ agree on all formulas in Γ_n, then they agree on ϕ. (7.18)

For then, ϕ will be equivalent to a boolean combination of formulas in Γ_n. To see this, reason as follows. The relation 'satisfies the same formulas in Γ_n' is an equivalence relation on the class of all models; as Γ_n is finite, there can only be finitely many equivalence classes. Choose representatives $(\mathfrak{M}_1, w_1), \ldots, (\mathfrak{M}_m, w_m)$, and for each i, with $1 \leq i \leq m$, let ψ_i be the conjunction of all formulas in Γ_n that are satisfied by $(\mathfrak{M}_i, w_i)$. Then ϕ is equivalent to $\bigvee\{\psi_i \mid (\mathfrak{M}_i, w_i) \Vdash_{\mathcal{L}} \phi\}$.

Now to conclude the proof of the theorem we need only observe that condition (7.18) is exactly the content of Lemma 7.59. $\dashv$

To conclude this section a few remarks are in order. First, the property of having a notion of finite degree can be characterized algebraically in terms of preservation under ultraproducts over the natural numbers; Theorem 7.60 can then be reformulated accordingly.

Second, in the proof of the Lindström Theorem the basic modal formula ψ that is found as the equivalent of the abstract modal formula ϕ is in the same vocabulary as ϕ. This means, for example, that the only abstract modal logic over a binary relation that has a notion of finite degree is the standard modal logic with a single modal operator $\Diamond$.

Here, we have only covered the modal logics as defined in Definition 1.12; in some cases extensions beyond this pattern can easily be obtained. As a first example, consider the basic temporal language with operators F and P, where $x \Vdash Fp$ ($x \Vdash Pp$) iff for some y, Rxy and $y \Vdash \phi$ (Ryx and $y \Vdash \phi$). Consider *temporal bisimulations* in which one not only looks forward along the binary relation, but also backward, and adopt the notion of height accordingly. Given the obvious definition of an *abstract temporal logic*, standard temporal logic is the only temporal logic over a single binary relation that has a notion of finite degree.

7.7 Summary of Chapter 7

▶ *Logical Modalities*: Logical modalities receive a fixed interpretation in every model. Simple examples are the past tense operator P, the global diamond E, and the difference operator D. As well as enhancing expressivity, some of them (notably P and D) make it possible to prove general completeness theorems using additional rules of proof.

▶ *Algebra of Diamonds*: Some modal languages offer not just a single logical modality but an entire algebra of diamonds. Good examples are PDL and BML.

▶ *Since and Until*: The since and until operators are interesting in applied logic because they enable us to specify guarantee properties. They are mathematically interesting because they are expressively complete over Dedekind complete total orders.

▶ *Completeness-via-Completeness*: While deductive completeness of since/until logic can be proved using standard modal techniques, for Dedekind complete total order there is an interesting alternative: taking a detour via expressive completeness.

▶ *Hybrid Logic*: The basic hybrid language lets us refer to states using nominals atomic symbols true at exactly one state in every model. Some stronger hybrid languages allow us to bind nominals.

▶ *Hybrid Proof Theory*: We can define a rule of proof called PASTE in the basic hybrid language. This rule is essentially a sequent rule lightly disguised. With its help, a frame completeness result covering all pure formulas can be proved fairly straightforwardly.

▶ *Guarded Fragment*: As the standard translation shows, modalities are essentially macros which permit restricted forms of quantification. Abstracting from this insight leads to the guarded fragment, a decidable fragment of first-order logic with the final model property.

▶ *Packed Fragment*: By taking this observation even further, and noting that the mosaic method suffices to prove decidability, it is possible to isolate an even larger decidable fragment of first-order logic: the packed fragment. This fragment also has the finite model property.

▶ *Multi-Dimensional Modal Logic*: Multi-dimensional modal logic is essentially modal logic in which evaluation is performed at a sequence of states, rather than at a single state. By viewing variable assignments as sequences of states, it is possible to view first-order logic itself as a multi-dimensional modal logic.

▶ *Lindström's Theorem*: Given a suitable (bisimulation centered) explication of what an abstract modal logic is, our Lindström Theorem for modal logic says that the general modal languages defined in Definition 1.12 are the strongest ones to have a notion of finite degree.

▶ *Extended Modal Logic*: In many ways, this chapter is badly named. Among other things, we have just seen that not only is it possible to introduce globality, more complex quantifier alternations in satisfaction definitions, names for states, and evaluation at sequences of states, but we can do so without losing the properties that made modal logic attractive in the first place. So forget the 'extended'. As we said in the Preface: *it is all just modal logic*!

Notes

A really serious guide to extended modal logic would have to cover the (vast) literature on temporal logics, fixed point logics, and variants of PDL discussed in the theoretical computer science literature, plus formalisms such as feature and description logic, and much else besides. We do not have space to do all that, and the following Notes stick to the six topics discussed in the text. Nonetheless, with the help of the following remarks (coupled with a little judicious reference chasing) the reader should be able to form a coherent map of territory.

Logical Modalities. It is hard to precise about when the idea of adding fixed interpretation operators to modal languages came to be seen as standard. Certainly the writings of Johan van Benthem (for example, his book on temporal logic, his 'manual' on intensional logic, and his influential survey of correspondence theory) played an important role. So did the new applications of modal logic, particularly computer science (once you have seen PDL it is hard to believe that the basic modal language is the be-all and end-all of modal logic). At any rate, by the end of the 1980s the idea that modal languages are abstract tools for talking about relational structures – tools that it was not only legitimate, but actually *interesting* to extend – was well established in both Amsterdam and Bulgaria. Nowadays this view is taken for granted by many (perhaps most) modal logicians, and given this perspective the use of logical modalities is as natural as breathing.

Of course, many of the operators we now call 'logical' have been around a lot longer than that. In a way, the global modality has always been there (after all its just a plain old **S5** operator). But when did it first emerge as an *additional* operator? We are not sure. Prior used it on a number of occasions (see, for example, [365, Appendix B4]), though sometimes Prior's global modality is actually the master modality [∗] discussed in Section 6.5 (that is, sometimes Prior views globality as the reflexive transitive closure of the underlying relation).

But it seems fair to say that it was the Bulgarian-school who first exploited it systematically: it is the Swiss Army knife underlying their investigation of BML, and their work on hybrid logic. Goranko and Passy [192] is a systematic study of the global modality as an additional operator, and is the source of Theorem 7.1, the Goldblatt-Thomason theorem for $ML(\Diamond, \mathrm{E})$. The operator has also been studied from an algebraic angle, being closely connected to the notion of a discriminator variety; these classes display nice algebraic behavior and have been intensively investigated in universal algebra. For, in the context of boolean algebra with operators, having the global modality is equivalent to having a so-called discriminator term; this is why in algebraic circles this modality is sometimes dubbed a 'unary discriminator term'; see Jipsen [247] for some information. The basic complexity results for the global modality were proved in Spaan's thesis [419]. Incidentally, the global modality is usually referred to as the 'universal' modality in the literature. However the word 'universal' suggests that we are working with a box, so we prefer the term 'global', which is appropriate for both boxes and diamonds.

The history of the difference operator is harder to untangle. It is probably due to von Wright [465] (who viewed it as a 'logic of elsewhere') and Segerberg gave an axiomatization in a *festschrift* for von Wright (see [406]). Segerberg's axiomatization, together with a more detailed completeness proof, was later published in [408]. But Segerberg treats D as an isolated modality. The use of D as an additional modality seems to have been proposed independently by Koymans [271] and Sain [397]. The difference operator is also discussed in Goranko [189]. For systematic investigation of D as an additional, logical modality, see de Rijke [378]. The D-Sahlqvist theorem in the text is due to Venema [444]. Theorem 7.8 is an unpublished result due to Szabolcs Mikulás.

BML is a Bulgarian school invention. The system is first described in Gargov, Passy and Tinchev [166] (as part of a wide ranging discussion of extended modal logic) and Gargov and Passy [165] concentrates on BML and gives proofs of the key completeness and decidability results. See also the results on modal definability in Goranko [189]. All these papers view modal languages as general tools for talking about structures, very much in the spirit of the present book. The window operator has an interesting independent history: van Benthem [38] used it as part of a logic of permissions and obligations, Goldblatt [176] used it to define negation in quantum logic, Humberstone [236] used it in a discussion of inaccessible worlds, while Gargov, Passy and Tinchev [166] view it as a 'logic of sufficiency' that balances the usual 'logic of necessity' provided by □. Complexity-theoretic aspects of BML have been studied and surveyed by Lutz and Sattler [303], while resolution-based decision procedures for extensions of BML and related languages are explored by Hustadt and Schmidt [238].

As we pointed out in the text, both BML and PDL are examples of modal languages equipped with highly structured collections of modal operators. The dynamic modal logic of de Rijke [384] is a further example, and many description logics allow for the construction of complex roles (that is, accessibility relations) by means of some or all of the booleans, converse, and sometimes even transitive closure and least fixed point constructors; see Donini *et al.* [115].

The algebraic counterparts of modal languages with structured collections of modal operators can best be phrased in terms of multi-sorted algebras, where the (algebraic counterparts of the) modal operators provide the links between the sorts. Kleene algebras [272] and Peirce algebras [382, 385] are two important examples. The former provide an algebraic semantics for PDL and consist of a boolean algebra and a regular algebra together with systematic links between them that are used to interpret the diamonds. The latter provide an algebraic semantics of dynamic modal logic and consist of a boolean and a relation algebra together with various links between them to interpret the modalities in the language.

Since and Until. The invention of since/until logic was a major breakthrough in the study of modal logic. Hans Kamp tells the story this way. In a semester-long course Arthur Prior gave on tense logic at UCLA in the fall of 1965, when Kamp had just started his PhD, Prior stressed that the P and F operators were strictly *topological*, and asked whether it was possible to develop some notion of *metric* time within the framework of tense logic. Now, a first requirement on such an enterprise is that it can express what it is for some proposition q to have been true since the last time some periodically true proposition p was true. Trying to find a genuinely topological tense logic in which these kinds of relations could be expressed lead Kamp to the definitions of since and until. As the technical interest of the new operators became clear, the original motivation seems to have been shelved (Kamp, personal communication, remarks that 'The question of how to embed a logic of metric temporal notions within a topological tense logic unfortunately never got properly off the ground.'). Kamp first showed that P and F cannot express since and until, and eventually succeeded in proving Theorem 7.12(i), the expressive completeness of the since and until operators over Dedekind complete total orders (see his thesis [258]). At that time, deductive completeness was the dominant interest in modal logic. Kamp's result showed that the neglected topic of modal expressivity deserved further attention, and can be regarded as a precursor to the study of correspondence theory that emerged in the 1970s.

The next step was taken by Dov Gabbay. Kamp's result was clearly important, but his direct proof was complex, and although Jonathan Stavi [422] succeeded in providing a direct proof of Theorem 7.12(ii), it was not obvious how to proceed. Matters were greatly simplified when Gabbay introduced the notion of *separability* (see [148, 150]). Roughly speaking, a language is separable over a class of models if every formula is equivalent to a boolean combination of atomic formulas, formulas that only talk about the past, and formulas that only talk about the future. This idea drastically simplifies the proofs of Theorem 7.12(i) and Theorem 7.12(ii), and opens the way to more general investigations. Nowadays a variety of techniques are used for proving expressive completeness results for modal (and other) languages; game-based approaches (see Immerman and Kozen [240]) have proved particularly useful. The best introduction to expressive completeness is the encyclopedic Gabbay, Hodkinson, and Reynolds [156]; both separability and game-based proofs are discussed, as well as many other results on since/until logic.

But what really made the until operator so popular is the simple observation made in the text: it offers precisely what is needed to express guarantee properties (this was first noted in Gabbay, Pnueli, Shelah, and Stavi [160]). Nowadays, until may well be the single best known modal operator (at least in computer science) and it occurs both in its original form, and in a number of variant forms in the study of linear and branching time temporal logics (see Clarke and Emerson [94] Goldblatt [177]).

Good discussions of step-by-step completeness proofs for since and until can be found in Burgess [78] and Xu [466]. The classification of properties of flow of time (in terms of safety, liveness, and guarantees) referred to in Section 7.: can be found in Manna and Pnueli's textbook [312] on using temporal logic fo specifying concurrent and reactive systems. Theorem 7.19 is due to Venema [443] the strategy of using expressive completeness to obtain axiomatic completenes results goes back at least to Gabbay and Hodkinson [155].

One final remark: in spite of the fact that its satisfaction definition makes use of more complex patterns of quantification, the since and until operators are genuinel modal. In particular, the notion of bisimulation can be adapted to these operator the only complication is that, instead of the simple 'complete the square' idea i lustrated in Figure 2.3 (65), bisimulations now need to match relational steps *pl intermediate intervals* in suitable ways. Kurtonina and de Rijke [288] contain solution to this issue as well as a survey of earlier proposals.

Hybrid Logic. Arthur Prior introduced and made systematic use of hybrid logic; see Prior [365] (in particular, Chapter 5 and Appendix B.3), several of the papers in Prior [366], and the posthumously published Prior and Fine [367]. Prior's systems typically allowed explicit quantification over states using $\forall$ and $\exists$, and contained the global modality. Technical aspects of such languages were explored in Bull [73], an important paper, which among other things notes that pure formulas give rise to easy frame completeness results. In the mid 1980s Passy and Tinchev independently reinvented the idea of 'names as formulas'. Their earliest paper [356] added nominals and the global modality to a rich version of PDL; in [357] they considered $\forall$ and $\exists$ (again in the setting of PDL); and [358], their beautiful essay on hybrid languages, remains one of the key papers on hybrid logic.

The subsequent history of hybrid logic revolves around attempts to find well-behaved sublanguages of such strong systems. The most obvious way to do this is the one explored in the text: treat nominals as names, rather than variables open to binding, and keep the underlying modal language relatively weak. Early papers which explore this option include Gargov and Goranko [164] (the basic modal language enriched with nominals and the global modality) and Blackburn [54] (the basic tense language enriched with nominals alone). The basic hybrid language discussed in the text can be viewed as an interesting compromise between simply adding nominals to the basic modal language (which makes the axiomatics messier, as Exercise 7.3.7 shows) and adding both nominals and the global modality (which raises the complexity to EXPTIME-complete). A proof of Theorem 7.21 (that the basic hybrid language has a PSPACE-complete satisfiability problem) can be found in Areces, Blackburn and Marx [14]. For a more detailed look at the complexity of hybrid logic, see [13] by the same authors. Theorem 7.29 is a modification of results proved in Blackburn and Tzakova [63]. It simplifies a similar result proved in Gargov and Goranko [164] with the aid of the global modality.

But the idea of binding variables to states turns out to be important. Binding admits a rich expressivity hierarchy. For a start, even if binding with $\forall$ and $\exists$ is allowed, when there are no satisfaction operators in the language, the resulting language does *not* have full first-order expressivity; see Blackburn and Seligman [59]. Moreover, as we mentioned in the text, the ↓ binder simply binds variables to the *current* state; in effect, it lets us create a name for the here-and-now (see Goranko [190], Blackburn and Seligman [59, 60], Blackburn and Tzakova [63]). If we enrich the basic hybrid language with the ↓ binder we obtain a hybrid language which corresponds to precisely the fragment of the first-order correspondence language which is invariant under generated submodels. This is proved in Areces, Blackburn and Marx [14] by isolating notions of bisimulation suitable for various hybrid languages and proving a characterization theorem. The paper also links these notions of bisimulation to restricted forms of Ehrenfeucht-Fraïssé games.

Hybrid logic provides a natural setting for modal proof theory. Seligman [411] is the pioneering paper here, and Seligman [412] discusses satisfaction operator based natural deduction and sequent systems. Blackburn [56] defines satisfaction operator driven tableaux and sequent systems and uses Hintikka sets to prove an analog of Theorem 7.29. Tzakova [439] combines the use of nominals with the prefix systems of Fitting [137]. Demri [107] defines a sequent calculus for the basic tense language enriched with nominals, and Demri and Goré [108] introduce a display calculus for the basic tense language enriched with nominals and D.

Hybrid logics turn up naturally in a number of applications. The AVMs used in computational linguistics (recall Example 1.17) can be viewed as modal logics: path re-entrancy tags are treated as nominals (see, for example, Blackburn and Spaan [61]). And while it has long been known that description logics are notational variants of modal logics, this relation only holds at the level of concepts. So-called A-Box (or assertional) reasoning – that is, reasoning about how concepts apply to particular individuals – corresponds to a restricted use of satisfaction operators, while the 'one-of' operators used in some versions of description logic are essentially disjunctions of nominals; see Blackburn and Tzakova [62], Areces and de Rijke [15], and Areces's PhD thesis [12]. Nominals also turn up in the Polish tradition of modal logics for information systems and rough-set theory: see Konikowska [269, 270]. They also provide a natural model of tense and other forms of temporal reference in natural language (see Blackburn [55]).

A final remark. The basic hybrid language shows that sorting is interesting in the setting of modal logic – so why not introduce further sorts? In fact, this step was already taken in Bull [73] who introduced a third sort of atomic symbol: *path nominals*, true at precisely the points belonging to some path through the model. For more information on hybrid logic, see the Hybrid Logic home page at www.hylo.net. For a recent 'manifesto' on hybrid logic that touches on most of the themes just mentioned, see Blackburn [57].

The Guarded Fragment. The guarded fragment was introduced by Andréka, van Benthem and Németi in 1994. The roots of the decidability proof date back to 1986, when Németi [340] showed that the equational theory of the class of so called relativized cylindric set algebras is decidable. The first-order counterpart of this result is that a certain subfragment of the guarded fragment is decidable.

The importance of this result for first-order logic was realized in 1994 when Andréka, van Benthem and Németi introduced the guarded fragment and showed that many nice properties of the basic modal system **K** generalize to it. In particular, the authors established a characterization in terms of guarded bisimulations, decidability and a kind of tree model property. The journal version of the paper is [9]. Some time later van Benthem was able to generalize some of the results, introducing the loosely guarded fragment in [49]. The slightly more general packed fragment was introduced in Marx [317] in order to give a semant characterization in terms of packed bisimulations. (An example of a packed sentence which is not equivalent to a loosely guarded sentence in the same signature is $\exists xyz\,(\exists wCxyw \wedge \exists w\,Cxzw \wedge \exists w\,Czyw \wedge \neg Cxyz)$.)

The mosaic based decision algorithms of Andréka, van Benthem and Németi were essentially optimal: a result established by Grädel [194]. In this paper, Grädel also defines and establishes the loose model property for the loosely guarded fragment. Our definition of a loose model is based on the definition of a tree model given there. Grädel and Walukiewicz [197] showed that the same bounds obtain when the guarded fragment is expanded with least and greatest fixed point operators. Marx, Mikulás and Schlobach [319] defined a PSPACE-complete guarded fragment (both locality principles) with the finite tree model property.

The finite model property for the guarded fragment, and several subfragments of the packed fragment, was established in an algebraic setting by Andréka, Hodkinson and Németi [7]. Grädel [194] provides a direct proof for the guarded fragment. The remaining open question for the full packed fragment was solved affirmatively by Hodkinson [231]. All these results are based on variants of a result due to Herwig [222]. The use of Herwig's Theorem to establish the finite model property and to eliminate the need for step-by-step constructions is due to Hirsch *et al.* [227].

Multi-Dimensional Modal Logic. The idea of evaluating modal languages at sequences of points, rather than at the points simpliciter, is extremely natural, so it is no surprise that over the years modal logicians with very diverse interests have devised multi-dimensional systems.

It seems that logicians interested in natural language were first off the mark. Natural language utterances are so context dependent, that evaluating at sequences of points (each coordinate modelling a different aspect of context) proved a useful idea. Evaluation at pairs of points is built into Montague's [336] general framework for natural language semantics. Kamp's [259] classic analysis of the word 'now' uses a second coordinate to keep track of utterance time. Vlach [452] provided an analysis of the word 'then', and in a series of papers, Åqvist and co-workers [11] developed rich multi-dimensional modal logics for analyzing temporal phenomena in natural language. Before long, such systems were subjected to rigorous logical investigation: see, for example, Segerberg's elegant decidability and completeness result in [405], and Gabbay's work on expressiveness and other topics (much of which reappeared in later work by Gabbay, Hodkinson and Reynolds [156]).

Somewhat later, a rich source of inspiration came from logic itself. Some work here, such as the sorted modal logic PREDBOX of Kuhn [286], fitted in the tradition of Quine-style first-order logic without variables, but most of it was linked, one way or another, with the algebraic logic framework of the Tarskian school (see the notes of Chapter 5). This certainly applies to the multi-dimensional logics that we presented in Section 7.5. Venema [445], from which our Theorem 7.51 originates, made the connection between modal logic and cylindric algebras. Subsequent research drew on existing ideas on relativized cylindric algebras (see Németi [340]) to use the modal framework to 'tame' first-order logic and its finite variable fragments (see our discussion of the abstract and relativized assignment frames in the text; more information on this program can be found in van Benthem [48] or Mikulás [329]). This line of work is closely related to arrow logic, which is a multi-dimensional modal logic in its own right (see Marx *et al.* [318] for more information) and in fact this strand of work ultimately lead to the isolation of the guarded fragment. All of these (and more) multi-dimensional modal logics are covered in the monograph Marx and Venema [320]; readers interested in complexity results should consult Marx [316].

Computer scientists have different motivations for studying multi-dimensional modal logics. In order to build formal models of an application domain, they need to take account of various features simultaneously. Of the wealth of literature on this topic we will just mention Fagin *et al.* [125], which concentrates on the combination of temporal and epistemic logics in the context of distributed systems. Such applications have led logicians to study various ways of constructing complex logics from relatively simple ones. A particularly interesting and mathematically non-trivial branch of multi-dimensional modal logics arises if one studies a modal language with various modal operators over a semantics in which the frames are cartesian products of frames for the individual operators. This area of so-called product logics, which has an early predecessor in Shehtman [413], has recently become very active; a monograph Gabbay *et al.* [159] is on its way.

Finally, multi-dimensional modal logic remains one of the most philosophically important branches of modal logic. Important references include Kaplan [264, 265], Stalnaker [421], and Chalmers [90].

The Lindström Theorem for Modal Logic. Theorem 7.60, a Lindström-type characterization of the modal languages defined in Definitions 1.9 and 1.12 is due to de Rijke [381]; the result was obtained as part of a general program to come up with modal counterparts of model-theoretic results in first-order logic [380]. The original first-order version of Lindström's Theorem was first presented in Lindström [302]. The original result states that, given a suitable explication of a 'classical logic', first-order logic is the strongest logic to possess the Compactness and Löwenheim-Skolem properties; it formed an important source of inspiration for the area of model-theoretic logics [26]. Definitions of the abstract notion of a logic can be found in Chang and Keisler [91] and in Barwise [25]. A very accessible presentation of Lindström's Theorem for first-order logic can be found in Doets [11, Chapter 4].

Appendix A

A Logical Toolkit

In this appendix we review basic first-order logic, define some model-theoretic concepts, introduce the ultraproduct construction, and briefly discuss second-order and infinitary logic. But it is not a self-contained introduction to first-order (or any other) logic: we assume the reader has had some prior exposure to formal logic and is comfortable with the basic ideas. If a significant portion of what follows is unfamiliar, you should consult an introduction to mathematical logic (Enderton [122] and Hodges [228] are good choices).

There are many sources for further information about model theory: Doets [111] is an approachable introductory text, Hodges [229] gives an encyclopedic view of the field, and Bell and Slomson [33] is a detailed guide to the ultraproduct construction. But the single most important source is probably Chang and Keisler [91]; our presentation of ultraproducts is largely based on theirs. For second-order logic, see Doets and van Benthem [112], Chapter 4 of Enderton [122], and Fitting [139]. For infinitary logic, see Keisler [267].

Languages, Models, and Satisfaction

A language of first-order logic is built from terms and formulas. *Terms* t are built from variables x, constants c, and function symbols f as follows:

$$t ::= x \mid c \mid f(t_1, \ldots, t_n).$$

The *atomic formulas* β are expressions of the form:

$$\beta ::= t_1 = t_2 \mid R(t_1, \ldots, t_n).$$

Here R is an n-ary relation symbol, and the t_i are terms of the language.

Arbitrary first-order *formulas* α are built from atomic formulas β using boolean operators and quantifiers as follows:

$$\alpha ::= \beta \mid \neg\alpha \mid \alpha \vee \alpha' \mid \exists x \, \alpha.$$

We define $\forall x\, \alpha$ to be $\neg \exists x\, \neg \alpha$. Other boolean connectives (in particular, $\wedge$, $\rightarrow$, $\leftrightarrow$, $\bot$, $\top$) can be defined in the standard way, as can the notions of *free* and *bound* variables of a formula. A first-order *sentence* is a formula without free variables.

Models for a first-order language $\mathcal{L}^1$ are tuples $\mathfrak{A} = (A, R, \ldots, f, \ldots, c, \ldots)$. Here A is a non-empty set (the *domain*, or *universe*), each R is a relation on the domain (a subset of A^n, for some n), each f is a function on the domain (a mapping from A^n to A, for some n) and each c is an element of the domain (the elements of A singled out in this way are often called *distinguished elements*). Models are essentially *relational structures* (see Definition 1.1), for n-place functions are $(n + 1)$-ary relations, and distinguished elements can be viewed as 0-ary functions.

The models used to interpret a language must 'match' it in certain obvious respects. First, the model must supply enough relations to interpret all the relation symbols of the language, enough functions to interpret all the function symbols of the language, and enough distinguished elements to interpret all the constants of the language. Furthermore, relations of arity n must be used to interpret n-ary relation symbols, and n-ary functions must be used to interpret n-ary function symbols. We will usually employ the same notation for a relation symbol and the actual relation interpreting this symbol, and likewise for function symbols and constants.

We need a mechanism to interpret free variables. Given a model $\mathfrak{A} = (A, R, \ldots, f, \ldots, c, \ldots)$, an *assignment* on $\mathfrak{A}$ is a function g that assigns an element of A to each variable of the language. Thus, given a model and an assignment we can interpret arbitrary terms through the obvious inductive definition: constants denote the corresponding distinguished elements, variables are interpreted using the assignment, and terms $f(t_1, \ldots, t_n)$ are interpreted by applying the interpretation of f to the interpretations of $t_1, \ldots, t_n$. We often call the interpretation of a term its *value*; $t^{\mathfrak{A}}[g]$ denotes the value of term t in model $\mathfrak{A}$ under assignment g.

The satisfaction definition is a 3-place relation $\models$ between a model, formula and assignment: $\mathfrak{A} \models \alpha[g]$ means that formula α is satisfied in model $\mathfrak{A}$ under assignment g. But before defining this relation, a notational point. Rather than using g, g', and so on to name arbitrary assignments, it is more common to use a notation that specifies which element has been assigned to which free variable. Thus if $\alpha(x_1, \ldots, x_n)$ is a formula in which $x_1, \ldots, x_n$ occur free, $\mathfrak{A}$ is a model and $a_1, \ldots, a_n$ are elements of $\mathfrak{A}$, then $\mathfrak{A} \models \alpha[a_1, \ldots, a_n]$ means that we are evaluating with respect to an assignment that assigns a_i to x_i (for $1 \leq i \leq n$). Sometimes this notation is convenient when defining the value of terms: in our discussion of ultraproducts we use expressions such as $t^{\mathfrak{A}}[a_1, \ldots, a_n]$ to denote the value of term t in model $\mathfrak{A}$ under an assignment that sends x_i to a_i, for $1 \leq i \leq n$.

The satisfaction relation is defined by induction on the structure of formula. The atomic cases are as follows: $\mathfrak{A} \models R(t_1, \ldots, t_n)[a_1, \ldots, a_n]$ iff the values of the terms $t_1, \ldots, t_n$ are R-related in the model $\mathfrak{A}$; and $\mathfrak{A} \models t_1 = t_2[a_1, \ldots, a_n]$ iff the values of terms t_1 and t_2 in $\mathfrak{A}$ are equal. The boolean cases are defined in the obvious way by $\mathfrak{A} \models \neg\alpha[a_1, \ldots, a_n]$ iff $\mathfrak{A} \not\models \alpha[a_1, \ldots, a_n]$; and $\mathfrak{A} \models \alpha \vee \alpha'[a_1, \ldots, a_n]$ iff either $\mathfrak{A} \models \alpha[a_1, \ldots, a_n]$ or $\mathfrak{A} \models \alpha'[a_1, \ldots, a_n]$. The case of the quantifier is defined as follows:

$$\mathfrak{A} \models \exists x\, \alpha[a_1, \ldots, a_n] \quad \text{iff} \quad \text{there exists } a \text{ in } \mathfrak{A} \text{ such that}$$
$$\mathfrak{A} \models \alpha[a_1, \ldots, a_n, x \mapsto a],$$

where $[a_1, \ldots, a_n, x \mapsto a]$ is the assignment that differs from $[a_1, \ldots, a_n]$, if at all, only in that the variable x is assigned the value a. Clearly, we have that $\mathfrak{A} \models \forall x\, \alpha[a_1, \ldots, a_n]$ iff for all a in $\mathfrak{A}$, $\mathfrak{A} \models \alpha[a_1, \ldots, a_n, x \mapsto a]$.

If $\mathfrak{A} \models \alpha[a_1, \ldots, a_n]$ then we say that the sequence $a_1, \ldots, a_n$ satisfies α in $\mathfrak{A}$.

It is easy to see that if a *sentence* is satisfied in a model under one assignment, then it is satisfied under all assignments; accordingly, if a sentence α is satisfied in a model $\mathfrak{A}$ under some assignment, then we write $\mathfrak{A} \models \alpha$ and say that α is *true* in $\mathfrak{A}$, or that $\mathfrak{A}$ is a *model* for α. Two models $\mathfrak{A}$ and $\mathfrak{B}$ are *elementarily equivalent* (notation: $\mathfrak{A} \equiv \mathfrak{B}$), if every sentence true in $\mathfrak{A}$ is true in $\mathfrak{B}$, and vice versa.

Definition A.1 (Validity and Semantic Consequence) A formula α is *valid* if for every model $\mathfrak{A}$ and every assignment, $\mathfrak{A} \models \alpha[a_1, \ldots, a_n]$.

Given a set of formulas Π (the premises) and a formula γ (the conclusion), we say that γ is a *semantic consequence* of Π (notation: $\Pi \models \gamma$), if for every model $\mathfrak{A}$ and every assignment

$$\mathfrak{A} \models \Pi[a_1, a_2, \ldots] \quad \text{implies} \quad \mathfrak{A} \models \gamma[a_1, a_2, \ldots].$$

That is, satisfiability of the premises (with respect to some assignment) guarantees satisfiability of the conclusion (with respect to the same assignment). ⊣

Basic Properties of First-Order Logic

First-order logic is *undecidable*: it is impossible to write a computer program that, given an arbitrary first-order formula as input, will stop after finitely many steps and (correctly) tell us whether the formula is valid or not. On the other hand, first-order validity is *recursively enumerable*. That is, it is possible to write a computer program that successively generates all valid formulas. (The relationship between decidability, undecidability, and recursive enumerability is discussed in Appendix C.)

The usual way of showing that the collection of first-order validities is recursively enumerable is to devise a (sound and complete) *proof system* for it. Many such systems are known (axiom systems, sequent systems, natural deduction systems) and though very different, they have one fundamental thing in common: they are purely *syntactic*. Proofs are essentially simple finite data structures (for example, lists or trees of formulas). Such a data structure is a proof of a formula if the symbols it contains fulfill certain (usually quite simple) syntactic criteria.

Let us write $\vdash \alpha$ to indicate that α is provable in a standard proof system. We would like $\vdash \alpha$ to hold if and only if $\models \alpha$, for if we could do this we would have reduced a complex semantic notion ($\models \alpha$ means satisfied in *all* models under all assignments, and there are lots of models and assignments) to a relatively simple syntactic one (patterns of symbols in finite data structures). It is not obvious that this can be done, but it can. Indeed, something better is the case. If we write $\Pi \vdash \gamma$ to indicate that the conclusion γ follows *syntactically* from the premises Π (this concept can be defined for any standard proof system) then we have:

Theorem A.2 (Soundness and Completeness) *Let Π be a set of first-order formulas, and α a first-order formula. Then $\Pi \vdash \alpha$ iff $\Pi \models \alpha$. When $\Pi = \varnothing$ we have as a special case that $\vdash \alpha$ iff $\models \alpha$.*

Proof. Completeness is the right to left implication: this assures us that the proof system captures *all* semantically correct inferences. Detailed proofs can be found in Chang and Keisler [91] and Enderton [122]. Hodges [228] has a good discussion of completeness proof strategies.

The left to right direction assures that the proof system does not produce semantic nonsense; this is called *soundness*. It is far easier to prove than completeness (typically, a fairly simple inductive argument suffices) and proofs for various types of proof system can be found in any standard text on mathematical logic.   ⊣

So first-order semantic consequence, and first-order validity, can be reduced to syntactic criteria on simple data structures. Thus first-order validities can be recursively enumerated: we merely write a program that systematically generates all finite data structures of the appropriate kind, and checks whether they fulfill the criteria demanded of proofs.

The Completeness Theorem is one of the fundamental theorems of first-order logic, but it plays a relatively modest role in this book. More important for us are the two theorems that follow:

Theorem A.3 (Compactness) *Let Σ be a set of first-order formulas. If each finite subset of Σ has a model, then Σ itself has a model.*

Theorem A.4 (Löwenheim-Skolem Theorem) *Let Σ be a set of first-order formulas. If Σ has a model, then it has a countable model.*

We will make use of both these results (actually, we will generally use a strong version of the Löwenheim-Skolem Theorem which is discussed below).

Both the Compactness and Löwenheim-Skolem Theorems are purely model theoretic: they make no reference to provability in some proof system. The Completeness, Compactness and Löwenheim-Skolem Theorems together characterize
first-order logic.

This completes our survey of the basics of first-order logic. If much of it was
unfamiliar, we suggest you consult Enderton [122] or Hodges [228].

Basic Model-Theoretic Concepts

Now for some basic model-theoretic concepts. First we need to know when two
models are *isomorphic*.

Definition A.5 (Isomorphism) Two models $\mathfrak{A}$ and $\mathfrak{A}'$ for the same first-order language are *isomorphic* if there is a bijective function f mapping A onto A' such
that

(i) For each n-place relation R of $\mathfrak{A}$ and the corresponding relation R' of $\mathfrak{A}'$,
$R(a_1,\ldots,a_n)$ iff $R'(fa_1,\ldots,fa_n)$.

(ii) For each m-place function F of $\mathfrak{A}$ and the corresponding function F' of $\mathfrak{A}'$,

$$f(F(a_1,\ldots,a_n)) = F'(fa_1,\ldots,fa_n).$$

(iii) For each distinguished element c of $\mathfrak{A}$ and the corresponding distinguished
element c' of $\mathfrak{A}'$, $f(c) = c'$.

A function f that satisfies these requirements is called an *isomorphism between* $\mathfrak{A}$
and $\mathfrak{A}'$. The notation $f : \mathfrak{A} \cong \mathfrak{A}'$ means that f is an isomorphism between $\mathfrak{A}$ and
$\mathfrak{A}'$. ⊣

In essence, isomorphic models are mathematically identical. Thus the following
proposition (which can be proved by induction on the structure of formulas) is
unsurprising:

Proposition A.6 *Let f be an isomorphism between $\mathfrak{A}$ and $\mathfrak{B}$. Then for all formulas $\alpha(x_1,\ldots,x_n)$ and n-tuples $a_1,\ldots,a_n \in A$, we have*

$$\mathfrak{A} \models \alpha[a_1,\ldots,a_n] \text{ iff } \mathfrak{B} \models \alpha[fa_1,\ldots,fa_n].$$

We are often confronted with situations in which a model we are interested in is
part of a larger one, or when we need to extend a given model to a bigger one. We
now define the basic model-theoretic notions useful in such cases.

Definition A.7 (Submodels and extensions) A model $\mathfrak{A}'$ is called a *submodel* of
$\mathfrak{A}$ (notation: $\mathfrak{A}' \subseteq \mathfrak{A}$), if $A' \subseteq A$ and

(i) Each n-place relation R' of $\mathfrak{A}'$ is the restriction to A' of the corresponding relation R of $\mathfrak{A}$.

(ii) Each n-place function f' of $\mathfrak{A}'$ is the restriction to A' of the corresponding function f of $\mathfrak{A}$.

(iii) Each distinguished element of $\mathfrak{A}'$ is the corresponding distinguished element of $\mathfrak{A}$.

If $\mathfrak{A}'$ is a submodel of $\mathfrak{A}$, then we say that $\mathfrak{A}$ is an *extension* of $\mathfrak{A}'$.   ⊣

That $\mathfrak{A}' \subseteq \mathfrak{A}$ is no guarantee that $\mathfrak{A}'$ and $\mathfrak{A}$ satisfy the same formulas – and in general, that is what we care about. This prompts the following definition:

Definition A.8 (Elementary Extension) $\mathfrak{B}$ is said to be an *elementary extension* of $\mathfrak{A}$, (notation: $\mathfrak{A} \preccurlyeq \mathfrak{B}$), if

(i) $\mathfrak{B}$ is an extension of $\mathfrak{A}$.

(ii) For any first-order formula $\alpha(x_1, \ldots, x_n)$ and any sequence $a_1, \ldots, a_n$ of elements in $\mathfrak{A}$, $a_1, \ldots, a_n$ satisfies α in $\mathfrak{A}$ iff it satisfies α in $\mathfrak{B}$.

When $\mathfrak{B}$ is an elementary extension of $\mathfrak{A}$ we also say that $\mathfrak{A}$ is an *elementary submodel* of $\mathfrak{B}$.

A mapping $f : A \to B$ is called an *elementary embedding* of $\mathfrak{A}$ into $\mathfrak{B}$, (notation: $f : \mathfrak{A} \preccurlyeq \mathfrak{B}$), if for all formulas $\alpha(x_1, \ldots, x_n)$ and n-tuples $a_1, \ldots, a_n \in A$, we have

$$\mathfrak{A} \models \alpha[a_1, \ldots, a_n] \text{ iff } \mathfrak{B} \models \alpha[fa_1, \ldots, fa_n].$$

That is, an elementary embedding of $\mathfrak{A}$ into $\mathfrak{B}$ is an isomorphism of $\mathfrak{A}$ onto an elementary submodel of $\mathfrak{B}$.   ⊣

When working with some model, it is often useful to move from the original language to a richer language in which every element on the model has a name:

Definition A.9 (Expansion) Let $\mathcal{L}$ be a first-order language, and $\mathfrak{A}$ a model for $\mathcal{L}$. We *expand* $\mathcal{L}$ to a new language $\mathcal{L}_A = \mathcal{L} \cup \{c_a \mid a \in A\}$ by adding a new constant symbol c_a for each element $a \in A$ (if $a \neq b$, then c_a and c_b are different symbols). We *expand* $\mathfrak{A}$ to the model $\mathfrak{A}_A = (\mathfrak{A}, a)_{a \in A}$ for $\mathcal{L}_A$ by stipulating that each element a of A is a distinguished element. Each new constant c_a of $\mathcal{L}_A$ is interpreted by the distinguished element a. If X is a subset of A, then $\mathcal{L}_X$ is the language $\mathcal{L} \cup \{c_a \mid a \in X\}$, and $\mathfrak{A}_X = (\mathfrak{A}, a)_{a \in X}$ is the obvious expansion of $\mathfrak{A}$ to a model for $\mathcal{L}_X$; that is, all the elements of X become distinguished elements.

Proposition A.10 $\mathfrak{B}$ *is an elementary extension of* $\mathfrak{A}$ *iff* $\mathfrak{A} \subseteq \mathfrak{B}$ *and*

$$(\mathfrak{A}, a)_{a \in A} \equiv (\mathfrak{B}, a)_{a \in A}.$$

We can now state the version of the Löwenheim-Skolem Theorem that will be most useful to us:

Theorem A.11 *Let $\mathfrak{A}$ be a model of cardinality α, and let the $|\mathcal{L}| \leq \beta \leq \alpha$, where $|\mathcal{L}|$ is the number of non-logical symbols in the language. Then $\mathfrak{A}$ has an elementary submodel of cardinality β. Furthermore, given any set $X \subseteq A$ of cardinality $\leq \beta$, $\mathfrak{A}$ has an elementary submodel of cardinality β which contains X.*

Actually, in two respects this result is more general than we need. First, we nearly always work with languages with at most countably many non-logical symbols. Secondly, we will always be interested in forming countable submodels (that is, we are interested in the case when β is $\aleph_0$). But the other two generalizations will be useful. First this form of the theorem guarantees that we can find not merely a submodel, but an *elementary* submodel. Second, we can select any (sufficiently small) subset of the original model we find interesting, and find an elementary submodel containing it.

Ultraproducts

This ultraproduct construction is an important tool for building new models out of old. Roughly speaking, it tells us how we can multiply together a collection of models to form a new model with the following property: any formula satisfied in most of the original models is satisfied in the new model, and vice versa. Mathematically, the notion 'most of' is cashed out with the aid of ultrafilters. Readers that are not familiar with this notion are advised to have a look at some of the exercises in Section 2.5.

Definition A.12 (Filters and Ultrafilters) Let W be a non-empty set. A *filter F over W* is a set $F \subseteq \mathcal{P}(W)$ such that

(i) $W \in F$.
(ii) If $X, Y \in F$, then $X \cap Y \in F$.
(iii) If $X \in F$ and $X \subseteq Z \subseteq W$, then $Z \in F$.

Obviously $\mathcal{P}(W)$ is itself a filter. A filter is called *proper* if it is distinct from $\mathcal{P}(W)$. An *ultrafilter over W* is a proper filter U such that for all $X \in \mathcal{P}(W)$, $X \in U$ if and only if $(W \setminus X) \notin U$. ⊣

A non-trivial example of a filter is the collection of all co-finite subsets of an infinite set. (A subset of an infinite set is co-finite if its complement is finite.) A large supply of filters is provided by the following definition.

Definition A.13 Let W be a non-empty set, and let E be a subset of $\mathcal{P}(W)$. By the *filter generated by* E we mean the intersection F of the collection of all filters over W which include E:

$$F = \bigcap\{G \mid E \subseteq G \text{ and } G \text{ is a filter over } W \,\}.$$

E has the *finite intersection property* if the intersection of any finite number of elements of E is non-empty.   ⊣

We have defined ultrafilters as a special kind of filters, satisfying an additional property. An alternative definition states that ultrafilters are *maximal* proper filters; that is, a filter is an ultrafilter if and only if it is proper but has no proper extensions. In many cases where we need to prove the existence of an ultrafilter containing a certain collection of sets, we apply the Ultrafilter Theorem.

Fact A.14 (Ultrafilter Theorem) *Fix a non-empty set* W. *Any proper filter over* W *can be extended to an ultrafilter over* W. *As a corollary, any subset of* $\mathcal{P}(W)$ *with the finite intersection property can be extended to an ultrafilter over* W.   ⊣

A special role is often played by the so-called *principal* ultrafilters.

Definition A.15 Let W be a non-empty set. Given an element $w \in W$, the *principal ultrafilter* π_w generated by w is the filter generated by the singleton set $\{w\}$. An equivalent definition would be to put $\pi_w = \{X \subseteq W \mid w \in X\}$.   ⊣

Are such sets really ultrafilters? Yes – see Exercise 2.5.2.
We are ready to introduce the ultraproduct construction. We first apply the construction to sets, and then to models. Suppose that U is an ultrafilter over a non-empty set I, and that for each $i \in I$, A_i is a non-empty set. Let $C = \prod_{i \in I} A_i$ be the cartesian product of those sets. That is: C is the set of all functions f with domain I such that for each $i \in I$, $f(i) \in A_i$. For two functions $f, g \in C$ we say that f and g are *U-equivalent* (notation $f \sim_U g$) if $\{i \in I \mid f(i) = g(i)\} \in U$.

Proposition A.16 *The relation* $\sim_U$ *is an equivalence relation on the set* C.

Definition A.17 Let f_U be the equivalence class of f modulo $\sim_U$, that is: $f_U = \{g \in C \mid g \sim_U f\}$. The *ultraproduct of* the sets A_i *modulo* U is the set of all equivalence classes of $\sim_U$. It is denoted by $\prod_U A_i$. So

$$\prod_U A_i = \{f_U \mid f \in \prod_{i \in I} A_i\}.$$

Let us now apply the same idea to models:

Definition A.18 Fix a first-order language $\mathcal{L}^1$, and let $\mathfrak{A}_i$ $(i \in I)$ be $\mathcal{L}^1$-models. The *ultraproduct* $\prod_U \mathfrak{A}_i$ *of* $\mathfrak{A}_i$ *modulo* U is the model described as follows:

(i) The universe A_U of $\prod_U \mathfrak{A}_i$ is the set $\prod_U A_i$, where A_i is the universe of $\mathfrak{A}_i$.

(ii) Let R be an n-place relation symbol, and R_i its interpretation in the model $\mathfrak{A}_i$. The relation R_U in $\prod_U \mathfrak{A}_i$ is given by

$$R_U f_U^1 \ldots f_U^n \text{ iff } \{i \in I \mid R_i f^1(i) \ldots f^n(i)\} \in U.$$

(iii) Let F be an n-place function symbol, and F_i its interpretation in $\mathfrak{A}_i$. The function F_U in $\prod_U \mathfrak{A}_i$ is given by

$$F_U(f_U^1, \ldots, f_U^n) = \{(i, F_i(f^1(i), \ldots, f^n(i))) \mid i \in I\}_U.$$

(iv) Let c be a constant, and a_i its interpretation in $\mathfrak{A}_i$. Then c is interpreted by the element $c' \in \prod_U A_i$ where $c' = \{(i, a_i) \mid i \in I\}_U$.

In the case where all the structures are the same, say, $\mathfrak{A}_i = \mathfrak{A}$ for all i, we speak of the *ultrapower* of $\mathfrak{A}$ modulo U, notation: $\prod_U \mathfrak{A}$. ⊣

To show that the above definition is coherent, we should check that the above clauses depend only on the equivalence classes $f_U^1, \ldots, f_U^{n+1}$. We leave this to the reader and go straight to the fundamental result.

Theorem A.19 (Łoś's Theorem) *Let U be an ultrafilter over a non-empty set I. For each $i \in I$, let $\mathfrak{A}_i$ be a model.*

(i) *For every term $t(x_1, \ldots, x_n)$ and all elements $f_U^1, \ldots, f_U^n$ of $\mathfrak{B} = \prod_U \mathfrak{A}_i$ we have*

$$t^{\mathfrak{B}}[x_1 \mapsto f_U^1, \ldots, x_n \mapsto f_U^n] = \{(i, t^{\mathfrak{A}_i}[f^1(i), \ldots, f^n(i)]) \mid i \in I\}_U.$$

(ii) *Given any first-order formula $\alpha(x_1, \ldots, x_n)$ in $\mathcal{L}_\tau^1$ and $f_U^1, \ldots, f_U^n$ in $\prod_U \mathfrak{A}_i$ we have*

$$\prod_U \mathfrak{A}_i \models \alpha[f_U^1, \ldots, f_U^n] \text{ iff} \tag{A.1}$$
$$\{i \in I \mid \mathfrak{A}_i \models \alpha[f^1(i), \ldots, f^n(i)]\} \in U.$$

Proof. We leave item (i) to the reader. To prove item (ii) we argue by induction on α. The atomic case holds by definition. Suppose that $\alpha \equiv \neg\beta(x_1, \ldots, x_n)$ and (A.1) holds for $\beta(x_1, \ldots, x_n)$. Then

$$\prod_U \mathfrak{A}_i \models \alpha[f_U^1 \ldots f_U^n] \quad \text{iff} \quad \text{not } \prod_U \mathfrak{A}_i \models \beta[f_U^1, \ldots, f_U^n]$$
$$\text{iff} \quad \{i \in I \mid \mathfrak{A}_i \models \beta[f_U^1, \ldots, f_U^n]\} \notin U$$
$$\text{iff} \quad \{i \in I \mid \mathfrak{A}_i \not\models \beta[f^1(i), \ldots, f^n(i)]\} \in U$$
$$\text{iff} \quad \{i \in I \mid \mathfrak{A}_i \models \alpha[f^1(i), \ldots, f^n(i)]\} \in U.$$

Here, the second equivalence follows from the inductive hypothesis, and the third from the fact that U is an ultrafilter.

Next we have to prove that if β and γ satisfy (A.1), then so does $\beta \wedge \gamma$. This uses the fact that filters are closed under intersections and supersets.

Finally, suppose that $\alpha(x_1, \ldots, x_n) \equiv \exists x_0 \, \beta(x_0, x_1, \ldots, x_n)$ and that (A.1) holds for β. Then the following are equivalent:

$$\prod_U \mathfrak{A}_i \models \alpha[f_U^1, \ldots, f_U^n] \text{ iff for some } f_U^0, \prod_U \mathfrak{A}_i \models \beta[f_U^0 f_U^1, \ldots, f_U^n]$$
$$\text{iff for some } f_U^0, \{i \in I \mid \mathfrak{A}_i \models \beta[f^0(i) f^1(i), \ldots, f^n(i)]\} \in U. \quad \text{(A.2)}$$

As $\mathfrak{A}_i \models \beta[f^0(i), \ldots, f^n(i)]$ implies $\mathfrak{A}_i \models \alpha[f^1(i), \ldots, f^n(i)]$, (A.2) implies

$$\{i \in I \mid \mathfrak{A}_i \models \alpha[f^1(i), \ldots, f^n(i)]\} \in U. \quad \text{(A.3)}$$

Conversely, if (A.3) holds, then we can easily select a function f^0 in $\prod_{i \in I} A_i$, where A_i is the universe of $\mathfrak{A}_i$, such that (A.2) holds. So (A.2) is equivalent to (A.3). $\dashv$

Corollary A.20 *Let $\prod_U \mathfrak{A}$ be an ultrapower of $\mathfrak{A}$. Then, for all first-order sentences α, $\mathfrak{A} \models \alpha$ iff $\prod_U \mathfrak{A} \models \alpha$.*

There is a natural embedding of a model $\mathfrak{A}$ in each of its ultrapowers. Define the *diagonal mapping* d of $\mathfrak{A}$ into $\prod_U \mathfrak{A}$ to be the function

$$a \mapsto (f_a)_U, \text{ where } f_a(i) = a, \text{ for all } i \in I.$$

Corollary A.21 *Let $\prod_U \mathfrak{A}$ be an ultrapower of $\mathfrak{A}$. Then the diagonal mapping of $\mathfrak{A}$ into $\prod_U \mathfrak{A}$ is an elementary embedding.*

Proof. Let $\alpha(x_1, \ldots, x_n)$ be a first-order formula, and $a_1, \ldots, a_n$ elements of $\mathfrak{A}$. By Theorem A.19 we have

$$\prod_U \mathfrak{A} \models \alpha[d(a_1), \ldots, d(a_n)] \text{ iff } \{i \in I \mid \mathfrak{A} \models \alpha[a_1, \ldots, a_n]\} \in U$$
$$\text{iff } \mathfrak{A} \models \alpha[a_1, \ldots, a_n]. \quad \dashv$$

The preceding results will be useful in our modal investigations. But ultraproducts can also be employed to characterize the expressive power of first-order languages and we will use this characterization on several occasions. First, we need to be precise about what it means to define a class of models in a first-order language.

Definition A.22 A class K of models for a fixed first-order language $\mathcal{L}^1$ is *defined by* a set Δ of $\mathcal{L}^1$-sentences if every model for the language is in K iff it is a model for Δ. A class of models is *elementary* if it is defined by some set of first-order sentences.

Theorem A.23 *A class of models K is definable by means of a set of first-order sentences iff it is closed under isomorphisms and ultraproducts, while its complement is closed under ultrapowers.*

Proof. See [91, Theorem 6.1.16]; a weaker version of the result states that K is elementary iff it is closed under ultraproducts and elementary equivalence [91, Theorem 4.1.12]. ⊣

In Lemma 2.73 we made use of the following result: *Let $\mathcal{L}$ be a countable first-order language, U a countably incomplete ultrafilter over a non-empty set I, and $\mathfrak{M}$ an $\mathcal{L}$-model. The ultrapower $\prod_U \mathfrak{M}$ is countably saturated.* A proof of this result can be found in [91, Theorem 6.1.1].

Extensions of First-Order Logic

We now briefly review two important extensions of first-order logic: second-order logic and infinitary logic.

In *second-order logic* quantification is allowed not only over individuals, as in first-order logic, but also over sets of individuals. That is, we can write expressions like

(*well-orderedness*) $\quad \forall X \, (\exists y \, Xy \rightarrow \exists y \, (Xy \wedge \neg \exists z \, Xz \wedge z < y))$

(*induction*) $\quad\quad\quad \forall X \, (X0 \wedge \forall n \, (Xn \rightarrow X(n+1)) \rightarrow \forall n \, Xn)$

Here the expression $\forall X$ is a second-order quantifier, and X is a variable over sets of individuals. Second-order formulas are interpreted on the same models as first-order formulas are, and second-order quantifiers have the obvious meaning (for example, $\forall X$ means 'for all subsets').

The two formulas just given are relatively simple second-order formulas. For a start, the only second-order quantifiers used are quantifiers over unary relations (that is, subsets), thus these formulas are what is known as *monadic* second-order formulas. Moreover, only universal quantifiers are used (indeed, in both examples only one such quantifier is used) and these stand right at the start of the formula, thus these formulas are examples of *universal* second-order formulas. In Chapter 3 we show that, when interpreted over frames, modal formulas are equivalent to universal second-order formulas.

As these examples make clear, second-order logic (indeed, even the universal monadic fragment of second-order logic) is far more expressive than first-order logic: neither well-orderedness nor induction is definable in first-order logic. But this increased expressive power comes at a price: many familiar results from first-order logic break down. For example, the validities of second-order logic are not recursively enumerable, and the Compactness and Löwenheim-Skolem theorems A.3 and A.4) do not hold for second-order logic. However there is a method due to Henkin [217] for 'taming' second-order logic. By working with a special class of non-standard models (usually called *generalized models* or *Henkin models*), it possible to obtain a first-order perspective on a useful fragment of second-order logic, and to prove a natural completeness theorem for this fragment. Good discussions of the method can be found in Doets and van Benthem [112], Chapter 4 of Enderton [122], and Fitting [139]. The modal analog of the method – the use of *general frames* – is discussed in detail in the text.

In *infinitary logic*, we are allowed to form infinitely long formulas. The infinitary logic $\mathcal{L}_{\omega_1\omega}$, for example, allows countably infinite conjunctions and disjunctions in addition to the usual first-order repertoire. At first glance, the idea of infinitely long formulas may seem bizarre – but in fact the the logic is a natural setting for formalizing many computational issues. For example, 'repeat the program α finitely many times' means the same as the infinite disjunction

skip, or do α once, or do α twice, or do α three times, or ... .

And likewise, the following infinitary formula expresses that S is the reflexive, transitive closure of R:

$$\forall xy \, (Sxy \leftrightarrow \bigvee_{i \geq 1} R^i xy),$$

where $R^0 xy := (x = y)$ and $R^{n+1} xy := \exists z \, (Rxz \wedge R^n zy)$. As we discuss in Chapter 2, when interpreted on models, propositional dynamic logic is a fragment of $\mathcal{L}_{\omega_1\omega}$.

Appendix B

An Algebraic Toolkit

In this appendix we review some basic (universal) algebraic notions used in Chapter 5. The first part deals with algebras and operations on (classes of) algebras, the second part is about algebraic model theory, and in the third part we discuss equational logic. Birkhoff's fundamental theorems are stated without proof.

For an introduction to universal algebra, see Burris and Sankappanavar [81] or Grätzer [198]; McKenzie, McNulty and Taylor [321] provide more comprehensive reading. Basic track readers may like the algebraic accounts of propositional logic given in Chapter 3 of Bell and Machover [32] and Chapters 1 and 2 of Bell and Slomson [33]. Many readers will find Davey and Priestly [105] useful supplementary reading.

Universal Algebra

An algebra is a set together with a collection of functions over the set; these functions are usually called *operations*. Algebras come in various *similarity types*, determined by the number and arity of the operations.

Definition B.1 (Similarity Type) An *algebraic similarity type* is an ordered pair $\mathcal{F} = (F, \rho)$ where F is a non-empty set and ρ is a function $F \to \mathbb{N}$. Elements of F are called *function symbols*; the function ρ assigns to each operator $f \in F$ a finite *arity* or *rank*, indicating the number of arguments that f can be applied to. Function symbols of rank zero are called *constants*. We will usually be sloppy in our notation and terminology and write $f \in \mathcal{F}$ instead of $f \in F$. ⊣

Given a similarity type, it is obvious what an algebra of this type should be.

Definition B.2 (Algebras) Let A be some set, and n a natural number; an n-*ary operation* on A is a function from A^n to A.

Let $\mathcal{F}$ be an algebraic similarity type. An *algebra* of type $\mathcal{F}$ is a pair $\mathfrak{A} = (A, I)$ where A is a non-empty set called the *carrier* of the algebra, and I is an *interpretation*, a function assigning, for every n, an n-ary operation $f_{\mathfrak{A}}$ on A to each function symbol f of rank n. We often use the notation $\mathfrak{A} = (A, f_{\mathfrak{A}})_{f \in F}$ for such an algebra. When no confusion is likely to arise, we omit the subscripts on the operations. ⊣

We now define the standard constructions for forming new algebras from old. First we define the natural notion of structure preserving maps between algebras.

Definition B.3 (Homomorphisms) Let $\mathfrak{A} = (A, f_{\mathfrak{A}})_{f \in F}$ and $\mathfrak{B} = (B, f_{\mathfrak{B}})_{f \in F}$ be two algebras of the same similarity type. A map $\eta : A \to B$ is a homomorphism if for all $f \in F$, and all $a_1, \ldots, a_n \in A$ (where n is the rank of f):

$$\eta(f_{\mathfrak{A}}(a_1, \ldots, a_n)) = f_{\mathfrak{B}}(\eta a_1, \ldots, \eta a_n). \tag{B.1}$$

(Here ηa_i is shorthand $\eta(a_i)$.) The special case for constants c is

$$\eta(c_{\mathfrak{A}}) = c_{\mathfrak{B}}.$$

The *kernel* of a homomorphism $f : \mathfrak{A} \to \mathfrak{B}$ is the relation $\ker f = \{(a, a') \in A^2 \mid f(a) = f(a')\}$. We say that $\mathfrak{B}$ is a *homomorphic image* of $\mathfrak{A}$ (notation: $\mathfrak{A} \twoheadrightarrow \mathfrak{B}$), if there is a surjective homomorphism from $\mathfrak{A}$ onto $\mathfrak{B}$. Given a class C of algebras, HC is the class of homomorphic images of algebras in C. ⊣

Definition B.4 (Isomorphisms) A bijective homomorphism is called an *isomorphism*. We say that two algebras are *isomorphic* if there is an isomorphism between them. Usually we do not distinguish isomorphic algebras, but if we do, we write IC for the class of isomorphic copies of algebras in C. ⊣

The second way of making new algebras from old is to find a small algebra inside a larger one.

Definition B.5 (Subalgebras) Let $\mathfrak{A}$ be an algebra, and B a subset of the carrier A. If B is closed under every operation $f_{\mathfrak{A}}$, then we call $\mathfrak{B} = (B, f_{\mathfrak{A}} \restriction B)_{f \in F}$ a *subalgebra* of $\mathfrak{A}$. We say that $\mathfrak{C}$ is *embeddable* in $\mathfrak{A}$ (notation: $\mathfrak{C} \rightarrowtail \mathfrak{A}$), if $\mathfrak{C}$ is isomorphic to a subalgebra of $\mathfrak{A}$; the isomorphism is called an *embedding*. Given a class C of algebras, SC denotes the class of isomorphic copies of subalgebras of algebras in C.

A third way of forming new algebras is to make a big algebra out of a collection of small ones.

Definition B.6 (Products) Let $(\mathfrak{A}_j)_{j \in J}$ be a family of algebras. We define the *product* $\prod_{j \in J} \mathfrak{A}_j$ of this family as the algebra $\mathfrak{A} = (A, f_{\mathfrak{A}})_{f \in F}$ where A is the cartesian product $\prod_{j \in J} A_j$ of the carriers A_j, and the operation $f_{\mathfrak{A}}$ is defined co-ordinatewise; that is, for elements $a_1, \ldots, a_n \in \prod_{j \in J} A_j$, $f_{\mathfrak{A}}(a_1, \ldots, a_n)$ is the element of $\prod_{j \in J} A_j$ given by:

$$f_{\mathfrak{A}}(a_1, \ldots, a_n)(j) = f_{\mathfrak{A}_j}(a_1(j), \ldots, a_n(j)).$$

When all the algebras $\mathfrak{A}_j$ are the same, say $\mathfrak{A}$, then we call $\prod_{j \in J} A$ a *power* of $\mathfrak{A}$, and write $\mathfrak{A}^J$ rather than $\prod_{j \in J} A$. Given a class C of algebras, **P**C denotes the class of isomorphic copies of products of algebras in C. ⊣

Suppose you are working with a class of algebras from which you cannot obtain new algebras by the three operations defined above. Intuitively, such a class is 'complete', for it is closed under the natural algebra-forming operations. Such classes play an important role in universal algebra. They are called *varieties*:

Definition B.7 (Varieties) A class of algebras is called a *variety* if it is closed under taking subalgebras, homomorphic images, and products. Given a class C of algebras, $\mathbb{V}$C denotes the variety generated by C; that is, the smallest variety containing C. ⊣

A well-known result in universal algebra states that $\mathbb{V}$C = **HSP**C. That is, in order to obtain the variety generated by C, you can start by taking products of algebras in C, then go on to take subalgebras, and finish off by forming homomorphic images. You do not need to do anything else: subsequent applications of any of these operations will not produce anything new.

Homomorphisms and Congruences

Homomorphisms on an algebra $\mathfrak{A}$ are closely related to special equivalence relations on the carrier of A.

Definition B.8 (Congruences) Let $\mathfrak{A}$ be an algebra for the similarity type $\mathcal{F}$. An equivalence relation $\sim$ on A (that is, a reflexive, symmetric and transitive relation) is a *congruence* if it satisfies, for all $f \in F$

$$\text{if } a_1 \sim b_1 \& \ldots \& a_n \sim b_n, \text{ then } f_{\mathfrak{A}}(a_1, \ldots, a_n) \sim f_{\mathfrak{A}}(b_1, \ldots, b_n), \quad \text{(B.2)}$$

where n is the rank of f. ⊣

The standard examples of congruences are the 'modulo' relations on the integers. Consider the algebra $\mathfrak{Z} = (\mathbb{Z}, +, *, 0, 1)$ of the integers under addition and multi-plication, and, for a positive integer n, let the relation $\equiv_n$ be defined by $z \equiv_n z'$ if n divides $z - z'$. We leave it to the reader to verify that these relations are all congruences.

The importance of congruences is that they are precisely the kind of equivalence relations that allow a natural algebraic structure to be defined on the collection of equivalence classes.

Definition B.9 (Quotient Algebras) Let $\mathfrak{A}$ be an $\mathcal{F}$-algebra, and $\sim$ a congruence on $\mathfrak{A}$. The *quotient algebra of* $\mathfrak{A}$ *by* $\sim$ is the algebra $\mathfrak{A}/\!\sim$ whose carrier is the set

$$A/\!\sim = \{[a] \mid a \in A\}$$

of equivalence classes of A under $\sim$, and whose operations are defined by

$$f_{\mathfrak{A}/\sim}([a_1], \dots, [a_n]) = [f_{\mathfrak{A}}(a_1, \dots, a_n)].$$

(This is well-defined by (B.2).) The function ν taking an element $a \in A$ to its equivalence class $[a]$ is called the *natural map* associated with the congruence. ⊣

As an example, taking the quotient of 3 under the relation $\equiv_n$ makes the algebra 3_n of arithmetic modulo n.

The close connection between homomorphisms and congruences is given by the following proposition (the proof of which we leave to the reader).

Proposition B.10 (Homomorphisms and Congruences) *Let $\mathfrak{A}$ be an $\mathcal{F}$-algebra. Then*

(i) *If $f : \mathfrak{A} \to \mathfrak{B}$ is a homomorphism, its kernel is a congruence on $\mathfrak{A}$.*
(ii) *Conversely, if $\sim$ is a congruence on $\mathfrak{A}$, its associated natural map is a surjective homomorphism from $\mathfrak{A}$ onto $\mathfrak{A}/\!\sim$.*

Algebraic Model Theory

Universal algebra can be seen as a branch of model theory in which one is only interested in structures where all relations are functions. The standard language for talking about such structures is *equational*, where an equation is a statement asserting that two *terms* denote the same element.

Definition B.11 (Terms and Equations) Given an algebraic similarity type $\mathcal{F}$ and a set X of elements called *variables*, we define the set $Ter_{\mathcal{F}}(X)$ of $\mathcal{F}$-terms over X inductively: it is the smallest set T containing all constants and all variables in X such that $f(t_1, \dots, t_n)$ is in T whenever $t_1, \dots, t_n$ are in T and f is a function symbol of rank n.

An *equation* is a pair of terms (s, t); the notation $s \approx t$ is usually used.

Having defined the algebraic language, we now consider the way it is interpreted in algebras. Obviously terms refer to elements of algebras, but in order to calculate the meaning of a term we need to know what the variables in the term stand for. This information is provided by an assignment.

Definition B.12 (**Algebraic Semantics**) Let $\mathcal{F}$ be an algebraic similarity type, X a set of variables, and $\mathfrak{A}$ an $\mathcal{F}$-algebra. An *assignment* on $\mathfrak{A}$ is a function $\theta : X \to A$ associating an element of A with each variable in X. Given such an assignment θ, we can calculate the *meaning* $\tilde{\theta}(t)$ of a term t in $Ter_{\mathcal{F}}(X)$ as follows:

$$\begin{aligned}
\tilde{\theta}(x) &= \theta(x), \\
\tilde{\theta}(c) &= c_{\mathfrak{A}}, \\
\tilde{\theta}(f(t_1, \ldots, t_n)) &= f_{\mathfrak{A}}(\tilde{\theta}(t_1), \ldots, \tilde{\theta}(t_n)).
\end{aligned}$$
$\dashv$

The last equality bears an obvious resemblance to the condition (B.1) defining homomorphisms. In fact, we can turn the meaning function into a genuine homomorphism by imposing a natural algebraic structure on the set $Ter_{\mathcal{F}}(X)$ of terms:

Definition B.13 (**Term Algebras**) Let $\mathcal{F}$ be an algebraic similarity type, and X a set of variables. The *term algebra of $\mathcal{F}$ over X* is the algebra $\mathfrak{Ter}_{\mathcal{F}}(X) = (Ter_{\mathcal{F}}(X), I)$ where every function symbol f is interpreted as the operation $I(f)$ on $Ter_{\mathcal{F}}(X)$ given by

$$I(f)(t_1, \ldots, t_n) = f(t_1, \ldots, t_n). \tag{B.3}$$

$\dashv$

In other words, the carrier of the term algebra over $\mathcal{F}$ is the set of $\mathcal{F}$-terms over the set of variables X, and the operation $I(f)$ or $f_{Ter_{\mathcal{F}}(X)}$ maps an n-tuple $t_1, \ldots, t_n$ of terms to the term $f(t_1, \ldots, t_n)$. Note the double role of f in (B.3): on the right-hand side, f denotes a 'static' *part* of the syntactic *term* $f(t_1, \ldots, t_n)$, while on the left-hand side $I(f)$ denotes a 'dynamic' *interpretation* of f as an *operation* on terms.

The perspective on $\mathcal{F}$-terms as constituting an $\mathcal{F}$-*algebra* is extremely useful. For example, we can view the operation of *substituting* terms for variables in terms as an *endomorphism* on the term algebra, that is, a homomorphism from an algebra to itself.

Definition B.14 Let $\mathcal{F}$ be a similarity type, and X a set of variables. A *substitution* is a map $\sigma : X \to Ter_{\mathcal{F}}(X)$ mapping variables to terms. Such a substitution can be extended to a map $\tilde{\sigma} : Ter_{\mathcal{F}}(X) \to Ter_{\mathcal{F}}(X)$ by the following inductive definition:

$$\begin{aligned}
\tilde{\sigma}(x) &:= \sigma(x), \\
\tilde{\sigma}(f(t_1, \ldots, t_n)) &:= f(\tilde{\sigma}(t_1), \ldots, \tilde{\sigma}(t_n)).
\end{aligned}$$

We sometimes use the word 'substitution' for a function mapping terms to terms that satisfies the second of the conditions above (that is, we sometimes call $\tilde{\sigma}$ a substitution). $\dashv$

Proposition B.15 *Let* σ : $Ter_{\mathcal{F}}(X) \rightarrow Ter_{\mathcal{F}}(X)$ *be a substitution. Then* σ : $\mathfrak{Ter}_{\mathcal{F}}(X) \rightarrow \mathfrak{Ter}_{\mathcal{F}}(X)$ *is a homomorphism.*

Moreover, the *meaning function* associated with an assignment θ is now a homomorphism:

Proposition B.16 *Given any assignment θ of variables X to elements of an algebra $\mathfrak{A}$, the corresponding meaning function $\tilde{\theta}$ is a homomorphism from $\mathfrak{Ter}_{\mathcal{F}}(X)$ to $\mathfrak{A}$.*

The standard way of making statements about algebras is to compare the meaning of two terms under the same valuation – that is, to use *equations*.

Definition B.17 (Truth and Validity) An equation $s \approx t$ *is true* or *holds* in an algebra $\mathfrak{A}$ (notation: $\mathfrak{A} \models s \approx t$), if for all assignments $\theta, \tilde{\theta}(s) = \tilde{\theta}(t)$.

A *set* E of equations holds in an algebra $\mathfrak{A}$ (notation: $\mathfrak{A} \models E$), if each equation in E holds in $\mathfrak{A}$. If $\mathfrak{A} \models s \approx t$ or $\mathfrak{A} \models E$ we will also say that $\mathfrak{A}$ is a *model* for $s \approx t$, or for E, respectively.

An equation $s \approx t$ is a *semantic consequence* of a set E of equations (notation: $E \models s \approx t$), if every model for E is a model for $s \approx t$.   ⊣

Algebraists are often interested in specific classes of algebras such as groups and boolean algebras. Such classes are usually defined by sets of equations.

Definition B.18 (Equational Class) A class C of algebras is equationally definable, or an *equational class*, if there is a set E of equations such that C contains precisely the models for E.   ⊣

The following theorem, due to Birkhoff, is one of the most fundamental results of universal algebra:

Theorem B.19 (Birkhoff) *A class of algebras is equationally definable if and only if it is a variety.*

Unfortunately, we do not have the space to prove this theorem here. The reader is advised to try proving the easy direction (that is, to show that any equationally definable class is closed under taking homomorphic images, subalgebras and direct products) for him- or herself.

Equational Logic

Equational logic arises when we formalize the rules that enable us to deduce new equations from old. Although we do not make direct use of equational logic in the text, it will be helpful if the reader is acquainted with it. Here is a fairly standard system.

Definition B.20 (Equational Logic) Let $\mathcal{F}$ be an algebraic similarity type, and E a set of equations. The set of equations that are *derivable* from E is inductively defined by the following schema:

(*axioms*)   The equations in E are derivable from E; they are called *axioms*.

(*reflexivity*)   Every equation $t \approx t$ is derivable from E.

(*symmetry*)   If $t_1 \approx t_2$ is derivable from E, then so is $t_2 \approx t_1$.

(*transitivity*)   If the equations $t_1 \approx t_2$ and $t_2 \approx t_3$ are derivable from E, then so is $t_1 \approx t_3$.

(*congruence*)   Suppose that all equations $t_1 \approx u_1, \dots, t_n \approx u_n$ are derivable from E, and that f is a function symbol of rank n. Then the equation $f(t_1, \dots, t_n) \approx f(u_1, \dots, u_n)$ is derivable from E as well. This schema is sometimes called *replacement*.

(*substitution*)   If $t_1 \approx t_2$ is derivable from E, then so is the equation $\sigma t_1 \approx \sigma t_2$, for every substitution σ.

The notation $E \vdash t_1 \approx t_2$ means that the equation $t_1 \approx t_2$ is derivable from E.

A *derivation* is a list of equations such that every element is either an axiom, or has the form $t \approx t$, or can be obtained from earlier elements of the list using the symmetry, transitivity, congruence/replacement, or substitution rules.   ⊣

A fundamental completeness result, also due to Birkhoff, links this deductive apparatus to the semantic consequence relation defined earlier.

Theorem B.21 *Let E be a set of equations for the algebraic similarity type $\mathcal{F}$. Then for all equations $s \approx t$, $E \models s \approx t$ iff $E \vdash s \approx t$.*

Appendix C

A Computational Toolkit

In this appendix we introduce the basic ideas of computability theory (the study of which problems are, and which problems are not, computationally solvable), and provide some background information on complexity theory (the study of the computational resources required to solve problems).

For detailed discussions of computability, see Rogers [391] or Odifreddi [343]. For accessible introductions to the subject, see Boolos and Jeffrey [70], or Cutland [103]. But the single most useful source is probably the (second edition of) Lewis and Papadimitriou [301]; this introduces computability theory, and then goes on to treat computational complexity. For more on computational complexity, try Garey and Johnson [163] and Papadimitriou [352]. Garey and Johnson's book is a source for information on NP-complete problems, but it discusses the basic ideas of computational complexity lucidly, and gives background information on other complexity classes. Papadimitriou's book is a well-written introduction to computational complexity covering far more than is needed to understand Chapter 6; if you want to go deeper into computational complexity, it is a good place to start.

Computability and Uncomputability

To prove theorems about computability – and in particular to prove that some problem is *not* computable – we need a robust mathematical model of computability. One of the most widely used models is the *Turing machine*. A Turing machine is a device which manipulates symbols written on a tape. The symbols are taken from some alphabet fixed in advance (often the alphabet simply consists of the two symbols 0 and 1). The tape is subdivided into squares, and only one symbol can be written on each square (squares containing no symbols are called blank). The tape is used to receive input, to present output, and acts as working memory. The tape is assumed to be infinitely long in both directions (so no finite upper bound on the amount of working memory is assumed).

Turing machines scan the squares of such tapes and act on the information the see; they can only scan one square at a time. A Turing machine has a finite number of internal states, and a finite number of rules which tell it what to do when it is in a certain state scanning a certain symbol. Turing machines can perform three basic actions: (1) move to the square immediately to the left of the square they are currently scanning, (2) move to the square immediately to the right of the square they are currently scanning, or (3) write a symbol (from the alphabet) on the square currently being scanned (thereby overwriting any symbol already written on that square). In addition to specifying which of these three actions will be performed, the rules also specify which internal state the Turing machine is to move into on completing the action. A Turing machine halts when (and if) it enters a special halting state. For some simple (and not so simple) examples of computations on Turing machines, see Chapter 4 of Lewis and Papadimitriou [301] and Chapters 1–3 of Boolos and Jeffrey [70].

The ideas just sketched can be made precise as follows:

Definition C.1 (Turing Machines) A *Turing machine* is a 5-tuple $(S, s, H, \Sigma, \delta)$ where S is a finite set of states, $s \in S$ is the initial state, $H \subseteq S$ is the set of halting states, Σ (the alphabet) is a finite set of symbols, and δ is a function from $(S \setminus H) \times \Sigma$ to $S \times (\Sigma \cup \{left, right\})$. ⊣

For example, the rule *if you are in state 57 scanning the symbol 1, move one square to the left and go into state 14* amounts to saying that $\delta(57, 1) = (14, left)$. The rule *if you are in state 30 scanning the symbol %, write the symbol 5 and go into state 12* means that $\delta(30, \%) = (12, 5)$. Since δ is a function, the action of such a machine is *deterministic*: when the machine is put in the initial state scanning some tape, what it does (if it does anything) is fixed.

Let f be a function, and suppose we have fixed some convention about how the elements of the domain and range of the function are to be represented. (For example, if f is a function from the natural numbers to the natural numbers, we might decide to use binary notation – that is, base 2 notation – to represent the numbers). Then f is *computable* (or *recursive*) if there is a Turing machine that when given (the representation of) an item x in the domain of f will halt after finitely many steps, leaving on an otherwise blank tape (the representation of) $f(x)$.

We can use Turing machines to provide yes/no answers to problems. Many logical problems – for example, is some formula ϕ satisfiable or not – are of this type. Suppose we have fixed the alphabet of a Turing machine, and have decided how we are going to represent the problems we are interested in (in Section 6.1, we discuss how to encode modal formulas and models as strings of 0s and 1s). Given our encoding conventions, some strings over the alphabet represent problem instances for which the answer is *yes*, while others represent problem instances for which the answer is *no*. A problem is *computable* (or *recursive*, or *decidable*) if there is a Turing machine which when given (the representation of) any instance of the problem, halts after finitely many steps leaving the (representation of) the correct answer on an otherwise blank tape.

Turing machines essentially provide answers to set membership problems: a _yes_ answer means that the input belongs to a set of interest (for example, the set of satisfiable formulas) while a _no_ means it does not. Thus it is common to talk of computable (or recursive, or decidable) _sets_. Another important notion is that of a _recursively enumerable_ set. A set is recursively enumerable (r.e.) if there is a Turing machine which successively writes, on an otherwise blank tape, all and only the members of the set. If the set is infinite, this listing process will never finish – but after some finite time, any given element of an r.e. set will eventually be listed.

All recursive sets are r.e., but there are r.e. sets that are not recursive. The best known example is the set of valid first-order formulas (in a sufficiently rich language). This set is not recursive, but (as we mentioned in Appendix A) it is recursively enumerable. Details for the following result may be found in Lewis and Papadimitriou [301, pp. 198–200, 267–273])

Proposition C.2 _A set is recursive iff both it and its complement are recursively enumerable_

Thus, from a computational perspective, the set of first-order formulas that are not valid is more complex than the set of valid formulas – the non-valid formulas cannot even be recursively enumerated.

It is common practice to identify problems with the set of those strings of symbols that provide the answer _yes_ to the problem. For once an alphabet has been fixed, each subset of the set of all finite strings over the alphabet can be regarded as the encoding of the problem. This abstract perspective is a convenient way of stating abstract computability and complexity results, and we adopt it later in this appendix; but in the text, when we apply these ideas to modal logic, we try to keep our statement of problems fairly concrete.

Because of its simplicity, the Turing machine model is widely used in theoretical computer science, particularly in complexity theory. But it is not a toy model of computation: experience has shown that it is remarkably robust and general. For example, we can allow Turing machines to have special read-only input tapes, special write-only output tapes, and allow them to access several working tapes independently – but none of these variations allows new functions to be computed or new problems solved. Moreover, we can move away from the Turing machine model in many different ways: for example, we can use Random Access Memory machines which model more directly the workings of a physical computer. Such variations make no difference: if a function is computable (or a problem decidable in one of these alternative models then it is also computable (decidable) on some Turing machine.

Another important variation is the use of *non-deterministic* Turing machines. The action of such a machine is not fixed by the symbol it is scanning and the state it is in: for any such combination, it may have a (finite) range of options. (Formally, we drop the requirement that the δ of Definition C.1 be a function and let it be an arbitrary relation.) We think of such a machine as following all the options allowed by δ simultaneously, and say that such a machine solves a problem if at least one such computation path halts leaving the correct answer on an otherwise blank tape.

The beauty of non-determinism is that it factors out search. Many problems require us to find a candidate solution and then see if it works, and if not, to look for another candidate, and so on. This process may be the major computational overhead. Non-deterministic machines abstract away from this: if there is a solution, a non-deterministic machine can find it far more efficiently (we will see a classic example of this when we discuss complexity theory). But as far as *computability* is concerned, non-determinism adds nothing: if a function (or problem) is computable using a non-deterministic Turing machine, it is also computable using a deterministic Turing machine, for we can (laboriously) work through all possible choices.

Such observations give rise to Church's thesis.

Thesis C.3 (Church's Thesis) A function is computable (a problem decidable) precisely when it can be computed (solved) using a Turing machine.

On the face of it, Church's Thesis just stipulates that the notion of computation defined by Turing machines is so robust that it makes sense to think of it as pinning down what we mean by computation. But its import is far wider: in essence it is an acknowledgment of the fact that all the general finitary models of computation that have been proposed (and there are probably several hundreds of these) have turned out to be equivalent. That is, Church's Thesis affirms that we *do* have a robust model of computation.

You may view Church's Thesis as saying that computable functions and problems are those which can be calculated/solved by writing a program in your favorite programming language when no limitations are placed on memory or execution time. In fact, in Chapter 6 we rarely talk explicitly of Turing machines: rather, we prove that problems are decidable by analyzing them till it becomes clear that any competent programmer could write a program that carries out the task.

The most important benefit of having a robust definition of computability is that it gives us a way of proving that some function or problem is *undecidable*. And many – indeed most – functions and problems are undecidable. For let M be a set of natural numbers. Is each such M decidable? A simple cardinality argument shows that the answer is *no*. Every Turing machine is a finite function over a finite set of states and a finite alphabet. It follows that there are only countably many Turing machines – but there are uncountably many M, so they cannot all be computable. It is not difficult to construct concrete examples of functions which no Turing machine can compute (see Lewis and Papadimitriou [301, Chapter 5], and Boolos and Jeffrey [70, Chapters 4, 5]).

But again, to prove undecidability it is not necessary to appeal to the definition of a Turing machine. It is usually easier to show problems are undecidable via reductions:

Definition C.4 Let Σ be an alphabet and let $L_1, L_2 \subseteq \Sigma^*$ be problems (note that we are adopting the abstract view of problems here). A *reduction* from L_1 to L_2 is a computable function $f : \Sigma^* \to \Sigma^*$ such that $s \in L_1$ iff $f(s) \in L_2$; here, Σ^* is simply the set of all finite strings over Σ. ⊣

Proposition C.5 *Let* $L_1, L_2 \subseteq \Sigma^*$ *be problems, and* f *be a reduction from* L_1 *to* L_2. *If* L_1 *is undecidable, then so is* L_2.

Proof. Easily established using a proof by contradiction, and the reader may like to try. You can find a detailed account on pages 254–258 of the second edition of Lewis and Papadimitriou [301]. ⊣

Nowadays a vast range of problems are known to be undecidable, and we can try to prove undecidability results by reduction from any one of these problems. We follow this strategy in Chapter 6.

One final remark: not all undecidable problems are alike. There is a precise sense in which some are worse than others. The key idea is to equip Turing machines with *oracles*. A Turing machine equipped with an oracle is allowed to temporarily halt in the middle of some computation, consult the oracle, and proceed with its computation taking the oracle's answer into account.

Oracles provide answers to *undecidable* problems (an oracle that provided answers to decidable problems would offer nothing new: it could always be replaced by a Turing machine). They are a mathematical abstraction which allow us to remove the limitation to finitary computation inherent in Church's thesis. It is common to specify what oracles can do in logical terms – for example, we might imagine we have a Turing machine hooked to an oracle that is able to determine whether an arbitrary second-order sentence has a model or not. It turns out that undecidable problems are not all the same: when we measure their difficulty with respect to the oracles required to solve them, there is a whole hierarchy of difficulty. A problem that is not merely undecidable, but requires the help of some such oracle to solve it, is called *highly undecidable*. In Section 6.5 we show that a certain modal satisfiability problem is highly undecidable, and in fact, Σ_1^1-hard

Roughly speaking, this means that the problem is as difficult as deciding whether a prenex formula in the *second-order* language of arithmetic, that begins with a block of existential quantifiers, is satisfiable on the natural numbers. Such formulas have immense expressive power, and Σ_1^1 problems are highly complex (certainly not recursively enumerable). Incidentally, 'ordinary' undecidable problems are often said to be Π_1^0-hard, and we use this terminology in Section 6.5 too. Roughly speaking, this means that such problems are 'only' as difficult as deciding whether a prenex formula (with what is known as a recursive matrix) in the *first-order* language of arithmetic, and with a quantifier prefix consisting of universal quantifiers only, is satisfiable on the natural numbers. Although such problems are undecidable, they are recursively enumerable. For more on highly undecidable problems, see Harel [208]. For precise definitions and further discussion of the classes Σ_1^1 and Π_1^0, see Odifreddi [343].

Complexity Theory

Complexity theory studies the computational resources required to solve (decidable) problems. The two main resources studied are *time* (the number of computation steps required) and *space* (the amount of memory required). Both time required and space required are measured as functions of the length of the input.

Ideally, complexity theory would give us a precise bound on the resources required to solve any problem that interested us. But this goal is far too ambitious. Instead, complexity theory classifies problems into various classes. In this book we mention the classes

$$P \subseteq NP \subseteq PSPACE \subseteq EXPTIME \subseteq NEXPTIME \qquad (C.1)$$

and we devote a lot of attention to NP, PSPACE, and EXPTIME.

Before defining these classes, some general remarks. It is currently unknown whether the inclusions in (C.1) are strict or not. It is widely conjectured that they are, but nobody has been able to prove (or disprove) any of these strict inclusions. All we know for sure is that $P \neq EXPTIME$.

Second, although a problem that belongs to any of these classes is decidable, P the class at the bottom of this putative hierarchy) is widely taken to be the class of problems that are *tractable*, or *efficiently solvable*.

Definition C.6 A deterministic Turing machine is *polynomially time bounded* if here is a polynomial $p(n)$ such that the machine always halts after at most $p(n)$ teps, where n is the length of the input. A problem is *solvable in polynomial time* a function f is *solvable in polynomial time*) if there is a polynomially bounded uring machine that solves it (that computes it). The class of all problems solvable polynomial time is called P. A problem is called *tractable* if it belongs to P. ⊣

One word of warning. When we solve a problem on a Turing machine, we choose a way of *representing* the problem (that is, encoding it in the symbols used by that machine). Needless to say, there are sensible ways of representing problems, and highly inefficient ways of doing so. If a sufficiently bad representation is chosen, this can give a completely misleading impression of the resources required to solve the problem. For example, a really bad representation could ensure that a problem solvable in polynomial time takes exponential time to compute.

Fortunately, for the complexity classes considered in this book, there is little to worry about. The main pitfall to be avoided concerns the representation of numbers: unary representations should be avoided as they are exponentially longer than binary (or higher base) representations. In the text, we assume we are working with binary representations of numbers. We discuss the representation of modal logical problems in Section 6.1.

Identifying tractable problems with those in P is not unproblematic, but it has proved useful. For a start, if a problem is solvable in polynomial time, then typically the polynomial is of low degree. Moreover, if a problem is *not* solvable in polynomial time, then (some instances of it) will be very hard to solve indeed. For example, if a problem requires resources exponential in the length of the input (for example, 2^n, where n is the length of the input), then no algorithm is going to solve all instances of the problem efficiently: on some input, even for quite small values of n, the computation will not halt within the expected lifetime of the universe.

Hardness and completeness

In order to define the complexity classes of interest, we need some additional concepts. We have already met the idea of reducing one problem to another (see Definition C.4). To make further progress, we need the notion of *tractably* reducing one problem to another. As we have identified 'tractable' with 'computable in polynomial time', the following notion is what we require.

Definition C.7 (Polytime Reduction) Let L_1, $L_2 \subseteq \Sigma^*$ be problems. A polynomial time computable function $f : \Sigma^* \to \Sigma^*$ is called a *polynomial time reduction* (or: a polytime reduction) from L_1 to L_2 if for each $s \in \Sigma^*$ we have that $s \in L_1$ iff $f(s) \in L_2$.

A polytime reduction from L_1 to L_2 is essentially a *tractable* way of compiling problem L_1 down to problem L_2. It follows that if L_2 is solvable in polynomial time (that is, if L_2 is tractable), then so is L_1: to test whether a string x is in L_1 simply compute $f(x)$ (this compilation step is polynomial time computable) and then test whether $f(x) \in L_2$ (which by assumption is polynomial time solvable). As $x \in L_1$ iff $f(x) \in L_2$, and as the composition of two polynomials is a polynomial, we have efficiently computed an answer to our original problem. On the other hand, if L_1 is *not* solvable in polynomial time, then neither is L_2, as the reader should verify. Summing up: if there is a polytime reduction from L_1 to L_2, then L_2 is *at least as hard* as L_1, and this observation leads us to the following fundamental definition:

Definition C.8 (Hardness and Completeness) Let C be a class of problems. A problem L is C-*hard* (with respect to polynomial time reductions) if every problem in C is polynomial time reducible to L; L is C-*complete* if it is C-hard and moreover $L \in C$. That is, the C-complete problems are the hardest problems in C. ⊣

The class P

This fundamental class does not play a direct role in the book, for it is widely believed that the problem of deciding whether a formula of classical propositional logic is satisfiable is *not* in P. (No proof of this is known – it is one of the best-known open problems in theoretical computer science.) As the modal logics discussed in this book contain classical propositional logic as a subpart, their satisfiability problems probably do not lie in P either.

The class NP

There are many naturally occurring problems which do not seem to belong to P but which can be solved efficiently using a non-deterministic Turing machine.

Definition C.9 A non-deterministic Turing machine is *polynomially time bounded* if there is a polynomial $p(n)$ such that no computation of the machine continues for more than $p(n)$ steps where n is the length of the input. NP is the class of all problems decided by a polynomially bounded non-deterministic machine. ⊣

The problems that seem *not* to be in P but which *are* in NP typically involve search. The classic example is the satisfiability problem for propositional logic: given a propositional formula ϕ, is there an assignment of truth values (0 and 1) to its proposition letters that makes the formula evaluate to 1? No deterministic polynomial time algorithm for propositional satisfiability is known, and it is widely believed that none exists.

But it is easy to design an NP algorithm to solve propositional satisfiability. When it is given as input the formula ϕ, the first step of the algorithm is to non-deterministically try out all possible combinations of truth values on the propositional variables in ϕ. If there is a solution, this is returned; if not, an arbitrary truth value assignment is returned instead. Either way, after one (non-deterministic) step we are given an assignment. We can then deterministically compute in polynomial time whether this assignment satisfies ϕ or not (all we have to do is perform one operation for each logical connective in ϕ). If ϕ evaluates to true, ϕ is satisfiable. On the other hand, if ϕ evaluates to false, it must be unsatisfiable, for the non-deterministic step would have returned a satisfying assignment had one existed. Does this mean that propositional satisfiability is really an easy problem to solve? Unfortunately, no. The only known way of implementing non-determinism is to simulate it on a deterministic Turing machine. All known simulations require exponential time to perform, and it is widely believed (though not proved) that no efficient simulation exists. Non-deterministic Turing machines are probably not a realistic model of efficient computation.

But non-determinism has proved to be a very useful way of thinking about problems consisting of a *search* for a solution, followed by a *verification* step that can be conducted in deterministic polynomial time. An extraordinary range of interesting problems have this general profile (see Garey and Johnson [163] for an extensive list), and by reducing the search to a single non-deterministic step, we see that such problems belong to NP.

Now for a more demanding question: are there any NP-hard problems? The celebrated Cook-Levine Theorem tells us that there are.

Theorem C.10 (Cook-Levine Theorem) *The propositional satisfiability problem is NP-complete.*

Proof. We have just given an informal argument showing that the propositional satisfiability problem is in NP. As for NP-hardness, we need to show (in accordance with Definition C.8) that *any* problem in NP whatsoever can be polytime reduced to the propositional satisfiability problem. It may seem that we do not have enough information to prove something this general – but amazingly, we do. An elegant proof is given by Lewis and Papadimitriou [301, pp. 309–317].

Once we have shown that one problem is NP-complete, it becomes much easier to show that other problems are NP-complete. Given a problem L which we suspect to be NP-complete, all we have to do is (i) show that it is in NP, and (ii) show that some problem known to be NP-hard is polynomial time reducible to L. In this book, showing point (ii) is trivial: all the logics we are interested in extend classical propositional logic, so NP-hardness is immediate by the Cook-Levine Theorem.

The classes NP and coNP (that is, the class of problems whose complement are in NP) seem to have very different complexity profiles. A classic problem in coNP is the *validity* problem for propositional calculus – the problem of deciding whether *all* assignments of truth values satisfy a propositional formula. The validity problem is widely believed not to be in P. However, (unlike the satisfiability problem) it does not seem to belong to NP either: because we need to consider a possible truth assignments, non-determinism does not seem to help us solve it. B e face another open problem here: although it is standardly conjectured that NP ≠ coNP, no-one has been able to prove or disprove it.

he class PSPACE

SPACE is the complexity class of most relevance to modal logic. It is defined in rms of *deterministic* Turing machines.

efinition C.11 A deterministic Turing machine is *polynomially space bounded* there is a polynomial $p(n)$ such that no computation of the machine scans more an $p(n)$ tape squares, where n is the length of the input. PSPACE is the class f all problems that are decided by a polynomially space bounded deterministic uring machine. ⊣

roposition C.12 *NP* ⊆ *PSPACE*.

roof. This is a special case of a more general result: see Papadimitriou [352, heorem 7.4(b)]. ⊣

/hat is the intuition behind this theorem? As we have remarked, a deterministc Turing machine can simulate a non-deterministic Turing machine, though it is idely believed that the simulation will in general run exponentially slower than e non-deterministic machine. Proposition C.12 tells us for any problem in NP, it always possible to carry out the simulation in such a way that there is no blow-) in *space* requirements. Roughly speaking, we work systematically through the :arch space, with the search for each item taking only polynomial space. When it time to search for the next item, we reuse the same squares. There is a bookkeep-ig overhead (we need to keep track of where we are in the search space) but (with areful management) this can be done using a relatively small number of squares. o while the simulation may take a long time, we do not need much memory.

It is widely conjectured that NP ⊂ PSPACE (that is, it is believed that there are :oblems in PSPACE that are not in NP) and indeed that coNP ⊂ PSPACE too. /e will now describe an important problem in PSPACE that seems to belong to ither NP nor coNP; we make use of this problem in Chapter 6.

The set of prenex quantified boolean formulas (QBFs) consists of expressions of e form

$$Q_1 p_1 \dots Q_m p_m \, \theta(p_1, \dots, p_m),$$

here each Q_i is either ∀ or ∃, and $\theta(p_1, \dots, p_m)$ is a formula of propositional lculus. The quantifiers range over the truth values 1 (true) and 0 (false), and a uantified boolean formula without free variables is *true* if and only if it evaluates 1; the *QBF-truth problem* is to determine whether such a formula is true or not.

This problem contains both the satisfiability and validity problems for propositional logic as special cases (let the quantifiers be all existential, or all universal, respectively). But the general problem of deciding the truth of QBF formulas seems to be harder than either of these. Nonetheless there is a certain modularity to deciding QBF-truth. We try out one sequence of truth value assignments: checking whether it works takes polynomial space. We record what we have done, and reuse the same space to check the next assignment sequence. In this way we check through all possible assignments, and each check is performed in the same working space. So it may take an awfully long time to solve the problem – but we do not have to use much memory. And in fact we have the following result.

Theorem C.13   *The QBF-truth problem is PSPACE-complete.*

Proof. See Papadimitriou [352, Theorem 19.1].   ⊣

PSPACE is defined in terms of *deterministic* Turing machines. NPSPACE, the class of problems computable by non-deterministic polynomial space bounded Turing machines, is defined by rephrasing the definition in terms of non-deterministic Turing machines. Intriguingly, NPSPACE contains nothing new:

Theorem C.14 *(*Savitch's Theorem*) PSPACE = NPSPACE.*

Proof. See Papadimitriou [352, Theorem 7.5].   ⊣

So if we want to show that a problem is in PSPACE, we can do so by showing that it is in NPSPACE, and we take advantage of this in the text.

Finally, PSPACE = coPSPACE. Why? Well, any deterministic Turing machine that decides a problem L in PSPACE can be converted to a machine that decides $\overline{L}$ simply by flipping *yes*s to *no*s and vice versa. This invariance under complementation has nothing much to do with PSPACE: for any time or space class C defined in terms of *deterministic* Turing machines, $C = coC$, as the 'switch the outputs' argument shows. (Note that this argument does not work with non-deterministic machines.)

The class EXPTIME

EXPTIME is defined in terms of deterministic Turing machines:

Definition C.15   A deterministic Turing machine is exponentially time bounded there is a polynomial $p(n)$ such that the machine always halts after at most $2^{p(n)}$ steps, where n is the length of the input.

A problem is solvable in exponential time if there is an exponentially time bounded Turing machine that solves it. The class of all problems solvable in exponential time is called EXPTIME.

Incidentally, the use of 2 in this definition is arbitrary. If we can show that a Turing machine is time bounded by a function of the form $c^{p(n)}$, where $c > 2$, then we can show that it is exponentially time bounded in the sense just defined: simply choose k such that $2^k > c$. Then the Turing machine is time bounded by $2^{k \cdot p(n)}$, for $c^{p(n)} < (2^k)^{p(n)} = 2^{k \cdot p(n)}$.

EXPTIME-hard problems are intractable. Some problems in EXPTIME are provably *outside* P (see Lewis and Papadimitriou [301, Theorem 6.1.2]), hence any EXPTIME-hard problem is at least as hard as such intractable problems.

The class NEXPTIME

NEXPTIME is the class of problems solvable using an exponentially bounded *non-deterministic* Turing machine. Like NP algorithms, NEXPTIME algorithms have a 'guess and check' profile. The crucial difference is that guessed information may be exponentially large in the size of the input, thus the deterministic checking that follows may take exponentially many steps in the size of the input.

We have not mentioned NEXPTIME much in Chapter 6, but it is implicitly present: when modal logics are proved decidable using the finite model property, a NEXPTIME algorithm is usually being employed.

$\mathcal{O}$ notation

In the text we try to do as little combinatorial analysis as possible, and we are often content simply to say that some procedure or other runs in polynomial, or exponential, time. But sometimes we state more precise bounds, and when we do, we use $\mathcal{O}$ notation. Basically, $\mathcal{O}$ notation is a way of stating bounds that ignores multiplicative constants and low order terms. For example, instead of saying that an algorithm runs in time $3n^2 + 2n + 7$ (where n is the size of the input) we would say that it runs in time $\mathcal{O}(n^2)$ (read this as: 'of the order n^2'). Roughly speaking, this means that for all sufficiently large n, the fact that we square the length of the input dominates all the other contributions. More precisely:

Definition C.16 Let f and g be functions from the natural numbers to the natural numbers. We say that $f = \mathcal{O}(g)$ if there are positive constants c and k such that for all $n \geq k$, $f(n) \leq c \cdot g(n)$. ⊣

Thus $3n^2 + 2n + 7 = \mathcal{O}(n^2)$, as $3n^2 + 2n + 7 \leq 6n^2$ for all $n \geq 2$.

Appendix D

A Guide to the Literature

Here we list and briefly describe a number of textbooks, survey articles, and more specialized books which the reader may find useful. We have not aimed for comprehensive coverage. Rather, we have commented on the sources the reader is most likely to run into, provided pointers to topics not discussed in this book (in particular, modal proof theory and theorem proving, and first-order modal logic) and drawn attention to some interesting emerging themes.

This is a good place to mention the *Advances in Modal Logic* initiative, which attempts to bring together scholars working in various areas of modal logic and its applications. You can find out more at: http://www.aiml.net. The collection *Advances in Modal Logic, Volume 1*, edited by Kracht *et al.* [281], contains a selection of papers from the first conference hosted by the initiative. Selections from later workshops have also been published; see *Advances in Modal Logic, Volume 2*, edited by Zakharyaschev *et al.* [469]; *Advances in Modal Logic, Volume 3*, edited by Wolter *et al.* [461]; and *Advances in Modal Logic, Volume 4*, edited by Balbiani *et al.* [18].

Textbooks on Modal Logic

To start, here is an annotated list of textbooks on modal logic.

o *A Manual of Intensional Logic*, van Benthem [44]. What is modal logic? What is not! This inspiring little book takes the reader on a whirlwind tour of the many faces of modal logic. The book is deceptively easy to read; alert readers will soon cotton onto the fact that the author indicates unexplored territory on practically every page.

o *The Logic of Provability*, Boolos [69]. Clear, up-to-date, introduction to provability logic.

o *Modal Logic*, Chagrov and Zakharyaschev [88]. A recent advanced textbook on modal logics in the basic modal language and their connections with superintuitionistic logics. Concentrates on the fine structure of the lattice of normal modal logics, using methods not covered by our book, to prove general results on various properties of logics.

○ *Modal Logic. An Introduction*, Chellas [92]. A readable introductory text which focuses on completeness-via-canonicity and decidability-via-filtration for the basic modal language. Also introduces neighborhood (or Montague-Scott) semantics, a tool for analyzing non-normal modal logics (Chellas calls neighborhood models minimal models).

○ *Proof Methods for Modal and Intuitionistic Logic*, Fitting [137]. This beautiful book explores in detail a number of proof methods (tableaux, sequent systems, natural deduction) for modal logic; we have no hesitation in recommending this classic work as a great source for finding out more about modal proof theory.

○ *Types, Tableaus, and Goedel's God*, Fitting [139]. An introduction to higher-order modal logic that ends by formalizing Goedel's ontological argument for the existence of God! Quite apart from anything else, it is one of the best introductions to ordinary (non-modal) higher-order logic around. Tableaux-based.

○ *First-Order Modal Logic*, Fitting and Mendelsohn [140]. An excellent introduction to first-order modal logic. Addresses both technical and philosophical aspects of quantification and equality. Provides both Hilbert-style and tableaux-based proof systems.

○ *Modal Logics and Philosophy*, Girle [172]. Readable introduction to modal logic, including first-order modal logic. Discusses temporal, dynamic, epistemic, and deontic interpretations, as well as the logic of necessity and possibility.

○ *Logics of Time and Computation*, Goldblatt [177]. Clearly written intermediate level text which focuses on completeness results for the basic modal and temporal languages, and a variety of extended modal languages, including until-based temporal languages, PDL, and first-order PDL. A useful book to have around.

○ *Mathematics of Modality*, Goldblatt [178]. This book brings together a number of the author's papers on modal logics. In particular, it contains his seminal PhD thesis which can be hard to obtain in its article version [184, 185].

○ *Dynamic Logic*, Harel, Kozen and Tiuryn [212]. A detailed, well written, introduction to propositional dynamic logic, and many of its extensions, including first-order dynamic logic. A good choice for readers wanting to find out more about this important branch of modal logic.

A Companion to Modal Logic, Hughes and Cresswell [234]. This was the first textbook to move beyond the staples of the classical period (relational semantics, canonical models, filtrations) and discuss distinctively modern topics (notably frame incompleteness). Short, accessible, and clear, it is still a valuable introductory text, though it has since been superseded by the next entry.

○ *A New Introduction to Modal Logic*, Hughes and Cresswell [235]. An admirably clear and wide ranging introductory text. Although many topics (such as the standard translation) are not discussed, it manages to at least mention many modern themes such as frame definability and extended modal languages. Contains a good discussion of first-order modal logic.

○ *Tools and Techniques in Modal Logic*, Kracht [279]. Advanced book on mathematical aspects of modal logic, taking a polymodal perspective. Contains an in-depth study of correspondence and completeness, duality theory, the lattice of modal logics, and transfer results from monomodal to polymodal logics.

○ *The 'Lemmon Notes': An Introduction to Modal Logic*, Lemmon and Scott [296]. All that exists of an unfinished monograph on modal logic (Lemmon's death in 1966 prevented its completion). The original source for work on filtrations and canonical models, for many years it was *the* definitive introduction to modal logic. Although out of date, its quality still shines through.

○ *Epistemic Logic for AI and Computer Science*, Meyer and van der Hoek [328]. An introductory text on epistemic logic. Covers the basic modal approach, as well as more advanced models and default reasoning.

○ *A Short Introduction to Modal Logic*, Mints [330]. This little book is a short introduction to Gentzen-style proof systems for **S5**, **S4** and **T**. A useful starting point in modal proof theory.

○ *First Steps in Modal Logic*, Popkorn [361]. If you find the present book too difficult and feel the need to consult something simpler, we suggest you try Popkorn's text. Like the present book it is semantically oriented and takes for granted that modal logic has more to offer than the basic modal language. It is clearly written, mathematically precise, and besides the present book, it is the only textbook we know that discusses bisimulations.

○ *Self-Reference and Modal Logic*, Smoryński [416]. The classic introduction to provability logic. Beautifully written. More demanding than the Boolos volume.

○ *From Modal Logic to Deductive Databases*, edited by Thayse [430]. Wide ranging introduction to modal logic. Discusses links with natural language, temporal reasoning, various forms of defeasible reasoning, and deductive databases.

Books in Other Languages

Next, here is a list of books in languages other than English, without comments.

○ *Essai de Logique Déontique*, Bailhache [17].

○ *Logicaboek*, Batens [30].

○ *Fondements Logiques du Raisonnement Contextuel. Une Etude sur les Logique des Conditionnels*, Crocco [101].

○ *Logica, Significato e Intelligenza Artificiale*, Frixione [143].

○ *La Logique du Temps*, Gardies [161].
○ *Essai sur les Logiques des Modalités*, Gardies [162].
○ *Logique. Volume 3. Méthodes pour l'intelligence artificielle*, Gochet, Gribomont, and Thayse [174].
○ *Una Introducción a la Lógica Modal*, Jansana [244].
○ *La Logique Déductive*, Kalinowski [257].
○ *Glauben, Wissen und Wahrscheinlichkeit. Systeme der Epistemischen Logik*, Lenzen [298].
○ *Pour une Logique du Sens*, Martin [315].
○ *Forcing et Sémantique de Kripke-Joyal*, Moens [331].
○ *Essais sur les Logiques non Chrysipiennes*, Moisil [332].
○ *Klassische und nichtklassische Aussagenlogik*, Rautenberg [373].

Survey Articles

Note that virtually all the survey articles listed below are drawn from the following sources:

○ *Handbook of Automated Reasoning*, edited by Robinson and Voronkov [388].
○ *Handbook of Logic and Language*, edited by van Benthem and ter Meulen [51].
○ *Handbook of Logic in Artificial Intelligence and Logic Programming*, edited by Gabbay, Hogger, and Robinson [157, 158].
○ *Handbook of Logic in Computer Science*, edited by Abramsky, Gabbay, and Maibaum [1].
○ *Handbook of Philosophical Logic*, edited by Gabbay and Guenthner [154].
○ *Handbook of Proof Theory*, edited by Buss [83].
○ *Handbook of Tableau Methods*, edited by D'Agostino, Gabbay, Hähnle, and Posegga [104].
○ *Handbook of Theoretical Computer Science*, edited by van Leeuwen [293].

In fact, these handbooks contain surveys of many other topics in, or related to, modal logic including: auto-epistemic logic, belief revision, combinations of tense and modality, computational treatments of time, conditional logic, decision procedures, deontic logic, description logics, epistemic aspects of databases, general decidable fragments, higher-order modal logic, logics of programs, non-monotonic temporal reasoning, philosophical perspectives on first-order modal logic, provability logic, reasoning about knowledge, time and change in AI.

Correspondence Theory, van Benthem [43]. For many years this was the only easily accessible general reference on correspondence theory, and it is still well worth reading. Lots of telling examples, and propelled by a clear vision of what the modal enterprise is all about.

o *Temporal Logic*, van Benthem [47]. A wide-ranging and thoughtful discussion of key themes in temporal logic.

o *Basic Modal Logic*, Bull and Segerberg [75]. An interesting survey, rich in historical detail, which provides a useful point of entry to a topic barely touched on in the present book: the fine structure of the lattice of normal modal logics in the basic modal language.

o *Basic Tense Logic*, Burgess [79]. An accessible survey, mostly devoted to the basic temporal language, but touching on until-based logics and multi-dimensional systems. Contains many useful examples of step-by-step completeness proofs.

o *Advanced Modal Logic*, Chagrov, Wolter and Zakharyaschev [87]. Takes up the story where the Bull and Segerberg survey leaves off. Strong on the fine structure of the lattice of normal modal logics.

o *Reasoning in Description Logics*, Donini, Lenzerini, Nardi, and Schaerf [115]. Approachable overview article which discusses four types of reasoning important in description logic. Both inference techniques and complexity results are covered.

o *Temporal and Modal Logic*, Emerson [121]. A detailed introduction to temporal logic from the perspective of theoretical computer science. Somewhat dated, but still a good introduction.

o *Basic Modal Logic*, Fitting [138]. An extremely clear introductory survey. Starts with a good discussion of the basic modal language (including Hilbert systems, natural deduction, tableaux methods, the standard translation, and alternative semantics), and then goes on to examine first-order modal logic.

o *Varieties of Complex Algebras* and *Algebraic Polymodal Logic*, Goldblatt [186, 181]. Both of these provide an introduction to the study of varieties of BAOs, emphasizing their connections with modal logics, and focusing on structural properties (such as canonicity) that are related to natural properties of modal logical systems.

o *Tableau Methods for Modal and Temporal Logics*, Goré [193]. Detailed survey of tableaux based proof methods for temporal languages.

o *Dynamic Logic*, Harel [209]. If you want to learn more about PDL, this is a good place to look, though it is more densely written than Harel, Kozen and Tiuryn [212]. Discusses a wide range of variants and extensions of PDL.

o *The Logic of Provability*, Japaridze and de Jongh [245]. A thorough overview of provability logic, covering propositional provability logic, interpretability logic and related areas, as well as predicate provability logic.

o *A Survey of Boolean Algebras with Operators*, Jónsson [253]. Gives an algebraic introduction to the theory of boolean algebras with operators. The papers by Goldblatt and this one by Jónsson are highly recommended to readers who want more on the algebraic side of modal logic than our Chapter 5 offers.

o *Logics of Programs*, Kozen and Tiuryn [274]. Essentially (though not entirely) an introduction to PDL. Useful, but Harel, Kozen and Tiuryn [212] is probably a better choice.

o *Resolution Decision Procedures*, Leitsch, Fermüller and Tammet [294]. An extensive survey of resolution-based decision procedures for fragments of first-order logic. Covers procedures for modal and guarded fragments and fragments corresponding to description logics.

o *Encoding Non-Classical Logics in Classical Logic*, Ohlbach, de Rijke, Nonnengart, and Gabbay [345]. A computationally oriented overview of translation methods for mapping modal and modal-like languages into first-order logic. Discusses various flavors of the standard (relational) translation and the functional translation, from the point of view of both expressive power and decidability.

o *Feature Logics*, Rounds [394]. Authoritative survey of feature logic. Devotes a lot of attention to the modal aspects of feature logic.

o *Modal and Temporal Logics*, Stirling [424]. An extremely useful, technically oriented article which will appeal to many readers of this volume. Starting with many of the same tools introduced in this book (transition systems, bisimulations, correspondence theory, unraveling), it goes on to discuss many other topics such as modal μ-calculi and links with automata theory.

o *An Overview of Interpretability Logic*, Visser [451]. An elegant overview of interpretability logic, an extension of provability logic which deals with interpretations between formal theories.

o *Canonical Formulas for Modal and Superintuitionistic Logics: A Short Outline*, Zakharyaschev [467]. Canonical formulas are an important frame-theoretic approach to the classification of modal formulas. This is an accessible survey, by the inventor of the method, with plenty of motivations, examples and definitions.

Other Books

o *Vicious Circles*, Barwise and Moss [27]. An introduction to non-well-founded set theory, a set theory where bisimulation rather than extensionality is the key concept. The authors advocate using infinitary modal logic to study such structures, and the area is becoming a research area in its own right.

> *Modal Logic and Classical Logic*, van Benthem [42]. This book (a reworking of the author's PhD thesis) is the primary source on what is now called correspondence theory. Although most of the book is devoted to frame definability, this is also where bisimulations (under the name p-relations) were introduced and shown to capture modal expressivity over models. It may be hard to get hold of, but it is well worth making the effort.

Logic of Time, van Benthem [45]. Explores temporal logic model-theoretically; modal logic is just one strand in the story. The book makes many interesting technical contributions (notably in the study of interval structures) but its real importance is methodological: it is by far the most convincing demonstration we know of the light model-theoretic methods can throw on difficult conceptual issues.

o *Reasoning about Knowledge*, Fagin, Halpern, Moses and Vardi [125]. Looks set to be the key reference on epistemic logic for quite some time to come. Treats a wide range of topics from a modern semantically-oriented perspective. Clearly written. Highly recommended.

o *Labelled Deductive Systems*, Gabbay [151]. This is not a book on modal logic, but an introduction to a proof theoretical methodology. But modal logics provide many of the nicest examples of the labeling method in action, and if you are interested in modal proof theory you need to know about this.

o *Temporal Logics. Mathematical Foundations and Computational Aspects. Volume 1*, Gabbay, Hodkinson and Reynolds [156]. The first of three projected volumes, this book discusses basic completeness and incompleteness, the temporal logics of special structures (such as $\mathbb{R}$), multi-dimensional temporal logic, fixed-point logic and propositional quantification, and much else besides. Contains a detailed discussion of expressive completeness.

o *Cylindric Algebras, Parts I & II*, by Henkin, Monk and Tarski [218]. The definitive work on cylindric algebras, it also contains a wealth of results on other branches of algebraic logic and on boolean algebras with operators in general.

o *Relation Algebras by Games*, by Hodkinson and Hirsch [226]. Still in manuscript form at the time of writing, this work promises to become a standard reference on relation algebras. Strong on the connections between relation algebras and model theory and a convincing argument for the application of games in algebraic logic.

o *The Temporal Logic of Reactive and Concurrent Systems. Volume 1: Specification*, Manna and Pnueli [312]. Textbook aimed at computer scientists interested in specifying reactive systems. All the necessary temporal logic is introduced and explained in the course of the book.

o *The Temporal Verification of Reactive Systems. Volume 2: Safety*, Manna and Pnueli [313]. Follow up to the previous volume, emphasizing safety issues.

o *Multidimensional Modal Logic*, Marx and Venema [320]. The key reference for work in multi-dimensional modal logic. Discusses two-dimensional modal logics, arrow logics, modal logics of intervals, modal logics of relations, and the problem of defining 'concrete' semantics (in the sense of Henkin) for arbitrary modal languages.

o *Temporal Logic*, Øhrstrøm and Hasle [347]. Introduction to temporal logic from a historical perspective. A good way of getting to grips with Prior's work.

o *Past, Present and Future*, Prior [365]. Arthur Prior seems to be one of those authors more often cited than read. This is pity: if you push on past his use

Polish notation you will discover a fascinating writer with a surprisingly up to date range of concerns.

o *Set Theory and the Continuum Problem*, Smullyan and Fitting [418]. So why is this one here? Because in Part III of the book the authors show how Cohen-style forcing arguments can be formulated using first-order **S4**. Roughly speaking, they prove independence results using sophisticated step-by-step arguments.

o *Proof Theory of Modal Logic*, Wansing [454]. A collection of papers devoted to proof-theoretical aspects of modal logic, covering many flavors of proof theory: sequents, resolution, tableaux, display calculi, and translation-based approaches.

o *Displaying Modal Logic*, Wansing [455]. A proof-theoretical monograph that shows in detail how to define display calculi for modal logics.

o *Modal Logic: The Lewis Modal Systems*, Zeman [474]. Provides sequent proof systems for the Lewis systems and some others. Written by a student of Arthur Prior, it draws on some of Prior's unpublished classroom material. Clearly written. Uses Polish notation.

The Ω-bibliography

Finally, a fairly comprehensive bibliography of books and articles on modal (and various non-classical) logics is provided by Volume II of the Ω-bibliography of mathematical logic. The references in this volume, edited by Rautenberg [375], stop somewhere in the 1980s; nevertheless, this work can be the starting point of a fascinating quest for treasures in the earlier literature of modal logic.

Bibliography

[1]   S. Abramsky, D.M. Gabbay, and T.S.E. Maibaum, editors. *Handbook of Logic in Computer Science*, volume 2. Clarendon Press, 1992.

[2]   P. Aczel. *Non-Well-Founded Sets*. CSLI Publications, 1988.

[3]   M. Aiello and J. van Benthem. A modal walk through space. *Journal of Applied Non-Classical Logics*, 12:319–364, 2002.

[4]   C.E. Alchourrón, P. Gärdenfors, and D. Makinson. On the logic of theory change: partial meet contraction and revision functions. *Journal of Symbolic Logic*, 50:510–530, 1985.

[5]   A. Andréka, A. Kurucz, I. Németi, and I. Sain. Applying algebraic logic to logic. In M. Nivat and M. Wirsing, editors, *Algebraic Methodology and Software Technology*, pages 201–221. Springer, 1994.

[6]   H. Andréka. Complexity of equations valid in algebras of relations. *Annals of Pure and Applied Logic*, 89:149–229, 1997.

[7]   H. Andréka, I.M. Hodkinson, and I. Németi. Finite algebras of relations are representable on finite sets. *Journal of Symbolic Logic*, 64:243–267, 1999.

[8]   H. Andréka, J.D. Monk, and I. Németi, editors. *Algebraic Logic. (Proceedings of the 1988 Budapest Conference)*, volume 54 of *Colloquia Mathematica Societatis János Bolyai*. North-Holland Publishing Company, 1991.

[9]   H. Andréka, J. van Benthem, and I. Németi. Modal languages and bounded fragments of predicate logic. *Journal of Philosophical Logic*, 27:217–274, 1998.

[10] I.H. Anellis and N. Houser. Nineteenth century roots of algebraic logic and universal algebra. In Andréka *et al.* [8], pages 1–36.

[11] L. Åqvist, F. Guenthner, and C. Rohrer. Definability in ITL of some subordinate temporal conjunctions in English. In F. Guenthner and C. Rohrer, editors, *Studies in Formal Semantics*, pages 201–221. North Holland, 1978.

[12] C. Areces. *Logic Engineering: The Case of Description and Hybrid Logics*. PhD thesis, ILLC, University of Amsterdam, 2000.

[13] C. Areces, P. Blackburn, and M. Marx. The computational complexity of hybrid temporal logics. *Logic Journal of the IGPL*, 8(5):653–679, 2000.

[14] C. Areces, P. Blackburn, and M. Marx. Hybrid logics: Characterization, interpolation and complexity. *Journal of Symbolic Logic*, 66:977–1010, 2001.

[15] C. Areces and M. de Rijke. Description and/or hybrid logic. In *Workshop Proceedings AiML-2000*, pages 1–14. Institut für Informatik, Universität Leipzig, 2000.

524

[16] F. Baader and K. Schulz, editors. *Frontiers of Combining Systems 1*. Kluwer Academic Publishers, 1996.

[17] P. Bailhache. *Essai de Logique Déontique*. Mathesis, 1991.

[18] P. Balbiani, N.-Y. Suzuki, F. Wolter, and M. Zakharyaschev, editors. *Advances in Modal Logic, Volume 4*. King's College London Publications, 2003.

[19] Ph. Balbiani, L. Fariñas del Cerro, T. Tinchev, and D. Vakarelov. Modal logics for incidence geometries. *Journal of Logic and Computation*, 7:59–78, 1997.

[20] A. Baltag. *STS: A Structural Theory of Sets*. PhD thesis, Indiana University, Bloomington, Indiana, 1998.

[21] A. Baltag. A logic for suspicious players: epistemic actions and belief updates in games. *Bulletin of Economic Research*, to appear, 2000. Paper presented at the Fourth Conference on *Logic and the Foundations of the Theory of Games and Decisions*.

[22] A. Baltag. STS: a structural theory of sets. In Zakharyaschev *et al.* [469].

[23] B. Banieqbal, H. Barringer, and A. Pnueli, editors. *Proc. Colloquium on Temporal Logic in Specification*, volume 398 of *LNCS*. Springer, 1989.

[24] H. Barendregt. *The Lambda Calculus: Its Syntax and Semantics*. North-Holland Publishing Company, 1984.

[25] J. Barwise. Model-theoretic logics: background and aims. In Barwise and Feferman [26], pages 3–23.

[26] J. Barwise and S. Feferman, editors. *Model-Theoretic Logics*. Springer, 1985.

[27] J. Barwise and L. Moss. *Vicious Circles*, volume 60 of *Lecture Notes*. CSLI Publications, 1996.

[28] J. Barwise and L.S. Moss. Modal correspondence for models. *Journal of Philosophical Logic*, 27:275–294, 1998.

[29] D. Basin and N. Klarlund. Automata based symbolic reasoning in hardware verification. *Journal of Formal Methods in Systems Design*, 13:255–288, 1998.

[30] D. Batens. *Logicaboek*. Garant Leven, Apeldoorn, 1991.

[31] P. Battigalli and G. Bonanno. Recent results on belief, knowledge and the foundations of game theory. *Research in Economics*, 53:149–225, 1999.

[32] J.L. Bell and M. Machover. *A Course in Mathematical Logic*. North-Holland Publishing Company, 1977.

[33] J.L. Bell and A.B. Slomson. *Models and Ultraproducts*. North-Holland Publishing Company, 1969.

[34] B. Bennet, C. Dixon, M. Fisher, E. Franconi, I. Horrocks, U. Hustadt, and M. de Rijke. Combinations of modal logic. *Journal of AI Reviews*, 17:1–20, 2002.

[35] J. van Benthem. A note on modal formulas and relational properties. *Journal of Symbolic Logic*, 40:85–88, 1975.

[36] J. van Benthem. *Modal Correspondence Theory*. PhD thesis, Mathematisch Instituut & Instituut voor Grondslagenonderzoek, University of Amsterdam, 1976.

[37] J. van Benthem. Two simple incomplete modal logics. *Theoria*, 44:25–37, 1978.

[38] J. van Benthem. Minimal deontic logics. *Bulletin of the Section of Logic*, 8:36–42, 1979.

[39] J. van Benthem. Syntactical aspects of modal incompleteness theorems. *Theoria*, 45:63–77, 1979.

[40] J. van Benthem. Canonical modal logics and ultrafilter extensions. *Journal of Symbolic Logic*, 44:1–8, 1980.

[41] J. van Benthem. Some kinds of modal completeness. *Studia Logica*, 39:125–141, 1980.

[42] J. van Benthem. *Modal Logic and Classical Logic*. Bibliopolis, 1983.

[43] J. van Benthem. Correspondence theory. In Gabbay and Guenthner [154], pages 167–247.

[44] J. van Benthem. *A Manual of Intensional Logic*, volume 1 of *Lecture Notes*. CSLI Publications, 1985.

[45] J. van Benthem. *The Logic of Time*. Kluwer Academic Publishers, second edition, 1991.

[46] J. van Benthem. Modal frame classes revisited. *Fundamenta Informaticae*, 18:307–317, 1993.

[47] J. van Benthem. Temporal logic. In Gabbay et al. [158], pages 241–351.

[48] J. van Benthem. *Exploring Logical Dynamics*. Studies in Logic, Language and Information. CSLI Publications, 1996.

[49] J. van Benthem. Dynamic bits and pieces. Technical Report LP-97-01, Institute for Language, Logic and Computation, 1997.

[50] J. van Benthem and W. Meyer Viol. Logical semantics of programming. Unpublished lecture notes. University of Amsterdam, 1993.

[51] J. van Benthem and A. ter Meulen, editors. *Handbook of Logic and Language*. Elsevier Science, 1997.

[52] R. Berger. The undecidability of the domino problem. Technical Report 66, Mem. Amer. Math. Soc., 1966.

[53] G. Birkhoff. On the structure of abstract algebras. *Proceedings of the Cambridge Philosophical Society*, 29:441–464, 1935.

[54] P. Blackburn. Nominal tense logic. *Notre Dame Journal of Formal Logic*, 34:56–83, 1993.

[55] P. Blackburn. Tense, temporal reference, and tense logic. *Journal of Semantics*, 11:83–101, 1994.

[56] P. Blackburn. Internalizing labelled deduction. *Journal of Logic and Computation*, 10:137–168, 2000.

[57] P. Blackburn. Representation, reasoning, and relational structures: a hybrid logic manifesto. *Logic Journal of the IGPL*, 8:339–365, 2000.

[58] P. Blackburn, C. Gardent, and W. Meyer-Viol. Talking about trees. In *Proceedings of the 6th Conference of the European Chapter of the Association for Computational Linguistics*, pages 21–29, 1993.

[59] P. Blackburn and J. Seligman. Hybrid languages. *Journal of Logic, Language and Information*, 4:251–272, 1995.

[60] P. Blackburn and J. Seligman. What are hybrid languages? In Kracht et al. [281] pages 41–62.

[61] P. Blackburn and E. Spaan. A modal perspective on the computational complexity o attribute value grammar. *Journal of Logic, Language and Information*, 2:129–169 1993.

[62] P. Blackburn and M. Tzakova. Hybridizing concept languages. *Annals of Mathe matics and Artificial Intelligence*, 24:23–49, 1998.

[63] P. Blackburn and M. Tzakova. Hybrid languages and temporal logic. *Logic Journal of the IGPL*, 7(1):27–54, 1999.

[64] W.J. Blok. An axiomatization of the modal theory of the veiled recession fram

Technical report, Department of Mathematics, University of Amsterdam, 1977.

[65] W.J. Blok. On the degree of incompleteness in modal logic and the covering relations in the lattice of modal logics. Technical Report 78–07, Department of Mathematics, University of Amsterdam, 1978.

[66] W.J. Blok. The lattice of modal algebras: An algebraic investigation. *Journal of Symbolic Logic*, 45:221–236, 1980.

[67] W.J. Blok and D. Pigozzi. Algebraizable logics. *Memoirs of the American Mathematical Society*, 77, 396, 1989.

[68] G. Boolos. *The Unprovability of Consistency*. Cambridge University Press, 1979.

[69] G. Boolos. *The Logic of Provability*. Cambridge University Press, 1993.

[70] G. Boolos and R. Jeffrey. *Computability and Logic*. Cambridge University Press, 1989.

[71] E. Börger, E. Grädel, and Y. Gurevich. *The Classical Decision Problem*. Springer, 1997.

[72] J.R. Büchi. On a decision method in restricted second order arithmetic. In *Proceedings International Congress on Logic, Methodology and Philosophy of Science 1960*. Stanford University Press, 1962.

[73] R. Bull. An approach to tense logic. *Theoria*, 36:282–300, 1970.

[74] R.A. Bull. That all normal extensions of S4.3 have the finite model property. *Zeitschrift für mathemathische Logik und Grundlagen der Mathematik*, 12:314–344, 1966.

[75] R.A. Bull and K. Segerberg. Basic modal logic. In Gabbay and Guenthner [154], pages 1–88.

[76] P. Buneman, S. Davidson, M. Fernandez, and D. Suciu. Adding structure to unstructured data. In *Proceedings ICDT'97*, 1997.

[77] J. Burgess. Decidability for branching time. *Studia logica*, 39:203–218, 1980.

[78] J. Burgess. Axioms for tense logic I: 'since' and 'until'. *Notre Dame Journal of Formal Logic*, 23:375–383, 1982.

[79] J.P. Burgess. Basic tense logic. In Gabbay and Guenthner [154], pages 89–133.

[80] J.P. Burgess and Y. Gurevich. The decision problem for linear temporal logic. *Notre Dame Journal of Formal Logic*, 26:115–128, 1985.

[81] S. Burris and H.P. Sankappanavar. *A Course in Universal Algebra*. Graduate Texts in Mathematics. Springer, 1981.

[82] R. Burstall. Program proving as hand simulation with a little induction. In *Information Processing '74*, pages 308–312. North-Holland Publishing Company, 1974.

[83] S.R. Buss, editor. *Handbook of Proof Theory*. Elsevier Science, 1998.

[84] D. Calvanese, G. De Giacomo, and M. Lenzerini. Modeling and querying semistructured data. *Networking and Information Systems*, pages 253–273, 1999.

[85] R. Carnap. Modalities and quantification. *Journal of Symbolic Logic*, 11:33–64, 1946.

[86] R. Carnap. *Meaning and Necessity*. University of Chicago Press, 1947.

[87] A. Chagrov, F. Wolter, and M. Zakharyaschev. Advanced modal logic. In *Handbook of Philosophical Logic*, volume 3, pages 83–266. Kluwer Academic Publishers, second edition, 2001.

[88] A. Chagrov and M. Zakharyaschev. *Modal Logic*, volume 35 of *Oxford Logic Guides*. Oxford University Press, 1997.

[89] L.A. Chagrova. An undecidable problem in correspondence theory. *Journal of*

Symbolic Logic, 56:1261–1272, 1991.

[90] D. Chalmers. *The Conscious Mind.* Oxford University Press, 1996.

[91] C.C. Chang and H.J. Keisler. *Model Theory.* North-Holland Publishing Company, Amsterdam, 1973.

[92] B.F. Chellas. *Modal Logic, an Introduction.* Cambridge University Press, 1980.

[93] B. Chlebus. Domino-tiling games. *Journal of Computer and System Sciences*, 32:374–392, 1986.

[94] E.M. Clarke and E.A. Emerson. Design and synthesis of synchronisation skeletons using branching time temporal logic. In D. Kozen, editor, *Logics of Programs*, pages 52–71. Springer, 1981.

[95] E.M. Clarke, O. Grumberg, and D.A. Peled. *Model Checking.* The MIT Press, 1999.

[96] E.M. Clarke and B.-H. Schlingloff. Model checking. In Robinson and Voronkov [388].

[97] B.J. Copeland, editor. *Logic and Reality. Essays on the Legacy of Arthur Prior.* Clarendon Press, 1996.

[98] W. Craig. On axiomatizability within a system. *Journal of Symbolic Logic*, 18:30–32, 1953.

[99] M.J. Cresswell. A Henkin completeness theorem for T. *Notre Dame Journal of Formal Logic*, 8:186–90, 1967.

[100] M.J. Cresswell. An incomplete decidable modal logic. *Journal of Symbolic Logic*, 49:520–527, 1984.

[101] G. Crocco. *Fondements Logiques du Raisonnement Contextuel. Une Etude sur les Logiques des Conditionnels.* Padova Unipress, 1996.

[102] L. Csirmaz, D.M. Gabbay, and M. de Rijke, editors. *Logic Colloquium '92*, number 1 in Studies in Logic, Language and Information. CSLI Publications, 1995.

[103] N. Cutland. *Computability. An Introduction to Recursive Function Theory.* Cambridge University Press, 1980.

[104] M. D'Agostino, D.M. Gabbay, R. Hähnle, and J. Posegga, editors. *Handbook of Tableau Methods.* Kluwer Academic Publishers, 1999.

[105] B.A. Davey and H.A. Priestly. *Introduction to Lattices and Order.* Cambridge University Press, 1990.

[106] G. De Giacomo. *Decidability of Class-Based Knowledge Representation Formalisms.* PhD thesis, Università di Roma "La Sapienza", 1995.

[107] S. Demri. Sequent calculi for nominal tense logics: a step towards mechanization? In Murray [339], pages 140–154.

[108] S. Demri and R. Goré. Cut-free display calculi for nominal tense logics. In Murray [339], pages 155–170.

[109] H. van Ditmarsch. *Knowledge Games.* PhD thesis, Department of Mathematics and Computer Science, Rijksuniversiteit Groningen, 2000.

[110] H.C. Doets. *Completeness and Definability: Applications of the Ehrenfeucht Game in Intensional and Second-Order Logic.* PhD thesis, Department of Mathematics and Computer Science, University of Amsterdam, 1987.

[111] K. Doets. *Basic Model Theory.* Studies in Logic, Language and Information. CSLI Publications, 1996.

[112] K. Doets and J. van Benthem. Higher-order logic. In Gabbay and Guenthner [153] pages 275–329.

[113] J. Doner. Tree acceptors and some of their applications. *Journal of Computer an*

System Sciences, 4:406–451, 1970.

[114] F.M. Donini, M. Lenzerini, D. Nardi, and W. Nutt. The complexity of concept languages. *Information and Computation*, 134:1–58, 1997.

[115] F.M. Donini, M. Lenzerini, D. Nardi, and A. Schaerf. Reasoning in description logics. In G. Brewka, editor, *Principles of Knowledge Representation*, Studies in Logic, Language and Information, pages 191–236. CSLI Publications, 1996.

[116] J. Dugundji. Note on a property of matrices for Lewis and Langford's calculi of propositions. *Journal of Symbolic Logic*, 5:150–151, 1940.

[117] M.A.E. Dummett and E.J. Lemmon. Modal logics between S4 and S5. *Zeitschrift für mathemathische Logik und Grundlagen der Mathematik*, 5:250–264, 1959.

[118] H.-D. Ebbinghaus and J. Flum. *Finite Model Theory*. Perspectives in Mathematical Logic. Springer, 1995.

[119] A. Ehrenfeucht. An application of games to the completeness problem for formalized theories. *Fundamenta Mathematicae*, 49:129–141, 1961.

[120] P. van Emde Boas. The convenience of tilings. Technical Report CT-96-01, ILLC, University of Amsterdam, 1996.

[121] E.A. Emerson. Temporal and modal logics. In van Leeuwen [293], pages 995–1072.

[122] H.B. Enderton. *A Mathematical Introduction to Logic*. Academic Press, New York, 1972.

[123] L.L. Esakia. Topological Kripke models. *Soviet Mathematics Doklady*, 15:147–151, 1974.

[124] L.L. Esakia and V.Yu. Meskhi. Five critical systems. *Theoria*, 40:52–60, 1977.

[125] R. Fagin, J.Y. Halpern, Y. Moses, and M.Y. Vardi. *Reasoning About Knowledge*. The MIT Press, 1995.

[126] T. Fernando. A modal logic for non-deterministic discourse processing. *Journal of Logic, Language and Information*, 8:455–468, 1999.

[127] K. Fine. Propositional quantifiers in modal logic. *Theoria*, 36:331–346, 1970.

[128] K. Fine. The logics containing $S4.3$. *Zeitschrift für mathemathische Logik und Grundlagen der Mathematik*, 17:371–376, 1971.

[129] K. Fine. An incomplete logic containing $S4$. *Theoria*, 40:23–29, 1974.

[130] K. Fine. Logics extending $K4$. Part I. *Journal of Symbolic Logic*, 39:31–42, 1974.

[131] K. Fine. Normal forms in modal logic. *Notre Dame Journal of Formal Logic*, 16:229–234, 1975.

[132] K. Fine. Some connections between elementary and modal logic. In Kanger [263].

[133] K. Fine. Modal logics containing $K4$. Part II. *Journal of Symbolic Logic*, 50:619–651, 1985.

[134] M. Finger. Handling database updates in two-dimensional temporal logic. *Journal of Applied Non-Classical Logics*, 2:201–224, 1992.

[135] M.J. Fischer and R.E. Ladner. Propositional dynamic logic of regular programs. *Journal of Computer and System Sciences*, 18:194–211, 1979.

[136] F. Fitch. A correlation between modal reduction principles and properties of relations. *Journal of Philosophical Logic*, 2:97–101, 1973.

[137] M. Fitting. *Proof Methods for Modal and Intuitionistic Logic*. Reidel, 1983.

[138] M. Fitting. Basic modal logic. In Gabbay *et al.* [157], pages 368–449.

[139] M. Fitting. *Types, Tableaus, and Goedel's God*. Kluwer Academic Publishers, 2002.

[140] M. Fitting and R.L. Mendelsohn. *First-Order Modal Logic*. Kluwer Academic Publishers, 1998.

[141] R. Fraïssé. Sur quelques classifications des systèmes de relations. *Publ. Sci. Univ. Alger.*, 1:35–182, 1954.

[142] N. Friedman and J.Y. Halpern. Modeling belief in dynamic systems, part i: Foundations. *Artificial Intelligence*, 95:257–316, 1997.

[143] M. Frixione. *Logica, Significato e Intelligenza Artificiale*. FrancoAngeli, Milano, 1994.

[144] A. Fuhrmann. On the modal logic of theory change. In A. Fuhrmann and M. Morreau, editors, *LNAI*, volume 465, pages 259–281. Springer, 1990.

[145] D.M. Gabbay. Decidability results in non-classical logics. *Annals of Mathematical Logic*, 10:237–285, 1971.

[146] D.M. Gabbay. On decidable, finitely axiomatizable modal and tense logics without the finite model property I. *Israel Journal of Mathematics*, 10:478–495, 1971.

[147] D.M. Gabbay. On decidable, finitely axiomatizable modal and tense logics without the finite model property II. *Israel Journal of Mathematics*, 10:496–503, 1971.

[148] D.M. Gabbay. The separation property of tense logics. Unpublished manuscript, September 1979.

[149] D.M. Gabbay. An irreflexivity lemma with applications to axiomatizations of conditions on linear frames. In U. Mönnich, editor, *Aspects of Philosophical Logic*, pages 67–89. Reidel, 1981.

[150] D.M. Gabbay. The declarative past and imperative future: executable temporal logic for interactive systems. In Banieqbal *et al.* [23], pages 431–448.

[151] D.M. Gabbay. *Labelled Deductive Systems*. Clarendon Press, Oxford, 1996.

[152] D.M. Gabbay and M. de Rijke, editors. *Frontiers of Combining Systems 2*. Research Studies Press, 2000.

[153] D.M. Gabbay and F. Guenthner, editors. *Handbook of Philosophical Logic*, volume 1. Reidel, 1983.

[154] D.M. Gabbay and F. Guenthner, editors. *Handbook of Philosophical Logic*, volume 2. Reidel, 1984.

[155] D.M. Gabbay and I.M. Hodkinson. An axiomatization of the temporal logic with Since and Until over the real numbers. *Journal of Logic and Computation*, 1:229–259, 1991.

[156] D.M. Gabbay, I.M. Hodkinson, and M. Reynolds. *Temporal Logic: Mathematical Foundations and Computational Aspects*. Oxford University Press, 1994.

[157] D.M. Gabbay, C.J. Hogger, and J.A. Robinson, editors. *Handbook of Logic in Artificial Intelligence and Logic Programming*, volume 1. Oxford University Press, 1993.

[158] D.M. Gabbay, C.J. Hogger, and J.A. Robinson, editors. *Handbook of Logic in Artificial Intelligence and Logic Programming*, volume 4. Oxford University Press, 1994.

[159] D.M. Gabbay, A. Kurucz, F. Wolter, and M. Zakharyaschev. *Many-Dimensional Modal Logics: Theory and Applications*, volume 146 of *Studies in Logic and the Foundations of Mathematics*. Elsevier, 2003.

[160] D.M. Gabbay, A. Pnueli, S. Shelah, and J. Stavi. On the temporal analysis of fairness. In *Proc. 7th ACM Symposium on Principles of Programming Language*, pages 163–173, 1980.

[161] J. Gardies. *La Logique du Temps*. Presses Universitaires de France, Paris, 1975.

[162] J. Gardies. *Essai sur les Logiques des Modalités*. Presses Universitaires de Franc

Paris, 1979.

[63] M.R. Garey and D.S. Johnson. *Computers and Intractibility. A Guide to the Theory of NP-Completeness*. W.H. Freeman, 1979.

[64] G. Gargov and V. Goranko. Modal logic with names. *Journal of Philosophical Logic*, 22:607–636, 1993.

[65] G. Gargov and S. Passy. A note on Boolean modal logic. In P.P. Petkov, editor, *Mathematical Logic. Proceedings of the 1988 Heyting Summerschool*, pages 311–321. Plenum Press, 1990.

[66] G. Gargov, S. Passy, and T. Tinchev. Modal environment for Boolean speculations. In D. Skordev, editor, *Mathematical Logic and its Applications*, pages 253–263. Plenum Press, 1987.

[67] F. Gecseg and M. Steinby. Tree languages. In Rozenberg and Salomaa [395], pages 1–68.

[68] M. Gehrke and J. Harding. Bounded lattice expansions. *Journal of Algebra*, 238:345–371, 2001.

[69] M. Gehrke and B. Jónsson. Bounded distributive lattices with operators. *Mathematica Japonica*, 40:207–215, 1994.

[70] J. Gerbrandy and W. Groeneveld. Reasoning about information change. *Journal of Logic, Language and Information*, 6:147–169, 1997.

[71] S. Ghilardi and G. Meloni. Constructive canonicity in non-classical logics. *Annals of Pure and Applied Logic*, 86:1–32, 1997.

[72] R. Girle. *Modal Logics and Philosophy*. Acumen, 2000.

[73] R. van Glabbeek. The linear time-branching time spectrum II; the semantics of sequential processes with silent moves. In *Proceedings CONCUR '93*, volume 715 of *LNCS*, pages 66–81. Springer, 1993.

[74] P. Gochet, P. Gribomont, and A. Thayse. *Logique. Volume 3. Méthodes pour l'intelligence artificielle*. Hermes, 2000.

[75] K. Gödel. Eine Interpretation des intuitionistischen Aussagenkalkülus. In *Ergebnisse eines mathematischen Kolloquiums 4*, pages 34–40, 1933.

[76] R. Goldblatt. Semantic analysis of orthologic. *Journal of Philosophical Logic*, 3:19–35, 1974.

[77] R. Goldblatt. *Logics of Time and Computation*, volume 7 of *Lecture Notes*. CSLI Publications, 1987.

[78] R. Goldblatt. *Mathematics of Modality*, volume 43 of *Lecture Notes*. CSLI Publications, 1993.

[79] R. Goldblatt. Saturation and the Hennessy-Milner property. In Ponse *et al.* [360].

[80] R. Goldblatt. Elementary generation and canonicity for varieties of boolean algebras with operators. *Algebra Universalis*, 34:551–607, 1995.

[81] R. Goldblatt. Algebraic polymodal logic: a survey. *Logic Journal of the IGPL*, 8:393–450, 2000.

[82] R. Goldblatt. Mathematical modal logic: a view of its evolution, 2000. To appear. Draft available at http://www.vuw.ac.nz/~rob.

[83] R.I. Goldblatt. First-order definability in modal logic. *Journal of Symbolic Logic*, 40:35–40, 1975.

[84] R.I. Goldblatt. Metamathematics of modal logic I. *Reports on Mathematical Logic*, 6:41–78, 1976.

[85] R.I. Goldblatt. Metamathematics of modal logic II. *Reports on Mathematical Logic*,

7:21–52, 1976.

[186] R.I. Goldblatt. Varieties of complex algebras. *Annals of Pure and Applied Logic*, 38:173–241, 1989.

[187] R.I. Goldblatt. The McKinsey axiom is not canonical. *Journal of Symbolic Logic*, 56:554–562, 1991.

[188] R.I. Goldblatt and S.K. Thomason. Axiomatic classes in propositional modal logic. In J. Crossley, editor, *Algebra and Logic*, pages 163–173. Springer, 1974.

[189] V. Goranko. Modal definability in enriched languages. *Notre Dame Journal of Formal Logic*, 31:81–105, 1990.

[190] V. Goranko. Hierarchies of modal and temporal logics with reference pointers. *Journal of Logic, Language and Information*, 5:1–24, 1996.

[191] V. Goranko and B. Kapron. The modal logic of the countable random frame. *Archive for Mathematical Logic*, 42:221–243, 2003.

[192] V. Goranko and S. Passy. Using the universal modality: Gains and questions. *Journal of Logic and Computation*, 2:5–30, 1992.

[193] R. Goré. Tableau methods for modal and temporal logics. In D'Agostino *et al.* [104].

[194] E. Grädel. On the restraining power of guards. *Journal of Symbolic Logic*, 64:1719–1742, 1999.

[195] E. Grädel, P. Kolaitis, and M.Y. Vardi. On the decision problem for two-variable first-order logic. *Bulletin of Symbolic Logic*, 3:53–69, 1997.

[196] E. Grädel, M. Otto, and E. Rosen. Two-variable logic with counting is decidable. In *Proceedings 12th IEEE Symposium on Logic in Computer Science LICS'97*, 1997.

[197] E. Grädel and I. Walukiewicz. Guarded fixed point logic. In *Proceedings 14th IEEE Symposium on Logic in Computer Science LICS'99*, 1999.

[198] G. Grätzer. *Universal Algebra*. Springer, 1979.

[199] A.J. Grove, J.Y. Halpern, and D. Koller. Asymptotic conditional probabilities: the non-unary case. *Journal of Symbolic Logic*, 61:250–275, 1996.

[200] A.J. Grove, J.Y. Halpern, and D. Koller. Asymptotic conditional probabilities: the unary case. *SIAM Journal on Computing*, 25:1–51, 1996.

[201] Y. Gurevich and S. Shelah. The decision problem for branching time logic. *Journal of Symbolic Logic*, 50:669–681, 1985.

[202] P. R. Halmos. *Algebraic Logic*. Chelsea Publishing Company, 1962.

[203] J.Y. Halpern. The effect of bounding the number of primitive propositions and the depth of nesting on the complexity of modal logic. *Artificial Intelligence*, 75:361–372, 1995.

[204] J.Y. Halpern and B.M. Kapron. Zero-one laws for modal logic. *Annals of Pure and Applied Logic*, 69:157–193, 1994.

[205] J.Y. Halpern and Y.O. Moses. A guide to the completeness and complexity for modal logics of knowledge and belief. *Artificial Intelligence*, 54:319–379, 1992.

[206] J.Y. Halpern and M.Y. Vardi. The complexity of reasoning about knowledge and time, I: Lower bounds. *Journal of Computer and System Sciences*, 38:195–237, 1989.

[207] J.Y. Halpern and M.Y. Vardi. Model checking vs. theorem proving: a manifesto. In J.A. Allen, R. Fikes, and E. Sandewall, editors, *Principles of Knowledge Representation and Reasoning: Proc. Second International Conference (KR'91)*, pages 325–334. Morgan Kaufmann, 1991.

[208] D. Harel. Recurring dominoes: making the highly undecidable highly understandable. In *Proc. of the Conference on Foundations of Computing Theory*, volume 158 of *LNCS*, pages 177–194. Springer, 1983.

[209] D. Harel. Dynamic logic. In Gabbay and Guenthner [154], pages 497–604.

[210] D. Harel. Recurring dominoes: making the highly undecidable highly understandable. *Annals of Discrete Mathematics*, 24:51–72, 1985.

[211] D. Harel. Effective transformations on infinite trees, with applications to high undecidability. *Journal of the ACM*, 33:224–248, 1986.

[212] D. Harel, D. Kozen, and J. Tiuryn. *Dynamic Logic*. The MIT Press, 2000.

[213] R. Harrop. On the existence of finite models and decision procedures for propositional calculi. *Proceedings of the Cambridge Philosophical Society*, 54:1–13, 1958.

[214] T. Hayashi. Finite automata on infinite objects. *Math. Res. Kyushu University*, 15:13–66, 1985.

[215] E. Hemaspaandra. The price of universality. *Notre Dame Journal of Formal Logic*, 37:174–203, 1996.

[216] L. Henkin. Logical systems containing only a finite number of symbols. Séminiare de Mathématique Supérieures 21, Les Presses de l'Université de Montréal, Montréal, 1967.

[217] L. Henkin. Completeness in the theory of types. *Journal of Symbolic Logic*, 15:81–91, 1950.

[218] L. Henkin, J.D. Monk, and A. Tarski. *Cylindric Algebras. Part 1. Part 2*. North-Holland Publishing Company, Amsterdam, 1971, 1985.

[219] M. Hennessy and R. Milner. Algebraic laws for indeterminism and concurrency. *Journal of the ACM*, 32:137–162, 1985.

[220] J.G. Henriksen, J. Jensen, M. Jørgensen, N. Klarlund, R. Paige, T. Rauhe, and A. Sandhol. MONA: Monadic second-order logic in practice. In *Proceedings TACAS'95*, LNCS, pages 479–506. Springer, 1995.

[221] M. Henzinger, T. Henzinger, and P. Kopke. Computing simulations on finite and infinite graphs. In *Proceedings 20th Symposium on Foundations of Computer Science*, pages 453–462, 1995.

[222] B. Herwig. Extending partial isomorphisms on finite structures. *Combinatorica*, 15:365–371, 1995.

[223] J. Hindley and J. Seldin. *Introduction to Combinators and the Lambda Calculus*. London Mathematical Society Student Texts vol. 1. Cambridge University Press, 1986.

[224] J. Hintikka. *Knowledge and Belief*. Cornell University Press, 1962.

[225] R. Hirsch and I.M. Hodkinson. Step by step — building representations in algebraic logic. *Journal of Symbolic Logic*, 62:225–279, 1997.

[226] R. Hirsch and I.M. Hodkinson. *Relation Algebras by Games*. Number 147 in Studies in Logic. Elsevier, Amsterdam, 2002.

[227] R. Hirsch, I.M. Hodkinson, M. Marx, Sz. Mikulás, and M. Reynolds. Mosaics and step-by-step. Remarks on 'A modal logic of relations'. In Orłowska [349], pages 158–167.

[228] W. Hodges. Elementary predicate logic. In Gabbay and Guenthner [153], pages 1–131.

[229] W. Hodges. *Model Theory*. Cambridge University Press, 1993.

[230] I.M. Hodkinson. Atom structures of cylindric algebras and relation algebras. *Annals*

of Pure and Applied Logic, 89:117–148, 1997.

[231] I.M. Hodkinson. Loosely guarded fragment has finite model property. *Studia Logica*, 70:205–240, 2002.

[232] M. Hollenberg. Safety for bisimulation in general modal logic. In *Proceedings 10th Amsterdam Colloquium*, 1996.

[233] M.J. Hollenberg. Hennessy-Milner classes and process algebra. In Ponse *et al.* [360].

[234] G. Hughes and M.J. Cresswell. *A Companion to Modal Logic*. Methuen, 1984.

[235] G. Hughes and M.J. Cresswell. *A New Introduction to Modal Logic*. Routledge, 1996.

[236] I. Humberstone. Inaccessible worlds. *Notre Dame Journal of Formal Logic*, 24:346–352, 1983.

[237] U. Hustadt. *Resolution-Based Decision Procedures for Subclasses of First-Order Logic*. PhD thesis, Universität des Saarlandes, Saarbrücken, Germany, 1999.

[238] U. Hustadt and R. A. Schmidt. Issues of decidability for description logics in the framework of resolution. In R. Caferra and G. Salzer, editors, *First-order Theorem Proving—FTP'98*, pages 152–161. Technical Report E1852-GS-981, Technische Universität Wien, 1998.

[239] M.R.A. Huth and M.D. Ryan. *Logic in Computer Science*. Cambridge University Press, 2000.

[240] N. Immerman and D. Kozen. Definability with bounded number of bound variables. In *Proceedings 4th IEEE Symposium on Logic in Computer Science LICS'87*. Computer Society Press, 1987.

[241] B. Jacobs. The temporal logic of coalgebras via Galois algebras. *Mathematical Structure in Computer Science*, 12:875–903, 2002.

[242] B. Jacobs and J. Rutten. A tutorial on (co)algebras and (co)induction. *Bulletin of the European Association for Theoretical Computer Science*, 62:222–259, 1997.

[243] D. Janin and I. Walukiewicz. On the expressive completeness of the propositional μ-calculus w.r.t. monadic second-order logic. In *Proceedings CONCUR '96*, 1996.

[244] R. Jansana. *Una Introducción a la Lógica Modal*. Editorial Tecnos, Madrid, 1990.

[245] G. Japaridze and D. de Jongh. The logic of provability. In Buss [83], pages 475–546.

[246] He Jifeng. Process simulation and refinement. *Formal Aspects of Computing,* 1:229–241, 1989.

[247] P. Jipsen. Discriminator varieties of boolean algebras with residuated operators. In C. Rauszer, editor, *Algebraic Methods in Logic and Computer Science*, volume 28 of *Banach Center Publications*, pages 239–252. Polish Academy of Sciences, 1993.

[248] P.J. Johnstone. *Stone Spaces*, volume 3 of *Cambridge Studies in Advanced Mathematics*. Cambridge University Press, Cambridge, 1982.

[249] D. de Jongh and A. Troelstra. On the connection between partially ordered sets and some pseudo-boolean algebras. *Indigationes Mathematicae*, 28:317–329, 1966.

[250] D. de Jongh and F. Veltman. Intensional logic, 1986. Course notes.

[251] B. Jónsson. Varieties of relation algebras. *Algebra Universalis*, 15:273–298, 1982.

[252] B. Jónsson. The theory of binary relations. In Andréka *et al.* [8], pages 241–292.

[253] B. Jónsson. A survey of boolean algebras with operators. In *Algebras and Order,* pages 239–286. Kluwer Academic Publishers, 1993.

[254] B. Jónsson. On the canonicity of Sahlqvist identities. *Studia Logica*, 4:473–49, 1994.

[255] B. Jónsson and A. Tarski. Boolean algebras with operators, Part I. *American Journal of Mathematics*, 73:891–939, 1952.

[256] B. Jónsson and A. Tarski. Boolean algebras with operators, Part II. *American Journal of Mathematics*, 74:127–162, 1952.

[257] G. Kalinowski. *La Logique Déductive*. Presses Universitaires de France, 1996.

[258] H. Kamp. *Tense Logic and the Theory of Linear Order*. PhD thesis, University of California, Los Angeles, 1968.

[259] H. Kamp. Formal properties of 'Now'. *Theoria*, 37:227–273, 1971.

[260] M. Kaneko and T. Nagashima. Game logic and its applications. *Studia Logica*, 57:325–354, 1998.

[261] S. Kanger. The morning star paradox. *Theoria*, pages 1–11, 1957.

[262] S. Kanger. *Provability in Logic*. Almqvist & Wiksell, 1957.

[263] S. Kanger, editor. *Proceedings of the Third Scandinavian Logic Symposium. Uppsala 1973*. North-Holland Publishing Company, 1975.

[264] D. Kaplan. Dthat. In P. Cole, editor, *Syntax and Semantics Volume 9*, pages 221–253. Academic Press, 1978.

[265] D. Kaplan. On the logic of demonstratives. *Journal of Philosophical Logic*, 8:81–98, 1978.

[266] R. Kasper and W. Rounds. The logic of unification in grammar. *Linguistics and Philosophy*, 13:33–58, 1990.

[267] H.J. Keisler. *Model Theory for Infinitary Logic*. North-Holland Publishing Company, 1971.

[268] H. Kirchner and C. Ringeissen, editors. *Frontiers of Combining Systems 3*. Springer, 2000.

[269] B. Konikowska. A formal language for reasoning about indiscernibility. *Bulletin of the Polish Academy of Sciences*, 35:239–249, 1987.

[270] B. Konikowska. A logic for reasoning about relative similarity. *Studia Logica*, 58:185–226, 1997.

[271] R. Koymans. *Specifying Message Passing and Time-Critical Systems with Temporal Logic*, volume 651 of *LNCS*. Springer, 1992.

[272] D. Kozen. A completeness theorem for Kleene algebras and the algebra of regular events. In *Proceedings 6th IEEE Symposium on Logic in Computer Science LICS'91*, pages 214–225, 1991.

[273] D. Kozen and R. Parikh. An elementary proof of the completeness of PDL. *Theoretical Computer Science*, 14:113–118, 1981.

[274] D. Kozen and J. Tiuryn. Logics of programs. In van Leeuwen [293], pages 789–840.

[275] M. Kracht. Even more about the lattice of tense logics. *Archive of Mathematical Logic*, 31:243–357, 1992.

[276] M. Kracht. How completeness and correspondence theory got married. In de Rijke [379], pages 175–214.

[277] M. Kracht. Splittings and the finite model property. *Journal of Symbolic Logic*, 58:139–157, 1993.

[278] M. Kracht. Lattices of modal logics and their groups of automorphisms. *Journal of Pure and Applied Logic*, 100:99–139, 1999.

[279] M. Kracht. *Tools and Techniques in Modal Logic*. Number 142 in Studies in Logic. Elsevier, Amsterdam, 1999.

[280] M. Kracht. Logic and syntax — a personal perspective. In Zakharyaschev *et al.*

[469], pages 355–384.

[281] M. Kracht, M. de Rijke, H. Wansing, and M. Zakharyaschev, editors. *Advances* i
Modal Logic, Volume 1, volume 87 of *Lecture Notes*. CSLI Publications, 1998.

[282] M. Kracht and F. Wolter. Simulation and transfer results in modal logic: A surve
Studia Logica, 59:149–177, 1997.

[283] S. Kripke. A completeness theorem in modal logic. *Journal of Symbolic Logi*
24:1–14, 1959.

[284] S. Kripke. Semantic analysis of modal logic I, normal propositional calcul
Zeitschrift für mathemathische Logik und Grundlagen der Mathematik, 9:67–9
1963.

[285] S. Kripke. Semantical considerations on modal logic. *Acta Philosophica Fennic*
16:83–94, 1963.

[286] S. Kuhn. Quantifiers as diamonds. *Studia Logica*, 39:173–195, 1980.

[287] N. Kurtonina. *Frames and Labels*. PhD thesis, OTS, Utrecht University, 1996.

[288] N. Kurtonina and M. de Rijke. Bisimulations for temporal logic. *Journal of Logi*
Language and Information, 6:403–425, 1997.

[289] N. Kurtonina and M. de Rijke. Expressiveness of concept expressions in first-ord
description logics. *Artificial Intelligence*, 107:303–333, 1999.

[290] A. Kurz. A co-variety-theorem for modal logic. In Zakharyaschev *et al.* [469].

[291] A.H. Lachlan. A note on Thomason's refined structures for tense logics. *Theori*
40:117–120, 1970.

[292] R. Ladner. The computational complexity of provability in systems of modal logi
SIAM Journal on Computing, 6:467–480, 1977.

[293] J. van Leeuwen, editor. *Handbook of Theoretical Computer Science*, volume I
Formal Models and Semantics. Elsevier, 1990.

[294] A. Leitsch, C. Fermüller, and T. Tammet. Resolution decision procedures. In Robi
son and Voronkov [388].

[295] E.J. Lemmon. Algebraic semantics for modal logics, Parts I & II. *Journal of Syn*
bolic Logic, pages 46–65 & 191–218, 1966.

[296] E.J. Lemmon and D.S. Scott. *The 'Lemmon Notes': An Introduction to Modal Logi*
Blackwell, 1977.

[297] O. Lemon and I. Pratt. On the incompleteness of modal logics of space: advancir
complete modal logics of space. In Kracht *et al.* [281].

[298] W. Lenzen. *Glauben, Wissen und Wahrscheinlichkeit. Systeme der Epistemisch*
Logik. Springer, 1980.

[299] C.I. Lewis. *A Survey of Symbolic Logic*. University of California Press, 1918.

[300] C.I. Lewis and C.H. Langford. *Symbolic Logic*. Dover, 1932.

[301] H.R. Lewis and C.H. Papadimitriou. *Elements of the Theory of Computatio*
Prentice-Hall, 1981.

[302] P. Lindström. On extensions of elementary logic. *Theoria*, 35:1–11, 1969.

[303] C. Lutz and U. Sattler. The complexity of reasoning with boolean modal logics.
Workshop Proceedings AiML-2000, pages 175–184. Institut für Informatik, Unive
sität Leipzig, 2000.

[304] R.C. Lyndon. The representation of relation algebras. *Annals of Mathematic*
51:707–729, 1950.

[305] H. MacColl. *Symbolic Logic and its Applications*. Longmans, Green, and Cc
London, 1906.

[306] R. Maddux. Introductory course on relation algebras, finite-dimensional cylindric algebras, and their interconnections. In Andréka *et al.* [8], pages 361–392.

[307] D.C. Makinson. On some completeness theorems in modal logic. *Zeitschrift für mathemathische Logik und Grundlagen der Mathematik*, 12:379–84, 1966.

[308] D.C. Makinson. A generalization of the concept of relational model. *Theoria*, pages 331–335, 1970.

[309] D.C. Makinson. Some embedding theorems for modal logic. *Notre Dame Journal of Formal Logic*, pages 252–254, 1971.

[310] L. Maksimova. Interpolation theorems in modal logic and amalgable varieties of topological boolean algebras. *Algebra and Logic*, 18:348–370, 1979.

[311] L.L. Maksimova. Pretabular extensions of Lewis S4. *Algebra and Logic*, 14:16–33, 1975.

[312] Z. Manna and A. Pnueli. *The Temporal Logic of Reactive and Concurrent Systems. Vol. 1 Specification.* Springer, 1992.

[313] Z. Manna and A. Pnueli. *Temporal Verification of Reactive Systems: Safety.* Springer, 1995.

[314] M. Manzano. *Extensions of First Order Logic*, volume 19 of *Tracts in Theoretical Computer Science*. Cambridge University Press, 1996.

[315] R. Martin. *Pour une Logique du Sens.* Presses Universitaires de France Paris, 1983.

[316] M. Marx. Complexity of modal logics of relations. Technical Report ML–97–02, Institute for Language, Logic and Computation, May 1997.

[317] M. Marx. Tolerance logic. *Journal of Logic, Language and Information*, 10:353–373, 2001.

[318] M. Marx, L. Pólos, and M. Masuch, editors. *Arrow Logic and Multi-Modal Logic.* Studies in Logic, Language and Information. CSLI Publications, 1996.

[319] M. Marx, S. Schlobach, and Sz. Mikulás. Labelled deduction for the guarded fragment. In D. Basin et al., editor, *Labelled Deduction*, Applied Logic Series, pages 193–214. Kluwer Academic Publishers, 2000.

[320] M. Marx and Y. Venema. *Multidimensional Modal Logic*, volume 4 of *Applied Logic Series*. Kluwer Academic Publishers, 1997.

[321] R. McKenzie, G. McNulty, and W. Taylor. *Algebras, Lattices, Varieties*, volume I. Wadsworth & Brooks/Cole, 1987.

[322] J.C.C. McKinsey. A solution to the decision problems for the Lewis systems S2 and S4 with an application to topology. *Journal of Symbolic Logic*, 6:117–134, 1941.

[323] J.C.C. McKinsey. On the syntactical construction of systems of modal logic. *Journal of Symbolic Logic*, 10:83–96, 1945.

[324] J.C.C. McKinsey and A. Tarski. The algebra of topology. *Annals of Mathematics*, pages 141–191, 1944.

[325] J.C.C. McKinsey and A. Tarski. Some theorems about the sentential calculi of Lewis and Heyting. *Journal of Symbolic Logic*, 13:1–15, 1948.

[326] K. McMillan. *Symbolic Model Checking.* Kluwer Academic Publishers, 1993.

[327] C. Meredith and A. Prior. Interpretations of different modal logics in the 'property calculus', 1956. Mimeographed manuscript. Philosophy Department, Canterbury University College.

[328] J.J.-Ch. Meyer and W. van der Hoek. *Epistemic Logic for AI and Computer Science.* Cambridge University Press, 1995.

[329] Sz. Mikulás. *Taming Logics.* PhD thesis, Institute for Language, Logic and Com-

putation, University of Amsterdam, 1995. ILLC Dissertation Series 95-12.

[330] G. Mints. *A Short Introduction to Modal Logic*, volume 30 of *Lecture Notes*. CSLI Publications, 1992.

[331] J.L. Moens. *Forcing et Sémantique de Kripke-Joyal*. Cabay, 1982.

[332] G. Moisil. *Essais sur les Logiques non Chrysipiennes*. Editions de l'Académie de la République Socialiste de Roumanie, 1972.

[333] F. Moller and A. Rabinovich. On the expressive power of CTL*. In *Proceedings 14th IEEE Symposium on Logic in Computer Science LICS'99*, 1999.

[334] J.D. Monk. On representable relation algebras. *Michigan Mathematical Journal*, 11:207–210, 1964.

[335] R. Montague. Logical necessity, physical necessity, ethics, and quantifiers. *Inquiry*, 4:259–269, 1960.

[336] R. Montague. Universal grammar. *Theoria*, 36:373–398, 1970.

[337] M. Mortimer. On languages with two variables. *Zeitschrift für mathemathische Logik und Grundlagen der Mathematik*, 21:135–140, 1975.

[338] D.E. Muller, A. Saoudi, and P.E. Schupp. Weak alternating automata give a simple explanation of why most temporal and dynamic logics are decidable in exponential time. In *Proceedings 3rd IEEE Symposium on Logic in Computer Science LICS'88*, pages 422–427, 1988.

[339] N. Murray, editor. *Conference on Tableaux Calculi and Related Methods (TABLEAUX), Saratoga Springs, USA*, volume 1617 of *LNAI*. Springer, 1999.

[340] I. Németi. Free algebras and decidability in algebraic logic, 1986. Thesis for D.Sc. (a post-habilitation degree) with Math. Inst. Hungar. Ac. Sci. Budapest. In Hungarian, the English version is [342].

[341] I. Németi. Algebraizations of quantifier logics: an overview. *Studia Logica*, 50:485–569, 1991.

[342] I. Németi. Decidability of weakened versions of first-order logic. In Csirmaz et al. [102], pages 177–242.

[343] P. Odifreddi. *Classical Recursion Theory*. North-Holland Publishing Company, 1989.

[344] H.J. Ohlbach, D.M. Gabbay, and D. Plaisted. Killer transformations. In Wansing [454].

[345] H.J. Ohlbach, A. Nonnengart, M. de Rijke, and D.M. Gabbay. Encoding non-classical logics in classical logic. In Robinson and Voronkov [388].

[346] H.J. Ohlbach and R.A. Schmidt. Functional translation and second-order frame properties of modal logics. *Journal of Logic and Computation*, 7:581–603, 1997.

[347] P. Øhrstrom and P Hasle. *Temporal Logic*. Kluwer Academic Publishers, 1995.

[348] H. Ono and A. Nakamura. On the size of refutation Kripke models for some linear modal and tense logics. *Studia Logica*, 39:325–333, 1980.

[349] E. Orłowska, editor. *Logic at Work: Essays Dedicated to the Memory of Elena Rasiowa*. Studies in Fuzziness and Soft Computing. Springer, 1999.

[350] M.J. Osborne and A. Rubinstein. *A Course in Game Theory*. The MIT Press, 1994.

[351] M. Otto. *Bounded Variable Logics and Counting — A Study in Finite Models*, volume 9 of *Lecture Notes in Logic*. Springer, 1997.

[352] C.H. Papadimitriou. *Computational Complexity*. Addison-Wesley, 1994.

[353] R. Parikh. The completeness of propositional dynamic logic. In *Mathematical Foundations of Computer Science 1978*, volume 51 of *LNCS*, pages 403–415. Springer.

1978.

54] D. Park. Concurrency and automata on infinite sequences. In *Proceedings 5th GI Conference*, pages 167–183. Springer, 1981.

55] W.T. Parry. The postulates for strict implication. *Mind*, 43:78–80, 1934.

56] S. Passy and T. Tinchev. PDL with data constants. *Information Processing Letters*, 20:35–41, 1985.

57] S. Passy and T. Tinchev. Quantifiers in combinatory PDL: completeness, definability, incompleteness. In *Fundamentals of Computation Theory FCT 85*, volume 199 of *LNCS*, pages 512–519. Springer, 1985.

58] S. Passy and T. Tinchev. An essay in combinatory dynamic logic. *Information and Computation*, 93:263–332, 1991.

59] A. Pnueli. The temporal logic of programs. In *Proc. 18th Symp. Foundations of Computer Science*, pages 46–57, 1977.

60] A. Ponse, M. de Rijke, and Y. Venema, editors. *Modal Logic and Process Algebra: A Bisimulation Perspective*, volume 53 of *Lecture Notes*. CSLI Publications, 1995.

61] S. Popkorn. *First Steps in Modal Logic*. Cambridge University Press, 1992.

62] V.R. Pratt. Semantical considerations on Floyd-Hoare logic. In *Proc. 17th IEEE Symposium on Computer Science*, pages 109–121, 1976.

63] V.R. Pratt. Models of program logics. In *Proc. 20th IEEE Symp. Foundations of Computer Science*, pages 115–222, 1979.

64] A.N. Prior. *Time and Modality*. Oxford University Press, 1957.

65] A.N. Prior. *Past, Present and Future*. Oxford University Press, 1967.

66] A.N. Prior. *Papers on Time and Tense*. Oxford University Press, New edition, 2003. Edited by Hasle, Øhrstrom, Braüner, and Copeland.

67] A.N. Prior and K. Fine. *Worlds, Times and Selves*. University of Massachusetts Press, 1977.

68] M.O. Rabin. Decidability of second-order theories and automata on infinite trees. *Transactions of the American Mathematical Society*, 141:1–35, 1969.

69] S. Rahman. Hugh MacColl: Eine bibliographische Erschließung seiner Hauptwerke und Notizen zu ihrer Rezeptionsgeschichte. *History and Philosophy of Logic*, 18:165–183, 1997.

70] H. Rasiowa and R. Sikorski. *The Mathematics of Metamathematics*. Polish Scientific Publishers, 1963.

71] H. Rasiowa and R. Sikorski. *An Algebraic Approach to Non-Classical Logics*. North Holland, 1974.

72] W. Rautenberg. Der Verband der normalen verzweigten Modallogiken. *Mathematische Zeitschrift*, 156:123–140, 1977.

73] W. Rautenberg. *Klassische und nichtklassische Aussagenlogik*. Vieweg & Sohn, 1979.

74] W. Rautenberg. Splitting lattices of logics. *Archiv für Mathematische Logik*, 20:155–159, 1980.

75] W. Rautenberg, editor. *Non-Classical Logics. Ω-bibliography of Mathematical Logic, Volume II*. Springer, 1987.

76] M. Reape. A feature value logic. In C. Rupp, M. Rosner, and R. Johnson, editors, *Constraints, Language and Computation*, Synthese Language Library, pages 77–110. Academic Press, 1994.

77] M. Reynolds. A decidable logic of parallelism. *Notre Dame Journal of Formal*

Logic, 38:419–436, 1997.

[378] M. de Rijke. The modal logic of inequality. Journal of Symbolic Logic, 57:566–58. 1992.

[379] M. de Rijke, editor. Diamonds and Defaults. Synthese Library vol. 229. Kluwe Academic Publishers, 1993.

[380] M de. Rijke. Extending Modal Logic. PhD thesis, ILLC, University of Amsterdan 1993.

[381] M. de Rijke. A Lindström theorem for modal logic. In Ponse et al. [360], page 217–230.

[382] M. de Rijke. The logic of Peirce algebras. Journal of Logic, Language and Infoi mation, 4:227–250, 1995.

[383] M. de Rijke, editor. Advances in Intensional Logic. Number 7 in Applied Log Series. Kluwer Academic Publishers, 1997.

[384] M. de Rijke. A system of dynamic modal logic. Journal of Philosophical Logi 27:109–142, 1998.

[385] M. de Rijke. A modal characterization of Peirce algebras. In Orłowska [349], page 109–123.

[386] M. de Rijke and H. Sturm. Global definability in basic modal logic. In H. Wansin; editor, Essays on Non-Classical Logic. King's College University Press, 2000.

[387] M. de Rijke and Y. Venema. Sahlqvist's theorem for Boolean algebras with opera tors. Studia Logica, 95:61–78, 1995.

[388] A. Robinson and A. Voronkov, editors. Handbook of Automated Reasoning. Elsevie Science Publishers, to appear.

[389] R.M. Robinson. Undecidability and nonperiodicity for tilings of the plane. Inven tiones Mathematicae, 12:177–209, 1971.

[390] P.H. Rodenburg. Intuitionistic Correspondence Theory. PhD thesis, University c Amsterdam, 1986.

[391] H. Rogers. Theory of Recursive Functions and Effective Computability. McGra Hill, 1967.

[392] E. Rosen. Modal logic over finite structures. Journal of Logic, Language ar Information, 6:427–439, 1997.

[393] M. Rößiger. Coalgebras and modal logic. Electronic Notes in Computer Scienc 33:299–320, 2000.

[394] W.C. Rounds. Feature logics. In van Benthem and ter Meulen [51].

[395] G. Rozenberg and A. Salomaa, editors. Handbook of Formal Languages, volume : Beyond Words. Springer, 1997.

[396] H. Sahlqvist. Completeness and correspondence in the first and second order se mantics for modal logic. In Kanger [263], pages 110–143.

[397] I. Sain. Is 'some-other-time' sometimes better than 'sometime' for proving parti correctness of programs? Studia Logica, 47:279–301, 1988.

[398] G. Sambin and V. Vaccaro. Topology and duality in modal logic. Annals of Pui and Applied Logic, 37:249–296, 1988.

[399] G. Sambin and V. Vaccaro. A topological proof of Sahlqvist's theorem. Journal c Symbolic Logic, 54:992–999, 1989.

[400] K. Schild. A correspondence theory for terminological logics. In Proc. 12th IJCA pages 466–471, 1990.

[401] D. Scott. A decision method for validity of sentences in two variables. Journal c

Symbolic Logic, 27:377, 1962.
[402] K. Segerberg. Decidability of S4.1. *Theoria*, 34:7–20, 1968.
[403] K. Segerberg. Modal logics with linear alternative relations. *Theoria*, 36:301–322, 1970.
[404] K. Segerberg. *An Essay in Classical Modal Logic*. Filosofiska Studier 13. University of Uppsala, 1971.
[405] K. Segerberg. Two-dimensional modal logics. *Journal of Philosophical Logic*, 2:77–96, 1973.
[406] K. Segerberg. 'Somewhere else' and 'Some other time'. In *Wright and Wrong*, pages 61–64, 1976.
[407] K. Segerberg. A completeness theorem in the modal logic of programs. *Notices of the American Mathematical Society*, 24:A–552, 1977.
[408] K. Segerberg. A note on the logic of elsewhere. *Theoria*, 46:183–187, 1980.
[409] K. Segerberg. A completeness theorem in the modal logic of programs. In T. Traczyk, editor, *Universal Algebra and Applications*, volume 9 of *Banach Centre Publications*, pages 31–46. PWN–Polish Scientific Publishers, 1982.
[410] K. Segerberg. Proposal for a theory of belief revision along the lines of Lindström and Rabinowicz. *Fundamenta Informaticae*, 32:183–191, 1997.
[411] J. Seligman. A cut-free sequent calculus for elementary situated reasoning. Technical Report HCRC-RP 22, HCRC, Edinburgh, 1991.
[412] J. Seligman. The logic of correct description. In de Rijke [383], pages 107–135.
[413] V. Shehtman. Two-dimensional modal logics. *Mathematical Notices of USSR Academy of Sciences*, 23:417–424, 1978.
[414] H. Simmons. The monotonous elimination of predicate variables. *Journal of Logic and Computation*, 4, 1994.
[415] A.P. Sistla and E.M. Clarke. The complexity of linear temporal logic. *Journal of the ACM*, 32:733–749, 1985.
[416] C. Smoryński. *Self-Reference and Modal Logic*. Springer, New York, 1985.
[417] R. Smullyan and M Fitting. *Set Theory and the Continuum Problem*. Clarendon Press, 1996.
[418] R.M. Smullyan and M. Fitting. *Set Theory and the Continuum Problem*. Oxford University Press, 1997.
[419] E. Spaan. *Complexity of Modal Logics*. PhD thesis, ILLC, University of Amsterdam, 1993.
[420] E. Spaan. The complexity of propositional tense logics. In de Rijke [379], pages 239–252.
[421] B. Stalnaker. Assertion. In P. Cole, editor, *Syntax and Semantics Volume 9*, pages 316–322. Academic Press, 1978.
[422] J. Stavi. Functional completeness over the rationals. Unpublished manuscript. Bar-Ilan University, Ramat-Gan, Israel, 1979.
[423] V. Stebletsova. *Algebras, Relations and Geometries*. PhD thesis, Zeno (The Leiden-Utrecht Research Institute of Philosophy), Utrecht, 2000.
[424] C. Stirling. Modal and temporal logics. In Abramsky *et al.* [1], pages 477–563.
[425] M.H. Stone. The theory of representations for boolean algebras. *Transactions of the American Mathematical Society*, 40:37–111, 1936.
[426] H. Sturm. *Modale Fragmente von $\mathcal{L}_{\omega\omega}$ und $\mathcal{L}_{\omega_1\omega}$*. PhD thesis, CIS, University of Munich, 1997.

[427] A. Tarski. On the calculus of relations. *Journal of Symbolic Logic*, 6:73–89, 1941.

[428] A. Tarski and S. Givant. *A Formalization of Set Theory without Variables*, volume 41. AMS Colloquium Publications, Providence, Rhode Island, 1987.

[429] J.W. Thatcher and J.B. Wright. Generalized finite automata theory with an application to a decision problem of second-order logic. *Mathematical Systems Theory*, 2:57–81, 1968.

[430] A. Thayse, editor. *From Modal Logic to Deductive Databases*. Wiley, 1989.

[431] W. Thomas. Automata on infinite objects. In van Leeuwen [293], pages 135–191.

[432] W. Thomas. Languages, automata and logic. In Rozenberg and Salomaa [395], pages 389–456.

[433] S.K. Thomason. Semantic analysis of tense logics. *Journal of Symbolic Logic*, 37:150–158, 1972.

[434] S.K. Thomason. An incompleteness theorem in modal logic. *Theoria*, 40:150–158, 1974.

[435] S.K. Thomason. Categories of frames for modal logics. *Journal of Symbolic Logic*, 40:439–442, 1975.

[436] S.K. Thomason. Reduction of second-order logic to modal logic. *Zeitschrift für mathemathische Logik und Grundlagen der Mathematik*, 21:107–114, 1975.

[437] S.K. Thomason. Reduction of tense logic to modal logic II. *Theoria*, 41:154–169, 1975.

[438] D. Toman and D. Niwiński. First-order queries over temporal databases inexpressible in temporal logic. Manuscript, 1997.

[439] M. Tzakova. Tableaux calculi for hybrid logics. In Murray [339], pages 278–292.

[440] A. Urquhart. Decidability and the finite model property. *Journal of Philosophical Logic*, 10:367–370, 1981.

[441] M.Y. Vardi. Why is modal logic so robustly decidable? In *DIMACS Series in Discrete Mathematics and Theoretical Computer Science 31*, pages 149–184. AMS, 1997.

[442] M.Y. Vardi and P. Wolper. Automata-theoretic techniques for modal logics of programs. *Journal of Computer and System Sciences*, 32:183–221, 1986.

[443] Y. Venema. Completeness via completeness: Since and Until. In de Rijke [379], pages 279–286.

[444] Y. Venema. Derivation rules as anti-axioms in modal logic. *Journal of Symbolic Logic*, 58:1003–1034, 1993.

[445] Y. Venema. Cylindric modal logic. *Journal of Symbolic Logic*, 60:591–623, 1995.

[446] Y. Venema. Atom structures and Sahlqvist equations. *Algebra Universalis*, 38:185–199, 1997.

[447] Y. Venema. Modal definability, purely modal. In J. Gerbrandy, M. Marx, M. de Rijke, and Y. Venema, editors, *JFAK. Essays Dedicated to Johan van Benthem on the Occasion of his 50th Birthday*. Vossiuspers AUP, Amsterdam, 1999.

[448] Y. Venema. Points, lines and diamonds: a two-sorted modal logic for projectiv geometry. *Journal of Logic and Computation*, 9:601–621, 1999.

[449] Y. Venema. Canonical pseudo-correspondence. In Zakharyaschev *et al.* [469], page 439–448.

[450] A. Visser. Modal logic and bisimulation. Tutorial for the workshop 'Three days c bisimulation', Amsterdam, 1994.

[451] A. Visser. An overview of interpretability logic. In Kracht *et al.* [281], pages 30⁷

359.

[452] H. Vlach. *'Now' and 'Then'*. PhD thesis, University of California, Los Angeles, 1973.

[453] H. Wang. Proving theorems by pattern recognition II. *Bell Systs. Tech. J*, 40:1–41, 1961.

[454] H. Wansing, editor. *Proof Theory of Modal Logic*. Kluwer Academic Publishers, 1996.

[455] H. Wansing. *Displaying Modal Logic*. Kluwer Academic Publishers, 1998.

[456] P. Wolper. Temporal logic can be more expressive. *Information and Control*, 56:72–93, 1983.

[457] F. Wolter. Tense logic without tense operators. *Mathematical Logic Quarterly*, 42:145–171, 1996.

[458] F. Wolter. Completeness and decidability of tense logics closely related to logics containing $K4$. *Journal of Symbolic Logic*, 62:131–158, 1997.

[459] F. Wolter. The structure of lattices of subframe logics. *Annals of Pure and Applied Logic*, 86:47–100, 1997.

[460] F. Wolter. The product of converse PDL and polymodal K. *Journal of Logic and Computation*, 10:223–251, 2000.

[461] F. Wolter, H. Wansing, M. de Rijke, and M. Zakharyaschev, editors. *Advances in Modal Logic, Volume 3*. World Scientific, 2002.

[462] F. Wolter and M. Zakharyaschev. Satisfiability problem in description logics with modal operators. In *Principles of Knowledge Representation and Reasoning: Proc. Sixth International Conference (KR'98)*, pages 512–523. Morgan Kaufmann, 1998.

[463] M. Wooldridge and N. Jennings. Intelligent agents: theory and practice. *Knowledge Engineering Review*, 10:115–152, 1995.

[464] G.H. von Wright. *An Essay in Modal Logic*. North-Holland Publishing Company, 1951.

[465] G.H. von Wright. A modal logic of place. In E. Sosa, editor, *The Philosophy of Nicholas Rescher*, pages 65–73. Publications of the Group in Logic and Methodology of Science of Real Finland, vol. 3, 1979.

[466] M. Xu. On some U,S-tense logics. *Journal of Philosophical Logic*, 17:181–202, 1988.

[467] M. Zakharyaschev. Canonical formulas for modal and superintuitionistic logics: a short outline. In de Rijke [383], pages 195–248.

[468] M. Zakharyaschev and A. Alekseev. All finitely axiomatizable normal extensions of K4.3 are decidable. *Mathematical Logic Quaterly*, 41:15–23, 1995.

[469] M. Zakharyaschev, K. Segerberg, M. de Rijke, and H. Wansing, editors. *Advances in Modal Logic, Volume 2*. CSLI Publications, 2000.

[470] M.V. Zakharyaschev. On intermediate logics. *Soviet Mathematics Doklady*, 27:274–277, 1983.

[471] M.V. Zakharyaschev. Normal modal logics containing $S4$. *Soviet Mathematics Doklady*, 28:252–255, 1984.

[472] M.V. Zakharyaschev. Syntax and semantics of modal logics containing $S4$. *Algebra and Logic*, 27:408–428, 1988.

[473] M.V. Zakharyaschev. Canonical formulas for $K4$. Part I: basic results. *Journal of Symbolic Logic*, 57:1377–1402, 1992.

[474] J. Zeman. *Modal Logic: The Lewis Modal Systems*. Oxford University Press, 1973.

List of Notation

Modalities

$\Diamond$, $\Box$, Definition 1.9
K, Example 1.10
$\triangle$, Definition 1.11
$\Diamond_a$, $\langle a \rangle$, Definition 1.11
∇, Definition 1.13
$\Box_a$, $[a]$, Definition 1.13
F, P, G, H, Example 1.14
$\langle \pi \rangle$, $[\pi]$, Example 1.15
$\circ$, $\otimes$, $1'$, Example 1.16
$\Diamond^0$, $\Diamond^1$, $\Diamond^2$, ..., Example 1.22
E, A, Example 2.4, page 415
$[*]$, page 371
D, $\overline{\text{D}}$, page 419
$\| \cdot \|$, page 424
S, U, page 427
S', U', Definition 429
@$_i$, page 436
$\downarrow$, page 444
Id_{ij}, Equation 7.11
$\bigcirc_\sigma$, page 461
$\iota\delta_{ij}$, Definition 7.41
$\bigcirc_{ij}$, E$_i^n$, D$_n$, page 467

Other syntax

$p, q, r, \ldots$, Definition 1.9
τ, Definition 1.11
ρ, Definition 1.11 and B.1
$\pi_1 \cup \pi_2, \pi_1 ; \pi_2, \pi^*, \pi_1 \cap \pi_2$, Example 1.15
$\phi?$, Example 1.15
$(\forall y \rhd x), (\exists y \rhd x)$, page 170
$\forall^r, \exists^r$, page 170
$\sim\phi$, Definition 4.79
$Bool$, Definition 5.1
$Ter_\tau(\Phi)$, Definition 5.18
i, j, k, page 435

Formulas and axioms

K, Dual, Definition 1.39
4, Example 3.6, page 207
T, 5, Example 3.6
M, Example 3.11
$\phi_{\mathfrak{F},w}$, page 143
REL, AT, POS, page 157

BOX-AT, page 162
B, D, .3, L, page 192
K$_\nabla^i$, Dual$_\nabla$, Definition 4.13
D$_r$, D$_l$, *den*, page 207
$.3_r$, $.3_l$, page 208
$.r_r$, D$_r$, L$_l$, page 212
$name(\phi)$, page 230
$\phi^{\mathcal{B}}(m)$, Table 6.5
$f_L(\beta)$, Table 6.6
(A1a)–(A7a), (A1b)–(A7b), Definition 7.13
D, L, W, N, Definition 7.13
$G(\overline{x}, \overline{y})$, page 446
$(CM1_i)$–$(CM8_i)$, Definition 7.48

Maps and operations on formulas

ϕ^σ, Definition 1.18
deg, Definition 2.28
ST_x, Definition 2.45
$c_\phi(x)$, Theorem 3.54
$\phi^\approx$, Definition 5.26
$Free(\phi)$, Definition 7.31
$(\cdot)^t$, Definition 7.42
$(\cdot)^\bullet$, page 466

Languages

τ, Definition 1.11
$\tau_\rightarrow$, Example 1.16
$ML(\tau, \Phi)$, Definition 1.12
PDL, Example 1.15
$\mathcal{L}_\tau^1(\Phi)$, Definition 2.44
$\mathcal{L}_\tau^2(\Phi)$, $\mathcal{L}_\tau^2$, page 126
SnS, Definition 6.17
KR, page 367
KRA, page 368
KRA$[*]$, page 371
$ML(\Diamond)$, page 415
$ML(\text{E})$, $ML(\Diamond, \text{E})$, page 415
$ML(\text{D})$, $ML(\Diamond, \text{D})$, page 419
MLR_n, Definition 7.41
$\mathcal{L}^1$, page 486
$\mathcal{L}_X$, Definition A.9
$\mathcal{L}_{\omega_1\omega}$, page 496

Sets of formulas

$Form(\tau, \Phi)$, Definition 1.12

544

Λ_F, Definition 1.28
$\mathrm{FL}(\Sigma)$, $\neg\mathrm{FL}(\Sigma)$, Definition 4.79
$\mathrm{Cl}\Gamma$, $\mathrm{Cl}(\phi)$, Definition 6.23
$\mathrm{Dem}(H, \Diamond\psi)$, Definition 6.43
PF, GF, PF_n, GF_n, Definition 7.31
L_n, $_n^r$, Definition 7.40
CML_n, Definition 7.41

Logics
Λ_F, Definition 1.28
K, Definition 1.42
Λ_S, $\Lambda_\mathfrak{S}$, Example 4.2
PC, page 190
K4, T, B, KD, S4, S5, K4.3, S4.3, KL, Table 4.1
$\mathbf{K}_\tau$, $\mathbf{K}_\tau\Gamma$, Definition 4.14
$\mathbf{K}_t$, Definition 4.33
$\mathbf{K}_t\mathbf{Q}$, Definition 4.40
$\Lambda_{\mathsf{F}t}$, Definition 4.32
KvB, Exercise 4.4.2
$\mathbf{K}_t\mathbf{Tho}$, $\mathbf{K}_t\mathbf{ThoM}$, page 212
$\mathbf{K}_t\mathbf{Q}^+$, Definition 4.66
PDL, Definition 4.78
$\mathbf{K}_n\mathbf{Alt}_1$, page 341
$\mathbf{K}_t\mathbf{N}$, page 360
$\mathbf{K}_g$, page 417
$\mathbf{K}_{td}\Sigma$, page 420
Λ_F^d, Theorem 7.8
B, BW, BN, page 431
$\mathbf{K}_h + \text{RULES}$, page 437 and 441
$\mathbf{K}_g$, Definition 7.23

Maps
m_R, Definition 1.30, 2.55
l_R, Definition 2.55
d, page 494

Operations on maps
$\tilde{\theta}$, Definition B.12, 5.23
θ^+, η_+, Definition 5.50

Relations
C, R, I, Example 1.8
R_1, page 424
R_a, R_b, ... Example 1.24(i)
R_Δ, Definition 1.23
R_π, Definition 1.26
$\leftrightharpoons$, Definition 2.16
$\leftrightharpoons_n$, Definition 2.30
R^Λ, Definition 4.18
R_P^Λ, R_F^Λ, Definition 4.34
$\equiv_C$, page 269
$\equiv_\Lambda$, Definition 5.29
Q_f, Definition 5.40
$\lessdot$, page 323
$\sim_U$, page 492
$\ker f$, Definition B.3
sub, Definition 4.92
$R_{-\alpha}$, page 424
$\equiv_i$, Equation 7.10
$\bowtie_\sigma$, Equation 7.13

elations between structures
$w \leftrightsquigarrow w'$, $\mathfrak{M} \leftrightsquigarrow \mathfrak{M}'$, Definition 2.1

$\mathfrak{M}' \rightarrowtail \mathfrak{M}$, Definition 2.5
$\mathfrak{M} \cong \mathfrak{M}'$, Definition 2.8, A.5
$\mathfrak{M} \twoheadrightarrow \mathfrak{M}'$, Definition 2.10
$Z : \mathfrak{M}, w \leftrightharpoons \mathfrak{M}', w'$ and $Z : \mathfrak{M} \leftrightharpoons \mathfrak{M}'$,
    Definition 2.16
$\mathfrak{M}, w \leftrightharpoons \mathfrak{M}', w'$ and $w \leftrightharpoons w'$, Definition 2.16
$\mathfrak{M} \leftrightharpoons \mathfrak{M}'$, Definition 2.16
$w \leftrightsquigarrow_\Sigma w'$, Definition 2.36
$\mathfrak{F}' \rightarrowtail \mathfrak{F}$, $\mathfrak{F} \twoheadrightarrow \mathfrak{F}'$, Definition 3.13
$f : \mathfrak{A} \preccurlyeq \mathfrak{B}$ and $\mathfrak{A} \preccurlyeq \mathfrak{B}$, Definition A.8
$\mathfrak{A} \equiv \mathfrak{B}$, page 487

Operations on relations
R^+, R^*, Example 1.6
$R^{\smile}$, Example 1.25
R^0, R^1, R^2, ..., Example 1.22(iii)
R^f, Definition 2.36
R^s, R^l, page 79
R^t, Definition 2.42
R^{ue}, Definition 2.57
$P?$, Example 2.80(ii)
$\sim R$, Example 2.80(iii)
R_β, R_ϵ, Convention 3.46

Truth, validity and consequence
$\mathfrak{M}, w \Vdash \phi$, Definition 1.20
$\mathfrak{M}, w \Vdash \Sigma$, Definition 1.20
$w \Vdash \phi$, page 18
$\mathfrak{M} \Vdash \phi$, Definition 1.21
$\mathfrak{F}, w \Vdash \phi$, Definition 1.28
$\mathfrak{F} \Vdash \phi$, Definition 1.28
$\mathsf{F} \Vdash \phi$, Definition 1.28
$\Sigma \Vdash_\mathsf{S} \phi$, Definition 1.35
$\Sigma \Vdash_\mathsf{S}^g \phi$, Definition 1.37
$(\mathfrak{F}, V)$, $\vec{s} \Vdash \rho$, Convention 5.88
$t^\mathfrak{A}[g]$, page 486
$t^\mathfrak{A}[a_1, \ldots, a_n]$, page 486
$\mathfrak{A} \models \alpha[a_1, \ldots, a_n]$, page 486
$\mathfrak{A} \models \alpha[a_1, \ldots, a_n, x \mapsto a]$, page 487
$\Pi \models \gamma$, page 487
$\mathfrak{A} \models s \approx t$, $\mathfrak{A} \models E$, Definition B.17

Structures
$(\mathbb{N}, <)$, $(\mathbb{Z}, <)$, $(\mathbb{Q}, <)$, $(\mathbb{R}, <)$, $(\omega, <)$,
    Example 1.2
$\mathfrak{S}_U$, Example 1.8
$\mathfrak{M}^\Lambda$, $\mathfrak{F}^\Lambda$, Definition 4.18 and 4.34
2, Definition 5.2
$\mathfrak{Form}(\Phi)$, Definition 5.3
$\mathfrak{P}(A)$, Definition 5.7
$\mathfrak{L}_C(\Phi)$, Definition 5.13
$\mathfrak{L}_\Lambda(\Phi)$, Definition 5.31
$\mathfrak{f}_\Lambda^c$, Example 5.61
$\mathfrak{N}_n$, Definition 6.17
$\mathfrak{C}_n(U)$, $\mathfrak{C}_n^W(U)$, Definition 7.45
$\mathfrak{A}_X$, Definition A.9
$\mathfrak{Ter}_\mathcal{F}(X)$, Definition B.13
$\mathfrak{Form}(\tau, \Phi)$, Definition 5.28

Operations on structures
$\biguplus_i \mathfrak{M}_i$, Definition 2.2
$\mathfrak{M} \upharpoonright k$, Definition 2.32
$\mathfrak{M}_\Sigma^f$, $\mathfrak{M}^f$, Definition 2.36

ue $\mathfrak{M}$, ue $\mathfrak{F}$, Definition 2.57
$\biguplus_i \mathfrak{F}_i$, Definition 3.13
$\mathfrak{F}_X$, $\mathfrak{F}_w$, Definition 3.13
$\mathfrak{M} = (\vec{W}, \vec{R}, \vec{V})$, Definition 4.51
$\mathfrak{F}^+$, Definition 5.21
$\mathfrak{A}_+$, Definition 5.40
$\mathfrak{Em}\mathfrak{A}_+$, Definition 5.40
$Uf\mathfrak{A}$, page 284
$\prod_U \mathfrak{A}_i$, $\prod_U \mathfrak{A}$, Definition A.18
$\prod_{j \in J} \mathfrak{A}_j$, $\mathfrak{A}^J$, Definition B.6
$\mathfrak{A}/\sim$, Definition B.9

$\approx$, Definition B.11
Π_1^0, Σ_1^1, page 508
Σ^*, Definition C.4
$\mathcal{O}$, page 515

Classes of structures
Fr_ϕ, Fr_Γ, Definition 3.1
Set, Definition 5.7
BA, Definition 5.10
F_1^n, page 341
C_n, R_n, Definition 7.45
HCF_n, page 468

Operations on classes
$\overline{\mathsf{K}}$, page 107
$\mathbf{Cm}$, Definition 5.21
$\mathbf{V}$, Definition 5.55
$\mathbf{H}$, Definition B.3
$\mathbf{I}$, Definition B.4
$\mathbf{S}$, Definition B.5
$\mathbf{P}$, Definition B.6
$\mathbb{V}$, Definition B.7

Special sets
Φ, Definition 1.9
V_m, Example 3.7
$At(\Sigma)$, Definition 4.80
W^Λ, Definition 4.18
T_n, Definition 6.17
$Ter_\mathcal{F}(X)$, Definition B.11

Operations on sets
$\mathcal{P}(W)$, Definition 1.19
$W \setminus X$, Definition 1.32
$Uf(W)$, Definition 2.57
P?, Example 2.80(ii)
$\widehat{A}$, page 242
$R[c]$, Proposition 5.83
$\prod_{i \in I} A_i$, page 492
$\prod_U A_i$, $\prod_U A$, Definition A.17
$Hin(\Sigma)$, Definition 6.53

Miscellaneous
$\vdash_\mathbf{K} \phi$, Definition 1.39
$|w|_\Sigma$, $|w|$, Definition 2.36
$\vdash_\Lambda \phi$, Definition 4.1
$[\phi]$, Definition 5.13
f_Δ, Definition 5.19
$\widehat{\phi}$, Example 5.61
$\vdash \alpha$, page 488
$\Pi \vdash \gamma$, page 488
π_w, Definition A.15
f_U, Definition A.17
f_a, f_w, page 494
$[a]$, Definition B.9

Index

abstract modal logic, 473
additivity, 275
admissible
  set, 29
  valuation, 29
algebra, 497
  as logic, 262
  of truth values, 265
algebraic interpretation of modal logic, 278
algebraic perspective, 39, 45
algebraizing
  modal axiomatics, 279
  modal semantics, 277
  propositional axiomatics, 268
  propositional semantics, 263
algorithm
  witness, 386
antisymmetry, 3, 217
arrow
  frame
    square, 8
  logic, 7, 16, 26, 30, 60
    and the three variable fragment, 91
    bisimilarity and squares, 72
    bounded morphism, 62
    finite model property, 83
    frame for, 8
    generated submodel, 57
    language of, 14
    model for, 23
    relativized square, 229
    square arrow frame, 8
    standard translation, 90
    ultrafilter extension, 99
  structure, 7
om, 241, 358
iom, 33, 192
  induction, 132
  Segerberg's, 132
iomatic system, see Hilbert system, normal modal
    logic
iomatizable, 342
iomatization, 195

k condition, 65

basic
  hybrid language, 435
  modal language, 9
    decidability, 340
    frame for, 16
    model for, 16
  temporal language
    bisimulation, 70, 72
    bounded morphism, 62
    completeness-via-canonicity, 204–209
    definability of bidirectionality, 26, 137
    definition of, 11
    dense frame, 129
    expressiveness, 63
    frame for, 21
    generated submodel, 57
    model for, 21
    over $(\mathbb{Q}, <)$, 224–237
    standard translation, 89
    strong completeness of $\mathbf{K}_t\mathbf{Q}^+$, 239
    transitive temporal filtration, 83
    ultrafilter extension, 99
    undefinability of progressive, 72
basic hybrid language, 435
basic temporal language, 137
bidirectional model and frame, 21
binary tree, 382–384
Birkhoff's Theorem, 502
bisimilar, 65
bisimilarity-somewhere-else, 98, 102
bisimulation, 64–73
  locality and computation, 67
  n-, 74
  safe for, 112
boolean
  algebra, 269
    and classical theoremhood, 269
  algebra with operators (BAO), 275
  homomorphism, 296
  modal logic, 424–425
bounded
  morphic image, 59
  morphism
    duals of, 313

for frames, 138
for general frames, 309
for models, 57–73
box
$\Box$, 9
$\Box_a$, 11
$[a]$, 11
boxed atom, 161
BR, 425
Bull's Theorem, 43, 247–252
bulldozing, 220–222

canonical
  class of algebras, 292
  embedding algebra, 288
  equation, 292
  formula, 203
  logic, 203
  model, 42
    arbitrary similarity type, 200
    basic modal language, 197
    basic temporal language, 205
    over a finite set of formulas, 243
  relation, 198
  valuation, 198
Canonical Model Theorem, 199
canonicity
  and d-persistence, 319
  and first-order definability, 215
  failure of, 211
  for a property, 204
Cantor's Theorem, 225, 229
carrier
  of an algebra, 497
Chagrova's Theorem, 167
characterization
  of frame classes, 125
Church's Thesis, 507
Church-Rosser property, 160
classical era, (1959–1972), 41
closed
  formula, 151
  subset of general frame, 315
closure
  Fischer-Ladner, 241
  transitive, 5
  under single negations, 241
  under subformulas, 77, 241
  universal, 445
cluster, 81
  and completeness proofs, 220
co-finite set, 30
cofinality, 213
compact
  logic, 212
compactness
  over models, 86
Compactness Theorem, 488
complete logic, 212
completely additive formula, 113
completeness
  strong, 194
    of $\mathbf{K}_t\mathbf{Q}$, 228

of $\mathbf{K}_t\mathbf{Q}^+$, 236
of $\mathbf{K}_t$, 206
of $\mathbf{K}_t$4.3, 220, 221
of $\mathbf{K}_t\mathbf{Q}$, 209
of $\mathbf{K}$4.3, 210
of $\mathbf{K}$4, 202
of $\mathbf{KB}$, 202
of $\mathbf{KD}$, 202
of $\mathbf{K}$, 199, 219
of $\mathbf{S}$4.3, 210
of $\mathbf{S}$4, 203, 219
of $\mathbf{S}$5, 203
of $\mathbf{T}$, 202
via completeness, 432
weak, 194
  via Stone's Theorem, 273
completeness-via-canonicity, 202
complex algebra, 277
composition relation, 7
computable, 505
confluence, *see* Church-Rosser property
congruence, 499
  natural map associated with, 500
connected, 3, 247
coNP, 512
consistency, 191, 194
  problem, 334
converse
  axioms
    canonicity of, 206
  of a relation, 22
Cook-Levine Theorem, 512
coPSPACE, 514
correspondence
  for models, 85
  global, for frames, 126
  language, 84
  local, for frames, 149
  theory, 86
corresponding algebraic similarity type, 275
Craig's Lemma, 342
cube
  over U, 465
  relativized, 465
current state, 18
cylindric modal logic, 462

d-persistence, 318
daughter-of, 6
dead-end, 19
decidability, 338, 505
  via finite models, 338–346
  via interpretations, 347–356
  via quasi-models and mosaics, 356–364
decidable, 338, 505
Dedekind complete, 429
deducibility, 190
definability
  of a property, 125
  of frame classes, 125
  of model classes, 88
  of models, 107–109
definable variant, 146

definably well-ordered, 431
degree, 74
density, 207
  canonicity of, 207
  definability of, 129
descriptive
  frames and BAOs, 311
  general frame, 306
Detour Lemma, 102
di-persistence, 318
diamond
  $\Diamond$, 9
  $\Diamond_a$, 11
  $\langle a \rangle$, 11
  saturated, 231
difference operator, 63, 238, 419–424
discriminator variety, 419
disjoint union
  of frames, 138
  of general frames, 310
  of models, 52–55
distinguishing model, 146
distribution axiom, *see* K axiom
domain, 2
downward monotone, 152
Dual axiom, 33, 191, 195
dual operator, 9
duality theory
  applications, 299–303
  basic ideas, 294–299
DUWTO-frame, 207

elementarily equivalent, 487
elementary
  class of models, 494
  extension, 490
  submodel, 490
elimination of Hintikka sets, 402–405
embedding, 58
endomorphism, 501
epistemic logic, 10, 19
equation, 500
equational
  class, 502
  logic, 502
equivalence
  elementary, 487
  expressive, 90
  first-order, 487
  modal, 52
euclidean, 128
existence Lemma
  basic modal language, 198
  for arbitrary programs in PDL, 245
  for basic programs in PDL, 243
  potentially deep models, 393–395
  pressive
  power, 43, 45, 73, 97
  pressive completeness, 430
  KPTIME, 514
  and modal logic, 393–406
  global modality, 422
  guarded fragment, 453

hardness via tiling, 395–402
extension
  elementary, 490
  ultrafilter, 138
extensions of **S4.3**
  Bull's Theorem, 247–252
  finite axiomatizability, 252–255
  Hemaspaandra's Theorem, 380
  negative characterization, 255–256

f.f.p., 335
f.m.p., 73, 336
false in a model, 18
falsifiable, 18
filter
  of a boolean algebra, 284
  over a set, 491
filtration, 77–82
  largest, 79
  natural map, 78
  smallest, 79
  transitive, 80
  transitive temporal, 83
Filtration Theorem, 79
finite
  character, 335
  frame property, 145–148
    general case, 335
  intersection property, 492
  meet property, 285
  model property
    for normal modal logics, 145
    for similarity types, 73
    general case, 336
    strong, 339
  models
    via filtrations, 77–82
    via selections, 74–77
  transitive frame, 143–144
  -variable fragment, 87
finitely
  axiomatizable, 252, 342
  based, 335
first-order logic
  assignment, 486
  basic properties, 487–489
  compactness, 488
  completeness, 488
  formula, 485
  Löwenheim-Skolem Theorem, 488
  modal fragment, 100
  model, 486
  model theory, 489–495
  satisfaction definition, 486
  sentence, 486
  term, 485
  ultraproducts, 492
  validity and semantic consequence, 487
Fischer-Ladner closure, 241
flat set of lists, 255
formula
  algebra
    modal logic, 280

propositional logic, 265
as term, 264
completely additive, 113
first-order, 485
Grzegorczyk, 256
Jankov-Fine, 143
Kracht, 170
Löb, 10, 130
McKinsey, 12, 30, 133
modal, 9, 11
monotone, 152
Sahlqvist
  general case, 164
  very simple, 156
Sahqlvist
  simple, 160
uniform, 151
universal second-order, 495
forth condition, 64
fragment
  guarded, 446
  packed, 446
  universal, 458
frame
  arbitrary similarity type, 20
  arrow logic, 8
  basic modal language, 16
  basic temporal language, 21
  connected, 247
  definability, 125
    relative, 125
  DUWTO, 207
  general, *see* general frame
  incompleteness, *see* incompleteness
  language
    first-order, 126
    second-order, 126
  of type τ, 20
  propositional dynamic logic, 22
  square, 229
  underlying a model, 17

general frame, 28–31
  algebraic perspective, 303–318
  and second-order logic, 136
  compact, 306
  descriptive, 306
  differentiated, 306
  discrete, 306
  full, 306
  general completeness result, 306
  refined, 306
  tight, 306
  topological aspects, 314–317
general ultrafilter frame, 310
generalization, 33, 191
generated
  point-, 56
  subframe, 138
  submodel, 55–57
global
  modality, 54, 367–371, 415–419
  truth, 18

Goldblatt-Thomason Theorem, 142, 178–181
  algebraic proof, 300–301
Grzegorczyk formula, 137, 256
guarantee properties, 427
guarded
  fragment, 446–458
  quantification, 448

height, 75
Hemaspaandra's Theorem, 380
Hennessy-Milner
  class, 92
  property, 92
  Theorem, 69
highly undecidable, 508
Hilbert systems and normal modal logic, 33–37
Hintikka set, 357
homomorphism, 57
  modal, 296
  of algebras, 498
hybrid logic, 434–444
hypercylindric frame, 468

identity arrow, 7
image-finite, 69
incompleteness, 44, 212
  first-order example, 216
  of $\mathbf{K}_t\mathbf{ThoM}$, 214
  of $\mathbf{KvB}$, 216
  ramifications of, 214–216
inconsistent
  formula, 191
  logic, 190
induction axiom, *see* Segerberg's axiom
infinitary logic, 496
instant, 2
internal, 18
interpretation
  in SnS, 347–355
invariance, 52
irreflexivity, 2, 217, 229
isomorphism
  of algebras, 498
  of first-order models, 489
  of modal models, 58

Jónsson-Tarski Theorem, 40
  and canonicity, 291–293
  statement and proof, 289–291
Jankov-Fine formula, 143

K axiom, 33, 191, 195
Kracht formula, 170
Kracht's Theorem, 171, 210
Kripke semantics, 42
Kruskal's Theorem, 253

Löb formula, 10, 130
Löwenheim-Skolem Theorem, 488
  for modal models, 86
labeled transition system, 3
lambda notation, 155
language

basic modal, 9
modal, 11
left-unboundedness, 207
Lindenbaum's Lemma, 197
Lindenbaum-Tarski algebra
modal logic, 281
propositional logic, 271
Lindström Theorem for modal logic, 470–476
linear order, 2
local, 18
frame correspondence, 149
logic
compact, 212
complete, 212
epistemic, 10, 19
equational, 502
infinitary, 496
modal, 189
normal modal, 191, 195
normal temporal, 205
of a class of frames, 24
provability, 10
second-order, 495
since/until, 426, 434
loose
model, 454–458
loosely guarded fragment, 458
Łoś's Theorem, 493

m-saturation, 91–93
master modality, 371–373
maximal consistent set, 196
McKinsey formula, 12, 30, 133
MCS, *see* maximal consistent set
modal
consequence relation, 31–32
constant, *see* nullary modality
homomorphism, 296
language of relations, 462
logic, 189
boolean, 424–425
multi-dimensional, 458
operator
arity, 11
modal logic
abstract, 473
modality
global, 406
master, 371–373
nullary, 11, 195, 201
substitution, 461
universal, 478
modally equivalent, 52
model
arrow logic, 23
based on a frame, 17
basic modal language, 16
basic temporal language, 21
canonical, 197
finitely based, 335
first-order logic, 486
loose, 454–458
named, 238

on a general frame, 29
pointed, 107
propositional dynamic logic, 22
saturated, 100
modern era, (1972–present), 44
modus ponens, 33, 189
monadic
second-order logic, 495
second-order theory of n successor functions, 348
monotone
downward, 152
formula, 152
upward, 152
monotonicity
in boolean algebras, 276
of formulas, 152
morphism, *see* bounded morphism
mosaics
for packed and guarded fragments, 450–453
for tense logic of naturals, 360–363
multi-dimensional modal logic, 458–470

n-bisimulation, 74
nabla, ∇, 11
named model, 238
natural valuation
on canonical model, 198
necessarily, 10, 19, 127
necessitation, *see* generalization
negation
single, 241, 356
negative occurrence, 151
network, 224, 232, 454
NEXPTIME, 515
node, 2
nominal, 238, 435
non-branching, 209
to the right, 193
normal modal logic, 33–37
alternative definition, 191
and finite frames, 145
arbitrary similarity type, 195
basic modal language, 191
basic temporal language, 205
generated by a set of formulas, 192
incomplete, 212
minimal, 192
propositional dynamic logic, 240
normality
algebraic definition, 275
NP, 511
and modal logic, 373–381
NPSPACE, 514
nullary modality, 11, 195, 201

operation, 497
operator
algebraic definition, 275
difference, 238
dual, 9
since, 43
until, 43, 72
oracle, 508

P, 511
packed
    fragment, 446–458
        universal, 458
    quantification, 448
partial order, 3, 217
PDL, *see* propositional dynamic logic
permissible, 16
persistence, 318
point, 2
point-generated
    frame, 139
    model, 56
pointed model, 107, 471
polynomial
    space, 513
    time, 509
        reduction, 510
polysize model property, 339
positive
    existential formula, 111
    occurrence, 151
possible world, 19
possibly, 10, 19, 127
preservation, 51
problem, 505
    as strings of symbols, 506
    completeness of, 511
    consistency, 334
    decidable, 505
    EXPTIME, 514
    hardness of, 511
    non-elementary, 373
    NP, 511
    P, 511
    provability, 334
    PSPACE, 513
    satisfiability, 333
    tiling, 366
    undecidable, 507
    validity, 333
product
    of algebras, 498
progressive operator, 72
proof
    in **K**, 33, 35
propositional dynamic logic, 16, 26
    and infinitary logic, 89
    as three variable fragment, 89
    axiomatization, 240
    bisimulation, 71
    bounded morphism, 62
    compactness failure, 240
    decidability of, 343
    definability of regular frames, 132
    EXPTIME algorithm for, 403
    EXPTIME hardness of, 397
    frame for, 22
    generated submodel, 57
    language, 12
    model for, 22
    regular, 13
    standard translation, 89–90

test, 13
    with intersection, 13, 367–373
provability
    logic, 10
    problem, 334
provable
    in **K**, 33
provable equivalence as a congruence
    modal logic, 281
    propositional logic, 270
PSPACE, 513
    algorithm for **K**, 384–389
    and modal logic, 381–393
    hybrid logic, 436
    Ladner's Theorem, 389–392

QBF, 389, 513
    truth problem, 513
    validity problem, 389
quantification
    guarded, 448
    packed, 448
quantified boolean formula, 389
quantifiers as modalities, 459–460
quotient algebra, 500

r-persistence, 318
Rabin's Theorem, 350
recursive, 505
recursively
    axiomatizable, 342
    enumerable, 506
    reduction, 508
reflexive
    and transitive tree, 7
    closure, 3, 8
    linear order, 3
    total order, 3
    transitive closure, 5
reflexivity, 3, 127
refutable, *see* falsifiable
refuted in a model, *see* false in a model
regular frames and models, 22
relational
    semantics, 41
    structure, 2
relativized
    cube over U, 465
    square, 229
reverse relation, 8
right-unboundedness, 193, 207
root
    of tree, 6
rooted, *see* point-generated
rules
    as sequents, 443
    for the undefinable
        IRR, 229–238

S4.3
    decidability of extensions, 345
    NP-completeness of extensions, 380
safety, 112

Sahlqvist
  Completeness Theorem, 210
  Correspondence Theorem, 165
  formula
    general case, 164
    simple, 160
    very simple, 156
Sahlqvist Completeness Theorem
  statement and proof, 322–325
Sahlqvist-van Benthem algorithm, 156–166
satisfaction operator, 238
satisfiability
  problem, 333
satisfiable, 18
  finitely, 92
satisfied in a model
  arbitrary similarity type, 20
  basic modal language, 17
  nullary modality, 20
saturated model, 100, 101
saturation
  diamond, 231
  m-, 91–93
Savitch's Theorem, 514
second-order
  logic, 495
  translation, 135
Segerberg's axiom, 13, 132
semantic consequence
  global, 32
  local, 31
set algebra, 267
similarity type
  algebraic, 497
  modal, 11
simulation, 110
since operator, 43
since/until logic, 426–434
situation, 2
SnS, 348
soundness, 193
square
  frame, 229
  over U, 8
  relativized, 229
standard translation, 83–91
  and second-order frame language, 135
  until, 429
state, 2
stavi connectives, 429
step-by-step method, 223–228
Stone Representation Theorem
  full statement and proof, 286
  short statement, 273
strict
  partial order, 2
  total order, 220
strong
  completeness, 194
    of K_tQ, 228
    of K_tQ^+, 236
    of K_t, 206
    of $K_t4.3$, 220, 221

    of K_tQ, 209
    of $K4.3$, 210
    of $K4$, 202
    of KB, 202
    of KD, 202
    of K, 199, 219
    of $S4.3$, 210
    of $S4$, 203, 219
    of $S5$, 203
    of T, 202
  finite model property, 339
  homomorphism, 58
subalgebra, 498
subformula closed, 77
submodel, 56
  elementary, 490
substitution, 15
  modality, 461
successor, 18
syntactic era, (1918–1959), 38
syntactically driven, 43

temporal, *see* basic temporal
tense logic, 40, 205
term, 485, 500
  algebra, 501
  as formula, 435
  theorem, 189
tile, 365
tiling
  game, 395
  problem
    $\mathbb{N} \times \mathbb{N}$, 366
    $\mathbb{N} \times \mathbb{N}$ recurrent, 366
times, 2
total order, 2
tractable, 509
transitive
  closure, 5
  tree, 7, 9
transitivity, 2
translation
  second-order, 135
  standard, *see* standard translation
tree, 6, 347, 348
  -like, 6
  binary, 382–384
  model, 62
  model property, 62
  reflexive and transitive, 7
  transitive, 7, 9
triangle, $\triangle$, 11
trichotomy, 2, 207
true in a model, *see* satisfied in a model
truth and validity in algebras, 502
Truth Lemma, 199
Turing machine, 504
  exponentially time bounded, 514
  non-deterministic, 507
  non-deterministic polynomially time bounded, 511
  polynomially space bounded, 513
  polynomially time bounded, 509
  with oracles, 508

ultrafilter
  countably incomplete, 106
  extension
    of a frame, 138
    of a model, 93–100
    temporal language, 99
    frame, 287
    general, 310
  of a boolean algebra, 284
  over a set, 491
  principal, 492
Ultrafilter Theorem, 285, 492
ultrapower, 493
ultraproduct, 492–493
  of modal models, 104–107
unbounded, 207
undecidability
  via tiling, 364–373
undecidable, 338, 507
unfolding, *see* unraveling
uniform
  formula, 151–155
  substitution, 33, 189
universal
  modality, 478
  second-order formula, 495
universally true, *see* global truth
universe, 2
unraveling, 63
  in completeness proofs, 218–220
until operator, 43, 72, 427
unwinding, *see* unraveling
upward monotone, 152

valid, 24
  at a state in a frame, 24
  in a frame, 24
  in a general frame, 29
  on a class of frames, 24
validity
  definition of, 24, 124
  problem, 333
valuation, 17
van Benthem Characterization Theorem, 100–104
variant
  definable, 146
variety, 499

weak
  completeness, 194
    of propositional dynamic logic, 239–246
    of **KL**, 212
    via Stone's Theorem, 273
  linearity, 207
  total order, 207
well-founded, 8
well-order
  definably, 431
well-ordered, 429
window, 424
witness
  algorithm, 386
  set, 385
world, 2

Printed in the United States
By Bookmasters